Neural and
Brain Modeling

NEUROSCIENCE

A Series of Monographs and Texts

edited by

Richard F. Thompson

University of Southern California
Los Angeles, California

Neural and Brain Modeling

Ronald J. MacGregor

Aerospace Engineering Sciences
University of Colorado
Boulder, Colorado

ACADEMIC PRESS, INC.
Harcourt Brace Jovanovich, Publishers
San Diego   New York   Berkeley   Boston
London   Sydney   Tokyo   Toronto

for Millie

ACADEMIC PRESS, INC.
1250 Sixth Avenue, San Diego, California 92101

United Kingdom Edition published by
ACADEMIC PRESS INC. (LONDON) LTD.
24–28 Oval Road, London NW1 7DX

Library of Congress Cataloging in Publication Data

MacGregor, Ronald J.
     Neural and brain modeling.

     (Neuroscience series)
     Includes index.
     1. Brain—Mathematical models. 2. Brain—Data
processing.   3. Nervous system—Mathematical models.
4. Nervous system—Data processing.   I. Title.
II. Series: Neuroscience series (San Diego, Cal.)
QP376.M18   1987   599'.0188   87-1157
ISBN 0—12—464260—8 (hardcover) (alk. paper)
ISBN 0—12—464261—6 (paperback) (alk. paper)

PRINTED IN THE UNITED STATES OF AMERICA

87 88 89 90   9 8 7 6 5 4 3 2 1

Contents

Preface

This book consists of two parts. Part I is a review of the literature in neural and brain modeling. There exists a tremendous volume of literature concerning the applications of theories and techniques of engineering modeling to the operations of functioning of the nervous system and brain. This literature is exceedingly rich in accomplishments to date, in currently energetically active areas, in opportunities for the future, and in significance for the overall development of neuroscience and brain science. The literature, however, is characterized by fragmentation and isolation with respect to both different subfields and related experimental work. One of the main goals of this review is to bring together under one cover, for what I believe is the first time, all theoretical and experimental approaches to neural and brain modeling. The review, which is as comprehensive and balanced as I can make it, surveys all those strands of engineering, mathematical, and computer-based modeling methodology as they have been directed toward neural and brain modeling.

Part II of this volume presents 54 computer programs in FORTRAN to simulate the dynamics of neurons and neuronal networks, together with instructions, examples, and exercises. It is my hope that these programs, widely used, will help bridge the gap in understanding between unit and systemic levels of nervous system functions.

Although the review in Part I is independent of the computer modeling of Part II, it provides a natural definition of the context within which that modeling has been undertaken. It should help to point out to students, scholars, and researchers the value, uses, and potential of modeling approaches so that they might see more clearly the relevance of modeling generally and of the modeling programs of Part II specifically.

xi

The computer programs contained in Part II were developed at the University of Colorado over the past 15 years. The final versions of the programs were written in the fall of 1984, and the text of Part II was written in the spring of 1985. The literature review in Part I was undertaken upon the recommendation of Richard Thompson and my editor at Academic Press and written in the months of October 1985–March 1986. A few minor modifications of the text and additions to the citation lists were made in September 1986.

The primary support for the most critical ingredients of the computer modeling of Part II was provided by NSF grant ESC 79-11794. Supplementary support was provided by the Computing Center of the University of Colorado and the Council on Research and Creative Work at the University of Colorado.

I thank Stephen Pilkenton for his invaluable assistance in refining the examples of Part II and for other hard and time-consuming programming work. I thank Paul Gaudiano for imaginative research on the embedding of memory traces. I thank other students whose enthusiasm and confidence in this work has helped tremendously in keeping me at it. I thank my colleagues Herb Alpern, Carol Barnes, Mark Dubin, Robert Eaton, Eva Fifkova, George Gerstein, Ted Lewis, Marvin Luttges, Steve Maier, Bruce McNaughton, John Meyer, Howard Wachtel, and my colleagues in Aerospace Engineering Sciences at the University of Colorado for various forms of support and encouragement.

I thank Photios Anninos, Jack Cowan, Robert Eaton, Sir John Eccles, Walter Freeman, George Gerstein, Stephen Grossberg, Phil Groves, E. Roy John, Ted Lewis, Bruce McNaughton, Andras Pellionisz, Don Perkel, Robert Plonsey, T. Poggio, Stephen Redman, Arthur Sanderson, Jose Segundo, Herbert Simon, Paul Smolensky, R. Stein, Richard Thompson, Roger Traub, Henry Tuckwell, Terry Winograd, and David Zipser for providing reprints and suggestions for the review.

I thank Lisa Reasoner for a fine job of drawing the figures. I thank my daughter Terri for her excellent typing of Part II of this strange, technical manuscript when she was much more interested in her upcoming wedding.

Finally, I thank my editor at Academic Press and Richard Thompson for their enthusiasm, encouragement, and help through the later stages of this work.

I

Models of the Nervous System: A Review of the Literature

The Brain—is wider than the Sky—
For—put them side by side—
The one the other will contain
With ease—and You—beside—

The Brain is deeper than the sea—
For—hold them—Blue to Blue—
The one the other will absorb—
As Sponges—Buckets—do—

The Brain is just the weight of God—
For—Heft them—Pound for Pound—
And they will differ—if they do—
As Syllable from Sound—

(Emily Dickinson)

Outline

2

T his review surveys all those strands of engineering, mathematical, and computer-based modeling methods as they have been directed toward neural and brain modeling. The models reviewed in this volume are categorized into three broad areas: models of generic mechanisms; models of specific neuronal systems; and models of generic operations, networks, and systems. The first section contains chapters on spike generation, dendritic trees, and interneuronal communication and biophysical processes. The second section contains chapters on models of simpler and invertebrate systems, lower-level vertebrate systems, limbic systems, and neocortical systems. This partitioning is based on the levels-of-construction approach to nervous system organization discussed in Chapter 8. Primary chapters of the last section are on neural networks and cognitive operations. This section also includes briefer chapters on clinical models and models at the outer limits of science.

The review discusses roots and central contributions and provides briefer specifics on work published in the past five years. Chapters 2, 3, 9, and 10 review what I consider to be the fundamental generic methods of neural and brain modeling. In these chapters I have tried to described (although briefly) fundamental theoretical basics as well as to review recent literature. Chapters 2, 3, and 9 are probably the strongest in the review. The material of Chapter 10 is somewhat outside my speciality, so it may be found wanting by specialists in that area. Nonetheless, I believe that the perspective offered here will be valuable. In Chapters 5, 6, 7, and 8 I catalogue and review models for specific neuronal systems as I have found them in the literature. I have been surprised, as I expect my colleagues will be, by the vast amount of literature in this area. Chapters 4, 11, and 12 are admittedly weaker than other chapters, as they reside at the peripheral margins of the focus of interest of this review. They are nonetheless necessary for completeness, and they contain valuable information and interesting models.

I believe the structure of this review, including as it does, the various subfields of neural and brain modeling in an intelligible cohesive framework, is in itself a valuable contribution. This framework easily houses the scope of activity and anticipated advances in the field. The bibliographies partition the field of neural and brain modeling into approximately 60 subdivisions and provide at a glance a thumbnail sketch of the development of each. Citations in each list are arranged chronologically, most recent first (alphabetically by author for each year). One might readily make interesting observations on the history and evolution of neural and brain modeling and some projections from these lists by plotting the number of publications against time.

I make no claim to have read the more than 1200 publications cited in this review. Moreover, I do not claim to be expert in many of the areas reviewed here. I am certain to have made some oversights and some errors of perspective. Still, I believe the ingredients of comprehensiveness and of a single unifying view to be important for the field at this time. Many areas of neural and brain modeling are developing very rapidly, and they will of course have generated a number of new and important publications before this review is published.

1

Introduction

Models of neural and brain operations are reviewed here in three broad sections: models of generic mechanisms; models of specific neuronal systems; and models of generic operations, networks, and systems. The first section contains chapters on spike generatrion, dendritic trees, and interneuronal communication and biophysical processes. The middle section contains chapters on models of simpler and invertebrate systems, lower-level vertebrate systems, limbic systems, and neocortical systems. This partitioning is based on the levels-of-construction approach to nervous system organization discussed in Chapter 8. The primary chapters of the last section are those on models of neural networks and of cognitive operations. This section also includes briefer chapters on clinical models and models at the outer limits of science.

Roots of Neural and Brain Modeling in Engineering

Table 1.1 identifies the central roots of engineering modeling methodology as relevant to neural and brain modeling. These seem to consist of three main groups: fundamental quantitative techniques developed in mathematics, engineering, and computer science; physical theories developed in physics, the engineering sciences, and mathematical physics; and the information processing paradigm and related concepts, which have emerged from computer science and artificial intelligence and have blossomed in the new field of cognitive science.

The quantitative techniques enumerated in Table 1.1 have been applied by theorists and modelers rudimentally throughout all three sections of the review (generic mechanisms, specific systems, and generic systems), but

Table 1.1
Roots of Neural and Brain Modeling in Engineering

Various quantitative techniques: (mathematics, engineering, and computation)
   1. Differential equations (ordinary, partial)
   2. Matrix algebra and tensor analysis
   3. Probability and statistics
   4. Set theory
   5. Fourier frequency analysis
   6. Digital and electronic simulation
   7. Discrete mathematics, Boolean algebra, etc.

Physical theories (engineering, physics)
   8. Feedback control system theory
   9. Classical mechanics and dynamical systems
   10. Continuum mechanics and conservation laws
   11. Nondimensional analysis and scaling
   12. Stochastic systems and information theory
   13. Statistical mechanics
   14. General systems theory

Information processing paradigm (computer science, engineering)
   15. Logic circuits, sequential machines
   16. Organization of computing systems
   17. Artificial intelligence

---

with particular success so far at the level of generic mechanisms of single neurons (Chapters 2 and 3). Ordinary and partial differential equations have been used to describe spike generation and processing in dendritic trees (Chapters 2 and 3). Matrix algebra has been used by Pellionisz to describe dynamic transformations at junctions in neural networks (Chapter 6). Probability and statistics have been applied to spike train analysis and, by Harth, to simple neural net modeling (Chapter 9). Set theory is used to characterize some models of knowledge structures (Chapter 10). Fourier analysis has been used to characterize processing in the visual system (Chapters 6 and 8) and to theorize on holographic-like operations of neural networks (Chapter 9). Digital and electronic simulation techniques are applied in all models in all sections of the review. Discrete mathematics (Boolean algebra, etc.) is used largely in conjunction with logical circuits and the sequential machine approach to neural networks (Chapter 9).

Physical theories from the engineering sciences and physics have been applied to neuronal networks (Chapter 9) and to a lesser extent to specific neuronal systems (Chapters 5–8). Control system theory, for example, with concepts of feedback, optimization, etc., has been applied ubiquitously to specific neuronal systems, particularly motor, autonomic, and appetitive systems. It has also been applied to spike generation mechanisms in single neurons. Dynamical system theory with concepts such as state space representation has been applied to neural networks (Chapter 9) and to models of spike generation (Chapter 2). The approaches of continuum mechanics and conservation laws (conservation mass, momentum, and energy), as exemplified in the partial differential equations of fluid mechanics and elasticity, have been applied to the dynamics of neuronal networks (Chapter 9). Concepts of nondimensional analysis and scaling taken from fluid mechanics have also been applied primarily to neural network modeling (Chapter 9). Stochastic systems theory and information theory have been applied primarily in general terms to neuronal networks and spike train analysis (Chapter 9) and to spike train generation (Chapter 2). Statistical mechanics, which deals with global properties of large numbers of interacting molecules, has been applied to neuronal networks (Chapter 9). General systems theory, which deals in broad terms with complex interrelated systems and their organizations and constraints, has been applied in considering the overall functional organiztion and operations of nervous systems, as discussed in Chapter 8.

The information processing paradigm has been applied primarily at the level of cognitive operations (Chapter 10), with some application to neural networks (Chapter 9). For example, the development of logic circuits, the theory of sequential machines, and automata theory have been advocated as models of neuronal networks (Chapter 9) in their performance of cognitive operations (Chapter 10); methodology regarding the organization of computers and computer systems has been applied implicity in terms of the information processing paradigm for cognitive operations (Chapter 10); fields of artificial intelligence have contributed largely to cognitive operations (Chapter 10), with some contributions (e.g., perceptrons) to neural networks (Chapter 9).

Stratification of Variables

The perspective of this review can be characterized in terms of the stratification of variables indicated in Table 1.2. This concept, introduced by MacGregor and Lewis in 1977, is useful for mapping and considering various theoretical and modeling efforts in neuroscience. The arrangement is useful primarily in making clear the polarization of bottom-up as opposed to top-down theorizing and modeling, elucidating certain principles of neural and brain modeling, mapping various disciplines and fields, and identifying particularly significant areas. A key ingredient in the stratification is the level of inferred theoretical constructs, level D, which lies between the levels of behavioral and experiential variables on the one hand and variables of electrical signaling on the other. Identification of this level clarifies the relations among certain fields critical to our focus of interest and shows the way to their eventual integration.

Table 1.2
Stratification of Variables

E. Behavior, experience, consciousness
(direct observations)

D. Theoretical inferred constructs
(superego, id, drives, cognitive maps, cell assemblies, statistical configurations)

C. Electrical signals
(field potentials, spike trains, generator potentials)

B. Membrane conductance modulations
(synaptic conductances, active dendritic conductances, action potential conductances)

A. Molecular and chemical processes
(chemical synaptic transmission, gating processes)

Several principles concerning modeling are suggested by this stratification of variables. (1) Models which relate variables of adjacent strata are the most powerful; there is a sense in which the behavior of variables at the upper level is explained in terms of the variables at the lower level. (2) Models which skip a level are difficult to test and generally low in believability. (3) Models which interrelate variables at a single level are weak in predictive power. (4) Models which relate variables of several strata are most broadly significant.

The fields of cognitive science, artificial intelligence, and psychology attempt to link levels D and E, that is, to explain behavior of variables at level E in terms of variables at level D. Classical neurophysiology and the field of neural modeling in a narrow sense have been concerned with linking variables of levels B and C. The domain of neural network theory is that of linking C and D.

In a review of this field published in 1977, MacGregor and Lewis concluded that the level of neuronal networks was most critically in need of development for the advancement of neuroscience. Since that time, as we will see in this review, there has been a tremendous amount of activity in this area, and very significant contributions have been made, both from the bottom up and from the top down.

Bibliography

General Neural Modeling
1983   Sanderson, A. C., and Peterka, R. J. Neural modeling and model identification. *CRC Crit. Rev. Biomed. Eng.* **12**, 237–309.
        Sanderson, A. C., and Zeevi, Y. Y. Neural and sensory information processing. *IEEE Trans. Syst., Man., Cybernet.* **13**, Spec. Issue.
1977   MacGregor, R. J., and Lewis, E. R., eds. *"Neural Modeling."* Plenum, New York.

1977   Scott, A. C., *"Neurophysics."* Wiley, New York.
1973   Leibovic, K. N., *"Nervous System Theory."* Academic Press, New York.
1970   Institute of Electrical and Electronics Engineers *Proc . IEEE,* Special Issue. June, 1968.
1967   Deutsch, S. *"Models of the Nervous System."* Wiley, New York.
        Harmon, L. D., and Lewis, E. R. Neural modeling. *Physiol. Rev.* **46**, 113–192.
1961   Wiener, N. "Cybernetics," 2nd ed. MIT Press, Cambridge Massachusetts.
1945   Wiener, N., "Cybernetics." 1st ed. MIT Press, Cambridge, Massachussetts.

Generic Mechanisms

Central accomplishments in twentieth century neuroscience have occurred at the levels of neurobiological mechanisms of electrical signal generation, information processing in single neurons, and the biochemical and electrical modes of communication among neurons. This section reviews the involvement and contributions of engineering, mathematics, and computer modeling in these areas. Chapter 2 discusses models for the generation of transmembrane electrical signals and for operational input–output dynamics in point neurons. Chapter 3 reviews models for passive and active dendritic trees, and Chapter 4 reviews models for intercellular communication and for biophysical and chemical foundations of control.

2

Single Neurons: Spike Generation

This chapter reviews models that describe the membrane processes of generating electrical signals in single neurons. We are primarily concerned with mechanisms that are crucial in effecting the overall transfer of information from input bombardment to output signaling in neurons. The primary focus is on models that predict and describe variations in transmembrane electrical potential in terms of input currents and membrane conductance modulations. All the models described and cited in this chapter are for isolated patches of neuronal membrane or for point neurons, except those describing conduction of action potentials in axons. Models describing spatial interactions in dendritic trees are described in Chapter 3. The rationale for considering point neurons is discussed in Chapter 14.

The fundamental guiding concept in neuron modeling is that the output signals of neurons consists of time series of action potentials and that these are produced in normal functioning of neurons by comparison of a graded generator potential to a threshold mechanism somewhere near the junction of the neuron's soma with its axon. A classic paper by Bishop on the functional significance of action potentials in nervous systems provides a valuable discussion of the validity and limitations of this point of view.

Single-Neuron Physiology

Perhaps the most significant neural models that have appeared thus far are those which describe the transmembrane electrical potentials in neurons in terms of their ionic basis. The fundamental picture is that of a relaxation circuit activated primarily by modulations in membrane conductance. The transmembrane potentials are maintained by differential concentrations of ions and served by ionic currents; the membrane conductance modulations are controlled by molecular gating processes active during synaptic activation, the generation of action potentials, and various other active membrane processes (often if not exclusively associated with calcium ions, as discussed in Chapter 3).

The central idea that neuronal membranes and ionic fluxes are germane to neuroelectric signaling was presented around the turn of the century by Bernstein and has been elaborated by a number of subsequent investigators.

Resting Potential: The Goldman Model

The primary model for the resting potential in neurons is expressed by the Goldman equation

$$V_G = \frac{kT}{q} \ln \frac{P_K[K]_o + P_{Na}[Na]_o + P_{Cl}[Cl]_i}{P_K[K]_i + P_{Na}[Na]_i + P_{Cl}[Cl]_o} \qquad (2.1)$$

where the P's are permeabilities of the membrane (proportional to conductances) to the different ionic species; the terms in brackets are concentrations of the different ionic species in the intracellular and extracellular fluid; k is Boltzmann's constant, T is the temperature, and q is the unitary electric charge. The idea underlying this equation is that when the transmembrane potential is fixed at its resting value, the total transmembrane current due to diffusion along concentration gradients and to constituent electric field gradients must sum to zero. In deriving this equation, Goldman built on Bernstein's fundamental concepts, on early work by Nernst, and on fundamental physics of ionic media as contributed to by Planck and Einstein.

The significance of this model is that it allows one to make precise, quantitative, causal predictions regarding modulations in transmembrane electrical potentials in neurons as they are influenced by variations in the constituency of intra- and extracellular fluids and by modulations in the permeabilities of the membrane to the various relevant ions. It embodies the fundamental causal mechanisms for electrical signaling in neurons and provides the scientific basis for interpreting and predicting how neuroelectric signals are modulated in normal functioning and can be altered experimentally.

The fundamental biophysical model embodied by the Goldman equation can be represented by a relatively simple equivalent circuit model, as discussed in Chapter 14.

Action Potential: The Hodgkin–Huxley Model

In 1952 Hodgkin and Huxley published the following equation system for the generation of single-action potentials in squid giant axons:

$$C \frac{dE}{dt} = GNO*m^3h(ENA - E) + GKO*n^4(EK - E)$$

$$+ GL*(EL - E) + SC) \tag{2.2a}$$

$$\frac{dn}{dt} = \alpha_n(1 - n) - \beta_n n$$

$$\frac{dm}{dt} = \alpha_m(1 - m) - \beta_m m \tag{2.2b}$$

$$\frac{dh}{dt} = \alpha_h(1 - h) - \beta_h h$$

$$\alpha_n = \frac{0.01(E + 10)}{e^{(E+10)/10} - 1}, \quad \alpha_m = \frac{0.1(E + 25)}{e^{(E+25)/10} - 1}, \quad \alpha_h = 0.07 \, e^{E/20} \tag{2.2c}$$

$$\beta_n = 0.125 \, e^{E/80}, \quad \beta_m = 4e^{E/18}, \quad \beta_h = \frac{1}{e^{(E+30)/10} + 1}$$

$$C = 1 \, \mu F/cm^2, \quad GNO = 120 \, \frac{m \cdot mho}{cm^2},$$

$$GKO = 36 \, \frac{m \cdot mho}{cm^2}, \quad GL = 0.3 \, \frac{m \cdot mho}{cm^2},$$

$$ENA = -115 \, mV, \quad EK = +12 \, mV, \quad EL = -10.6 \, mV \tag{2.2d}$$

$$n_0 = 0.31767691, \quad m_0 = 0.05293249; \quad h_0 = 0.59612075$$

In this model the fundamental equation for the transmembrane potential E is Eq. (2.2a), which describes E in terms of the resting and active membrane conductances in accordance with the equivalent circuit model discussed in Chapter 14. The active conductances for sodium, G_{Na}, and potassium, G_K, are defined in terms of hypothetical continuous functions n, m, and h as shown in Eq. (2.2b); n, m, and h in turn are given by rate equations guided by α and β parameters as shown in Eq. (2.2c). Finally, these rate-governing α and β parameters, are represented by experimentally determined functions of E as shown in Eq. (2.2d). In conjunction with this model, Hodgkin and Huxley also described a large number of carefully controlled experiments on transmembrane currents, conductances, and potentials which quantitatively matched the predictions of the equation system. The work is very significant because (1) it established that large, brief excursions in conductances to sodium and potassium are the fundamental physical processes underlying action potentials in neurons; (2) it helped firmly establish the ionic picture of the basis of transmembrane potentials in neurons; and (3) it held out by example the hope that significant neuroelectric events could be precisely and quantitatively modeled by mathematical and engineering techniques.

The Hodgkin–Huxley model has at the same time strong and elegant features and fundamental weaknesses. On the one hand, and most significantly, it accurately describes the transmembrane potential in the course of single-action potentials in terms of underlying membrane conductance modulations. It embodies and describes quantitatively the causal mechanisms underlying the generation of action potentials. On the other hand, the model does not embody a satisfactory picture of the biophysical basis of the membrane conductance modulations in terms of more fundamental molecular events. For example, one hypothetical interpretation of the Hodgkin–Huxley equation system, as presented by Hodgkin and Huxley themselves, would be that the functions n, m, and h represent the relative concentrations of three "gating" molecules in the intra- or extracellular fluid adjacent to the membrane. Then, $1 - n$, $1 - m$, and $1 - h$ would represent the concentrations of these molecules on the other side of the membrane. The rate equations for n, m, and h would then be considered as representing diffusion of these molecules across the membrane. The α's then would represent the momentary ease of influx of these molecules and the β's would represent the momentary ease of efflux (or vice versa). Then the form of the governing equations for the conductances G_{Na} and G_K in terms of n, m, and h might reflect that opening a channel for a sodium ion would require the confluence of one m molecule and three h molecules on one side of the membrane, whereas gating a potassium channel would require the confluence of four n molecules. That is, the likelihood of gating a channel would be proportional to the probability of finding the right combination of gating molecules at hand, which in turn would be proportional to the product of all the relevant concentrations.

Hodgkin and Huxley presented this equation system with the full understanding that it was merely a conceptual guide and probably did not accurately represent the underlying processes. Again, the fundamental contribution of the model is that it relates the level of electrical signals to the level of membrane conductances, not that it describes the molecular events at still lower levels. However, the precise form of the equation system is, in fact, quite arbitrary. The fundamental equation for the transmembrane potential E is based on the equivalent circuit model and is therefore reliable, but the functional forms of the other equations are somewhat suspect. Specifically, Eqs. (2.2a), which relate G_{Na} and G_K to n, m, and h, and Eqs. (2.2b), the rate equations for n, m, and h themselves, are without satisfactory physical foundation. Equations (2.2c) for the α's and β's are experimentally reliable since they are based on curve-fitting experimental data from squid giant axons, but they are only shorthand representations of data—that is, they do not represent any physical principles in themselves, and moreover the α's and β's themselves are hypothetical constructs defined in the weakly based equations for *n, m,* and *h.* Therefore, although the equation system does give a satisfactory quantitative description of the generation of action potentials and the time courses of the sodium and potassium conductances, one does not feel totally confident in the functional form of the equations or their biophysical basis. Therefore, one cannot reliably predict with the model how fundamental changes at the molecular and biochemical level will influence the behavior of the system, or how a neuronal tissue other than the squid giant axon may differ in behavior on the basis of known biophysical differences.

The model also has strengths and weaknesses as a model of information processing operations in neurons. It is accurate in terms of describing events leading up to the generation of single-action potentials and the shapes of single-action potentials in the squid axon. It exhibits accommodation to activating currents in ways consistent with detailed studies of spinal motoneurons (see the model of Hill described later). On the other hand, the model does not accurately describe the generation of two or more action potentials even in the squid axon. Within an information processing orientation one is most interested in ongoing repetitive activity in neurons. The mechanisms of interest are primarily those that determine the modulations of the ongoing graded generator potential, the threshold, and their interaction, rather than the detailed shape of the action potential once triggering has occurred. Therefore, this is a significant limitation of the model.

The Hodgkin–Huxley model has spawned a very large number of attempts to model neuroelectric activity in a wide variety of contexts. First, there have been attempts to extend and modify the Hodgkin–Huxley parameters in application of the equation system to other tissue. For example, Lewis constructed an elegant analog simulation model wherein the parameters of the Hodgkin–Huxley equation system could be adjusted manually by knobs on a control panel to produce a wide variety of neuronal activity. Frankenhauser showed how to adjust the parameters to describe action potentials in frogs. Nobel presented a number of publications showing how to extend the equation system to describe the generation of contractions in heart muscle fibers. Second, Connor and Stevens attempted to extend the equation system to describe repetitive firing. Third, a large number of mathematical and computer modeling studies have been presented which elaborate the properties of the Hodgkin–Huxley model in various contexts. These studies are cited later in this chapter.

Other researchers interested in ongoing information processing in single neurons and in neuronal networks and systems have found it more fruitful to go back to the basic equivalent circuit model and to proceed in slightly different ways than have Hodgkin and Huxley.

In the Hodgkin–Huxley model, the excursion in potassium conductance exhibits a relatively long tail beyond the time required to bring the potential down to resting level. This tail contributes to an afterhyperpolarization in the potential and moreover contributes to refractoriness in the cell by its propensity to leak off any excitatory current. Kernell presented a model which accounts nicely for the adaptive behavior and refractory properties exhibited by spinal motoneurons in response to applied step currents wherein the elevated postfiring potassium conductances decay exponentially in time and accumulate from successive output spikes. This model can be represented by Eq. (2.3):

$$\frac{dg_K}{dt} = \frac{-g_K + bS}{T_K} \qquad (2.3)$$

Here g_K is potassium conductance above resting level; S is 1 when the cell is firing and 0 otherwise; b determines the magnitude of the postfiring potassium increment; and T_K is the time constant of decay of g_K.

Circuit Model for Synaptic Activation: The Eccles Model

The first major accomplishment of twentieth century neuroscience has been to establish the fundamental biophysical model of the ionic basis of neuroelectric activity. The second major accomplishment has been to elucidate the modes and characteristics of interaction and communication between and among neurons. This picture is still unfolding. At this stage four things at least are clear: (1) The various modes of interaction and communication are the essential ingredients in producing the overall functional organization of the brain and its operatons. (2) There is a rich variety of modes of naturally occurring communication among neurons, including both synaptic and nonsynaptic interaction and both chemical and electrical modes of interaction. (3) A central and widely used form of intercellular communication is direct chemically mediated synaptic activation between the axon terminal of the sending cell and the dendritic or somatic regions of the receiving cell. (4) Such direct chemical synapses are highly malleable and plastic: they depend on complex physiology and biochemistry, they exhibit short-term plasticity in the forms of habituation and sensitization that are certainly important in ongoing dynamic operations, and they exhibit long-term changes in efficacy which are probably related to the embedding of memory traces and the formation of cognitive structures.

Sir John Eccles contributed centrally to the understanding of synaptic interaction among neurons by elucidating the electrophysiological mechanisms of direct chemically mediated synaptic activation, and integrating these within the ionically mediated relaxation circuit model to produce a dynamic model for information processing in neurons and, by implication, for neuronal networks. A simple summary of Eccles' basic picture for synaptic activation at chemical synapses is shown in Fig. 2.1. In this model postsynaptic responses are produced by selective alterations in conductance

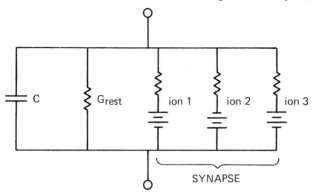

Fig. 2.1. Eccles equivalent circuit model for synaptic activation.

to particular ionic species in the subsynaptic membrane. This in turn selectively encourages the influx or efflux of the appropriate ionic species constituting electrical currents and corresponding modulations in transmembrane potential across the subsynaptic membrane. Eccles posited that the most common synapses in spinal motoneurons are excitatory synapses, whose transmitters open pores in the subsynaptic membrane to all operational ionic species and thereby tend to drive potentials toward a completely depolarized state, and inhibitory synapses, whose transmitters open pores selectively to potassium and chloride ions and thereby tend to drive membrane potentials to about 10 mV below resting level. In such a model postsynaptic responses exhibit transient excursions away from resting level while the synapse is activated and decay passively back toward resting level with the time constant of the membrane when the synapse is no longer active. The fundamental implication is that the ongoing continuous generator potentials in the somadendritic regions of neurons consist of summations of large numbers of excitatory and inhibitory postsynaptic potentials (PSPs), corresponding to the arrival input action potentials at such chemically mediated synapses. The mechanisms of interaction of the PSPs are determined by those of the equivalent circuit model shown in Fig. 2.1.

Eccles spelled out many significant and interesting properties of postsynaptic potentials and their interactions in a large number of experiments conducted over three decades. These included the observations that the integrative properties of neurons would include spatial and temporal summation, in that PSPs outlived the action potentials which triggered them and summed with other PSPs over a time frame determined by the time constant of decay to produce responses in receiving neurons. Temporal summation refers to the summation of PSPs in a single input channel by a burst of input action potentials; spatial summation refers to the summation of PSPs from separate input synapses. Eccles also pointed out that inhibitory synapses would produce significantly larger inhibitory responses when the existing generator potential in the receiving cell was depolarized close to threshold than in the resting state. He pointed out that inhibitory synapses would be much more effective when placed at or near the soma rather than in peripheral regions of dendrites and stated that this seemed to be the case as far as the anatomy and physiology of the day could determine.

Several significant postulates building on Eccles' basic findings were widely taken up. These include the following: all interactions between neurons are mediated by direct chemical synaptic activation; all synapses are either excitatory or inhibitory; a given neuron projects either all excitatory or all inhibitory synapses to all its target cells.

We know now that none of these postulates is universally true; that there are significant and widespread violations or exceptions to each. Nonetheless, this picture was almost universally adopted in modeling studies of neuronal networks, as we will see in Chapter 9, and probably does represent some significant portion of neural operations.

Threshold Variation: The Hill Model

The simplest neuromines use a fixed threshold for comparison with a graded generator potential to determine firing. However, in real neurons excitability and thresholds do indeed vary and a number of models of such variations have been presented. In accordance with the Hodgkin–Huxley model, the triggering of action potentials is caused by the excursion of membrane conductance to sodium. Therefore, fundamental understanding of threshold variation will be dependent on understanding of processes at the molecular level which gate the operations of these channels. These are being increasingly elucidated, but are not yet sufficiently understood to allow quantitative modeling of their effects at the electrical signal level. At the electrical signal level various modeling stratagems involving threshold modulations have been suggested to produce various electrophysiological effects. For example, one can produce refractoriness in neuron models easily by simply positing that the threshold is infinitely high for a short time immediately following firing (absolute refractoriness) and then decays gradually to resting level (relative refractoriness).

Short-term plastic effects in neuronal firing such as accommodation and postinhibitory rebound (PIRB) are often modeled using variations of threshold. The first such model was presented in 1933 by Rashevsky. In 1936 A. V. Hill presented an improved version of this model and reports of experiments corroborating its predictions. The basic phenomenon in accommodation is that the threshold required for firing tends to increase as the neuron is accumulatively stimulated at the subthreshold level. Hill posited a two-state variable model for this effect, as shown in Eq. (2.4).

$$\frac{dE}{dt} = -\frac{E}{\tau} + I$$

$$\frac{dU}{dt} = \frac{-(U - U_o) + CE}{T_U} \tag{2.4}$$

The first of these equations is for the transmembrane potential E as determined by the conventional equivalent circuit model. The second equation is for the threshold U considered as continuous function of time. The essential mechanism is that the threshold is driven away from its resting level according to the transmembrane potential, and relaxes or decays exponentially back to its resting level with time constant T_U. The deflection of threshold depends on the integral of the membrane potential and is driven upward by depolarization and downward by hyperpolarizations. The response of threshold to potential can be quick or sluggish depending on the value assigned to its time constant of response, T_U. With this model Hill was able to demonstrate accommodation and successfully match the output of the model with experimental data from some spinal motoneurons. Hill showed that the model and some neurons exhibited a critical minimum level of scope of applied ramp currents, such that currents with slope less than this minimum level produced no firing in cells if left on for long times so that the input current attained damagingly high levels. Hill also demonstrated what was then called "anode break" behavior, which is now called postinhibitory rebound. This is the property of neurons to fire action potentials to an input that is normally subthreshold or indeed to no input at all during recovery from intense inhibition. In the model it results from the accumulated depression of threshold sufficiently below normal level to allow firing to a normally subthreshold stimulus.

Hill discussed Eq. (2.4) with the sensitivity parameter c equal to 1. Bradley and Somjen and MacGregor extended this model to include values of c in the range 0 to 1, which allows representation of "ceilings" and degrees of accommodation. The Hodgkin–Huxley model exhibits accommodation by desensitizing the activation of the sodium conductance channel by the m and h functions. The accommodative properties of the Hodgkin–Huxley equation system to ramp currents are matched nicely by Hill's model with the parameter values $c = 0.85$ and $T_U = 6.25$ msec.

Postinhibitory rebound has been modeled by Andersen using ideas essentially similar to Hill's in studies of the generation of the alpha rhythm, and by a number of other neuron modelers. Various other *ad hoc* modulations of threshold are also often used in neuron models.

Digital Computer Neuromimes

Neuromimes are devices which mimic neural function. They are considered here in the narrower sense of mathematical and computational devices which produce output signals according to rules for input interactions in single-point neurons. They can be conveniently partitioned into four classes: digital computer neuromimes, electronic analog neuromimes, continuous mathematical neuromimes, and stochastic neuromimes.

Of these, the digital computer neuromimes are probably the most practical, being the most readily accessible to the widest range of researchers, most flexible, most adaptable, and most easily absorbed into large-scale representations of larger systems.

The simplest neuromimes compare an instantaneous combination of arriving excitatory and inhibitory signals on multiple input channels in a point neuron with a fixed threshold to determine whether the model neuron fires at a particular instant in time. The specific rules for combining inputs may vary from model to model: the algebraic sum of the number of excitatory (plus) and inhibitory (minus) inputs might be used; individual input channels might be given variable weights; inhibitory inputs might be considered so overwelmingly powerful as to block all excitation; and so on.

The seminal model of this type was presented in 1943 by McCulloch and Pitts, who created simple logic circuits composed of interconnected neuronlike elements. The units exhibit only binary activity (on or off) and interact through time in discrete steps corresponding to the interunit conduction time. McCulloch and Pitts were more concerned with the possible cognitive and logical operations of their circuits than with electrical activity of the nervous system per se, and they stressed a conformity between the all-or-none law of neuronal action potentials and the logic of propositions. (Their only literature citations in this work are to Carnap, Russell and Whitehead, and Hilbert and Ackerman.) They felt that much of psychology could be reduced to neurophysiology through the medium of their logic circuits.

This work has been a seminal starting point for a vast amount of subsequent work and advances in areas of computer science and artificial intelligence, spawning such subdisciplines as automata theory, the theory of sequential machines, and logic circuits. Associated with these fields was a considerable amount of disscussion in the 1950s as to the applicability of discrete mathematics and Boolean algebra to nervous system function, the relevance of digital versus analog processing, and so on. The field has produced a great many network models, many of which have possible relevance to nervous system operations. The field continues to produce important theoretical ideas, including contemporary discussions of sequential as opposed to parallel processing in neuronal networks. The approach is discussed further in Chapters 9 and 10.

Another seminal starting point is this kind of modeling was presented in 1961 by Caianiello, who defined the mechanics of neuronlike networks by Eq. (2.5).

$$u_h(t + T) = 1[a^r_{hk}u_k(t - rT) - s_h]$$ (2.5)

Here 1 is binary function whose value may be positive (1) or negative (0), t is time measured in discrete steps, T is the elemental time step, s is the threshold, and a^r_{hk} are coupling coefficients which express influences on neuron h from neuron k with time lag r. Many physiological effects such as synaptic facilitation or depression, refractoriness, and long-lasting transmitter action can be heuristically represented by appropriate definitions of these coefficients. Caianiello went on to show how one can represent learning in this model by allowing these coefficients to depend on the firing histories of the neurons. Caianiello has maintained a steady and influential production of publications over the past 25 years elaborating dynamic properties of these nets. These studies are cited in Chapter 9.

Despite the attractiveness and theoretical parsimony of the approach to information processing in neuromimes and neuronal networks initiated by McCulloch and Pitts and by Caianiello, neurobiologists and many neurobiologically oriented modelers have felt the need for more refined descriptions of the graded generator potentials in neurons and the processes underlying them, and more explicit attention to the level of electrical signaling in neuronal networks intermediate between the level of neural mechanisms and the level of psychological function. More realistic digital computer neuromimes began to appear in the 1950s and 1960s, those presented by Perkel and MacGregor being perhaps most representative.

Perkel's neuromimes, as first described in 1963–1965 and with elaborations in 1976, determine the occurrence of output spikes on the basis of the comparison of a graded generator potential with a threshold. The generator potential is determined by the linear algebraic sum of excitatory and inhibitory PSPs. The arrival of an input PSP is signaled by an abrupt rise or drop in the generator potential and a subsequent exponential decay. Threshold is infinitely high for a short period following firing to represent absolute refractoriness and then decays exponentially to a resting level to represent relative refractoriness. The model can be driven to exhibit spontaneous pacemaker firing by setting the resting value of the threshold below the resting value of the generator potential. The method of operation of the computer program is based on the concept called "interesting times," wherein the computer projects forward in time to the earliest of the next arrival times of an input PSP or the next project occurrence of a spontaneous spike according to the pacemaker mechanism described previously. Perkel has effected a number of refinements in this neuromime since it was first published in 1964. He used the neuromime to explore many questions concerning the relative timing of spike potentials in single and interconnected neurons. He bombarded the neuromime with various stochastic input trains and examined many auto- and cross-correlation histograms. Much of this work was carried out in collaboration with George Moore, George Gerstein, Jose Segundo, and others with particular reference to experimental observations in *Aplysia*. The main application of the neuromime was to the timings of firings of the pacemaker neurons in *Aplysia*. Segundo and Kohn and some others have continued the study of pacemaker interactions with this neuromime. Torras i Ginis extended the model to study learning in pacemaker neurons in *Aplysia* and crayfish. Sugano adapted Perkel's neuromime to study the effects of doublet impulse sequences in model neurons. Beginning as early as 1962 and 1963, Perkel used the neuromime in computer models of neuronal networks, as will be discussed in Chapter 9. More recently, Perkel has moved on to modeling the dendritic trees of neurons.

A representative contemporary digital computer neuromime was presented in 1974 by MacGregor. The model operates in terms of four state variables: the transmembrane potential, the threshold, an active (supraresting level) conductance to potassium, and a spiking variable. Firing is determined by comparison of the transmembrane potential with the threshold. The transmembrane potential is driven with nonlinear interactions of input conductance changes according to the model of Eccles. The threshold exhibits accommodative and rebound properties according to the model of Hill. Post-firing modulations of potassium conductance represent the properties of relative refractoriness and adaptation as described by Kernell. Four parameters determine the relative strengths of the accommodative and adaptive properties in the particular neuron being stimulated. The model produces reliable and realistic behavior with integration steps of 1 msec. It is therefore very computer-efficient.

The primary application of this neuromime has been to the fundamental generic mechanisms of spike production, accommodation, and adaptation in the triggering sections of neurons. It has been further used in conjuction with models of dendritic regions to produce realistic models of entire neurons, as will be discussed in Chapter 3, and in simulations of large interconnected networks and systems, as will be discussed in Chapter 9. The specific equations and detailed neural modeling methodology associated with this and related neural models are discussed at length in Part II.

Electronic Analog Neuromimes

Electronic analog neuromimes have been increasingly outdistanced in practicality by the digital computer neuromimes, the former being much less accessible to people outside the electrical engineering profession. Their contemporary use is restricted largely to special research projects in specially equipped laboratories and to pedagogical or historical interest. Many analog models are presented in the current literature in application to specific neural systems, as will be reviewed in Chapters 5-8 on specific neuronal systems. Moreover, the specific neuronal systems considered in those chapters are often modeled by electronic analog simulation models that are similar in methodology to these electronic neuromimes but are intended to represent entire systemic behavior rather than specific processes of pulse generation in single neurons.

The first modern computational neuromime was an electric analog published by Schmitt in 1937. Faber, Taylor, Harmon, Lewis, and MacGregor have constructed generic electronic neuromimes, as have many others in the past and present.

Harmon's electronic model simulates the essential nonlinear interactions of input synaptic conductance changes and exhibits refractoriness by modeling the threshold. Accommodation is approximated by supposing that excitation of the model produces a long slow hyperpolarizing process in addition to the faster depolarizing process. Harmon's neuromime and a similar early electronic neuromime by Lewis have been used by a number of workers to investigate basic mechanisms of neuronal interaction. For example, Harmon early applied his neuromimes to mutually inhibiting pairs of neurons. He showed that with fatique operating in the individual neurons, such a mutually inhibiting pair tend to exhibit oscillations in dominance in that there is a rhythmic oscillation in firings of bursts between the two. Wilson and Waldron used neuromimes provided by Lewis to simulate the generation of flight patterns of motoneurons of locusts. They used the concept of recurrent excitation to produce accelerated bursts and mutual inhibition as described by Harmon to produce alternations in firing by depressor and elevator motoneuron groups. Fernald, working with Gerstein, used Lewis's neuromimes to simulate auditory neurons in the cochlear nucleus of the cat. Various other applications and extensions of these early neuromime efforts continue to appear in the literature, as do presentations of similar neuromimes.

Lewis developed an extremely elaborate neuromime to simulate the Hodgkin–Huxley equations and to allow further modulation of the values of most of the critical parameters in the equations by knobs on a control panel. This neuromime was intended to expedite the study of the generality of the Hodgkin–Huxley model and its relation to fundamental membrane properties in other nerve cells besides the squid giant axon. This modeling study did show how to change parameters in the Hodgkin–Huxley model to match dynamics in several other species but ran into difficulty when trying to simulate accurately the ongoing group activity of a number of interconnected neurons in an invertebrate cardiac ganglion.

MacGregor's electronic neuromimes were intended to help visualize ongoing activity in small pools of neurons. The neuromimes, of which there were 50, were plugged simultaneously into a 4-ft-square power board. The neuromimes consisted of an input triggering section and an output axon-delay line element. The axon elements exhibited green or red lights depending on whether they were excitatory or inhibitory. All elements were equipped with patches of conducting sponge rubber by means of which functional connections could be effected by metal-tipped wires. One could then stand in front of the board and adjust the circuitry being simulated by plugging in or rearranging wires in the various elements. This neuromime system was used in studies attempting to visually observe coherent patterns in internally sustained activity in recurrently connected pools of neurons. The elements simulated Eq. (14.1) of Part II.

On balance, although the electronic neuromime approach has generated some understanding at the level of generic unitary interactions, it seems at present that digital simulation offers prospective modelers much greater availability, accessibility, flexibility, and adaptability.

Continuous Mathematical Neuromimes

Mathematical neuromimes are generally less accessible to less mathematically oriented researchers. They are also less flexible and less adaptable than digital computer neuromimes. Nonetheless, mathematical neuromimes provide austere delineation of fundamental principles of operation which are often buried and lost in the detail of brute computer simulation. The first mathematical neuromime was published by Rashevsky in 1933. This stream of work continues to the present day and has included notable contributions by Hill, Hodgkin and Huxley, and Kernell, as we have seen, and by Bonhoffer, Fitzhugh, Tuckwell, Holden, and others.

Rashevsky's model published in 1933 is very similar to the model presented for accommodation in 1936 by Hill. The only significant difference is that in Rashevsky's model the threshold is driven up by the applied current rather than by the potential as in Hill's model.

After Hill, the next significant model in this stream was that presented by Hodgkin and Huxley as discussed previously. A number of modelers have performed various mathematical manipulations based on the Hodgkin–Huxley model. Fitzhugh reformulated the Hodgkin–Huxley model in terms of a dynamical systems theory approach based on a state-space representation of two variables, "excitation" and " recovery," in analogy to a van der Pol oscillator. Fitzhugh described the characteristics of recurrent oscillations and stability of this model.

Game has extended the Fitzhugh model to include a mechanism of inactivation and compared his results with the Hodgkin–Huxley and Fitzhugh models. In addition, Holden, Yoga, Hayden, and Winlow have continued mathematical exploration of the Hodgkin–Huxley model and its generalizations. The focus is on the response of the membrane potential to activation and inactivation gating variables (analogous to the *n, m,* and *h* functions in the Hodgkin–Huxley model). These researchers describe the dynamic properties of their model in terms of concepts from chaos theory (see Chapter 9) such as multiple equilibria, "erratic behavior," endogenous paroxysmal depolarizing shifts, and complex multiple spikes, in addition to stable and unstable oscillations and repetitive discharge properties. Holden and colleagues emphasize that much intraneuronal information processing is the result of membrane conductance properties rather than electronic spatial properties.

Stochastic Neuromimes

An intriguing subpopulation of neuron models has attempted to describe the characteristics of spike generation in neurons in terms of concepts from stochastic systems theory. The earliest such model was presented in 1964 by Gerstein, and another seminal contribution by Stein appeared in 1965. Subsequent significant contributions have been made by Fineberg, Gluss, Ricciardi, Cowan, Wilbur and Rinzel, Tuckwell, Smith and Smith, and others.

A central goal of this field is to predict interspike interval distributions in terms of parameters characterizing input interactions and membrane dynamics so as, for example, to be able to make inferences concerning these parameters in specific neurons from their observed interspike interval histograms. Gerstein's model, which is based on a simplified random walk analog for spike generation, exhibits considerable success in this direction by showing that the interspike interval distribution might be approximated by

$$I(a, b, t) = z \frac{e^{-[(a/t)+bt)]}}{t^{3/2}} \tag{2.6}$$

In this model, *a* is associated with the threshold and *b* with the difference in rates between excitatory and inhibitory PSPs (EPSPs and IPSPs). The fundamental concepts in this field are clean and elegantly illustrated by Gerstein's model. However, the accurate mathematics are complicated and have not yet lent themselves to clean general solution. Researchers have focused on various limiting cases.

The models discussed in this chapter, which focus on the mechanisms of generation of transmembrane signals and action potentials in single patches of membrane or point neurons, cannot be complete within themselves. Thus, the information processing characteristics of a neuron can be satisfactorily considered only in terms of the networks and systems of which that neuron is a part. The concept of massively parallel processing in neuronal networks, as discussed in Chapter 9, is bringing into question the significance of temporal firing patterns in neurons as considered in terms of the networks of which they are constituent elements. Even more limiting, the information processing characteristics of any given neuron are influenced, often to a large degree, by the spatiotemporal processing in its dendritic trees. We turn in Chapter 3 to models describing these processes.

Bibliography

Models of Single-Neuron Physiology
1986   Smith, C. E., and Goldberg, J. M. A stochastic afterhyperpolarization model of repetitive activity in vestibular afferents. *Biol. Cybernet.* **54**, 41–51.
1979   Baldissera, F., and Parmiggiani, F. Afterhyperpolarization conductance time-course and repetitive firing in a motoneurone model with early inactivation of the slow potassium conductance system. *Biol. Cybernet.* **34**, 233–240.
1975   Jack, J. J. B., Noble, D., and Tsien, R. W. "Electric Current Flow in Excitable Cells." Oxford University Press (Claredon), London and New York.
1974   MacGregor, R. J., and Oliver, R, M. A model for repetitive firing in neurons. *Biol. Cybernet.* **16**, 53–64.
1972   Noble, D. Conductance mechanisms in excitable cells. *Biomembranes* **3**, 427–447.
1971   Connor, J. A., and Stevens, C. F. Prediction of repetitive firing behavior from voltage clamp data on an isolated neuron soma. *J. Physiol. (London)* **213**, 31–54.
1970   Sabah, N. H., and Spangler, R. A. Repetitive response of the Hodgkin–Huxley model for the squid giant axon. *J. Theor. Biol.* **29**, 155–172.
1968   Kernell, D. The repetitive impulse discharge of a simple neuron model compared to that of spinal motoneurons. *Brain Res.* **11**, 685–687.
        Kernell, D., and Sojoholm, H. Repetitive impulse firing comparisons between neurone models based on 'voltage clamp equation' and spinal motoneurones. *Acta Physiol. Scand.* **74**, 319–330.
1966   Noble, D. Applications of Hodgkin–Huxley equations to excitable tissues. *Physiol. Rev.* **46**, 1–50.
1965   Frankenhauser, B., and Valbo, H. Accommodation in myelinated nerve fibers of *Xenopus laevis* as computed on the basis of voltage clamp data. *Acta Physiol. Scand.* **63**, 1–20.
        Lewis, E. R. Neuroelectric potentials derived from an extended version of the Hodgkin–Huxley model. *J. Theor. Biol.* **10**, 125–158.
1964   Eccles, J. C. "The Physiology of Synapses." Academic Press, New York.
1963   Frankenhauser, B. A quantitative description of potassium currents in myelinated nerve fibers of *Xenopus laevis. J. Physiol. (London)* **169**, 424–430.
1962   Noble, D. A modification of the Hodgkin–Huxley equations applicable to Purkinje fiber action and pace-maker potentials. *J. Physiol. (London)* **160**, 317–352.
1961   Bradley, K., and Somjen, G. G. Accommodation in motoneurons of the rat and the cat. *J. Physiol. (London)* **156**, 75–92.
1960   Frankenhauser, B. Sodium permeability in a toad nerve and in squid nerve. *J. Physiol. (London)* **152**, 159–166.
1957   Eccles, J. C. "The Physiology of Nerve Cells." Johns Hopkins Press, Baltimore, Maryland.
1956   Bishop, G. H. Natural history of the nerve impulse. *Physiol. Rev.* **36**, 376–399.

1952   Hodgkin, A. L., and Huxley, A. F. A quantitative description of membrane current and its application to conduction and excitation in nerve. *J. Physiol.* (London) **117**, 500–544.
1950   Goldman, D. E. Potential, impedance, and rectification in membranes. *J. Gen. Physiol.* **27**, 37–60.
1936   Hill, A. V. Excitation and accommodation in nerve. *Proc. R. Soc. London, Ser. B* **119**, 305–355.
1912   Bernstein, J. "Elecktrobiologie." Fraiedr. Vieweg, Braunschweig.
1908   Nernst, W. Zur Theorie des elecktrischen Reizes. *Arch. Gesamte Physiol. Menschen Tiere* **122**, 275–314.

Simplest Neuromimes

1961   Caianiello, E. R. Outline of a theory of thought-processes and thinking machines. *J. Theor. Biol.* **2**, 204–235.
1945   McCulloch, W. S., and Pitts, W. A logical calculus of the ideas immanent in nervous activity. *Bull. Math. Biophys.* **5**, 115–133.

Digital Computer Neuromimes

1985   Torras i Genis, C. Pacemaker neuron model with plastic firing rate: Entrainment and learning ranges. *Biol. Cybernet.* **52**, 79–91.
1984   Sugano, N. Effect of doublet impulse sequences on the transient and steady responses in the computer-simulated nerve cell. *Biol. Cybernet.* **50**, 123–128.
1983   deBruin, G. Ypey, D. L., and Meerwijk, W. P. M. Synchronization in chains of pacemaker cells by phase resetting action potential effects. *Biol. Cybernet.* **48**, 175–186.
1981   Mulloney, B., Perkel, D. H., and Budelli, R. W. Motor-pattern production: Interaction of chemical and electrical synapses. *Brain Res.* **229**, 25–33.
        Sequndo, J. P., and Kohn, A. F. A Model of excitatory synaptic interactions between pacemakers. Its reality, it generality, and the principles involved. *Biol. Cybernet.* **40**, 113–126.
1974   MacGregor, R. J., and Oliver, R. M. A model for repetitive firing in neurons. *Biol. Cybernet.* **16**, 53–64.
1965   MacGregor, R. J. "Input-Output Relations for Axo-Somatic Activation in a Neuron Model," P-3672. The RAND Corp., Santa Monica, California.
        Perkel, D. H. Applications of a digital computer simulation of a neural network. *In "Biophysics and Cybernetic Systems"* (M. Maxfield, A. Callahan, and L. J. Fogel, eds.) Spartan Books, Washington, D.C.
1964   Perkel, D. H. "A Digital Computer Model of Nerve-cell Functioning," RM-4132-NIH. The RAND Corp., Santa Monica, California.
        Perkel, D. H., Schulman, J. H., Bullock, T. H., Moore, G. P., and Segundo, J. P., Pacemaker neurons: Effects of regularly spaced synaptic input. *Science* **145**, 61–63.

Electronic Neuromime Models

1983   Bruckstein, A. M., and Zeevi, Y. Y. Demodulation methods for an adaptive neural encoder model. *Biol. Cybernet.* **49**, 45–53.
1982   Yoshizawa, S., Osada, H., and Nagumo, J. Pulse sequences generated by a degenerate analog neuron model. *Biol. Cybernet.* **45**, 23–33.
1981   Mitchell, C. E., and Friesen, W. O. A neuromime system for neural circuit analysis. *Biol. Cybernet.* **40**, 127–137.
        Zeevi, Y. Y., and Bruckstein, A. M. Adaptive neural encoder model with selfinhibition and threshold control. *Biol. Cybernet.* **40**, 79–92.
1977   MacGregor, R. J. *In "Neural Modeling"* (R. J. MacGregor and E. R. Lewis, eds.), pp. 220–224. Plenum, New York.
        Scott, P. D. A simple hardware model neuron. *J. Physiol. (London)* **269**, 19–20.

1973   MacGregor, R. J., and Oliver, R. M. A general-purpose electronic model for arbitrary configurations of neurons. *J. Theor. Biol.* **38,** 527–538.

1972   Nagumo, J., and Sato, S. On a response characteristic of a mathematical neuron model. *Biol. Cybernet.* **10,** 155–164.

Rosen, M. J. A theoretical neural integrator. *IEEE Trans. Biomed. Eng.* **BME-19,** 362–367.

1971   Fernald, R. A neuron model with spatially distributed synaptic input. *Biophys. J.* **11,** 323.

1968   Lewis, E. R. An electronic model of neuroelectric point processes. *Biol. Cybernet.* **5,** 30–46.

Lewis, E. R. Using electronic circuits to model simple neuroelectric interactions. *Proc. IEEE* **56,** 931–949.

Wilson, D. M., and Waldron, I. Models for the generation of the motor output pattern in flying locusts. *Proc. IEEE* **56,** 1058–1064.

1964   Harmon, L. D. Neuromines: Action of a reciprocally inhibitory pair. *Science* **146,** 1323–1325.

1962   Nagumo, J., Arimoto, S., and Yoshizawa, S. An active pulse trasmission line simulating nerve axon. *Proc. I.R.E.* **50,** 2061–2070.

1961   Harmon, L. D. Artificial neuron. *Science* **129,** 962–963.

Harmon, L. D. Studies with artificial neurons. I. Properties and functions of an artificial neuron. *Biol. Cybernet.* **1,** 89–101.

1940   Fabre, P. Retour sur un modéle du nerf. 1. *Arch. Intern. Physiol..* **1,** 12–32.

Fabre, P. Retour sur un modèle du nerf. 2. *Arch. Intern. Physiol.* **1** 185–196.

1937   Schmitt, O. H. An electrical theory of nerve impulse propagation. *Am. J. Physiol.* **119,** 399.

Schmitt, O. H. Mechanical solution of the equations of nerve impulse propagation. *Am. J. Physiol.* **119,** 399–400.

Mathematical Neuromimes: Continuous

1984   Chay, T. R. Abnormal discharges and Chaos in a neuronal model system. *Biol. Cybernet.* **50,** 301–311.

Luzader, S. Multiple pulse propagation in a Fitzhugh–Nagumo nerve model. *J. Theor. Neurobiol.* **3,** 25–40.

1983   Holden, A. V., Haydon, P. G., and Winlow, W. Multiple equilibria and exotic behavior in excitable membranes. *Biol. Cybernet.* **46,** 167–172.

Matsumoto, G., and Shimizu, H. Spatial coherence and formation of collectively-coupled local nonlinear oscillators in squid giant axons. *J. Theor. Neurobiol.* **2,** 29–46.

Mironov, S. L. Theoretical model of slow-wave membrane potential oscillations in molluscan neurons. *Neuroscience* **10,** 899–905.

1982   Game, C. J. A. BVP models: An adjustment to express a mechanism of inactivation. *Biol. Cybernet.* **44,** 223–229.

1981   Holden, A. V., and Yoda, M. Ionic channel density of excitable membranes can act as a bifurcation parameter. *Biol. Cybernet.* **42,** 29–38.

Holden, A. V., and Yoda, M. The effects of ionic channel density on neuronal function. *J. Theor. Neurobiol.* **1,** 60–81.

Plant, R. E. Bifurcation and resonance in a model for bursting nerve cells. *J. Math. Biol.* **15,** 15–32.

1980   Holden, A. V. Autorhythmicity and entrainment in excitable membranes. *Biol. Cybernet.* **38,** 1–8.

Rinzel, J., and Miller, R. N. Numerical calculation of stable and unstable periodic solutions to the Hodgkin–Huxley equations. *Math. Biosci.* **9,** 27–59.

1979   Bell, J., and Cook, L. P. A model of the nerve action potential. *Math. Biosci.* **46,** 11–36.

1978   Hassard, B. Bifurcation of periodic solutions of the Hodgkin-Huxley model for the squid giant axon. *J. Theor. Biol.* **71**, 401-420.

Reauch, J., and Smollier, J. Qualitative theory of the Fitzhugh-Nagumo equations. *Adv. Math.* **27**, 12-44.

Rinzel, J. On repetitive activity in nerve. *Fed. Proc., Fed. Am. Soc. Exp. Biol.* **37**, 2793-2802.

1977   Carpenter, G. A. Periodic solutions of nerve impulse equation. *Math. Anal. Appl.* **58**, 152-173.

Sirovitch, L., and Knight, B. On subthreshold solutions of the Hodgkin-Huxley equations. *Proc. Natl. Acad. Sci. U.S.A.* **74**, 5199-5202.

1976   derHeiden, U. Existance of periodic solutions of a nerve equation. *Biol. Cybernet.* **21**, 37-39.

Parnas, I, Hochstein, S., and Parnas, H. Theoretical analysis of parameters leading to frequency modulation along an inhomogeneous axon. *J. Neurophysiol.* **39**, 909-923.

Plant, R. E., and Kim, M. Mathematical description of a bursting neuron by a modification of the Hodgkin-Huxley equations. *Biophys. J.* **16**, 227-244.

Spira, M. E., Yarom, Y., and Parnas, I. Modulation of spike frequency by regions of special axonal geometry and by synaptic inputs. *J. Neurophysiol.* **39**, 882-899.

Troy, W. C. Bifurcation phenomena in Fitzhugh's nerve conduction equations. *J. Math. Anal. Appl.* **54**, 678-690.

1975   Adelman, W. J., and Fitzhugh, R. Solutions of the Hodgkin-Huxley equation modified for potassium accumulation in a periaxonal space. *Fed. Proc., Fed Am. Soc. Exp. Biol.* **34**, 1322-1329.

1974   Goldstein, S. S., and Rall, W. Changes of action potential shape and velocity for changing ore conductor geometry. *Biophys. J.* **14**, 731-757.

Moore, J. W., and Ramon, F. On numerical integration of the Hodgkin and Huxley equations for a membrane action potential. *J. Theor. Biol.* **45**, 249-273.

Troy, W. C. Oscillation phenomena in the Hodgkin-Huxley equations. *Proc.-Soc. Edinburgh,* **74**, 299-310.

1973   Rinzel, J., and Keller, J. B. Travelling wave solutions of a nerve conduction equation. *Biophys. J.* **13**, 1313-1337.

1972   Sabah, N. H., and Leibovic, K. N. The effect of membrane parameters on the properties of the nerve impulse. *Biophys. J.* **12**, 1132-1143.

1970   McKean, H. P. Nagumo's equation. *Adv. Math* **4**, 209-223.

1969   Sabah, N. H., and Leibovic, K. N. Subthreshold oscillatory response of the Hodgkin-Huxley cable model for the squid giant axon. *Biophys. J.* **9**, 1206-1222.

1966   Cooley, J., and Dodge, F. Digital computer solutions for excitation and propagation of the nerve impulse. *Biophys. J.* **6**, 583-599.

1965   Cooley, J., Dodge, F., Cohen, H. Digital computer solutions for excitable membrane models. *J. Cell. Comp. Physiol.* **66**, Suppl. 2, 99-110.

1962   Fitzhugh, R. Computation of impulse initiation and saltatory conduction in a myelinated nerve fiber. *Biophys. J.* **2**, 11-21.

1961   Fitzhugh, R. Impulses and physiological states in theoretical models of nerve membrane. *Biophys. J.* **1**, 445-466.

1960   Fitzhugh, R. Thresholds and plateaus in the Hodgkin-Huxley nerve equations. *J. Gen. Physiol.* **43**, 867-896.

Rashevsky, N. (1960). "Mathematical Biophysics: Physico-Mathematical Foundations of biology," 3rd ed. Dover, New York.

1955   Fitzhugh, R. Mathematical models of threshold phenomena in the nerve membrane. *Bull. Math. Biophys.* **17**, 257-278.

1952   Hodgkin, A. L., and Huxley, A. F. (1952). A quantitative description of membrane current and its application to conduction and excitation in nerve. *J. Physiol. (London)* **117**, 500–544.

1948   Bonhoeffer, K. F. Activation of passive iron as a model for excitation of nerve. *J. Gen. Physiol.* **32**, 69–91.

1938   Rashevsky, N. Mathematical Biophysics: Physico-Mathematical Foundations of Biology, 1st ed. Dover, New York.

1926   Van der Pol, B. On relaxation oscillations. *Philos. Mag. [7]* **2**, 978.

Stochastic Models of Single Neurons

1984   Smith, C. E., and Smith, M. V. Moments of voltage trajectories for Stein's model with synaptic reversal potentials. *J. Theor. Neurobiol.* **3**, 67–77.

Tuckwell, H. C. Neuronal response to stochastic stimulation. *IEEE Trans. Syst., Mem, Cybernet.***SMC-14**, 464–469.

Tuckwell, H. C., Wan, F. Y. M., and Wong, Y. S. The interspike interval of a cable model neuron with white noise input. *Biol Cybernet.* **49**, 155–167.

1983   Hanson, F. B., and Tuckwell, H. C. Diffusion approximations for neuronal activity including synaptic reversal potentials. *J. Theor. Neurobiol.* **2**, 127–153.

Lansky, P. Selective interaction models of evoked neuronal activity. *J. Theor. Neurobiol.* **2**, 173–183.

Tuckwell, H. C., and Walsh, J. B. Random currents through nerve membranes. *Biol. Cybernet.* **49**, 99–110.

Wilbur, W. J., and Rinzel, J. A theoretical basis for large coefficient of variation and bimodality in neuronal interspike interval distributions. *J. Theor. Biol.* **105**, 345–368.

1982   Vasudevan, R., and Vittal, P. R. Interspike interval density of the leaky integrator model neuron with a Pareto distribution of PSP amplitudes and with simulation of refractory time. *J. Theor. Neurobiol.* **1**, 219–227.

Wan, Y. M., and Tuckwell, H. C. Neuronal firing and input variability. *J. Theor. Neurobiol.* **1**, 197–218.

Wilbur, W. J., and Rinzel, J. An analysis of Stein's model for stochastic neuronal excitation. *Biol. Cybernet.* **45**, 107–114.

1981   Thompson, R. S., and Gibson, W. G. Neural model with probabilistic firing behavior. I. General considerations. *Math. Biosci.* **56**, 239–253.

Thompson, R. S., and Gibson, W. G. Neural model with probabilistic firing behavior. II. One- and two-neuron networks. *Math. Biosci.* **56**, 255–285.

Walsh, J. B. A stochastic model of neuronal response. *Adv. Appl. Prob.* **13**, 231–281.

1979   Skaugen, E., and Walloe, L. Firing behavior in a stochastic nerve membrane model based on the Hodgkin–Huxley equations. *Acta Physiol. Scand.* **107**, 343–363.

Tuckwell, H. C. Synaptic transmission in a model for stochastic neural activity. *J. Theor. Biol.* **77**, 65–81.

Wan, F. Y. M., and Tuckwell, H. C. The response of a spatially distributed neuron to white noise current injection. *Biol. Cybernet.* **33**, 36–59.

White, B. C., and Ellias, S. A stochastic model for neuronal spike generation. *SIAM J. Appl. Math.* **37**, 206–233.

1978   Tuckwell, H. C., and Richter, W. Neuronal interspike time distributions and the estimation of neurophysiological and neuroanatomical parameters. *J. Theor. Biol.* **71**, 167–183.

1977   Levine, M. W., and Schefner, J. M. A model for the variability of interspike intervals during sustained firing of a retinal neuron. *Biophys. J.* **19**, 241–252.

Sampath, G., and Srinivasan, S. K. "Stochastic Models for Spike Trains of Single Neurons," Lect. Notes in Biomath. No. 16. Springer-Verlag, Berlin and New York.

1976   Clay, J. R. A stochastic analysis of the graded excitatory response of nerve membrane. *J. Theor. Biol.* **59**, 141–158.

Holden, A. V. "Models of the Stochastic Activity of Neurones," Lect. Notes Biomath. No. 12. Springer, New York.

Srinivasan, S. K., and Sampath, G. On a stochastic model for the firing sequence of a neuron. *Math. Biosci.* **30**, 305-323.

1973   Clay, J. R., and Goel, N. S. Diffusion models for firing of a neuron with varying threshold. *J. Theor. Biol.* **39**, 633-644.

1972   Feinberg, S. E. Stochastic models for single neurone firing trains: A survey. *Biometrics* **30**, 399-427.

Feinberg, S. E., and Hochman, H. G. Modal analysis of renewal models for spontaneous single neuron discharges. *Biol. Cybernet.* **11**, 201-207.

Skaugen, E. Firing behavior in stochastic nerve membrane models with different pore densities. *Acta Physiol. Scand.* **108**, 49-60.

1970   Feinberg, S. E. A note of the diffusion approximation for single neuron firing problems. *Biol. Cybernet.* **6**, 227-229.

1967   Gluss, B. A model for neuron firing with exponential decay of potential resulting in diffusion equations for probability density. *Bull. Math. Biophys.* **29**, 233-243.

Johannesma, P. I. M. Diffusion models for the stochastic activity of neurons. *In "Neural Networks"* (E. R. Caianiello, ed.), pp. 116-144. Springer-Verlag, Berlin and New York.

Stein, R. B. Some models of neuronal variability. *Biophys. J.* **7**, 38-68.

1966   Geisler, C. D., and Goldberg, I. J. M. A stochastic model of the repetitive activity of neurons. *Biophys. J.* **6**, 53-69.

1965   Stein, R. B. A theoretical analysis of neuronal variability. *Biophys. J.* **5**, 173-194.

1964   Gerstein, G. L., and Mandelbrot, B. Random walk models for the spike activity of a single neuron. *Biophys. J.* **4**, 41-68.

3

Single Neurons: Dendritic Trees

This chapter is concerned with the influence of spatial characteristics on information processing in single neurons. In order to obtain a realistic representation and understanding of the dynamics of entire neurons it is necessary to give accurate descriptions of the interactions and propagations of signals in dendritic trees and axonal regions. Realistic models of entire neurons must make use of both material of this chapter and material of Chapter 2. The fundamental approach discussed in this chapter is applicable to both dendrites and axons but is discussed primarily in application to dendrites.

The basic modern approach to the influence of spatial characteristics in single neurons is that of the core conductor theory, wherein one considers a neuron, aside from the soma, to consist of an interconnected collection of individual segments each of which may be considered as a longitudinal element whose length is considerably greater than its radius. In neuronal dendrites and axons electrical current tends to flow inside the core parallel to the cylinder axis for considerable distances before a significant fraction can leak out across the cell membrane. This is because the electrical resistance to current flow for reasonable longitudinal distances is much less than the resistance to electrical current flow transversely across the membrane. The great advantage of this characteristic is that it permits reliable modeling of conduction processes in dendrites and axons by considering only one spatial dimension, namely distance longitudinally along the core conductor. Theoretical studies by Rall and Shepherd of three-dimensional effects in neurons have shown that the one-dimensional cable theory approach to core conduction provides excellent accuracy for most situations of interest. However, significant three-dimensional effects can occur close to point sources of current and in extracellular volumes where the distribution of potential is of primary importance.

In the second century Galen postulated that a force was transmitted through nerves, but he was uncertain whether the force was material or immaterial and whether the nerves were hollow or solid. In the seventeenth century Descartes suggested that nerves were hollow tubes which conducted fluids or vapors from the brain, which in turn constituted a sort of central reservoir. In the eighteenth century Haller proposed that a nerve might be composed of a long row of spheres each in contact with its neighbors so that a sharp rap on the first sphere would be conducted to the last sphere almost immediately. Newton proposed that nerves were translucent and propagated signals as optical vibrations. The modern idea that nerves operate in terms of electrical signals was established over the period from the late eighteenth century to the mid-nineteenth century, with contributions from Lord Cavendish, John Walsh, Galvani, Volta, and duBois-Reymond. duBois-Reymond theorized that signals were conducted in nerves by polarizable elements ("Molenken") that aligned themselves longitudinally under the influence of applied longitudinal current. This was thought to be analogous to the magnetization of magnetic materials by externally applied magnetic fields.

The modern concept of core conduction in nerves was first proposed and demonstrated by Matteucci in 1863. The idea was further developed and advocated by Herman from 1870 through 1900. The core conductor concept has been applied in this century to axonal conduction by Lillie (the "iron-wire" model), by Bonhoffer and Fitzhugh (state-variable models), and in the current literature by Tuckwell, Holden, and others, as discussed in Chapter 2.

The second and more widely influential stream of core conductor application can be labeled the cable theory stream. This approach has been applied largely although not exclusively to dendrites and dendritic regions and is generally based on rather rigorous mathematical formalism. The first contriubtion in this field was that of Weber, who applied classical mathematical physics (application of Bessel functions for the steady flow of electricity in a cylindrical coordinate system) to a three-dimensional representation of the core conductor model expounded at that time by his colleague, Herman. In 1855 Thomson (later to become Lord Kelvin) made public his exposition of cable theory, wherein longitudinal conduction in thin cylindrical elements was described in terms of one spatial dimension and time. Thomson discussed the application of this theory to underwater transatlantic cables. Around 1900 several investigators recognized the applicability of Thomsons's theory to core conduction in neurons. Virtually all of the important contributions to the understanding of electrical signal processing in dendrites in the twentieth century have been based on this approach. Important reviews of these contributions are provided by Davis and Lorente de No, Taylor, Harmon and Lewis, and Rall. Wilfred Rall, in particular, has contributed greatly to this development in the past quarter century.

Cable Theory Model of Core Conduction

The fundamental mathematical statement of cable theory as applied to neuronal dendrites and axons is shown in Eq. (3.1)

$$\frac{1}{2\pi r(R_i + R_e)} \frac{\partial^2 E}{\partial x^2} = C \frac{\partial E}{\partial t} + GE + \sum g_i (E - E_i) + I \qquad (3.1)$$

Here E is the transmembrane potential, x the longitudinal distance, and t the time, r the radius of the core conductor, R_e and R_i the external and internal resistances per unit length, C and G, respectively, the capacitance and conductance per unit area of the membrane surrounding the core conductor, g_i any conductances per unit area active in the area in addition to the resting conductance G, E_i the equilibrium potentials associated with the g_i, hr and I the current applied to the segment by an external source. Physically, Eq. (3.1) states that the change in longitudinal electrical current is equal to whatever current leaks out of the membrane in that differential length, given by the conventional capacitive and conductance terms, plus whatever current is applied externally in that segment. (This can readily be seen by recognizing that since the first longitudinal derivative of E is proportional to the longitudinal current, the longitudinal derivative of this represents the change in that longitudinal current.) The spatiotemporal development of the transmembrane potential E in a given single cylindrical segment is then analyzed by aplying Eq. (3.1) to that segment in conjunction with the appropriate values for the potential and current at the ends of the segment and the initial distribution of the potential in the segment—the so-called boundary and initial conditions. To obtain the solution for groups of segments which interconnect to form a dendritic or axonal tree, one solves Eq. (3.1) for the individual segments and matches the currents and potential values appropriately at junctions (currents are proportional to the first spatial derivatives of E) in addition to satisfying the initial and external boundary conditions.

As long as the active membrane conductances g_i are zero and the other parameters in the equation are constant, Eq. (3.1) can be analyzed with relative ease by mathematical techniques established in the eighteenth and nineteenth centuries. Indeed, in the engineering sciences and physics there are many applications of this and similar equations to diffusion, heat conduction, electrical and magnetic field distributions, etc. and volumes of well-established techniques and solutions for such boundary value problems.

The special difficulty in applying Eq. (3.1) to neurons is that the assumptions of passivity—that is, constant parameter values and zero values of the g_i—are virtually never satisfied in any situation of real interest. Thus, synaptic activation of dendritic regions involves active excursions of transmembrane conductances at the locations under the synapses ($g_i \neq 0$). Moreover, as shown by contemporary experimental research, many if not most neurons exhibit very significant and often spatially localized active membrane conductance modulations in ongoing normal function. Third, to represent ongoing information processing in an entire neuron, it is necessary to combine models of regions that are described by Eq. (3.1) with models of other regions that include the active spike-generating mechanisms, thought to reside primarily at the soma or axon hillock regions of neurons.

In general, then, comprehensive applications of cable theory to information processing in neurons have often included both mathematical analyses based on Eq. (3.1) and computer simulations which take into account the active processes that are more difficult to explicate mathematically. This combined mathematical and computational approach provides considerable insight into the understanding of electrical signal processing and information processing characteristics of dendritic regions, which are generally difficult to explore by direct experimentation.

Passive Dendritic Trees: The Models and Theories of Wilfred Rall

In the 1960s and 1970s Rall made a substantial contribution to our understanding of the dynamics of information processing in single neurons by applying mathematical and computational formulations of cable theory to passive dendritic trees. This modeling effort neglected spike-triggering regions of neurons and the possibility of active membrane conductances in dendritic regions, and assumed rather that dendritic trees exhibited passive membranes except under synapses. Rall's lead was followed by a large number of mathematically oriented modelers.

Basic Electrotonic Analysis: The Equivalent Cylinder Model

One of Rall's concepts that has been widely followed by experimentalists has been that of reducing an entire dendritic tree to an equivalent cylinder. Specifically, Rall showed that the spatiotemporal development of signals in a passive dendritic tree would be the same as that in a single equivalent cylinder provided the following conditions are satisfied: (1) at each bifurcation the 3/2 power of the diameter of the proximal cylinder equals the sum of the 3/2 powers of the diameters of the distal cylinders; (2) all dendritic terminals are at equal electrotonic distances from the origin; (3) the input current density divided by the 3/2 power of the diameter is equal at all points which are the same electrotonic distance from the soma; (4) all the terminals have the same boundary conditions.

This concept of a single equivalent cylinder is useful in predicting the responses of neurons to experimentally applied input currents. For example, if one applies a step current to the soma of a neuron and records the transmembrane potential at the soma, one can make certain quantitative inferences regarding the dendritic tree on the basis of the equivalent cylinder model. Rall showed mathematically in 1960 that for a soma attached to a single dendrite, the response in the soma to a step current applied in the soma would exhibit a time constant different from that of the neural membrane and showed experimentalists how to estimate the membrane time constant and size of the dendritic tree from such experiments.

Later Rall showed that the solution for a finite length of cable with sealed ends can be more conveniently expressed by the infinite series shown in Eqs. (3.2). The response at the soma ($x = 0$) in this case reduces to that shown in Eq. (3.2b), where the effective time constant is given by Eq. (3.2):

$$E(x, t) = \sum_{n=0}^{\infty} B_n \cos\left(n\pi \frac{x}{L}\right) \exp\left\{-\left[1 + \left(\frac{n\pi}{L}\right)^2\right]t\right\}$$

(3.2a)

$$E = C_0 e^{-t/\tau_0} + C_1 e^{-t/\tau_1} + C_2 e^{-t/\tau_2} + \ldots$$

(3.2b)

$$\tau_n = \frac{C/G}{1 + (n\pi/L)^2}$$

(3.2c)

$$C_n = B_n \cos(n\pi x/L)$$

(3.2d)

(sealed ends)

One can also estimate the effective electrotonic length of the equivalent cylinder model, which is the same as the electrotonic length of the idealized dendritic tree, by matching the observed response to the theoretical response in Eqs. (3.2). Several investigations using this approach indicate that many mammalian neurons exhibit equivalent length constants in the range of about 1.0 to 1.5. That is, the effective length of the idealized dendritic tree is a little longer then one electrotonic length constant. The quantity ρ is defined as the ratio of the combined dendritic input conductance to the soma membrane conductance. This can also be estimated for given neurons on the basis of the equivalent cylinder model. Reported values for ρ for mammalian neurons range from about 1.4 to as high as 20. Application of Rall's method to mammalian neurons has generally resulted in estimates of membrane time constants in the range of 5–20 msec.

These estimates for particular observed neurons based on the equivalent cylinder model give valuable aids to intuitive considerations of the character of electrical signals and their interactions in otherwise unobservable dendritic regions. These considerations are further enhanced by computational simulations of synaptic activations in dendritic regions.

Spatiotemporal Propagation in Passive Dendritic Trees

Rall performed simple computer experiments based on the cable equation to provide reliable quantitative estimates of the decrements in magnitude and changes in shape experienced by postsynaptic potentials (PSPs) as they propagate through passive dendritic regions. These fundamental results showed that potentials tend to exhibit slower rise and fall times and smaller peak amplitudes at locations progressively removed from their point or origin. In later simulations from the same model, Rall presented data which show investigators how to infer the dendritic origin of a PSP observed at the soma.

Rall also showed with this model that different spatiotemporal sequences of input activation could produce markedly different responses at the soma. A sequence of synaptic inputs which activates distal regions first tends to produce a noticeably higher peak and more temporally localized response at the soma than an otherwise equivalent input sequence which activates more proximal regions first.

Also, since the magnitudes of PSPs in dendritic regions tend to approach sizable fractions of the equilibrium potential of the excitatory synapse, nonlinear interactions among excitatory postsynaptic potentials (EPSPs) in dendritic regions should be much more prominent than those experienced by axosomatic excitatory synapses, where the amplitude of the ongoing generator potential is a much smaller fraction of the excitatory synaptic equilibrium potential. One implication of this is that the effects of dendritic inhibitory postsynaptic potentials (PSPs) on dendritic EPSPs and of occlusion among multiple EPSPs are much more marked when those PSP groups are activated in the same or locally neighboring regions of dendritic trees than when they are spread out in electronically separate regions of the dendritic tree.

Rall's early work then did a great deal to facilitate the understanding of interactions and processing in dendritic regions. It led to what might now be considered a classic picture of the integrative properties of passive dendritic regions in neurons. Basically, this picture suggests that in passive dendritic regions much of axodendritic synaptic activation serves primarily as a background signal from which fine-grained discriminability is absent or minimal, whereas axosomatic synaptic activation can more readily sustain such spatial and temporal discriminability. Axodendritic activation may modulate the sensitivity of the neuron to the spatiotemporal patterns applied at either excitatory of inhibitory axosomatic synapses.

Many investigators including Rall and his many collaborators have continued to refine and amplify the formal cable theory approach to electrical signal processing and information processing in dendritic regions of neurons. Representative methodological contributions are those provided by Brown, Butz, Cowan, Jack, Johnston, Norman, Redman, Rinzel, Tuckwell, Turner, and others. Fox has shown how to infer the location of synaptic input conductances from somatic observations based on cable theory. Barnes and McNaughton showed that senescent rats exhibit dendritic atrophy but that the physiological effects on PSPs seem insignificant.

Rall's methodology has been applied to particular neuron types by a number of investigators, as may be seen in the contemporary pages of many leading neuroscience journals. Some of these studies on specific neuronal systems will be referred to in Chapters 5–8. The most significant contemporary applications of cable theory are to active conductances in dendritic regions and to dendritic spines.

Models of Active Conductances in Dendrites

It was recognized as early as 1960 that vertebrate neuronal membrane exhibits active processes beyond those involved in action potential generationn and chemical synaptic transmisison. A great deal of experimental work largely in the past five years has established and elaborated these processes. The bulk of the studies have been on pyramidal neurons of the hippocampus, but similar processes have been observed in other neurons as well, and indeed such processes may be widespread throughout the nervous system.

The complete picture involving these active membrane conductances is not yet established. There is a good review of the current status of the experimental work by Crill and Schwindt. Essentially, there appears to be a commonly occurring calcium conductance change in localized regions and an associated calcium-mediated long-lasting increase in potassium conductance. A simplified picture that is perhaps not too far wrong is that in the dendritic regions of some neurons there occur actively mediated pseudo-action potentials which are fundamentally similar to the Hodgkin–Huxley action potentials mediated by the axonal regions. They are similar in that there is a threshold level of potential required for their activation and that they are mediated by intrinsic transitory excitatory and inhibitory conductance changes sequenced to produce a transient depolarization followed by a transient hyperpolarization. The pseudo-action potentials differ from Hodgkin–Huxley action potentials in that they are slower (the depolarizing wave lasts from 5 to 15 msec; the afterhyperpolarization lasts from 50 to

150 msec), and the excitatory conductance change is specific for calcium ions rather than sodium ions (in both types of action potentials potassium ions are the hyperpolarizing ingredient). A primary manifestation of these active membrane conductances is the occurrence of bursts of fast Hodgkin–Huxley action potentials observed at the soma. An interpretation is that the slow pseudo-action potential invades the soma from the dendritic region and causes the triggering of multiple fast Hodgkin–Huxley action potentials for propagation out of the axon. It seems that the production of such slow pseudo-action potentials or calcium-mediated waves in dendritic regions permits integration within local neighborhoods of dendritic trees. The occurrence of such a slow action potential and its transmission to the soma would signal to the soma that a significantly large confluence of excitatory input had occurred within this region at this time. This would ensure that action potentials were fired at the soma in response to this event and moreover the transmission of normal Hodgkin–Huxley action potentials would communicate this information to the target cells of the neuron.

There have been two attempts to model these active membrane processes within the framework of information processing models of neuronal dynamics. An elegant model based on the Hodgkin–Huxley approach has been presented by Traub. In this model the fundamental new ingredients are an active conductance for calcium and an active slow conductance for potassium. These are mediated by new state variables s, r, and q, which are in turn governed by rate equations of the type governing the n, m, and h functions of the Hodgkin–Huxley model. Traub has taken great pains to adjust the rate functions in terms of their hypothetical dependence on membrane potential and calcium conductance in order to match the electrophysiological intricacies of CA3 neurons in the hippocampus. He has also adjusted the Hodgkin–Huxley potassium conductance term by the inclusion of the state variable y, which serves to deactivate this fast potassium conductance to match the observed properties of afterdepolarizing potentials in these neurons. Finally, he has adjusted the s and r dependences on calcium to terminate the calcium spike as observed experimentally. This representation of active conductance changes is used by Traub in a 28-compartment model for CA3 pyramidal cells. The model accurately reproduces bursts generated in either the soma or the apical dendrites by sets of conductances localized in the same respective membrane region. Traub has applied this and similar techniques to model a number of features of single-neuron and local network activity in hippocampal regions associated with epileptic seizures, rhythmicity, and ongoing information processing. We will refer again to this work in Chapters 7 and 8.

A more simplified model of these active conductances is presented in detail in Chapter 15. This model describes the activity of a patch of neuronal membrane in terms of four state variables: the transmembrane potential, the membrane active conductance to calcium, the intracellular concentration of calcium, and the membrane active conductance to potassium. An active conductance to calcium is triggered when the potential in the compartment exceeds a given threshold level. This tends to drive the potential toward the calcium equilibrium potential of about $+50$ mV and also allows the internal calcium concentration to increase. When the internal calcium concentration rises above a certain threshold value, an active hyperpolarizing conductance to potassium is activated. The various parameters in the system are adjustable to produce slow calcium-mediated depolarizing waves followed by the long-lasting potassium-mediated afterhyperpolarization. Properties of this model and its inclusion in network simulation programs are discussed in Part II.

Traub's model provides an elegant and penetrating description of the fundamental physiological properties and their quantitative characteristics in CA3 pyramidal neurons. It can be valuable in helping delineate the fundamental physiological gating processes. On the other hand, since it is based on the Hodgkin–Huxley equation system, it is subject to the weaknesses of that model at the biophysical level as discussed in Chapter 2. Morever, Traub's model requires considerable computational power, updating approximately 22 variables at 200 times per simulated millisecond. By contrast, our model provides a satisfactory description of the generic mechanisms involved, is readily transportable to other neuron types by adjusting a few parameters, and is several thousand times more computer-efficient, updating four state variables once per millisecond.

Koch has presented a mathematical approach to modeling active membrane processes wherein the active process are linearized and Laplace transforms are used.

Responses to Activation in Dendritic Spines

Many if not the majority of synapses in the central nervous system terminate on the heads of spines. Geometrically, spines can be approximated by a cylindrical shaft (neck) capped by a spherical bulb (head) on which input synapses are effected. The great interest in the dynamics of spines stems from the observation that the amplitude of responses in spines should be a strong function of the spine geometry (in particular the length and diameter of the neck), so possible modulations in spine geometry during normal function should serve to modulate synaptic strength and could therefore serve as a primary locus for the physical basis of memory in the brain. Indeed, a number of experiments have shown that spine neck geometry tends to change in association with electrophysiological activity. It appears that the change in geometry consists of an expansion of the spherical head and a corresponding shortening of the neck with the total amount of membrane in the spine remaining constant. Crick has suggested that spines might effect such structural changes very quickly and utilize them in short-term memory rather than or in addition to using them in long-term memory.

A number of mathematical and computer modeling studies of spines based on passive cable theory have confirmed and elaborated these speculations. For passive dendritic spines and attached dendrites, PSP amplitude is indeed a function of neck size and can increase significantly if the spine neck is shortened or enlarged in radius. The high input resistance associated with narrow spine necks tends to cause particularly large amplitude depolarizations in the spine heads and also tends to isolate spines electrotonically from one another. Perkel has shown that the transient peaks of PSPs in particular tend to attenuate dramatically from the heads of spines to the contiguous parent dendrite.

Interpretation of these modeling studies is complicated by the recent realization that active membrane conductances are probably ubiquitous in dendritic regions of central nervous system neurons. Perkel and Perkel have modeled the responses to transient input conductance changes in a dendritic spine containing active calcium-mediated conductances. They find that the active membrane processes in the spine can amplify the signal from spine head to parent dendrite rather than attenuate it as in passive spines. Moreover, this amplification is a nonmonotonic function of the neck size (there is an optimum neck size for maximum amplification), so that changing neck size can either increase or decrease response amplitude. It has been noted that the calcium influx associated with these active conductances can regulate enzymatic activity in the spine head that could affect phosphorylation of cytoskeletal proteins which maintain spine shape. Thus, the electrical and structural features of spines could be intimately and directly linked.

Contemporary Simulation Models of Overall Information Processing in Neurons with Dendritic Trees

The models of Traub and those presented in Chapter 21 are representative of the current state of the art in simulating overall information processing in neurons including passive or active dendritic trees. These models are based on the compartmentalized approach to the core conductor theory and contain a number of adjustable parameters so that one can produce most of the observed characteristics of dendritic regions relatively accurately. They are somewhat complicated, sometimes a bit difficult to tune, and require a considerable amount of computer power. Traub's model, as indicated above, updates approximately 22 variables 200 times per millisecond in each compartment simulated. MacGregor's model (Chapter 21) updates four state variables every millisecond from every compartment simulated. General approaches to compartmentalized modeling of dendritic trees are discussed by Perkel, Mulloney, and Budelli and in Chapter 21 of this book.

Chapters 14 and 15 will introduce a less accurate, highly simplified simulation model for overall electrical signal processing in neurons with dendritic trees which is much more computer-efficient than these more elaborate and accurate models. This model replaces core conductance through passive resistance by weighted averaging of potentials in adjacent compartments. The dendritic compartments are connected to a somatic triggering section. The model includes representation of active calcium-mediated conductance changes in the individual compartments.

Bibliography

Historical

1977   MacGregor, R. J., and Lewis, E. R., "Neural Modeling." Plenum, New York.
       Rall, W. Core conductor theory and cable properties of neurons. (J. M. Brookhart,
           V. B Mountcastle, and E. R. Kandel, eds.) Rev. ed., Sect. 1, Vol. 1, pp. 39–97. In
           "Handbook of Physiology" Am. Physiol. Soc., Bethesda, Maryland.
1966   Harmon, L. D. and Lewis, E. R., Neural modeling. *Physiol. Rev.* **46**, 513–591.
       Lewis, E. R. "A Brief review of Neural Modeling," Tech Rep. General Precision, Inc.,
           Librascope Group, Los Angeles.
1963   Taylor, R. E. Cable theory. In "Physical Techniques in Biological Research" (W. L.
           Nastuk, ed.), Vol 6, Part B, pp. 219–262. Academic Press, New York.
1947   Davis, L., and Lorente de No, R. Contribution to the methematical theory of the
           electrotonus. *Stud. Rockefeller Inst. Med. Res.* **131**, 442–496.
1936   Lillie, R. S. The passive iron wire model of protoplasmic and nervous transmission
           and its physiological analogues. *Biol. Rev. Cambridge Philos. Soc.* **11**, 181–209.
1914   Lillie, R. S. The conditions determining the rate of conduction in irritable tissues and
           especially in nerve. *Am J. Physiol.* **34**, 414–445.
1899   Hermann, L. Zur Theorie der Erregungsleitung und der elektrischen Erregung. *Arch.
           Gesamte Physiol. Menschen Tiere* **75**, 574–590.
1873   Weber, H. Ueber die stationaren Stromungen der Elektricitat in Cylindern. *J. Reine
           Angew. Math.* **76**, 1–20.
1872   Hermann, L. Ueber eine wirkung galvanischer Strome auf Muskeln und Nerven.
           Arch. Gesamte Physiol. Menschen Tiere **6**, 312–360.
1868   Matteucci, C. Recherches physico-chimiques appliquées à l'éctro-physiologie. *Hebd.
           Seances Acad. Sci.* **66**, 580–585.
1863   Matteucci, C. Sur le pouvoir électromoteur secondaire des nerfs, et son application
           à l'électro-physiologie, *Hebd. Seances Acad. Sci.* 56, 760–764.
1852   duBois-Reymond, E. "On Animal Electricity" (H. Bence-Jones, trans). Churchill,
           London.
1848   duBois-Reymond, E. "Untersuchungen über thierische elektricitat," Vol. 1. Reimer,
           Berlin.
1794   Galvani, L. "Dell'uso e dell'attivita dell'arco Connduttore nelle Contrazione dei
           Muscoli." Thommaso D'Aquino, Bologna.
1791   Galvani, L. "De Viribus Electricitatus in Motu Musculari, Commentarius. Ex Typo-
           graphia Instituti Scientarium, Bologna.

1776   Cavendish, H. An account of some attempts to imitate the effects of the torpedo by electricity. *Trans. R. Soc. London* **66**, 196–225.

1680   Borelli, A. "De Motu Animalium," Vols. 1 and 2. Bernado, Rome.

Passive Dendritic Trees

1985   Fox, S. E. Location of membrane conductance changes by analysis of the input impedance of neurons. I. Theory. *J. Neurophsiol.* **54**, 1578–1593.

Shelton, D. P. Membrane resistivity estimated for the Purkinje neuron by means of a passive computer model. *Neuroscience* **14**, 111–131.

Tuckwell, H. C. "Some Aspects of Cable Theory with Synaptic Reversal Potentials," Res. Rep. No. 121, Dept. Math., Monash University, Clayton, Australia.

Waldrop, B., and Glantz, R. M. Synaptic mechanisms of a tonic EPSP in crustacean visual interneurons: Analysis and simulation. *J. Neurophysiol.* **54**, 636–650.

Walsh, J. B., and Tuckwell, H. C. Determination of the electrical potential over dendritic trees by mapping onto a nerve cylinder. *J. Theor. Neurobiol.* **4**, 27–46.

1984   Wilson C. J. Passive cable properties of dendritic spines and spiny neutrons. *J. Neurol. Sci.* **4**, 281–297.

1983   Horwitz, B. Unequal diameters and their effects on time-varying voltages in branched neurons. *Biophys. J.* **41**, 51–66.

Johnston, D., and Brown, H. Interpretation of voltage-clamp measurements in hippocampal neurons. *J. Neurophysiol.* **50**, 464–486.

Koch, C., Poggio, T., and Torre, V. Non-linear interactions in a dendritic tree: localization, timing and role in information processing. *Proc. Natl. Acad. Sci. U.S.A.* **80**, 2799–2802.

Lev-Tov, A., Miller, J. P., Burke, R. E., and Rall, W. Factors that control amplitude of EPSPs in dendritic neurons. *J. Neurophysiol.* **50**, 399–412.

1981   Brown, T. H., Fricke, R. A., and Perkel, D. H. Passive electrical constants in three classes of hippocampal neurons. *J. Neurophysiol.* **46**, 812–827.

Horwitz, B. An analytical method for investigating transient potentials in neurons with branching dendritic trees. *Biophys. J.* **36**, 155–192.

Johnston, D. Passive cable properties of hippocampal CA3 pyramidal neurons *Cell Mol. Neurobiol.* **1**, 41–55.

1980   Turner, D. A., and Schwartrzkroin, P. A. Steady-state electrotonic analysis of intracellularly stained hippocampal neurons *J. Neurophysiol.* **44**, 184–199.

1979   Barnes, C. A., and McNaughton, B. L. Neurophysiological comparison of dendritic cable properties in adolescent, middle-aged, and senescent rats. *Exp. Aging Res.* **5**, 195–206.

Jack, J. (1979). An introduction to linear cable theory. *In* "The Neurosciences: Fourth Study Program" (F.O. Schmitt and F. G. Worden, eds.), pp. 324–437. MIT Press, Cambridge, Massachusetts.

1978   Perkel, D. H., and Mulloney, B. Electrotonic properties of neurons: Steady-state compartmental Model. *J. Neurophysiol.* **41**, 621–639.

Perkel, D. H., and Mulloney, B. Calibrating compartmental models of neurons. *Am. J. Physiol.* **235**, R93–R98.

Rinzel, J. Integration and propagation of neuroelectric signals. *In* "Studies in Mathematical Biology" (S. A. Levin, ed), pp. 1–66. Math. Assoc. Am., Washington, D. C.

1977   McCabe, G. P., and Samuels, M. L. Random censoring and dendritic trees. *Biometrics* **33**, 69–84.

Rall, W. Core conductor theory and cable properties of neurons. *In* "Handbook of Physiology" (J. M. Brookhart, V. B. Mountcastle, and E. R. Kandel, eds.), Rev. ed., Sect. 1, Vol. 1, pp. 39–97. Am. Physiol. Soc., Baltimore, Maryland.

1976   Berry, M., and Bradley, P. The application of network analysis to the study of branching patterns of large dendritic fields. *Brain Res.* **109**, 111–132.

Redman, S. J. A quantitative approach to integrative function of dendrites. *Int. Rev. Physiol.* **10**, 1–35.

Schmitt, F. O., Dev, P., and Smith, B. H. Electrotonic processing of information by brain cells, *Science* **193**, 114–120.

1975   Hollingsworth, T., and Berry, M. Network analysis of dendritic fields of pyramidal cells in neocortex and Purkinje cells in the cerebellum of the rat. *Philos. Trans. Soc. London, Ser. B* **270**, 227–264.

Jack, J. J. B., Noble, D., and Tsein, R. W. "Electric Current Flow in Excitable Cells." Oxford Univ. Press, London, and New York.

1974   Butz, E. G., and Cowan, J. D. (1974). Transient potentials in dendritic systems of arbitrary geometry. *Biophys. J.* **14**, 661–689.

Lewis, E. R. A note on transfer and driving point functions of iterated ladder networks. *IEEE Trans. Circuits Syst.* **21** (1) 321–352.

1973   Rall, W., and Rinzel, J. Branch input resistance and steady attenuation for input to one branch of a dendritic neuron model. *Biophys. J.* **13**, 648–688.

Redman, S. J. The attenuation of passively propagating dendritic potentials in a motoneurone cable model. *J. Physiol. (London)* **234**, 637–664.

1972   Norman, R. S. Cable theory for finite length dendritic cylinders with initial and boundary conditions. *Biophys. J.* **121**, 25–45.

1971   Jack, J. J. B., and Redman, S. J. An electrical description of the motoneurone and its application to the analysis of synaptic potentials. *J. Physiol. (London)* **215**, 321–352.

1969   Rall, W. Time constants and electrotonic length of membrane cylinders and neurons. *Biophys. J.* **9**, 1483–1508.

Rall, W. Distributions of potential in cylindrical coordinates and time constants for a membrane cylinder. *Biophys. J.* **9**, 1509–1541.

1968   MacGregor, R. J. A model for responses to activation by axodendritic synapses. *Biophys. J.* **8**, 305–318.

Rall, W., and Shepherd, G. M. Theoretical reconstruction of field potentials and dendrodendritic synaptic interactions in olfactory bulb. *J. Neurophysiol.* **31**, 884–915.

1967   Rall, W. Distinguishing theoretical synaptic potentials computed for different somadendritic distributions of synaptic input. *J. Neurophysiol.* **30**, 1138–1168.

1962   Rall, W. Electrophysiology of a dendritic neuron model. *Biophys. J.* **2**, 145–167.

Ramon Moliner, E. An attempt at classifying nerve cells on the basis of their dendritic patterns. *J. Comp. Neurol.* **119**, 211–227.

Rall, W. Theory of physiological properties of dendrites. *Ann. N.Y. Acad. Sci.* **96**, 1961/62, 1071–1092.

1960   Rall, W. Membrane potential transients and membrane time constant of motoneurons. *Exp. Neurol.* **2**, 503,532.

1911   Ramón y Cajál, S. (1911). "Histologie du Système Nerveux de l'Homme et des Vertèbres" (L. Asoulay, trans.). Maloine, Paris.

Active Conductances

1987   MacGregor, R. J. "Neural and Brain Modeling," Part II. Academic Press, Orlando, Florida.

1984   Koch, C. Cable theory in neurons with active, linearized membranes. *Biol. Cybernet.* **50**, 15–33.

1983   Crill, W. E., and Scwhindt, P. C. Active currents in mammalian central neurons. *Trends Neurosci.* **6**, 236–240.

1982   Traub, R. D. Simulation of intrinsic bursting in CA3 hippocampal neurons. *Neuroscience* **7**, 1233–1242.

1981   Perkel, D. H., Mulloney, B., and Budelli, R. W. Quantitative methods for predicting neuronal behavior. *Neuroscience* **6**, 823–837.

Dendritic Spines

1985   Coss, R. G., and Perkel, D. H. The function of dendritic spines: A review of theoretical issues. *Behav. Neural Bio.* **44**, 151–185.

Perkel, D. H., and Perkel, D. J. Dendritic spines: Role of active membrane in modulating synaptic efficacy. *Brain Res.* **325**, 331–335.

Pongracz, F. The function of dendritic spines: A theoretical study. *Neurscience* **15**, 933–946.

Shepherd, G. M., Brayton, R. K., Miller, J. P., Segev, I., Rinzel, J., and Rall, W. Signal enhancement in distal cortical dendrites by means of interactions between active dendritic spines. *Proc. Natl. Acad. Sci. U.S.A.* **82**, 2192–2195.

1984   Kawato, M., Hamaguchi, T., Murakami, F., and Tsukahara, N., Quantitative analysis of electrical properties of dendritic spines. *Biol. Cybernet.* **50**, 447–454.

Miller, J. P., Rall, W., and Rinzel, J. Synaptic amplification by active membrane in dendritic spines. *Brain Res.*

Turner, D. A. Conductance transients onto dendritic spines in a segmental cable model of hippocampal neurons. *Biphys. J.* **46**, 85–96.

Wilson, C. J. Passive cable properties of dendritic spines and spiny neurons. *J. Neurosci.* **4**, 281–297.

1983   Kawato, M., and Tsukahara, N. Theoretical study on electrical properties of dendritic spines. *J. Theor. Biol.* **103**, 507–522.

Koch, D., and Poggio, T. A theoretical analysis of electrical properties of spines. *Proc. R. Soc. London, Ser. B* **218**, 455–477.

Koch, C., and Poggio, T. Electrical properties of dendritic spines. *Trends Neurosci.* **6**, 80–83.

Perkel, D. H. Functional role of dendritic spines. *J. Physiol. (Paris)* **78**, 695–699.

Turner, D. A., and Schwaratzkroin, P. A. (1983). Electrical characteristics of dendrites and dendritic spines in intracellularly-stained CA3 and dentate hippocampal neurons. *Neuroscience* **3**, 2381–2394.

Wilson, C. J., Groves, P. M., Kitai, S. T., and Linder, J. C. (1983). Three-dimensional structure of dendritic spines in the rat neostriatum. *J. Neurosci.* **3**, 383–398.

1982   Boycott, B. B. Some further comments concerning dendritic spines. *Trends Neurosci.* **5**, 328–329.

Crick, T. Do dendritic spines twitch? *Trends Neursci.* **5**, 44–46.

1981   Swindale, N. V. Dendritic spines only connect. *Trends Neurosci.* **4**, 240–241.

1978   Rall, W. Dendritic spines and synaptic potency. *In* "Studies in Neurophysiology" (R. Porter, ed.), pp. 203–209. Cambridge Univ. Press, London and New York.

1974   Purpura, D. P. Dendritic spine dysgenesis and mental retardation. *Science* **86**, 1126–1128.

1970   Dimaond, J., Grey, E. G., and Yasargil, G. M. (1970). The function of the dendritic spine: An hypothesis. *In* "Excitatory Synaptic Mechanisms" (P. Anderson and J. K. S. Jansen, eds). pp. 213–222. Universitetsforlag, Oslo.

1979   Valverde, F., and Ruiz-Marcos, Dendritic spines in the visual cortex of the mouse: Introduction to a mathematical model. *Exp. Brain Res.* **8**, 269–283.

4

Biophysical and Chemical Levels of Control and Intercellular Communication

Biophysical and Chemical Control

A number of significant discoveries have markedly changed our concepts of the biophysical and chemical levels of control of neuroelectric activity.

Metabolic Control of Ionic Channels

As we have seen in Chapters 2 and 3, neuroelectric activity in individual neurons is controlled primarily by modulations of ionic membrane conductances. These are exemplified by the models of Goldman for the resting potential, of Hodgkin and Huxley for the action potential, of Eccles for chemical synaptic transmission, and of active calcium-related conductances in dendritic regions. In the classical picture, which includes all but the last-named of these effects, metabolism is not directly involved in the mediation of ionic currents but only creates the prerequisites for these currents by forming transmembrane electrochemical gradients by active ion transport and maintaining the integrity of the intra- and extracellular fluid. Much recent evidence however, supports the generalization that ionic channels whose activity is connected directly or indirectly with the transfer of calcium ions into the cell are under direct metabolic control and that this control is mediated via phosphorylation of protein components. In one case such phosphorylation seems to act as a gating mechanism by transferring the channel from the closed to the open state, whereas in another case it only creates conditions necessary for the proper gating mechanism of the channel. The calcium is critical because it is involved both in the mediation of membrane conductance modulations related to neuronal electrical

48

signaling and in the regulation of practically all intracellular structural and metabolic processes including synthesis of substances, intracellular transport, construction of cellular structures, and release of substances. Modulation of calcium concentration in neurons then can be seen as a factor directly coupling membrane electrical signaling and neuronal structural modification. We have seen reference to this in the modeling of calcium-related active conductances in dendritic spines by Perkel and Perkel and will refer to it again in Chapter 9 on models of learning.

Informational Molecular Regulators

Schmitt has documented the many new informational substances in the nervous system that are being discovered by the application of DNA technology. In addition to transmitters, these include peptides, hormones, factors, and various proteins, the most salient of which are indicated in Table 4.1. Schmitt hypothesizes that these constitute a parasynaptic system of control in parallel with the synapse-linked circuitry in the nervous system. In this parasynaptic system informational substances reach specific target cell receptors by diffusion from release points through extracellular fluids. The system is postulated to exhibit more versatility and plasticity than the synaptic circuitry but the same degree of selectivity. Moreover, it can modulate and influence the synaptic system. Such parasynaptic informational regulators might be particularly involved in the generation of global characteristics such as mood, emotion, and state of arousal. As far as we know, no quantitative models of these concepts have yet appeared.

Chemical Synaptic Transmission

The most widely studied and presumably preeminent mode of intercellular communication in the vertebrate nervous system is direct chemical synaptic activation of one neuron by another. The fundamental concept is that a given gating molecule from a presynaptic terminal of a sender neuron diffuses across a junction, combines with a specific type of receptor complex on the membrane of a given receiving cell, and causes a selective pattern of conductance changes in local subsynaptic regions of the membrane of the receiving neuron. The effects of a given transmitter may vary depending on the specific receptors with which it interacts.

The 1950s and 1960s saw detailed elaboration of the properties of chemically mediated synaptic transmission, as in the model of Eccles discussed in Chapter 2 and in the binomial quantal release model discussed below. Classical chemical transmitters discovered in these years as enumerated by Schmitt are acetylcholine, γ-aminobutyrate, aspartate, dopamine, epinephrine, glutamate, glycine, histamine, norepinephrine, and

Table 4.1
Molecular Regulators[a]

Neuropeptides	Neurohormones
Angiotensins I, II, III	Adrenocorticotropic hormonee
Bombesin	Corticotropin-releasing hormone
Bradykinin	β-Endorphin
Calcitonin	Glucagon
Carnosine	Gonadotropin-releasing hormone
Cholecystokinin	Growth hormone
Corticotropin-releasing factor	Hypothalamic-releasing hormone
Ependymin (β, γ)	Insulin
β-Endorphin	Luteinizing hormone-releasing hormone
Gastrin	Melanocyte-stimulating hormone
Glucagon	Melanocyte-stimulating hormone-releasing hormone
Gastrointestinal polypeptide	Neurohypophysial hormone
Insulin	Somatomedins
Leu- and Met-Enkephalin	Thyroid hormone
Melatonin	Thyroid-stimulating hormone
α- and β-Melanocyte-stimulationg hormone	Thyrotropin-releasing hormone
Motilin	Estrogens (female):
Neurophysin	estradiol, estrone, progesterone
Neuropeptide Y	Androgen (male): testosterone
Neurotensin	Glucocorticoids: cortisone
Oxytocin	Mineralocorticoids: aldosterone
Pancreatic polypeptides	
Physalaemin	
Pituitary peptides	
Proctolin	
Prolactin	
Secretin	
Somatomedin	
Somatostatin	
Substance P	
Thyrotropin	
Arginine vasopressin	
Lysine vasopressin	
Vasotocin	
Vasointestinal polypeptide	

[a]Adapted from F. O. Schmitt (1984). *Neuroscience* **13**. Copyright 1984, Pergamon Press, Ltd.

serotonin. It has been estimated that there may be as many as 50 classical chemical transmitters. Corresponding to this increased understanding of the identity, distribution, and modes of action of chemical transmitters has been considerable development in neuropharmacology and in the treatments of various mental conditions and diseases by pharacological agents. Also, the actions of many psychoactive drugs can be understood in terms of chemically mediated synaptic transmission. These effects can be quantitatively modeled by incorporating the postulated chemical controls or influences in the equations for the binomial quantal release model, the corresponding changes in postsynaptic membrane conductance, and the resulting influences on postsynaptic potentials as determined by cable theory and equivalent circuits as discussed in Chapters 2–4. The remainder of this section considers models for describing the postsynaptic conductance change in terms of presynaptic events.

Time Course

The time course of a postsynaptic conductance change at chemically mediated synapses is often modeled by

$$g = kte^{-\alpha t} \tag{4.1}$$

In this expression g is the postsynaptic conductance charge, α determines its time course, and k scales its amplitude. Often in discrete time simulations of neuronal activity, postsynaptic conductance change is taken as a pulse function, being on at some specified level during the occurrence of presynaptic action potentials and zero at other times.

Quantal Model

In the 1950s Katz, with collaborators Fatt and later del Castillo, established that chemical synaptic transmission in the neuromuscular junction of the frog is a stochastic phenomenon wherein clusters of individual quanta of transmitter are released by the presynaptic terminal. Further investigations (of the spinal cord of cats by Kuno and Martin and of many other preparations, including other neuromuscular junctions in vertebrates and invertebrates, certain peripheral synapses in the sympathetic ganglia, central synapses of the spinal cord of the frog and the cat, the squid, and *Aplysia*) have firmly established the stochastic quantal model of transmitter release. In this model n quanta of transmitter in the presynaptic terminal are available for release. Spontaneous release of transmitter may occur in the absence of electrical activity in the presynaptic terminal. When an action potential arrives at the presynaptic terminal, molecular mechanisms (presumably involving extracellular calcium) tend to release each of the transmitter quanta with the same probability p. Thus, the number of quanta released in response to a given action potential is a random variable x, where the probability distribution of x is given by

$$f(x) = \frac{x!}{n!(n-x)!} p^x (1-p)^{n-x}, \qquad x = 0, 1, 2, ..., n \tag{4.2}$$

$$m = np$$

$$\sigma = \sqrt{np(1-p)}$$

This is the standard binomial distribution which results from n independent Bernoulli trials, each with probability of success p. The mean of the distribution is $m = np$ and the standard deviation is σ.

In applying this model to electrophysiological experiments, one often refers to the mean number, m, of quanta released by presynaptic action potentials as the "quantal content." The peak amplitude of the postsynaptic potential associated with m is sometimes called the "mean quantal amplitude." Observed values for m and p vary considerably from different types of synapses. For example, m varies from about 1–3 in the synapses of the spinal cord and sympathetic ganglia of vertebrates to about 100–300 in vertebrate neuromuscular synapses, squid giant synapses, and *Aplysia* central synapses; n varies from about 3 at single terminals of the crayfish to about 1000 at vertebrate neuromuscular synapses; and p ranges from about 0.15 to 0.5. When n is large and p is very small, as is sometimes the case, the binomial model can be approximated by the Poisson model, which has one parameter, m, as shown in Eq. (4.3)

$$f(x) = (1/m)\,e^{-x/m} \tag{4.3}$$

The binomial model appears to be an accurate representation of chemical synaptic transmission and is widely used. Perkel and Feldman have generalized this model by deriving probability generator functions for the case where the assumptions of uniformity and stationarity are relaxed.

Synaptic Plasticity

A significant and widespread characteristic of chemical synaptic transmission is plasticity—the tendency of chemical synapses to facilitate or depress during repetitive firing. This is a ubiquitous phenomenon and dramatically influences information transfer and processing in the nervous system. Facilitation and depression have been observed and studied experimentally by many investigators in modern electrophysiology. Representative and thorough contemporary descriptions of synaptic plasticity are presented by McNaughton for neurons in the hippocampus and by Mendel for neurons in the spinal cord.

A simple model for synaptic plasticity, presented in Chapter 15, posits that the peak amplitude of a PSP (or, equivalently, the quantal component m) has a resting level toward which it decays at any instant with a given time constant and experiences an increment whenever the synapse is fired, as described in program PTNRN 15 in Chapter 15. This model describes some basic features of short-term plasticity as reported by Mendel for motoneurons in the spinal cord, but is inadequate to describe the full range of synaptic plasticity. McNaughton, for example, has constructed a more elaborate model, wherein plastic effects over several time spans are compounded, to describe his experimental observations on hipppocampal neurons. Earlier, Lara *et al.* presented models for synaptic plasticity.

The Variety of Synaptic Types

The direct chemical synapse described above (which may be axodendritic or axosomatic) is just one of a number of types of functional junctions between neurons. Several other varieties have been described, but modeled less thoroughly.

Chemical Axoaxonic Synapses (Presynaptic Inhibition)

Eccles described the properties of synapses on synapses wherein the presynaptic terminal B of a synapse effected from B to C itself experiences an input from a presynaptic terminal A. Eccles showed that firing an input pulse in terminal A diminished the effectiveness of synaptic transmission from B to C. Apparently this system has not been seriously modeled aside from the empirical data provided by Eccles. One could include such an effect in existing modeling techniques by using input activity in A to produce a conductance g_B in element B and a corresponding change in potential in B, and then letting the potential E_B diminish the strength of transfer from B to C, thereby diminishing the postsynaptic conductance change in C, g_C, in accordance with the experimental data provided by Eccles.

Dendrodendritic Synapses

Many investigators, notably Gordon Shepherd, have described chemical synapses between dendrites of neighboring cells throughout the central nervous system. The characteristics of these synapses and their functional significance are largely unknown. Apparently no serious modeling studies of the action of these synapses have been reported. One could initiate simple modeling of such effects by driving a postsynaptic conductance change g_B in element B as a function of the continuous graded potential E_A in the presynaptic element A at the point of the synapse. The situation is complicated by the fact that such junctions often tend to exhibit reciprocal or even triadic chemical synapses localized contiguously. Preliminary modeling of these more complicated situations could be undertaken along the lines indicated.

Electric Gap Synapses

Electric gap synapses have been increasingly observed, particularly in prevertebrate systems, although they occur in vertebrate nervous systems as well. They have been modeled according to the assumption of passive coupling between the pre- and postsynaptic elements; that is, one supposes that the current driven into the receiving cell is proportional to the difference in electrical potential between the two elements. If the gap junction is a two-way conductor, the current flows in and out of each cell depending on which cell has the higher potential at a given time. If it is a rectifying junction, conduction occurs in one way only.

Bidirectional Chemical Synapses

Andersen has described what appear to be bidirectional chemical synapses in the vertebrate central nervous system. As fas as we know these have not been modeled, and their properties are unknown.

Interneuronal Communication by Electric Currents and Fields

Interaction and communication among neurons can also occur in terms of more global effects such as extracellular electric or magnetic fields. The extracellular fluid surrounding neurons is volume conductor and the ionic currents that mediate neuroelectric activity are closed loops that traverse the membrane at two points, exhibiting both an intracellular largely longitudinal component and an extracellular component. These current loops are inextricably coupled with electric field gradients in both the intra- and extracellular fluids they traverse. The extracellular currents and fields may range for variable distances from the parent neuron and may or may not penetrate or influence neighboring neurons, depending on local geometric and packing configurations. It is generally thought that such extracellular electric currents and fields mediate electrical interaction among neurons under two conditions: (1) when the surrounding fluid is densely packed with dendrites and axons of neurons and (2) when there is synchronous firing of large number of local neurons such that ionic currents from large numbers of neighboring cells tend to coalesce into particularly large and coordinated extracellular effects. The latter case is thought to underly various macroscopic effects observed in electroencephalogram (EEG) signals, for example, the prominent alpha or theta rhythms.

The foundations of theoretical modeling of extracellular currents and potentials in volume conductors are discussed by Plonsey. A detailed analysis of local extracellular fields surrounding a single neuron and of more global electrical fields surrounding a population is provided by Rall and Shepherd.

It was observed some years ago that propagation of an action potential along an axon could be influenced by the occurrence of action potentials in neighboring axons. This effect was called "ephaptic" transmission. A representative contemporary study of coupling among neurons by extracellular fields is provided by Dudek and Taylor for hippocampal neurons. Traub and Dudek have presented a model to simulate these interactions. Programs POOL 12 and POOL 20 in Chapter 16 incorporate primitive representations of coupling of activity among neurons by extracellular fields.

Extracellular Magnetic Fields

Since the ionic currents which underly neuroelectric signals are closed loops, they generate magnetic fields perpendicular to the planes of the loops according to Ampere's law. Roth and Wikswo have computed the extracellular magnetic field associated with a single action potential in an axon and compared this theoretical curve with the experimentally measured magnetic fields surrounding giant axons in the crayfish. Presumable the magnitudes of magnetic fields associated with the larger and longer-lasting synaptic current loops should be proportionately larger. As far as we know, a direct influence of small magnetic fields on neuroelectric activity has not been demonstrated or modeled. Nonetheless, the possibility that global and perhaps even small local magnetic fields from individual neurons influence electrical activity in other neurons in normal function is intriguing and does not seem to be ruled out by what is known at the present time.

The analogy of this kind of picture to the classical picture of magnetic materials in physics is remarkable. In the latter picture the properties of a magnetic material are imagined to be due to the magnetic fields produced by the current loops of the electrons in the atoms of the material as those electrons move in their orbits around the nuclei. In nonmagnetic materials, that is, most materials, the atoms are arranged more or less at random, so the electronic orbits are oriented at random and the different magnetic fields tend to cancel each other out. In magnetic materials it is imagined that the atoms are all aligned relative to one another with the electron orbits in the same plane with essentially the same orientation so that the magnetic fields produced by the individual atoms are all aligned in the same direction. Their sum in this case amounts to a measurable macroscopic entity. Is it not possible that a similar thing might happen in different regions of the brain? In this case the driving currents would be the current loops associated with synaptic activation primarily in regions of the brain that are highly regular in structure, of which there are many (reticular formation, cerebellum, hippocampus, cerebral cortex, etc.). It seems likely that under certain conditions of stimulation many of the synaptic current loops in such regular structures would tend to be aligned in space. Moreover, if the input to the system were synchronous in time, many of those current loops might start to circulate simultaneously, so there could be both a temporal and a spatial confluence of these fundamental current loops. Not only might these currents sum to form macroscopically significant currents and associated electrical fields, but also the resulting composite magnetic fields associated with these loops might have some functional and behavioral significance. Such fields could conceivably mediate control over various local operations on the basis of the global distribution of activity through a region.

A number of investigators have developed the magnetoencephalogram, wherein magnetic fields surrounding brains are measured and analyzed. It has been discovered that such fields exhibit regular characteristics and moreover exhibit variations which can be correlated with cognitive operations of the brain. These efforts are discussed in Chapter 8.

Models of Growth of Neurons and Neuronal Networks

Two general concepts have been proposed to account for the growth of connections in neuronal networks. One concept posits that early in development there is promiscuous, nondirected projection of neurites and indiscriminate connections, with overproduction of neurons and connections. The neurons and connections which are insufficiently activated die. Thus selection of permanent networks is effected by neuronal activity. In the other concept, growing axons are directed to their target point for synapsing by several guiding mechanisms, including electric fields, mechanical guidance, guidance by diffusion, guidance by extracellular gradients of recognition molecules, extracellular concentrations of synaptic transmitters, and possible molecular cues on cell surfaces. A mathematical model for the development of projections in the retinotectal junctions of fish, amphibians, and birds adapting the second point of view has been presented by Gierer. Von der Malsburg and Cowan presented a model for the development of iso-orientation domains in visual cortex wherein interconnections reorganize themselves to restrict activity to families of patterns in a previsual period and later develop a one-to-one mapping of each stimulus orientation to one of the cortical patterns. A mathematical model for the growth of dendritic trees has been presented by Berry, McConnel, and Seivers.

Bibliography

Biophysical Levels

1986   Ashida, H. Stochastic description of the three-state model of the Na channel. *J. Theor. Biol.* **121**, 45–57.
   Goldfinger, M. D. Poisson process stimulation of an excitable membrane cable model. *Biophys. J.* **50**, 27–40.
1985   Fenstermacher, J. D. Current models of blood–brain transfer. *Trends Neurosci.* **8**, 449–453.
   Fohlmeister and Adelman on axons, gating. *Biophys. J.*

1984   Horn, R., and Vandenberg, C. A. Statistical properties of single sodium channels. *J. Gen Physiol* **84**, 505-534.

Kostyuk, P. G. Metabolic control of ionic channels in the neuronal membrane. *Neuroscience* **13**, 983-989.

Schmitt, F. O. Molecular regulators of brain function: A new view. *Neuroscience.* **13**, 991-1001.

1979   Llinas, R. The role of calcium in neuronal function. *In* "The Neurosciences: Fourth Study Program" (F. O. Schmitt and F. G. Worden, eds.) pp. 557-571. MIT Press, Cambridge, Massachusetts.

1976   Gillespie, C. J. Towards a molecular theory of the nerve membrane: Inactivation. *J. Theor. Biol.* **60**, 19-35.

1968   Cole, K. S. "Membrances, Ions, and Impulses" Univ. of California Press, Berkley.

Chemical Synaptic Transmission

1986   Bradford, H. F. "Chemical Neurobiology: An Introduction to Neurochemistry." Freeman, New York.

Lustig, C., Parnas, H., and Segel, L.A. On the quantal hypothesis of neurotransmitter release: An explanation for the calcium dependence of the binomial parameters. *J. Theor. Biol.* **120**, 205-213.

1984   Chesselet, M. F. Presynaptic regulation of neurotransmitter release in the brain: Facts and hypothesis. *Neuroscience.* **12**, 347-375.

1983   Nelson, P. G., Marshall, K. C., Pun, R. Y. K., Christian, C. N., Sheriff, W. H., MacDonald, R. L., and Neale, E. A. Synaptic interactions between mammalian central neurons in cell culture. II. Quantal analysis of EPSPs. *J. Neurophysiol.* **49**, 1442-1458.

1982   Faber, D. S., and Korn, H. Transmission at a central inhibitory synpapse. I. Magnitude of unitary postsynaptic conductance change and kinetic channel activations. *J. Neurophysiol.* **48**, 654-707.

1981   Carlen, P. L., and Durand, D. Modeling the postsynaptic location and magnitude of tonic conductance changes resulting from neurotransmitters or drugs. *Neuroscience* **6**, 839-846.

McNaughton, B. L., Barnes, C. A., and Andersen, P. Synaptic efficacy and EPSP summation in granule cells of rat fascia dentata studied in vitro. *J. Neurophysiol.* **46**, 952-966.

Marshall, K. C., Engberg, I., and Nelson, P. G. Studies of EPSP mechanisms in spinal neurons. *Adv. Physiol. Sci.* **1**, 100-104.

1978   McLachlan, E. M. The statistics of transmitter release at chemical synapses. *Int. Rev. Physiol.* **17**, 49-117.

1977   Martin, A. R. Junctional transmission. II. Pre-synaptic mechanisms. *In* "Handbook of Physiology" (J. M. Brookhart, V. B. Mountcastle, and E. R. Kandel, eds.), Rev. ed., Sect. 1, Vol. 1, pp. 329-355. Am. Phyiol. Soc., Bethesda, Maryland.

1976   Kandel, E. R. "Cellular Basis of Behavior." Freeman, San Francisco, California.

1969   Kuno, M., and Miyahara, J. T. Analysis of synaptic efficacy in spinal motoneurones from 'quantum' aspects. *J. Physiol. (London)* **201**, 479-493.

1966   Martin, A. R. Quantal nature of synaptic transmittsion. *Physiol. Rev.* **46**, 51-66.

1964   Kuno, M. Quantal components of excitatory synaptic potentials in spinal motoneurones. *J. Physiol.* **175**, 81-99.

Kuno, M. Mechanism of facilitation and depression of the excitatory synaptic potential in spinal motoneurones. *J. Physiol (London)* **175**, 100-112.

1963   Katz, B., and Miledi, R. A study of spontaneous miniature potentials in spinal motoneurones. *J. Physiol. (London).* **168**, 389-422.

1954   del Castillo, J., and Katz, B. The effect of magnesium on the activity of motor nerve endings. *J. Physiol. (London)* **124**, 553–559.

del Castillo, J., and Katz, B. Quantal components of the end-plate potential. *J. Physiol. (London)* **124**, 560–573.

del Castillo, J., and Katz, B. Statistical factors involved in neuromuscular facilitation tation and depression. *J. Physiol. (London)* **124**, 574–585.

del Castillo, J., and Katz, B. Changes in end–plate activity produced by pre-synaptic polarization. *J. Physicol. (London)* **124**, 586–604.

1952   Fatt, P., and Katz, B. Spontaneous subthreshold activity at motor nerve endings. *J. Physiol. (London)* **128**, 109–128.

1951   Fatt, P., and Katz, B. An analysis of the end-plate potential recorded with an intracellular electrode. *J. Physiol. (London)* **115**, 320–370.

Synaptic Plasticity

1986   McNaughton, B. L., Barnes, C. A., Rao, G., Baldwin, J., and Rasmussen, M. Long-term enhancement of hippocampal synaptic transmission and the acquisition of spatial information. *J. Neurosci.* **6**, 563–571.

1980   Lara, R., Tapia, R., Cervantes, F., Moreno, A., and Trujillo, H. Mathematical models of synaptic plasticity. II. Habituation. *J. Neurol. Res.* **2**, 1–18.

Variety of Synaptic Types

1985   Andersen, P. A. Physiology of a bidirectional excitatory chemical synapse. *J. Neurophysiol.* **53**, 821–835.

1983   Vizi, E. S., Gyires, K., Somogyi, G. T., and Ungvary, G. Evidence that transmitter can be released from regions of the nerve cell other than presynaptic axon terminal: Axonal release of acetylcholine without modulation. *Neuroscience* **10**, 967–972.

1982   Carnevale, N. T., and Johnston, D. Electrophysiological characterization of remote chemical synapses. *J. Neurophysiol.* **47**, 606–619.

1974   Shepherd, G. M. "The Synaptic Organization of the Brain" Oxford Univ. Press, London and New York.

1964   Eccles, J. C. "The Physiology of Synapses" Academic Press, New York.

Extracellular Currents and Electric Fields

1986   Yim, C. C., Krnjevic, K., and Dalkara, T. Ephaptically generated potentials in CA1 neurons of rat's hippocampus in situ. *J. Neurophysiol.* **56**, 99–122.

1985   Miller-Larsson, A., An analysis of extracellular single muscle fibre action potential field—modeling results. *Biol. Cybernet.* **51**, 271–284.

Taylor, C. P., and Dudek, F. E., Excitation and hippocampal pyramidal cells by an electrical field effect. *J. Neurophysiol.* **52**, 126–142.

Traub, R., and Dudek, F. E. Simulation of hippocampal afterdischarges synchronized by electrical interactions. *Neuroscience* **14**, 1033–1038.

1984   Feenstra, B. W. A., Hofman, F., and van Leeuwen, J. J. Syntheses of spinal cord field potentials in the terrapin. *Biol. Cybernet.* **50**, 409–418.

Plonsey, R. "Quantitative formulations of electrophysiological sources of potential fields in volume conductors. *IEEE Trans. Biomed. Eng.* **BME-13**, 868–872.

Taylor, C. P., and Dudek, F. E. Synchronization without active chemical synapses during hipcampal afterdischarges. *J. Neurophysiol.* **52**, 143–155.

1982   Holsheimer, J., Boer, J., Lopes da Silva, F. H., and van Rotterdam, A. The double dipole model of theta rythm generation: Simulation of laminar field potential profiles in dorsal hippocampus of the rat. *Brain Res.* **235**, 31–50.

1980   van Rotterdam, A. A computer system for the analysis and synthesis of field potentials. *Biol. Cybernet.* **37**, 33–39.

1979   Erulkar, S. D., and Soller, R. W. Neuronal interactions in a central nervous system model. *In* "Origin of Cerebral Field Potentials" (E. J. Speckmann and H. Caspers, eds.), pp. 13-20. Thieme, Stuttgart.

Kawato, M., Sokabe, M., and Suzuki, R. Synergism and antagonism of neurons caused by an electrical synapse. *Biol. Cybernet.* **34**, 81-89.

1978   van Rotterdam, A. A one dimensional formalism for the computation of extracellular potentials: Linear systems analysis applied to volume conduction. *Prog. Rep. Inst. Med. Phys., TNO* **PRG** 115-122.

1976   Schmitt, F. O., Dev, P., and Smith, B. H. Electronic processing of information by brain cells. *Science* **193**, 114-120.

1971   Pickard, W. F. Electrotonus on a cell of finite dimensions. *Math. Biosci.* **10**. 201-213.

1970   Eisenberg, R. S., and Johnson, E. A. (1970) Three dimensional electrical field problems in physiology. *Prog. Biophys.* **20**, 1-65.

1969   Plonsey, R. "Bioelectric Phenomena." McGraw-Hill, New York.

Rall, W. Distributions of potential in cylindrical coordinates and time constants for a membrane cylinder. *Biophys. J.* **9**, 1509-1541.

1968   Clarke, J., and Plonsey, R. The extracellular potential field of the single active nerve fiber in a volume conductor. *Biophys. J.* **8**, 842-864.

Hellerstein, D. Passive membrane potentials: A generalization of the theory of electrotonus. *Biophys. J.* **8**, 358-379.

Pickard, W. F. A contribution to the electromagnetic theory of the unmyelinated axon. *Math. Biosci.* **2**, 111-121.

Rall, W., and Shepherd, G. M. (1968) Theoretical reconstruction of field potentials and dendrodendritic synaptic interactions in olfactory bulb. *J. Neurophysiol.* **31**, 884-914.

1964   Plonsey, R. Volume conductor fields of action currents. *Biophys. J.* **4**, 317-328.

1941   Weinberg, A. M. Weber's theory of the Kernleiter. *Bull. Math. Biophys.* **3**, 39-55.

Extracellular Magnetic Fields

1985   Roth, B. J., and Wikswo, J. P. The magnetic field of a single axon. *Biophys. J.* **48**, 93-109.

Roth, B. J., and Wikswo, J. P. The magnetic field of a single axon: A volume conductor. *Math. Biosci.* **76**, 37-58.

Woosley, J., Roth, B. J., and Wikswo, J. P. The electrical potential and magnetic field of an axon in a nerve bundle. *Math. Biosci.* **76**, 1-36.

1983   Wikswo, J. P. Cellular magnetism: Theory, experiments, and applications. *IEEE Front. Eng. Comp. Health Care*, p. 432.

1982   Copson, D. A. "Informational Bioelectromagnetics". Matrix Publ., Beaverton, Oregon.

1981   Adey, W. R. Tissue interactions with nonionizing electromagnetic fields. *Physiol. Rev.* **61**, 435-514.

1979   Adey, W. R. Neurophysiologic effects of radiofrequency and microwave radiation. *Bull. N.Y. Acad. Med.* **55**, 1079-1093.

Models of Growth of Neurons and Neural Networks

1986   Liestol, K., Maehlen, J., and Nja, A. Selective synaptic connections: Significance of recognition and competition in mature sympathetic ganglia. *Trends Neurosci.* **9**, 21-24.

1985   Kater, S., and Letourneau, P. "Biology of the Nerve Growth Cone". Alan R. Liss, New York.

1984   Erdi, P. System-theoretical approach to the neural organization: Feed-forward control of the ontogenetic development. *In* (R. Trapp, ed.), "Cybernetics and System Research" Vol. 2, pp. 229-235. Elsevier/North-Holland, Amsterdam.

Erdi, P., and Barna, G. Self-organizing mechanism for the formation of ordered neural mappings. *Biol. Cybernet.* **51**, 93–101.

1983   Szentágothai, J. The modular architectonic principle of neural centers. *Rev. Physiol. Biochem. Pharmacol.* **98**, 11–61.

1982   Overton, K. J., and Arbib, M. A. The extended branch-arrow model of the formation of retino-tectal connections. *Biol. Cybernet.* **45**, 157–175.

Trisler, D. Are molecular markers of cell position involved in the formation of neural circuits? *Trends Neurosci.* **5**, 306–310.

von der Malsbur, C., and Cowan, J. D. Outline of a theory for the ontogenesis of iso-orientation domains in visual cortex. *Biol. Cybernet.* **45**, 49–56.

1981   Gierer, A. Development of projections between areas of the nervous system. *Biol. Cybernet.* **42**, 69–78.

Wolpert, L. Positional information and pattern formation. *Philos. Trans. R. Soc. B* **295**, 441–450.

1980   Berry, M., McConnell, P., and Seivers, J. Dendritic growth and the control of neuronal form. *Curr. Top. Biol.* **15**, 67–101.

Bonhoeffer, F., and Huf, J. Recognition of cell types by axonal growth cones *in vitro*. *Nature (London)* **288**, 162–164.

Hirai, Y. A new hypothesis for synaptic modification: An interactive process between postsynaptic competition and presynaptic regulation.*Biol. Cybernet.* **36**, 41–50.

1978   Bunge, R. A., Johnson, M., and Ross, C. D. Nature and nurture in development of the autonomic neuron. *Science* **199**, 1409–1416.

Cowan, J. D. Aspects of neural development. *Int. Rev. Physiol. Neurophysiol. III*, **17**, 149–189.

Meinhardt, H. Models for the ontogenetic development of higher organisms. *Rev. Physiol. Biochem. Pharmacol.* **80**, 47–104.

Szentágothai, J. Specificity versus (quasi-) randomness in cortical connectivity. *In* "Architectonics of the Cerebral Cortex" (M. A. B. Brazier and H. Petsche, eds.). Raven Press, New York.

1976   Changeux, J. P., and Danchin, A. Selective stabilization of developing synapses as a mechanism for the specification of neural networks. *Nature (London)* **264**, 705–712.

1974   Meinhardt, H. Applications of a theory of biological pattern formation based on lateral inhibition. *J. Cell Sci.* **15**. 321–346.

1973   Changeux, J. P., Courrege, P., and Danchin, A. A theory of the epigenesis of neural networks by selective stabilization of synapses. *Proc. Natl. Acad. Sci. U.S.A.* **70**, 2974–2978.

1972   Gierer, A., and Meinhardt, H. A theory of biological pattern formation. *Biol. Cybernet.* **12**, 30–39.

1971   Glansdorff, P., and Prigogine, J. "Thermodynamic Theory of Structure, Stability, and Fluctuation." Wiley, London.

1970   Crick, F. H. C. Diffusion in embryogenesis. *Nature (London)*. **225**, 420–422.

1968   Goodwin, B. C., and Cohen, M. H. A phase shift model for the spatial and temporal organization of developing systems. *J. Theor. Biol.* **29**, 99–107.

1963   Sperry, R. W. Chemoaffinity in the orderly growth of nerve fiber patterns and connections. *Proc. Natl. Acad. Sci. U.S.A.* **50**, 703–710.

Models of Specific Neuronal Systems and Networks

This section of the book reviews models applicable to specific networks and subsystems throughout the nervous system. Chapter 5 reviews models for invertebrates and "simpler" systems, and Chapters 6, 7 and 8 deal with the vertebrate nervous system. The categorization of vertebrate systems as reflected in the chapter titles is conceived within the levels-of-construction point of view of Hughlings Jackson and elaborated in the triune brain theory of Paul MacLean. Thus, Chapter 6 considers systems whose rudimentary forms at least are operative in and characteristic of a reptilian level of development (although certain suprareptilian elaborations are also included when they do not indicate a higher level of central organization—as in elaborated peripheral sensory processing, for example). Chapter 7 considers systems of the limbic level of organization, thought to be characteristic of the earliest mammals. Chapter 8 considers models of systems at the neocortical and thalamocortical level of organization.

This section of the book makes no attempt to discuss thoroughly all the models relevant to these specific subsystems. I do, however, attempt to cite the significant references in each category and to comment briefly on recent and particularly significant or representative model studies in each area. I hope that specialists in these areas forgive the inevitable oversights.

5

"Simpler Systems" and Invertebrate Systems

Experimentally Based Network Wiring Models for "Simpler" Neuronal Systems

In the early 1960s a number of investigators began sustained studies of a number of invertebrate nervous systems, guided by the idea that since these systems are considerably smaller in size, scope, and complexity than their counterparts in the vertebrate nervous system, it should be correspondingly easier to understand their dynamic activity and functioning in relation to anatomy, neurochemisty, neurophysiology, functional physiology, and behavior. Pioneers in this area were Kandel, Kennedy, Willows, and a number of others who worked on such animals as *Aplysia*, Tritonia, and leeches. The work has continued steadily in many laboratories and is reflected in a large number of contemporary publications in leading experimental journals. The work ususaly focuses on neuronal pools or ganglia and includes some related vertebrate research.

The guiding thrust in this work is to identify key neurons whose electrophysiological signals correlate directly with observable features of behavior or physiology, and then to further identify neuronal circuit connectivity patterns which underlie these signals. The behaviors are usually rather simple—regular rythmic patterns in swimming, walking, or flying, respiration or heartbeat, simple escape or startle behaviors, release of ink, and so forth. The commonly observed eletrophysiological features include endogenous activity, often rhythmic or bursting, and modulations of activity during normal or experimentally driven functioning. Such modulations include augmentation or diminution of bursts (which may be accelerating or decelerating); elevation or depression of firing correlated with particular behavioral or physiological events; positive or negative correlations between activities in neuron pairs or among activities in neuron groups, sometimes giving evidence of direct synaptic excitation or inhibition; and coordinated spreading "waves" of activity through neuron groups, corresponding, for example, to wavelike motions in swimming.

Key theoretical concepts in the driving and functional organization in these hypothetical neuronal circuits include the ideas of "command neurons" and "pattern generators." Command neurons are neurons whose activity is necessary and sufficient to release a given behavioral pattern. Pattern generators are single neurons or small clusters of neurons whose intrinsic dynamics or interconnection patterns reliably and repeatedly generate the key dynamic patterns required by the particular behavioral or physiological event under study. Neurons in these systems tend to be spontaneously (endogenously) active and to exhibit pacemaker or bursty activity. They fatigue and exhibit postinhibitory rebound. There are mixed synapses which exert both excitatory and inhibitory effects. There are electrical junctions between neurons. Activity patterns are largely sculptured by inhibition; that is endogenous mechanisms and sensory drive produce reverberating excitations which must be controlled and channeled by inhibitory processes. "Disinhibition" is a common controlling mechanism. The energy and flavor of work in this area of "simpler systems" are indicated by 10 papers appearing in the 1985 issues of the *Journal of Neurophysiology*, of which 9 deal with mechanisms of rhythmic movements (swimming, flight, and feeding).

This work is pushing closer to the ideal of providing a comprehensive understanding of the global activity of entire functional subsystems in relation to their functioning in an intact animal. However, that understanding has not yet been attained. There are considerable methodological difficulties and controversy in this area. One problem is that the contextual relations of the abstracted neuronal circuit diagrams to the remainder of the nervous system are not usually clear. One usually abstracts a few neurons from a ganglion or pool containing several hundred to several thousand neurons. The ganglion or pool usually deals with a number of functions, including but not restricted to the one under consideration. Many of the neurons in the hypothetical circuit diagram participate in other functions as well as one under consideration. The general assumption is that other neurons in the ganglion or pool are not particularly significant with regard to the behavior under consideration. Some key neurons in the neuronal circuit have preeminence because their activities are found to be necessary and sufficient for the observed behavior. On the other hand, the participation of other neurons in the behavior may be clearly significant but neither necessary nor sufficient. Thus at the outer limits of these circuit diagrams inclusion or noninclusion of neurons becomes to some extent a matter of perspective and is sometimes arbitrary. Often the degreee of convergence of inputs on particular constituent neurons of the circuit diagrams are not explicitly considered; nor are the overall degrees of convergence and divergence of interconnections throughout the neuronal circuitry of the ganglion or the model. Thus, the attempt to understand with the model all features of observed activity relevant to the behavior at hand is usually limited by the fact that the simplified circuit diagrams themselves do not include the entire range of possibilities in the actual ganglia or pools. Thus a global view of the dynamics of the ganglion and of the function or circuit model within this overall view has largely not yet been elucidated in this work.

Key investigators in this area have come up against fundamental theoretical difficulties involving the relation of function to structure in neuronal systems. Again, the concept of the dynamics of parts in relation to the operations of wholes is questioned. See Eaton and DiDomenico for an excellent review of this topic.

Another limitation of this work is that not many of the hypothetical neuronal circuit connectivity diagrams have been modeled quantitatively or simulated with computer models. This is important, because the dynamics of neuronal systems are very complex; the intricacies of dynamic activity for neurons in a particular (even "simple") interconnection pattern cannot be reliably predicted without such computational assistance. Such modeling techniques as described in Part II of this book can readily be applied to many of these neuronal circuit diagrams, in constructive ways.

Simulation Models of "Simpler" Systems

Selverston has reviewed models of central pattern generators. Salient models include endogenous rhythmic bursters, mutually inhibiting centers with fatigue (Harmon), recurrent cyclic inhibition (Szekely, Freisen, and Stent), bursting by positive feedback through electrical connections (Willows), and postinhibitory rebound in mutual inhibition (Satterlie).

As touched on in Chapter 2, various digital computer and elctronic analog neuromimes have been used to model many basic generic behaviours as observed in these simple-systems studies. Endogenous rhythmic oscillations (pacemaker acitivity) have been simulated in neuromimes with leaky membranes wherein the resting potential is higher than the resting threshold. A number of modeling studies have shown how the phase settings and periods of rhythms and rhythmic bursts in endogenous pacemakers, recurrently connected rings, and mutually inhibiting groups can be manipulated by various forms of excitatory or inhibitory input bombardment. Recurrent inhibition of projection cells by local inhibitory cells can produce rhythmic firing in the projection cells when they are bombarded with tonic input (these have been called "ring" networks). These rhythms have periods determined by the time constant of the recurrent IPSP or by the refractory properties of the cells. Decelerating transient bursts are modeled readily in neuromimes with accommodation and its mirror image, postinhibitory rebound, wherein threshold is driven up or down by excitatory or inhibitory potential excursions, respectively. Accelerating bursts are readily produced in neuron models with recurrent excitatory connections. Rhythmic alternations are readily produced between pairs of neurons or pairs of groups of neurons which mutually inhibit. This acitivity can be enhanced by fatigue and by postinhibitory rebound. The duration of the bursts and the interburst period can be varied by adjusting the parameters in the model.

Palvidas has shown that introduction of randomly fluctuating variables can produce such effects as rhythm splitting, spontaneous changes in free-running period with loss and recovery of rhythmicity, and various aftereffects. All of the effects described in this section so far can be obtained with single neurons and pairs of single neurons. Friesen and Stent have shown that neurons connected sequentially in small rings provide another source of rhythm generation which they call recurrent cyclic inhibition. Such linkages as these can be shown to produce multiphasic activity rhythms. Palvidas and Enright have modeled rhythmic oscillations with populations of small numbers of loosely coupled interacting oscillators. Christensen and Lewis have simulated circadian local motor rhythm in an insect with a control system representation of a population of weakly coupled feedback oscillators. This model successfully simulates entrainment, phase response curves, temperature compensation, Ascolt's rule for activity rhythms, rhythm splitting, and various spontaneous changes and aftereffects consistent with experimental data. Segundo and collaborators have modeled dynamic mechanisms of crayfish stretch receptors and dynamic interactions among pacemakers.

In 1974 MacGregor and Palasek showed that tonic bombardment of simulated pools of 50–100 interconnected neuromimes with recurrent excitation readily produced rhythmic coordinated firing patterns in which the period of the rhythm was produced by the refractoriness of the constituent neurons. Such populations tended to produce longitudinally propagating waves of rhythmically coordinated firing when the constituent neurons exhibited longitudinal organiation, in the sense that recurrent excitatory connections tended to fall with higher probability on neighboring cells (exponential distribution). In 1978 MacGregor and McMullen simulated interactions of groups of neuron pools with 50–100 neurons per pool interconnected by diffuse convergent–divergent junctions. Such systems are remarkably sensitive to synchronized clusters in the input firing pattern, and recurrent excitation tends to remarkably enhance coordination of firing among constituent neurons.

Probably the most representative modeling study in this area is still Wilson and Waldron's study of the neuronal dynamics involved in the control of flight patterns in locusts, which was published in 1968. It seems clear that the level of development of conceptual models and biological experimentation justifies a considerably higher level of quantitative modeling studies than has yet been seen in this area. With the availability of methodologies and simulation programs such as those described in Part II of this book, there is no reason that such modeling cannot be undertaken at this time.

Models of Sensory Information Processing in Prevertebrates

One of the most significant modeling studies of prevertebrates was the study of visual information processing in the eye of the horseshoe crab *Limulus* by Hartline and various collaborators including Ratliff. They discovered and modeled the mechanism of lateral inhibition and showed that this mechanism could account for a variety of fundamental features of spatiotemporal sculpturing of information, including the phenomenon of boundary enhancement and, by inference, feature extraction. Since lateral inhibition is universally active in vertebrate nervous systems it will be discussed in Chapter 6.

Reichardt, Poggio, and Hausen have modeled the visual system of the fly. In one study the three-dimensional trajectory of a fly chasing another fly is modeled by two simple control systems for fixation and tracking. Other studies address the algorithms used by the fly to separate figure from ground on the basis of motion information. Models for the neuronal circuitry underlying these processes are presented.

Stavenga and Beersma have presented a vector and matrix description of the connectivity in the fly visual system and discussed the applicability of this formalism to other visual systems.

Bibliography

Experimentally Based Studies of Simpler Systems

Miall, R. C., Simple or complex systems? *Behav. Brain Sci.* (in press).
1986   Bassler, U. On the definition of central pattern generator and its sensory control. *Biol. Cybernet.* **54**, 65–69.
Reichert, H., and Rowell, C. H. F. Neuronal circuits controlling flight in the locust: How sensory information is processed for motor control. *Trends Neurosci.* **9**, 281–283.
Zill, S. N. A model of pattern generation of cockroach walking reconsidered. *J. Neurobiol.* **17**, 317–328.
1985   Eaton, R. C. ed. "Neural Mechanisms of Startle Behavior." Plenum, New York.

Eaton, R., C., and DiDomenico, R. Command and the neural causation of behavior. *Brain Behav. Evol.* 27, 132–364.

Egelhaaf, M. On the neuronal basis of figure–ground discrimination by relative motion in the visual system of the fly. III. Possible input circuitries and behavioural significance of the FD-cells. *Biol. Cybernet* 52, 267–280.

Grillner, S., and Wallen, P. (1985). Central pattern generators for locomotion, with special reference to vertebrates. *Annu. Rev. Neurosci.* 8, 233–261.

1983   Christensen, N. D., and Lewis, R. D. The circadian locomotor rhythm of *Hemideina thoracica*: A population of weakly coupled feedback oscillators as a model of the underlying pacemaker. *Biol. Cybernet.* 47, 165–172.

Eaton, R. C. Is the Mauthner cell a vertebrate command neuron? A neuroethological perspective on an evolving concept. "In Advances in Vertebrate Neuroethology" (J. Ewert, R. R. Capranica, and Ingle, D. J. eds.) pp. 629–636. Plenum, New York.

1982   Heetderks, W. J., and Batruni, R. Multivariate statistical analysis of the responses of the cockroach giant interneuron system to wind puffs. *Biol. Cybernet.* 43, 1–11.

Kemmerling, S., and Vajur, D. (1982). Regulation of the body–substrate–distance in the stick insect: Step responses and modelling the control system. *Biol. Cybernet.* 44, 59–66.

Mial, R. C. (1982). Central organization of crustacean abdominal posture motoneurons: Connectivity and command fiber inputs. *J. Exp. Zool.* 224, 45–56.

Thompson, R. S. A model for basic pattern generating mechanisms in the lobster stomatogastric ganglion. *Biol. Cybernet.* 43, 71–78.

1981   van Dongaen, P. A. M., and van den Bercken, J. H. L. (1981). Structure and function in neurobiology: A conceptual framework and the localization of functions. *Int. J. Neurosci.* 15, 49–68.

1980   Selverston, A. I. Are central generators understandable? *Behav. Brain Sci.* 3, 535–571.

1979   Kandel, E. R. Small systems of neurons. *Sci. Am.* 241 66–76.

1978   Fowler, C. A., and Turvey, M. T. The concept of 'command neurons' in explananations of behavior. *Behav. Brain Sci.* 1, 20–22.

Kupfermann, I., and Weiss, K. R. The command neuron concept. *Behav. Brain Sci.* 1, 3–39.

Russell, D. F., and Hartline, D. K. Bursting neural networks. A re-examination. *Science* 200, 453–456.

1976   Bullock, T. H. In search of principles in neural integration. *In* "Simple Networks and Behavior" (J. D. Fentress, ed.), pp. 52–60. Sinauer, Sunderland, Massachusetts.

Davis, W. J. Organizational concepts in the central motor networks of invertebrates. *In* "Neural Control of Locomotion"(R. M. Herman, S. Grillner, P. S. G. Stein, and D. G. Stuart eds.), pp. 265–292. Plenum, New York.

Fentress, J. D., ed. "Simpler Networks and Behavior." Sinauer, Sunderland, Massachusetts.

Herman, R. M., Grillner, S., Stein, P. S. G., and Stuart, D. G., eds. "Neural Control of Locomotion." Plenum, New York.

Kennedy, D. Neural elements in relation to network function. *In* "Simple Networks and Behavior" (J. D. Fentress, ed.), pp. 65–81. Sinauer, Sunderland, Massachusetts.

Kristan, W. B., Jr., and Calabrese, R. L. Rhythmic swimming activity in neurons of the isolated nerve cord of the leech. *J. Exp. Biol.* 63, 643–666.

Williams, W. J., and Porter, W. A. Some system considerations of neuron pools with feedback. *Biol. Cybernet.* 21, 79–83.

1973   Willows, A. O. D., Dorsett, D. A., and Hoyle, G. The neuronal basis of behavior in Tritonia. III. Neuronal mechanism of a fixed action pattern. *J. Neurobiol.* **4**, 255–85.

1972   Maynard, D. M. Simpler networks. *Ann. N.Y. Acad. Sci.* **193**, 59–72.

1971   Kennedy, D. Nerve cells and behavior. *Sci. Am.* **59**, 36–42.

1969   Willows, A. O. D., and Hoyle, G. Neuronal network triggering a fixed action pattern. *Science* **166**, 1549–1551.

1961   Bullock, T. H. The origins of patterned nervous discharge. *Behaviour* **17**, 48–59.

Simulation Models of/for Simpler Systems

1985   Satterlie, R. A. Reciprocal inhibition and postinhibitory rebound produce reverbation in a locomotor pattern generator. *Science* **229**, 402–404.

Segundo, J. P., and Martinez, O. D. Dynamic and static hysteresis in crayfish stretch receptors. *Biol. Cybernet.* **52**, 291–296.

Vibert, J. F., Caille, D., and Segundo, J. P. Examination with a computer of how parameter changes and variabilities influence a model of oscillator entrainment. *Biol. Cybernet.* **53**, 79–91.

1984   Kumar, P., Gallagher, R. R., and Kammer, A. E. A neuronal model of the crayfish escape response. *Eng. Med. Biol.* **37**, 29.7, 190.

Oguztorelli, M. N. Some problems concerning nonlinear oscillators in neural networks. *In* "Trends in Theory and Practice of Nonlinear Differential Equations" (V. Lakshmikantham, ed.), pp. 425–433. Dekker, New York.

Tsutsumi, K., and Matsumoto, H. A synaptic modification algorithm in consideration of the generation of rhythmic oscillation in a ring neural network. *Biol. Cybernet.* **50**, 419–430.

Tsutsumi, K., and Matsumoto, H. Ring neural network qua a generator of rhymic oscillation with period control mechanism. *Biol. Cybernet.* **51**, 181–194.

1983   R. Grissell, Hodgson, J. P. E., and Vanowich, M., eds. "Oscillations in Mathemathical Biology." Springer-Verlag, Berlin and New York.

Martinez, O. D., and Segundo, J. P. Behavior of a single neuron in a recurrent excitatory loop. *Biol. Cybernet.* **47**, 33–41.

1982   Bardakjian, B. L., El-Sharkawy, T. Y., and Diamant, N. E. On a multiport synthesized relaxation oscillator representing a bioelectric rhythm. *IEEE Front. Eng. Comp. Health Care*, p. 433.

Kaswahara, K. and Mori, S. A two compartment model of the stepping generator: Analysis of the roles of a stage-setter and a rhythm generator. *Biol. Cybernet.* **43**, 225–230.

1981   Kohn, A. F., da Rocha, A. F., and Segundo, J. P. Presynaptic irregularity and pacemaker inhibition. *Biol. Cybernet.* **41**, 5–18.

Segundo, J. P., and Kohn, A. F. A model of excitatory synaptic interactions between pacemakers. Its reality, its generality, and the principles involved. *Biol. Cybernet.* **40**, 113–126.

1980   Kawahara, T. Coupled van der Pol oscillators—a model of excitatory and inhibitory neural interactions. *Biol. Cybernet* **39**, 37–43.

1979   Grasman, J., and Jansen, M. J. W. Mutually sunchronized relaxation oscillators as prototypes of oscillating systems in biology. *J. Math. Biol.* **7**, 171–197.

1978   Glass, L., and Pasternack, J. S. Prediction of limit cycles in mathematical models of biological oscillations. *Bull. Math. Biol.* **40**, 27–44.

MacGregor, R. J., and McMullen, T. Computer simulation of diffusely-connected neuronal populations. *Biol. Cybernet.* **28**, 121–127.

Pavlidas, T. What do mathematical models tell us about circadian clocks? *Bull. Math. Biol.* **40**, 625–635.

Pavlidas, T. Quantitative similarities between the behaviour of coupled oscillators and circadian rhythms. *Bull. Math. Biol.* **40**, 675–692.

Stein, P. S. G. Motor systems, with specific reference to the control of locomotion. *Annu. Rev. Neursci.* **1**, 61–81.

1977   Friesen, W. O., and Stent, G. S. Generation of a locomotory rhythm by a neural network with recurrent cyclic inhibition. *Biol. Cybernet.* **28**, 27–40.

MacGregor, R. J., and Lewis, E. R., "Neural Modeling." Plenum, New York.

1976   Friesen, W. O., Poon, M., and Stent, G. S. An oscillatory neuronal circuit generating a locomotory rhythm. *Proc. Natl. Acad Sci. U.S.A.* **73**, 3734–3738.

Herman, R. M., Grillner, S., Stein, P. S. G., and Stuart, D. G., eds. "Neural Control of Locomotion." Plenum, New York.

Pavlidas, T. Spatial and temporal organization of populations of interacting oscillators. In "The Molecular Basis of Circadian Rhythms" (J. W. Hastings and H. G. Schweiger, eds.). Dahlem Konferenzen, Berlin.

Segundo, J. P., Tolkunov, B. F., and Wolfe, G. E. Relation between trains of action potentials across an inhibitory synapse. Influence of presynaptic irregularity. *Biol. Cybernet.* **24**, 169–179.

Selverston, A. I. Neuronal mechanisms for rhythmic motor patterns generation in a simple system. In *"Neural Control of Locomotion"* (R.M. Herman, S. Grillner, P. S. G. Stein, and D. G. Stuart, eds.), pp. 377–400. Plenum, New York.

Williams, W. J., and Porter, W. A. Some system considerations of neuron pools with feedback. *Biol. Cybernet.* **21**, 79–83.

1975   Perkel, D. H., and Mulloney, B. Motor pattern production in reciprocally inhibitory neurons exhibiting postinhibitory rebound. *Science,* **185**, 181–183.

Stein, R. B., Leung, K. V., Mangeron, D., and Oguztoreli, M. N. Improved neuronal models for studying neural networks. *Biol. Cybernet.* **15**, 1–9.

1972   Johnsson, A., and Karlsson, H. G. A feedback model for biological rhythms. I. Mathematical description and basic properties of the model. *J. Theor. Biol.* **36**, 153–174.

1971   Pavlidas, T. Populations of biochemical oscillators as circadian clocks. *J. Theor. Biol.* **33**, 319–338.

1970   Dunnin-Barkovskii, V. L. Fluctuations in the level of activity in simple closed neurone chains. *Biofizika* **15**, 374–378.

Pozin, N. V., and Shyulpin, Y. A. Analysis of the work of auto-oscillatory neurone junctions. *Biofizika* **15**, 156–163.

1968   Adams, A. Simulation of rhythmic nervous activites. II. Mathematical models for the function of networks with cyclic inhibition. *Biol. Cybernet.* **5**, 103–109.

Kling, V., and Szekely, G. Simulation of rhythmic nervous activities. I. Function of networks with cyclic inhibitions. *Biol. Cybernet.* **5**, 89–102.

Lewis, E. R. (1968). Using electronic circuits to model simple neuroelectric interactions. *Proc. IEEE* **56**, 931–949.

Wilson, D. M., and Waldron, I. (1968). Models for the generation of the motor output pattern in flying locusts. *Proc. IEEE* **56**, 1058–1064.

1967   Winfree, A. T. (1967). Biological rhythms and the behavior of populations of coupled oscillators. *J. Theor. Biol.* **16**, 15–42.

Sensory Processing in Prevertebrates

1986   von der Malsburg, C. and Schneider, W. A neural cocktail-party processor. *Biol. Cybernet.* **54**, 29–40.

1985   Egelhaaf, M. On the neuronal basis of figure–ground discrimination by relative motion in the visual system of the fly. III. Possible input circuitries and behavioural significance of the FD-cells. *Biol. Cybernet.* **52**, 267–280.

1983   Reichardt, W., Poggio, T., and Hausen, K. Figure–ground discrimination by relative movement in the visual system of the fly. II. Towards the neural circuitry. *Biol. Cybernet.* **46** Suppl., 1–30.

1981   Poggio, T., and Reichardt, W. (1981). Visual fixation and tracking by flies: Mathematical properties of simple control systems. *Biol. Cybernet.* **40**, 101–112.

Reichardt, W., and Poggio, T., eds. (1981). "Theoretical Approaches in Neurobiology." MIT Press, Cambridge, Massachusetts.

1975   Stavenga, D. G., and Beersma, D. G. M. (1975). Formalism for the neural network of visual systems. *Biol. Cybernet.* **19**, 75–81.

6

Early and Lower-Level
Vertebrate Systems

Many features of autonomic, motor, and sensorimotor function lend themselves to modeling by engineering control system methods. Such systems are often clearly defined by single variables, which are in turn readily measured by single magnitudes; feedback control loops are ubiquitous; temporal variations such as oscillations and transients seem to be significant features, as they are in engineering control systems.

Autonomic Systems

Many models of autonomic function have focused on rhythmic oscillations, and of these many have found Harmon's concept of rhythmicity generated by mutually inhibiting centers driven by tonic drive to be useful. For example, Carpenter and Grossberg present a model for the generation of circadian rhythms by suprachiasmatic nuclei in the mammalian hypothalamus wherein each nucleus generates a rhythm by gating of positive feedback signals by slowly accumulating chemical transmitters, and the nuclei coordinate according to mutual inhibition.

The control of rhythmicity in respiration is somewhat more complicated. The standard picture is that at least three factors contribute fundamentally to the rhythm. First is the likely mutual inhibition between inspiratory and expiratory centers. Second, the excitation of stretch receptors by expansion of the lungs triggers shutting down of inspiratory neurons (the so-called Herring–Beurer reflex) and possibly also excitation of the respiratory neurons. Third, there is a so-called pneumotaxic center which is driven by inspiratory neurons and which presumably, after an accumulation of input from the inspiratory neurons, tends to both shut down inspiratory and perhaps excite expiratory neurons. A model of the respiratory oscillator consisting of four neuron pools inconnected to produce coordinated oscillations was presented in 1976 by German and Miller. Van Dooren and Vis reinterpret parameters in the German–Miller model to correctly predict afterdischarges. This model produces a variety of features similar to those observed in respiration.

A definitive experimental study of the neural control of respiration has been under way by Morton Cohen for the past 10 or 15 years wih various collaborators, including most recently Jack Feldman. Cohen presents a model in which phase switching from inspiratory to expiratory activity is a key ingredient in the generation and maintenance of rhythmicity. Feldman and Cowan applied a mathematical theory for large neural nets (see Chapter 9) to coupled nets of excitatory and inhibitory cells to represent respiratory oscillations and rapid switching from inspiratory to expiratory phases.

Other models have focused on the generation of autonomic rhythmicities at the cellular level. For example, Noble extended the Hodgkin–Huxley model to produce rhythmic generation of single Hodgkin–Huxley-like axon potentials (and therefore muscle contractions) in single Purkinje fibers of the heart muscle. This model is simulated in Chapter 22 of this book.

Vibert, Caille, and Segundo modeled the driving of respiratory rhythms by simulating a postsynaptic pacemaker subjected to inhibition by a presynaptic pacemaker.

Motor and Sensorimotor Systems

A large number of penetrating and useful models have appeared to describe various levels of operation in motor and sensorimotor systems.

Spinal Muscle Circuits

A very useful modeling study of spinal muscle motor operations has been provided by Stein and Oguztoreli. This model includes representation of single muscles by nonlinear mechanical and viscoelastic elements, interactions between antagonistic muscles, and driving and modulation of muscle function by both alpha and gamma motoneuron types. Stein and Oguztoreli elaborated the dynamic properties of this model with particular relevance to the influnce of the various components on stability and oscillatory tendencies. They also applied optimization theory to investigate the minimization of various "costs" associted with time, energy, reliability, and so forth.

A number of other investigators have presented models to describe detailed time courses of responses to stretches in muscle fibers: Hatze simulated such responses in FORTRAN; Hasan included a property akin to friction in a nonlinear differential equation system to predict responses to ramp and hold stretch; and Daunicht modeled muscle spindles with one position-dependent and three velocity-dependent components. All these studies have been quite successful in describing the peculiar temporal features of stretch responses in spindles.

A number of other useful models have described responses in motor units: Christakos modeled the properties of muscle force waveform and the electromyogram in terms of the asynchronous activity of muscle units modeled by a number of linear systems in parallel; Ducati *et al.* represented posttetanic potentiation in single motor units by a simple viscoelastic model which successfully describes potentiation effects by using a prolonged decay time of the active state; Inbar and Ginat showed that overall linearity and stability result in a model neuromuscular system with a large number of motor units.

Miller-Larsson presented a quantitative model for the extracellular field for an action potential in a single muscle fiber. Feenstra *et al.* presented compartmental model for an amphibian motor neuron to produce simulated extracellular field potentials in the spinal cord so as to discriminate between different activation hypotheses for the spinal cord potential. They found that simulated excitatory input gives more realistic output when distributed over the entire dorsal dendritic tree. Gouze *et al.* presented a theoretical model for the development of innervation of muscles in which a number of candidate nerve terminals compete for a postsynaptic retrograde factor and losers disappear. The model predicts that only one nerve terminal becomes stabilized per muscle fiber.

Cleveland *et al.* presented an electric circuit analog model of the Renshaw cell membrane to model the response of Renshaw cells to bombardment in alpha motoneuron axons. In this model saturation at high frequency is attributed to nonlinear occlusion among simultaneously active excitatory synapses. Windhorst and Koehler simulated the alpha motoneuron–Renshaw cell system with three principal negative feedback loops interconnected via cross-feedback pathways. The dynamics of this system are very sensitive to the parameters of recurrent inhibition and exhibit a number of potentially significant temporal effects. Small changes in inhibitory strength, for example, can markedly influence phase relationships. The inhomogeneity of recurrent inhibition helps to prevent a strong phase separation.

Kawahara and Mori presented a model to simulate the interaction of the spinal "stepping generator" and the postural control system in the mesencephalic cat. This model succeeds in simulating a variety of observed local motor patterns including stepping automatism.

Corner has reviewed early spontaneous motor rhythms and motility throughout the animal kingdom and suggests that neural systems generating these most primitive of behavior patterns persist into later life and again become active during sleep.

Agarwal *et al.* have presented a time series model of the dynamic response of the human ankle joint to torques. Transfer function techniques are used to describe the torque and electromyographic activity in input–output pairs. Gielen and van Zuylen have applied Pellionisz's tensorial theory (see below) to model the recruitment threshold in the activation of arm muscles during flexion/extension and supination/pronation. Hemani and Stokes present a model of four interrelated neural circuits to control ballistic-type biped movement.

Brain Stem Sensorimotor Systems

A number of sensorimotor control systems utilize nuclei embedded in the brain stem.

Pupil Size   Stark and various collaborators have presented a number of control system models to represent dynamics of control of pupil size.

Oculomotor Systems   Control system models have been used extensively to describe saccadic eye movements and optokinetic reflexes. Jurgens, Becker, and Kornhuber present a model which suggests that saccadic innervation in response to perceived target eccentricity is controlled by a local feedback loop. Tweed and Vilis describe a model for saccade generation involving multiple comparators and two independent burst generators. They suggest that the visual system stores the target location in inertial coordinates but the feedback loop which guides saccades and the motor error signal work in retinotopic coordinates. Bohmer and Allum and Gillis *et al.* present feedback control models to represent the dynamics of optokinetic reflexes. Bahill and McDonald present a model to represent human target tracking; they state that this control model can overcome an inherent time delay and produce zero-latency tracking as humans are known to do. Cannon *et al.* present a lateral inhibitory network model to generate eye position for motor neurons from eye velocity signals in oculomotor integration.

Arbib and Lara describe computer simulation of interactions of a linear array of tectal columns to simulate prey-catching behavior in amphibians. The model details inhibitory and excitatory interactions related to such functions as increased acuity in the direction of the prey, facilitation to moving stimuli, and preference for the head of the stimulus. They have extended this model to include stimulus-specific habituation. In this model prey–predator recognition is performed by command units as a result of retina–tectum–pretectum interactions.

Vestibulo-Oculomotor Interactions   In a number of publications Pellionisz and collaborators have described the application of a tensorial approach to transformations in neuronal networks to the neural control of gaze by the oculomotor and vestibular systems. In this approach junctions between neuronal populations are represented by matrices, and tensorial transformation theory is applied to elucidate properties of their connectivity and dynamics. The approach helps to identify multidimensional natural coordinate systems in neuronal systems. Feldman presents a model in which horizonal eye movements in pursuit and saccades are controlled by a superposition of reciprocal and unidirectional central commands of agonist and antagonist muscles. Segal and Outerbridge present a detailed model for nonlinear dynamics in primary afferents of the semicircular canal.

Buizza and Schmid apply control system techniques to model visual–vestibular interactions in patients with labyrinthine and cerebellar pathologies. The model estimates the state of the system in such patients and greatly assists diagnosis.

Schor presents a model system which allows construction of mathematical models for arbitrary combinations of linear cascades, parallel pathways, and feedback loops and applies this general system to responses to tilt in cat vestibular nuclei.,

Head Movements   A sixth-order nonlinear model for horizontal head rotations in humans is presented by Zangemeister, Lehman, and Stark. This model fits common head acceleration types and also less common dynamic overshoot trajectories and various salient features of neck muscle electromyograms (EMGs).

Cerebellum

The cerebellum is prominent and seems to function similarly and toward similar purposes in all vertebrates. It appears to be involved in higher-order modulations or fine tuning in sensorimotor integration. Its inputs and outputs appear to be concerned with information on proprioception, balance, and muscle tone and to project to and from sensorimotor centers in the spinal cord, brian stem, and basal ganglia and motor regions of the thalamus and cerebral cortex. The cerebellum is massive (approximately one-third of the brain in humans) but is highly regular and appears to consist largely of repetitions of a single five-cell unit which appears to be relatively simply constructed. Of all parts of the brain the cerebellum might be most readily likened to a computer.

In the 1970s Pellionisz, first alone and then with Llinas and Perkel, constructed several large-scale computer simulation models of cerebellar networks and cells. In work published in 1970 Pellionisz simulated 64,000 neurons representative of the neuronal circuitry in the cerebellar cortex. Visual displays based on this model provide insight into the spatiotemporal coordination of activity in this structure. Later, Pellionisz and Llinas presented a mathematical computer model of frog Purkinje cells, using passive cable equations and the Hodgkin–Huxley equations in 62 compartments. Responses to climbing fiber activation and to anti- and orthodromic invasion in parallel fibers were studied; most of the electrical properties of Purkinje cells were demonstrated. In an elaborate digital computer simulation of the cerebellar cortex of the frog Pellionisz, Llinas, and Perkel simulated 1.68 million granule cells, 8285 Purkinje cells, and 16,820 mossy fibers. A stochastic net was used in which the connectivity was specified in terms of probability density functions. The model was used to display spatiotemporal patterns of activity in response to various input configurations. The authors emphasize the functional specificity inherent in the net even though the net is based on stochastic interconnection functions. They speculate that a limited amount of genetic specificity may be necessary to generate functional specificity in ceratin neuronal networks.

A wide range of functions have been attributed to the cerebellum, including coordination of interrelated muscle activities, fine control of reflexes, timing of motor events, and learning of motor skills. These have been tabulated and discussed by Pellionisz. Most contemporary models of cerebellar function have been greatly influenced by one of two points of view. The first is that the primary role of the cerebellum is as the principal agent in learning of (largely voluntary) motor skills. This idea was first modeled by Brindley in 1964, elaborated by Marr in 1969, and contributed to by a number of people including Albus, Gilbert, Dunin-Barkowski, and Grossberg. The essential picture here is quite simple. The chief neurons of the cerebellar cortex (the Purkinje cells) learn to recognize a pattern over many of their input fibers (the parallel fibers) if the pattern coincides with the deep depolarization of the Purkinje cell evoked by the dense arbor of climbing fibers which innervates it. Specifically, appropriately timed signals to a Purkinje cell from the climbing fibers lead to long-term modification of the synaptic efficacy of junctions to the same cell from the parallel fibers. In Albus's theory the important distinction is that the efficacy of the parallel fiber–Purkinje cell synapses is reduced rather than increased in the learning. In this theory each inferior olivary cell signals a cerebral instruction for an elemental movement such as bending a single finger joint. Contextual information regarding proprioception and balance is also projected to the Purkinje cells.

A number of quantitative models of this theory have been presented. For example, Fujita reformulated this model with linear system analysis into an "adaptive linear filter" model of the cerebellum. This model functions as a phase lead or lag compensator with learning capability and is applied to the vestibulo-ocular reflex. Melkonian *et al.* simulated synaptic modification in cerebellar networks and explored learning processes in the context of the classical conditioning paradigm. Dunin-Barkowski and Larionova simulated a network consisting of one Purkinje cell and 20,000 granule cells. The demonstrated associative information recall and a storage capacity of about 0.6 bit per binary memorizing synapse.

In an expanded view in this stream of theorizing, Ito hypothesized that the cerebellum serves as an active coordinator of reflex action in the nervous system. As one example of this level of control, Ito hypothesized that the cerebellum monitors the vestibulo-ocular reflex which produces eye movements to compensate for head movements and prevent images in the visual field from moving across the retina when the head moves. In Ito's theory, when the reflex is inadequate, error signals which indicate "slip" of images across the retina activate Purkinje cells in the flocculus of the cerebellum via climbing fiber inputs to produce a correction of the vestibulo-ocular reflex performance.

The second stream of theorizing, due largely to the computer science-oriented theorists, recommends that the function of the cerebellum can be best conceived within the entire motor system of which it is a part. Arbib advocated this view some time ago. More recently it has been championed vociferously by Pellionisz. Pellionisz has also presented a powerful and coherent model of overall cerebellar function framed in the language of his tensor transformation model of neuronal junctions and circuits. In this theory the nervous system represents external space–time events in the coordinated activity in neuronal networks, which may be represented as continua in reference frames, and transforms these events among and between networks in ways that can be modeled with tensorlike transformations, while the cerebellum acts to coordinate movements by establishing coincidences of goal-directed movements of limbs in space–time with external targets. In this theoretical view learning is seen as a working part of cerebellar function but not as its salient operational feature. Pellionisz applies this model to the modification of vestibulo-ocular reflexes by the cerebellum. He and Ito both see the action of the cerebellum on these reflexes as a late addition in evolution, a "side path," a supplementary refinement of sensorimotor control.

Thompson has elaborated a model of neuronal substrates of basic associative learning and conditioning as exemplified in the nictitating membrane and eyelid response. One implication of this theory as discussed by McCormack, Thompson, and colleagues is that the engram for this response in the rabbit is laid down at least in part in the cerebellum.

Torioka presents a random neural net model for cerebellar cortex in which the significance of inhibitory connections and numbers of connections for pattern separation are emphasized. Nagana and Ohmi present a model for Golgi cells in the cerebellar cortex in which feed-forward inputs to Golgi cells keep the firing rates of granule cells approximately constant over wide variations in mossy fiber activity, while feedback inputs produce oscillations in granule cell activity. Licata *et al.* present a simulation model in which the interpositus nucleus, with the help of negative feedback from muscles and acting through the rubrospinal tract, serves as an interface between programming and executing motor structures.

Basal Ganglia

The basal ganglia (consisting of the neostriatum–caudate nucleus and putamen, globus pallidus, and sometimes other brain stem centers such as the substantia nigra, red nucleus, subthalamic nucleus, claustrum, and amygdala) have long been thought to play important roles in sensorimotor integration and perhaps preimmanent coordinating functions in sensorimotor integration in lower vertebrates such as reptiles. For example, degeneration of neurons within the neostriatum is well known in Huntington's disease, and degeneration of dopaminergic input to the neostriatum occurs in Parkinson's disease. Until recently, however, the specific operations and functions of these regions were largely unknown. Current studies of the neostriatum tend to emphasize its intregral functioning within a feedback loop involving the globus pallidus, the ventroanterior nucleus of the thalamus, and the cerebral cortex. Groves, for example, has advocated a theory in which the neostriatum is involved in preparation for and execution and guidance of voluntary movement in response to the analysis and volition of the cerebral cortex. In Groves' model the neostriatum receives its commands from widespread areas of the cerebral cortex and the brain stem and exerts its influence on voluntary movement through the thalamocortical projections and their impact on the motor cortex. Groves further models the neuronal networks in neostriatum in two functional cell systems: a lateral inhibitory network comprising the common Spiny I neurons which inibit their targets and a Spiny II cell cluster consisting of one excitatory efferent neuron and three interneurons.

Wilson has also theorized on neostriatal circuitry. So far no mathematical or computer models of this system have appeared.

Behavioral

The explosively expanding field of robotics is creating a multidisciplinary mileu for advances in and reconceptualization of motor and sensorimotor behavior involving contributions from behavioral psychology, neurobiology, mechanical engineering, control system theory, optimization theory, and computer science. The few entries described here may be taken as representative of work in this area but certainly not as comprehensive or exhaustive.

Meyer, Smith, and Wright discuss models for the speed and accuracy of limb movements. The show that a model wherein a limb is driven by product of a force parameter and a time function can account for two fundamental relations. The first is Fitt's law, which states that movement time is a logarithmic function of the movement distance divided by the width of the target toward which the movement proceeds. In the second relation movement error (as measured by deviation from the target center) is a linear function of the movement distance divided by the movement time. The model assumes summetry, curvilinearity, and force–time rescalabilty and may be realized in terms of underlying biomechanical (mass–spring) or neurophysiological (EMG) mechanisms.

Nelson asserts that skilled movements appear to be generally constrained toward the objectives of "ease," economy of effort, or efficiency as well as toward specific task-oriented objectives. He applies optimization theory to optimize such ease in terms of specific costs related to movement time, distance, peak velocity, energy, peak acceleration, and rate of change of acceleration (jerk).

A number of persons have applied conventional control theory to modeling the behavior of sensorimotor systems. A representative model of this type is presented by Powers. The essential feature of this model is that the motor system is conceived as a hierarchically arranged array of control systems in which a comparator at each level compares a set of signals at its own level with corresponding target levels for those signals provided by the next higher level and, on the basis of this comparison, produces an adjustment in the target values for the signals of the next lower level. (See further discussion in Chapter 8.)

Other investigators have begun to consider the structure and interrelations of various internalized motor programs. For example, Carter and Shapiro postulate the existance of internal generalized motor programs which consists of variant and invariant features rather than a set of separate motor programs for individual actions. In their concept overall movement spread is a variant feature, whereas the proportion of total time allocated to each segment in a compound movement is an invariant feature of the program.

In an overall article comparing robotics with biological motor systems, Hollerbach suggests that a hierarchical movement plan is developed at three levels of abstraction: an object level, where a task command such as "pick up the cup" is converted into a planned trajectory a joint level, where the object trajectory is converted to coordinated control of multiple joints; and an actuator level, where the joint movements are converted to appropriate motor or muscle activations. Hollerbach points out that in robotics a modular structure for planning and control has emerged in which modules for trajectory planning, compliance to environmental constraints, dynamics, and feedback control are operative at each of the three levels of movement planning. Hollerbach suggests that this modular hypothesis may also be representative of the organization of planning and control in biological motor processes in the nervous system.

Sensory Systems

Many features of the functions of sensory systems as classified, for example, by Shepherd (see Table 6.1) may be considered more satisfactorily in the framework of more general functional systems and will therefore be considered in other sections of this book. For example, receptors related to muscle and balance can be considered in the framework of sensorimotor or motor systems, and receptors associated with the chemical senses can be considered within autonomic systems. Further, the remaining "classical sensory" modes (vision, audition, somatosensation, taste, and smell) all have

Table 6.1
Classification of Sensory Modalities[a]

Modality	Receptor organ (cell)
Chemical	
Common chemicals	Various (free nerve endings)
Arterial oxygen	Carotid body (cells and nerve endings)
Toxins (vomiting)	Medulla (chemoreceptor cells)
Osmotic pressure	Hypothalamus (osmoreceptors)
Glucose	Hypothalamus (glucoreceptors)
pH (cerebrospinal fluid)	Medulla (ventricle cells)
Taste	Tongue and pharynx (taste bud cells)
Smell	Nose (olfactory receptors)
Somatosensory	
Touch	Skin (nerve terminals)
Pressure	Skin and deep tissue (encapsulated nerve endings)
Temperature	Skin and hypothalamus (nerve terminals and central neurons)
Pain	Skin and various organs (nerve terminals)
Muscle	
Vascular pressure	Blood vessels (nerve terminals)
Muscle stretch	Muscle spindle (nerve terminals)
Muscle tension	Tendon organ (nerve terminals)
Joint position	Joint capsule and ligaments (nerve terminals)
Balance	
Linear acceleration (gravity)	Vestibular organ (hair cells)
Angular acceleration	Vestibular organ (hair cells)
Hearing	Cochlea (hair cells)
Vision	Retina (photoreceptor)

[a]Adapted from "Neurobiology" by G. Shepherd, Oxford University Press, 1983.

fundamental linkages and contributions to sensorimotor systems, limbic systems, and neocortical and thalamocortical systems. Thus, this section will review quantitative models for what might be called "preproccessing in peripheral regions of classical sensory channels."

Vision

Feature Extraction   The central concept in studies of preprocessing in peripheral regions of the visual system (and in more central regions of the visual system, as we will see in Chapter 8) has been that of "feature extraction." This fundamental concept has been elaborated in two main streams. In a classical study in the first stream published in 1959, Lettvin, Maturana, McCulloch, and Pitts showed that retinal ganglion cells in the frog can be classified into four types: those that respond selectively to boundaries, moving curvatures, changing contrasts, and local dimming. The point of view of these authors is that the activites of these neurons are signaling features in the external environment which are particulary significant to the animal in obtaining prey or avoiding predators and that the retinal circuitry therefore extracts this information from the incident visual field. Hubel and Weisel showed in extensive studies that various neurons in the cat visual system respond selectively to borders, orientation, motion, length of line, and the like. The underlying assumption in this work is that neuronal circuitry abstracts features relevant to objects and their characteristics in the external field; the concept of figure–ground from the gestalt era is relevant here.

The second stream was initiated by Campbell and Robson, who showed that different neurons in the visual system respond preferentially to different characteristic lengths in the visual field. This can be conceptualized and investigated in terms of Fourier decomposition of visual fields into spatial frequencies. This point of view tends to emphasize more global features of the visual field than the first strean of work, to lead one to think in terms of texture more readily than in terms of objects, and to be related to holographic views of visual perception and brain function, as will be discussed in Chapter 9. The first point of view leads one to think of the brain as operating in terms of an internal model of the world, whereas the second lends itself more readily to alternative views of the interaction of the brain with externals.

Lateral Inhibition   The processes of feature extraction are brought about by local and largely lateral excitatory and inhibitory interactions in the circuitry of the visual system, of which lateral inhibition is the best known and most throughly studied. This mechanism was first elaborated by Hartline and Ratliff for the lateral connections in the eye of the horseshoe crab *Limulus* and subsequently for networks throughout the visual system and, with good reason, thought to operate ubiquitously throughout the central nervous system. Lateral inhibition emphasizes responses around boundaries in the visual field. Regions internal to uniformly illuminated zones (activated by comparable levels of excitation and inhibition) exhibit response levels intermediate between the high and low levels at boundaries. The mechanism of lateral inhibition can also produce sensitivity to temporal contrast if the time required to conduct the signals laterally is significant with respect to the operating times of the network.

The basic quantitative model for steady-state and time-varying lateral inhibition in the eye of *Limulus* was presented by Ratliff. Vertebrates tend to mediate lateral inhibition both peripherally and more centrally through separate local inhibitory cells. The dynamic and functional properties are similar, however, and Ratliff discussed many implications of this laterally mediated sensitivity to contrast and change in space and time for the psychophysics of vision in humans and animals.

A number of models have been presented to simulate preprocessing in retinal circuits making fundamental use of the properties of lateral inhibition. For example, Runge *et al.* in 1968 constructed an electronic analog simulation of the bird retina containing 145 cone cells, 58 horizontal cells, 80 bipolar cells, 88 amacrine cells, and 7 ganglion cells. Varju, and later Morishita and Yajima, used matrix algebra to describe the interactions in laterally inhibiting networks. Coleman and Renninger extended the theory for *Limulus* in spatially uniform fields, describing "burst" and "rest" periods of activity. More recently, retinal circuitry has been elaborately modeled by Siminov, who presents detailed electronic circuits for individual cones, horizontal cells, bipolar cells, amacrine cells, and ganglion cells and their interrelations. This modeling has been applied specifically to catfish and turtle retina. Electrical coupling of photoreceptors in the retina has been modeled by Marcelja. Van Ouwerkerk *et al.* studied the stability of the sombrero sensitivity distribution in the retina by Fourier analysis. Pinter showed how nonlinearities in lateral inhibition can affect the shape of the spatial-modulation transfer functions in the retina. Oguztorelli modeled the retinal circuitry with 12 model neurons. Richter and Ullman modeled the X and Y types of ganglion cells in the primate retina. In this model the X-type temporal response is determined primarily by the delay between center and surround contributions, whereas the Y-type response is generated in the inner plexiform layer by a derivativelike operation (mediated by recurrent inhibition in the dyad synaptic structure) on the bipolar cells' input followed by a rectification in the convergence of these inputs onto the Y ganglion cell. Fukurotani and Hara modeled receptive fields of the L cells in the carp retina with forward and feedback loops. Van Doorn and Koenderink presented a network model for the inner plexiform layer which is sensitive to the direction of moving patterns. Fromel presented a system-theoretical description of the cat's superior colliculus which exhibits velocity-dependent responses and direction specificity. Koch *et al.* modeled dynamic pattern generation in receptor fields by "jump"- or "flux"-like transitions.

Tensorial Approach   McCollum, Pellionisz, and Llinas applied the tensorial method of modeling neuronal networks to color vision. They assume that color vision is produced by the transformation of activity photoreceptors representing covariant components into a contravariant perception which we know as color at the bipolar level. This model includes activity vectors for the receptors, horizontial cells, and bipolar cells and matrices for transitions in activity between cell types as mediated by local neuronal circuitry. The model describes many basic features of color perception including the complete range of colors by particular mixtures of white and pink light as discovered by Land.

Audition

Representative contemporary models of peripheral preprocessing in the auditory system are thsoe of McMullen, Lewis, and Mountain. McMullen and Mountain match the summating potential of the cochlea with a feedback model including voltage-dependent cilia stiffness and nonlinear transducer channel resistance. Lewis models high-frequency rolloff and its attenuating influence on traveling waves in the cochlea. Lewis and Narins have also modeled extraordinary sensitivity in the ear to seismic vibrations.

Rozsypal has presented a computer simulation of an ideal lateral inhibition function for the auditory channel. He emphsizes that this function can compensate for the frequency-desharpening effect of the mechanical properties of the basilar membrane and thereby permit sharp frequency resolution. Nomoto has presented a digital computer simulation model of information processing in auditory networks. This model displays sharpening of spatial patterns by lateral inhibition which is augumented by spontaneous activity and random extrinsic input.

Models of peripheral preprocessing in the auditory system are published largely in specialized audition rather than mainstream neuroscience journals. Access to this considerable literature may be obtained through the reference list at the end of this chapter.

Somatosensation

A mathematical model of the pacinian corpuscle that accounts for receptor potentials and spikes has been presented by Grandori and Pedotti. Another model for mechanoreceptors is presented by Chorzempa. Licata *et al.* describe a computer simulation model of thalamic ventrobasal neurons during sleep and waking based on responses to three independent Poisson input pulse trains.

Olfaction

Freeman describes in numerous publications an elegant theoretical model for the mechanics of spatiotemporal propagation of activity in the olfactory bulb. A key element in this model is the distinction between interacting wave and pulse densities. Freeman shows that this model matches experimental observations.

Gustation

I have found no recent models of taste processes.

Central Control

The levels-of-construction point of view suggests that central control at the reptilian level of development is mediated primarily by sublimbic structures, which would imply primarily the basal ganglia (including the amygdaloid region) and/or the reticular formation and related centers and regions. Satisfactory elucidation of this level of control is significant in the levels-of-construction orientation because the nature of central control developed in more advanced species would be thought to reflect at least in part their original interfacing with these lower levels of central control. For example, Bishop and Penfield championed the notion of "centrencephalic" organization, in which many of the information-processing properties of more recently evolved neocortical systems are envisioned as highly skilled subroutines operating under the control of more primitive central regions in the reticular formation, basal, ganglia, and limbic system.

Reticular Formation

The reticular formation as an anatomical structure extends throughout the core of the brain stem, caudally throughout the core of the spinal cord, and rostrally into the limbic system, particularly the hypothalamus and the medical forebrain bundle, and into the thalamus. It is characterized by cells with rather sparse unbranching dendrites and many diffuse connections effected by the many axon collaterals emanating from each axon.

The reticular formation as a unitary functional concept, however, is not very clear. Early theorizing was based largely on the idea of the ascending reticular activating system (ARAS). This sytem was thought to correspond to an "arousal" dimension in mediating a continuum of states ranging from death through sleep, quiet waking, and on to a highly aroused state. The ARAS was pictured as being responsive to all sensory modalities, modulating the tone of motoneurons, being solely responsible for behavioral arousal, and in some way mediating attention. The network was thought to be highly multisypnatic, nonspecific, and essentially gland- or organlike in structural organization and function.

Subsequent work showed that, although the reticular formation is a major integrative center vitally involved in states of consciousness and arousal, the one-dimensional glandlike picture is an oversimplification of brain stem central control in many ways. Two models attempted to incorporate multiplicity of functioning in the reticular formation. Kilmer presented a model in which the function of the reticular formation is to commit an animal to one or another mode of behavior among approximately 25 modes such as sleep, eat, drink, fight, flee, hunt, search, mate, and groom. According to Kilmer *et al.* "the primary problem of the RF core is how, in a fraction of a second, can a million or more neurons reach a workable consensus as to the proper mode of total commitment." This model emphasizes parallel (as opposed to sequential) processing, which is gaining favor today as discussed in Chapter 9. Kilmer then constructed a computational model to effect such mode selection. In subsequent developments of the model, capabilities for development, generalization, discrimination, habituation, classical conditioning, extinction, avoidance conditioning, trial-and-error learning, and the like are included.

In a modeling study by MacGregor the reticular formation is viewed as consisting of a large number of quasi-independent "ladder nets." Each subnet signals a unique internal state of affairs according to its unique distribution of fibers. Individual nets are activated preferentially by different patterns of activity in the input fibers. The modeling study spells out various dynamic features of ladder nets such as their tendency to fire collectively in groups, to exhibit characteristic responses called "recruitment fuses" (wherein many constituent neurons contribute in sequence one action potential to a coordinated wavelike response), and to respond selectively to input fibers which conduct at the same speed as the fibers of the net. The last property would render reticular networks particularly sensitive to slowly conducting pain fibers.

Brain Stem Control of States

McCarley has presented an interesting representation of the control of states of consciousness by brain stem centers by modeling the interactions between neuron groups in the locus coeruleus (LC) and the nucleus gigantocellularis (GC) in terms of a predator–prey model of the Lotka–Volterra type. In this model GC cells are excitatory and correspond to prey; LC cells are inhibitory and correspond to predators. Parameters in the predator–prey interpretation correspond respectively to reproduction, being eaten, competing for prey, and eating. The cyclic oscillations in population level in the predator–prey model correspond to the cyclic state oscillations during sleep in the brain stem model. High levels of activity in the GC cells and corresponding low levels of activity in LC cells correspond to desynchronized sleep.

Basal Ganglia

The functions of the basal ganglia in central control have remained largely obscure and, as far as we know, no overall models for their role in central control have appeared.

Bibliography

Autonomic Functions

1986   Bates, J. H. T. The minimization of muscular energy expenditure during inspiration in linear models of the respiratory system. *Biol. Cybernet.* **54**, 195-200.

1984   Cohen, M. I., and Feldman, J. L. Discharge properties of dorsal medullary inspiratory neurons: Relation to pulmonary afferent and phrenic efferent discharge. *J. Neurophysiol.* **51**, 753-776.

1983   Carpenter, G. A., and Grossberg, S. A neural theory of circadian rhythms: The gated pacemaker. *Biol. Cybernet.* **48**, 35-59.

1983   Carpenter, G.A., and Grossberg, S. Dynamic models of neural systems: Propagated signals, photoreceptor transduction, and circadian rhythms. *In* "Oscillations in Mathematical Biology" (R. Grissell, J. P. E. Hodgson, and M. Yanowich, eds.). Springer-Verlag, Berlin and New York.

        Feldman, J. L., and Speck, D. F. Interactions among inspiratory neurons in dorsal and ventral respiratory groups in cat medulla. *J. Neurophysiol.* **49**, 472-490.

1982   Kronauer, R. E., Czeiler, C. A., Pilato, S. F., Moore-Ede, M. C., and Weitzman, E. D. Mathematical model of the human circadian system with two interacting oscillators. *Am. J. Physiol.* **242**, R3-R17.

        van Dooren, J. A. H. Q., and Vis, A. (1982). A reinvestigation of the Geman-Miller respiratory oscillator model. *Biol Cybernet.* **44**, 205-210.

        Cohen, M. I. (1981). Central determinants of respiratory rhythm. *Annu. Rev. Physiol.* **43**, 91-104.

1981   Vilbert, J. F., Caille, D., Segundo, J. P. Respiratory oscillator entrainment by periodic vagal afferents: An experimental test of a model. *Biol. Cybernet.* **41**, 119-130.

1980   Kawato, M., and Suzuki, R. Two coupled neural oscillators as a model of the circadian pacemaker. *J. Theor. Biol.* **86**, 547-575.

1979   Cohen, M. I. Neurogenesis of respiratory rhythm in the mammal. *Physiol. Rev.* **59**, 1105-1173.

1977   Beeler, G. W., and Reuter, H. Reconstruction of the action potential of ventricular myocardial fibers. *J. Physiol. (London)* **268**, 177-210.

        Cohen, M. I., and Feldman, J. L. Models of respiratory phase-switching. *Fed. Proc. Fed. Am. Soc. Exp. Biol.* **37**, 2367-2374.

        Wyman, R. J. Neural generation of the breathing rhythm. *Annu. Rev. Physiol.* **39**, 417-448.

1976   Feldman, J. L. (1976). A network model for control of inspiratory cutoff by the pneumotaxic center with supportive experimental data in cats. *Biol. Cybernet.* **21**, 131–138.

German, S, and Miller, M. Computer simulation of brainstem respiratory activity. *J. Appl. Physiol.* **41**, 931–938.

1975   McAllister, R. E., Noble, D., and Tsien, R. W. (1975). Reconstruction of the electrical activity of cardiac Purkinje fibres. *J. Physiol. (London)* **251**, 1–59.

Mitchell, R. A., and Berger, A. J. (1975). Neural regulation of respiration. *Am. Rev. Respir. Dis.* **11**, 206–224.

1972   Noble, D. Conductance mechanisms in excitable cells. *Biomembranes* **3**, 427–447.

Rubro, J. E. A new model of the repiratory centre. *Bull. Math. Biophys.* **34**, 467–481.

1966   Noble, D. Applications of Hodgkin–Huxley equations to excitable tissues. *Physiol. Rev.* **46**, 1–50.

1962   Noble, D. A modification of the Hodgkin–Huxley equations applicable to Purkinje fibre action and pacemaker potentials. *J. Physiol. (London)* **160**, 317–352.

1932   Cannon, W. B. "The Wisdom of the Body." Norton, New York.

Spinal Muscle Circuits

1986   Gielen, C. C. A. M., and van Zuylen, E. J. Coordination of arm muscles during flexion and supination: Application of the tensor analysis approach. *Neoscience* **17**, 527–540.

1985   Koehler, W., and Windhorst, U. Responses of the spinal alpha–motoneurone–Renshaw cell system to various differentially distributed segmental afferent and descending inputs. *Biol. Cybernet.* **51**, 417–426.

Miller-Larsson, A. An analysis of extracellular single muscle fibre action potential field—modeling results. *Biol. Cybernet.* **51**, 271–284.

1984   Agarwal, G. C., Goodarzi, S. M., O'Neill, W. D., and Gottlieb, G. L. (1984). Time series modeling of neuromuscular system. *Biol. Cybernet.* **49**, 103–112.

Feenstra, B. W. A., Hofman, F., and van Leeuwen, J. J. Syntheses of spinal cord field potentials in the terrapin. *Biol. Cybernet.* **50**, 409–418.

Hogan, N. Adaptive control of mechanical impedance by coactivation of antagonist muscles. *IEE Trans. Autom. Control.* **29**, 681–690.

Moty, E. A., and Khalil, T. M. The application of information theory in EMG processing. *Eng. Med. Biol.* **37**, 11.7, 73.

Stein, R. B., and Oguztoreli, M. N. Modification of muscle responses by spinal circuitry. *Neuroscience* **11**, 231–240.

1983   Daunicht, W. J. Re-examination of a linear systems approach to the behavior of mammialian muscle spindles. *Biol. Cybernet.* **48**, 85–90.

Gouze, J., Lasry, J., and Changeux, J. Selective stabilization of muscle innervation during development. *Biol. Cybernet.* **46**, 207–216.

Hasan, Z. A model of spindle afferent response to muscle stretch. *J. Neurophysiol.* **49**, 989–1106.

Hemani, H., and Stokes, B. T. Four neural circuit models and their role in the organization of voluntary movement. *Biol. Cybernet.* **49**, 69–77.

Inbar, G. F., and Ginat, T. Effects of muscle model parameter dispersion and multiloop segmental interaction on the neuromuscular system performance. *Biol. Cybernet.* **48**, 69–83.

Oguztoreli, M. N., and Stein, R. B. Optimal control of antagonistic muscles. *Biol. Cybernet.* **48**, 91–99.

Oguztoreli, M. N., and Stein, R. B. A model for the spinal control of antagonistic muscles. *J. Theor. Biol.* **2**, 81–100.

Windhorst, U, and Koehler, W. Dynamic behavior of alpha motoneurone sub-pools subjected to inhomogeneous Renshaw cell inhibition. *Biol Cybernet.* **46**, 217–228.

1982   Agarwal, G. C., and Gottlieb, G. L. Mathematical modeling and simulation of the postural control loop. Part I *CRC Crit. Rev. Biomed. Eng.* **8**, 93–134.

Christakos, C. N. A study of the electromyogram using a population stochastic model of skeletal muscle. *Biol. Cybernet.* **45**, 5–12.

Christakos, C. N. A linear stochastic model of the single motor unit. *Biol. cybernet.* **44**, 79–89.

Christakos, C. N. A study of the muscle force waveform using a population stochastic model of skeletal muscle. *Biol. Cybernet.* **44**, 91–106.

Ducati, A., Parmiggiani, F., and Schieppati, M. Simulation of post-tetanic potentiation and fatique in muscle using a visco-elastic model. *Biol. Cybernet.* **44**, 129–133.

Kawahara, K., and Mori, S. A two compartment model of the stepping generator: Analysis of the roles of a stage-setter and a rhythm generator. *Biol. Cybernet.* **43**, 225–230.

Oguztoreli, M. N., and Stein, R. B. Analysis of a model for antagonistic muscles. *Biol. Cybernet.* **45**, 177–186.

Stein, R. B., and Oguztoreli, M. N. A model of whole muscles incorporating functionally important nonlinearities. *In* "Nonlinear Phenomena in Mathematical Sciences" (V. Lakshimikantham, ed.), pp. 749–766. Academic Press, New York.

1981   Cleveland, S., Kuschmierz, A., and Ross, H. G. Static input–output relations in the spinal recurrent inhibitory pathway. *Biol Cybernet.* **40**, 223–231.

Hatze, H. "Myocybernetic Control Models of Skeletal Muscle—Characteristics and Applications." Univ. of South Africa Press, Pretoria.

Hatze, H. Analysis of stretch responses of a myocybernetic model muscle fibre. *Biol. Cybernet.* **39**, 165–170.

Stein, R. B., and Oguztoreli, M. N. The role of gamma-motoneurons in mammalian reflex systems. *Biol. Cybernet.* **39**, 171–179.

1980   Agarwal, G. C., and Gottlieb, G. L. Mathematical modeling and simulation of the postural control loop. Part I. *CRC Crit. Rev. Biomed. Eng.* **8**, 93–134.

Hatze, H. Optimal process of neuro-musculo-skeletal control systems. *Biomathem.* **33**, 19–39.

Hatze, H. Neuromusculoskeletal control system modeling—a critical survey of recent developments. *IEE Trans. Autom. Control,* **25**, 375–385.

1979   Oguztoreli, M. N., and Stein, R. B. Interactions between centrally and peripherally generated neuromuscular oscillations. *J. Math. Biol.* **7**, 1–30.

1978   Hatze, H. A general myocybernetic control model of skeletal muscle. *Biol. Cybernet.* **28**, 143–157.

Hemani, H. Reduced order models for biped locomotion. *IEEE Trans. Syst., Man., Cybernet.* **SMC-8**, 321–325.

Menzies, H. E., Albert, C. P., and Jordan, L. M. Testing a model for the spinal locomotor generator. *Soc. Neurosci. Abstr.* **4**, 1219.

1977   Cleveland, S., and Ross, H. G. Dynamic properties of Renshaw cells: Frequency response characteristics. *Biol. Cybernet.* **27**, 175–186.

Hatze, H. A myocybernetic model of skeletal muscle. *Biol. Cybernet.* **25**, 103–119.

Hatze, H., and Buys, J. D. Energy-optimal controls in the mammalian neuromuscular system. *Biol.Cybernet.* **27**, 9–20.

Traub, R. D. Repetitive firing of Renshaw spinal interneurons. *Biol Cybernet.* **27**, 71–76.

1976   Bawa, P., Mannard, A., and Stein, R. B. Predictions and experimental tests of a visco-elastic muscle model using elastic and inertial loads. *Biol. Cybernet.* **22**, 139–145.

Oguztoreli, M. N., and Stein, R. B. The effects of multiple reflex pathways on the oscillations in neuro-muscular systems. *J. Math. Biol.* **3**, 87–101.

Stein, R. B., and Oguztoreli, M. N. Tremor and other oscillations in neuro-muscular systems. *Biol. Cybernet.* **22**, 147–157.

1975   Oguztoreli, M. N., and Stein, R. B. An analysis of oscillations in neuromuscular systems. *J. Math. Biol.* **2**, 87–105.

1974   Reitz, R. R., and Stiles, R. N. A viscoelastic-mass mechanism as a basis for normal postural tremor. *J. Appl. Physiol.* **37**, 852–860.

1973   Hatze, H. A theory of contraction and a mathematical model of striated muscle. *J. Theor. Biol.* **40**, 219–246.

1970   Coggshall, J. C., and Bekey, G. A. A stochastic model of skeletal muscle based on motor unit properties. *Math. Biosci.* **7**, 405–419.

1962   Reiss, R. F. A theory and simulation of rhythmic behavior due to reciprocal inhibition in small nerve nets. 1962 Spring Joint Computer Conference . *AFIPS Proc.* **21**, 171–194.

1938   Hill, A. V. (1938). The heat of shortening and the dynamic constants of muscle. *Proc. R. Soc. London, Ser, B.* **126**, 136–195.

Brain Stem Sensorimotor Systems

1986   Schmid, R., and Ron, S. A model of eye tracking of periodic square wave target motion. *Biol. Cybernet.* **54**, 179–187.

1985   Krenz, W. C., and Stark, L. W. Systems model for pupil size effect. II. Feedback model. *Biol. Cybernet.* **51**, 391–397.

Lara, R., and Arbib, M. A. A model of the neural mechanisms responsible for pattern recognition and stimulus specific habituation in toads. *Biol Cybernet.* **51**, 223–237.

Ostriker, G., Pellionisz, A., and Llinas, R. Tensorial computer model of gaze. I. Oculomotor activity is expressed in non-orthogonal natural coordinates. *Neuroscience* **14**, 483–500.

Pellionisz, A., and Llinas, R. Tensor network theory of the metaorganization of functional geometries in the central nervous system. *Neuroscience* **16**, 245–270.

Pellionisz, A. Tensorial aspects of the multidimensional approach to the vestibulo-oculomotor reflex and gaze. *In* ''Adaptive Mechanisms in Gaze Control: Facts and Theories'' (B. Jones and M. Jones, eds.). Elsevier, Amsterdam.

Schor, R. H. Design and fitting of neural network transfer functions. *Biol. Cybernet.* **51**, 357–362.

Tweed, D., and Vilis, T. A two dimensional model for saccade generation. *Biol. Cybernet.* **52**, 219–227.

1984   Ezure, K., and Graf, W. A quantitative analysis of the spatial organization of the vestibulo-ocular reflexes in lateral- and frontal-eyed animals. II.Neuronal networks underlying vestibulo-oculomotor coordination. *Neuroscience* **12**, 95–109.

Gillis, G. L. Godaux, E., Beaufays, F., and Henri, V. P. The optokinetic reflex in the cat: Modeling and computer simulation. *Biol. Cybernet.* **50**, 135–141.

Ostriker, G., Pellionisz, A., and Llinas, R. Tensorial computer movie display of the metaorganization of oculomotor metric network. *Soc. Neurosci. Abstr.* **10**, 162.

1983   Bahill, A. T., and McDonald, J. D. Model emulates human smooth pursuit system producing zero-latency target tracking. *Biol. Cybernet.* **48**, 213–222.

Cannon, S. C., Robinson, D. A., and Shamma, S. A proposed neural network for the integrator of the oculomotor system. *Biol. Cybernet.* **49**, 127–136.

Sun, F., Krenz, W. C., and Stark, L. W. A systems model for the pupil size effect. I. Transient data. *Biol Cybernet.* **48**, 101–108.

1982   Arbib, M. A., and Lara, R. A neural model of the role of the tectum in prey-catching behavior. *Biol. Cybernet.* **44,** 58–68.

Arbib, M. A., and Lara, R. A neural model of the interaction of tectal columns in prey-catching behavior. *Biol. Cybernet.* **44,** 185–196.

Buizza, A., and Schmid, R. Visual–vestibular interaction in the control of eye movement: Mathematical modeling and computer simulation. *Biol. Cybernet.* **43,** 209–223.

Lara, R., and Arbib, M. A. A neural model of interaction between pretectum and tectum in prey selection. *Cognit. Brain Theory* **5,** 149–171.

Lara, R., Arbib, M. A., and Cromarty, A. S. The role of the tectal column in facilitation of amphibian prey-catching behavior: A neural model. *J. Neurosci.* **2,** 521–530.

Segal, B. N., and Outerbridge, J. S. A nonlinear model of semicircular canal primary afferents in bullfrog. *J. Neurophysiol.* **47,** 563–578.

Usi, S., and Stark, L. A model for nonlinear stochastic behavior of the pupil. *Biol. Cybernet.* **45,** 13–21.

1981   Bohmer, A., and Allum J. H. J. Human optokinetic responses under quasi-open and closed loop conditions. *Biol. Cybernet.* **40,** 233–238.

Eckmiller, R. A model of the neural network controlling foveal pursuit eye movements. *In* "Progress in Oculomotor Research " (A. F. Fuch and W. Becker, eds.), pp. 541–550. Am. Elsevier, New York.

Feldman, A. G. The composition of central programs subserving horizontal eye movements in man. *Biol. Cybernet.* **42,** 107–116.

Jurgens, R., Becker, W., Kornhuber, H. H. Natural and drug-induced variations of velocity and duration of human saccadic eye movements: Evidence for a control of the neural pulse generator by local feedback. *Biol. Cybernet.* **39,** 87–96.

Zangemeister, W. H., Lehman, S., and Stark, L. Simulation of head movement trajectories: Model and fit to main sequence. *Biol. Cybernet.* **41,** 19–32.

Zangemeister, W. H., Lehman, S., and Stark, L. Sensitivity analysis and optimization for a head movement model. *Biol. Cybernet.* **41,** 33–45.

1980   Bahill, A. T., Latimer, J. R., and Troost, B. T. Linear homeomorphic model for human movement. *IEEE Trans. Biomed. Eng.* **BME-27,** 631–639.

Schmid, R., Buizza, and Zambarbieri D. A non-linear model for visual–vestibular interaction during body rotation in man. *Biol. Cybernet.* **36,** 143–151.

Schmid, R., Zambarbieri, D., and Sardi, R. A mathematical model of the optokinetic reflex. *Biol. Cybernet.* **34,** 215–225.

1978   Abel, L. A., Dell'Osso, L. F., and Daroff, R. B. Analog model for gaze-evoked nystagmus. *IEEE Trans. Biomed. Eng.* **BME-25,** 71–75.

Bock, O., and Zangemeister, W. H. A mathematical model of air and water caloric nystagmus. *Biol. Cybernet.* **31,** 91–95.

1977   Semmlow, J., and Chen, D. A simulation model of the human pupil light reflex. *Math. Biol.* **33,** 5–24.

1976   Davies, P., and Jones, G. M. An adaptive neural model compatible with plastic changes induced in the human vestibulo-ocular reflex by prolonged optical reversal of vision. *Brain Res.* **103,** 546–550.

Didday, R. L. A model of visuomotor mechanisms in the frog optic tectum. *Math. Biosci.* **30,** 169–180.

Kamath, B. Y., and Keller, E. L. A neurological integrator for the oculomotor control system. *Math. Biosci.* **30,** 34–352.

1972   Collewijn, H. An analog model of the rabbit's optokinetic system. *Brain Res.* **36,** 71–88.

1970   Sugie, N., and Wakakuwa, M. Visual target tracking with active head rotation. *IEE Trans. Syst., Man, Cybernet.* **SMC-6,** 103–109.

1968   Stark, L. "Neurological Control Systems." Plenum, New York.

1959   Lettvin, J. Y., Maturana, R., McCulloch, W. S., and Pitts, W. H. What the frog's eye tells the frog' brain. *Proc. IRE* **47,** 1940–1951.

Cerebellum

1986   Mauk, M. D., Steinmetz, J. E., and Thompson, R. F. Classical conditioning using stimulation of the inferior olive as the unconditioned stimulus. *Proc. Natl. Acad. Sci. U. S. A.* **83,** 5349–5353.

1985   Dunin-Barkowski, W. L., and Larionova, N. P. Computer simulation of a cerebellar cortex compartment. I. General principles and properties and properties of a neural net. *Biol. Cybernet.* **51,** 399–406.

Dunin-Barkowski, W. L., and Larionova, N. P. Computer simulation of a cerebellar cortex compartment. II. An information learning and its recall in the Marr's memory unit. *Biol. Cybernet.* **51,** 407–415.

Pellionisz, A. David Marr's theory of the cerebellar cortex; a model in brain theory for the Galilean combination of simplication, unification and mathematization. *In* "Brain Theory" (G. Palm and A. Aertsen, eds.). Spring-Verlag, Berlin and New York.

1984   Ito, M. "The Cerebellum and Neural Control." Raven Press, New York.

McCormick, D. A., and Thompson, R. F. Cerebellum: Essential involvement in the classically conditioned eyelid response. *Science* **223,** 296–299.

Pellionisz, A. (1984). Coordination: A vector–matrix description of transformations of overcomplete CNS coordinates and tensorial solution using the Moore–Penrose generalized inverse. *J. Theor. Biol.* **110,** 353–375.

Pellionisz, A. (1984). Tensorial brain theory in cerebellar modeling. *In* "Cerebellar Functions" J. Bloedel *et al.,* eds.), pp. 201–229. Springer-Verlag, Berlin and New York.

Thompson, R. F., Clark, G. A., Donegan, N. H., Levond, D. G., Madden, IV, J. Mamounas, L A., Mauk, M. D., and McCormick, D. A. Neuronal substrates of basic associative learning. *In* "Neuropsychology of Memory" (L. R. Squire and N. Butters, eds.), pp. 424–442. Guilford Press, New York.

1983   Pellionisz, A. Brain theory: Connecting neurobiology to robotics. Tensor analysis: Utilizing intrinsic coordinates to describe, understand and engineer functional geometries of intelligent organisms. *J. Theor. Neurobiol.* **2,** 185–211.

1982   Andersen, P. Cerebellar synaptic plasticity—putting theories to the test. *Trends Neurosci.* **5,** 324–325.

Fujita, M. Adaptive filter model of the cerebellum. *Biol. Cybernet.* **45,** 195–206.

Fujita, M. Simulation of adaptive modification of the vestibulo-ocular reflex with an adaptive filter model of the cerebellum. *Biol. Cybernet.* **45,** 207–214.

Melkonian, D. S., Mkrtchian, H. H., and Fanardjian, V. V. Simulation of learning processes in neuronal networks of the cerebellum. *Biol. Cybernet.* **45,** 79–88.

Pellionisz, A., and Llinas, R. Space–time representation in the brain; the cerebellum as a predictive space–time metric tensor. *Neuroscience* **7,** 2949–2970.

1981   Linas, R. Cerebellar modeling. *Nature (London)* **291,** 279–180.

McCormick, D. A., Lavond, D. G., Clark, G. A., Kettner, R. E., Rising, C. E., and Thompson, R. F. The engram found? Role of the cerebellum in classical conditioning of nictitating membrane and eyelid responses. *Bull. Psychon. Soc.* **18,** 103–105.

1980   Pellionisz, A., and Llinas, R. Tensorial approach to the geometry of brain function: Cerebellar coodination via a metric tensor. *Neuroscience* **5,** 1125–1136.

1979   Licata, F., Perciavalle, V., Sapienza, S., Urbano, A., and Viscuso, A. A computer model of intermediate cerebellum dynamic operations in motor control. *Biol. Cybernet.* **35**, 137–144.

Pellionisz, A., and Llinas, R. Brain modeling by tensor network theory and computer simulation. The cerebellum: Distributed processor for predictive coordination. *Neuroscience* **4**, 323–348.

1978   Nagano, T., and Ohmi, O. Plausible function of Golgi cells in the cerebellar cortex. *Biol. Cybernet.* **29**, 75–82.

Pellionisz, A. Synthesis of fragmented data on neuronal sytstems: A computer model of cerebellum. *Prog. Cybernet. Syst. Res.* **3**, 411–427.

Torioka, T. Pattern separability and the effect of the number of connections in a random neural net with inhibitory connections. *Biol. Cybernet.* **31**, 27–35.

1977   Eccles, J. C. An instruction–selection theory of learning in the cerebellar cortex. *Brain Res.* **127**, 327–352.

Pellionisz, A., and Llinas, R. A computer model of cerebellar Purkinje cells. *Neuroscience* **2**, 37–48.

Pellionisz, A., Llinas, R., and Perkel, D. H. A computer model of the cerebellar cortex of the frog. *Neuroscience* **2**, 19–36.

1975   Llinas, R. The cortex of the cerebellum. *Sci. Am.* **232**, 56–71.

1974   Gilbert, P. F. C. A theory of memory that explains the function and structure of the cerebellum. *Brain Res.* **69**, 1–20.

1972   Calvert, T. W., and Meno, F. Neural systems modeling applied to the cerebellum. *IEEE Trans. Syst., Man. Cybernet.* **2**, 363–374.

Ito, M. Neural design of the cerebellum motor control system. *Brain Res.* **40**, 81–84.

1971   Albus, J. S. A theory of cerebellar function. *Math. Biosci.* **10**, 25–61.

Llinas, R. Frog cerebellum: Biological basis for a computer model. *Math. Biosci.* **11**, 137–151.

1970   Pellionisz, A. Computer simulation of the pattern transfer of large cerebellar neuronal fields. *Acta Biochem. Biophys, Acad. Sci. Hung.* **5**, 71–79.

1969   de Callatay, A. Cerebellum and Cerebrum model with periodic processing, neurotubules conduction hypothesis. *Curr. Mod. Biol.* **3**, 45–61.

Marr, D. A theory of cerebellar cortex. *J. Physiol. (London)* **202**, 437–470.

1968   Szentágothai, J. Structuro-functional considerations of the cerebellar neuron network. *Proc. IEEE* **56**, 960–968.

1967   Eccles, J. C., Ito, M., and Szentágothai, J. "The cerbellum as as Neuronal Machine." Springer-Verlag, Berlin and New York.

1965   Eccles, J. C. Functional meaning of the patterns of synaptic connections in the cerebellum. *Perspect. Biol. Med.* **8**, 289–310.

1962   Marr, D. A theory of the cerebellar cortex. *J. Physiol. (London)* **202**, 437–470.

Basal Ganglia

Wilson, C. J., Groves, P. M., Kitai, S. T., and Linder, J. C. Three-dimensional structure of dendritic spines in the rat neostriatum, *J. Neurosci.,* (in press).

1983   Groves, P. M. A theory of the functional organization of the neostriatum and neostriatal control of voluntary movement. *Brain Res. Rev.* **5**, 109–132.

1982   Wilson, C. J., Chang, H. T., and Kitai, S. T. Origins of postsynaptic potentials evoked in identified rat neostriatal neurons by stimulation in substantia nigra, *Exp. Brain Res.,* **45**, 157–167.

1971   Kornhuber, H. H. Motor function of the cerebellum and basal ganglia. *Biol. Cybernet.* **8**, 157–162.

Behavioral and Robotics

1985   Asimov, I., and Frenkel, K. A. "Robots: Machines in Man's Image." Harmony
        Books, New York.
        Hunt, D. "Smart Robots: A Handbook of Intelligent Robotic Systems." Chapman &
        Hall, London.
1984   Carter, M. C., and Shapiro, D. C. Control of sequential movements: Evidence for
        generialized motor programs. *J. Neurophysiol.* **52**, 787–796.
1983   Nelson, W. L. Physical principles for economies of skilled movements. *Biol. Cybernet.*
        46, 135–147.
1982   Hollerbach, J. M. Computers, brains, and the control of movement. *Trends
        Neurosci.* **4**, 189–192.
        Meyer, D. E., Smith, J. E. K., and Wright, C. E. Models for the speed and accuracy
        of aimed movements. *Psychol. Rev.* **89**, 449–482.
1981   Albus, J. "Brains, Behavior, and Robotics." McGraw-Hill, New York.

Vision

1985   Oguztorelli, M. N., Caelli, T. M., and Steil, G. Information processing in vertebrate
        retina. *In* "Trends in the Theory and Practice of Nonlinear Analysis " (V.
        Lakshmikantham, ed.), pp. 345–356. North-Holland Publ., Amsterdam.
        Pinter, R. B. Adaptation of spatial modulation transfer functions via nonlinear
        lateral inhibition. *Biol. Cybernet.* **51**, 285–291.
        Siminoff, R. Model of the cone–horazontal cell circuit in the catfish retina. *Biol.
        Cybernet.* **51**, 363–374.
        Siminoff, R. Dynamics of the cone–horazontal cell circuit in the turtle retina. *Biol.
        Cybernet.* **52**, 1–14.
        Siminoff, R. Modeling effects of a negative feedback circuit from horizontal cells to
        cones on the impulse response of cones and horizontal cells in the catfish retina.
        Biol. Cybernet. **52**, 307–313.
1984   Pinter, R. B. On feedback and feedforward lateral in neural networks of visual
        systems. *In* "Encyclopedia of Systems and Control" (M. Singh, ed.). Pergamon,
        Oxford.
        Siminoff, R. Electronic simulation of cones, horizontal cells, and bipolar cells of
        generalized vertebrate cone retina. *Biol. Cybernet.* **50**, 173–192.
        Siminoff, R. Electronic simulation of ganglion cells of generalized verebrate cone
        retina. *Biol. Cybernet.* **50**, 193–211.
        Siminoff, R. Influence of amacrine cells of receptive field organization of ganglion
        cells of the generalized vertebrate cone retina: Eletronic simulation. *Biol. Cybernet.*
        50, 213–234.
1983   Edwards, D. H. Response vs. excitation in response-dependent and stimulus-
        dependent inhibitory networks. *Vision Res.* **23**, 469–472.
        McCollum, G., Pellionisz, A., and Llinas, R. Tensorial approach to color vision. *J.
        Theor. Neurobiol.* **2**, 23–28.
        Niznik, C. A. Robotic vision analogies for the chemistry of human vision. *Eng. Med.
        Biol.* **36**, 14.2, 62.
        Oguztoreli, M. N. Modeling and simulation of vertebrate primary visual system:
        Basic network. *IEEE Trans. Syst., Man, Cybernet.* **13**, 764–781.
        Pinter, R. B. Product term nonlinear lateral inhibition enhances visual selectivity for
        small objects or edges. *J. Theor. Biol.* **100**, 525–531.
        Pinter, R. B. The electrophysiological bases for linear and for monlinear product-
        term lateral inhibition and the consequences for wide-field textured stimuli. *J.
        Theor. Biol.* **105**, 233–243.

Siminoff, R. An analogue model of the luminosity-channel in the vertebrate cone retina. I. *Biol. Cybernet.* **46**, 101–110.

1982   Oguztoreli, M. N. Response analysis of vertebrate retina. *Biol. Cybernet.* **44**, 1–8.

Richter, J., and Ullman, S. A model for the temporal organization of X- and Y-type receptive fields in the primate retina. *Biol. Cybernet.* **43**, 127–145.

Srinisvasan, M. V.. Laughlin, S. B., and Dubs, A. Predictive coding: A fresh view of inhibition in the retina. *Proc. R. Soc. London, Ser. B* **216**, 427–459.

1981   Carpenter, G. A., and Grossberg, S. Adaptation and transmitter gating in vertebrate photoreceptors. *J. Theor. Neurobiol.* **1**, 1–42.

Koch, A. S., Nienhaus, R., Lautsch, M., and Lukovits, I. An advanced version of the dynamic receptor pattern generation model: The flux model. *Bio. Cybernet.* **39**, 105–109.

Siminoff, R. Modeling of the vertebrate visual system. III. Topological analysis of the cone mosaic. *J. Theor. Biol.* **91**, 435–476.

1980   Chorzempa, A. A neuron-like model for sensory illusions produced by two-point stimulation and the contrast phenomenon. *Math. Biosci.* **51**, 43–70.

Lennie, P. Parallel visual pathways: A review. *Vision Res.* **20**, 561–594.

Marcelja, S. Electrical coupling of photoreceptors in retinal network models. *Biol. Cybernet.* **39**, 15–220.

Oguztoreli, M. N. Modelling and simulation of vertebrate retina. *Biol. Cybernet.* **37**, 53–61.

Oguztoreli, M. N., and O'Mara, K. S. Modelling and simulation of vertebrate retina: Extended network. *Biol. Cybernet.* **38**, 9–17.

Siminoff, R. Modeling of the vertebrate visual system. I. Wiring diagram of the cone retina. *J. Theor. Biol.* **86**, 673–708.

Siminoff, R. Modeling of the vertebrate visual system. II. Application of the turtle cone retina. *J. Theor. Biol.* **87**, 307–331.

van Ouwerkerk, H. J., Tulp, J. H., Piceni, H. A. L., Roufs, J. A. F., and Blommaert, F. J. J. Instabilities in a continuous medium model for the retina. *Biol. Cybernet.* **39**, 11–14.

1979   Koch, A. S., Feher, G., and Lukovits, I. A simple model of dynamic receptor pattern generation. *Biol. Cybernet.* **32**, 125–138.

Rodieck, R. W. Visual pathways. *Annu. Rev. Neurosci.* **2**, 193–225.

1978   Berman, S. M., and Stewart, A. L. Laterally induced impedance effects in vision. *J. Math. Physiol.* **18**, 73–99.

1977   Fromel, G. Neuronal network characteristics in the cat superior colliculus. *Biol. Cybernet.* **28**, 15–26.

1976   van Doorn, A. J., and Koenderink, J. J. A directionally sensitive network. *Biol. Cybernet.* **21**, 161–170.

1975   Coleman, B. D. Consequences of delayed lateral inhibition in the retina of *Limulus*. I. Elementary theory of spatially uniform fields. *J. Theor. Biol.* **51**, 243–265.

Eckmiller, R. Electronic simulation of the vertebrate retina. *IEEE Trans. Biomed. Eng.* **BME-22**, 305–311.

Fukurotani, K., and Hara, K. I. A dynamic model of the receptive field of L-cells in the carp retina. *Biol. Cybernet.* **20**, 1–8.

Kelly, D. H. Spatial frequency selectivity in the retina. *Vision Res.* **15**, 665–672.

Stavenga, D. G., and Beersma, D. G. M. Formalism for the neural network of visual systems. *Biol. Cybernet.* **19**, 75–81.

1972   Eckmiller, R., and Grusser, O. J. Electronic simulation of the velocity function of movement-detecting neurons. *Bibl. Ophthalmol.* **82**, 274–279.

1969   Ratliff, F., Knight, B. W., and Graham, N. On tuning and amplification by lateral inhibition. *Proc. Natl. Acad. Sci. U.S.A.* **62**, 733–740.

1966   Lange, D., Hartline, H. K., and Ratliff, F. The dynamics of lateral inhibition in the compound eye of *Limulus*. II. *In* "The Functional Organizational of the Compound Eye" (C. G. Bernhard, ed.), pp. 425–449. Pergamon, New York.

1965   Ratliff, F. "Mach Bands: Quantitative Studies on Neuronal Networks in the Retina." Holden-Day, San Francisco, California.

        Varju, D. On the theory of lateral inhibition. *In* "Cybernetic of Neural Processes" (E. R. Caianiello, ed.), pp. 291–316. CNR, Rome.

1963   Ratliff, F., Hartline, H. K., Miller, W. H. Spatial and temporal aspects of retinal inhibitory interaction. *J. Opt. Soc. Am.* **53,** 110–120.

Audition

1985   Lewis, E. R., and Narins, P. M. Do frogs communicate with seismic signals? *Science* **227,** 187–189.

        McMullen, T. A., and Mountain, D. C. Model of d.c. potentials in the cohlea: Effects of voltage-dependent cilia stiffness. *Hear. Res.* **17,** 127–141.

        Rozsypal, A. J. Computer simulation of an ideal lateral inhibition function. *Biol. Cybernet.* **52,** 15–22.

1984   Lewis, E. R. Inertial motion sensors. *In* "Comparitive Physiology of Sensory Systems" (L. Bolis, R. D. Keynes, and S. H. P. Maddrell, eds.), pp. 587–610. Cambridge Univ. Press, London and New York.

        Lewis, E. R. High-frequency rolloff in a cochlear model wthout critical-layer resonance. *J. Acoust. Soc. Am.* **76,** 779–786.

1983   Hudspeth, A. J., and Corey, D. Analysis of the microphonic potential of the bullfrog's sacculus. *J. Neurosci.* **3,** 942–976.

        Mountain, D. C., Hubbard, A. E., and McMullen, T. A. Electromechanical processes in the cochlea. *In* "Mechanics of Hearing" (E. deBoer and M. A. Viergever, eds.), Delft Univ. Press, Delft.

1982   Hudspeth, A. J. Extracellular curent flow and the site of transduction by vertebrate hair cells. *J. Neurosci.* **2,** 1–10.

        Weiss, T. F. Bidirectional transduction in vertebrate hair cells: A mechanism for coupling mechanical and electrical processes. *Hear. Res.* **7,** 353–360.

1981   Crawford, A. C., and Fettiplace, R. An electrical tuning mechanism in turtle cochlear hair cells. *J. Physiol. (London)* **312,** 377–422.

        Crawford, A. C., and Fettiplace, R. Non-linearities in the responses of turtle hair cells. *J. Physiol. (London)* **315,** 317–338.

        Dallos, P. Cochlear physiology. *Annu. Rev. Psychol.* **32,** 153–190.

        Steele, C. R., and Taber, L. A., Three-dimensional model calculations for the guinea pig cochlea. *J. Acoust. Soc. Am.* **69,** 1107–1111.

        Taber, L. A., and Steele, C. R. Cochlear model including three-dimensional fluid and four modes of partition flexibility. *J. Acoust. Soc. Am.* **70,** 426–436.

1980   Crawford, A. C., and Fettiplace, R. The frequency selectivity of auditory nerve fibres and hair cells in the cochlea of the turtle. *J. Physiol. (London)* **306,** 79–125.

        Fettiplace, R., and Crawford, A. C. The origin of tuning in turtle cochlear hair cells. *Hear. Res.* **2,** 447–454.

        Hall, J. L. Cochlear models: Two-tone suppression and the second filter. *J. Acoust. Soc. Am.* **67,** 1722–1728.

        Wilson, J. P. Model for cochlear echoes and tinnitus based on an observed electrical correlate. *Hear. Res.* **2,** 527–532.

1979   Lutkenhoner, B., and Hoke, M. A new model of an auditory nerve fibre. *Scand. Audiol., Suppl.* **9,** 93–107.

        Nomoto, M. Spatial firing patterns of auditory neuron network modeling by computer simulation. *Biol Cybernet.* **32,** 227–237.

Smith, R. L. Adaptation, saturation, and physiological masking in single auditory-nerve fibers. *J. Acoust. Soc. Am.* **65**, 166–178.

Steele, C. R., and Taber, L. A. Comparison of WKB and finite difference calculations for a two-dimensional cochlear model. *J. Acoust. Soc. Am.* **65**, 1001–1006.

Steele, C. R., and Taber, L. A. Comparison of WKB calculations and experimental results for three-dimensional cochlear models. *J. Acoust. Soc. Am.* **65**, 1007–1018.

1978   Dallos, P. Biophysics of the cochlea *In* "Handbook of Perception" (E. C. Carterette, and M. P. Fiedman, eds.), Vol. 4, pp. 125–162. Academic Press, New York.

1977   Allen, J. B. Two-dimensional cochlear fluid model: New results. *J. Acoust. Soc. Am.* **61**, 110–119.

1976   Geisler, C. D. Mathematical models of the mechanics of the inner ear. *In* "Handbook of Sensory Physiology" (W. D. Keidel and W. D. Neff, eds.), Vol. 5, pp. 391–415. Spring-Verlag, Berlin and New York.

Steele, C. R. Cochlear mechanics. *In* "Handbook of Sensory Physiology" (W.D Keidel and W. D. Neff, eds.), Vol. 5, pp. 443–478. Springer-Verlag, Berlin and New York.

Wiederhold, M. L. Mechanosensory transduction in 'sensory' and 'motile' cilia. *Annu. Rev. Biophys. Bioeng.* **5**, 39–62.

1975   Schroeder, M. R. Models of hearing. *Proc. IEEE* **63**, 1332–1350.

1974   Siebert, W. M. Rancke revisited—a simple short-wave cochlear model. *J. Acoust. Soc. Am.* **56**, 594–600.

1973   Dallos, P. "The Auditory Periphery." Academic Press, New York.

Schroeder, M. R. An integrable model for the basilar membrane. *J. Acoust. Soc. Am.* **53**, 429–434.

1970   Pfeiffer, R. R. A model for two-tone inhibition of single cochlear nerve fibers. *J. Acoust. Soc. Am.* **48**, 1373–1378.

1968   Engerbretson, A. M., and Eldredge, D. H. Model for the nonlinear characteristics of cochlear potentials. *J. Acoust. Soc. Am.* **44**, 548–554.

1966   Johnstone, J. R., and Johnstone, B. M. Origin of summating potential. *J. Acoust. Soc. Am.* **47**, 504–509.

1965   Davis, H. A model for transducer action in the cochlea. *Cold Spring Harbor Symp. Quant. Biol.* **30**, 181–189.

Whitfield, I. C., and Ross, H. F. Cochlear-microphonic and summating potentials and the outputs of individual hair cell generators. *J. Acoust. Soc. Am.* **38**, 126–131.

1963   Glaesser, E., Caldwell, W. F., and Stewart, J. L. "An Electronic Analog of the Ear," Tech. Rep. AMRL-TDR-63-60, pp. 1–66. Biophys. Lab., Wright-Patterson AFB, Ohio.

1950   Peterson, L. C., and Bogert, B. P. A dynamical theory of the cochlea. *J. Acoust. Soc. Am.* **22**, 369–381.

Rancke, O. F. Theory of operation of the cochlea: A contribution to the hydrodynamics of the cochlea. *J. Acoust. Soc. Am.* **22**, 772–777.

Somatosensation

1982   Grandori, F., and Pedotti, A. A mathematical model of he Pacinian corpuscle. *Biol-Cybernet.* **46**, 7–16.

1981   Chorzempa, A. An attempt to determine the structure of the nervous system serving mechanoreceptors. *Biol. Cybernet.* **42**, 51–56.

1980   Grandori, F., and Pedotti, A. Theoretical analysis of mechano-to-neural transduction in pacinian corpuscles. *IEEE Trans. Biomed. Eng.* **BME-27**, 559–565.

Halvorsen, O., and Walloe, L. Summation of excitation and inhibition in a second order sensory neuron investigated by computer simulation. *Biol. Cybernet.* **36**, 153–158.

1978   Licata, F., Percialvalle, V. Sapienza, S., Urbano, A., and Viscuso, A. Computer
        simulated discharges of thalamic ventrobasal neurones during sleep and
        wakefulness. *Biol. Cybernet.* **31,** 55–62.
1971   Deutsch, S. A model of sensory receptor tranducer. *TIT J. Life Sci.* **1,** 29–40.
1966   Loewenstein, W. R., and Skalak, R. Mechanical transmission in a pacinian corpus-
        cle: An analysis and a theory. *J. Physiol. (London)* **182,** 346–378.

Olfaction

1987   Freeman, W. J. "Spatial EEG apterns, nonlinear dynamics, and perception: The
        neo-Sherringtonian view. *Brain Res. Rev.* (in press).
1986   Schild, D. "System analysis of the goldfish olfactory bulb: Spatio-temporal transfer
        properties of the mitral cell granule cell complex. *Biol. Cybernet.* **54,** 9–19.
        de Prisco, G. V., and Freeman, W. J. (1985). Odor-related bulbar EEG Spatial pat-
        tern analysis during appetitive conditioning in rabbits. *Behav. Neurosci.* **99,**
        964–978.
1985   Freeman, W. J. Analytic techiques used in the search for the physiological basis of
        the EEG. *In* "Handbook of EEG and Clinical Neurophysiology"(A. Gevins and
        A. Remond, eds.), Vol. 3A, Chapter 18. Elsevier, Amsterdam.
1981   Freeman, W. J. A neural mechanism for generalization over equivalent stimuli in the
        olfactory system. *SIAM-AMS Proc.* **13,** 25–38.
1980   Freeman, W. J. Use of spatial deconvolution to compensate for distortion of EEG by
        volume conduction. *IEEE Trans. Biomed. Eng.* **BME-27,** 421–429.
1979   Freeman, W. J. Nonlinear dynamics of paleocortex manifested in the olfactory EEG.
        Biol. Cybernet. **35,** 21–37.
        Freeman, W. J. EEG analyses give model of neuronal template-matching mechanism
        for sensory search with olfactory bulb. *Biol. Cybernet.* **35,** 221–234.
1978   Freeman, W. J. Models of the dynamics of neuronal populations. *Elec-
        troencephalogr. Clin. Neurophysiol. Suppl.* **34,** 9–18.
1975   Ahn, S. M., and Freeman, W. J. Neural dynamics under noise in the olfactory
        system. *Biol. Cybernet.* **17,** 165–168.
        Freeman, W. J. "Mass Action in the Nervous System." Academic Press, New York.

Gustation

1954   Beidler, L. M. A theory of taste stimulation. *J. Gen. Physiol..* **38,** 133–139.

Central Control

1986   Fiedler, K., and Stroehm, W. What kind of mood influences what kind of memory:
        The role of arousal and information structure. *Mem. Cognit.* **14,** 181–188.
1978   Morse, A., and Kilmer, W. A neural net capable of competitive and cooperative com-
        putation. *Biol. Cybernet.* **30,** 1–6.
1975   McCarley, R. W., and Hobson, J. A. Neuronal excitability modulation over the sleep
        cycle: A structural and mathematical model. *Science.* **189,** 58–60.
        Penfield, W. "The Mystery of the Mind." Princton Univ. Press, Princton, New
        Jersey.
1974   Pittman, J. C., and Feeney, D. M. Modulation of recurrent inhibition in cat associa-
        tion cortex by reticulocortical arousal. *Exp. Neurol.* **44,** 160–170.
1972   MacGregor, R. J. A model for reticular-like networks: Ladder nets, recruitment
        fuses, and sustained responses. *Brain Res.* **41,** 345–363.
1969   Kilmer, W. L., McCulloch, W. S., and Blum, J. A model of the vertebrate central

command system. *Int. J. Man-Mach. Stud.* **1,** 279–309.

1968   Kilmer, W. L., McCulloch, W. S., and Blum, J. An embodiment of some vertebrate command and control principles. *Curr. Mod. Biol.* **2,** 81–97.

Bishop, G. H. (1957). The place of a cortex in a reticular system. *In* "The Return Formation of the Brain Stem" (H. H. Jaspers, ed.), pp. 413–421. Little, Brown, Boston, Massachusetts.

7

Limbic Systems

Appetitive/Motivational/Reward/Emotional

Functional models for the biological and homeostatic level of motivation have usually centered around the concept of drive reduction. Salient representative models are those based on mutually exciting and inhibiting centers in the medial hypothalamus for hunger and for thirst. A representative overall control system model of this type as presented by Lindsay and Norman is shown in Fig. 7.1. In this model an imbalance between target state and state monitor triggers drive or arousal, which in turn initiates seeking behavior and primes certain target images. When the seeking behavior produces a sensory image which matches the target image, the state goal comparator triggers consummatory behavior provided other contingent circumstances are compatible with this behavior. Consummatory behavior readjusts the internal biochemical state in the direction of the target state and thereby reduces drive and motivational level. The model is fundamentally similar to the concept of homeostasis and the maintenance of the internal milieu as originated by Cannon for the autonomic system. The concept of arousal as mediated by the reticular formation is compatible with this model and easily finds a place within it.

Pleasure or reward centers are thought to be triggered by the combination of an "incentive stimulus" and an "action contingency situation." (Winnie the Pooh opines that although eating honey is quite nice, there is a moment just before the eating that is a little nicer but he doesn't know exactly what that is called.) The concepts of pleasure and pain centers are compatible with the model of Lindsay and Norman for biological homeostatic level of motivation.

Gallistel, Shizgal, and Yeomans review the electrophysiological properties of the neural system mediating reinforcement and motivation effects of the

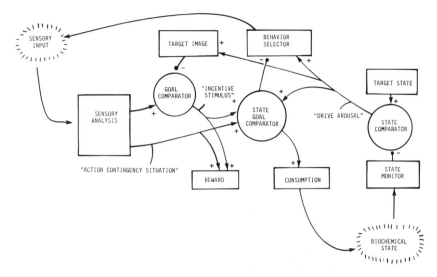

Fig. 7.1. Model of appetitive/motivational/reward system. (Adapted from "Human Information Processing" by P. H. Lindsay and D. A. Norman, copyright © 1972 by Harcourt Brace Jovanovich, Inc. Reprinted by permission of the publisher.)

medial forebrain bundle (the primary locus of pleasure/reward centers) to try to define more precisely the neurophysiological basis of conditioning and motivation. They note that the first-order neurons integrate over long time periods, accommodate slowly, and exhibit postinhibitory rebound. The second-order neurons exhibit surprisingly simple spatial and temporal integrative properties.

Clynes has described individually recurring temporal forms of finger pressure ("sentic" forms) associated individually with emotions of anger, hate, grief, love, sex, joy, and reverence. He has also shown that these emotions tend to be individually identified in acoustic sinusoidal carrier tones when these are frequency- and amplitude-modulated according to the same corresponding time forms.

The pleasure and pain centers are thought to serve as links to higher-order cognitive processes involving reinforcement and conditioning. Along these lines, Buck has claimed an increasing recognition in the psychological community that behavior can be inititaed by external stimuli and cognitive processes as well as by the biological deficits that have served as the cornerstone of the drive reduction model. Expectation of reward and adaptive behavior that anticipates homeostatic deficits before they occur seem to imply substantial cognitive participation. Buck goes on to postulate a theory of hierarchically organized primary motivational/emotional systems which he calls "primes." He suggests that emotion corresponds to the readout of emotional potential inherent in parts of these systems and can occur at various levels including adaptive, homeostatic, outward expressive, and direct experiential.

Conditioning and Paleocognition—The Hippocampus

The hippocampus is fundamentally involved in cognition, particularly in the learning and storing of memory traces. Many theories and models over the last 20–30 years have linked hippocampal function to simple classical conditioning and viewed the hippocampus as operating primarily conjointly with the motivational/reward system. This point of view is compatible with the view of the limbic system as a wholistic functional unit. In this view one looks to the hippocampus as the physiological substratum of classical conditioning concepts such as unconditioned and conditioned stimuli, reinforcements, habits, and drives. More recent findings and theories have tended to emphasize the more purely cognitive features of hippocampus function in learning. The salient concept is that of the hippocampus as a cognitive map.

A possible theoretical framework for these various points of view has been elaborated by Thompson on the basis of ideas originally presented by Squire. In this theory the hippocampus is the substratum of declarative learning (learning "what"), whereas procedural learning (learning "how") resides in other regions, most likely the cerebellum. The theory allows the possibility that declarative memory developed in the hippocampus from more ancient procedural learning operations, the latter capacity then moving to the cerebellum. In this view simple classical conditoning, which is largely procedural learning, is operative primarily in cerebellar rather than hippocampal circuitry in higher forms. On the other hand, when conditioning paradigms become complicated enough to involve declarative learning, hippocampal circuitry is necessarily involved. This theory accounts for the fundamental experimental observations that unit activity in the hippocampus is markedly engaged during conditioning procedures, the hippocampus is not necessary for the learning of simple conditioned responses and the hippocampus is necessary when greater demands are placed on the system as in latent inhibiting discrimination reversal and trace conditioning. Thompson, Berger, and various colleagues have elaborated this general theory for the neuronal substrates of associative learning and classical conditioning primarily in the hippocampus and cerebellum as exemplified in the nictitating membrane and eyelid response. Gluck and Thompson have presented a computational circuit model to accompany this theory, with specific reference to elementary associative learning in *Aplysia*.

In 1971 Marr presented a model for the hippocampus which limits its capabilities to memorizing (as opposed, for example, to neocortex, to which he attributes the ability to classify as well as memorize). He also suggested that very little information about a single learned event is necessary to provoke its recall and that the granule and pyramidal cells are memorizing cells. Marr arrived at these conclusions from consideration of the numerical constraints imposed by hippocampal anatomy and elaborated the operations of his model in terms of the hippocampal circuitry.

A model which stresses the participation of cognition in motivational behavior is presented by Grossberg. In this model the operations of conditioning-related concepts such as reinforcement, drive, incentive motivation, and habit are delineated in terms of hypothetical neural networks and related to cognitive concepts such as expectancy, competition, and resonance. The model emphasizes mechanisms of how short-term and long-term memory work together to control a shifting balance in processing of expected and unexpected events. Drive, reinforcer, and arousal inputs work together to regulate motivational baseline and use the hippocampus as a final common path. Grossberg criticizes O'Keefe and Nadel's idea of a cognitive map on philosophical and scientific grounds and suggests an alternative notion of a bilaterally organized motor map of approach and avoidance gradients wherein the bilateral asymmetry of the map's activity pattern at any time controls the network's approach and avoidance tendencies at that time. O'Keefe and Nadel's data on place learning are seen as probing properties of the latter pathway.

Grossberg has elaborated this model in great detail, linking many experimental findings and theoretical concepts in psychology to imaginatively predicted dynamic operations of hypothetical neuronal networks and then further mapping these neuronal network predictions speculatively onto likely key areas of the brain. In principle, his work allows the possibility of linking specific observed neuronal activity in neurons and populations through the level of inferred constructs (as discussed in Chapter 1) to the level of behavioral and experiential variables all within the framework of cohesive operational theories.

The central idea in the cognitive map theory of O'Keefe and Nadel is that the hippocampus is structured to produce multiple sets or groups of neurons which fire specifically in association with certain sets of related features in the animal's environment. That is, the animal houses within its hippocampus representations of different regions of the external world and operates cognitively with reference to these internal representations. The mapping occurs in three stages. The first stage is the organizing of environmental inputs into schema within the dendate gyrus; the second stage occurs in CA3, where the initial part of the map is made by representing places in an environment and the connections between those places; the third stage, which occurs in CA1, continues the map and contains the "misplace" system, which signals the presence of something new in a place or the absence of something old.

Rawlins presents an alternative theoretical model in which the hippocampus is seen as a high-capacity, intermediate-term memory store which processes stimuli of all kinds but only when there is a need to associate those stimuli with other events that are temporally discontiguous.

Stanley has modeled the dendate gyrus region of the hippocampus as a memory network capable of learning sequences of inputs separated by various time intervals and of repeating those sequences when cued by their initial portions.

Zipser has presented a model to show how neuronal networks might operate dynamically in constructing place fields within the framework of the O'Keefe and Nadel cognitive map theory and has also presented a theoretical model of classical conditioning of pyramidal cells of the hippocampus. His model uses a presynaptic learning rule to explain neuron recruitment and a tapped delay line to account for the temporal learning properties of the system and the spectrum of individual unit responses.

McDowell, Bass, and Kessell derive a rate equation for reinforcement effects which includes interreinforcement and interresponse intervals and avoids the assumption of optimal benefit which characterizes most such theories.

Hippocampal Anatomy and Electrophysiology

Teyler and DiScenna review anatomical interconnections of the hippocampus and conclude that the topological attributes of the hippocampus represent a four-dimensional array with the dimensions being the afferent input to laminated dendritic zones, the intrinsic trisynaptic hippocampal circuit, the longitudinal and commisural association systems, and time. They note that the hippocampal–neocortical pathways maintain topological specificity and suggest that the hippocampus represents a coordinate system which is capable of reciprocally addressing neocortical loci in space and time and is thereby related to the role of the hippocampus in the storage and recall of information.

Traub, with several collaborators, has described computer simulations of epileptic seizures in neuronal networks of the hippocampal slice. These simulations involve 100 interrelated, multicompartmental neurons which are individually modeled with the extended Hodgkin–Huxley equations to include active calcium-mediated conductances as described in Chapter 3. The model includes direct synaptic interaction between neurons, the influence of extracellular potassium ion concentration, the possible coupling of neighboring neurons by extracellular electric fields, and electrotonic junctions between neurons.

Leung has presented a control system model for local interactions within the CA1 region, detailing the mathematical transfer functions for each junction in the model and showing how it accounts for various features of CA1 electrical activity including its tendency to oscillate in various frequency regions. Leung earlier presented a linear, lumped model for responses in CA1 to afferent input.

The most elaborate model for single hippocampal neurons is that presented by Traub and discussed in Chapter 3. This model includes representation of active calcium-mediated conductances and produces such salient activity as complex spikes and bursting. Chapters 15 and 21 of Part II of this book describe models of single hippocampal neurons containing passive and active dendritic regions and dendritic spines. These models show that increasing spine neck diameter by 30% can produce increases of 15-20% in peaks of PSPs observed at the soma. Brown, Frick, and Perkel have modeled hippocampal pyramidal cells in terms of the equivalent cylinder model of Rall as discussed in Chapter 3 and find that for these neurons $L = 0.9$; the equivalent time constants are 11.6 (granule), 15.3 (CA1 pyramidal), and 19.3 (CA3 pyramidal) ms, and the ratio of dendritic to somatic conductance is 1.4. Turner and Schwartzkroin have applied passive cable theory modeling to hippocampal pyramidal cells. Leung has modeled a CA1 pyramidal cell with a series of passive compartments representing the soma and different parts of the dendritic tree. He also computes a simulated extracellular field potential and explores the spatial properties of this potential when pyramidal cells are driven at the theta frequency at different sites of the dendritic tree by simulated septal input. McNaughton has presented a model for short-term plasticity in hippocampal synapses as indicated in Chapter. 4. Stanley's model also represents habituation in hippocampal circuits.

Bibliography

Appetitive/Motivational/Reward/Emotional

1986   Ono, T., Nakamura, K., Nishijo, H., and Fukada, M. Hypothalamic neuron involvement in integration of reward, aversion, and cue signals. *J. Neurophysiol.* **56**, 63–79.

1985   Buck R. Prime theory: An integrated view of motivation and emotion. *Psychol. Rev.* **92**, 389–413.

1981   Gallistel, C. R. Shizgal, P., and Yeomans, J. S. A portrait of the substrate for self-stimulation. *Psychol. Rev.* **88**, 228–273.

1980   Clynes, M. Generation of emotionally expressive sounds from the dynamic forms of expressive finger pressure. *Conf. Am. Acoust. Soc., 1980.*

        Clynes, M., and Walker, J. Sound patterns and movements: An introduction to a neurophysiologically based theory of musical rhythm. *Int. Cong. Acoust., 10th, 1980.*

1977   Olds, J. "Drives and Reinforcements: Behavioral Studies of Hypothalamic Functions." Raven Press, New York.

1972   Lindsay, P. H., and Norman, D. A. "Human Information Processing." Academic Press, New York.

Functional Models for the Hippocampus

1987   Gluck, M. A., and Thompson, R. F. Modeling the neural substrates of associative learning and memory: A computational approach. *Psychol. Rev.* **94**, 176–191.
1985   Rawlins, J. N. P. Associations across time: The hippocampus as a temporary memory. *Behav. Brain Sci.* **8**, 479–528.
1984   Teyler, T. J., and diScenna, P. LTP as a candidate mnemonic device. *Brain Res. Rev.* **5**, 15–28.
        Zipser, D. "A Theoretical Model of Hippocampal Learning during Classical Conditioning," *ICS Rep 8408*. University of California, San Diego.
1983   Baerger, T. W. Rinaldi, P. C., Weisz, D. J., and Thompson, R. F. Single-unit analysis of different hippocampal cell types during classical conditioning of rabbit nictitating membrane response. *J. Neurophysiol.* **50**, 1197–1219.
        McDowell, J. J., Bass, R., and Kessel, R. Variable-interval rate equations and reinforcement and response distributions. *Psychol. Rev.* **90**, 364–375.
        Thompson, R. F. Neuronal substrates of simple associative learning: Classical conditioning *Trends Neurosci.* **6**, 270–275.
        Thompson, R. F., Berger, T. W., and Madden, J. IV, Cellular processes of learning and memory in the mammalian CNS *Annu. Rev. Neursci.* **6**, 447–491.
        Zipser, D. "A computational model of hippocampus place-fields," *Tech. Rep. I.C.S.* University of California, San Diego.
1982   Gray, J. A. "The Neuropsychology of Anxiety: An Enquiry into the Functions of the Septo-Hippocampal System." Oxford Univ. Press, London and New York; see also multiple reviews of this book in *Behav. Brain Sci.* **5**, 469–534 (1982).
        Grossberg, S. A psychophysiological theory of reinforcement, drive, motivation, and attention. *J. Theor. Neurobiol.* **1**, 286–369.
        Grossberg, S. Processing of expected and unexpected events during conditioning and attention: A psychophysiological theory. *Psychol. Rev.* **89**, 529–572.
        Grossberg, S. Some psychophysiological and pharmacological correlates of a developmental, cognitive, and motivational theory. *In* "Cognition and Brain Activity" (J. Cohen, R. Karrer, and P. Tueting, eds.). N.Y. Acad. Sci. New York.
        Squire, L. R. The neuropsychology of human memory. *Annu. Rev. Neurosci.* **5**, 241–273.
1979   Olton, D. S., Becker, J. T., and Handelmann, G. E., Hippocampus, space, and memory. *Behav. Brain Sci.* **2**, 313–365.
1978   O'Keefe, J., and Nadel, L. "The Hippocampus as a Cognitive Map." Oxford Univ. Press (Clarendon), London and New York.
1976   Stanley, J. C. Simulation studies of a temporal sequence memory model. *Biol. Cybernet.* **24**, 121–137.
1971   Grossberg, S. Pavlovian pattern learning by nonlinear neural networks. *Proc. Natl. Acad. Sci. U.S.A.* **68**, 828–831.
        Marr, D. Simple memory: A theory for archicortex. *Philos. Trans. R. Soc. London, Ser. B* **262**, 23–81.
1968   Kimble, D. P. Hippocampus and internal inhibition. *Psychol. Bull.* **70**, 285–295.
1967   Douglas, R. J. The hippocampus and behavior. *Psychol. Bull.* **67**, 416–442.
1948   Tolman, E. C. Cognitive maps in rats and men. *Psychol. Rev.* **55**, 189–208.

Electrophysiology of the Hippocampus

1985   Knowles, W. D., Traub, R. D., Wong, R. K. S., and Miles, R. Properties of neural networks: Experimentation and modeling of the epileptic hippocampal slice. *Trends Neurosci.* **8**, 61–67.

Traub, R. D., Dudek, F. E., Taylor, C. P., and Knowles, W. D. Simulation of hippocampal afterdischarges synchronized by electrical interactions. *Neuroscience* **14**, 1033-1038.

Traub, R. D., Dudek, F. E., Snow, R. W., and Knowles, W. D. Computer simulations indicate that electrical field effects contribute to the shape of the epileptiform field potential. *Neuroscience* **15**, 947-958.

1984   Fricke, R. A., and Prince, D. A. Electrophysiology of dentate gyrus granule cells. *J. Neurophysiol.* **52**, 195-209.

Leung, L. S. Model of gradual phase shift of theta rhythm in the rat. *J. Neurophysiol.* **42**, 1051-1065.

Miles, R., Wong, R. K. S., and Traub, R. D., (1984). Synchronized afterdischarges in the hippocampus: Contribution of local synaptic interactions. *Neuroscience* **12**, 1179-1189.

Teyler, T. J., and DiScenna, P. The topological anatomy of the hippocampus: A clue to its function. *Brain Res. Bull.* **12**, 711-719.

Traub, R. D., Knowles, W. D., Miles, R., and Wong, R. K. S. Synchronized afterdischarges in the hippocampus: Simulation studies of the cellular mechanism. *Neuroscience* **12**, 1191-1200.

1983   Brown, T. H., and Johnston, D. Voltage-clamp analysis of mossy fiber synaptic input to hippocampal neurons. *J. Neurophysiol.* **50**, 487-506.

Johnston, D. and Brown, T. H. Interpretation of voltage-clamp measurements in hippocampal neurons. *J. Neurophysiol.* **50**, 464-486.

Traub, R. D., and Wong, R. K. S. Synaptic mechanisms underlying interictal spike initiation in a hippocampal network. *Neurology* **33**, 257-266.

Traub, R. D., and Wong, R. K. S. Synchronized burst discharge in disinhibited hippocampal slice. II. Model of cellular mechanism. *J. Neurophysiol.* **49**, 459-471.

Turner, D. A., and Schwartzkroin, P. Electrical characteristics of dendrites and dendritic spines in intracellularly stained CA3 and dentate hippocampal neurons. *J. Neurosci.* **3**, 2381-2394.

1982   Leung, L. S. Nonlinear feedback model of neuronal populations in hippocampal CA1 region. *J. Neurophysiol.* **47**, 845-868.

Traub, R. D. Simulation of intrinsic bursting in CA3 hippocampal neurons. *Neuroscience* **7**, 1233-1242.

Traub, R. D., and Wong, R. K. S. Cellular mechanism of neuronal synchronization in epilepsy. *Science* **216**, 745-747.

1981   Brown, T. H., Fricke, R. A., and Perkel, D. H. Passive electrical constants in three classes of hippocampal neurons. *J. Neurophysiol.* **46**, 812-827.

Johnston, D. Passive cable properties of hippocampal CA3 pyramidal neurons. *Cell Mol. Neurobiol.* **1**, 41-55.

1980   Turner, D. A., and Schwartzkroin, P. A. Steady-state electrotonic analysis of intracellularly-stained hippocampal neurons. *J. Neurophysiol.* **44**, 184-199.

1979   Traub, R. B., and Llinas, R. Hippocampal pyramidal cells: Significance of dendritic ionic conductances for neuronal function and epileptogenesis. *L. Neurophysiol.* **42**, 476-496.

1978   Leung, L. S. Dynamic model of neuronal response in the rat hippocampus. *Biol. Cybernet.* **31**, 219-230.

1976   Mates, J. W. B., and Horowitz, J. M. (1976). Instability in a hippocampal neuronal network. *Comput. Programs Biomed.* **6**, 74-84.

Stanley, J. C. Simulation studies of a temporal sequence memory model. *Biol. Cybernet.* **24**, 121-137.

8

Neocortical and Thalamocortical Systems

Perhaps the broadest theoretical principle applicable to the organization of neocortical systems is the principle of levels of construction, advocated initially by Hughlings Jackson. In this view nervous systems are organized according to their phylogenetic evolution, and the development of new functional systems at any stage builds on and makes use of earlier developed functional systems. Thus, one may identify various more or less stable stages of functional organization (construction) in nervous systems corresponding to more or less stable stages of organic functioning. In this view function is the primary determinant of brain organization and anatomy is secondary.

MacLean has elaborated Jackson's view in a so-called "triune brain theory" wherein three stable levels of development are posited: a neomammalian level characterized by neocortical control; a paleomammalian level characterized by limbic system control; and a reptilian level characterized by basal ganglia and brain stem control. MacLean's view has been popularized by Sagan, who associates aggression and hostility primarily with the reptilian level, appetitive behavior and pleasure/pain primarily with the paleomammalian level, and objective cognition primarily with the neomammalian level.

Brown asserts that the emergence of neocortical left–right asymmetry in humans provides a quantal leap in cognitive ability and extends MacLean's theory to recognize two levels of neomammalian development: a lower level characteristic of dogs, horses, and apes and a higher level unique to human beings. Brown speculates freely on the characteristics of behavior and consciousness at each of these levels. Humans are distinguished cognitively from lower forms primarily by the exhibition of vast amounts of symbolic representation and processing, which are served by both a large degree of noncommitted cortical tissue and the functional asymmetry of the two hemispheres. The latter feature in particular allows a constant and continual interplay between analysis, comparison, sharpening, identification, content-based activity on the one hand and synthetic, relational, context-based activity on the other. Consciousness is asserted by Brown to exist in all these forms and to evolve from dim and body-centered at the reptilian level, through dreamlike at the paleomammalian level, and through "brighter" degrees at the neomammalian level.

Jaynes has also presented a theory for the development of consciousness in which consciousness is centrally dependent on neocortical asymmetry and restricted to modern human beings.

Penfield, on the other hand, argues that the brain is organized in a centrencephalic manner and that neocortical regions are largely subroutines for specialized operations and are under the overall functional control of brain stem and limbic regions. This view is consistent with the concept of the brain stem reticular arousal system which dominated much of brain research in the 1960s and early 1970s.

Luria has presented a theory for the functional organization of the human brain including both a theoretical orientation, involving the integration of a number of concepts from engineering systems theory, and a thorough and detailed elaboration of the identity and operations of specific functional systems in the brain. In his broad view the organization of the brain is best considered in terms of its functional systems and their dynamics and only secondarily in terms of its anatomy. He points out, for example, that functional loci move around within the brain with development and age. Luria also emphasizes that the organization of the brain should be considered within the levels-of-construction theory of Hughlings Jackson. Narrow views of localization of function are usually misleading: within localized regions of functional systems there exist equipotentiality and mass action, as enunciated by Lashley, and there is redundancy of information storage and representation. The nervous system is stochastic and noisy. In Luria's view the operations and organization of the nervous system are better considered from a top-down "systems" point of view rather than a bottom-up mechanism point of view. He further suggests that any form of higher activity involves the coordination of many brain parts and that focusing on any particular one of these may be misleading. Finally, he points out that the operation of the brain on information in any individual is heavily dependent on the idiosyncratic history of that individual.

Luria's specific model for brain organization partitions the neocortical level of brain organization into three functional systems: a unit for regulating and maintaining tone housed in the limbic system and brain stem; a unit for receiving and processing information occupying the posterior regions of the neocortex; and a unit for programming and overall regulation of function occupying the anterior regions of the neocortex.

The unit for receiving, analyzing, and processing information occupies the lateral and posterior surfaces of the cerebral cortex, comprising the parietal, temporal, and occipital lobes. The unit is concerned primarily with processing somatosensory, auditory, and visual information and integrating it with various internal schemes. In Luria's model this system consists of a hierarchy of three subunits: primary, secondary, and tertiary zones. Primary zones are the primary sensory receiving areas (17, 41, and 3). These zones are modality-specific and exhibit wired-in spatiotopic or tonotopic organization and high specificity in feature extraction. The secondary zones are the secondary sensory zones (18, 19, 21, 22, 1, 3, 5, and part of 40). These zones are modality-specific and synthesize input from primary areas into functional patterns. The tertiary zones are the posterior association areas of the cortex. They synthesize information from the various modalities to produce representations of "objects." They perform spatial organization of inputs, convert from successive stimuli to simultaneously processed groups, transfer to symbolic processing, and effect the memorizing of organized experience into "internal schemes." The primary zones have highly elaborated layer IV neurons, whereas the secondary and tertiary zones have amplified layers II and III and a great many short association neurons.

Three operational principles apply to the functioning of this unit according to Luria. The first is the law of hierarchical structuring, which posits that in youth the organization is from below upward, whereas in maturity it is from top downward. That is, in youth attention is paid to specific units of experience as they occur, whereas in maturity organization is dictated by the schema within which they are categorized. The second is the law of diminishing specificity from bottom to top, wherein the higher subunits deal with progressively generalized features. The third is the law of progressive lateralization, wherein the two hemispheres are very similar in the primary zones, differ somewhat in the secondary zones, and differ markedly in the tertiary zones. The dominant hemisphere becomes specialized for analytic, verbal, sequental, linear content and the other hemisphere becomes specialized for spatial, melodic, parallel, global, wholistic context.

In Luria's model the anterior unit for programming and regulation lies in the frontal lobe anterior to the precentral gyrus and contains the motor cortex and the frontal association areas. This unit serves will, volition, organization of conscious activity, planning, anticipation, worry, posture, and motor activity. The unit has three zones: tertiary (frontal association areas), secondary (premotor areas), and primary (primary motor areas). Plans and programs are formed in the tertiary and secondary areas and pass through the primary area for final preparation in conjunction with lower motor regions prior to transmission to muscle effector systems. The principles of hierarchical organization, lateralization, and diminishing specificity as described above for the posterior cortical system apply to this system also. (Luria's concept of the organization of language operations in this schema is discussed in Chapter 10.)

Eccles and Szentàgothai have also presented models for the global operations of neocortical systems.

Some years ago Barlow presented the idea that single neocortical neurons might individually represent percepts. This idea has not found favor in comparison with views that attribute percepts to ensembles of neurons or properties of ensembles of neurons such as Hebb's cell assemblies and John's statistical configurations, which are discussed in Chapter 9.

A description of properties of association areas of the cerebral cortex has been presented by Pandya and Sulter. Many of the generic and conceptual models discussed in Chapters 9 and 10 have direct relevance to the neocortical networks, particularly those of the posterior regions.

Anterior Unit for Programming and Regulation

Powers has presented a model which views the nervous system as a nested hierarchy of control systems in which the set points at any one level are determined by comparisons at the next higher level. The essential ingredient of the model is the idea that neural circuitry at a given level produces in a given situation a system of targets or set points for variables at the next lower level. These set points are compared at a given time with the actual values of variables in the lower level to compute a corresponding set of error signals for the lower level. The error signals in turn are converted by the neural circuitry of the lower level to produce sets of targets or set points for the variables of a yet lower level. And so on down through the system. Also essential to this model is the hierarchical stratification to a set of operations thought to be germane to purposive behavior. These include (from the bottom up) intensities, sensations, configurations, transitions, sequences, relationships, programs, principles, whole-system concepts, and extrasystem or ultimate influences.

The operations of the model can be read in two ways. First, from the bottom up it is imagined that the values of variables at a given level are determined by combinations of inputs from the next lower level. Second, from the top down it is imagined that behavior at the one level is governed by targets set by the next higher level. The model conceives that in purposive behavior one's perceptions are being controlled so as to be consistent with the cascaded sequence of desiderata as mapped onto the systems of set points throughout the system.

The main trigger at any stage is disharmony, measured as the sets of mismatches between inputs and desired set points. Disharmony at any stage triggers activity; the highest state is an equilibrium state in which all variables at all levels match their corresponding targets and correspondingly all error signals are zero. There is a rich multiplicity of variables at all levels and particularly at the lower levels. One implication of this is that the times required for error reduction are much longer for the higher levels, since matching variables at any level with the targets for that level requires multiple hierarchically nested looping through all lower subsystems.

Posterior Unit for Receiving and Processing Information

Association Cortex

In 1970 Marr presented a model showing how the networks of the cerebral coretx could use experience to produce "classificatory units" which could be used to interpret subsequent experience. Marr imagines that the pyramidal cells of the posterior cerebral cortex correspond to his classificatory units and that, when a given input event is presented to the brain, all the classificatory units to which it belongs are thrown into action so that the resulting distribution of activity over those cells signals the set of classes to which it belongs. Marr defines the essential diagnosis and interpretation operations which his model must perform in terms of probability theory and spells out interpretations of network connections and interactions in terms of the coefficients of this probability model.

Visual

Of all the operations of the posterior cortical unit, the functions of the visual system have been modeled most widely and thoroughly. The two central theoretical concepts behind contemporary modeling of visual function are feature extraction, which is associated with the identification of forms or objects or figure–ground discrimination, and spatial decomposition, which is often associated with textural analysis and more global views of visual operation.

Feature Extraction: Form, Object, Figure–Ground Discrimination   Models and concepts here are based fundamentally on the picture of visual system operation as evolved by Hubel and Wiesel. An early and influential model of this type was presented by Marr. In this model visual information processing is pictured as occurring in four stages. The first stage corresponds to the creation of a so-called primary sketch in which features such as edges, lines, shadings, and their various properties such as position, orientation, termination points, size and fuzziness are defined. The second stage corresponds to the grouping of contents in ways that are appropriate for later recognition. It corresponds essentially to the separation of figure from ground. Marr notes that since most of this separation can be carried out by techniques that do not depend on the particular images under consideration, figure–ground separation can normally precede the description of the shape of the abstracted form. He further notes that higher-level knowledge and purpose should influence overall control of rather than participation in this figure–ground discrimination. The third stage consists of processes which derive shape descriptions. Marr hypothesizes that these must involve means for identifying natural axes of a shape in its image and a mechanism for transforming viewer-centered axis specificaitons to specifications in an object-centered coordinate system. The fourth stage is shape recogniton. This involves a collection of stored shape descriptions and some means of associating newly derived descriptions from stage three processing with these stored descriptions. Marr goes on to elaborate hypothetical schemes whereby such abstractions and transformations might be effected.

A model similar to Marr's, called the "modulo system," has been presented by Fuhrmann. In this concept visual pattern recognition is effected by a network which extracts in successive functional layers contours, line segments, directions, vertices, etc. In the sixth and final functional layer of this model, the simplest known pattern which has the observed processed features is associated with the input field.

Stephen Grossberg with Mingolla and Cohen presented a model for visual form, brightness, and color estimation based on two parallel processes, a boundary contour process and a feature contour process, both of which are attributed to area 17 of the visual cortex.

Harth developed a model for visual perception based on an "alopex" principle in which positive feedback is emphasized and causes feature-specific enhancement of input. In this model "percepts" are produced by feature extraction using spatial derivatives and positive feedback. Borello *et al.* extended this model to produce "percepts" by feature extraction involving spatial derivatives and positive feedback. Foster presented a modeling approach to elucidate the natural structure (topological, metric, differentiable) of visual space.

Spatial Decomposition and Texture   In 1962 Julesz initiated an approach to visual perception based on generating stochastic texture pairs with idential second-order statistics and studying their effortless (preattentive) discrimination. Subsequent work, some in collaboration with Calli, showed that effortless discrimination must be the result of local features, whereas higher integrative processes seem adequately defined in terms of dipole autocorrelation or amplitude spectra of the textural domains. Local texture discriminatinon has been reduced to two features by Julesz.

Models in this stream are largely based on original experimental investigations by Robson, Campbell, and various colleagues, who showed that neurons throughout the visual system are particularly sensitive to various spatial frequences in the visual field, so that the visual system as a whole might be pictured as operating in terms of spatial frequency decomposition of the visual field.

In a theoretical model based on spatial decomposition, Schwartz notes that cortical columns could produce "difference maps" based on comparing spatial frequency decompositions of two input channels. Such difference maps from left and right retinal views could provide information on binocular disparity of objects in the scene. Physical objects located at a common distance would possess common textural values according to their manifestation of similar spatial frequency distributions. In this model the difference map is carried by a spatial frequency modulation determined by the period of columnar interlacing. He suggests that difference mapping may reflect a general synergistic mechanism relating topographic mapping and columnar architecture, which reduces the problem of feature extraction and segmentation for depth and color opponent channels to a single textural mechanism.

Earlier, Gafni and Zeevi presented a two-dimensional Fourier transform model for visual signal processing in which cortical cells sensitive to moving gratings are interpreted as narrowband spatial–temporal frequency filters. Glezer and Cooperman presented a similar model in which the transform is performed over the logarithm contrast function.

Legendy interprets the Campbell and Robson data in terms of conventinal receptive field organizations without the assumptions of Fourier analysis.

Models of Operations   Reitboeck and Altman have presented a model for size- and rotation-invariant pattern processing in the visual system. They note that the mapping of retinal space onto the striate cortex of some mammals can be appproximated by a log-polar function and that an exact log-polar transform converts centered scaling and rotation into translations. Therefore, any subsequent transformation which exhibits translation invariance will produce overall size and rotation invariance in the visual processing. They present a model for producing translation invariance in simple neural networks and show with several examples that this coupled with the log-polar mapping can produce object recognition independent of size, orientation, and position.

Mallot has explored mathematical functions which model the retinotopic mapping onto the cat's cortical areas 17, 18, and 19. All three mappings are found to be simple modifications of a complex power function with an exponent of 0.43. This function is further decomposed to give an intermediate stage which is common to all three mappings and can be regarded as a model of the retinotopic mapping onto the lateral geniculate nucleu (LGN).

Cowan has presented equations for a conformational projection of the visual field onto the visual cortex in terms of retinal polar coordinates and cortical rectangular coordinates. Various natural and drug-induced hallucination patterns are describable in terms of this conformable mapping.

Bennett and Hoffman have presented a mathematical model for the computation of three-dimensional structure and motion from changing retinal images. They show that objects whose textural elements rotate about a common axis but at varying angular velocities can in principle be computed from three successive retinal images of four texture elements or from four successive images of two texture elements.

Lehky and Sugie independently presented models for the binocular interaction of left and right channels relevant to brightness perception, Fechner's paradox, and retinal rivalry. Lehky's model involves nonlinear reciprocal feedback inhibition followed by linear summation between channels. Montalvo extends a self-organizing feature-extracting network model of Malsburg to two feature dimensions to encompass line orientation and color and applies it to contingent aftereffects. Hirai and Fukushima apply a multilayered network model of analog threshold elements to describe binocular parallax.

Burt and Sperling have presented a model for visual apparent motion based on the concept of path selection wherein the path with the greatest strength becomes the dominant path. In this model motion perception is based on the outputs of elementary detectors which are scaled replicas of each other, all having the same geometry and time delays and differing only in size and orientation. Caelli and Dodwell, on the other hand, claim that perceived motion paths cannot be explained solely in terms of feature-specific analyzers and present a dynamic network model with simple filtering and summation properties to predict the geometric paths of apparent motion.

Fisher presents a model of visual search incorporating serial scanning mechanisms and capacity limitations framed as a continuous time Markov process. Hamada presents a mathematical multistage model for border contrast wherein gradients of convolved luminance function are used to detect inflection points and averaged.

Wilson has presented a model of interacting analog elements for direction selectivity and threshold motion perception. A mathematical model by Marko describes threshold modulation for a variety of spatiotemporal patterns.

Cortical Feature Extraction   Nielsen has presented a model in which the neuronal networks housing the simple cells of the visual cortex are activated by inputs from center/surround cells assumed to be located in the LGN. The model produces receptive fields containing four or five consecutively antagonistic subfields and is able to account for orientation selectivity, spatial frequency filtering, phase relationships, subfield orientation selectivity, and slight end inhibition. The receptive fields of the simple cells closely resemble Gabor functions.

Nagana, Fujiwara, and Kurata have presented a model for the development of complex cells. Complex cell responses are learned in the model by means of a modifiable inhibitory synapse. Computer simulations of the model explore its orientation and direction selectivity and responses to random dot patterns.

Cooper, Liberman, and Oja, and earlier Nass and Cooper, also presented models for the development of specificity of feature extraction in cortical neurons. Cortical neurons become committed to arbitrary but repeated external patterns.

Finette, Harth, and Csermely present a computer simulation model with 3000 neurons of neural circuitry representing underlying orientation sensitivity and shape selectivity in visual cortex.

Spitzer and Hochstein present a model of the receptive fields of cortical complex cells in which the receptive field is built from two subunits, each of which performs a linear spatial summation, while summation between subunits is nonlinear. In an earlier paper these authors reported evidence of visual cortical receptive fields which resemble the "zero-crossings" requested by Marr for detection of light intensity edges.

Hartmann discusses the implications of recursive structures in the formation of receptive fields of visual neurons. He states that a recursive system would provide the same information to the visual cortex as a set of n virtual retinas with receptive fields of 2^n-fold diameter. Hence the different spatial frequency channels of the visual cortex would receive information from different virtual retinas. The neuronal circuitry, however, would be identical.

Braccini *et al.* present a model for extrafoveal processing in X cells of the retina and LGN. They show that the linear increase of receptive field diameter with eccentricity produces center frequencies that decrease from the center toward the periphery of the visual field and explore topological transformations, and the types of filtering taking place between the retina and the visual cortex.

Kulikowski *et al.* present mathematical inequalities relating the size of receptive field to corresponding spatial frequency and orientation tuning characteristics in cells of the striate cortex.

A model by Krone *et al.* represents the algorithms used by the cat in detecting simple patterns. Pattern recognition is described in two steps: (1) extraction of features and (2) classification.

Tolkmitt suggests that the rod system serving peripheral vision acts primarily to guide orienting behavior and to trigger more fine-grained sensory operations by the foveal system. He presents a computer simulation model representing receptive, retinal, geniculate, and cortical operations of the foveal system.

Jensen presents a model for the ananotimal organization of the visual cortex in the cat and a system-theoretic analysis of the Clare Bishop (CB) area. According to the model, this area is involved in the task of orientation with certain similarities to the superior colliculus. Preceeding stages provide an input consisting of images transformed in various ways. The CB cells have a generalizing function which takes place after previous specification. The cells function selectively with respect to the relation between object and background.

Dinse and von Seelen present a model which interprets area 18 as a system parallel to area 17 with respect to velocity and spatial frequency, although the function of feature extraction is the same in both systems. They show that pattern distortion and shift due to motion are eliminated by a spatial asymmetry of coupling in specific combinations of on–off systems so that extraction of features despite pattern movement is possible in area 18. They suggest that hypercomplex cell systems probably originate in recurrent inhibition and lead to differentiation of pattern along contour lines. In earlier papers, von Seelen and Hoffmann modeled the effect of noise in visual processing. In their model type I complex cells extract moving deterministic signals, whereas type II complex cells respond primarily to the stochastic input.

Vaitkevicius *et al.* present a model for monocular line perception in which the line inclination is coded by maximally excited orientation detector cells, inclination is represented by an excitation vector in subjective space, and differential sensitivity is constant over the whole range of line inclinations.

Sakitt and Barlow present a theoretical analysis of the first stage of cortical image processing based on the principle of minimization of requisite channels.

Audition

Itoh presents a model shown to represent auditory masking and unmasking processes. Masking is effected by forward and backward inhibitions on the auditory memory of signals by noise; unmasking is a result of disinhibition.

Shepard notes that increasingly accurate accounts of musical pitch are provided by increasingly generalized geometrically regular helical structures and presents a two-dimensional melodic map of these double helical structures which provides optimally compact representations of musical scales and melodies. Krumhansl and Kessler similarly produce a four-dimensional spatial map of the distances between keys as located on the surface of a torus. Deutch and Feroe, on the other hand, present a model for pitch sequences as retained in hierarchical networks.

Fedor presents a self-learning neuronal network model for the processing of simple melodies in the auditory system. The model represents neuronal trees that have the ability to specialize so that every melody can be recognized after a learning period. Longuet-Higgins presents a computer program intended to represent the perception of rhythmic and tonal relationships of classical melodies by Western musicians.

Gibson and Hirsch present a two-channel model for the perception of loudness in the presence of noise.

Somatosensation

I have discovered no mathematical models of neocortical somatosensation.

Global Electrophysiology

New Measures of Global Activity

Three new experimental techniques have emerged which promise to greatly expand our understanding of the global operations of the brain. These are positron emission tomography (PET), which traces radioactive isotopes tied to metabolically active species suh as sugar, oxygen, carbon, or nitrogen; the measurement of regional cerebral blood flow (rCBF) in different types of brain activity; and the measurement of global magnetic fields surrounding the brain during different kinds of activity. PET has been used by Reivich, Kuhl, Sokoloff, Wolf, and others to show that schizophrenia and manic-depressive illnesses are associated with localized metabolic changes in specific brain areas; that seizure-producing sites are less active than neighboring regions between seizures and more active during seizures; that solving spatial problems increases metabolic activity on the right side of the brain in right-handed subjects, while solving verbal problems increases activity on the left side; and that brain metabolism is more widely affected in stroke victims than was previously thought.

Roland and Friberg have shown that rCBF increases in the superior prefrontal cortex during three types of thinking, which they attribute to involvement of this area in overall organization of thinking. They found that different types of thinking increase rCBF in different cortical areas as follows: starting from 50 and continuously subtracting 3 from the result increased rCBF in the angular cortex; jumping every second word in a nine-word circular jingle increased rCBF exclusively in the right midtemporal cortex; and thinking about route finding increased rCBF in a number of areas in the occipital, temporal, and parietal lobes.

Reite, Zimmerman, and others have shown that global magnetic fields can be recorded around regions of the cerebral cortex in response to visual, auditory, or somatosensory stimultion. Current sources in the primary receiving cortices seem to underlie these fields. Models of the biophysical generation of magnetic fields by neural events are discussed in Chapter 4.

Event-Related Potentials

Event-related potentials have been studied by a number of investigators, including E. Roy John and Grey Walter. Walter has identified a contingent negative variation (cnv), which is a potential in the frontal regions associated with anticipation of movement by an individual.

Models of the Electroencephalogram

The most fundamental model of the mechanisms underlying the EEG was discussed by Elul in 1972. In this model the EEG reflects primarily the aggregate of large current loops associated with synaptic activation. Fundamental biophysical models for these kinds of events were presented by Plonsey and others, as discussed in Chapter 4.

A number of years ago Nuñez presented a mathematical description of the EEG in terms of Fourier frequency components. Anninos has presented several discussions of EEG characteristics on the basis of comptuer simulations of neuronal networks. He finds, for example, that some EEG activity is not affected by statistical properties of single elements in a neural network, even in networks containing as few as 200 elements. On the other hand, global activity of the net is highly sensitive to the connectivity among elements. Global activity is approximately Gaussian when connectivity is low but becomes less so when connectivity is increased. In 1975 Anninos and Raman presented a diffusion equation for the EEG and a Fourier series solution to that equation. A number of other theorists have presented partial differential equations for the spatiotemporal propagation of activity in neural networks (discussed in Chapter 9) which can be applied to the EEG.

Wright and Kydd model global electrocortical activity with a mass of linked oscillators that generate activity with a number of resonant modes that are subject to damping by an external fiber system, which they attribute to the lateral hypothalamus. They use this model to predict left and right electrocortical power spectra following unilateral lesions of the lateral hypothalamus and exlore implications such as the prediction that fixed phase velocities should be associated with each wavelength.

Epilepsy

Anninos and Cyrulnik in 1976 and Kokkinidis and Anninos in 1985 published computer simulation studies of neural networks exhibiting seizure activity. The seizure behavior of the nets is determined by a combination of certain parametrers in a phase diagram. Several mechanisms, including the number of spontaneously firing neruons beyond a certain threshold, produce seizures. Traub and a number of colleagues have presented several papers describing seizure activity in model hippocampal networks. The single-neuron model used in these simulations, as discussed in Chapter 2, is highly realistic, including features such as active calcium-related conductances in dendritic trees, complex spiking, and electrical interactions among neighboring units. In 1976 Kaczmarek showed that the dynamics of a simple model neuron with excitatory and inhibitory feedback produced epilepticlike firing patterns. The firing patterns could be varied by varying parameters descriptive of the external potassium concentraton, which suggests that the accumulation of extracellular potassium during seizures may be responsible for progressive changes observed in firing patterns in seizures.

Kindling is a lasting change in brain function which results from repeated focal stimulation and leads to the appearance of behavioral automatisms and convulsions as responses to brief electrical stimulation. This model has been widely used as a source of hypotheses concerning clinical epilepsy. Goddard and McNamara *et al.* have reviewed the kindling model of epilepsy. Johnston and Brown reported experiments confirming that a paroxysmal depolarizing shift in epilepsy is mediated by a giant excitatory postsynaptic potential (EPSP).

Lieblich and Amari and later Alonso-deFlorida and Minzone presented continuous ordinary differential equation models for the kindling model of epilepsy. In Leiblich and Amari's model the amygdaloid–pyriform complex is a self-stimulating excitatory aggregate, the hypothalamus an excitatory aggregate, and the hippocampal–septal–preoptic complex an inhibitory aggregate.

Petsche, Vollmer, and various colleagues have provided detailed quantitative graphic descriptions of the spatiotemporal development of seizures in rabbit cortex.

Spreading Depression

Spreading cortical depression is a complex wave phenomenon that moves slowly across cortical tissue at about 3 mm/min. It was first described by Leao in 1944 for the rabbit brain. It was early theorized that spreading depression is associated with the discharge of potassium into extracellular space by the tail of action potentials, bringing about sufficient depolarization to fire neighboring cells, which in turn generate more extracellular posassium, firing yet other cells, and so on. This firing would be brought to a halt only by an inactivation mechanism in indiviual neurons. An early model by Reshodko and Bures simulated spreading depression in computer networks of 1000–7000 neurons. Subsequent studies showed the importance of other substances such as glutamate and enkephalins in spreading depression. Tuckwell and Hermansen present an elegant mathematical model for the movements of basic ions (potassium, calcium, sodium, and chlorine) and two neurotransmitters (which might be taken as glutamate and γ-aminobutyric acid, GABA) to describe the characteristics of spreading depression.

Global Interconnections

Cook presents a model to explain the linguistic and paralinguistic specializations of the left and right hemispheres according to homotopic inhibition by the corpus collosum in interhemispheric communication. This mechanism produces patterns of cortical activity in each hemisphere which are the mirror images and photographic negatives of the patterns in the contralateral hemisphere. Applied to language generation and understanding, for example, excitation predominantly in the left hemisphere produces on the right inhibition of language-related neurons as well as excitation of all surrounding contextual neurons.

Bibliography

Neocortical: General

See also Chapters 9 and 10 on general neural network models and cognitive operations.
1986   Massing, W. The brain, time reversal and Libet's concept of antedating. *J. Theor. Biol.* **120**, 443–446.

Sejnowski, T. J. Open questions about computation in cerebral cortex. *In* "Parallel Distributed Processing: Explorations in the Microstructure of Cognition" (J. L. McClelland and D. E. Rumelhart, eds.), Vol. 2. MIT Press, Cambridge, Massachusetts.

1985   Peters, A., and Jones, E. G., eds. "Cerebral Cortex." Plenum, New York.
1983   Szetnágothai, J. The modular architectonic principle of neural centers. *Rev. Physiol. Biochem. Pharmacol.* **98**, 11–61.
1982   Abeles, M. "Local Cortical Circuits." Springer-Verlag, Berlin and New York.

Eccles, J. C. How the self acts on the brain. *Psychoneuroendocrinology* **7**, 271–283.

Pandya, D. N., and Seltzer, B. Association areas of the cerebral cortex. *Trends Neurosci.* **5**, 386–390.

1981   Eccles, J. C. The modular operation of the cerebral neocortex considered as the material basis of mental events *Neuroscience* **6**, 1839–1856.
1978   Creutzfeldt, O. D. The neocortical link: Thoughts on the generality of structure and function of the neocortex. *In* "Architectonics of the Cerebral Cortex" (M. A. B. Brazier and H. Petche, eds.), pp. 357–383. Raven Press, New York.

Szentágothai, J. The neuron network of the cerebral cortex: A functional interpretation. *Proc. R. Soc. London, Ser. B* **201**, 219–248.

1977   Brown, J. "Mind, Brain and Consciousness." Academic Press, New York.

Popper, K. R., and Eccles, J. C. "The Self and its Brain." Springer-Verlag, Berlin and New York.

Sagan, C. "The Dragons of Eden." Ballantine, New York.

1976   Jaynes, J. "The Origins of Consciousness in the Breakdown of the Bicameral Mind." Houghton Mifflin, Boston, Massachuetts.

1975   MacLean, P. D. On the evolution of three mentalities. *Man–Environ. Syst.* **5**, 213–224.

Penfield, W. "The Mystery of the Mind." Princeton Univ. Press, Princeton, New Jersey.

Szentágothai, J. The 'module-concept' in cerebral cortex architecture. *Brain Res.* **95**, 475–496.

1973   Luria, A. R. "The Working Brain." Basic books, New York.

1972   Barlow, H. A single neuron doctrine of perception. *Perception* **1**, 371–380.

1970   Eccles, J. C. "Facing Reality." Springer-Verlag, Berlin and New York.

MacLean, P. D. The triune brain, emotion, and scientific bias. *In* "The Neurosciences: Second Study Program" (F. O. Schmitt, ed.), pp. 336–349. Rockefeller Univ. Press, New York.

Marr, D. A theory of cerebral neocortex. *Proc. R. Soc., London, Ser. B* **176**, 161–234.

1969   Sperry, R. A modified concept of consciousness. *Psychol. Rev.* **76**, 532–536.

1968   MacLean, P. Forward in "The Structure of Ammon's Horn," S. Ramón y Cajál (L. M. Kraft, transl.). Thomas, Springfield, Illinois.

Riss, W. An overview of the design of the central nervous system. *Brain, Behav. Evol.* **1**, 124–131.

1967   Freeman, W. F. Analysis of function of cerebral cortex by use of control system theory. *Logistics Rev.* **3**, 5–40.

1966   Luria, A. R. "Higher Cortical Functions in Man." Basic Books, New York.

1951   Sherrington, C. S. "Man on His Nature." Cambridge Univ. Press, New York.

1932   Jackson, H. L. "Selected Writings of John Hughlings Jackson." Hodder & Stoughton, London.

1929   Sherrington, C. S. Some functional problems attaching to convergence. *Proc. R. Soc. London, Ser. B* **105**, 332–362.

1906   Sherrington, C. S. "The Integrative Activity of the Nervous System." Yale Univ. Press, New Haven, Connecticut.

Anterior Regions

1986   Reiner, A. Is prefrontal cortex found only in mammals? *Trends Neurosci.* **9**, 298.

1984   Tucker, D. M., and Williamson, P. A. Asymmetric neural control systems in human self-regulation. *Psychol. Rev.* **91**, 185–215.

1980   John, E. R. A neurophysiological model of purposive behavior. *In* "Neural Mechanisms of Goal-Directed Behavior and Learning" (R. F. Thompson, L. H. Hicks, and V. B. Shvyrkov, eds.). Academic Press, New York.

Pontius, A. A., and Yudowitz, B. S. Frontal lobe system dysfunction in some criminal actions as shown in the narratives test. *J. Nerv. Ment. Dis.* **168**, 111–117.

Ronald, P. E., Larsen, B., Lassen, N. A., and Shinhoe, E. Supplementary motor area and other cortical areas in organization of voluntary movements in man. *J. Neurophysiol.* **43**, 118–136.

1979   Albus, J. S. Mechanisms of planning and problem solving in the brain. *Math. Biosci.* **45**, 247–293.

1976   Deecke, L. Grozinger, B., and Kornhuber, H. H. Voluntary finger movement in man: Cerebral potentials and theory. *Biol. Cybernet.* **23**, 99–119.

1973   Luria, A. R. "The Working Brain." Basic Books, New York.

Powers, W. T. (1973). "Behavior: The Control of Perception." Aldine, Chicago, Illinois.

1948   Tolman, E. C. Cognitive maps in rats and men. *Psychol. Rev.* **55**, 189–208.

Visual: Feature Extraction, Form–Object, Figure–Ground Discrimination

1986   Konishi, M. Centrally synthesized maps of sensory space. *Trends Neurosci.* **9**, 163–166.

1985   Feldman, J. A. Connectionist models and parallelism in high level vision. *Comp. Vision Gr. Im. Proc.* **31**, 178–200.

Grossberg, S. Neural dynamics of form perception: Boundary completion, illusory figures, and neon color spreading. *Psychol. Rev.* **92**, 173–211.

Grossberg, S., and Mingolla, E. Neural dynamics of perceptual grouping: Textures, boundaries, and emergent segmentations. *Percept. & Psychophys.* **38**, 141–171.

1984   Cohen, M. A., and Grossberg, S. Neural dynamics of brightness perception: Features, boundaries, diffusion, and resonance. *Percept. & Psychophys.* **36**, 428–456.

1983   Grossberg, S. The quantized geometry of visual space: The coherent computation of depth, form, and brightness. *Behav. Brain Sci.* **6**, 625–692.

1982   Marr, D. "Vision. A Computational Investigation into the Human Representation and Processing of Visual Information." Freeman, San Francisco, California.

Schectman(1982). *Proc. R. Soc. London, Ser. B* **219**, 471–473.

1981   Borello, L. Ferraro, M., Penego, P., and Rossotti, M. L. A model of visual perception. *Biol. Cybernet.* **39**, 79–85.

Fuhrman, G. Modeling the visual cortex with modulo system concept. *Biol. Cybernet.* **40**, 39–48.

1980   Marr, D., and Hildreth, E. Theory of edge detection. *Proc. R. Soc. London, Ser. B* **207**, 187–217.

1979   Gibson, J. J. "The Ecological Approach to Visual Perception." Houghton-Mifflin, Boston, Massachusetts.

Hubel, D. H., and Wiesel, T. N. Brain mechanisms of vision. *Sci. Am.* **241**, 150–163.

Marr, D., and Poggio, T. (1979). A computational theory of human stereo vision. *Proc. R. Soc. London, Ser. B* **204**, 301–328.

Tzanakou, E., Michalak, R., and Harth, E. (1979). The apolex process: Visual receptive fields by response feedback. *Biol. Cybernet.* **35**, 161–174.

1978   Marr, D. Representation and recognition of the spatial organization of three-dimensional shapes. *Proc. R. Soc. London, Ser. B* **200**, 269–294.

1976   Harth, E. Visual perception: A dynamic theory. *Biol. Cybernet.* **22**, 169–180.

Marr, D. Early processing of visual information. *Philos. Trans. R. Soc. London, Ser. B* **275**, 483–519.

Marr, D. Analyzing natural images: A computational theory of texture vision. *Cold Spring Harbor Symp. Quant. Biol.* **40**, 647–662.

1975   Foster, D. H. An approach to the analysis of the underlying structure of visual space using a generalized notion of visual pattern recognition. *Biol. Cybernet.* **17**, 77–79.

Minsky, M. A framework for representing knowledge. *In* "The Psychology of Computer Vision" (P. H. Winston, ed.), pp. 211–277. McGraw-Hill, New York.

1974   Harth, E., and Tzanakou, E. Alopex: A stochastic method for determining visual receptive fields. *Vision Res.* **14**, 1475–1482.

1972   Barlow, H. B., Narasimhan, R., and Rosenfeld, A. Visual pattern analysis in machines and animals. *Science* **177**, 567–575.
1971   Binford, O. B. Visual perception by computer. *IEEE Conf Syst., Man, Cybernet.*
1966   Gibson, J. J. "The Senses Considered as Perceptual Systems." Houghton Mifflin, Boston, Massachusetts.
1962   Hubel, D. H., and Wiesel, T. N. Receptive fields, binocular interaction, and functional architecture in the cat's visual cortex. *J. Physiol. (London)* **160**, 106–154.

Visual: Spatial Decomposition and Texture

1984   Julesz, B. A brief outline of the texton theory of human vision. *Trends Neurosci.* **7**, 41–45.
1982   Schwartz, E. L. Columnar architecture and computational anatomy in primate visual cortex: Segmentation and feature extraction via spatial frequency coded difference mapping. *Biol. Cybernet.* **42**, 157–168.
1981   Julesz, B. A theory of preattentive texture discrimination based on first-order statistics of textons. *Biol. Cybernet.* **41**, 131–138.
        Kulikowski, J. J., and Bishop, P. O. Fourier analysis and spatial representation in the visual cortex. *Experientia* **37**, 160–163.
        Schwartz, E. L. Cortical anatomy, size invariance, and spatial frequency analysis. *Perception* **10**, 455–468.
1980   Caelli, T. M. Facilitative and inhibitory factors in visual texture discrimination. *Biol. Cybernet.* **39**, 21–26.
        Conners, R. W., and Harlow, C. A. (1980). A theoretical comparison of texture algorithms. *IEEE Trans. Pattern Anal.* **PA-1**, 204–222.
        Julesz, B. (1980). Spatial nonlinearities in the instantaneous perception of textures with identical power spectra. *Philos. Trans. R. Soc. London, Ser.* **290**, 91–97.
        Schwartz, E. L. Computational anatomy and functional architecture of striate cortex: A spatial mapping approach to perceptual coding. *Vision Res.* **20**, 645–669.
1979   Caelli, T. M. and Julesz, B. Pschophysical evidence for global feature processing in visual texture discrimination. *J. Opt. Soc. Am.* **69**, 675–678.
        Julesz, B., and Caelli, T. On the limits of Fournier decomposition in visual texture perception. *Perception* **8**, 69–73.
        Stone, J., Dreher, B., and Leventhal, A. Hierarchical and parallel mechanisms in the organization of visual cortex. *Brain Res. Rev.* **1**, 345–394.
1978   Caelli, T., and Julesz, B. On perceptual analyzers underlying visual texture discrimination. I. *Biol. Cybernet.* **28**, 167–175.
        Caelli, T. M., and Julesz, B. On perceptual analyzers underlying visual texture discrimination. II. *Biol. Cybernet.* **29**, 201–214.
1977   Gafni, H., and Zeevi, Y. Y. A model for separation of spatial and temporal information in the visual system. *Biol. Cynernet.* **28**, 73–82.
        Glezer, V. D., and Cooperman, A. M. Local spectral analysis in the visual cortex. *Biol. Cybernet.* **28**, 101–108.
        Schwartz, E. L. A quantitative model of the functional architecture of human striate cortex with application to visual illusion and cortical texture analysis. *Biol. Cynerbet.* **28**, 1–14.
1975   Foster, D. H. An approach to the analysis of the underlying structure of visual space using a generalized notion of visual pattern recognition. *Biol. Cybernet.* **17**, 77–79.
        Julesz, B. Experiments in the visual perception of texture. *Sci. Am.* **232**, 34–43.
        Legendy, C. R. Can the data of Campbell and Robson be explained without assuming Fourier analysis? *Biol. Cybernet.* **17**, 157–163.
        Wilson, H. R. (1975). A synaptic model for spatial frequency adaptation. *J. Theor. Biol.* **50**, 327–352.
1973   Harmon, L. D., and Julesz, B. Masking in visual recognition: Effects of two-

dimensional filtered noise. *Science* **180**, 1194–1197.

1971   Julesz, B."Foundations of Cyclopean Perception." Univ. of Chicago Press, Chicago, Illinois.

1970   Campbell, F. W., Nachmias, J., and Jukes, J. Spatial frequency discrimination in human vision. *J. Opt. Soc. Am.* **60**, 555–559.

1969   Campbell, R. W. C., and Robson, J. Applications of Fourier analysis to the visibility of gratings. *J. Physiol. (London)* **197**, 551–566.

Visual: Operations

1987   Cowan, J. D. What do drug-induced visual hallucinations tell us about the brain? Unpublished manuscript.

Cowan, J. D. Brain mechanisms underlying visual hallucinations. Unpublished manuscript.

1985   Bennet, B. M., and Hoffman, D. D. The computation of structure from fixed-axis motion: Nonrigid structures. *Biol. Cybernet.* **51**, 293–300.

Mallot, H. A. An overall description of retinotopic mapping in the cat's visual cortex areas 17, 18, and 19. *Biol. Cybernet.* **52**, 45–51.

Wilson, H. R. A model for direction selectivity in threshold motion perception. *Biol. Cybernet.* **51**, 213–222.

1984   Hamanda, J. A multi-stage model for border contrast. *Biol. Cybernet.* **51**, 65–70.

Reitboeck, H. J., and Altman, J. A model for size- and rotation-invariant pattern processing in the visual system. *Biol. Cybernet.* **51**, 113–121.

1983   Adelson, E. H., and Bergen, J. R. Spatiotemporal energy models for the perception of motion. *Meet. Opt. Soc. Am., 1983. Conf. Opt. Soc. Am.*

Hoffman, D. D. The interpretation of visual illusions. *Sci. Am.* **249**, 154–162.

Lehky, S. R. A model of binocular brightness and binaural loudness perception in humans with general applications to nonlinear summation of sensory inputs. *Biol. Cybernet.* **49**, 89–97.

Poggio, T. Visual algorithms. *In* "Physical and Biological Processing of Images" (O. J. Braddick and A. C. Sleigh, eds.), pp. 128–153. Springer-Verlag, Berlin and New York.

1982   Fisher, D. L. Limited-channel models of automatic detection: Capacity and scanning in visual search. *Psychol. Rev.* **89**, 1662–692.

Sugie, N. Neural models of brightness perception and retinal rivalry in binocular vision. *Biol. Cybernet.* **43**, 13–21.

1981   Burt, P., and Sperling, G. Time, distance, and feature trade-offs in visual apparent motion. *Psychol. Rev.* **88**, 171–195.

Marko, H. The z-model—a proposal for spatial and temporal modeling of visual threshold perception. *Biol. Cybernet.* **39**, 111–123.

1980   Caelli, T. M., and Dodwell, P. C. On the contours of apparent motion: A new perspective on visual space–time. *Biol. Cybernet.* **39**, 27–35.

Hildreth, E. A computer implementation of a theory of edge detection. *MIT At. Lab Rep.*, p. 579.

1979   Ermentrout, G. B., and Cowan, J. D. A mathematical theory of visual hallucination patterns. *Biol. Cybernet.* **334**, 137–150.

Wilson, H. R., and Bergen, J. R. A four mechanism model for spatial vision. *Vision Res.* **19**, 19–32.

1978   Caelli, T., Preston, G., and Howell, E. Implications of spatial summation models for processes of contour perception: A geometric perspective. *Vison Res.* **18**, 723–734.

Schwartz, E. L. Spatial mapping in the visual system. *J. Opt. Soc. Am.* **68**, 1371.

Schwartz, E. L. Spatial mapping in the primate sensory projection: Analytic structure and relevance to perception. *Biol. Cybernet.* **25**, 181–194.

Schwartz, E. L. Afferent geometry in the primate visual cortex and the generation of neuronal trigger features. *Biol. Cybernet.* **28**, 1–14.

Siegle, R. K. Hallucinations. *Sci. Am.* **237**, 132–140.

Tolkmitt, F. J. A computer simulation model of the afferent part of the visual foveation system. *Biol. Cybernet.* **15**, 195–203.

1976   Montalvo, F. S. A neural network model of the McCollough effect. *Biol. Cybernet.* **25**, 49–56.

1975   Hirai, Y., and Fukushima, K. A model of neural network extracting binocular parallax. *Biol. Cybernet.* **18**, 19–29.

1974   Quick, R. F. A vector-magnitude model of contrast detection. *Biol. Cybernet.* **16**, 65–67.

1973   Poggio, T., and Reichardt, W. Considerations on models of movement detection. *Biol. Cybernet.* **13**, 223–227.

1972   Luce, R., and Green, D. A neural timing theory for response times and the psychophysics of intensity. *Psychol. Rev.* **70**, 14–57.

1971   Foster, D. H. A model of the human visual system in its responses to certain classes of moving stimuli. *Biol. Cybernet.* **8**, 69–84.

1967   Wathen-Dunn, W. ed. "Models for the Perecption of Speech and Visual Form." MIT Press, Cambridge, Massachusetts.

Visual: Cortical Feature Extraction

1986   Okajima, F. K. A mathematical model of the primary visual cortex and hypercolumn. *Biol. Cybernet.* **54**, 107–114.

1985   Spitzer, H., and Hochstein, S. A complex-cell receptive-field model. *J. Neurophysiol.* **53**, 1266–1286.

1984   Hochstein, S. Zero-crossing detectors in primary visual cortex. *Biol. Cybernet.* **51**, 195–199.

1983   Krone, G., Kunz, D., and von Seelen, W. On the analysis of the cat's pattern recognition system. *Biol. Cybernet.* **48**, 115–124.

Nielsen, D. E. A functional model of the wiring of the simple cells of visual cortex. *Biol. Cybernet.* **47**, 213–222.

Vaitkevicius, H., Karalius, M., Meskauskas, A., Sinius, J., and Sokolov, E. A model for the monocular line orientation analyzer. *Biol. Cybernet.* **48**, 139–147.

1982   Braccini, C., Gambardella, G., Sandini, G., and Tagliasco, V. A model of the early stages of the human visual system: Functional and topological transformations performed in the peripheral visual field. *Biol. Cybernet.* **44**, 47–58.

Hartmann, G. Recursive features of circular receptive fields. *Biol. Cybernet.* **43**, 199–208.

Kulikowski, J. J., Marcelja, S., and Bishop, P. O. Theory of spatial position and spatial frequency relations in the receptive fields of simple cells in the visual cortex. *Biol. Cybernet.* **43**, 187–198.

Richter, J. and Ullman, S. A model for the temporal organization of X- and Y-type receptive fields in the primate retina. *Biol. Cybernet.* **43**, 127–145.

Sakitt, B., and Barlow, H. B. A model for the economical encoding of the visual image in cerebral cortex *Biol. Cybernet.* **43**, 97–108.

1981   Deutsch, S. A simplified model of the input layers of the visual cortex. *IEEE Front. Eng. Comp. Health Care*, pp. 60–63.

Dinse, H. R. O., and von Seelen, W. On the function of cell systems in area 18.1 *Biol. Cybernet.* **41**, 47–57.

Dinse, H. R. O., and von Seelen, W. On the function of cell systems in area 18.11. *Biol. Cybernet.* **41**, 59–69.

Nagano, T., and Kurata, K. A self-organizing neural network model for the development of complex cells. *Biol. Cybernet.* **40**, 195–200.

1980   Amari, S. Topographic organization of nerve fields. *Bull. Math. Biol.* **42**, 339–364.

Jensen, H. J. System-theoretical analysis of the Clare Bishop area in the cat. *Biol. Cybernet.* **39**, 53–66.

Nagano, T. and Kurata, K. A model of the complex cell based on recent neurophysiological findings. *Biol. Cybernet.* **38**, 103–105.

1979   Braitenberg, V., and Braitenberg, C. Geometry of orientation columns in the visual cortex. *Biol. Cybernet.* **33**, 179–186.

Cooper, L. N., Liberman, F., and Oja, E. A theory for the acquisition and loss of neuron specificity in visual cortex. *Biol. Cybernet.* **33**, 9–28.

Nagano,T., and Fujiwara, M. A neural network model for the development of direction selectivity in the visual cortex. *Biol. Cybernet.* **32**, 1–8.

1978   Amari, S., and Takeuchi, A. A mathematical theory on formation of category detecting nerve cells. *Biol. Cybernet.* **29**, 127–136.

Finette, S., Harth, E., and Csermely, T. J. Anisotropic connectivity and cooperative phenomena as a basis for orientation sensitivity in the visual cortex. *Biol. Cybernet.* **30**, 231–240.

Hoffman, K. P., and von Seelen, W. Analysis of neuronal networks in the visual system of the cat using statistical signals—simple and complex cells. II. *Biol. Cybernet.* **31**, 175–185.

1977   Amari, S. Neural theory of association and concept formation. *Biol. Cybernet.* **26**, 175–185.

Nagano, T. A model of visual development. *Biol. Cybernet.* **26**, 45–52.

Tolkmitt, F. J. A computer simulation model of the afferent part of the visual foveation system. *Biol. Cybernet.* **25**, 195–203.

1976   Schiller, P. H., Finlay, B. L., and Volman, S. F. Quantitative studies of single-cell properties in monkey striate cortex. V. Multivariate statistical analysis and models. *J. Neurophysiol.* **39**, 1362–1374.

von Seleelen, W., and Hoffman, K. P. Analysis of neuronal networks in the visual system of the cat using statistical signals. *Biol. Cybernet.* **22**, 7–20.

1975   Nass, M. and Cooper, L. N. A theory for the development of feature detecting cells in visual cortex. *Biol. Cybernet.* **19**, 1–18.

Perez, R., Glass, L., and Schlaer, R. Development of specificity in the cat visual cortex. *J. Math. Biol.* **1**, 275–288.

1973   von der Malsburg, C. Self-organization of orientation selective cells in the striate cortex. *Biol. Cybernet.* **14**, 85–100.

Audition

1985   Itoh, D. A neuro-synaptic model of the auditory masking and unmasking process. *Biol. Cybernet.* **52**, 229–235.

1982   Krumhansl, C. L., and Kessler, E. J. Tracing the dynamic changes in perceived tonal organization in a spatial representation of musical keys. *Psychol. Rev.* **89**, 334–368.

Shepard, R. N. Geometrical approximations to the structure of a musical pitch. *Psychol. Rev.* **89**, 305–333.

1981   Deutsch, D., and Feroe, J. The internal representation of pitch sequences in tonal music. *Psychol. Rev.* **88**, 503–522.

1977   Fedor, P. Principles of the design of D-neuronal networks. I. A neural model for pragmatic analysis of simple melodies. *Biol. Cybernet.* **27**, 129–146.

Gates, A., and Bradshaw, J. L. The role of the cerebral hemispheres in music. *Brain Lang.* **6**, 403–431.

1976   Longuet-Higgins, H. C. Perception of melodies. *Nature (London)* **263**, 646–653.

1975   Gibson, J. M., and Hirsch, H. R. Psychoneural models of the auditory masking process. *J. Theor. Biol.* **51**, 135–147.

1974   Clynes, M. The pulse of musical genius. *Psych. Today*, July, pp. 51–55.

Blood Flow Imaging and Positron Emission Tomography

1986   Stahl, S. M., Leenders, K. L., and Bowery, N. G. Imaging neurotransmitters and their receptors in living human brain by positron emission tomography. *Trends Neurosci.* **9**, 241–245.

1985   Roland, P. E. Application of brain blood flow imaging in behavioral neurophysiology: The cortical field activation hypothesis. *In* "Brain Imaging and Brain Function" (L. Sokoloff, ed.), pp. 89–106. Raven Press, New York.

Roland, P. E., and Friberg, L. Localization of cortical areas activated by thinking. *J. Neurophysiol.* **53**, 1219–1243.

1984   Roland, P. E. Metabolic measurements of the working frontal cortex in man. *Trends Neurosci.* **7**, 430–435.

1983   Roland, P. E., and Friberg, L., Are cortical rCBF increases during brain work in man due to synaptic excitation or inhibition? *J. Cereb. Blood Flow Metab.* **3**, Suppl. 1, 244–254.

1982   Roland, P. E. Cortical regulation of selective attention in man. *J. Neurophysiol.* **48**,1059,1078.

1975   Moskalenko, V. V. Regional cerebral blood flow and its control at rest and during increased functional activity. *In* "Brain Work" (H. D. Ingvar and N. A. Lassen, eds.) pp. 343–351. Munksgaard, Copenhagen.

Magnetic Fields

1981   Reite, M., Zimmerman, J. T., and Zimmerman, S. E. Magnetic auditory evoked fields: Interhemispheric asymmetry. *Clin. Neurophysiol. Electroencephalogr.* **51**, 388–392.

Williamson, S. J., and Kaufman, L. Magnetic fields of the cerebral cortex. *In* "Biomagnetism" (S. N. Erne, H. D. Hahlbohm, and H. Lubbig, eds.). deGruyter, Berlin.

Zimmerman, D. Seeing the brain at work. *Mosaic (Greenwich, Conn.)*, May/June, pp. 9–14.

1979   Cohen, D. Magnetic measurement and display of current generators in the brain. I. *Proc. Int. Congr. Med. Biol. Eng., 12th, 1979*, pp. 14–15.

1978   Brenner, D., Lipton, J., Kaufman, L., and Willismson, S. J. Somatically evoked magnetic field of the human brain. *Science* **199**, 81–83.

Reite, M., and Zimmerman, J. E. Magnetic phenomena of the central nervous system. *Annu. Rev. Biophys. Bioeng.* **7**, 167–188.

1975   Rush, S. On the independence of magnetic and electric body surface recordings. *IEEE Trans. Biomed. Eng.* **BME-22**, 157–167.

EEG

1985   Al-Nashi, H., Lee, H. C., Caines, P. E., and Gotman, J. Estimation of EEG evoked potential via Kalman filtering. *Eng. Med. Biol.* **38**,7.1,42.

Taylor, C. P., and Dudek, F. E. Excitation of hippocampal pyramidal cells by an electrical field effect. *N. Neurophysiol.* **52**, 126–142.

1984   Petsche, H., Pockberger, H., and Rappelsberger, P. On the search for the sources of the electroencephalogram. *Neuroscience* **11**, 1–27.

Taylor, C. P., and Dudek, F. E. Synchronization without active chemical synapses during hippocampal afterdischarges. *J. Neurophysiol.* **52**, 143–155.

1983   Anninos, P., Zenone, S., and Elul, R. Artificial neural nets: Dependence of the EEG amplitude's probability distribution on statistical parameters. *J. Theor. Biol.* **103**, 339–348.

1982   Holsheimer, J., Boer, J., Lopes da Silva, F. H., and van Rotterdam, A. The double dipole model of theta rhythm generation: Simulation of laminar field potential profiles in dorsal hippocampus of the rat. *Brain Res.* **235**, 31–50.

1981   Gevins, A. S. Dynamic brain electrical patterns of cognition. *IEEE Front. Eng. Comp. Health Care*, pp. 174–181.

Nuñez, P. L. "Electric Fields of the Brain. The Neurophysics of the EEG." Oxford Univ. Press, London and New York.

Sanderson, A. C. Hierarchical approaches to modeling EEG and evoked potentials. *IEEE Front. Eng. Comp. Health Care*, pp. 201–207.

1980   van Rotterdam, A. A computer system for the analysis and synthesis of field potentials. *Biol. Cybernet.* **37**, 33–39.

1979   Gevins, A. S., Zeitlin, G. M., Doyle, J. C., Kingling, C. D., Schafer, R. E., Callaway, E., and Keager, C. L. Electroencephalogram correlates of higher cortical functions. *Science* **203**, 665–667.

1978   Bohdanecky, Z., Lansky, P., Indra, M., and Radil-Weiss, T. EEG alpha and non-alpha intervals alternation. *Biol. Cybernet.* **30**,109,113.

van Rotterdam, A. A one dimensional formalism for the computation of extracellular potentials: Linear systems analysis applied to volume conduction. *Prog. Rep. Inst. Med. Phys., TNO* **PR6**, 115–122.

1977   Kaiser, F. Limit cycle model for brain waves. *Biol. Cybernet.* **27**, 155–163.

1976   Kawabata, N. Test of statistical stability of the electroencephalogram. *Biol. Cybernet.* **22**, 235–238.

Nogawa, T., Karayama, K., Tabata, Y., Ohshio, T., and Kawahara, T. The brain wave equation. *Dig. Int. Conf. Med. Biol. Eng., 11th, 1976*, pp. 546–547.

Schmitt, F. O., Dev, P., and Smith, B. H. Electrotonic processing of information by brain cells. *Science* **193**, 114–120.

1975   Anninos, P. A., and Raman, S. Derivation of a mathematical equation for the EEG and the general solution within the brain and in space. *Int. J. Theor. Phys.* **12**, 1–9.

Basar, E., Gonder, A., Ozesmi, C., and Ungan, P. Dynamics of brain rhythmic and evoked potentials. I. *Biol. Cybernet.* **20**, 137–143.

1974   Lopes da Silva, F. H., Hoeks, A., Smits, H., and Zetterberg, L. H. Model of brain rhythmic activity. The alpha rhythm of the thalamus. *Biol. Cybernet.* **15**, 27–37.

Nuñez, P. L. Wavelike properties of the alpha rhythm. *IEEE Trans. Biomed. Eng.* **BME-21**, 473–482.

Nuñez, P. L. The brain wave equation: A model for the EEG. *Math. Biosci.* **21**, 279–297.

1972   Elul, R. The genesis of the EEG. *Int. Rev. Neurobiol.* **15**, 227–272.

1968   Andersen, P., and Andersson, S. A. "Physiological Basis of the Alpha Rhythm." Appleton, New York.

1957   Mundy-Castle, A. C. The electroencepahlogram and mental activity. *Electroencephalogr. Clin. Neurophysiol.* **9**, 643–655.

Epilepsy

1986   Alonso-deFlorida, F., Minzoni, A. A., and Morales, M. A. A synaptic model for the kindling effect. *J. Theor. Biol.* **120**, 285–302.

1985   Kokkindis, M., and Anninos, P. Noisy neural nets exhibiting epileptic features. *J. Theor. Biol.* **113**, 559–588.

Traub, R. D., Dudek, F. E., Snow, R. W., and Knowles, W. D. Computer simulations indicate that electrical field effects contribute to the shape of the epileptiform field potential. *Neuroscience* **15**, 947–958.

1983   Goddard, G. V. The kindling model of epilepsy. *Trends Neurosci.* **6**, 275–278.

1981   Johnston, D., and Brown, T. H. Giant synaptic potential hypothesis for epileptiform activity. *Science* **211**, 294–297.
1980   Alonso-deFlorida, F., and Minzoni, A. A. A nonlinear network oscillator model for kindling and a dysrhythmic cerebral state. *In* "Limbic Epilepsy and the Dyscontrol Syndrome" (M. Girgis and L. G. Kiloh, eds.), pp. 63–74. Elsevier/North-Holland, Amsterdam.

McNamara, J. O., Byrne, M. C., Dasheiff, R. M., and Fitz, J. G. The kindling model of epilepsy: A review. *Prog. Neurobiol.* **15**, 139–159.
1978   Lieblich, I., and Amari, S. An extended first approximation model for the amygdaloid kindling phenomenon. *Biol. Cybernet.* **28**, 129–135.

Prince, P. A. Neurophysiology of epilepsy. *Annu. Rev. Neurosci.* **1**, 395–415.

Vollmer, R., Petsche, H., Pockberger, H., Prohaska, O., and Rappelsberger, P. Spatiotemporal analysis of cortical seizure activities in a homogenous cytoarchitectronic region. *In* "Architectonics of the Cerebral Cortex" (M. A. B. Brazier and H. Petsche, eds.), pp. 281–305. Raven Press, New York.
1977   Anninos, P. A., and Cyrulnik, R. A neural net model for epilepsy. *J. Theor. Biol.* **66**, 695–709.
1976   Kaczmarek, L. K. A model of cell firing patterns during epileptic seizures. *Biol. Cybernet.* **22**, 229–234.
1975   Petsche, H., Nagypal, T., Prohaska, O., Rappelsberger, P., and Vollmer, R. Approaches to the spatio-temporal analysis of seizure patterns. *In* "Computerized EEG Analysis" (G. Fischer, ed.), pp. 111–126. Thieme, Stuttgart.
1972   Purpura, D. P., Penry, J. K., Tower, D., Woodbury, D. M., and Walter, R., eds. "Experimental Models of Epilepsy." Raven Press, New York.
1968   Jasper, H. H., Ward, A. A., and Pope, A., eds. "Basic Mechanisms of the Epilipsies." Little, Brown, Boston, Massachusetts.

Spreading Depression

1981   Sprick, U., Oitzl, M. S., Ornstein, K., and Huston, J. P. Spreading depression induced by microinjection of enkephalins into the hippocampus and neocortex. *Brain Res.* **210**, 243–252.

Tuckwell, H. C. Simplified reaction–diffusion equations for potatssium and calcium ion concentrations during spreading cortical depression. *Int. J. Neurosci.* **12**, 95–107.

Tuckwell, H. C., and Hermansen, C. L. Ion and transmitter movements during spreading cortical depression. *Int. J. Neurosci.* **12**, 109–135.
1980   Tuckwell, H. C. Predictions and properties of a model of potassium and calcium ion movements during spreading cortical depression. *Int. J. Neurosci.* **10**, 145–164.
1978   Tuckwell, H. C., and Miura, R. M. A mathematical model for spreading cortical depression. *Biophys. J.* **23**, 257–276.
1975   Reshodko, L. V., and Bures, J. Computer simulations of reverberating spreading depression in a network of cell automata. *Biol. Cybernet.* **18**, 181–189.
1974   Bures, J., Buresova, O., and Krivánek, J. "The Mechanism and Applications of Leao's Spreading Depression of Electroencephalographic Activity." Academic Press, New York.
1962   Ochs, S. The nature of spreading depression in neural networks. *Int. Rev. Neurobiol.* **4**, 1–69.
1959   Marshall, W. H. Spreading cortical depression of Leao. *Physiol. Rev.* **39**, 239–279.
1956   Grafstein, B. Mechanism of spreading cortical depression. *J. Neurophysiol.* **19**, 154–171.

Miscellaneous Global Electrophysiology

1984   Cook, N. D. Callosal inhibition: The key to the brain code. *Behav. Sci.* **29**, 98–110.

.   Wright, J. J., and Kydd, R. R. A linear theory for global electrocortical activity and its control by the lateral hypothalamus. *Biol. Cybernet.* **50**, 75–82.

Wright, J. J., and Kydd, R. R. A test for constant natural frequencies in electrocortical activity under lateral hypothalamic control. *Biol. Cybernet.* **50**, 83–88.

Wright, J. J., and Kydd, R. R. Inference of a stable dispersion relation for electrocortical activity controlled by the lateral hypothalamus *Biol. Cybernet.* **50**, 89–94.

1981   Adey, W. R. Tissue interactions with nonionizing electromagnetic fields. *Physiol. Rev.* **61**, 435–514.

Pirch, J. H., and Peterson, S. L. Event-related slow potentials and activity of single neurons in rat frontal cortex. *Int. J. Neurosci.* **154**, 141–146.

Wright, J. J., and Ihaka, G. R. A preliminary mathematical model for lateral hypthalamic regulation of electrocortical activity. *Electroencephalogr. Clin. Neurophysiol.* **52**, 107–115.

1979   Adey, W. R. Neurophysiologic effects of radiofrequency and microwave radiation. *Bull. N.Y. Acad. Med.* [2] **55**, 1079–1093.

1976   Deecke, L., Grozinger, B., and Kornhuber, H. H. Voluntary finger movement in man: Cerebral potentials and theory. *Biol. Cybernet.* **23**, 99–119.

1974   Katchalsky, A., Rowland, V., and Blumenthal, R. "Dynamic Patterns of Brain Cell Assemblies." MIT Press, Cambridge, Massachusetts.

See also references in Chapter 4 on extracellular electric and magnetic fields.

Generic Operations, Systems, and Networks

The last major frontier of brain research, now on the horizon and beckoning us to privileged glimpses of its secrets in the first half of the 21st century, consists of those nether reaches where mind meets brain: where conscious experience, feeling, volition, and cognition come together with global patterns of electrochemical activity in the vast, intricate neuronal networks of the brain. It is at this level of operations where neurobiology has most need for guidance from theory and models and where the abilities of theory and modeling to deal with complexity are most sorely challenged. It is, in short, the primary testing ground where engineering- and computer-based theory and modeling should make their maximum and unique contribution to brain science. In the later chapters of the preceding section we surveyed some models relevant to these levels of functioning and operations in specific neuronal systems and networks. The present section considers general theoretical and modeling approaches to these higher-level functions. Chapter 9 considers models and theories for the dynamics and operations of neuronal networks; Chapter 10, models and theories for cognitive operations and structures; Chapter 11, models of clinical conditions, and Chapter 12, models at the outer limits of brain and mind function.

9

Neural Networks

This chapter reviews general theories and models for the dynamic operations of neuronal networks. Network models of specific networks are reviewed in the preceding section on the specific neuronal systems.

Systems Theoretical and Computer Simulation Models

Many of the earliest theoretical papers on neuronal networks appearing from the 1940s into the 1970s dealt with random networks. Typically, theorists were concerned with the dynamics of the overall activity level of random networks and the relation of dynamics characteristics to the underlying connections. For example, nets with only recurrent excitatory connections were shown to exhibit an "ignition" phenomenon in that activity either died out or became all encompassing in a neuron net, depending on its initial level. Investigators characterized activity levels of random nets as oscillatory, monostable, bistable, etc. and investigated stability criteria. A simple and elegant model of this type was presented in 1970 by Harth and Anninos, who characterized the conditions requisite for stable internally sustained activity in randomly connected nets with recurrent excitation and recurrent inhibition. (The particulars of this model are discussed in Chapter 18.) Stubbs and Good discussed features of connectivity of brain networks under the assumption that they are randomly connected. As experimental research in the 1960s and 1970s began to establish that actual neural networks are under highly specific genetic control and not randomly connected, interest in this line of modeling greatly diminished even though many of the basic findings are adaptable to nonrandom networks.

A number of models of lateral inhibition in neural networks appeared around 1970. These models exhibited boundary enhancement, pattern sharpening, etc., as described earlier by Ratliff and Hartline. More recently, as we saw in earlier chapters, Pellionisz described matrix-based, structure-oriented models for junctions between neuronal populations.

Much of the computer- and system-oriented modeling of neural networks from the 1960s to the present has been carried out or inspired by Caianiello and his school, based on his "neuronic" equations discussed in Chapter 2. A large number of papers in this area have described transients, oscillations, stability, limit cycles, equilibrium states, reverberations, cycling modes, synchronous, or asynchronous oscillations, traveling waves, diffuse reverberations, etc. in networks of threshold elements which are often characterized as switching nets, automata nets, or Boolean nets. In addition to Caianiello's work, representative studies of this type have been presented by Anninos, who in 1972 described self-maintaining cycling modes in simulated nets and their dependence on parameters of the network structure. Anninos showed that many properties of these models are dependent on overall global parameters of networks and independent of the detailed connectivity structure of the net. Subsequenlty, Anninos published several papers describing the properties of so-called multiple memory domains, which are generated in artificial neural networks wherein the connections are set up by means of chemical markers.

A refreshingly unusual early theoretical model of neural networks was presented by Freeman, who focused on the relation between a continuous graded measure of local activity density and constituent neuronal spike trains in terms of a theory of neural masses.

Recent computer simulation and system theoretic modeling of neuronal networks has tended increasingly to deal with state space or chaos theoretical descriptions of neural activity, more realistic computer simulation of neural net activity, or the properties of learning and memory in adaptive neural networks.

State Space Representations

Smolensky has presented a nice interpretation of conceptual processing and neural processing in parallel distributed systems. In this model a lower-level description of the systems' neural processing is given in terms of evolution equations of the Caianiello type. Conceptual or mental processing is described in terms of the collective activity of the lower level by defining new variables from many old ones and substituting the new variables for the old in the evolution equations. The fundamental equation supposed to be applicable to the neural level of processing lends itself readily to linear algebraic manipulation and to display in vector spaces.

Anderson has presented a similar state space approach to cognitive and psychological operations with neural models. Large patterns of activities of groups of neurons are represented by state vectors and rules of operations prescribed for these state vectors so as to produce learning and manipulation of information based on categorization.

Earlier papers by Cull, Josin, and Guilfind and Walker discuss measures of distance between states and self-control in binary switching nets.

Chaos Theoretical Concepts

Computer simulation studies have shown that neuronal networks are often prone to erratic activity, and that their large-scale dynamic behavior is often highly sensitive to small perturbations. Intermittency, abrupt transitions, compound rhythmicities, periodic doubling, and the like are all features of dynamic activity which have been subsumed under the title of chaos theory and which can under various conditions be observed in artificial and probably real neural networks.

Guevara *et al.* reviewed the concept of chaos in neurobiology and pointed to recurrent inhibition, periodic forcing of neural oscillators, and systems operating in ranges of parameters that lead to abnormal dynamics as likely commonly occurring sources of chaotic neuroelectric activity.

Choi and Huberman showed that nonlinear networks made of synchronous threshold elements exhibiting both excitatory and inhibitory couplings display collective behavior which can be described as either multiple periodic or deterministic chaotic. They discussed these properties in terms of statistical mechanical model for spin glass and the like, discussed later in this chapter.

Grondin *et al.* performed a modeling study of synchronism in systems of threshold elements and discussed their properties, including limit cycles and multifrequency oscillations, in terms of the concepts of chaos theory.

Holden also characterized irregular bursting activity in single molluscan pacemaker neurons in terms of chaos theoritical concepts.

Realistic Computer Simulation

Most of the computer modeling of large-scale networks of neuronlike elements has focused primarily on the global properties of such networks and has not attempted to incorporate a high degree of verisimilitude in the individual model neurons. Perkel and MacGregor both attempted to incorporate relatively realistic neuronal dynamics in large-scale simulation models for neuronal networks. Perkel's models were described in Chapter 2. MacGregor's models are described in Part II of this book. Traub also developed a highly realistic model of neuroelectric activity in networks of the hippocampus, as described in Chapters 3 and 7.

Stochastic Systems Theoretic Models

Stochastic systems theory generates models such as random walk, Markovian processes, and maximization of entropy, to describe underlying causal processes in indeterminate probabilistic systems. We discussed such models for events in single neurons in Chapter 2 and for characteristics of neuronal networks earlier in this chapter.

Stochastic systems theory also characterizes the active variables of indeterminate probabilistic systems in terms of probability density functions, correlations, measures of informations, and the like. These techniques have been applied to the nervous system by a number of investigators to provide information theoretic measures of activity in neurons and neuron populations, operational descriptions and analysis of single neuronal spike trains, and inferences concerning interconnections and emergent properties in neuronal networks on the basis of activity correlations among constituent units.

Information Theoretic Measures of Activity in Neuronal Systems

In 1949 Shannon and Weaver published a now classical characterization of the information in stochastic systems. The essential idea is that the observed occurrence of a specific event from a milieu of possibilities in an indeterminate system constitutes a specification of information. Shannon and Weaver showed how one could quantify the amount of information in such situations. Qualitatively, the less likely the event, the more information is associated with its occurrence. They further characterize the binary decision as the most elementary informational element and label the amount of information associated with a binary decision as a "bit". More recently, Harary and Batell characterized the introduction of confusion by means of additional alternatives as negative information.

In the mid-1960s, Perkel, Gerstein, Moore, Sequndo, and various colleages introduced operational techniques for characterizing the dynamic patterns of neuronal spike trains. These techniques were widely adopted by many practicing electrophysiologists throughout the 1960s and 1970s and are used as a matter of course to the present day. The fundamental character of this approach is adequately described in several places and will not be reviewed here.

A number of investigators in the early 1970s discussed the applications of information theoretic ideas to neuronal spike trains. Nakahama and various collaborators defined the concept of "dependency" for discrete variables in a population in terms of Shannon's entropy and the mth order conditional entropy. This measure provides sufficient descriptions of the degree of the first-order Markovian characteristics of neuronal spike trains and is claimed by the authors to be the most concise measure for expressing higher-order properties of time series and to be superior to correlation or spectral measures.

Palm has extended the information theoretic approach to provide quantitative measures of "evidence" and "surprise" in probabilistic neuronal systems. Palm also applied information theory to the storage capacity of neural associated memory with randomly distributed storage elements. He estimated that at least 0.05 bit per storage element can be stored in such systems.

Windhorst and Schultens present a method for calculating the information transfer or "transinformation" in multiple input/single output neuronal systems. This is an extention of an approach introduced by Eckhorn and Poeple in 1974. Tsukada, Ishii, and Sato discuss Shannon's entropy concept with regard to some properties of temporal pattern discrimination in neurons.

Koenderink has presented a model based on the idea that functional order in nervous nets may be signaled through the total covariances of the activities carried by the members of the collection. These covariances are available to the system itself, whereas spatial relations as signaled by somatopy in individual neurons, for example, are not available to the system and are irrelevant to the organization of the system as an abstract machine. Kohenderink elaborates this model system for the retina.

Sejnowski also discusses the significance of correlations with regard to information processing in nervous systems. Borisyuk and varous colleagues from Moscow proposed a new method for analyzing several simultaneously recorded spike trains to identify and evaluate interconnections. Epping *et al.* described a technique called the "neurochrome" for displaying neural activity patterns.

Stochastic Characterization of Multiple Neuronal Spike Train and Emergent Properties in Neuronal Networks

George Gerstein with Perkel and other collaborators extensively explored the interrelations in firings of up to 20 simultaneously recorded spike trains from various neuronal populations and made inferences regarding the patterns of activity sustained by the network as a whole. These studies revealed the following. (1) Information about spatial direction of a sound source is available in the near-coincident firings of neurons even though it is not present in the individual spike trains of the neurons. (2). The effective connectivity within networks as shown by the prominent correlations can change spontaneously and can change in response to stimulation. (3). The dynamic network patterns, again as measured by prominent correlations among constituents, varies in response to a given tone depending on the position of that tone in a sequence. There is no evidence that the single neuron components individually exhibit melody discrimination in this sense. These examples illustrate what Gerstein characterizes as "emergent" properties of the neuronal population in that they signal characteristics not found explicitly in the individual constitutent neurons.

Gerstein and collaborators developed a striking visual display of the coordinated activity of a neuronal network based on a state space approach of classical mechanics. The basic concept is to transform n simultaneously recorded spike trains into an n-body problem in an n-dimensional space and to map correlated firing between units into gravititational attraction between the corresponding bodies in the n-space. Thus, units that tend to fire together correspond to bodies that tend to come together in the n-space. With movies based on this approach and snapshots from these movies, Gerstein showed a striking visual representation of dynamic correlated firing in clusters of neurons. Kruger and Bach described a similar multiunit approach to neuronal ensembles in visual cortex.

Statistical Mechanics

One of the most intriguing ideas in theoretical brain science is that it should be possible to some extent to describe the global electrical activity of neuronal networks in terms of theoretical concepts similar to those used to describe the global properties of materials in statistical mechanics. This idea was discussed as early as 1948 by Norbert Wiener in his classic book "Cybernetics." In the statistical mechanics of gases, for example, the properties of central interest are global and often relatively simple (such as pressure–volume–temperature relations) even though they are determined by complicated dynamic interactions of vast numbers of constituent molecules. The theoretical science of statistical mechanics makes inferences concerning global properties and constraints based on the aggregate of the physical rules describing the individual molecular interactions. In various forms of such theorizing in classical mechanics and the thermodynamics of solids and fluids, intermediary global functions such as the Lagrangian, the Hamiltonian, the total energy, the action, and the entropy are defined. The highest and most powerful principles in these approaches involve statements that the systems tend toward extrema of these functions—for example, minimization of total energy or maximization of entropy. The attraction of these theoretical approaches for some brain theorists resides in their ability to pinpoint the essential global behavior of rich and complicated systems by a single global function and a single principle. In application to brain function one is led by this kind of thinking, for example, to a sympathy with the theory of "cognitive dissonance," wherein opinions and attitudes are formed within a cognitive structure to minimize the overall dissonance of that structure. In the same vein, one is led to imagine that subjective experiences of the levels of truth and beauty of a set of elements may correspond to the minimization of some global measure of those elements, such as minimization of contradiction for truth and minimization of some measure of disharmony for beauty.

Although Wiener pointed out the possible analogy between vast numbers of interacting neurons in a neural network and vast numbers of interacting molecules in a solid or gas some 40 years ago, and a few relevant papers on this topic appeared in the 1960s and 1970s, only in the past 4 or 5 years has intensive study of this approach been undertaken by a number of investigators in the engineering and computer sciences.

Energy, Entropy, and the "Boltzmann Machine"

In 1972 Bergstrom and Nevanlinna postulated that the state of a neural system could be described by its total neural energy E and its entropy distribution H. Two governing global principles for such a neural network are that the total energy remains constant and that the system always moves to maximize its entropy. They suggested that the entropy of such a system could be given in terms of its total energy, number of elements n, and number of connections N according to

$$H(E, n, N) = \frac{4E}{nN} \left(1 - \frac{E}{nN} \right) (n - 1) \log(N + 1) \qquad (9.1)$$

In 1972 Pfaffelhuber described learning as a process in which entropy decreases in time. In 1975 Takatsuji also discussed the application of the maximum entropy principle to systems of interacting elements.

Hinton has constructed a computational device (containing both hardware and software features) which, given a signal and a random milieu, performs cognitive operations by systematically changing states according to a principle of minimizing the global energy of the system. The system produces best interpretations of given input signals in terms of stored memories by this procedure. To avoid settling at false or local minima, Hinton's program introduces random fluctuations ("jiggling").

Hamiltonians

In classical mechanics the Hamiltonian H is a single global function of a dynamic system determined by the degrees of freedom of the system and the associated momenta such that the differential equations of motion of the system are obtained by differentiating with respect to the independent variable associated with each degree of freedom. In 1967 Cowan defined a

Hamiltonian and found a corresponding invariant for the dynamics of a single two-cell loop as described by

$$H = \sum_r B_r [1 + q_r(e^{v_v} - v_v)]$$

$$\text{Inv} = \frac{e^{-\alpha H}}{\pi_r \int_{-\infty}^{\infty} e^{-\alpha H_r}} \, dv_v$$

(9.2)

Hopfield presented a Hamiltonian-like model for cognitive operations in neuronal networks. In this model, Hopfield establishes a correspondence between the dynamics of a neuron model and the Ising spin model for glass and writes the Hamiltonian shown in Eq. (9.3) for this system in terms of its synaptic coefficients c_{ij} and its internal state.

$$h_i = \sum_j \frac{C_{ij}}{2} \sigma_j + \left[\sum_j \frac{C_{ij}}{2} \pm \theta_i \right]$$

$$h_i \sigma_i \geq 0$$

(9.3)

$$H(I) = - \sum_i h_i \sigma_i$$

Equilibrium states for this system are then determined as the minima of the function H. These equilibrium states in turn are imagined to represent the activation of certain learned mental states corresponding to the sets of synaptic strengths c_{ij} associated with those learned mental states. The Hamiltonian defined by Hopfield is intriguing because its minimization corresponds to the minimization of dissonance or conflict among synaptic input influences distributed throughout the network. Thus the function h_i defined in Eq. (9.3) represents the contribution by the individual units to the global Hamiltonian H. The function h_i depends on the distribution of input activity to the ith unit as follows. The connection coefficients c_{ij} are positive for excitatory synapses and negative for inhibitory synapses. Thus the state function is $+1$ if the cell is on hand -1 if the cell is off. One can see that from Eq. (9.3), then, that h becomes large positive if its strong excitatory inputs are on and its strong inhibitory inputs are off. Conversely, h becomes large negative when its strong inhibitory inputs are on and its strong excitatory inputs off. Since $h_i \sigma_i$ is always positive, either of these conditions contributes a maximum negative increment to H. Mixed input configurations to individual units, wherein both excitatory and inhibitory inputs are on, tend to give values of h_i closer to zero and therefore to give lesser increments toward minimizing H. Both conditions which contribute to minimizing H correspond to consonance or agreement among the synaptic influences throughout the network. Many cells are receiving locally agreeing signals on whether to become active. Larger values of H correspond to smaller absolute values of the h_i and reflect mixed or conflicting synaptic messages being distributed on individual constituent cells. A central focus of continuing work in this area consists of trying to show that patterns of activity corresponding to the activation of memory traces embedded by Hebbian learning do indeed correspond to minimization of the function H as defined by Hopfield. Subsequent work will probably show how to modify that definition to achieve this goal.

The modeling study of Hopfield is reviewed and extended by Peretto to include noisy processes. Peretto also discusses a similar model by Little based on Markovian evolution equation and state transition probability functions.

Fluid Mechanical Models

Dynamic Similarity and Scaling

In fluid mechanics the concepts of dynamic similarity and associated nondimensional analysis identify certain critical nondimensional numbers which are used to great advantage in characterizing the overall global features of a flow field. Such numbers provide broad theoretical characterization of the nature of a flow field, guidance in designing a scaled-down model version of the flow field, and compact representation of the influence of a number of different parameters on the flow field. For example, the Reynolds number, which can be considered as the ratio of the inertial forces to the viscous forces, is a critical determinant of the global behavior of fluids. One of the broadest generalizations in fluid mechanics is that when this ratio is very high, so that inertial forces dominate viscous forces, the fluid flow is turbulent, whereas if this number is smaller, the viscous forces damp out inertial osciallations and the flow is laminar or smooth. For most fluid situations there is a relatively narrow band of values for the Reynolds number in which the onset of turbulence occurs. Moreover, in making a model representation of a flow situation (say flow over the wing of an airplane or around a bend of a river), one of the first constraints imposed on the laboratory setup is that the Reynolds number in the model situation be the same as that in the situation of practical interest. Finally, in doing a comprehensive experimental study of a phenomenon one maps the characteristics of observed behavior throughout a wide range of Reynolds numbers.

Clearly, it would be useful to identify for neuronal networks various nondimensional numbers such as the Reynolds number which would allow similar characterization of their dynamic neuroelectric activity. For example, it would be useful to be able to predict tendencies to stable internally sustained activity as opposed to tendencies to oscillations in a particular neuronal population on the basis of the ratio of recurrent excitation to recurrent inhibition in that population. Moreover, it would be useful to have theoretical guidance in constructing scaled-down simulation models of neuronal populations, which typically contain many more neurons than are simulated in their corresponding model representations. Currently, the numbers of neurons and connections used in such models are chosen quite arbitrarily.

In Chapter 17 we present a definition of such nondimensional numbers which are representative of a convergent–divergent junction between two neuronal populations. One immediate result of this analysis is that in scaled-down models one must choose between verisimilitude in terms of network connectivity and intercohesion and verisimilitude at the level of individual neuronal dynamics.

Partial Differential Equations Representing Spatiotemporal Spread of Neuroelectric Activity

In fluid mechanics the detailed spatiotemporal structure of a flow field is described in terms of the spatiotemporal distributions of the velocity components and, if appropriate, thermodynamic properties such as pressure, temperature, and density throughout the flow field. The distributions of these functions are obtained from overall governing partial differential equations and corresponding boundary conditions. The differential equations in turn represent the physical constraints of conservation of mass, momentum, and energy applied to each point in the fluid. Under certain conditions such systems of differential equations can be reduced to a single governing partial differential equation for a "potential" function, from which the velocity components may be obtained. The governing differential equations may be derived from either the "continuum" point of view, wherein the fluid is conceived as a continuous medium, or the "kinetic theory" point of view, wherein the fluid is pictured as consisting of large numbers of interacting particles. (The latter point of view is related to the statistical mechanical point of view discussed above.)

Several attempts have been made to develop similar governing partial differential equations to describe the spatiotemporal development of electrical activity in neuronal populations. In these theories one considers the overall mean level of activity at a given point in space rather than the firing rate in any specific neuron.

One of the first models of this type was published by Beurle in 1956. Essential assumptions of Beurle's model are shown in Eq. (9.4)

Mean rate of arrival of impulses at x and $t + \tau$

$$= \int_{-\infty}^{\infty} F(X, t)\xi(x - X)\, dX \qquad (9.4a)$$

Mean value of integrated excitation at x and $t + \tau$

$$= \overline{N}(x, t) = \int_{-\infty}^{0} \int_{-\infty}^{\infty} F(X, T)\xi(x - X)\chi(t - T) \, dx \, dT \qquad (9.4b)$$

Proportion of cells at x which have total excitation of $q - 1$ at $t + \tau$

$$= p_{q-1} = e^{-N} \frac{N^{q-1}}{(q - 1)!} \qquad (9.4c)$$

In these expressions $F(x, t)$ represents the number of excited cells at position x at time t, $\xi(x - y)$ the mean number of connections between cells separated by distance $x - y$, $\chi(T)$ the portion of excitation produced by a given primary cell which remains at time T after it has become active, and N the number of elements in the population. Beurle derived

$$F(x, t + \tau) = RP_{q-1} \int_{-\infty}^{\infty} F(X, t)\xi(x - X) \, dX \qquad (9.5a)$$

$$F(t + \tau) = F(t) \qquad (9.5b)$$

$$1 - R = Fr \qquad (9.5c)$$

$$e^{-Fks} \frac{(Fks)^{q-1}}{(q - 1)!} (k - Fks_s') = 1 \qquad (9.5d)$$

as governing continuous random activity in a neuronal network. In these equations R is the proportion of cells which are nonrefractory and q is taken as the threshold. Equations (9.5b) and (9.5c) are conditions Beurle imposed to sustain constant activity at a particular level. Equation (9.5d) is the condition derived for sustained steady firing under these constraints. Here F is the level of sustained activity, k the integral of X over its entire range, and S the time interval over which excitation is integrated by the cells and Fks the total integrated excitation. Beurle also used this methodology to describe propagation of waves through neural networks. The governing equation for this propagation is shown is shown in Eq.(9.6a), which is derived from the constraints shown in Eqs.(9.6b) and (9.6c)

$$d^2R/dt^2 + (1 - R\phi/F) \, dR/dt = 0 \qquad (9.6a)$$

$$dF/dt = R\phi - F \qquad (9.6b)$$

$$dR/dt = -F \qquad (9.6c)$$

$$R = \frac{1}{m} - \frac{2B}{m} \tanh B(t - t_0) \qquad (9.6d)$$

$$F = \frac{2B^2}{m} \frac{1}{\cosh^2 B(t - t_0)} \qquad (9.6e)$$

where ϕ is the probability that a sensitive cell will be excited above its threshold in unit time. The solutions to this equation shown in Eqs. (9.6d) and (9.6e), represent traveling waves; B is a constant which determines the rate of rise and fall of activity, and M is a constant representing the ratio of ϕ to F.

In 1963 Griffith published a theory of spatiotemporal propagation of neuroelectric activity which is summarized here in Eqs. (9.7).

$$H\psi = kF \tag{9.7a}$$

$$F(x_i, t) = f[\phi(x_i, t)] \tag{9.7b}$$

where

$$\phi(x_i, t) = \int_{-\infty}^{\infty} \psi(x_i, t + \epsilon)i(\epsilon)\, d\epsilon \quad \begin{cases} i(\epsilon) = 0 & \epsilon > -\tau \\ i(\epsilon) \geqslant 0 & \epsilon < -\tau \end{cases}$$

$$\nabla^2\psi = \tfrac{1}{4}\beta^2\psi + \left(\frac{\beta}{v} - 4\pi A_1\right)\frac{\partial\psi}{\partial t} + \left(\frac{1}{v^2} - 4\pi A_2\right)\frac{\partial^2\psi}{\partial t^2} - 4\pi F(\psi)$$

$$\tag{9.7c}$$

$$\nabla^2\psi = \tfrac{1}{4}\beta^2\psi + \gamma\frac{\partial\psi}{\partial t} - 4\pi F(\psi) \tag{9.7d}$$

In these expression, ψ represents the overall excitation of the nervous network and F the activity of the soma of the neurons, "Excitation is regarded as being carried by a continual shuttling between sources and field, F creates psi, and so on." Griffith stated that H is an undefined operator, and k is a constant. Griffith used various theoretical arguments to derive Eq.(9.7c) from Eq.(9.7a). Equation (9.7c) is the main governing differential equation for the spatiotemporal distribution of excitation. The further simplfication shown in Eq.(9.7d) assumes that conduction velocity is infinitely large and neglects the second time derivative.

Okuda et al. extended Griffith's approach to show that a spreading active region tends to be disk-shaped in two dimensions and sphere-shaped in three dimensions. These workers also showed that the velocity of the boundary between a large active region and a large resting region is given by Eq.(9.8) and showed the influence of fatigue on this velocity.

$$u = (1 - 2a^2\tau\theta/k)v \tag{9.8}$$

In 1973 Wilson and Cowan presented Eq.(9.9) to describe the spatiotemporal development of E, the proportion of excitatory cells becoming active per unit time in a neuronal network.

$$E(\chi, t + \tau)\rho_e\delta\chi st = [1 - \int_{-t-r_e}^{1} E(\chi, T)\, dT]\, \rho_e\delta\chi$$

$$\times L_e\left\{\int_{-\infty}^{1}\left[\int_{-\infty}^{\infty} \rho_e E\left(X, T - \frac{|\chi - X|}{v_i}\right)\beta_{ee}(\chi - X)\, dX\right.\right.$$

$$- \int_{-\infty}^{\infty} \rho_i I\left(X, T - \frac{|\chi - X|}{v_i}\right)\beta_{ie}(\chi - X)\, dX$$

$$\pm P(z, T)\Big]\alpha(t - T)\, dT\Big\}\delta t \tag{9.9}$$

Here e_e and e are the surface density of the excitatory and inhibitory neurons respectivel, L_e is the expected proportion of excitatory neurons which receive at least treshold excitation per unit time, the beta's are interconnection functions, P represents afferent excitation applied from outside the net, and alpha is an excitatory postsynaptic potential, and I is the proportion of inhibitory cells becoming active at a point and is governed by an analogous equation. Wilson and Cowan derived coarse-grained versions of these governing equations which are shown in Eqs.(9.10).

$$\mu\,\frac{\partial E}{\partial t} + E = (1 - r_e E)S_e(N_e) \tag{9.10a}$$

$$\mu\,\frac{\partial I}{\partial t} + I = (1 - r_i I)S_i(N_i). \tag{9.10b}$$

with

$$N_e(x, t) = \alpha\mu\left[\int_{-\infty}^{\infty} q_e E(X, t - |x - X|/v_e)\,\beta_{ee}(x - X)\, dX\right.$$

$$- \int_{-\infty}^{\infty} q_i I(X, t - |x - X|/v_i)$$

$$\beta_{ie}(x - X)\, dY + P_e(x, t)\Big] \tag{9.10c}$$

$$N_i(x, t) = \alpha\mu\left[\int_{-\infty}^{\infty} q_e E(X, t - |x - X|/v_e)\,\beta_{ei}(x - X)\, dX\right.$$

$$- \int_{-\infty}^{\infty} q_i I(X, t - |x - X|/v_i)$$

$$\times \beta_{ii}(x - X)\, dX + P_i(x, t)\Big] \tag{9.10d}$$

These equations involve convolutions. Kawahara has simplified these equations of Wilson and Cowan and applied them to oscillations in coupled Van der Pohl oscillators and to the description of action potentials as in the Fitzhugh–Naguma equations.

Oguztoreli also presented a nonlinear integro–partial differential equation for the spatiotemporal spread of neuroelectric activity.

Ventriglia has presented Eqs. (9.11) to describe the spatiotemporal evolution of activity in neuronal populations.

$$\left[\frac{\partial}{\partial t} + \mathbf{v} \cdot \frac{\partial f}{\partial \mathbf{r}} \right] f(\mathbf{r}, \mathbf{v}, t)$$

$$= - \sigma \psi v f(\mathbf{r}, \mathbf{v}, t) + \lambda \sigma a f_r^*(\mathbf{v}) \int_{\cdot}^1 de\, \phi(\mathbf{r}, e, t)$$

$$\times \iint_{-\infty}^{\infty} d\mathbf{v}\, v f(\mathbf{r}, \mathbf{v}, t - t_0) + S(\mathbf{r}, \mathbf{v}, t) \qquad (9.11a)$$

$$\left[\frac{\partial}{\partial t} + \mu\, (e_r - e) \frac{\partial}{\partial e} \right] \phi(\mathbf{r}, e, t)$$

$$= \sigma \iint_{-\infty}^{\infty} d\mathbf{v}\, v f(\mathbf{r}, \mathbf{v}, t - t_0)[a\phi(\mathbf{r}, e - i, t) + (1 - a)\phi(\mathbf{r}, e + i, t)$$

$$- \phi(\mathbf{r}, e, t)] + \delta(e - e_r)\phi_d(\mathbf{r}, t - \tau) \qquad (9.11b)$$

$$\frac{\partial \phi_d(\mathbf{r}, t)}{\partial t} = - \phi_d(\mathbf{r}, t - \tau) + \sigma a \iint_{-\infty}^{\infty} d\mathbf{v}\, v f(\mathbf{r}, \mathbf{v}, t - T_0)$$

$$\times \int_{\cdot}^1 de\, \phi(\mathbf{r}, e, t) \qquad (9.11c)$$

where ϕ is the internal excitation of neurons, ϕ the proportion of neurons in a refractory state, f the velocity of impulses, ψ the neuron density, a the percentage of excitatory neurons, s the synaptic density, axonic branching, r position, $\mathbf{v}$ velocity, and δ the Dirac delta funtion. The local density of impulses, r and the local mean subthreshold excitation E are determined by f and ϕ according to Eq. (9.11b). Ventriglila applies this equation system to study the propagation of informational waves, dynamic activities, and some memory effects.

Models of Learning and Memory in Neural Networks

Learning and memory are components of a vast field of neural plasticity which merges at its peripheral limits with development and genetics and comprises many types of changes in neural structure and operation such as habituation, sensitization, adaptation, short-term memory, and long-term memory. In this section we consider models of long-term memory, that is, memory which is relatively permanent and whose specific content is not genetically prescribed. Such memories can be sensory, motor, cognitive, experiential, or volitional. Tulving partitioned such memories into episodic (sequential, historical, subjective, biographical) and structured (semantic, asequential, objective). Structured memories have been considered as the result of conditioning or association or alternatively as cognitive maps. Squire partioned associational memories into procedural (how) and declarative (what). Richard Thompson and several colleagues presented a theory of the neuronal substrates of associative learning with primary emphasis on conditioning and on the hippocampus and cerebellum. Citations for this work are given in Chapters 6 and 7.

It is almost universally assumed that long-term memories are defined in the nervous system in terms of variations in synaptic efficiency and that biochemical changes associated with learning and memory act primarily by mediating such changes in synpatic efficiency. The neurobiological particulars are largely unknown. Important contributions include experimental evidence of conjunctive or associative "enhancement" of synapses (i.e., changes in strength in specific synapses dependent on combined activity of multiple inputs) in hippocampal neurons, speculation and evidence that changes in the morphology of dendritic spines contribute substantially to learning and memory in central neurons, and speculation that calcium ions mediate changes in synaptic efficiency by contributing to increases in postsynaptic receptors, protein synthesis involved in spine swelling, transport of dendritic microtubules, and output of presynaptic transmitter.

This section reviews "connectionist" models, which explore the dynamics of neural networks due to hypothetical modulations of synaptic efficiency; "supraconnectionist" models, which consider more global features of neuroelectric activity in neuronal networks relevant to learning and memory; and models in the formation processing approach to neural function, whose primary focus in on the cognitive or psychological functions rather than electrical activity, per se. A number of models of learning and memory in specific neuronal systems have been considered in Chapters 5–8.

Connectionist Models of Learning and Memory

Virtually all connectionist models of the embedding of memory traces in neural networks refer to the seminal work of Hebb published in 1949. Hebb imagined that memory traces were embedded in neuronal networks by virtue of distributions of enhanced synaptic strengths between specific input and receiving neurons; that such systems of varied synaptic effectiveness are embedded in the network in the course of its operations according to its experience; that the individual neurons of specific networks were thereby arranged into coordinated clusters or "cell assemblies," activation of any one of which corresponded to a given cognitive or psychological event—that is, a given memory trace; and that the cognitive or functional operations of the networks of the brain consist of sequential grouping of the activation of cell assemblies in "phase sequences." The widely held contemporary view—that short-term memories are served by reverberating electrical activity patterns in specific neuronal pathways, with permanent and long-term modifications occurring in the synapses of those pathways to mediate long-term memories—is a component of this point of view. Hebb further imagined that the intrinsic original connectivity structure of the network was of secondary importance and might for theoretical purposes be considered as random or homogeneous and that the original network might be considered akin to the *tabula rasa* or blank tablet of Locke.

Hebb speculated that the significant synaptic changes in strength occurred incrementally on those clusters of input synapses on a cell whose coincident activation succeeded in producing an output spike in the postsynaptic or receiving cell. This allowed for the fundamental ingredient of associational learning: after adequate enhancement of all the synapses in a given input pattern had occurred, it was possible to successfully fire the postsynaptic cell by activation involving only various subsets of the original requisite input group. Thus a given neuron, and by implication the cell assembly of which it was part, could respond with a total pattern (cat) in response to a partial (tail) stimulus.

A number of investigators, including Farley and Clarke, Rochester, and Smith and Davidson, constructed early computer simulation models of Hebb's basic ideas. Using models involving hundreds of interconnected neurons, they discovered diffuse reverberations and the emergence of coordinated firing in specific collections of cells which could readily be called "assemblies" in the sense of Hebb. Inhibitory cells in particular seemed to be critical in partitioning cells into assemblies and in maintaining interassembly distinctions.

Caianiello (1961) also incorporated Hebb's ideas into his "adiabatic learning hypothesis." Caianiello's school has focused primarily on dynamics of neuronal networks, with minimal attention to learning, but has from time to time consistently made use of this learning rule.

Another early study involving Hebbian concepts that has been influential in pattern recognition and artifical intelligence was carried out by Rosenblatt. Rosenblatt constructed electronic computational units, which he called "perceptrons," whose primary purpose was to explore learning and cognitive operations in artifical devices with brainlike mechanisms (threshold elements and Hebbian learning). A notable feature of Rosenblatt's perceptrons was the inclusion of a reinforcement control system or "teacher" which exerted external control over the learning operations of his networks. Rosenblatt's perceptrons were best at pattern recognition and not very successful even in this area. Rosenblatt presented his research in formal mathematical terms with many precise definitions, theorems, and corollaries. His work was a seminal effort in pattern recognition and artificial intelligence. Pappert and Minsky published a thorough critique of this approach in that field in 1969.

Also in 1969 Brindley published a theoretical paper elaborating Hebb's ideas with direct relevance to neuronal networks. Brindley pointed out that within Hebb's general conceptual framework one could identify a number of specific plausible synaptic learning rules. About this time Marr presented notable theoretical papers involving learning in the networks of the cerebellum, hippocampus, and neocortex, as discussed in Chapters 6–8.

In the past 15 years or so a large number of computational models incorporating Hebbian ideas have appeared in the literature. The models of Kohonen, Fukushima, and Amari are particularly notable. Kohonen simulated associative learning in two-dimensional networks of about 3000 threshold elements which exhibit one-to-one connections with an input field, short-range lateral inhibition, and longer-range recurrent excitatory connections that mediate the associative learning. In this model Kohonen and collaborators demonstrated the selective recall of signal patterns corresponding individually to 500 distinct photographs of faces. The signals could be recalled with a fragment of the corresponding photographs. Kohonen discovered that such model systems can automatically form one- or two-dimensional maps of features which are present in sets of input signals, with the responses acquiring the same topological order as their corresponding stimulus events, and analyzed this process mathematically.

Fukushima noted that the capabilities of Rosenblatt's perceptron model might be enlarged if the number of neural layers was increased. With a number of collaborators, he constructed a multilayer "cognitron" which contains modifiable feedback connections from the last layer cells to the front layer cells as well as conventional feed-forward connections and showed that the cognitron acquires selective responses to individual stimulus patterns as predicted by Hebb. In this model a signal keeps circulating in the network in response to a given stimulus even after termination of the stimulus, and the characteristic response pattern gradually emerges during a "self-organization" period. In later work Fukishima added modifiable inhibitory feedback connections as well as conventional modifiable excitatory feed-forward connections in the cognitron. He showed that such a network can be made to respond only to novel stimuli and to perform both associative memory and pattern recognition. Such pattern recognition can be unaffected by shifts in position. Fukishima's cognitron has several hundred elements laid out in a two-dimensional array. Deutsch presented a simplified version of Fukishima's cognitron with a 10-element one-dimensional field. This model demonstrates primitive pattern recognition with invariance to lateral shift.

Amari also studied orthogonal and covariance learning in self-organizing neural-like networks and their activities in forming associations and representation of concepts. Amari's models have been applied primarily to formation of category-detecting cells and topographic maps, as indicated in Chapter 8. Silverstein and Yeshurun and Richter-Dyn presented theoretical discussion of Hebbian learning in neural networks. Willwacher demonstrated selective signaling of input patterns and selective associative recall in a 100-element model of Hebbian learning on a UNIVAC computer. Easton and Gordon showed that stability can be mediated on Hebbian nets by modifiable recurrent inhibition while other simple inhibitory configurations do not mediate stability. They also theoretically analyzed for the occurrence of cell assemblies with a linearization assumption. Reilly, Cooper, and Elbaum presented a model for learning of pattern categories in which the concept of a pattern class develops from storing in memory a limited number of class elements in a Hebbian-type model involving supervised learning (i.e., learning with a "teacher"). Bobrowski presented an algorithm for perceptronlike learning in multilayer threshold nets without feedback, using a teacher's decision. In later work Bobrowski discussed two kinds of neurons involved in unsupervised learning: a neuron that passes signals most frequently occurring in the learning sequence, and another neuron that detects rareness and speculates on the role of such formal neurons in receptive field dynamics. Hampson and Kibler presented a multilayer neuron assembly model which learns arbitrary Boolean functions and allows multiple output systems to share a common memory. Nelson presented a network model consisting of several layers of cells capable of pattern recognition, concept formation, and recognition of patterns of events in time. Nelson stressed the ability of his model to recognize two or more patterns in a context-dependent manner, to permit the recognition of logical relations among patterns, and to separate the processes of association and recognition. In his model association corresponds to forward flow of information and recognition to reverse flow of information.

Shaw presents a model of Hebbian learning in neuronal networks which specifies the consequences for correlated firing among constituent neurons in a cell assembly that should be readily testable by experiments with multiple extracellular microelectrodes. Shaw discusses the Hebbian cell assembly with respect to the theories of E. Roy John (see below) and suggests a modified hypothesis for Hebbian learning.

Shiozaki points out that the ability to activate assemblies selectively in correlation matrix models such as those discussed in this chapter is greatly dependent on the orthogonality (separability) of the correponding stimulus patterns. He notes that the vast majority of these modeling efforts have involved two-dimensional models. He constructs a three-dimensional Hebbian learning model and shows that the selective recallability is enhanced for overlapping or nonorthogonal stimuli.

Palm has applied graph theory to determine the number and size of cell assemblies available in simplified Hebbian networks. Homogeneous two-dimensional nets have at least $2^n + 2^m - 4$ assemblies, where n and m are the dimensions of the net. Palm and Bonhoeffer discussed the limitations of certain ways of simulating computation in neural networks on computers.

Munroe suggested two alternative postsynaptic mechanisms for learning in single neurons which could produce neurons that specialize respectively in specification or generalization, and studied a module consisting of several cells of these two types in relation to visual processing.

Fedor and Majernik proposed a model for the molecular mechanisms of synaptic memory in which the receptors of the neuron surface are divided into functional independent fields. Recording in memory corresponds to cooperative transitions of protein subunits; discrimination corresponds to time-controlled transport of ions; and accommodation is served by changes in concentrations of protein complex subunits.

Sutton and Barto assert that modeling studies of Hebbian learning in neural networks have had only moderate success in the recognition, processing, and associative storage and retrieval of spatial patterns, and have shown a conspicuous absence of significant processing of temporal patterns. They speculate that a narrow interpretation of Hebb's rule for synaptic learning does not allow for the richness of learning processes as evidenced in classical conditioning studies or psychological learning theory. They suggest that an extended rule contributed to by Rescorla and Wagner and by Widrow and Hoff, which incorporates stimulus context effects and adjusts for stability and saturation difficulties, provides a more valid model of classical conditioning. They present a further extension of this rule which incorporates the predictive nature of classical conditioning. In this model a temporal period of eligibility for modification is triggered whenever a presynaptic signal occurs; the extent to which an eligible synapase is modified depends on the reinforcement level during the period of eligibility.

In another paper these authors describe an associative search network for reinforcement learning which does not require a "teacher" but rather searches for an ouput pattern that optimizes an external reinforcement signal for any given input pattern. Barto and Sutton apply this model to the control of locomotion in terms of landmarks in a spatial environment containing various types of olfactory gradients. In a subsequent publication they show that a layered extension of this network model can control movement in a spatial environment by forming associations between optimal directions of movement and stimulus patterns determined by its position with respect to landmarks.

Carpenter, Cohen, and Grossberg have presented theoretical papers on the properties and constraints of Hebbian-type learning in neural networks. They include topics such as stability, category learning, common recognition, attention, memory consolidation, retention, and amnesia.

Chapters 20 and 22 of this book present computer simulation programs in FORTRAN to simulate Hebbian learning in neuronal networks of arbitrary structure.

Theodorescu presents a model for stochastic learning in neuronal networks wherein the transformation from one state of the network to the next state is accomplished by means of an optimization operator. Each active step is followed by a relaxation step so that a fundamental relation between optimization and learning occurs as the system converges in response to a repeated stimulus pattern. Pfaffelhuber has also discussed learning in relation to stochastic optimization and information theory in general.

Supraconnectionist Models of Learning and Memory in Neural Networks

A smaller number of theorists have considered the more global dimensions of embedding memory traces in neuronal systems. Contributions in this area include Lashley's finding that the ability of regions of the cortical networks to embed memory traces seems to be governed by the rules of equipotentiality and mass action; the apparent stochastic or probabilistic nature of neuroelectric activity; and the idea from general systems theory that the nervous system might operate as a globally coordinated system with a certain amount of security as to the functioning of any of its constitutent parts.

Some 20 years ago Ross Adey presented several theoretical discussions emphasizing the global nature of neuroelectric signaling and its dependence on the biochemical constituency and processes of the extracellular and intracellular fluids in which the nervous system is embedded. Adey discussed the significance of the stochastic nature of the nervous system, the likelihood of Fourier representation in neuronal processes, and the importance of calcium as a global unifier in nervous processes.

One of the most notable supraconnectionist models is the "statistical configuration theory" of E. Roy John. In this model memory traces are represented in internally coherent spatiotemporal patterns of electrical activity which move through the neural networks that support them with a large degree of apparent independence of the detailed structure of the underlying networks. In this concept physiology rather than anatomy is the primary coordinating ingredient. John further imagines that these spatiotemporal patterns are stochastic and may occur redundantly in separate regions of the nervous system. John describes numerous experiments in cats which tend to support the statistical configuration concept.

MacGregor has shown in computer simulations that coherent patterns of activity which tend to move relatively freely across two-dimensional model networks can arise in response to particular stimulus patterns learned in interconnected networks according to changing synpaptic efficiency. In these simulations the movement of patterns seems to be determined primarily by the confluence of postsynaptic potentials (PSPs) as projected forward through the network by the current locus of the physiological activity and only secondarily by the substratum of synaptic strengths. MacGregor has also constructed a theoretical model wherein the dynamic manifestations of memory traces could exhibit the characteristics of statistical configurations although they are embedded in a set of increased synaptic efficiencies in specific interconnections as described by Hebb.

Karl Pribram has described a theory of brain operations which stresses the global character of brain activity. In this model the brain operates primarily in terms of spatial and temporal frequency signals rather than explicit representation of objects. Pribram points to the representation of the visual system as a Fourier analyzer as presented by Robson and Campbell and to the similarities of the brain's function in learning and memory to holographic systems, as suggested by Lashley's work. This view lends itself more readily to the abilities of the brain to resonate with various environmental signals and to Eastern mystical views of consciousness than to the brain's operating in terms of an internal model of the objects of the external world.

Edelman has developed a theory of learning, memory, and mental operations which links contemporary neurobiological genetics, Hebbian synaptic learning theory, and global brain operations. In this theory mental operations are dependent on three phases. In the first (prenatal) phase, control of neural development by cell adhesion molecules (CAM's) result in the emergence of collections of "neuronal groups" within developing neuronal networks. These neuronal groups are defined by their patterns of synaptic interconnections. Since the formation of barriers and specific regions within developing tissue depends on variables beyond the control of cell adhesion molecules, the patterns of neuronal groups are dependent on individual history and are idiosyncratic to a given individual. Edelman stresses that even idential twins will have different neuronal groupings at this early stage. In the second phase of development the neuronal groups further develop to form a "secondary repertoire" within a given brain region, which is also later referred to as a "map." A secondary repertoire is a pattern of interconnected neuronal groups that have come to respond better to certain specific stimuli on the basis of Hebbian strengthening of synapses. Secondary repertoires develop in the early stages of life after birth. This is seen as a process of selection, somewhat similar to Darwinian selection, wherein various stimulus patterns complete for the neuronal groups. A particular secondary repertoire in a given brain region consists of a large number of interrelated neuronal groups, and the activation of each set correponds to the particular stimulus configuration. Different stimuli activate different interconnected subsets of neuronal groups, In the third phase, the various secondary repertoires in various brain regions constitute sets of maps, and any significant mental operation involves the simultaneous activation of large numbers of maps and their interactions through reciprocal anatomical pathways between the brain regions. For example, in the representation of a perceptual event, sensory stimuli would be expected to have activated maps in the primary and secondary areas of auditory and visual cortex and in posterior association cortex on both sides of the brain, and various associated patterns in limbic and frontal areas would be active as well. The conscious and unconscious mental operations associated with this image would involve the global pattern of activity over the entire brain and in particular the cross-referencing and interactions among these maps. Edeleman further theorizes that mental operations such as categorization and imagination consist of the processes and properties of these intermap interactions.

Significant features of Edelman's theory are that individual brains are seen as entirely unique and idiosyncratic; there is an element of Darwinian-like competition and selection application to brain structure; fundamental significance is attached to global activity patterns and interactions (as in Luria's model); and the nature of mental life is highly dependent on processes such as imagination and creative thought which involve the global processes of intermap interaction. Thus in this model an individual's mental life is unique, idiosyncratic, and continually developing in time on the basis of continued cultivation and interactions involving its current structure and activation patterns.

Models of Neural Networks as Cognitive or Functional Operators

An important stream of modeling has attempted to construct network models of neuronlike elements for hypothetical functional and cognitive operations with little interest in the detailed properties of individual units or their correspondence to specific electrical signals in the brain. This point of view is justifiable because the primary focus of these models is at the level of cognitive and functional variables, neuronal systems are extremely complicated and some simplification is necessary, and it is often not desirable to specify ahead of time exactly what neuronal components are repesented by the individual elements—the individual elements may represent parts of neurons, single neurons, clusters of neurons, networks, functional subsystems, etc.

The automaton model published in 1943 by McCulloch and Pitts, as discussed in Chapter 2, is the starting point for this work. Basic neuronal operators implied by the McCulloch–Pitts approach are illustrated in Fig. 9.1. McCulloch and Pitts showed how such elements could be used to construct sequential machines to perform combinations of logical operations of any degree of complexity. Implicit is the idea that neuronal activity is interpretable in terms of propositions. They state that "the all-or-none law of nervous activity is sufficient to insure that the activity of any neuron may be represented by a proposition."

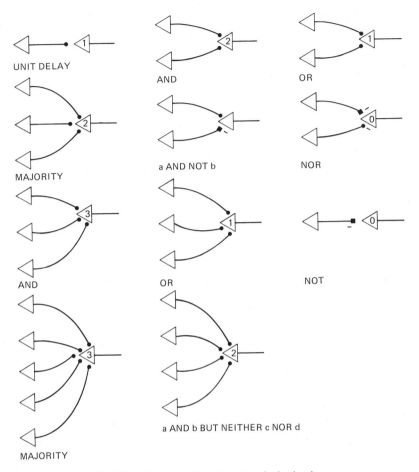

Fig. 9.1. Representative elementary logic circuits.

In a later paper McCulloch and Pitts showed how a network operating according to these concepts could abstract universals in sensory input streams—chord and timbre independent of pitch, for example.

A number of workers in the fields of logic circuits, theory of automata (or sequential machines), and related areas have elaborated the cognitive and functional properties of neuronal networks of McCulloch–Pitts-like elements. A good representative summary is given by Minsky, who shows how simple networks of, say, 3 to 12 McCulloch–Pitts-like elements serve to produce delays, gating, controlled switching or routing of information, short-term memory, scaling, addition, subtraction, and various encoding and decoding procedures, including interchanges between parallel and serial representations of information and between representations of space and time.

Since it can be shown that any finite state machine (e.g., a Turing machine) can be represented by a McCulloch–Pitts-like nerve network, virtually all the theoretical conclusions in the field of sequential machines are applicable to McCulloch–Pitts nerve networks. For example, Goedel's theorem that no closed cognitive system can be both consistent and complete would apply to McCulloch–Pitts nerve networks for which the constraints of the Turing system apply.

Continued work in this field is vigorously advocated by Feldman, who has presented a refreshingly new theoretical concept of operations in neuronal networks based on the importance of parallel rather than serial processing. Feldman claims that brains are inherently massively parallel in anatomy and function and that our thinking has been significantly off-course in conceiving neural net operations as sequential. He states that the fundamental premise of connectionism is that individual neurons do not transmit large amounts of symbolic information. Instead they compute by being appropriately connnected to large numbers of similar units. He points out that it is unreasonable to imagine the nervous system operating in terms of trains of transmitted sequential signals when entire complex behaviors are carried out in a few hundred milliseconds, implying a comparable number of time steps. This, coupled with the observation that significant central neurons experience thousands, tens of thousands, and even upward of a hundred thousand input or output synapses, seems to establish parallel processing as a central concept in our understanding of the functioning of neuronal networks.

A number of other investigators have advocated and employed highly parallel neural models. These include Killmer, Arbib, Anderson, Ratcliff, McClellan, Hinton and Hopfield.

Feldman also spells out a formal model for the elements of such functional nets and continues the elaboration of their functional properties in the manner of McCulloch and Pitts and Minsky. In Feldman's model a unit has associated with it an input, a potential, a state, and an output. Inputs are prescribed external to the unit, whereas the temporal evolutions of potential, state, and output are each determined by appropriate functions of input, potential, and state. This general unit may represent parts of neurons, entire neurons, clusters of neurons, systems of neurons, etc., depending on the context. Feldman imagines that distribution of activity in his units maps uniquely onto functional and operational variables; for example, when activity in a given unit is high, the cognitive or functional variable associated with that unit is activated. He constructs simple networks of these units which can cause motion to turn left or to turn right, perform multiplications, modify associations, make decisions ("winner take all," "coalition") and do functional decomposition, limited-precision computation, coarse and coarse-fine coding, tuning, spatial coherence, object recognition, time and sequence operations, space-time interconversion, and the like.

Stephen Grossberg has made a significant theoretical effort over about 18 years to understand the functional operations of nervous systems on the basis of network models of neuronlike elements. Like the other investigators discussed in this section, he is not interested in the dynamic behavior of the elements themselves but rather in the emergent properties of the networks of which they are parts. Grossberg, with various collaborators, has modeled autonomic functions (Chapter 5), conditioning processes (see Chapter 7), and numerous features of cognitive processes. This work is an interesting example of the top-down approach to brain function: Grossberg has offered numerous interpretations of the organization of neuronal networks of the brain on the basis of thought experiments concerning their psychological operations.

McClelland and Rumelhart have edited a two-volume book entitled "Parallel Distributed Processing: Explorations of the Microstructure of Cognition" which contains many contributions representative of the state of the art in the approach to neural networks as cognitive operators. It has appeared too late, however, to be fully reviewed here.

Bibliography

General Neural Networks

1977   MacGregor, R. J., and Lewis, E. R., eds."Neural Modeling", Plenum, New York.
        Metzler, J., ed."Systems Neuroscience"Academic Press, New York.
1975   Szentágothai, J., and Arbib, M.A. " Conceptual Models of Neural Organization."
        MIT Press, Cambridge, Massachusetts.
        Weinberg, G. M. "An Introduction to General Systems Thinking." Wiley, New York.
1974   Katchalsky, A. L., Rowland, V., and Blumenthal, R. "Dynamic Patterns of Brain Cell Assemblies." MIT Press, Cambridge, Massachusetts.
1973   Leibovic, K. N. "Nervous System Theory; An Introductory Study." Academic Press, New York.
1972   Laszlo, E. "The Systems View of the World." Braziller, New York.
1971   Gawronski, R. "Bionics: The Nervous Systems as a Control Sytem." Am. Elsevier, New York.
1969   Klir, G. J. "An Approach to General Systems Theory." Van Nostrand-Reinhold, Princeton, New Jersey.
1968   von Bertalanffy, L. "General System Theory," rev. ed. Braziller, New York.
1967   Deutsch, S. "Models of the Nervous System." Wiley, New York.
1966   Milsum, J. H. "Biological Control Systems Analysis." McGraw-Hill, New York.
1961   Wiener, N. "Cybernetics: Control and Communication in the Animal and the Machine," 2nd ed. MIT Press, Cambridge, Massachusetts.

Computer Simulation and System Theoretic Descriptions

1984   Annison, P.A. and Kokkinidis, M. A neural net model for multiple memory domains. *J. Theor. Biol.* **109**, 95-110.

1986   Oguztoreli, M. N., Steil, G. M. and Caelli, T. M. Control mechanisms of a neural network. *Biol. Cybernet.* **54**, 21–28.
1984   Anninos, P. A., Argyrakis, P., and Skouras, A. A computer model for learning processes and the role of the cerebral commissures. *Biol. Cybernet.* **50**, 329–336.
　　　　Anninos, P. A., Kokkinidis, M., and Skouras, A. Noisy neural nets exhibiting memory domains. *J. Theor. Biol.* **109**, 581–594.
　　　　Glass, L. Boolean and topological models for physiological dynamics. *Eng. Med. Biol.* **37**, 20.3, 124.
1983   Anninos, P. A., and Argyrakis, P. A mathematical model for the decay of short-term memory with age. *J. Theor. Biol.* **102**,191–197.
1982   Gelfand, A. E., and Walker, C. C. On the character of and distance between states in a binary switching net. *Biol. Cybernet.* **43**, 79–86.
1980   Amari, S. Topographic organization of nerve fields. *Bull. Math. Biol.* **42**, 339–364.
1979   Erulkar, S. D., and Soller, R. W. Neuronal interactions in a central nervous system model. *In* "Origin of Cerebral Field Potentials" (E. J. Speckmann and H. Caspers, eds.), Thieme, Stuttgart.
　　　　Oguztoreli, M. N. Activity analysis of neural networks. *Biol. Cybernet.* **34**, 159–169.
　　　　Torioka, T. Pattern separability in a random neural net with inhibitory connections. *Biol. Cybernet.* **34**, 53–62.
1978   Amari, S., and Takeuchi, A. A mathematical theory on formation of category detecting nerve cells. *Biol. Cybernet.* **29**, 127–136.
　　　　Ermentrout, G. B., and Cowan, J. D. Temporal oscillations in neural nets. *J. Math. Biol.* **12**, 43–51.
　　　　MacGregor, R. J., and McMullen, T. (1978) Computer simulation of diffusely-connected neuronal populations. *Biol. Cybernet.* **20**, 1–7.
　　　　Siljak, D. D. "Large-Scale Dynamic Systems." North Holland Publ., New York.
　　　　Zetterberg, L. H., Kristiansson, L., and Mossberg, K. Performance of a model for a local neuron population. *Biol. Cybernet.* **31**, 15–26.
1977   Amari, S., Dynamics of pattern formation in lateral-inhibition type neural fields. *Biol. Cybernet.* **27**, 77–87.
　　　　Amari, S. Neural theory of association and concept formation. *Biol. Cybernet.* **26**, 175–185.
　　　　Josin, G. M. Self-control in neural nets. *Biol. Cybernet.* **27**, 185–187.
　　　　MacGregor, R. J., and McMullen, T. Theory of monosynaptic transfer between neuron populations. *Behav. Sci.* **22**, 207–217.
　　　　Tokura, T., and Morishita, I. Analysis and simulation of double-layer neural networks with mutually inhibiting interconnections. *Biol. Cybernet.* **25**, 83–92.
　　　　Torioka, T., and Koga, K. Pattern separability of a three-layered random nerve network with inhibitory connections. *Electron. Commun. Jpn.* **60A**, 34–41.
1976   Caianiello, E. R. and Grimson, W. E. L. Methods of analysis of neural nets. *Biol. Cybernet.* **22**, 1–6.
　　　　Kurokawa, T. On reverberatory processes in homogeneous neuronal spaces. *Biol. Cybernet.* **21**, 139–144.
　　　　Leake, B., and Anninos, P. Effect of connectivity on the activity of neuronal net models. *J. Theor. Biol.* **58**, 337–363.
　　　　Lopes da Silva, F. H., van Rotterdam, A., Barts, P., and van Heuden, E. Models of neuronal populations: The basic mechanisms of rhythmicity. *Prog. Brain Res.* **45**, 281–308.
1975   Amari, S. Homogeneous nets of neuron-like elements. *Biol. Cybernet.* **17**, 211–220.
　　　　Caianiello, E. R., and Grimson, W. E. L. Synthesis of Boolean nets and time behavior of a general mathematical neuron. *Biol. Cybernet.* **18**, 11–117.
　　　　Cull, P. Control of switching nets. *Biol. Cybernet.* **19**, 137–145.

Freeman, W. J. "Mass Action in the Nervous System." Academic Press, New York.

Stubbs, D. F. Connectivity and the brain. Unpublished manuscript.

Stubbs, D. F., and Good, P. I. Connectivity in random networks. Unpublished manuscript.

1974   Amari, S. A method of statistical neurodynamics. *Biol. Cybernet.* **14**, 201–215.

Kurokawa, R., and Tamura, H. Networks of neural nuclei. *Biol. Cybernet.* **16**, 69–77.

MacGregor, R. J., and Palasak, R. Computer simulation of rhythmic oscillations in a neuron pool. *Biol. Cybernet.* **16**, 79–86.

Sait, N. Localized oscillations in neural networks. I. *J. Theor. Biol.* **47**, 421–437.

Yoshizawa, S. Some properties of randomly connected networks of neuron-like elements with refractory. *Biol. Cybernet.* **16**, 173–182.

1973   Anninos, P. A. Evoked potentials in artificial neural nets. *Biol. Cybernet.* **13**, 24–29.

Findlay, J. M., and Daniell, G. J. A model for pattern recognition by cell networks. *J. Theor. Biol.* **38**, 641–645.

Wong, R., and Harth, E. M. Stationary states and transients in neural populations. *J. Theor. Biol.* **40**, 77–106.

1972   Amari, S. Characteristics of random nets of analog neuron-like elements. *IEEE Trans. Syst., Man., Cybernet.* **2**, 643–657.

Anninos, P.A. Cyclic modes in artifical neural nets. *Biol. Cybernet.* **11**, 5–14.

Freeman, W. J. Linear analysis of the dynamics of neural masses. *Annu. Rev. Biophys. Bioeng.* **1**, 225–226.

McMurtie, R. E. Determinants of stability of large randomly connected systems. *J. Theor. Biol.* **35**, 227–232.

Morshita, I., and Yajima, A. Analysis and simulation of networks of mutually inhibiting neurons. *Biol. Cybernet.* **11**, 154–165.

1971   Amari, S. Characteristics of randonly connected threshold-element networks and network systems. *Proc. IEEE* **59**, 35–47.

deLuca, A. On some dynamical properties of linear and affine networks. *Biol. Cybernet.* **8**, 123–127.

Meno, F. "Neural Systems Modeling," Tech. Rep. Carnegie-Mellon University, Pittsburgh, Pennsylvania.

1970   Aiello, A., Burattini, E., and Caianiello, E. R. Synthesis of reverberating neural networks. *Biol. Cybernet.* **7**, 191–195.

Anninos, P. A., Beek, B., Csermely, T. G., Harth, E. M., and Pertile, G. Dynamics of neural structures. *J. Theor. Biol.* **26**, 121–148.

Gardner, N. R., and Ashby, W. R. Connectance of large dynamic (cybernetic) systems: Critical values for stability. *Nature (London)* **228**, 784.

Harth, E. M. Csermely, T. J., Beek, B., and Lindsay, R. D. Brain functions and neural dynamics. *J. Theor. Biol.* **26**, 93–120.

Mowle, F. Controllability of nonlinear sequential networks. *J. Assoc. Comput. Mach.* **17**, 518–524.

Rosen, J. "Dynamical System Theory." Wiley, New York.

1969   Averbukh, D. Ya. Random nets of analog neurons. *Avtom. Telemekh.* **10**, 116–123.

deLuca, A., and Ricciardi, R. M. Decomposition of a matrix as the sum of direct products. Applications to neural networks. *Calcolo* **6**, 225–231.

Kauffman, S. A. Metabolic stability and epigenesis in randomly constructed genetic nets. *J. Theor. Biol.* **22**, 437–467.

Rozonor, L. I. Random logical nets. I. *Avtom. Telemekh.* **5**, 137–147.

Rozoner, L. I. Random logical nets. II. *Avtom. Telemekh.* **6**, 99–109.

Rozonoer, L. I. Random logical nets. III. *Avtom. Telemekh.* **7**, 127–136.

1968   Klemera, P. Random network as a model of associative region of cerebral cortex. *Sb. Ved. Pr. Lek. Fak. Univ. Karlov Hradci Kralove*, **11**, 559–565.

1967   Caianiello, E. R., deLuca, A., and Ricciardi, L. M. Reverberations and control
        of neural networks. *Biol. Cyberenet.* **4**, 10–18.
        Chuang, Y. H., Bell, N. R., and Stacy, R. W. An automaton analysis approach to
        the study of neural nets. *Comp. Biomed. Red.* **1**, 173–186.
1966   Caianiello, E. R. (1966). Decision equations and reverberations. *Biol. Cybernet.*
        3, 998–100.
1963   Dusheck, G. J. A flexible neural logic network. *IEEE Trans. Mil. Electron.* **7**,
        208–213.
1962   Ashby, W. R., von Foerster, H., and Walker, C. C. Instability of pulse activity in a
        network with threshold. *Nature (London)* **196**, 561–562.
        Cohn, M. Controllability in linear sequential networks. *IRE Trans. Circuit Theory*
        9, 74–78.
1961   Caianiello, E. R. Outline of a theory of thought-processes and thinking machines.
        J. Theor. Biol. **1**, 209–235.
1956   Allanson, J. T., Some properties of randomly connected neural networks. *In*
        "Information Theory" (C. Cherry, ed.), pp. 303–313. Butterworth, London.
        Ashby, W. R. "An Introduction to Cybernetics." Chapman & Hall, London.
        von Neumann, J. Probabilistic logic and the synthesis of reliable organisms from
        unreliable components. In "Automata Studies." (C.E. Shannon and J. McCarthy,
        eds.). Princeton Univ. Press, Princeton, New Jersey.
1952   Landau, H. G. On some problems of random nets. *Bull. Math. Biophys.* **14**, 203–212.
        Rapoport, A. (1952). Ignition phenomena in random nets. *Bull. Math. Biophys.* **14**,
        35–44.
1948   Rapoport, A. Cycle distribution in random nets. *Bull Math. Biophys.* **10**, 145–157.

Chaos Theoretic Descriptions

1984   Chay, T. R. Abnormal discharges and chaos in a neuronal model system. *Bio.
        Cybernet.* **50**, 301–311.
        King, R., Barchas, J. D., and Huberman, B. A. Chaotic behavior in dopamine
        neurodynamics. *Proc. Natl. Acad. Sci. U.S.A.* **81**, 1244–1247.
1983   Choi, M. Y., and Huberman, B. A. Dynamic behavior of nonlinear networks. *Phys.
        Rev. A* **28**, 1204–1206.
        Grondin, R. O., Porod, W., Loeffler, C. M., and Ferry, D. K. Sychronous and
        asynchronous systems of threshold elements. *Biol. Cybernet.* **49**, 1–7.
        Guevara, M. R., Glass, L., Mackey, M. C., and Shrier, A. (1983) Chaos in Neuro-
        biology. *IEEE Trans. Syst. Man. Cybernet.* **13**, 790–798.
1982   Holden, A. V., Winlow, W., and Haydon, P. G. "The induction of periodic and
        chaotic activity in a mulluscan neurone. *Biol. Cybernet.* **43**, 169–173.
1981   Haken, H., ed. "Chaos and Order in Nature." Springer-Verlag, Berlin and New
        York.
        Hofstader, D. R. Metamagical themas: Strange attractors: Mathematical patterns
        delicately posed between order and chaos. *Sci. Am.* **245**. 22–43.
        Holden, A. V. Membrane current fluctuations and neuronal information processing.
        Adv. Physiol. Sci. **30**, 23–42.
        Ott, E. (1981). Strange attractors and chaotic motions of dynamical system. *Rev.
        Mod. Phys.* **53**, 655–671.
1976   Cronin, J. Mathematical aspects of periodic catatonic schizophrenia. *Bull. Math.
        Biol.* **39**, 187–200.

Realistic Computer Simulations

1987   MacGregor, R. J. "Neural and Brain Modeling," Part II. Academic Press. Orlando,
        Florida.

1985   Knowles, W. D., Traub, R. D., Wong, R. K. S., and Miles, R. Properties of neural networks: Experimentation and modeling of the epileptic hippocampal slice. *Trends Neurosci.* **8**, 73–79.

1982   MacGregor, R. J., and Gochis, P. A computer software system in FORTRAN for simulating large nerve networks. II Application. *J. Theor. Neurobiol.* **1**, 135–149.

Traub, D. R. Simulation of intrinsic bursting in CA3 hippocampal neurons. *Neuroscience* **7**, 1233–1242.

1981   MacGregor, R. J., A computer software system in FORTRAN for simulating large nerve networks. I. The system. *J. Theor. Neurobiol.* **1**, 82–104.

1977   MacGregor, R. J., and Lewis, E. R., "Neural Modeling." Plenum, New York.

1976   Perkel, D. H. A computer program for simulating a network of interacting neurons. I. Organization and physiological assumptions. *Comput. Biomed. Res.* **9**, 31–44.

Perkel, D. H. A computer program for simulating a network of interacting neurons. III. Applications. *Comput. Biomed. Res.* **9**, 67–74.

Perkel, D. H., and Smith, M. S. A Computer program for simulating a network for interacting neurons. II. Programming aspects. *Comput. Biomed. Res.* **9**, 45–66.

1968   Dill, J. C., Randall, D. L., and Richer, I. PLEXUS—an on-line system for modeling neural networks. *Commun. ACM* **11**, 622–640.

1966   Harmon, L. D., and Lewis, E. R. Neural modeling. *Physiol. Rev.* **46**, 113–192.

1965   Perkel, D. H. Applications of a digital-computer simulation of a neural network. *In* "Biophysics and Cybernetic Systems" (M. Maxfield, A. Callahan, and L. J. Fogel, eds.), pp. 37–51. Spartan Books, Washington, D.C.

Information Theoretic Approaches

1985   Nakao, M., Hara, K., Kimura, M., and Sato, R. Identification and estimation algorithm for stochastic neural system. II. *Biol. Cybernet.* **52**, 71–78.

1984   Moty, E. A., and Khalil, T. M. The application of information theory in EMG processing. *Eng. Med. Biol.* **37**, 11.7, 73.

1983   Bennet, C. H., and Landauer, R. The fundamental physical limits of computation. *Sci. Amer.* 48–56.

Nakahama, H., Yamamoto, M., Aya, K., Shima, K., and Fujii, H. Markov dependency based on Shannon's entropy and its application to neural spike trains. *IEEE Trans. Syst., Man, Cybernet.* **13**, 692–701.

1982   Windhorst, U., and Schultens, H. A. Measures of transinformation for multiple input/single output neuronal systems. *Biol. Cybernet.* **45**, 57–61.

1981   Palm G. Evidence, information, and surprise. *Biol. Cybernet.* **42**, 57–68.

Palm, G. On the storage capacity of an associative memory with randomly distributed storage elements. *Biol. Cybernet.* **39**, 125–127.

1979   Palm, G. Entropy for dynamical systems. *Fund. Univ. Bras., Trab. Mat.* **151**, 1–22.

Uttley, A. M. "Information Transmission in the Nervous System." Academic Press. London.

1978   Harary, F., and Batell, M. F. The concept of negative information. *Behav. Sci.* **23**, 264–270.

Landolt, J. P., and Correia, M. H. Neuromathematical concepts of point process theory. *IEEE Trans. Biomed. Eng.* **BMI 25**, 1–12.

Usami, H., Massaki, S., and Sato, R. The stochastic properties of the basic neuron populations as information processing system. *Biol. Cybernet.* **29**, 167–173.

1977   Nakahama, H., Yamamoto, M., Fujii, H., Aya, K., and Tani, Y. Dependency representing Markov properties of spike trains recorded from central single neurons. *Tohoku J. Ex. Med.* **22**, 99–111.

1976   Eckhorn, R., Gruesser, O. J., Kroeller, U., Pellnitz, K., and Poeple, B. (1976).

Efficiency of different neuronal codes: Information transfer calculations for three different neuronal systems. *Biol. Cybernet.* **22**, 49–60.

1975   Laki, K., An attempt to measure 'disorder' in mental disorders. *Perspect. Biol. Med.* **15**, 157–161.

Takatsuji, M. An information-theoretical approach to a system of interacting elements. *Biolg. Cybernet.* **17**, 208–210.

Tsukada, M., Ishii, N., and Sato, R. Temporal pattern discrimination of impulse sequences in the computer-simulated nerve cells. *Biol. Cybernet.* **17**, 19–28.

1974   Eckhorn, R., and Poepel, B. Rigorous and extended application for information theory to the afferent visual system of the cat. I. Basic concepts. *Biol. Cybernet.* **16**, 191–200.

1973   Bergstrom, R. M. (1973). Neural micro- and macro-states. *In* "Advances in Psychobiology" (G. Newton and A. Riesen, eds.). Wiley, New York.

1972   Bergstrom, R. M. Information transfer and neurophysiology. *Int. J. Neurosci.* **4**. 167–169.

Bergstrom, R. M., and Nevanlinna, O. An entropy model of primitive neural systems. *Int. J. Neurosci.* **4**, 171–173.

Grusser, O. J. Informationstheorie und die Signalverarbeitung in den Sinnesorganen und im Nervensystem. *Naturwissenschaften.* **59**, 436–447.

Guttinger, W. Problem of information processing in the nervous system. *Int. J. Neursci.* **3**, 61–66.

Pfaffelhuber, E. Learning and information theory. *Int. Rev. Neurosci.* **3**, 83–88.

1970   Shrader-Frechette, K. S. "Information—Theoretic Descriptions and Physicalistic Accounts of Mind." University of Louisville, Kentucky.

1969   Bergstrom, R. M. An Entropy model of the developing brain. *Dev. Psychobiol.* **2**, 139–152.

1967   Bergstrom, R. M. Neural macrostates. *Synthese* **17**, 425–443.

1965   Billingsley, P. "Ergodic Theory and Information." Wiley, New York.

1963   Winograd, S., and Cowan, J. D. "Reliable Computation in the Presence of Noise." MIT Press, Cambridge, Massachusetts.

1962   Reichardt, W. Auto-correlation, a principle for the evaluation of sensory information by the central nervous system. *In* "Sensory Communication" (W. Rosenblith, ed.), pp. 303–317. MIT Press, Cambridge, Massachusetts.

1954   Uttley, A. M. The classification of signals in the nervous system. *Electroencephalogr. Clin. Neurphysiol.* **6**, 479–494.

1949   Shannon, C. E., and Weaver, W. "The Mathematical Theory of Communication." Univ. of Illinois Press, Urbana.

1946   Gabor, D. Theory of communication. *J. Inst. Electr. Eng.* **93**, 429–457.

Multiple Neuronal Spike Train Approaches

1986   Gerstein, G., G., Aertsen, A., Bloom, M., Espinosa, I., Evanczuk, S., and Turner, M. Multi-neuron experiments: Observation of state in neural nets. In "Synergetics" (H. Haken, ed.), pp. 58–70. Springer-Verlag, Berlin and New York.

Marmarelis, V. Z., Citron, M. C., and Vivo, C. P. Minimum-order Wiener modelling of spike-output systems. *Biol. Cybernet.* **54**, 115–123.

1985   Aertsen, A., and Gerstein, G. L. Evaluation of neuronal connectivity: Sensitivity of cross correlation. *Brain Res.* **340**, 341–354.

Aertsen, A., Gerstein, G. L., and Johannesma, P. From neuron to assembly: Neuronal organization and stimulus representation. *In* "Brain Theory" (A. Aertsen and G. Palm, eds.). Springer-Verlag, Berlin and New York.

Borisyuk, G. N., Borisyuk, R. M., Kirillov, A. B., Kovalenko, E. I., and Kryukov, V. I. A new statistical method for identifying interconnections between neuronal network elements. *Biol. Cybernet.* **52**. 301–306.

Gerstien, G. L., and Aertsen, A. Representation of cooperative firing activity among simultaneously recorded neurons *J. Neurophysiol.* **54**, 1513–1528.

Gerstein, G. L. Perkel, D. H., and Dayhoff, J. E. Cooperative firing activity in simultaneously recorded populations of neurons: Detections and mesurement. *J. Neurosci.* **5**, 881–889.

1984   Epping, W., van den Boogaard, H., Aertsen, A., Eggermont, J., and Johannesma, P. The neurochrome. An identity preserving representation of activity patterns from neural populations. *Biol. Cybernet.* **50**, 235–240.

Koenderink, J. J. Simultaenous order in nervous nets from a functional standpoint. *Biol. Cybernet.* **50**, 35–41.

Koenderink, J. J. Geometrical structures determined by the functional order in nervous net. *Biol. Cybernet.* **50**, 43–50.

1983   Dayhoff, J. E., and Gerstein, G. L. Favored patterns in spike trains. I. Detection. *J. Neurophysiol.* **49**, 1334–1348.

Dayhoff, J. E., and Gerstein, G. L. Favored patterns in spike trains, II. Application. *J. Neurophysiol.* **49**, 1349–1363.

Gerstein, G. L., Bloom, M., Espinosa, I., Evanczuk, S., and Turner, M. "Design of a laboratory for multi-neuron studies. *IEEE Trans. Syst., Man, Cybernet.* **13**, 668–676.

Schneider, J., Eckhorn, R., and Reitbock, H. Evaluation of neuronal coupling dynamics. *Biol. Cybernet.* **46**, 129–134

1982   Shaw, G. L., Harth, E., and Scheibel, A. B. Cooperativity in brain function: Assemblies of approximately 30 neurons. *Exp. Neurol.* **77**, 324–357.

1981   Kruger, J., and Bach, M. Simultaneous recording with 30 microelectrodes in monkey visual cortex. *Exp. Brain Res.* **41**, 191–194.

1979   Roberts, W. M. Optimal recognition of neuronal waveforms. *Biol. Cybernet.* **35**, 73–80.

1978   Gerstein, G. L., Perkel, D. H., and Subramanian, K. N. Identification of functionally related neural assemblies. *Brain Res.* **140**, 43–62.

1976   Sejnowski, T. J. On global properties of neuronal interaction. *Biol. Cybernet.* **22**, 85–95.

Sejnowski, T. J. On the stochastic dynamics of neuronal interaction. *Biol. Cybernet.* **22**, 203–211.

1960   Gerstein, G. L., and Perkel, D. H. Simultaneously recorded trains of action potentials: Analysis and functional interpretation. *Science* **164**, 828–830.

1969   Verzeano, M., and Negishi, K. Neuronal activity in cortical and thalamic networks. *J. Gen. Physiol.* **43**, 177–195.

Statistical Mechanics Approaches

1986   Peretto, P., and Niez, J. J. Stochastic dynamics of neural networks. *IEEE Trans. Syst., Man, Cybernet.* **16**. 73–83.

Smolenski, P. Neural and conceptual interpretations of parallel distributed processing models. *In* "Parallel Distributed Processing: Explorations in the Microstructure of Cognition." (J. L. McClelland and D. E. Rumelhart, eds.), Vol. 2, Chapter 22. MIT Press, Cambridge, Massachusetts.

Toulouse, G., Dehaene, S., and Changeux, J. A spin glass model of learning by selection. *Conf. Neural Networks, 1986.*

1985   Ackley, D. H., Hinton, G. E., and Sejnowski, T. J. A learning algorithm for Boltzmann machines. *Cognit. Sci.* **9**, 147.

Weisbuch, G., and Fogelman-Souilie, F. Scaling laws for the attractors of Hopfield networks. *J. Phys. Lett.* **46**, 623–630.

1984   Hinton, G. E., Sejnowski, T. J., and Ackley, D. H. Boltzmann machines:

Constraint Satisfaction Networks that Learn," Tech. Rep. SMU-CS-84-119. Carnegie-Mellon, University Pittsburgh, Pennsylvania.

Hopfield, J. J. Neurons with graded response have collective properties like those of two-state neurons. *Proc. Natl. Acad. Sci. U.S.A.* **81**, 3088-3092.

Peretto, P. Collective properties of neural networks: A statistical physics approach. *Biol. Cybernet.* **50**, 51-62.

1983   Anderson, J. A. Cognitive and psychological computation with neural models. *IEEE Trans. SCM-Syst., Man, Cybernet.* **13**, 799-815.

Ingber, L. Statistical mechanics of neocortical interactions. Dynamics of synaptic modifications. *Phys. Rev. A* **8**, 395-416.

Ventriglia, F. Kinetic theory of neural systems: Analysis of the activity of the two-dimensional model. *Biol. Cybernet.* **46**, 93-99.

1982   Hopfield, J. J. Neural networks and physical systems with emergent collective computational abilities. *Proc. Natl. Acad. Sci. U.S.A.* **79**, 2554-2558.

Ingber, L. Statistical mechanics of neocortical interactions. I. Basic formulation. *Physica D* **5D** 83-107.

van Hemmen, J. L. Classical spin-glass model. *Phys. Rev. Lett.* **49**, 409-412.

1981   Hinton, G. E. Implementing semantic networks in parallel hardware. In "Parallel Models of Associative Memory" (G. E. Hinton and J. A. Anderson, eds.), pp. 161-187. Erlbaum, Hillsdale, New Jersey.

1978   Little, W. A., and Shaw, G. L. Analytic study of the memory storage capacity of a neural network. *Math. Biosci.* **39**, 281-290.

Pastur, L. A., and Figotin, A. L. Theory of disordered spin systems. *Teor. Mat. Fiz.* **35**, 193-210.

1977   Amari, S., Yoshida, K., and Kanatani, K. (1977). A mathematical foundation for statistical neurodynamics. *SIAM J. Appl. Math.* **33**, 95-126.

Anderson, J. A., Silverstein, J. W., Ritz, S. A., and Jones, R. S. Distinctive features, categorical perception, and probability learning: Some applications of a neural model. *Phsychol. Rev.* **84**, 413-451.

1976   Sejnowski, T. J. On global properties of neuronal interactions. *Biol. Cybernet.* **22**, 85-95.

1975   Little, W. A., and Shaw, G. L. A statistical theory of short and long term memory. *Behav. Biol.* **14**, 115-133.

Takatsuji, M. An information-theoretical approach to a system of interacting elements. *Biol. Cybernet.* **17**, 207-210.

1974   Amari, S. A method of statistical neurodynamics. *Biol. Cybernet.* **14**, 201-215.

Little, W. A. The existence of persistent states in the brain. *Math. Biosci.* **19**, 101-120.

1972   Bergstrom, R. M., and Nevalinna, O. An entropy model of primitive neural systems. *Int. J. Neurosci.* **4**, 171-173.

Pfaffelhuber, E. Learning and information theory. *Int. Rev. Neurosci.* **3**, 83-88.

1968   Cowan, J. D. Statistical mechanics of nervous nets. In "Neural Networks: Proceedings of the School on Neural Networks, June, 1967, in Ravello" (E.R. Caianiello, ed.), pp. 182-188. Springer-Verlag, Berlin and New York.

1963   Glauber, R. J. Time-dependent statistics of the Ising model. *Phys. Rev.* **4**, 294-307.

1961   Wiener, N. "Cybernetics: Control and Communication in the Animal and the Machine," 2nd ed. MIT Press, Cambridge, Massachusetts.

1959   Goldstein, H. "Classical Mechanics." Addison-Wesley, Reading, Massachusetts.

Dynamic Similarity and Scaling

1987   MacGregor, R. J. "Neural and Brain Modelings," Part II, Chapter 14. Academic Press, Orlando, Florida.

1979   Schlichting, H. "Boundary Layer Theory," 7th ed. (J. Kestin, transl.). McGraw-Hill, New York.

1963   Fitzhugh, R. Dimensional analysis of nerve models. *J. Theor. Biol.* **40**, 517-541.

Partial Differential Equations

1983   Kawahara, T., Katayama, K., and Nogawa, T. Nonlinear equations of reaction-diffusion type for neural populations. *Biol. Cybernet.* **48**, 19-25.

Ventriglia, F. Kinetic theory of neural systems. Analysis of the activity of the two-dimensional model. *Biol. Cybernet.* **46**, 93-99.

1980   Ermantrout, G. B., and Cowan, J. D. Large scale spatially organized activity in neural nets. *SIAM J. Appl. Math.* **38**, 1-21.

Kawahara, T. Coupled van der Pol oscillators—a model of excitatory and inhibitory neural interactions. *Biolg. Cybernet.* **39**, 37-43.

Ventriglia, F. Numerical investigations for one-dimensional neural systems. *Biol. Cybernet.* **36**, 125-130.

1978   Ventriglia, F. Propagation of excitation in a model of neural systems. *Biol. Cybernet.* **30**, 75-79.

1977   Nogawa, T., Karayama, K., Tabata, Y., Ohshio, T., and Kawahara, T. Simplified model equations and envelope equation for neural populations. *Proc. Annu. Conf. Eng. Med. Biol. 80.* **30**.

1976   Nogawa, T., Karayama, K., Tabata, Y., Oshio, T., and Kawahara, T. The brain wave equation. *Dig. Int. Conf. Med. Biol. Eng., 11th, 1976,* pp. 546-547.

1975   Feldman, J. L. and Cowan, J. D. Large-scale activity in neural nets. I. *Biol. Cybernet.* **17**, 29-38.

Oguztoreli, M. N. On the activities in a continuous neural network. *Biol. Cybernet.* **18**, 41-48.

1974   Fujita, S., and Okuda, M. Investigation of two- and three-dimensional behaviours of activity by neural field equations. *Mem. Fac. Eng., Osaka City Univ.* **15**, 143-147.

Okuda, M., Yoshida, A., and Takahashi, K. A dynamical behaviour of active regions in randomly connected neural networks. *J. Theor. Biol.* **48**, 51-73.

Ventriglia, F. Kinetic approach to neural systems. I. *Bull. Math. Biol.* **36**, 534-544.

1973   Okuda, M. Growth of an active region in neural networks. *J. Phys. Soc.* **35**, 316.

Ventriglia, F. Kinetic approach to neural systems. *Int. J. Neurosci.* **6**, 29-30.

Wilson, H. R., and Cowan, J. D. A mathematical theory of the functional dynamics of cortical and thalamic nervous tissue. *Biol. Cybernet.* **3**, 55-80.

1972   Wilson, H. R. and Cowan, J. D. Excitatory and inhibitory interactions in localized populations of model neurons. *Biophys. J.* **12**, 1-24.

1971   Griffith, J. S. "Mathematical Neurobiology." Academic Press, New York.

1969   Lauria, F. E. Mathematical approach to the study of a cerebral cortex. II. *J. Theor. Biol.* **23**, 72-86.

1965   Griffith, J. S. A field theory of neural nets. II. *Bull. Math. Biophys.* **27**, 187-195.

Lauria, F. E. Mathematical approach to the study of a cerebral cortex. I. *J. Theor. Biol.* **8**, 54-68.

1963   Griffith, J. S. A field theory of neural nets. I. *Bull. Math. Biophys.* **25**, 11-120.

1956   Beurle, R. L. Properties of a mass of cells capable of regenerating pulses. *Philos. Trans. Soc. London, Ser. A* **240**, 55-94.

Learning and Memory in Neural Networks: General

1984   Hawkins, R. D., and Kandel, E. R. Is there a cell-biological alphabet for simple forms of learning? *Psychol. Rev.* **91**, 375-391.

Lynch, G., and Baudry, M. The biochemistry of memory: A new and specific hypothesis. *Science* **224**, 1057–1063.

983   Eccles, J. C. Calcium in long-term potentiation as a model for memory. *Neuroscience* **10**, 1071–1081.

Harley, C. B. Learning—an evolutionary approach. *Trends Neurosci.* **6**, 204–208.

Morris, R. Modelling amnesia and the study of memory in animals. *Trends Neursci.* **6**, 479–483.

Thompson, R. F. Neuronal substrates of simple associative learning: Classical conditioning. *Trends Neurosci.* **6**, 270–275.

1982   Goldmeier, E. "The Memory Trace: Its Formation and its Fate." Erlbaum, Hillsdale, New Jersey.

Kety, S. S. The evolution of concepts of memory. In "The Neural Basis of Behavior" (A.L. Beckman, ed.), pp. 95–102. SP Med. & Sci. Books, New York.

Morris, R. New approaches to learning and memory. *Trends Neurosci.* **5**, 3–4.

Squire, L. R. The neuropsychology of human memory. *Annu. Rev. Neurosci.* **5**, 241–273.

Woody, C. D. "Memory, Learning, and Higher Function: A Cellular View." Springer-Verlag, Berlin and New York.

1981   deLorenzo, R. J. The calmodulin hypothesis of neurotransmission. *Cell Calcium* **2**, 365–385.

Martinez, J. L., Jr., Jensen, R. A., Messing, R. B., McGaugh, J. L., and Rigter, H., eds. "Endogenous Peptides and Learning and Memory Processes." Academic Press, New York.

1976   Berger, T. W., Alger, B., and Thompson, R. F. Neuronal substrate of classical conditioning in the hippocampus. *Science* **192**, 483–485.

1974   Cooper, L. N. A possible organization of animal learning and memory. In "Collective Properties of Physical Systems" (B. Lundqvist & S Lundqvist, eds.), pp. 252–264. Academic Press, New York.

Mark, R. "Memory and Nerve Cell Connections: Criticisms and Contributions from Developmental Neurophysiology." (Clarendon), Oxford Univ. Press, London and New York.

1972   Thompson, R. F., Patterson, M. M., and Teyler, T. J. The neurophysiology of learning. *Annu. Rev. Psychol.* **23** 73–104.

1966   Griffith, J. S. A theory of the nature of memory. *Nature (London)* **211**, 1160–1163.

Connectionist Models of Learning and Memory

1986   Gelperin, A. Complex associative learning in small neural networks. Trends Neurosci. **9**, 325–329.

Kelso, S. R., Ganong, A. H., and Brown, T. H. Hebbian synapses in hippocampus. *Proc. Natl. Acad. Sci. U.S.A.* **83**, 5326–5330.

Liestol, K., Maethlen, J., and Nja, A. Selective synaptic connections: Significance of recognition and competition in mature sympathetic ganglia. *Trends Neurosci.* **9**, 21–24.

Lynch, G., Synapses, circuits, and the beginnings of memory. In "Cognitive Neurobiology" (M. Gazzaniga, ed.). MIT Press, Cambridge, Massachusetts.

Pereto, P. and Wiez, J. J. Long term memory storage capacity of multiconnected neural networks. *Biol. Cybernet.* **54**, 53–63.

Ritter, H. and Schulten K. On the stationary state of Kohonen's self-organizing sensory mapping. *Biol. Cybernet.* **54**, 99–106.

1985   Carpeneter, G. A., and Grossberg, S. Neural dynamics of category learning and recognition: Attention, memory consolidation, and amnesia. In "Brain Structure, Learning, and Memory," (J. Davis, R. Newburgh, and E. Wegman, eds.). Am. Assoc. Adv. Sci., Washington, D.C.

Sahley, C. Co-activation, cell assemblies, and learning. *Trends Neurosci.* **8**, 423–424.

1984   Easton, P., and Gordon, P. E. Stabilization of Hebbian neural nets by inhibitory learning. *Biol. Cybernet.* **51**, 1–9.

Fukushima, K. A hierarchical neural network model for associative memory. *Biol. Cybernet.* **50**, 105–113.

Kohonen, T. "Self-Organization and Associative Memory." Springer-Verlag, Berlin and New York.

Miyake, S., and Fukushima, K. A neural network model for the mechanism of feature extraction. *Biol. Cybernet.* **50**, 377–384.

Munro, P. W. A model for generalization and specification by single neurons. *Biol. Cybernet.* **51**, 169–179.

Palm, G. Local synaptic modification can lead to organized connectivity patterns in associative memory. *In* "Synergetics from Microscopic to Macroscopic Order," (E. Frehland, ed.). Springer-Verlag, Berlin and New York.

Palm, G. and Bonhoeffer, T. Parallel processing for associative and neuronal networks. *Biol. Cybernet.* **51**, 201–204.

Shiozaki, A., Recollection ability of three-dimensional correlation matrix associative memory. *Biol. Cybernet.* **50**, 337–342.

Theodorescu, D. Stochastic processes with optimization—a model of learning in higher systems. *Biol. Cybernet.* **50**, 313–327.

1983   Cohen, M. A., and Grossberg, S. Absolute stability of global pattern for formation and parallel memory storage by competitive neural networks. *IEEE Trans. Syst., Man, Cybernet.* **SMC 13**, 815–826.

Hampson, S., and Kibler, K. A Boolean complete neural model of adaptive behavior. *Biol. Cybernet.* **49**, 9–19.

Hopfield, J. J., Feinstein, D. I., and Palmer, R. G. Unlearning has a stabilizing effect in collective memories. *Nature (London)* **304**, 158–159.

McNaughton, B. L. Activity-dependent modulation of hippocampal synaptic efficiency: Some implications for memory processes. In "Neurobiology of the Hippocampus" W. Seifert, ed.), pp. 233–249. Academic Press, New York.

Nelson, T. J. A neural network model for cognitive activity. *Biolg. Cybernet.* **49**, 79–88.

1982   Barto, A. G., Anderson, C. W., and Sutton, R. S. Synthesis of nonlinear control surfaces by a layered associative search network. *Biol. Cybernet.* **43**, 175–185.

Bobrowski, L. Rules of forming receptive fields of formal neurons during unsupervised learning processes. *Biol. Cybernet.* **43**, 23–28.

Edelman, G. M., and Reeke, G. N. Jr., Selective networks capable of representative transformations, limited generalizations, and associative memory. *Proc. Natl. Acd. Sci. U.S.A.* **79**, 2091–2095.

Fukushima, K., and Miyake, S. Neocognitron: A new algorithm for pattern recognition tolerant of deformations and shifts in position. *Pattern Recognition.* **15**, 455–469.

Hirai, Y. A learning network resolving multiple match in associative memory. *Proc. Int. Conf. Pattern Recognition, 6th, 1982*, pp. 1949–1052.

Klopf, A. H. "The Hedonistic Neuron: A Theory of Memory, Learning, and Intelligence." Hemisphere, Washington, D.C.

Kohonen, T. Self-organized formation of topologically correct feature maps. *Biol. Cybernet.* **43**, 59–69.

Kohonen, T. Analysis of a simple self-organizing process. *Biol. Cybernet.* **44**, 135–140.

McNaughton, B. L. (1982). Long-term synaptic enhancement and short-term potentiation in rat fascia dentata act through different mechanisms. *J. Physiol. (London)* **324**, 249–262.

Melkonian, D. S., Mkartchian, H. H., and Fanardjian, V. V. Simulation of learning processes in neural networks of the cerebellum. *Biol. Cybernet.* **45**, 79–88.

Palm, G. Rules for synaptic changes and their relevance for the storage of information in the brain. *In* "Cybernetics and Systems Research" (R. Trappl, ed.). Elsevier, Amsterdam.

Palm, G. "Neural Assemblies: An Alternative Approach to Artificial Intelligence." Springer-Verlag, Berlin and New York.

Reilly, D. L., Cooper, L. N., and Elbaum, C. A neural model for category learning. *Biol. Cybernet.* **45**, 35–41.

Wigstrom, H., McNaughton, B. L., and Barnes, C. A. Long-term synaptic enhancement in hippocampus is not regulated by postsynaptic membrane potential. *Brain Res.* **233**, 195–199.

Willwacher, G. Storage of a temporal pattern sequence in a network. *Biol. Cybernet.* **43**, 115–126.

1981 Barto, A. G., and Sutton, R. S. Landmark learning: An illustration of associative search. *Biol. Cyberenet.* **42**, 1–8.

Barto, A. G., Sutton, R. S., Brouwer, P. S. Associative search network: A reinforcement learning associative memory. *Biol. Cybernet.* **40**, 201–211.

Deutsch, S. A simplified version of Kunihiko Fukushima's neocognitron. *Biol. Cybernet.* **42**, 17–21.

Hinton, G. E. and Anderson, J. A. "Parallel Models of Associative Memory." Erlbaum, Hillsdale, New Jersey.

McNaughton, B. L., Barnes, C. A., and Andersen, P. Synaptic efficacy and EPSP summation in granule cells of rat fascia dentata studied in vitro. *J. Neurophysiol.* **46**, 952–966.

Nagano, T. and Kurata, K. A self-organizing neural network model for the development of complex cells. *Biol. Cybernet.* **40**, 195–200.

Palm, G. Towards a theory of cell assemblies. *Biol. Cybernet.* **39**, 181–194.

Sutton, R. S., and Barto, A. G. (1981). Toward a modern theory of adaptive networks: Expectation and prediction. *Psychol. Rev.* **88**, 135–170.

1980 Amari, S. Topographic organization of nerve fields. *Bull. Math. Biol.* **42**, 339–364.

Barnes, C. A., and McNaughton, B. L. (1980). Physiological compensation for loss of afferent synapses in rat hippocampal granule cells during senescence. *J. Physiol. (London)* **309**, 473–485.

Barnes, C. A., and McNaughton, B. L. Spatial memory and hippocampal synaptic plasticity in senescent and middle-aged rats. *In* "Psychology of Aging" (D. Stein, ed.), pp. 253–274. Elsevier, Amsterdam.

Bobrowski, L., and Caianiello, E. R. Comparison of two unsupervised learning algorithms. *Biol. Cybernet.* **37**, 1–7.

Fukushima, K. Neocognitron: A self-organizing neural network model for a mechanism of pattern recognition unaffected by shift in position. *Biol. Cybernet.* **36**, 193–202.

Hirai, Y. A new hypothesis for synaptic modification: An interactive process between postsynaptic competition and presynaptic regulation. *Biol. Cybernet.* **36**, 41–50.

Kohonen, T. "Content-Addressable Memories." Springer-Verlag, Berlin and New York.

Lara, R., Tapia, R., Cervantes, F., Moreno, A., and Trujillo, H. Mathematical models of synaptic plasticity. II. Habituations. *J. Neurol. Res.* **2**, 1–18.

1979   Bechtereva, N. P. Biolelectrical expression of long-term memory activation and its possible mechanism. *Brain Res. Monogr.* **4**, 311-327.

Cooper, L. N., Liberman, F., and Oja, E. A theory for the acquisition and loss of neural specificity in visual cortex. *Beiol. Cybernet.* **33**, 9-28.

Kohonen, T. Representation and processing of associations using vector space operations. In "Dynamic Systems and Pattern Recognition" (H. Haken, ed.), pp. 199-207. Springer-Verlag, Berlin and New York.

Levy, W. B., and Steward, O. (1979). Synapses as associative memory elements in the hippocampal formation. *Brain Res.* **175**, 233-245.

Takeuchi, A., and Amari, S. Formation of topographic maps and columnar microstructers in nerve fields. *Biolg. Cybernet.* **35**, 63-72.

Yeshurun, Y., and Richter-Dyn, N. Analysis of recall and recognition in a certain class of adaptive networks. *Biol. Cybernet.* **32**, 35-40.

1978   Amari, S., and Takeuchi, A. A mathematical theory on formation of category detecting nerve cells. *Biol. Cybernet.* **29**, 127-136.

Anderson, J. A., and Cooper, L. N. "Les modèles mathématiques de l'organization biologique de la mémoire," Pluriscience 168-175. Encyclopaedia Universalis, Paris.

Bobrowski, L. Learning processes in multilayer threshold nets. *Biol. Cybernet.* **31**, 1-6.

Braitenberg, V. Cell assemblies in the cerebral cortex. In "Theoreteical Approaches to Complex Systems" (R. Heim and G. Palm eds.), p. 171. Springer-Verlag, Berlin and New York.

Fukushima, K., and Miyake, S. A self-organizing neural network with a function of associative memory:Feedback-type cognitron. *Biol. Cybernet.* **28**, 201-208.

Kohonen, T. "Associative Memory—a System—Theoretical Approach," 2nd ed. Springer-Verlag, Berlin and New York.

Lehtio, P., and Kohonen, T. Associative memory and pattern recognition. *Med. Biol.* **56**, 110-116.

Little, W. A., and Shaw, G. L. Analytic study of the memory storage capacity of a neural network. *Math. Biosci.* **39**, 281-290.

McNaughton, B. L., Douglas, R. M., and Goddard, G. V. Synaptic enhancement in fascia dentata: Cooperativity among coactive afferents. *Brain Res.* **157**, 277-293.

Shaw, G. L. Space–time correlations for neuronal firing related to memory storage capacity. *Brain Res. Bull.* **3**, 107-113.

1977   Amari, S. I. Neural theory of association and concept formation. *Biol. Cybernet.* **26**, 175-185.

Anderson, J. A., Silverman, J. W., Ritz, S. A., and Jones, R. S. Distinctive features, categorical perception, and probability learning: Some applications of a neural model. *Psychol. Rev.* **85**, 413-451.

Caianiello, E. R. Some remarks on organization and structure. *Biol. Cybernet.* **26**, 151-158.

Conrad, M. Principle of superposition-free memory. *J. Theor. Biol.* **67**, 213-219.

Eccles, J. C. An instruction–selection theory of learning in the cerebellar cortex. *Brain Res.* **127**, 327-352.

Fedor, P., and Majernik, V. A neuron model as an universal element of self-learning networks for pattern recognition. *Biol. Cybernet.* **26**, 25-33.

Kohonen, T. "Associative Memory." Springer-Verlag, Berlin and New York.

Kohonen, T., Lehtio, P., Rovamo, J., Hyvarinen, J., Bry, K., and Vainio, L. A principle of neural associative memory. *Neuroscience.* **2**, 1065-1076.

Nagano, T. A model of visual development. *Biol. Cybernet.* **26**, 45-52.

Teodorescu, D. Learning and optimization in biological systems—a model of sequential actions. *Biol. Cybernet.* **28**, 83–93.

Tokura, T., and Morishita, I. Analysis and simulation of double-layer neural networks with mutually inhibiting interconnections. *Biol. Cybernet.* **25**, 83–92.

Wigstrom, H. Spatial propagation of associations in a cortex-like neural network model. *J. Neurosci. Res.* **3**, 301–319.

1976   Arbib, M. A., Kilmer, W. L., and Spinelli, D. N. Neural models and memory. In "Neural Mechanisms and Memory" (M. R. Rosenzweig, and E. L. Bennet, eds.). MIT Press, Cambridge, Massachusetts.

Changeux, J. P. and Danchin, A. Selective stabilisation of developing synapses as a mechanism for the specification of neural networks. *Nature (London)* **264**, 705–712.

Kohonen, T., and Oja, E. Fast adaptive formation of orthogonalizing filters and associative memory in recurrent networks of neuron-like elements. *Biol. Cybernet.* **21**, 85–95.

Kohonen, T. E., Reuhkale, E., Makisara, K., and Vaino, L. Associative recall of images. *Biol. Cybernet.* **22**, 159–168.

Silverstein, J. W. Asymptotics applied to a neural network. *Biol. Cybernet.* **22** 73–84.

Stanley, J. C. Simulation studies of a temporal sequence memory model. *Biol. Cybernet.* **24**, 121–137.

Woody, C. D., Buerger, A. A., Ungar, R. A., and Levine, D. S. Modeling aspects of learning by altering biophysical properties of a simulated neuron. *Biol. Cybernet.* **23**, 73–82.

1975   Fukushima, K. Cognitron: A self-organizing multilayered neural network. *Biol. Cybernet.* **20**, 121–136.

Little, W. A., and Shaw, G. L. A statistical theory of short and long term memory. *Behav. Biol.* **14**, 115–133.

Pfaffelhuber, E. Correlation memory models—a first approximation in a general learning scheme. *Biol. Cybernet.* **18**, 217–223.

Wigstrom, H. Associative recall and formulation of stable modes of activity in neural network models. *J. Neurosci. Res.* **1**, 287–313.

Wilson, H. R. A synaptic model for spatial frequency adaptation. *J. Theor. Biol.* **50**, 327–352.

1974   Grossberg, S. Classical and instrumental learning by neural networks. In "Progress in Theoretical Biology" (R. Rosen and F. Snell, eds.) Academic Press, New York.

Kohonen, T. An adaptive associative memory principle. *IEE Trans. Comput.* **23**, 444–445.

Little, W. A. The existence of persistent states in the brain. *Math. Biosci.* **19**, 101–120.

Shaw, G. L. and Vasudevan, R. Persistent states of neural networks and the random nature of synaptic transmission. *Math. Biosci.* **21**, 207–218.

Wigstrom, H. A neuron model for a neural network with recurrent inhibition. *Biol. Cybernet.* **16**, 103–112.

1973   Fukushima, K. A model of associative memory in the brain. *Biol. Cybernet.* **12**, 58–63.

Huttunen, M. O. General model for the molecular events in synapses during learning. *Perspect. Biol. Med.* **17**, 103–108.

Kohonen, T. A new model for randomly organized associative memory. *Int. J. Neurosci.* **5**, 27–29.

Kohonen, T., and Ruohonen, M. Representation of associated data by matrix operators. *IEEE Trans. Comput.* **7**, 701–702.

Pfaffelhuber, E. "A model for learning and imprinting with finite and infinite memory range", *Biol. Cybernet.* **12**, 229–236.

Pfaffelhuber, E., and Damle, P. S. Learning and imprinting in stationary and non-stationary environment. *Biol. Cybernet.* **13**, 229–237.

Stent, G. S. A physiological mechanisms for Hebb's postulate of learning. *Proc. Natl. Acad. Sci. U.S.A.* **70**, 997–1001.

Tsypkin, Y.Z. "Foundations of the Theory of Learning Systems." Academic Press, New York.

Wigstrom, H. A neuron model with learning capability and its relation to mechanisms of association. *Biol. Cybernet.* **12**, 204–215.

1975   Amari, S. Learning patterns and pattern sequences by self-organizing nets of threshold elements, *IEEE Trans. Comput.* **21**, 1197–1206.

Amari, S. I. Characteristics of random nets of analog neuron-like elements. *IEEE Trans. Comput.*

Anderson, J. A. A simple neural network generating an interactive memory. *Math. Biosci.* **14**, 197–220.

Kohonen, T. Correlation matrix memories. *IEEE Trans. Comput.*, **21**, 353–359.

Kohonen, T. Computers and association. *Data (Copenhagen)* **6**, 17–21.

Minsky, M., and Papert, S. "Perceptrons." MIT Press Cambridge, Massachusetts.

Nakano, K. Associatron: A model of associative memory. *IEET Trans. Syst. Man. Cybernet.* **SMC 2**, 380–388.

Pfaffelhuber, E. Learning and information theory. *Int. J. Neurosci.* **3**, 83–88.

1971   Marr, D. Simple memory: A theory for archicortex. *Philos. Trans. Soc. London, Ser. B* **262**, 23–81.

1970   Anderson, J. A. Two models for memory organization using interacting traces. *Math. Biosci.* **8**, 1137–160.

Fukushima, K. A feature extractor for curvilinear patterns: A design suggested by the mammalian visual system. *Biol. Cybernet.* **7**, 153–160.

Marr, D. A theory of cerebral neocortex. *Proc. R. Soc. London, Ser. B* **176**, 161–234.

Uttley, A. M. The informon: A network for adaptive pattern recognition. *J. Theor. Biol.* **27**, 31–67.

1969   Brindley, G. S. Nerve net models of plausible size that perform many simple learning tasks. *Proc. R. Soc. London* **174**, 173–191.

Marr, D. A theory of cerebellar cortex. *J. Physiol. (London)* **202**, 437–470.

Smith, D. R., and Davidson, C. H. Maintained activity in neural nets. *J. Assoc. Comput. Mach.* **9**, 268–279.

1968   Anderson, J. A. A memory storage model utilizing spatial correlation functions. *Biol. Cybernet.* **5**, 113–119.

Legendy, C. R. How large are Hebb's cell assemblies? *In* "Cybernetic Problems in Bionics" (H.L. Oestreicher and D. R. Moore, eds.). Gordon & Breach, New York.

1967   Brindley, G. S. The classification of modifiable synapses and their use in models for conditioning. *Proc. R. Soc. London, Ser. B* **168**, 361–376.

Legendy, C. R. On the scheme by which the human brain stores information. *Math. Biosci.* **1**, 55–597.

1962   Block, H. D. The perceptron: A model for brain functioning. I. *Rev. Mod. Phys.* **34**, 123–135.

Block, H. D., Knight, B. W., and Rosenblatt, F. Analysis of a four layer series coupled perceptron. *Rev. Mod. Phys.* **34**, 135–142.

Greene, P. H. On looking for neural networks and 'cellular assemblies' that underlie behavior. I. Mathematical model. *Bull. Math. Biophys.* **24**, 247–275.

Rosenblatt, F. "Principles of Neurdynamics." Spartan Books, Washington, D.C.

1961   Farley, B. G., and Clark, W. A. Activity in networks of neuron-like elements. *In*

"Information Theory" (C. Cherry, ed.), Butterworth, London.

Keller, H. B. Finite automata, pattern recognition and perceptrons. *J. Assoc. Comput. Mach.* **8**, 1-20.

1960   Zadeh, L. A. Time-varying networks. *I. Proc. IRE*, **49**, 1488-1503.

Rosenblatt, F. Perceptron simulation experiments. *Proc. IRE* **48**, 301-309.

1957   Milner, P. M. The cell assembly: Mark II. *Psychol. Rev.* **64**, 242-252.

1956   Rochester, N., Holland, J. H., Haibt, L. H. and Duda, W. L. Tests on a cell assembly theory of the action of the brain, using a large digital computer. *IRE Trans. Circuit Theory* **2**, 80-93.

1955   Clark, W. A., and Farley, B. G. Generalization of pattern recognition in a self-organizing system. *Proc. West. Comput. Conf., 1955*, pp. 86-91.

1954   Farley, B. G., and Clark, W. A. Simulation of self-organizing systems by digital computer. *IRE Trans. Circuit Theory* **4**, 76-84.

1949   Hebb, D. O. "Organization of Behavior." McGraw-Hill, New York.

Supraconnectionist Models of Learning and Memory

1985   Finkel, L. H., and Edelman, G. M. Interactions of synaptic modification rules within populations of neurons. *Proc. Natl. Acad. Sci. U.S.A.* **82**, 1291-1295.

Reeke, G. N., and Edelman, G. M. Selective networks and recognition automata, *Ann. N.Y. Acad. Sci.* **82**, 181-201.

1984   Edelman, G. M. Expression of cell adhesion molecules during embryogenesis and regeneration. *Exp. Cell Res.* **161**, 1-16.

1983   Edelman, G. M. Cell adhesion molecules. *Science*, **219**, 450-457.

Edelman, G. M., and Finkel, L. H. Neuronal group selection in the cerebral cortex. *In* "Dynamic Aspects of Neocortical Function" (G. M. Edelman, W. E. Gall, and W. M. Cowan, eds.), pp. 653-695. Wiley, New York.

Edelman, G. M. Group selection as the basis for higher brain function. *In* "The Organization of the Cerebral Cortex" (F.O. Schmitt, and Bloom, F. E. eds.), pp. 535-563. MIT Press, Cambridge, Massachusetts.

Edelman, G. M. Group selection and phasic reentrant signaling: A theory of higher brain function. *In* "The Mindful Brain" (G. M. Edelman and V. B. Mountcastle, eds.), pp. 51-100. MIT Press, Cambridge, Massachusetts.

1982   Edelman, G. M. Neural Darwinism: Population thinking and higher brain funciton. *In* "How We Know" (M. Shafto, ed.), pp. 1-30, Harper & Row, New York.

Edleman, G. M. Through a computer darkly: Group selection and higher brain function. *Bull. Am. Acad. Arts Sci.* **36**, 20-48.

MacGregor, R. J., and Gochis, P. A computer software system in FORTRAN for simulating large nerve networks, II. Application. *J. Theor. Neurobiol.* **1**, 135-149.

1981   MacGregor, R. J., A Theory of the Representation of Memory Schemata in Nerve Networks". University of Colorado, Boulder.

1978   John, E. R., and Schwartz, E. L. The neurophysiology of information processing and cognition. *Annu. Rev. Psychol.* **29**, 1-29.

Mikhaltsev, I. E., On the physics of CNS memory. *J. Theor. Biol.* **70**, 33-49.

Wood, C. C. Variations on a theme by Lashley: Lesion experiments on the neural model of Anderson, Silverstein, Ritz, and Jones. *Psychol. Rev.* **85**, 582-591.

1977   Wess, O., and Roder, U. A holographic model for associative memory chains. *Biol. Cybernet.* **27**, 89-98.

1976   Ramos, A., Schwartz, E., and John, E. R. An examination of the participation of neurons in readout from memory. *Brain Res. Bull.* **1**, 77-86.

1973   Bartlett, F., and John, E. R. Equipotentiality quantified: The anatomical distribution of the engram. *Science* **181**, 764-767.

John, E. R., Bartlett, F., Shimokochi, M., and Kleinman, D. Neural readout from memory. *J. Neurophysiol.* **36**, 893–924.

1972   Adey, W. R. Organization of brain tissues: Is the brain a noisy processor? *Int. J. Neurosci.* **3**, 271–284.

John, E. R. Switchboard versus statistical theories of learning and memory. *Science.* **177**, 850–864.

Adey, W. R. Evidence for cerebral membrane effects of calcium, derived from DC gradient, impedance, and intracellular records. *Exp. Neurol.* **30**, 78–102.

1971   Pribram, K. H. "Languages of the Brain." Prentice-Hall, Englewood Cliffs, New Jersey.

1970   Adey, W. R. Spectral analysis of EEG data from animals and man during alerting, orienting, and discriminative responses. *In* "Attention in Neurophysiology" (T. Mulholland and C. Evans, ed.), pp. 194–229. Butterworth, London.

Westlake, P. R. The possibilities of neural holographic processes within the brain. *Biol. Cybernet.* **7**, 129–153.

1969   Adey, W. R. Neural information processing; windows without and the citadel within. *In* "Biocybernetics of the Central Nervous System" (L. D. Proctor, ed.), pp. 1–27. Little, Brown, Boston, Massachusetts.

Gabor, G. Associative holographic memories. *IBM J. Res. Dev.* **13**, 156–159.

Willshaw, D. J., Buneman, O. P., and Longuet-Higgins, H.C. Non-holographic associative memory. *Nature (London)* **222**, 960–962.

1968   Gabor, D. Holographic model of temporal recall. *Nature (London)* **217**, 584.

Longuet-Higgins, H. C. Holographic model of temporal recall. *Nature (London)* **217**, 104.

1967   Adey, W. R. Intrinsic organization of cerebral tissue in alerting, orienting, and discriminative responses. *In* "The Neurosciences: A Study Program" (G.C. Quarton, T. Melnechuk, and F. O. Schmitt, eds.), pp. 615–633.

Elazar, Z., and Adey, W. R. Spectral analysis of low frequency components in the electrical activity of the hippocampus during learning. *Electroencephalogr. Clin. Neurophysiol.* **23**, 225–240.

John, E. R. "Mechanisms of Memory." Academic Press, New York.

1966   Adey, W. R., Kado, R. T., McIlwain, J. T., and Walter, D. O. The role of neuronal elements in regional cerebral impedance changes in alerting, orienting, and discriminative responses. *Exp. Neurol.* **15**, 490–510.

1963   Adey, W. R., Kado, R. T., Didio, J., and Schindler, W. J. Impedance changes in cerebral tissue accompanying a learned discriminative performance in the cat. *Exp. Neurol.* **7**, 259–281.

1960   Lashley, K. S., *In* "The Neurophsychology of Lashley" (F. A. Beach, D. O. Hebb, C. T. Morgan, and H. W. Nissen, eds.). McGraw-Hill, New York.

1950   Lashley, K. S. In search of the engram. *Symp. Soc. Exp. Biol.* **4**, 454–482.

Neural Networks as Cognitive Operators

1986   Golden, R. M. The 'brain-state-in-a-box' neural model is a gradient descent algorithm. *J. Math. Psychol.* **30**, 73–80.

McClelland, J. L., and Rumelhart, D. E., eds. (1986). "Parallel Distributed Processing: Explorations in the Microstructure of Cognition." MIT Press Cambridge, Massachusetts.

985   Feldman, J. A., ed. Special issue on Connectionist models and their applications. *Conit. Sci.* **9**, No. 1.

Hopfield, J. J., and Tank, D. W. 'Neural' computation of decisions in optimization problems. *Biol. Cybernet.* **52**, 141–152.

Kawamoto, A., and Anderson, J. A. A Neural network model of multistable perception. *Acta Psychol.* **59**, 35–65.

1983   Anderson, J. A. Cognitive and psychological computation with neural models. *IEEE Trans. Syst., Man, Cybernet.* **SMC 13**, 799–815.

Grossberg, S. The quantized geometry of visual space: The coherent computation of depth, form, and brightness. *Behav. Brain Sci.* **6**, 625–692.

1982   Feldman, J. A., Dynamic connections in neural networks. *Biol. Cybernet.* **46**, 27–39.

Feldman, J. A. and Ballard, D. H. Connectionist models and their properties. *Cognit. Sci.* **6**, 205–254.

Grossberg, S. "Studies of Mind and Brain: Neural Principles of Learning, Perception, Development, Cognition, and Motor Control." Reidel, Boston, Massachusetts.

Oja, E. A simplified neuron model as a principal component analyser. *J. Math. Biol.* **15**, 267–273.

Palm, G. "Neural Assemblies: An Alternative Approach to Artificial Intelligence," Stud. Brain Funct. No. 7. Springer-Verlag, Berlin and New York.

1981   Hinton, G. E., and Anderson, J. A., eds. "Parallel Models of Associative Memory." Erlbaum, Hillsdale, New Jersey.

McClelland, J. L., and Rumelhart, D. W. An interactive activation model of the effect of context in perception. I. *Psychol. Rev.* **88**, 375–407.

1980   Grossberg, S. How does a brain build a cognitive code? *Psychol. Rev.* **87**, 1–51.

1979   Dreyfus, H. L. "What Computers Can't Do: The Limits of Artificial Intelligence," 2nd ed. Harper, New York.

1978   Grossberg, S. Competition, decision, and consensus. *J. Math. Anal. Appl.* **66**, 470–493.

Ratcliff, R. A theory of memory retrieval. *Psychol. Rev.* **85**, 59–108.

1977   Anderson, J. A., Silverstein, J. W., Ritz, S. A., and Jones, R. S. Distinctive features, categorical perception, and probability learning: Some applications of a neural model. *Psychol. Rev.* **84**, 413–451.

Weber, C. "Propositional Logic Performed by Neural Structures," term paper. University of Colorado, Boulder.

1975   Ellias, S. A., and Grossberg, S. Pattern formation, contrast control, and oscillations in the short term of shunting on-center off-surround networks. *Biol. Cybernet.* **20**, 69–98.

1974   Blomfield, S. Arithmetic operations performed by nerve cells. *Brain Res.* **69**, 115–124.

1973   Arbib, M. A. Automata theory in the context of theoretical neurophysiology. *In* "Foundations of Mathematical Biology" (R. Rosen, ed.), Vol. 3. Academic Press, New York.

1972   Arbib, M. B. "The Metaphorical Brain." Wiley, New York,

1969   Moreno-Diaz, R., and McCulloch, W. S. Circularities in nets and the concept of functional matrices. *In* "Biocybernetics of the Central Nervous System." (R. Proctor, ed.), pp. 145–152. Little, Brown, Boston, Massachusetts.

1968   Griffith, J. S. The unification of neural activity. *Proc. R. Soc. London, Ser. B* **171**, 353a–359.

1967   Minsky, M. L. "Computation: Finite and Infinite Machines." Prentice-Hall, Englewood Cliffs, New Jersey.

von Bertalanffy, L. "Robots, Men, and Minds." Braziller, New York.

von Foerster, G. Computation in neural nets. *Curr. Mod. Biol.* **1**, 47–93.

1965   Rapoport, A. Net theory as a tool in the study of gross properties of nervous systems. *Perspect. Biol. Med.* **4**, 142–164.

1964   Arbib, M. A. "Brains, Machines, and Mathematics." McGraw-Hill, New York.

1963   Perles, M., Rabin, M. O., and Shamir, E. The theory of definite automata. *IEEE Trans. Electron. Comput.* **12**, 233–243.

1962   Culbertson, J. T. Nerve net theory. *Comput. Appl. Behav. Sci.*, pp. 468–489.

1961   Ashby, W. R. "An Introduction to Cybernetics." Wiley, New York.

1958   von Neuman, J. "The Computer and the Brain." Yale Univ. Press, New Haven, Connecticut.

1956   Kleene, S. C. Representation of events in nerve nets and finite automata. In "Automata Studies." (C. E. Shannon and J. McCarthy, eds.), pp. 3–41. Princeton Univ. Press Princeton, New Jersey.

Moore, E. F. Gedanken experiments on sequential machines. In "Automata Studies" (C. E. Shannon and J. McCarthy, eds.). Princeton Univ. Press, Princeton, New Jersey.

1951   von Neuman, J. The general logical theory of automata. In "Cerebral Mechanisms in Behavior—The Hixon Symposium" (L.A. Jeffries, ed.). Wiley, New York.

1947   McCulloch, W. S., and Pitts, W. How we know universals. The perception of auditory and visual forms. *Bull Math. Biophys.* **9**, 127–147.

1943   McCulloch, W. S., and Pitts, W. A logical calculus of the ideas immanent in nervous activity. *Bull. Math. Biophys.* **5**, 115–137.

1930   Kubie, L. S. A theoretical application to some neurological problems of the properties of excitation waves which move in closed circuits. *Brain* **53**, 166–177.

10

Cognitive Operations and Structures

The fields of computer engineering, computer science, artificial intelligence, and cognitive science are providing an increasingly rich fund of ideas and findings potentially relevant to brain and nervous system function in areas such as logic circuits and the theory of automata or sequential machines, organization and control of computers and computer systems, natural and computer languages, programming and operations (algorithms, . . .), various computer-oriented mathematical techniques, adaptive or learning systems, systems for knowledge representation (data structures, . . .), representation of images, problem solving (expert systems, . . .), natural and formal logics, development, creativity, and fundamental philosophical questions concerning the relation of computers to brains and minds and the nature of intelligence and indeed of mind itself. At present, the approach of this body of study to neuroscience is characterized by a great deal of activity, energy, and optimism. Viewing at least certain aspects of brain and nervous system function from this perspective allows one to set up a context wherein the brain in these functions is simply one of a potentially infinite number of cognitive or intelligent systems and its properties, characteristics, and limitations in these dimensions are brought out in bold relief by comparison to other imaginary, manmade, or otherwise alternative intelligent systems. This kind of comparison provides a powerful stage for prodding advances in both understanding of brain and nervous system function and development of new generations of artificially intelligent devices.

From this massive body of study and information we will remark on the following areas as particularly relevant to brain and nervous system function: (1) basic orientation toward brain functions as information-processing operations (discussed immediately below); (2) logic circuits, automata, and sequential machines as models of neuronal networks (see Chapter 9); (3) the information-processing paradigm in learning and memory (discussed in this chapter); (4) recent literature concerning cognitive operations and structures utilized in brain functioning (discussed in this chapter); and (5) basic concepts concerning brain operations in language (discussed in this chapter). We have also seen examples of contributions from this milieu to modeling of specific neuronal systems in Chapters 5–8 (e.g., in Marr's models of neuronal networks and of visual processing) and to basic theoretical concepts in modeling neuronal networks generally (e.g., in the statistical mechanical models of Hinton and Hopfield). Excellent introductory summaries of fundamentals in the broad computer science-related fields are provided by Gardner, Hunt, Winston, Booth, and others indicated in the bibliography.

The Information Processing Paradigm of Brain Function

The essential features of the information processing paradigm are that the primary purpose of the system is to support signals which are in turn characterizable in terms of the information they carry and that the system is organized in terms of (1) paths along which the information is channeled through the system (and which may be represented by an operational flowchart and (2) operations which are performed on the signals in the component parts of the system (and which can be represented by algorithms). Different investigators present slightly different definitions of operations and corresponding algorithms and slightly different overall flowcharts for the paths and channels in the system, but all share a common philosophical orientation. Research is directed toward sharpening and elaborating the definitons of the operations, paths, and channels and their interrelations.

The Information Processing Paradigm in Learning and Memory

The information processing paradigm has been applied primarily to the level of cognitive operations and indeed primarily to fields which incorporate memory structures and learning processes. Watkins has reviewed human memory and the infromation-processing metaphor and states that "research [in the psychology of memory] not conceptualized in information processing terms is all but extinct."

Representative overall flowcharts for the processes of learning and memory as presented by Klatsky and by Hunt are shown in Fig. 10.1. Another general model of this type was presented by Kessner, who suggests

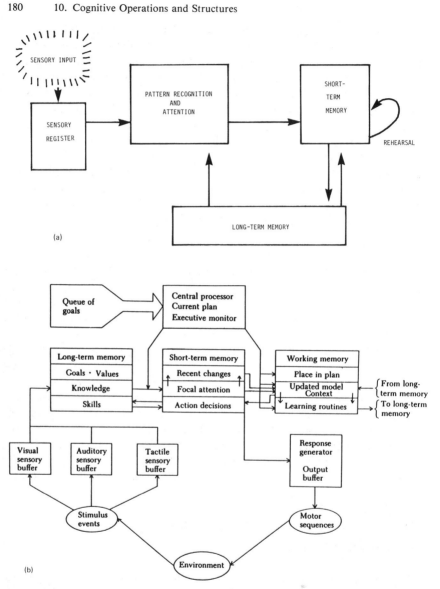

Fig. 10.1. (a) Flowchart for human information processing. (Adapted from "Human Memory: Structures and Processes" by R. L. Klatzky. Copyright © 1975, 1980. W. H. Freeman and Company. Used by permission.) (b) Flow diagram for human information processing. (From "The Universe Within" by M. Hunt. Copyright © 1982. Reprinted by permission of Simon & Schuster, Inc.)

that memory from an information processing viewpoint can be subdivided into cue access, short-term memory, and long-term memory storage and retrievable systems and that these in turn are controlled by operations such as match–mismatch, decay, selective attention, expectancy, rehearsal, arousal, consolidation, and readout processes. Kessner goes on to suggest that match–mismatch and the cue access system occur primarily in sensory systems and the cerebral cortex, short-term memory processes occur primarily in the midbrain reticular formation and association cortex, and consolidation and readout of the long-term memory system are effected by the hippocampus.

In addition to the universally accepted distinction between short-term memory and long-term memory, a permanent "recognition" memory is often associated with pattern recognition and assumed to be separate from but related to the long-term memory system; the concept of "working memory" is sometimes considered in conjunction with the label short-term memory and sometimes separately; and Tulving's distinction between episodic and semantic types of long-term memory is widely accepted. In this section we will briefly review models for memory in pattern recognition, semantic long-term memory, episodic long-term memory, short-term memory, and general learning operations.

Pattern Recognition

A classical model for pattern recognition called "pandemonium" was presented by Selfridge in 1959. In this model patterns are recognized in stages, the first of which is feature extraction in primary sensory modes; elements responding to particular features in a given stimulus field respond at higher levels of activity ("yell louder") than other elements. The second stage consists of cognitive elements which respond to lists or sets of features; the cognitive elements optimally activated by feature lists in a given stimulus field respond at higher levels than their neighbors. The last stage consists of a decision element or elements which determine which cognitive element is most active and thereby recognize the pattern.

In the early 1960s Feigenbaum and Simon developed a very influential model for the learning of pattern recognition called EPAM whose primary operative structure is a discrimination net. The discrimination net grows from a single input node and branch through multiple *n*-ary branchings to terminate in recognizable images. The images consist of lists of parts or features, as do objects in sensory fields presented to the net. Nodes in the net represent test points where lists of the object are compared to corresponding lists associated with the particular branches of the tree feeding the node at hand; channeling of the recognition process or definition of new tree structures is determined by the outcomes of these tests. This model is very powerful and describes many features of human verbal learning and pattern recognition.

Barsalou and Bower criticized the applicability of discrimination net models to pattern recognition in humans on the grounds that they suffer from a high degree of test contingency, which does not plague alternative models such as the pandemonium model. Feigenbaum and Simon presented a vigorous reply to this criticism.

The connectionist neural network perceptron model presented by Rosenblatt and discussed in Chapter 9 has also been influential in pattern recognition.

Semantic Long-Term Memory

The most widely influential model for the structure of long-term memory is the concept of an associative semantic network, first presented in the late 1960s by Quillian and then elaborated in 1973 by Anderson and Bower (under the name of HAM for human associative memory). In this concept familiar categories of objects are placed at the nodes of a network, and the nodes are interconnected according to the semantic relationships between the corresponding concepts. Categories are conceived in terms of sets and subsets, and the arrangement of interconnections among the nodes reflects the set relations and sizes of the categories. If, for example, two categories have minimal overlap or have very different sizes they would tend to be far apart in the network. Links between nodes may be of different types and may provide differential accessibility by means of different strength levels or travel times. In 1975 Collins and Loftus elaborated spreading activation as a mechanism for various operations in such a semantic net, including searching and retrieval. In 1976 Wicklegren elaborated the notion of network strength theory to describe retrieval dynamics in associative semantic networks.

In 1981 Raaijmakers and Shriffrin presented a model for seach of associative memory (SAM) for retreival from long-term memory, combining features of associative network models and cue-dependent random search models. Gillund and Shiffrin have extended this model to include a familiarity process in exploring the distinction between recall and recognition.

Maida and Shapiro have argued that nodes in propositional semantic networks represent only intentions and that extentionality is only minimally represented in such networks. They further argue that many psychologically motivated network models are restricted to assertional information and do not represent structural of definitional information. Such networks are said to be prone to memory confusions about knowing unless augmented by domain-specific inference processes or by structural information.

Several authors have advocated distributed models as alternatives to semantic associative network models. Eich and Murdock presented a holographic-based model for formation, storage, and retrieval of associations in semantic memory. Items are represented as sets of features and are associatively stored in composite memory by convolution, which interacts all items of one entry with all items of another. Retrieval is effected by correlation. Items are stochastic and retrieval is intrinsically noisy and perhaps ambiguous. This model dispenses with the need for search. The model is akin to the supraconnectionist models discussed in Chapter 9.

Pike has also advocated the distributed approach to memory systems and argued that a matrix model for associations is superior to the convolution approach on the grounds of neural plausibility, complexity, and storage economy. Pike relates his matrix approach to the connectionist models of Kohonen and others discussed in Chapter 9.

Some years ago Ratcliff and Murdock reviewed theoretical approaches to retrieval processes in long-term semantic memory. Ratcliff presented a continuous random walk (or diffusion) model for memory retrieval wherein access to memory traces is viewed in terms of a resonance metaphor. Ratcliff asserted that this model can be interfaced with neural network models with little difficulty. Ratcliff has applied retrieval theory to the ordering of relations in perceptual matching. This theory uses a memory model wherein representation of items is distributed over position and the comparison process assesses the amount over overlap. The comparison process is modeled with a random walk or diffusion process. Ratcliff and McKoon presented a theoretical discussion of the concept of spreading activation in the retrieval of information in associative semantic networks of the type advocated by Quillian and by Anderson and Bower.

Other investigators have emphasized memory structure more than processes. For example, Tulving and Watkins some years ago described a theory of memory traces as bundles of attributes or features wherein the valences of retrieval cues are emphasized. Morton and collaborators propose that memory is made up of individual unconnected records labeled with a heading and that retrieval can be accomplished only by addressing the heading. Brachman and Schmolze have described a system called KL-ONE for representing knowledge in artificial intelligence programs. Wittlin finds that overload obstructs human learning performance, probably through undesirable autonomic changes, and can be prevented by changing the structure of inputs.

Williams and Hollan present a model for recall of names of acquaintances which views retrieval as a problem-solving process. Baddeley discusses four applied problem areas in long-term memory in terms of domains of processing, distinctions between interactive and independent contexts, and active processes of search and evaluation. Brown *et al.* discuss retrieval of subjective dates of natural events in terms of an accessibility principle which states that the more known the more recent the event will seem. In several papers Osherson, Stob, and Weinstein discuss a highly rigorous formal learning theory which they conceive as a means of relating theories of comparative grammar to studies of linguistic development. They suggest that the search for ideal learning machines in this formal sense is not a fruitful strategy for artificial intelligence or cognitive science and that one might rather investigate the limitations inherent in various kinds of learning devices. They thus search for constraints on linguistic theory and probe such questions as, how many natural languages are there?

Episodic Long-Term Memory

Kolodner presents a computer model for episodic memory based on reconstructive processes. The model has been programmed to store events in the lives of famous people and to answer questions posed in English concerning that information. It includes algorithms for reconstruction-based retrieval and for regorganization of memory as new unanticipated items are added.

Reiser *et al.* present a model in which autobiographical events are organized in memory by the knowledge structure that guided comprehension and planning during the experience. Retreival of individual experience is effected by accessing knowledge structures used to encode the event and then using information in these structures to predict features of the target event.

Walker and Kintsch suggest that retrieval from natural categories as in real-world knowledge can be modeled by a loosely interassociated memory network of the type suggested by Raaijmakers and Shiffron but must be extended to a more complex control structure wherein retrieveal cues are primarily episodic rather than semantic.

Short-Term Memory

Thirty years ago Miller asserted that the capacity of short-term memory was limited to about seven "chunks" of information. More recently, Baddeley suggested that short-term memory capacity should be measured in terms of the number of symbols that can be reproduced in a limiting time period (about 2 sec). Simon with various collaborators, in trying to reconcile these points of view, identified a visual or semantic component of short-term memory, whose capacity is approximately three chunks, in addition to the acoustical short-term memory, whose capacity is approximately 7 chunks but can be expressed better as a weighted sum of chunks and syllables. The latter relation can be interpreted in terms of Baddeley's articulatory loop.

The long-held view that short-term memory and information processing call on a common pool of scarce resources has also come under attack. Brainerd and Kingma performed experiments which show that the resources supporting short-term memory and information processing do not overlap but rather develop independently. Baddeley and collaborators have been forced to reject the idea of a single hypothetical short-term memory structure, which plays a crucial role in many information processing tasks, and have moved toward a more functional analysis based on the concept of working memory. Brainerd has described a working memory system comprising four types of storage operations and three types of processing operations to describe probability judgments in children.

General Learning Operations

In 1977 Schneider and Shiffrin presented a theory of automatic and controlled processing in memory-related tasks. Automatic processing is characterized as fast, parallel, effortless, not under subjective control, not limited by short-term memory capacity, and applicable primarily to well-developed skill behaviors, whereas controlled processing is characterized as slow, serial, effortful, capacity-limited, subject-regulated, and applicable to novel or inconsistent information. This theory has been criticized by Ryan and defended by Shiffrin and Schneider. Keil has theorized that cognitive development is guided by complex sets of constraints which are specifically tailored to particular cognitive domains and which sharply limit the class of naturally learnable structures in each domain. He discusses this theory with respect to ontological knowledge, number concepts, deductive reasoning, and natural language syntax. Lawler proposes a theory for the processes of learning and structure of knowledge wherein cognitive structures are highly specific to limited domains and new cognitive structures develop from and call upon existing structures. Anderson discusses a theory for skill acquisition wherein declarative and procedural stages are distinguished.

Cognitive Operations and Structures

The exciting and powerful idea that the instancing of concepts by particular events may be graded rather than all-or-none, embodied in the theory of "fuzzy sets" as developed by Zadeh, has become very influential in cognitive science. Osherson and Smith elaborated this concept as the so-called prototype theory and pointed out that fuzzy set theory does not provide a totally satisfactory model for the theory. Zadeh provides an alternative definition of a prototype for which the theory of fuzzy sets is appropriate. Jones proposes yet another version of prototype theory in which conceptual gradedness is represented ordinally rather than by fuzzy sets. Cohen and Murphy present a knowledge representation model of prototype theory. Smith and Osherson continue to find difficulties with the fuzzy set theory approach and argue for an approach to conceptual combination that starts with the prototype representations themselves.

Pendse, applying information theory to perceptual categorization, theorizes that the number of categories into which a person breaks down information input is a function of the signal-to-noise ratio. He further theorizes that when information is to be retransmitted the optimal number of categories decreases and the information undergoes verbalization. This suggests that the verbalization hemisphere (dominant, usually left) has a smaller optimal number of categories than the nonverbal hemisphere. Murphy and Medin address the question of what makes the members of a concept form a comprehensible class. They conclude that similarity, feature correlations, and various theories of categorization provide inadequate accounts of such conceptual coherence. They propose the alternative that concepts are coherent to the extent that they fit people's background knowledge or naive theories about the world. McNamara presents data suggesting that the mental representation of spatial relations is consistent with partially hierarchical theories rather than strongly hierarchical or nonhierarchical theories. That is, spatial relations are redundantly represented in that many spatial relations that could be computed are also stored explicitly.

A number of papers discuss mental imagery and spatial relations. Kosslyn posits that visual mental images are transitory data structures that occur in an analog spatial medium and that these "surface" representations are generated from more abstract "deep" representations in long-term memory. Once formed, such representations can be operated on in various ways. Farah presents a componential information processing model of imagery. She concludes that there is a region in the posterior left hemisphere which is critical for the image generation process of her model, that long-term visual memories used in imagery are also used in recognition, and that dreaming and waking visual imagery share some underlying processes. Kosslyn and Shwartz earlier presented a computer simulation model of mental image processing whose central assumption is the existence of a fixed resolution analog spatial medium. Koenderink suggests that visual perception treats images on several levels of resolution simultaneously and presents a model wherein any image can be embedded in a one-parameter family of derived images with resolution as the parameter. In this model any image can be described as a juxtaposed and nested set of light and dark blobs and each blob has a limited range of resolution in which it manifests itself. Leyton conceives perceptual organization as nested control. In this model perceptual sequences are split into two subsequences; one can change under external action and the other describes the way the set is perceived internally. He applies this theory to shape perception, gestalt grouping, an orientation and form problem, and motion perception. Leyton suggests that planning hierarchies have the same dynamically structured algebraic sequences and are thus formally equivalent to perceptual organization. Kuipers has criticized the "map in the head" metaphor for the representation of spatial knowledge. He argues that a person's cognitive map is built up from observations obtained as the person travels through an environment and acts as problem solver to find routes and relative positions. He proposes an alternative metaphor for the representation of spatial knowledge which includes separate metrical and topological components and is based on computational structures.

Strelow points out that in both mobility (traveling through a spatial environment) and reading considerable research has been performed on component tasks but very little attention has been paid to theories of overall psychological operations. Gibson has made major statements on visually guided mobility by focuses on visual processes. Strelow presents a theory wherein mobility can be directed by nonvisual control stimuli and processes of spatial learning, including stimulus–regroup (S–R) rote learning, motor plans, schemas, and cognitive maps, as well as by visual control stimuli. This theory is applicable to blind as well as to seeing subjects.

Wilensky discusses a model for plannng and metaplanning wherein a level of metaplanning is involved in understanding someone else's plan and creating plans for one's own use and rests on knowledge about the world and about the planning process itself. Knowledge about the planning process consists of metagoals and metaplans which can be processed in the same ways that goals and plans are processed at a lower level of planning processes. This theory is embodied in PAM, a story understanding program, and PANDORA, a problem-solving and planning system. Wilks and Bien present a model of beliefs for computer understanding of natural language in which the beliefs of A about B are computed from the knowledge structures about A and B separately. The model includes nestings of beliefs, belief percolation, and a least-effort hypothesis for mental functioning. Schoenfeld discusses the way in which belief systems shape people's behavior as they solve problems. Gentner presents a structure mapping theory for analogy wherein relations between objects rather than attributes of objects are mapped. Kahneman and Tversky discuss mechanisms underlying intuitions and judgments under uncertainty. This model is criticized by Evans.

Yates theorizes that awareness is a unique independent component of cognition whose properties may be different from those of other components and suggests that the contents of awareness may be characterized as a model of the world capable of simulating future events, anticipating present events, and thereby formulating appropriate actions. Sternberg discusses a comprehensive componential metatheory for human intelligence based on higher-order control processes used for executive processes and decision making, lower-order processes used in the execution of various strategies for task performance, knowledge-acquisition components involved in learning new information and storing it in memory, and their interactions. Fodor presents a theory in which language and perception are mediated by systems of modules which represent subsystems in the brain. These modules are "informationally encapsulated" in that their data processing is limited to inputs from the body's transducers and information stored in the module itself, exhibit a number of other simplifying constraints, and are associated with fixed neural architecture. Posner also considers the relation of some models of human performance and cognition to underlying neural systems. Fleischaker compares the "traditional" model for perception and the theory of knowledge with the information-based model of perception by Gibson and the "organizational closure" model of perception by Maturana and Varela. He concludes that the traditional model and Gibson's model, as a variant of it, exhibit problems, whereas the organizational closure model is free of these problems.

Brain Operations in Language

A fundamental and widely influential concept of the function of the brain in language has been developed over the years by Chomsky. Chomsky asserts that there exists a specialized center in the brain called a language acquisition device (LAD) which contains the essential relationships among "agents," "actions," and "objects," representations of which are common to all languages. The LAD is innate in all human brains. Different languages originate culturally *ad hoc* with various arbitrary sounds mapped through phonemes and words onto this LAD. This LAD embodies a "deep structure" which defines and constrains our linguistic mappings of the world. Particular phonemes and words in specific languages are represented in more superficial levels of brain function.

Over the years Luria has developed an elaborate model for the areas serving language activity which explains much of normal language function and clinical language disturbances. This model consists of five main operative regions. (1) Centers in the anterior tertiary zones house the intention, plans, and programming of speech and are involved in active searching, internal speech, predicative structure, linear structure of the sentence, etc. (2) Secondary zones of the anterior unit contain articulemes, that is, speech–muscle combinations for elemental units of speech, and regions for switching among articulemes. (The latter area corresponds to the classical Broca area.) (3) Tertiary zones of the posterior unit house semantic association nets for words. Words are conceived as multidimensional matrices of connections, including connections to acoustic areas (secondary zones in the auditory parts of the posterior unit), morphological areas (secondary zones in the visual part of the nondominant side of the posterior unit), lexical areas (secondary zones in the visual part of the dominant side of the posterior unit), and semantic areas (frontal fields, limbic fields, etc.).

Logical schema are housed in this area. (4) Secondary zones of the auditory parts of the posterior unit house phonemes (the fundamental units of word sounds) and speech memory (which contains series of word sounds). (This is the classical Wernicke area.) (5) Secondary areas of the visual parts of the posterior unit house either lexical features (dominant side) or morphological features (nondominant side). Two fundamental interconnecting tracts in this model are (1) the tract from the word sound system (secondary zones of auditory part of posterior unit) to the motor speech system (secondary zones of anterior unit) (this tract is the arcuate fasciculus) and (2) the tract from the phoneme areas to both the secondary visual zones and the posterior tertiary semantic association nets.

The operations of this system in expressive speech might include several stages. One follows from some deep impression originating in unknown parts of the nervous system, perhaps the limbic system or frontal cortex, from which some sense of direction is formed, largely in the tertiary zones of the anterior unit. This sense of direction is translated into a speech scheme in the inferior part of the tertiary zone of the anterior unit. This in turn activates association nets and word matrices involving the tertiary and secondary zones of the posterior unit and triggers word sounds in the secondary auditory areas. The latter, in turn, trigger articulemes in the secondary zones of the anterior unit by way of the arcuate fasciculus. Activation of the articulemes then projects outward through the motor systems to the speech muscles to produce speech. Notice that in this model there is an "internal speech" wherein the intention of producing certain speech sounds activates the sensory phoneme circuits in the auditory cortex before the words are spoken; indeed, it is the activation of phoneme circuits which drives subsequent activation of the frontal articuleme activity.

McClelland has developed a model for speech perception based on the concept of massively parallel processing in neural networks, discussed in Chapter 9. This model operates on the principle of interactive activation, in which large numbers of excitatory and inhibitory processing units work continuously to update their individual activation on the basis of the activations of other units to which they are connected. McClelland and Elman described two computer simulation programs called TRACE (I and II) which effect this processing and include features such as identifying phonemes in context and interactions of phonemes, words, and perception.

Liberman and Mattingly have presented a revised motor of speech perception wherein phonetic information is perceived in a biologically distinct system that is specialized to detect the intended gestures of speakers, which in turn form the basis for phonetic categories. In this system relations between gestures and acoustic patterns are built in and allow direct perception of phonetic structure from preliminary auditory impressions without intermediary translations.

Grossberg has presented an intriguing model for recognition and recall of words and nonwords using a real-time network processing theory which he calls adaptive resonance theory. The approach emphasizes the moment-by-moment dynamical interactions that control language. Cohen and Grossberg present a model of the translation of a speech stream into context-sensitive representations in language.

The considerable literature in linguistics is beyond the scope of this review.

Bibliography

Introductions to Artificial Intelligence and Cognitive Science

1986   Leyton, M. A theory of information structure. I. General principles. *J. Math. Psychol.* **30**, 103–160.
1985   Gardner, H. The Mind's New Science." Basic Books, New York.
1984   Winston, P. H. "Artificial intelligence," 2nd ed. Addison-Wesley, Reading, Massachusetts.
1982   Hunt, M. "The Universe Within." Simon & Schuster, New York.
1981   Bernstein, J. Profiles: Marvin Minsky. *New Yorker* Dec 14, pp. 50–126.
        Haugeland, J. ed. "Mind Design." Bradford, Montgomery, Vermont.
1974   Carney, J. D., and Scheer, R. K. "Fundamentals of Logic," 2nd ed. Macmillan, New York.
1972   Newell, A., and Simon, H. A. "Human Problem Solving." Prentice-Hall, Englewood Cliffs, New Jersey.
1971   Nilsson, N. J. "Problem-Solving Methods in Artificial Intelligence." McGraw-Hill, New York.
1968   von Bertalanffy, L. "General System Theory," rev. ed. Braziller, New York.
1967   Booth, T. L. "Sequential Machines and Automata Theory." Wiley, New York.
1966   Ashby, W. R. Mathematical models and computer analysis of the functions of the central nervous system. *Ann. Rev. Physiol.* **28**, 89–106.
1962   Ashby, W. R. Simulation of a brain. *In* "Computer Applications in the Behavioral Sciences" (H. Borko, ed.), pp. 452–467. Prentice-Hall, Englewood Cliffs, New Jersey.
1960   Miller, G. A., Galanter, E. H., and Pribram, K. H. "Plans and the Structure of Behavior." Holt, New York.

The Information Processing Paradigm in Learning and Memory

1986   Atlan, H., Ben-Ezro, E., Fogelman-Soulie, F., Pellegrin, D., and Weisbuck, G. Emergence of classification procedures in automata networks as a model for functional self-organization. *J. Theor. Biol.* **120**, 371–380.
        Reeke, G. N., and Edelman, G. M. Selective networks and recognition automata. *Conf. Neural Networks, 1986.*
        Shastri, L., and Feldman, J. A. Evidential reasoning in semantic networks: A formal theory. *Conf. Neural Networks, 1986.*
1985   Brachman, R. J., and Schmolze, J. G. An overview of the KL-ONE knowledge representation system. *Cognit. Sci.* **9**, 171–216.
        Brainerd, C. J., and Kingma, J. On the independence of short-term memory and working memory in cognitive development. *Cognit. Psychol.* **17**, 210–247.

Brown, N. R., Rips, L. J., and Shevell, S. K. The subjective dates of natural events in very-long-term memory. *Cognit. Psychol,* **17,** 139–177.

Kolodner, J. L. "Retrieval and Organizational Stragtegies in Conceptual Memory: A Computer Model." Erlbaum, Hillsdale, New Jersey.

Morton, J., Hammersley, R. H., and Bekerian, D. A. Headed records: A model for memory and its failures. *Cognition* **20,** 1–23.

Reiser, B. J., Black, J. B., and Abelson, R. P. Knowledge structures in the organization and retrieval of autobiographical memories. *Cognit. Psychol.* **17,** 89–137.

Walker, W. H., and Kintsch, W. Automatic and strategic aspects of knowledge retrieval. *Cognit. Sci.* **9,** 261–283.

Yu, B., Zhang, W., Jing, Q., Peng, R., Zhang, G., and Simon, H. A. STM capacity for Chinese and English language materials. *Mem. & Cognit.* **13,** 202–207.

Zhang, G., and Simon, H. A. STM capacity for Chinese words and idioms: Chunking and acoustical loop hypotheses. *Mem. & Cognit.* **13,** 193–201.

1984   Baddeley, A. D. Neuropsychological evidence and the semantic/episodic distinction. *Behav. Brain Sci.* **7,** 238–239.

Barsalou, L. W., and Bower, G. H. Discrimination nets as psychological models. *Cognit. Sci.* **8,** 1–26.

Feigenbaum, E. A., and Simon, H. A. EPAM-like models of recognition and learning. *Cognit. Sci.* **8,** 305–336.

Gillund, G., and Shiffrin, R. M. A retrieval model for both recognition and recall. *Psychol. Rev.* **91,** 1–67.

Osherson, D. N., Stob, M., and Weinstein, S. Learning theory and natural language. *Cognition* **17,** 1–28.

Pike, R. Comparison of convolution and matrix distributed memory systems for associative recall and recognition. *Psychol. Rev.* **91,** 281–294.

Shiffrin, R. M., and Schneider, W. Automatic and controlled processing revisited. *Psychol. Rev.* **91,** 269–276.

Wittlin, A. S. Are the limits of the mind expandable? *Behav. Sci.* **51-60.**

Yokoi, H., and M. Saito, A mathematical model of human memory. *Eng. Med. Biol.* **29**(8),191.

1983   Anderson, J. R. "The Architecture of Cognition." Harvard Univ. Press, Cambridge, Massachusetts.

Baddeley, A. D. Working memory. *Philos. Trans. R. Soc. London, Ser. B* **302,** 311–324.

Kolodner, J. L. Maintaining organization in a dynamic long-term memory. *Cognit. Sci.* **7,** 243–280.

Kolodner, J. L. Reconstructive memory: A computer model. *Cognit. Sci.* **7,** 281–328.

Ryan, C. Reassessing the automaticity–control distinction: Item recognition as a paradigm case. *Psychol. Rev.* **90,** 171–178.

1982   Anderson, J. R. Acquisition of cognitive skill. *Psychol. Rev.* **89,** 369–406.

Baddeley, A. D. Domains of recollection. *Psychol. Rev.* **89,** 708–729.

Eich, J. M. A composite holographic associative recall model. *Psychol. Rev.* **89,** 627–661.

Grossberg, S. "Studies of Mind and Brain: Neural Principles of Learning, Perception, Development, Cognition, and Motor Control." Reidel, Boston, Massachusetts.

Hunt, M. "The Universe Within." Simon & Schuster, New York.

Maida, A. S., and Shapiro, S. C. Intensional concepts in propositional semantic networks. *Cognit. Sci.* **6,** 291–330.

Murdock, B. B. A theory for the storage and retrieval of item and associative information. *Psychol. Rev.* **89,** 609–626.

Osherson, D. N., and Weinstein, A note on formal learning theory. *Cognition* **11,** 77–88.

Osherson, D. N., Stob, M., and Weinstein, S. Ideal learning machines. *Cognit. Sci.* **6**, 277–290.

Rumelhart, D. E., and McClelland, J. L. An interactive activation model of context effects in letter perception. II. The contexual enhancement effect and some tests and extensions of the model. *Psychol. Rev.* **89**, 60–94.

1981   Baddeley, A. D. The concept of working memory: A view of its current state and probable future development. *Cognition* **10**, 17–23.

Bernstein, J. Profiles: Marvin Minsky. *New Yorker Mag.* Dec. 14, pp. 50–126.

Brainerd, C. J. Working memory and the developmental analysis of probability judgement. *Psychol. Rev.* **88**, 463–502.

Keil, F. C. Constraints on knowledge and cognitive development. *Psychol. Rev.* **88**, 197–227.

Lawler, R. W. The progressive construction of mind. *Cognit. Sci.* **5**, 1–30.

McClelland, J. L, and D. E. Rumelhart, An interactive model of context effects in letter perception. I. An account of basic findings. *Psychol. Rev.* **88**, 375–407.

Raaijmakers, J. G. W., and Shiffrin, R. M. Search of associative memory. *Psychol. Rev.* **88**, 93–134.

Ratcliff, R. A theory of order relations in perceptual matching. *Psychol. Rev.* **88**, 552–572.

Ratcliff, R., and McKoon, G. Does activation really spread? *Psychol. Rev.* **88**, 454–462.

Watkins, M. J. Human memory and the information-processing metaphor. *Cognition* **10**, 331–336.

Williams, M. D., and Hollan, J. D. The process of retrieval from very long-term memory. *Cognit. Sci.* **5**, 87–119.

1979   Simon, H. A. Information processing models of cognition. *Annu. Rev. Psychol.* **30**, 363–396.

Wickelgren, W. A. Chunking and consolidation: A theoretical synthesis of semantic networks, configuring in conditioning. S-R versus cognitive learning, normal forgetting, the amnesic syndrome, and the hippocampal arousal system. *Psychol. Rev.* **86**, 44–60.

1978   Frey, P. W., and Sears, R. J. Model of conditioning incorporating the Rescorla-Wagner associative axiom, a dynamic attention process, and a catastrophe rule. *Psychol. Rev.* **85**, 321–340.

Medlin, D. L., and Schaffer, M. M. Context theory of classification learning. *Psychol. Rev.* **85**, 207–238.

Ratcliff, R. A theory of memory retrieval. *Psychol. Rev.* **85**, 59–108.

1977   Schneider, W., and Shiffrin, R. M. Controlled and automatic human information processing. I. Detection, search, and attention. *Psychol. Rev.* **84**, 1–66.

Shiffrin, R. M., and Schneider, W. Controlled and automatic human information processing. II. Perceptual learning, automatic attending and a general theory. *Psychol. Rev.* **84**, 127–189.

1976   Anderson, J. A. Neural models with cognitive implications. *In* "Basic Processes in Reading: Perception and Comprehension" (D. La Berge and S. J. Samuels, eds.). Erlbaum, Hillsdale, New Jersey.

Cofer, C. N., ed. "The Structure of Human Memory." Freeman, San Francisco, California.

Meyer, D. E., and Schvaneveldt, R. W. Meaning, memory structure, and mental processes. *Science* **192**, 27–33.

Ratcliff, R., and Murdock, B. B. Retrieval processes in recognition memory. *Psychol. Rev.* **83**, 190–214.

Uttley, A. M. A two-pathway informon theory of conditioning and adoptive pattern recognition. *Brain Res.* **102**, 23–35.

Uttley, A. M. Simulation studies of learning in an information network. *Brain Res.* **102**, 37–53.

Uttley, A. M. Neurophysiological predictions of a two-pathway informon theory of neural conditioning. *Brain Res.* **102**, 55–70.

Wicklegren, W. A. Network strength theory of storage and retrieval dynamics. *Psychol. Rev.* **83**, 466–478.

1975   Collins, A. M., and Loftus, E. F. A. A spreading–activation theory of semantic processing. *Psychol. Rev.* **82**, 407–428.

Hollan, J. D. Features and semantic memory: Set-theoretic or network model? *Psychol. Rev.* **82**, 154–155.

Klatzky, R. L. "Human Memory." Freeman, San Francisco, California.

Rips, L. J, Shoben, E. J., and Smith, E. E. Set-theoretic and network models reconsidered: A comment on Hollan's 'Features and semantic memory.' *Psychol. Rev.* **82**, 156–157.

Smith, E. E., Shoben, E. J., and Rips, L. J. Structure and process in semantic memory: A featural model for semantic decisions. *Psychol. Rev.* **8**, 214–241.

Tulving, E., and Watkins, M. J. Structure of memory traces. *Psychol. Rev.* **82**, 261–275.

Uttley, A. M. The information in classical conditioning. *J. Theor. Biol.* **49**, 355–376.

1974   Narendra, K. S. and Thathachar, M. A. L. Learning automata—a survey. *IEEE Trans. Syst., Man, Cybernet.* **SMC-4**, 323–334.

Simon, H. A. How big is a chunk? *Science* **183**, 482–488.

Smith, E. E., Shoben, E. J., and Rips, L. J. Structure and process in semantic memory: A featural model for semantic decision. *Psychol. Rev.* **81**, 214–241.

Wicklegren, W. A. Single-trace fragility theory of memory dynamics. *Mem. & Cognit.* **2**, 775–780.

1973   Kessner, R. A neural system analysis of memory storage and retrieval. *Psychol. Bull.* **80**, 177–203.

Tsetlin, M. L. "Automaton Theory and Modeling of Biological Systems." Academic Press, New York.

1972   Anderson, J. R. FRAN: A simulation model of free recall. *In* "The Psychology of Learning and Motivation: Advances in Research and Theory" (G. H. Bower, ed.), Vol. 5, pp. 315–378. Academic Press, New York.

Gregg, L. W., ed. "Cognition in Learning and Memory." Wiley, New York.

Murdock, B. B., and Dufty, P. O. Strength theory and recognition memory. *J. Exp. Psychol.* **94**, 284–290.

Rumelhart, D. E., Lindsay, P. H., and Norman, D. A. A process model for long-term memory. *In* "Organization of Memory" (E. Tulving and W. Donaldson, eds.), pp. 197–246. Academic Press, New York.

Tulving, E. Episodic and semantic memory. *In* "Organization of Memory" (E. Tulving and W. Donaldson, eds.), pp. 381–404. Academic Press, New York.

1971   Grossberg, S. Pavlovian pattern learning by nonlinear neural networks. *Proc. Natl. Acad. Sci. U.S.A.* **68**, 828–831.

1970   Kintsch, W. Models for free recall and recognition. *In* (D. A. Norman, ed.), "Models of Human Memory" Academic Press, New York.

Lockhart, R. S., and Murdock, B. B. Memory and the theory of signal detection. *Psychol. Bull.* **74**, 100–109.

Mendel, J. M., and Fu, K. S. "Adaptive Learning, and Pattern Recognition Systems: Theory and Applications." Academic Press, New York.

Morton, J. A functional model for memory. *In* "Models of Human Memory" (D. A. Norman, ed.), pp. 203–254. Academic Press, New York.

Simon, H. A. "Sciences of the Artificial." MIT Press, Cambridge, Massachusetts.

Spinelli, D. N. OCCAM: A computer model for a content addressable memory in the central nervous system. *In* "Biology of Memory" (K. Pribram and D. Broadbent, eds.), pp. 293-306. Academic Press, New York.

1969   Klopf, A. H., and Gose, E. An evolutionary pattern recognition network. *IEEE Trans. Syst., Man, Cybernet.* **SMC-5**, 247-250.

1968   Atkinson, R. C., and Shiffrin, R. M. Human memory: A proposed system and its control processes. *In* "The Psychology of Learning and Motivation: Advances in Research and Theory" (K. W. Spence and J. T. Spence, eds.), Vol. 2, pp. 90-197. Academic Press, New York.

Quillian, M. R. Semantic memory. *In* "Semantic Information Processing" (M. Minsky, ed.), MIT Press, Cambridge, Massachusetts.

Weizenbaum, J. Contextual understanding by computers. *In* "Recognizing Patterns" (P.A. Kolers and M. Eden, eds.), pp. 170-193. MIT Press, Cambridge, Massachusetts.

1967   Booth, T. L. "Sequential Machines and Automata Theory." Wiley, New York.

Bower, G. H. "A multicomponent theory of the memory trace." *In* "The Psychology of Learning and Motivation: Advances in Research and Theory" (K. W. Spence and J. T. Spence, eds.), Vol. 1, pp 230-327. Academic Press, New York.

Feigenbaum, E. A. Information processing and memory. *Proc. Berkeley Symp. Math. Stat. Prob., 5th, 1967,* Vol. 4, pp. 37-51.

Quillian, M. R. Word concepts: A theory and simulation of some basic semantic capabilities. *Behav. Sci.* **12**, 410-430.

1965   Nilsson, N. J. "Learning Machines." McGraw-Hill, New York.

Waugh, N. C., and Norman, D. A. Primary memory. *Psychol. Rev.* **72**, 89-104.

1964   Simon, H. A., and Feigenbaum, E. A. An information-processing theory of some effects of similarity, familiarization, and meaningfulness in verbal learning. *J. Verb. Learn. Behav.* **3**, 385-396.

1963   Giulano, V. E. Analog networks for word association. *IEEE Trans. Mil. Electron.* **7**, 221-234.

1962   Feigenbaum, E. A., and Simon, H. A. A theory of the serial position effect. *Br. J. Psychiatry* **53**, 307-320.

Rosenblatt, F. "Principles of Neurodynamics." Spartan Books, Washington, D. C.

1961   Feigenbaum, E. A. The simulation of verbal learning behavior. *Proc. West. J. Comput. Conf.* **19**, 121-132.

Feigenbaum, E. A., and Simon, H. A. Forgetting in an associative memory. *Proc. Assoc. Natl. Comput. Mach. Conf.* **16**, 202-205.

Zadeh, L. A. Time-varying networks. I. *Proc. IRE* **49**, 1488-1503.

1959   Estes, W. K. The statistical approach to learning theory. *In* "Psychology: A Study of a Science" (S. Koch, ed.), Vol. 2, 380-491. McGraw-Hill, New York.

Feigenbaum, E. A. "An Information Processing Theory of Verbal Learning." The RAND Corp., Santa Monica, California.

Selfridge, O. G. Pandemonium: A paradigm for learning. *In* "The Mechanization of Thought Processes," pp. 511-534. H. M. Stationery Office, London.

1956   Miller, G. A. The magical number seven, plus or minus two. Psychol. Rev. **63**, 81-97.

1955   Estes, W. K. Statistical theory of spontaneous recovery and regression. *Psychol. Rev.* **62**, 145-154.

Cognitive Operations

See also references above in "Information Processing Paradigm in Learning and Memory."

1986   McNamara, T. P. Mental representations of spatial relations. *Cognit. Psychol.* **18**, 87-121.

1985   Murphy, G. L., and Medin, D. L. The role of theories in conceptual coherence. *Psychol. Rev.* **92**, 289-316.

Strelow, E. R. What is needed for a theory of mobility: Direct perception and cognitive maps—lessons from the blind. *Psychol. Rev.* **92**, 226-248.

Yates, J. The content of awareness is a model of the world. *Psychol. Rev.* **92**, 249-284.

1984   Cohen, B., and Murphy, G. L. Models of concepts. *Cognit. Sci.* **8**, 27-58.

Farah, M. J. The neurological basis of mental imagery: A componential analysis. *Cognition* **18**, 245-272.

Fleischaker, G. R. The traditional model for perception and theory of knowledge: Its metaphor and two recent alternatives. *Behav. Sci.* **29**, 40-50.

Koenderink, J. J. The structure of images. *Biol. Cybernet.* **50**, 363-370.

Leyton, M. Perceptual Organization as nested control. *Biol. Cybernet.* **51**, 141-153.

Putnam, H. Models and modules. *Cognition* **17**, 253-264.

Smith, E. E., and D. N. Osherson. Conceptual combination with prototype concepts. *Cognit. Sci.* **8**, 337-361.

1983   Anderson, J. R. "The Architecture of Cognition." Harvard Univ. Press, Cambridge, Massachusetts.

Armstrong, S. L., Gleitman, L. R., and Gleitman, H. What some concepts might not be. *Cognition* **13**, 263-308.

Fodor, J. "The Modularity of Mind." Bradford Books, Montgomery, Vermont.

Genter, D. Structure-mapping: A theoretical framework for analogy. *Cognit. Sci.* **7**, 155-170.

Schoenfeld, A. H. Beyond the purely cognitive: Belief systems, social cognitions and metacognitions as driving forces in intellectual performance. *Cognit. Sci.* **7**, 329-363.

Sternberg, R. J., Components of human intelligence. *Cognition* **15**, 1-48.

Wilks, Y., and Bien, J. Beliefs, points of view, and multiple environments. *Cognit. Sci.* **7**, 95-119.

1982   Evans, J. St. B. T. On statistical intuitions and interential rules: A discussion of Kahneman and Tversky. *Cognition* **12**, 319-323.

Eysenck, H. "A Model of Intelligence." MIT Press, Cambridge, Massachusetts.

Jones, G. V. Stacks not fuzzy sets: An ordinal basis for protoype theory of concepts. *Cognition* **12**, 281-290.

Kahneman, D., and Tversky, A. On the study of statistical intuitions. *Cognition* **11**, 123-141.

Kuipers, B. The 'map in the head' metaphor. *Environ. Behav.* **14**, 202-220.

Zadeh, L. A. A note on prototype theory and fuzzy sets. *Cognition* **12**, 291-297.

1981   Burt, P. J., Hong, T. H., and Rosenfeld, A. Segmentation and estimation of image region properties through cooperative hierarchical computation. *IEEE Trans. Syst., Man, Cybernet.* **SMC-11**, 802-825.

Farah, M. J., and Kosslyn, S. M. Structure and strategy in image generation. *Cognit. Sci.* **4**, 371-383.

Kosslyn, S. M. The medium and the message in mental imagery: A theory. *Psychol. Rev.* **88**, 46-66.

Mamdani, E. H., and Gaines, B. R. eds. "Fuzzy Reasoning and Its Applications." Academic Press, New York.

Mervis, C. B., and Rosch, E. Categorization of natural objects. *Annu. Rev. Psychol.* **32**, 89-115.

Osherson, D. N., and Smith, E. E. On the adequacy of prototype theory as a theory of concepts. *Cognition* **9**, 35-58.

Posner, M. I. Cognition and neural systems. *Cognition* **10**, 361-366.

Pslyshyn, Z. W. The imagery debate: Analogue media versus tacit knowledge. *Psychol. Rev.* **87**, 16–45.

Smith, E. E., and Medin, D. L. "Categories and Concepts." Harvard Univ. Press, Cambridge, Massachusetts.

Wilensky, R. Meta-planning: Representing and using knowledge about planning in problem solving and natural language understanding. *Cognit. Sci.* **5**, 197–233.

1980   Dubois, D., and Padre, H. "Fuzzy Sets and Systems: Theory and Applications." Academic Press, New York.

Johnson-Laird, P. N. Mental models in cognitive science. *Cognit. Sci.* **4**, 71–115.

Kosslyn, S. M. "Image and Mind." Harvard Univ. Press, Cambridge, Massachusetts.

Triesman, A. M., and Gelade, G. A feature-integration theory of attention. *Cognit. Psychol.* **12**, 97–136.

1979   Fahlman, S. E. "NETL: A System for Representing and Using Real-World Knowledge." MIT Press, Cambridge, Massachusetts.

Gibson, J. J. "The Ecological Approach to Visual Perception." Houghton-Mifflin, Boston, Massachusetts.

Klopf, A. H. Goal-seeking systems from goal-seeking components. Implications for AI. *Cognit. Brain. Theor. Newsl.* **3**, 54–62.

1978   Brownell, H. H., and Carmazza, A. Categorizing with overlapping categories. *Mem. & Cognit.* **6**, 481–490.

Kuipers,. B. Modeling spatial knowledge. *Cognit. Sci.* **2**, 129–153.

Pendse, S. G. Category perception, language and brain hemispheres: An information transmission approach. *Behav. Sci.* **23**, 421–427.

Rosch, E., and Lloyd, B. B. "Cognition and Categorization." Erlbaum, Hillsdale, New Jersey.

Zadeh, L. PRUF. A meaning representation language for natural languages. *Int. J. Man-Mach. Stud.* **10**, 395–460.

1977   Kosslyn, S. M., and Shwartz, S. P. A simulation of visual imagery. *Cognit. Sci.* **1**, 265–295.

Ogden, G. C. Integration of fuzzy logical information. *J. Exp. Psychol., Hum. Percept. Perform.* **3**, 565–575.

1976   Hersch, H. M., and Caramazza, A. A fuzzy set approach to modifiers and vagueness in natural language. *J. Exp. Psychol. Gen.* **105**, 254–276.

1975   Kosslyn, S. M., Pick, H. L., and Fariello, G. R. Cognitive maps in children and men. *Chld. Dev.* **45**, 707–716.

Zadeh, L. A., Fu, K.-S., Tanaka, K., and Shimura, M., eds. "Fuzzy sets and Their Applications to Cognitive and Decision Processes." Academic Press, New York.

1969   Gougen, J. A. The logic of inexact concepts. *Synthese* **19**, 325–373.

1966   Gibson, J. J. "The Senses Considered as a Perceptual System." Houghton-Mifflin, Boston, Massachusetts.

1965   Zadeh, L. Fuzzy sets. *Inf. Cont.* **8**, 338–353.

1958   Gibson, J. J. Visually controlled locomotion and visual orientation in animals. *Br. J. Psychiatry* **49**, 182–194.

1950   Gibson, J. J. "The Perception of the Visual World." Houghton Mifflin, Boston, Massachusetts.

1948   Tolman, E. C. (1948). Cognitive maps in rats and men. *Psychol. Rev.* **55**, 189–208.

Brain Operations in Language

1986   Dell, G. S. A spreading–activation theory of retrieval in sentence production. *Psychol. Rev.* **93**, 283–321.

Grossberg, S., and Stone, G. Neural dynamics of word recognition and recall: Attentional priming, learning, and resonance. *Psychol. Rev.* **93**, 46–74.

McClelland, J. L., and Elman, J. L. The TRACE model of speech perception. *Cognit. Psychol.* **18**, 1–86.

1985   Golden, R. M. A developmental neural model of word perception. *Proc. Conf. Cognit. Sci. Soc., 1985.*

Liberman, A. M., and Mattingly, I. G. The motor theory of speech perception revised. *Cognition* **21**, 1–36.

1973   Luria, A. R. "The Working Brain." Basic Books, New York.

1972   Chomsky, N. "Language and Mind." Harcourt, Brace, Jovanovich, New York.

Greene, J. "Psycholinguistics." Penguin, Baltimore, Maryland.

1962   Halle, M. and Stevens, K. N. Speech recognition: A model and a program for research. *IRE Trans. Inf. Theory* **8**, 155–159.

11

Clinical Models

This chapter is only an entrance to the literature. I have found a few interesting clinically oriented engineering models of cardiac action potentials and conduction, conduction in axons, global electric signals, schizophrenia, stuttering, and neurological function. The citations are representative of current efforts in these areas, but are not exhaustive.

Bibliography

Cardiac Function

1983   Hafner, D. Hodkin–Huxley-model analysis of cardiac action potentials. *Eng. Med. Biol.* **36**, 36.3, 178.
        Trujillo, V. A., and Baumgartner, D. N. Primitive communication element for cardiac conduction model. *Eng. Med. Biol.* **36**, 18.3, 83.
        See also Noble in Chapter 2.

Conduction in Axons

1984   Restivo, M. A., Li, J. K.-J., Welkowitz, W., and El-Sherif, N. Extracellular potential fields at sites of conduction block. *Eng. Med. Biol.* **37**, 29.4, 187.
1983   Wood, S. L., and Waxman, S. G. Computer simulation of conduction in myelinated and demyelinated axons. A review. *IEEE Front. Eng. Comp. Health Care,* pp. 425–431.
1982   Wood, S. L., and Waxman, S. G. Conduction in demyelined nerve fibers: Computer simulations of the effects of variation in voltage-sensitive ionic conductances. *IEEE Front. Eng. Health Care,* pp. 424–428.
1981   Wood, S. L., and Cummins, K. L. Bidirectional nerve refractory characteristics: Simulations of conduction resulting from direct and remote stimulation. *IEEE Front. Eng. Health Care,* pp. 419–423.
        See also Chapters 2 and 22.

Global Electric Fields

1985   Schoonhoven, R., Stegman, D. F., and van Oosterom, A. Volume conductor modeling in electroneurography. *Eng. Med. Biol.* **38**, 9.21, 53.
1984   Rusinko, J. B., Sepulveda, N. G., and Walker, C. F. 3-dimensional numerical solution of field distribution. *Eng. Med. Biol.* **37**, 42.1, 280.
        Sherif, M. H., and Gregor, R. J. Modeling myoelectric interference patterns. *Eng. Med. Biol.* **37**, 11.2, 68.
        Kim, Y., Tompkins, W. J., and Webster, J. G. Development of a modifiable computer body model. *IEEE Front. Eng. Comp. Health Care,* pp. 45–50.
1983   de Weerd, J. P. C. Modeling and analysis in electroneurography. *IEEE Front. Eng. Comp. Health Care,* p. 423.
        Doslak, M. J., and Hsu, P. C. A model of the ERG and the effect of vitreous hemorrhage. *IEEE Front. Eng. Comp. Health Care,* p. 623.
1982   Rush, S., Budd, R., and Baldwin, A. F. III, (1982). Model study of effects of local change of resting potential on injury currents and ECG. *IEEE Front. Eng. Comp. Health Care,* p. 418.
1981   Rusinko, J. B., Walker, C. F., and Sepulveda, N. G. Finite element modeling of potentials within the human thoracic spinal cord due to applied electrical stimulation. *IEEE Front. Eng. Comp. Health Care,* p. 76.
        See also EEG in Chapter 8.

Schizophrenia

1983   Schmolling, P. A systems model of schizophrenic dysfunction. *Behav. Sci.* **28**, 253–267.
1976   Colby, K. M. Clinical implications of a simulation model of paranoid processes. *Arch. Gen. Psychiatry* **33**, 854–857.
        Cronin, J. Mathematical aspects of periodic catatonic schizophrenia. *Bull. Math, Biol.* **39**, 187–200.
1976   Hartmann, E. Schizophrenia: A theory. *Psychopharmacology (Berlin)* **49**, 1–15.
1975   Laki, K. An Attempt to measure 'disorder' in mental disorders. *Perspect. Biol. Med.* **15**, 157–161.

Stuttering

1985   Nudelman, H. B., Herbrich, K. E., Hoyt, B. D., and Rosenfield, D. B. Modeling for classification and therapy in stuttering. *Eng. Med. Biol.* **38**, 10.2, 58.

Neurological Quantification

1981   Kondraske, G. V., Potvin, A. R., Tourtellotte, W. W., and Syndulko, K. Computer-based quantification of neurological function. *IEEE Front. Eng. Comp. Health Care,* pp. 404–408.
        McDowell, F. H. Quantitative testing of neurological deficits. *IEEE Front. Eng. Comp. Health Care,* p. 403.
        Syndulko, K., Potvin, A. R., Tourtellotte, W. W., and Potvin, J. H. Multisystem quantitative assessment of neurologic function for clinical trials. *IEEE Front. Eng. Comp. Health Care,* pp. 409–414.

12

Outer Limits

I am not sufficiently conversant with the literature of this area to comment on it, and the reference list included here is certainly not comprehensive. Nonetheless, the list does offer some intriguing entries and I include it both as a reminder that no real comprehensive treatment of neural and brain modeling can be complete without consideration of these limits and as a goad to many of us to expand our understanding at this level. I have made no attempt to discriminate "respectable" from "unrespectable" entries in this list.

Bibliography

Specific Models

1986   Denker, M. W., Achenback, K. E., and Keller, D. M. Computer simulation of Freud's counterwill theory: Extension to elementary social behavior. *Behav. Sci.* **31**, 103–141.
1983   Newcomb, R. W. Psychic fields and neural networks. *Eng. Med. Biol.* **36**, 14.1,61.
1980   Newcomb, R. W. Neural-type microsystems: Some circuits and considerations. *Proc. IEEE Int. Conf. Circuits and Comput.* **2**, 1072–1074.
1978   Greenwald, A. G., and Ronis, D. L. Twenty years of cognitive dissonance: Case study of the evolution of a theory. *Psychol. Rev.* **85**, 53–57.
1977   Newcomb, R. W. "Psychic Fields: Basics and Main Ideas," Microsyst. Gen. Networks Rep. University of Maryland, Baltimore.
        Wegman, C. A computer simulation of Freud's counterwill theory. *Behav. Sci.* **22**, 218–233.
1976   John, E. R. A model of consciousness. *In* "Consciousness and Self-Regulation" (G. E. Schwartz and D. Shapiro, eds.). Plenum, New York.
1975   MacLean, P. D. On the evolution of three mentalities. *Man-Environ. Syst.* **5**, 213–224.

1972   Moser, U., von Zepplin, I., and Schneider, W. Reply to Blackmore: Some comments on computer simulation of a model of neurotic defense processes. *Behav. Sci.* **17**, 232-234.

1970   Moser, U., von Zeppelin, I., and Schneider, W. Computer simulation of a model of neurotic defense processes. *Behav. Sci.* **15**, 194-202.

1968   Moser, U., von Zepplin, I., and Schneider, W. Computer simulation of a model of neurotic defense mechanisms (clinical part). *Bull. Psychol. Inst. Univ. Zurich* **2**, 1-76.

1966   Freud, S. Project for a scientific psychology. *In* "The Complete Psychological Works of Sigmund Freud" (J. Strachey, ed. and transl.), Vol. 1, pp. 283-397. Hogarth Press, London (Originally Published 1893).

1957   Festinger, L. (1957). "A Theory of Cognitive Dissonance." Stanford Univ. Press, Stanford, California.

Outer Limits of Brain

1982   MacLean, P. D. Evolution of the psychencephalon. *Zygon* **17**, 187-211.

1979   Crick, F. H. C. Thinking about the brain. *Sci. Am.* **241**, 219-232.

1977   Sagan, C. "The Dragons of Eden." Ballantine, New York.

1975   Penfield, W. "The Mystery of the Mind." Princeton Univ. Press, Princeton, New Jersey.

1971   Sinsheimer, R. L. The brain of Pooh: An essay on the limits of mind. *Am. Sci.* **59**, 20-28.

1970   Eccles, J. C. "Facing Reality." Springer-Verlag, Berlin and New York.

Brains, Minds, and Computers

1986   Denning, P. J. Will machines ever think? *Am. Sci.* **74**, 344-346.
        Kugel, P. When is a computer not a computer? *Cognition* **23**, 89-94.
        Sternberg, R. J. Inside intelligence. *Am. Sci.* **74**,137-143.
        Winograd, T., and Flores, F. "Understanding Computers and Cognition: A New Foundation for Design." Ablex, New York.

1985   Haddon, R. C., and Lamola, A. A. The molecular electronic device and the biochip computer: Present status. *Proc. Natl. Acad. Sci. U.S.A.* **82**, 1874-1878.
        Johnson-Laird, P. N. Human and computer reasoning. *Trends Neurosci.* **8**, 54-57.

1983   Johnson-Laird, P. N. "Mental Models." Cambridge Univ. Press, London and New York.
        Thompson, I. D. Thinking about thinking. *Trends Neurosci.* **6**, 161-162.

1982   Hunt, M. "The Universe Within." Simon & Schuster, New York.
        Michie, D. 'Mind-like' capabilities in computers: A note on computer induction. *Cognition* **12**, 97-108.

1981   Johnson-Laird, P. N. Cognition, computers, and mental models. *Cognition* **10**, 139-143.
        Longuet-Higgins, H. C. Artificial intelligence—a new theoretical psychology? *Cognition* **10**, 197-200.

1979   Dreyfus, H. L. "What Computer's Can't Do. The Limits of Artificial Intelligence." Harper & Row, New York (previously published 1972).

1978   Braine, M. D. S. On the relation between the natural logic of reasoning and standard logic. *Psychol. Rev.* **85**, 1-21.

1974   Zettler, F. Uber die logische struktur eines nervensystem. *J. Comput. Physiol.* **95**, 123-167.

1973   Conrad, M. Is the brain an effective computer? *Int. J. Neurosci.* **5**, 167-170.

1970   Jensen, R. E. Brain-computer relationships. *Prog. Brain Res.* **33**, 1-8.

1967   von Bertalanffy, L. "Robots, Men, and Minds." Braziller, New York.

Philosophy of Mind, Mysticism, Philosophy of Science

1985   Feinberg, F. "Solid Clues: Quantum Physics, Molecular Biology, and the Future of Science." Simon & Schuster, New York.

Winson, J. "Brain and Psyche: The Biology of the Unconscious." Anchor/Double-day, New York.

1984   Carr, T. H., Brown, T. L., and Sudevan, P. Straw men and glass houses? A reply to Churchland on reductionism. *Neuroscience* **13**, 1397–1400.

Churchland, P. S. "Matter and Consciousness." Bradford/MIT, Cambridge, Massachusetts.

Churchland, P. S. Psychology and the study of the mind–brain: A reply to Carr, Brown, and Sudevan. *Neuroscience* **13**, 1401–1404.

Griffin, D. R. "Animal Thinking." Harvard Univ. Press, Cambridge, Massachusetts.

Reiser, M. F. "Mind, Brain, Body: Toward a Convergence of Psychoanalysis and Neurobiology," Basic Books, New York.

1983   Manicus, P. T., and Secord, P. F. Implications for psychology of the new philosophy of science. *Am. Psychol.* **38**, 399–413.

Walker, F. "Animal Thought." Routledge & Kegan Paul, Boston, Massachusetts.

Wilson, R. A. "Prometheus Rising." Falcon Press, Phoenix, Arizona.

1982   Churchland, P. M. Mind–brain reduction: New light from the philosophy of science. *Neuroscience* **7**, 1041–1047.

Leary, T. "Changing My Mind, Among Others." Prentice-Hall, Englewood Cliffs, New Jersey.

1981   Gregory, R. L. "Mind in Science." Weidenfeld & Nicolson, London.

1980   Brown, B. "Supermind." Harper & Row, New York.

Chapman, A. J., and Jones, D. M. "Models of Man." Br. Psychol. Soc., Leicester.

Churchland, P. M. A perspective on mind–brain research. *J. Philos.* **77**, 185–207.

Talbot, M. "Mysticism and the New Physics." Bantam Books, New York.

1979   Bateson, G. "Mind and Nature." E. P. Dutton, New York.

1978   Goodman, N. "Ways of World Making." Hackett Publ., Cambridge, Massachusetts.

1977   Brown, J. "Mind, Brain and Consciousness." Academic Press, New York.

Weinberg, A. M. The limits of science and trans-science. *Int. Sci. Rev.* **2**, 337–342.

Weisskopf, V. F. The frontiers and limits of science. *Am. Sci.* **65**, 405–411.

Wilson, "Cosmic Trigger." Pocket Books, New York.

1976   Diamond, S. J. Brain circuits for consciousness. *Brain Behav. Evol.* **13**, 376–395.

Jaynes, J. "The Origin of Consciousness in the Breakdown of the Bicameral Mind." Houghton Mifflin, Boston, Massachusetts.

Sperry, R. W. Changing concepts of consciousness and free will. *Perspect. Biol. Med.* **20**, 9–19.

Zangwill, O. L. Thought and the brain. *Br. J. Psychiatry* **67**, 301–314.

1975   Bhaskar, R. "A Realistic Theory of Science." Leeds Books, Leeds, England.

Capra, F. "The Tao of Physics." Shambhala, Boulder, Colorado.

Stent, G. S. Limits to the scientific understanding of man. *Science* **187**, 1052–1057.

1972   Bateson, G. (1972) "Towards an Ecology of Mind." Ballantine Books, New York.

Lilly, J. "The Center of the Cyclone." Bantam Books, New York.

Lilly, J. "The Human Biocomputer." Bantam Books, New York.

1971   Pribram, K. H. "Languages of the Brain." Prentice-Hall, Englewood Cliffs, New Jersey.

1969   Sperry, R. W. A modified concept of consciousness. *Psychol. Rev.* **76**, 532–536.

1968   Popper, K. R. On the theory of the objective mind. *Proc. Int. Congr. Philos., 14th, 1968,* Vol. 1.

1966   Bronowski, J. The logic of the mind. *Am. Sci.* **21**, 1–14.

1965   Bronowski, "The Identity of Man." Nat. Hist. Press, New York.
1959   Popper, K. R. (1959). "The Logic of Scientific Discovery." Hutchinson, London.
1956   Boulding, K. (1956). "The Image." Ann Arbor Press, Ann Arbor, Michigan.

Epilogue

The review in the preceding chapters shows clearly that a rich variety of techniques and methods within engineering and computer approaches to neural and brain modeling has helped to deal with the complexities of neuroscience and brain science. There is a vast literature (over 1200 publications cited here) in many different subdivisions of neural and brain modeling. Approximately 60 separate bibliographic lists on individual subareas within neural and brain modeling have been presented.

The area of neuronal networks is now, as it was 10 years ago, a critical level in the integration and forward movement of brain science. Moreover, as Lewis and I projected in 1977, this area has been the focus of a tremendous amount of theoretical study and modeling. In my opinion, the three most significant developments related to neuronal networks are (1) the theoretical recognition that their dynamic patterns are likely to be "supraconnectionist" in the sense of being constrained by anatomy only within limits, (2) the recognition that their essential connectivity patterns are primarily massively parallel rather than sequential, and (3) the demonstration that some of their significant global operations may be understood theoretically on the basis of minimization (or maximization) principles in the manner of statistical mechanics.

The last of these three developments may be in the broadest sense the most significant guide for further theorizing and modeling in brain science. So far in neural and brain modeling we have seen extensive application of fundamental mathematical and computational techniques, and application of computer science methodology (the information processing paradigm). Both of these approaches have made and will continue to make substantial contributions, but now we are on the verge of deeper fundamental premises and orientations. There are a number of fundamental engineering and physical sciences such as statistical mechanics, thermodynamics, fluid mechanics, electromagnetic theory, dynamic systems theory, quantum mechanics, and relativity theory whose underlying premises are based on physical laws describing essential operating principles of physical systems. Such sciences may serve as better guides and models for theorists trying to develop a fundamental understanding of brain operations than can straightforwardly applied mathematical and computational procedures. In coming years we should see increased application of deeper premises and starting points similar to those of fundamental engineering and physical sciences by brain theorists and modelers.

A fundamental split revealed by this review is that between the computer science and neural modeling approaches. These two fields should find ways of working together to bring about progress in this area. The computer science approach has been top-down, concerned with linking variables at levels D and E and concerned primarily with cognitive operations. It has represented a total restructuring of the top-down approach to modeling brain/mind functioning in terms of the information processing paradigm and, moreover, has been very helpful in providing a number of models and theories of cognitive contexts for the operations of neuronal networks. In this last endeavor, however, it has often been cavalier in its disregard of neuroscience.

The neural modeling approach has been bottom-up, dealing primarily with linking variables at levels B and C and focusing on generic neuronal mechanisms and models of the electrical activity of neural networks. It has made substantial inroads in providing techniques for exploring electrical activity patterns in large interconnected neural networks, as evidenced by the computer programs in Part II of this book. On the other hand, with its concern for mechanism, it has sometimes been slow in dealing with the overall functional properties of networks as coordinated wholes. A main goal of the review in Part I and of the programs in Part II of this book is to help remedy these deficiences and promote a balanced understanding of neuronal networks in which their global properties are seen in perspective in relation to both the behavioral and experiential operations they serve and the underlying mechanisms that produce them.

In the last 10 to 15 years neuroscience generally has seen a shift in focus from the systematic point of view that was prevalent in the late 1960s to great interest in the molecular and chemical levels. It is time, I believe, to emphasize the systemic level again and to recognize the contributions that can be made at that level by the theoretical and modeling approaches discussed in this review.

Experimentally oriented researchers should seek guidance by theory through the complexities of brain function and should recognize that possible models for such guidance exist in the engineering, physical, and computer sciences. They should recognize that such modeling and theoretical efforts can play leading roles and are not merely devices for explaining data that are produced in this or that laboratory, although such purposes should indeed be served as well.

II

Computer Simulation of Neuronal Systems

This is my letter to the World,
That never wrote to Me—
The simple News that Nature told—
With tender Majesty.

Her Message is committed
To Hands I cannot see—
For love of Her—Sweet—countrymen—
Judge tenderly—of Me

(Emily Dickinson)

Outline

Several factors motivated the writing of this part of the book. First, the need for an easily available methodology for the simulation of nervous system activity seems obvious given the tremendous and continuing increase in computer power and availability and the corresponding emergence of new generations of computer-literate young people in all fields. Realistic neuronal simulation models should help provide advances in experimental neuroscience research, in clinical studies and applications, in instructional contexts, and in speculative brain/mind theory.

Second, although neural modeling studies appear in the literature, descriptions of the methodology and techniques of neuronal modeling and simulation are not easily or widely available to the many students, researchers, and scholars who might find good use for them. This book contains packaged programs in FORTRAN for simulating an extremely broad range of neural activity, ranging from simple point neurons to neurons that may exhibit elaborate dendritic trees and active conductance modulations or Hodgkin–Huxley action potentials, to neuronal junctions, pools, and interconnected networks, to junctions that may exhibit Hebbian learning, to systems of multiple populations whose individual populations may exhibit dendritic trees and whose junctions may exhibit Hebbian learning.

Third and finally, the initial and driving impetus for the creation of this book was my sense of a fundamental need in my own research area. It has been my impression and growing belief for nearly 20 years that a central ingredient in the advance of neuroscience must be some means of bridging the chasm between the molecular level (physiological and anatomical rules of interaction in single and small groups of neurons), on the one hand, and the global, macroscopic level (level of systemic neuronal dynamics: cell assemblies? statistical configurations? massively parallel operations?) on the other—some means of showing what specific global outcomes are produced from various specific sets of determinants at the molecular level; for example, what global dynamic patterns are produced by different specific neurophysiological mechanisms in neuronal networks and systems in various structural arrangements. Showing this is particularly essential for an understanding of the nervous system, where neuronal networks, as coordinated wholes, are so refractory to experimental investigation. One wants, for example, to study the set and distribution of multiple neuronal signals from vast numbers of constituent neurons simultaneously—not just from single EEG signals. This kind of capability, it seems to me, will greatly enhance our ability to speculate intelligently about the overall dynamic properties of neuronal networks and systems as wholes and about the possible sets of principles and constraints that govern the behavior of activity of these levels.

On this basis, then, I have undertaken for the past 15 years or so the work that has at last resulted in this book. I believe that the computer programs contained in this book do finally give us the capability to make reasonable projected inferences about the overall dynamic properties of neuronal systems—how they depend on various properties and parameters of the systems from which they emerge, how they are responsive or not responsive to various input stimuli, how sensitive they are to various parameters or features of the system, what sorts of internal logic or coherence they have, what sort of overall dynamic constraints they must adhere to. The next generation of neuroscientists will likely take realistic, large-scale computer simulation of neuronal systems as much for granted as a component tool of scientific study of nervous system dynamics as this generation has taken the microelectrode for granted. Realistic computer simulation merely provides grist for the mill (raw data comparable to that produced by experimental preparations themselves) from which the higher-level work of conceptualizing theoretical understanding can be made, and I believe it is at this level of understanding that some of the most far-reaching and profound mysteries concerning the brain and mind will be understood. In the past 15 years, neuroscience has made tremendous advances at the molecular and neurochemical levels, but it has experienced corresponding diminution of progress and loss of focus at the systemic and information-processing levels. The programs of this book can be used to make significant contributions to a renaissance of systemic and information-processing neuroscience, which, of necessity, must occur sooner or later.

I envision a period of 10 to 20 years in which programs such as those in this book will produce dynamic predictions on many specific neuronal systems of ever-increasing reliability, accuracy, and significance. I envision this development occurring at the hands of many investigators in many laboratories with many foci of interest and points of view. If this book contributes to only some small fraction of the realization of this vision, I will be content with the effort that went into its creation.

13

Introduction

This book presents packaged programs and techniques for simulating the electrical activity of neurons and neuronal systems. The programs range from simple models for the dynamics of point neurons, through interacting pools, junctions, and networks of neurons, to arbitrarily arranged systems of neuronal populations which may exhibit Hebbian associational learning at any prescribed junctions and which may contain neuronal assemblies with dendritic trees. The approach of the book is "bottom-up" in the sense that neuronal potentials are purported to be determined relatively accurately by the underlying anatomy and physiology.

The central contribution of this book is to translate the physiological principles underlying neuroelectric activity into mathematical physics by virture of an equivalent circuit model for neuronal membrane and then to translate this in turn into structured computational algorithms and programs in representing anatomical arrangements of neurons and neuron populations. We have made no attempt to describe programming techniques for various graphical displays of simulation output or techniques for analysis of such output. These problems are idiosyncratic from laboratory to laboratory and should be within the fundamental programming capabilities of technicians and researchers in using their own systems. What the programs in this book do is to produce relatively realistic numerical evaluations of ongoing electrical signals in neurons and neuron populations in response to prescribed input stimulation and endogenous activity, as mediated by internal connections and various pertinent parameters under the control of the user.

Chapters 14, 15, 21, and 22 deal with simulation programs and methods for single neurons. Those in Chapters 14 and 15 are simplified approximations but are computationally quite efficient. Those in Chapters 21 and 22 are more accurate representations of neuronal processes but require considerably more computation. Chapters 16, 17, and 18 deal with individual neuronal assemblies: junctions, pools, and networks, respectively. Chapters 19 and 20 deal with systems of multiple, interconnected neuronal populations and the possible occurence of Hebbian associational learning at prescribed junctions within the systems. For clarity, all programs in Chapters 16 through 20 make use of a relatively simple point neuron model, which is introduced in Chapters 14 and 15. Chapter 23 extends the most fundamental and most general of the programs of Chapters 16 through 20 to include representation of dendritic trees on the neurons in the populations in the programs. Some of the neuron models in Chapters 14, 15, 20, and 23 include a first-pass representation of active calcium-related conductances in dendritic regions. This model produces relatively realistic representations of complex spikes, bursting, and the mechanisms thought to underlie them, but can be only approximate at this stage since the actual properties of these underlying mechanisms are only now being investigated experimentally. The programs of Chapters 14 and 15 should be very useful in quantitative modeling of overall input-output dynamics in response to particular prescribed patterns of synaptic bombardment in neurons. The models in Chapter 20 should be of value in assisting experimental studies which penetrate more deeply into the structure and significance of dendritic trees in both passive electrotonic behavior and ongoing input-output behavior. Although these models do not include detailed representation of events within dendritic spines, they do include representation of a theoretical relationship which accounts for the influence of the spine by placing an equivalent synapse with a modified conductance amplitude on the main stem of the dendrite.

The programs of Chapters 16, 17, and 18 can be of use in exploring many questions involving the sculpturing and processing of information in local networks of neurons—for example, the filtering or integrating properties of convergent-divergent junctions, the effecting of feature extraction or boundary enhancement by lateral inhibition, the effects of coupling of adjacent neurons by extracellular fields and perhaps the genesis and propagation of synchronized activity through pools, the maintenance of internally sustained activity in recurrently connected networks by appropriate balance between recurrent excitation and recurrent inhibition, and the tendency for recurrently mediated activity in two-dimensional networks to remain constrained in space or to migrate through populations depending on the relative lengths of spread of recurrent excitation and recurrent inhibition. Examples relevant to each of these phenomena are presented in Chapters 16 through 18. More important, the range of properties of information coding in neuronal assemblies is perhaps only hinted at by this brief inventory of mechanisms. Computer programs in these chapters should assist subsequent creative exploration in this field.

The programs in Chapter 19 can be used to simulate ongoing electrical activity in arbitrarily structured neuronal populations. These models can be applied to an extremely wide range of functional neuronal systems. A primitive example presented in Chapter 19 represents the circuitry of the hippocampus with six neuronal populations. This example focuses on the propensity of these individual regions to generate rhythms of 4 to 11 per second (theta rhythm) and the concept that the rhythmic firings in the individual regions are brought into phase by coincident activation from the medial septum. The models of this chapter can be used to explore and illustrate fundamental concepts of organization in many neuronal systems, including the visual system, the motor control system of the cerebellum, and auditory information processing in the auditory system, to name just a few.

The programs can be used as they stand to illustrate and explore guiding principles of organization in these systems. Alternatively, they can easily be extended and refined to assist more penetrating evaluation of behavior and concepts with respect to hard experimental findings in various regions.

The programs of Chapter 20 can be used to simulate the embedding of memory traces in neuronal systems. Concepts such as the cell assembly from Hebb's theories or the statistical configuration concept of Roy John can be elucidated and clarified with the simulation models of this chapter. Indeed, it would seem that computational devices such as these programs are essential to bridge the findings of experimental neuroscience to such global theoretical concepts. We include in Chapter 20 one or two preliminary examples of the nature of dynamic memory traces as mediated by these mechanisms. However, thorough exploration of these phenomena awaits future effort. The programs of Chapter 22 simulate the generation and propagation of action potentials according to the Hodgkin-Huxley equation system developed for the giant axon of the squid. Use of these programs will be somewhat limited, primarily because the equation system is known not to describe accurately the generation, shape, and propagation of mulitple action potentials close together in time. However, they can be useful as an instructional tool and may be applicable to a wide variety of questions related to single action potentials. Moreover, they may form the basis for subsequent generalization or modification to equation systems that do a better job with repetitive firing. For example, Chapter 22 includes one program which extends the Hodgkin-Huxley model to describe the electodynamics of contraction of heart muscles.

Finally, the programs of Chapter 23 allow one to simulate the dynamics and possible associational learning within systems of multiple neuronal populations which may be arbitrarily interconnected, wherein the neuronal populations contain neurons with dendritic trees. The programs of this chapter are the most general in the book and may be the place where the most significant application of computer neural modeling occurs. Again, the most meaningful and significant applications of these programs in the understanding of neuronal dynamics await future effort, ideally by a number of investigators in a number of different contexts. The appendices give listings of the programs and summarize the symbols used throughout the book.

Each of the technical chapters—that is, Chapters 14 through 23—contains four distinct sections. The first sections (sections A) present input-output flowcharts for the programs and discuss application of the programs from the standpoint of a user who is not interested in the details of the FORTRAN programming. These sections are particularly appropriate for those who are interested primarily in using the programs as they stand to illustrate and explore various concepts of neuronal dynamics, and they can be read independently of the other parts of the book. The programs can be typed directly from the listings into the user's computing system. The second sections in the chapters (sections B) discuss the specific FORTRAN listings for the programs. Users and readers who are particularly interested in the programming methods of this book can study sections B for this information. Sections C of these chapters contain discussion of the theoretical background for the modeling procedures developed in the book. It is in sections C that mathematical physics meets neurophysiology. Readers who are particularly interested in the theoretical foundations and limitations as well as the significance of various modeling techniques used in this book should read these sections and the references cited in them. Finally, sections D of these technical chapters contain various exercises which illustrate the properties of the programs. The exercises include both applications where the programs can be used "as is" and example exercises where minor modifications of existing programs are required to effect various idiosyncratic applications. The latter exercises should be appropriate for user-oriented courses or instructional programs.

In a sense, this book is three separate books: sections A comprise a collection of packaged simulation programs and instructions on how to use them; sections B constitute an instructional course in constructing FORTRAN programs to simulate neuronal activity; sections C comprise a book on neural modeling.

The book is organized in terms of progression from simple to complex. Substantially, the first program represents a simple point neuron, and the last programs represent arbitrary systems of neurons which exhibit dendritic trees and effect Hebbian learning. The partitioning of chapters and the sequencing of programs within chapters have been structured primarily with the student of sections B in mind—that is, on the basis of pedagogical presentation of new programming devices and techniques.

All the programs are written in standard FORTRAN. FORTRAN 77 is required to make use of all the programs in the later chapters. However, most of the programs in Chapters 14 through 18 can be implemented with a FORTRAN IV compiler. Virtually all the development of these programs was done on the Cyber system of the Computing Center at the University of Colorado, utilizing 143,000 words of central memory and a dual CDC 6700 central processing unit. This is easily adequate for most uses of programs in the earlier chapters of the book, say up through Chapter 18. For neuronal systems involving more than 3000 to 5000 neurons or networks of neurons with dendritic trees, it is desirable to have more central memory than this. Simulation, of course, slows down when one makes use of memory stores outside central memory. For many applications, particularly in research contexts on laboratory computers, this slowdown in simulation time on dedicated computers should not be a major problem. On the other hand, for the most intensive and serious use of the most advanced programs in the book, say those from Chapter 23, it is desirable to have access to supercomputer capabilities. For example, with the most primitive point neurons and making use of data packing, it is possible to simulate 20,000 or 30,000 neurons with central memory of about 140,000 words. However, if one wants to simulate tens of thousands of cells realistically or to represent, say, 5000 cells with dendritic trees realistically and to do this while having an effective ongoing interaction with the simulation, it is necessary to have access to a central memory of ½ million words or more and computational power like that available on the CRAY and Cyber 205 supercomputers.

14

Simplified Models of Single Neurons

Computational efficiency in realistic representation of fundamental neuronal processes is the main accomplishment of this chapter. The chapter presents computer programs (which are located in Appendix 2) that simulate neuronal processes with integration step sizes of about 1 msec in model neuron time. This represents an approximately 500- to 1000-fold increase in computer efficiency over techniques which focus on mathematical exactness or very fine-grained descriptions of precise temporal time courses. The significance of this increased efficiency becomes clear when one considers that the equations and algorithms presented for single neurons in this chapter are to be used repeatedly in larger hierarchical programs representing networks and systems of tens of thousands or more individual neurons, all with realistic properties. The neuronal processes simulated include a state variable model that describes the fundamental processes of accommodation and repetitive firing of trains of all-or-none action potentials; a state variable model of active calcium-related conductances including representation of active calcium spikes, complex spikes, and bursting activity; and a model of a neuron involving the above processes with a compartmentalized dendritic tree. The dendritic tree model in particular represents an approximately 1000-fold increase in computer efficiency in integration over time and a 25-fold increase in integration over space, representing an approximately 25,000-fold increase in efficiency over more exact programs presented in Chapter 22. The programs in this chapter include activation of these models only by simulated experimentally applied currents; synaptic bombardment of these models is presented in Chapter 15. Part A of this chapter shows how to use the programs, gives a nonmathematical discussion of their behavior, and presents basic examples. Part B presents the equation systems used in the models and the computer listing of the programs. Part C discusses the theoretical background of the models. Part D suggests various exercises in using the programs.

A. Using the Programs: Input–Output Charts and Illustrative Examples

1. State-Variable Point Model for Repetitive Firing in Neurons (PTNRN10)

This model produces relatively realistic ongoing input-output dynamics for a neuron with basic accommodative properties. For example, on-responses, off-responses, and representative intensity-duration curves for responses to ramp currents are produced by it. Figure 14.1 shows the input-output flowchart for this program. The program is activated by a single input function SCN, which represents the stimulating current from an experimentally applied electrode as a function of time. The model produces four main state variables or output functions: the transmembrane potential E, the time-varying threshold TH, a spiking variable S, and the potassium conductance above resting level, GK. The function P(t) is a cosmetic output function that represents the combination of the transmembrane potential E and the spiking variable S; P looks like the record that would be measured with an intracellular microelectrode. Figure 14.1 also shows the various parameters to be specified in using the model. STEP is the step size of the numerical integration scheme; typically it is taken as 1.0, representing 1 msec. EK is the equilibrium potential of the potassium conductance, typically taken as -10, representing 10 mV below resting potential. LTSTOP is the duration of the particular simulation, e.g., 60 msec. SC, SL, SCSTRT, and SCSTP are parameters defining the time shape of the input current. SC is the amplitude of a step current, SL the slope of a ramp current, SCSTRT the time the stimulation starts, and SCSTP the time the simulation stops. C, TTH, B, TGK, T0, TH0, and TMEM are parameters describing the point neuron. TH0 is the resting threshold of the cell, typically taken to be about 10 mV. TMEM is the membrane time constant, typically about 5 msec. C and TTH are parameters describing the rise of the threshold used to simulate accommodation in the cell; C varies from 0 to 1, and TTH is a time constant in milliseconds, usually taken to be about 25. B and TGK are parameters determining the refractoriness of the cell— in particular the postfiring increase in potassium conductance of the membrane. B, typically taken to be about 20, represents the amplitude of the postfiring potassium conductance decay, typically taken to be about 5 msec.

The essential behavior of this state variable model for repetitive firing is shown in Fig. 14.2. If one, for example, applies a steady current of

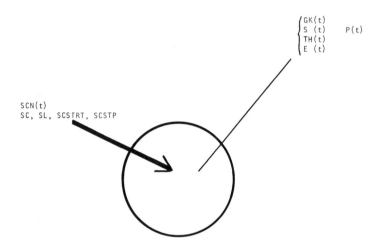

PARAMETERS

STEP, EK
C, TTH, B, TGK, THO, TMEM
LTSTOP

INPUT FUNCTIONS   STATE AND OUTPUT FUNCTIONS

$\begin{cases} GK(t) \\ S\ (t) \\ TH(t) \\ E\ (t) \end{cases}$ P(t)

SCN(t)
SC, SL, SCSTRT, SCSTP

Fig. 14.1. This program simulates the responses of a four-state-variable model of a single point neuron to applied step and ramp input currents.

amplitude SC, as shown in the figure, the transmembrane potential E rises exponentially as shown. As a result of the rise in membrane potential, the threshold TH rises in the model in proportion to the parameter C, as shown in Fig. 14.2 and with the time constant TTH. If the potential E at any time exceeds the threshold, a spike is generated, indicated in Fig. 14.2 by the variable S going from 0 to 1. When an action potential is elicited, the conductance to potassium GK increases by an amount proportional to B. The potassium conductance then decays exponentially back toward 0 with a time constant of TKG. While the potassium conductance is in an elevated state it causes the transmembrane potential E to course back downward toward the equilibrium potential, typically − 10mV. As the potassium conductance weakens with time, however, the applied current again predominates, sending the potential back up, and it may fire again as indicated by another excursion of S from 0 to 1 and another increase in GK. Finally, the second GK brings the potential back down. Again, as GK decreases the potential goes back up, and if the threshold has now risen to a level higher than the

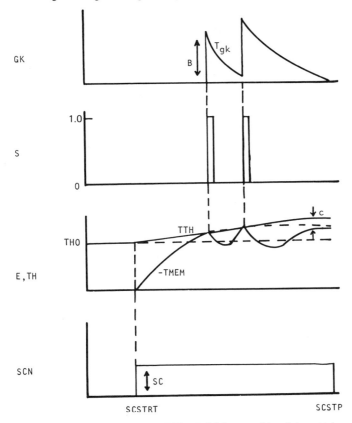

Fig. 14.2. Essential behavior of state variable model for repetitive firing at trigger sections (PTNRN10, …).

equilibrium value of the membrane potential associated with the value of SC, no further firing ensues. Figure 14.2 thus illustrates the so-called on-response to a steady step current. Figure 14.2 also shows clearly the influence of the various parameters in the model. TMEM indicates the sluggishness with which the potential responds to the stimulus SC. C represents the amount of threshold rise associated with a given level of E. TTH denotes the time constant rise of the threshold. B is related to the amplitude of the potassium conductance rise after firing, and TGK represents the time constant of decay of the potassium conductance.

In essence, then, this model produces the response of four state variables—transmembrane potential, threshold, spiking variable, and postfiring potassium conductance—to a single stimulating current SCN. The several parameters allow one to create different kinds of firing propensity in the individual neurons simulated. For example, with C = 0, no accommodation, one produces a neuron that tends to fire tonically or at steady rates with given levels of current. On the other hand, with C between 0.5 and 1, one creates a cell that tends to fire transiently at the onset or perhaps offset of various stimuli. With B and TGK at relatively low levels, one creates model neurons that are able to recover from firing relatively quickly and thereby fire at relatively high rates for a given level of input stimulation; with B and TGK larger, on the other hand, one creates cells that have large refractory periods, respond rather sluggishly, and fire at lower rates.

Examples of Behavior from PTNRN10   Figure 14.3a maps the basic character of the output response of PTNRN10 in terms of tonic (steady) firing or phasic (transient) firing against its basic accommodative parameters C and TTH for different levels of SC. Figure 14.3b shows a representative train of action potentials produced by PTNRN10 in response to a steady applied current. Phasic responses are obtained with larger values of C and

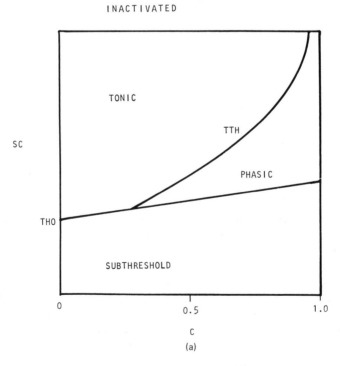

Fig. 14.3.   (a) Parameter space for PTNRN10; (b) generator potential from neuron; (c) response of model to step current; (d) comparison of model response to data from a spinal motoneuron (λ = 6.91; c = 0.521); (e) accommodation by the model.

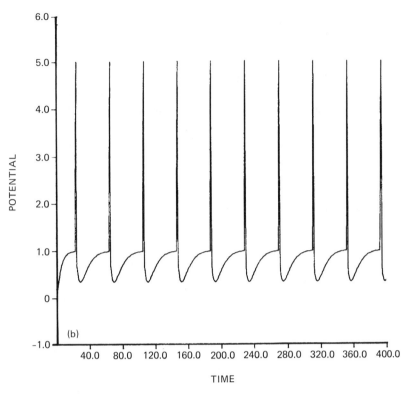

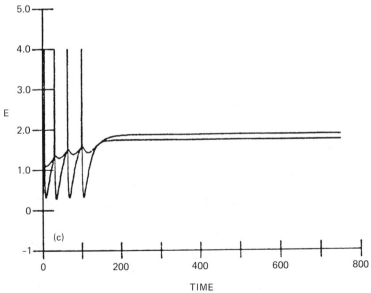

Fig. 14.3 (*Continues on next page*)

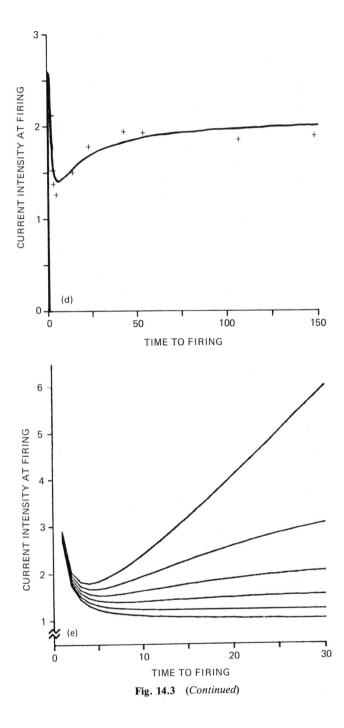

Fig. 14.3 (*Continued*)

larger values of TTH. Figure 14.3c shows an on-response to a step current obtained from PTNRN10 by increasing the value of C to about 0.75. The second curve in Fig. 14.3c is the threshold. These accommodative properties can be explored more quantitatively with ramp currents. The intensity of current at firing is plotted versus the latency to firing to obtain a so-called intensity-latency curve. The behavior or the model is matched to data from a spinal motoneuron as shown in Fig. 14.3d. More generally, the accommodative properties of the model are shown in Fig. 14.3e. If C is 0 there is no accommodation and the intensity-duration curve decreases monotonically to a plateau called the rheobase. If C = 1 the curve comes back up and approaches asymptotically a line with slope 1/TTH. At C between 0 and 1 the curve approaches various ceilings as indicated in the figure. The example shown in Fig. 14.3d, for example, represents a typical "high-ceiling curve" from the data of Bradely and Somjen. See Chapter 2 for citations on these properties. This state variable model is used to represent triggering sections of neurons in virtually all programs of this book except where the Hodgkin-Huxley model is used in Chapter 22.

2. Active Calcium-Related Conductances in a "Two-Point" Model Neuron (PTNRN20)

Figure 14.4 shows the input-output flowchart for program PTNRN20. This program simulates a neuron with two points: one representing the soma, which is in turn simulated by the state variable model used in PTNRN10, and the other representing the entire dendritic tree. Quantitatively, each compartment in the model is represented by an equivalent circuit and the two circuits are allowed to conduct a current back and forth between them in proportion to their potential difference. The model is activated by two input functions: SCS, which represents stimulating current applied to the soma, and SCD, which represents stimulating current applied to the dendritic region. The model produces eight state and output variables. First are ES, THS, S, and GKS, representing respectively the membrane potential, threshold, spiking variable, and potassium conductance of the soma region. Four state variables describing the dynamics of the dendritic compartment are ED, GCA, CA, and GKD, representing respectively the potential, the conductance to calcium, the calcium concentration, and the conductance to potassium in the dendrites. Collectively these last four functions constitute a state variable representation of active conductances producing calcium spikes in dendritic regions. PS and PD are cosmetic output functions that represent essentially what would be observed if an intracellular electrode was stuck in the soma or dendritic region, respectively.

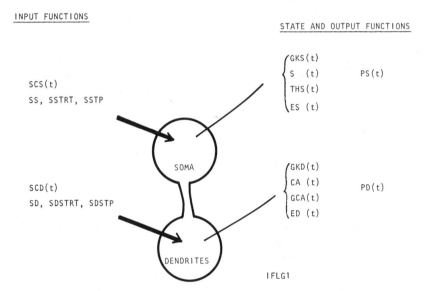

PARAMETERS

STEP, NSTEPS, EK, ECA

TS, GDS, THO, C, TTH, B, TGK

TD, GSD, THD, D, TGC, A, TCA, CAO, BD, TGKD

LTSTOP

INPUT FUNCTIONS

STATE AND OUTPUT FUNCTIONS

SCS(t)

SS, SSTRT, SSTP

GKS(t)

S (t)   PS(t)

THS(t)

ES (t)

SOMA

SCD(t)

SD, SDSTRT, SDSTP

GKD(t)

CA (t)   PD(t)

GCA(t)

ED (t)

DENDRITES

IFLG1

Fig. 14.4. Program PTNRN20. This program simulates the responses of a two-point model of a single neuron with active calcium-related conductances to step currents applied in either of the two regions.

Figure 14.4 also shows the parameters that must be specified to operate PTNRN20. STEP is the essential integration step size, again taken to be 1.0 for 1 msec. NSTEPS must typically be set at about 10 or 20 for this model and represents a more fine-grained temporal integration. NSTEPS represents the number of temporal integration steps per STEP; it is necessary in PTNRN20 because of the exchange of current between the two compartments in the model. EK is again the potassium equilibrium potential, taken to be −10 mV; ECA is the equilibrium potential of calcium, taken to be about 50 mV above resting level. LTSTOP is again the simulation time, taken to be, say, 100 msec. IFLG1 is a parameter which, if set greater than 0, allows the program to print out instantaneous values of all eight state variables and, if set equal to 0 suppresses this printout. SS, SSTRT, and SSTP are the amplitude and on/off times of the soma current; SD, SDSTRT, and SDSTP are the amplitude and on/off times of the dendritic current. TS is the time constant of the soma membrane, GDS the conductance from the dendrite to the soma. Again, TH0, C, TTH, B, and TGK are parameters for the soma triggering model as discussed for PTNRN10. TD is the time constant of the dendritic compartment. Typically one would expect the time constant of the membrane in a dendritic region to be the same as that of the membrane in the soma region. However, when representing a perhaps large dendritic tree with a single compartment, it is more appropriate to make the effective time constant larger than the membrane time constant. GSD is the conductance from the soma to the dendrite. Typically, if the soma area is imagined to be bigger than the dendritic area then GSD should be proportionally bigger than GDS, and vice versa. Nominal values of GDS and GSD of 1 to about 10 are representative. THD is the threshold in the dendrites to conductance to calcium, D the amplitude of calcium conductance in response to membrane potential, TGC the time constant of rise of dendritic calcium conductance, A the amplitude of calcium concentration rise in proportion to calcium conductance, TCA the time constant of calcium concentration rise, CA0 the threshold level of calcium concentration required to trigger potassium conductance in the dendrites, BD the amplitude of potassium conductance in the dendrites in response to calcium concentration, and TGKD the time constant of decay of potassium conductance in the dendrites. Representative values for these parameters are given in Section B and in Appendix 1.

The essential dynamic features of the state variable model for active calcium-related conductances are illustrated in Fig. 14.5. If one applies a steady current of amplitude SD in a dendritic compartment, the potential in the compartment goes up essentially exponentially with the time constant TD. As illustrated in Fig. 14.5, when the potential ED exceeds its threshold THD, it causes an increase in conductance to calcium in the membrane. The parameter D determines the amplitude of this calcium conductance. The parameter TGC is the time constant with which it rises. As the calcium conductance goes up, it causes, (1) current to flow into the cell, driving the potential ED up toward the calcium equilibrium potential of + 50, and (2) an increase in calcium concentration inside the cell above resting level, indicated by the variable CA in Fig. 14.5. The parameter A determines the amplitude of the calcium concentration increase in response to an increase in calcium conductance. TCA is the time constant of increase of calcium. When the calcium concentration exceeds its threshold level, CA0, as indicated in Fig. 14.5, it causes increases in conductance to potassium in the dendritic membrane. BD is the parameter that determines the amount of potassium conductance increase corresponding to a given level of CA. In this model, GKD is imagined to increase incrementally (in proportion to BD) as long as CA is above CA0. Now, when GKD rises, it allows current to

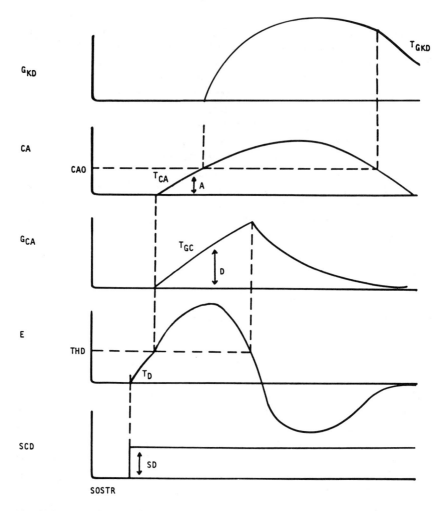

Fig. 14.5. Essential behavior of active calcium-related conductances in dendritic regions (PTNRN20, NEURN30, ...).

leak out of the model neuron, tending to drive the potential of the model neuron down toward its equilibrium potential of -10 mV. The resulting drop in ED is shown in Fig. 14.5. When ED drops again below its threshold THD, the calcium conductance is no longer driven upward and tends to decay with time constant TGC down to zero. As GCA goes down to zero, the calcium concentration again begins to drop. When it drops below its threshold CA0, then GKD is no longer being augmented and it in turn begins to decay exponentially back down to zero. The net result of this mechanism is a relatively slow calcium spike of the order of 3 to 5 msec and a relatively long-lasting afterhyperpolarization, both of which are shown in Fig. 14.5.

Examples of Behavior from PTNRN20   The essential results of the active calcium-related conductances in this model are to produce waves of "calcium spikes" through dendritic regions and to alter the firing patterns of neurons from trains of relatively widely spaced single spikes to clusters and groups of bursts of spikes. Figure 14.6a shows the response of PTNRN20 to a step current applied at the soma when the parameter D is set equal to zero, thereby disengaging all the active calcium conductance changes. The repetitive firing pattern is very similar to that obtained from PTNRN10, where individual spikes are produced by the threshold rule in the soma only. Figure 14.6b shows the response of PTNRN20 where all parameters are identical to those used for Fig. 14.6a except that the parameter D is set to 2.2, bringing in the active calcium-related conductance changes. In this simulation, potential at the soma rose to threshold, triggering a single action potential. This somatic potential propagated out to the dendritic region and was above its threshold, causing the active calcium spike to be initiated. This wave in turn propagated back to the soma and caused the soma potential to be above its threshold, lasting until the strong hyperpolarizing potassium conductance in the dendrites was activated, which in turn finally brought the potential back down toward resting level in the soma. The result then represents the slow calcium wave, and the corresponding burst of spikes produced at the soma triggering section represents a so-called complex spike. Figure 14.6c, taken from the same simulation as 14.6b, shows the essential dendritic functions underlying these processes. The first somatic spike triggered the first sawtooth increase in the somatic potassium conductance GKS. Very shortly after this, and reflecting the somatic potential propagating into the dendrites and the dendritic potential increasing its threshold, the conductance to calcium GCA is seen to rise. Immediately following this rise one sees the calcium concentration rise, peaking at about 19 msec. Someplace along its time course the calcium concentration has exceeded CA0 and this has triggered GKD, the potassium conductance in the dendrites. It in turn rises to about time 28 and then decays exponentially.

The parameter listings in Fig. 14.4 show one how to use program PTNRN20 to explore more thoroughly the characteristics and properties of active calcium-related conductances as represented in this state variable model. Again, one can create artificial neurons with different dynamic properties and characteristics by varying the pertinent parameters. For example, we have already seen how to produce cells that are essentially tonic or phasic by varying C and TTH and how to produce cells that are essentially bursty or

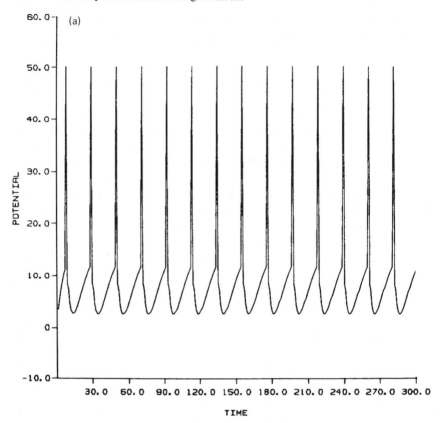

Fig. 14.6. Response of PTNRN20 to step current illustrating the influence of active conductances; (a) without active calcium conductance changes; (b) with active calcium conductance charges; (c) underlying dendritic functions.

nonbursty in firings of individual spikes by adjusting the calcium conductance parameter D. One can further influence the duration of the calcium spike by varying TGC, for example, and the relative strength of the dendritic as opposed to the somatic potassium conductance change by varying BD and TGKD. This state variable model for active calcium-related conductances and dendrites is used in both the simplified conpartmentalized dendritic tree model, NEURN30 and NEURN31, discussed immediately below, and the more elaborate and more accurate dendritic series discussed in Chapter 21.

3. Model for Repetitive Firing in Neurons with Compartmentalized Dendritic Trees (NEURN30)

The simplified model described in this section, NERUN30, represents propagation and interaction of signals in dendritic trees of neurons. Its

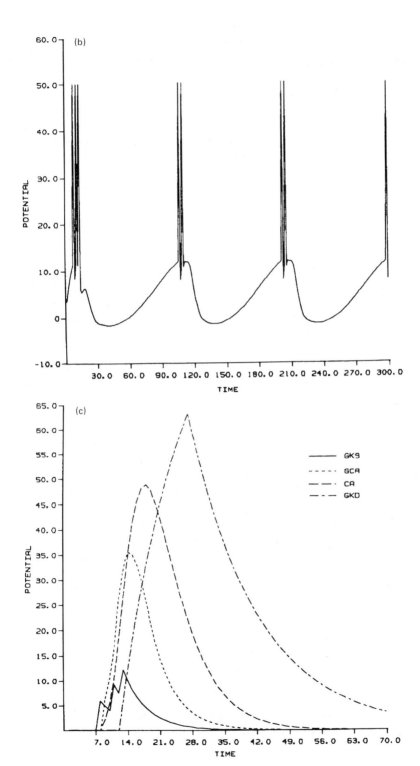

Fig. 14.6 (*Continued*)

essential feature is that it represents propagation along dendrites by a simple averaging (with a variable parameter) of potentials from neighboring compartments in lieu of more accurate representations of propagation by intercompartmental resistances or by appropriate partial differential equations, as is done in Chapter 22. The model also includes in each of the dendritic compartments the active calcium-related conductances as described for PTNRN20 and the state variable model for repetitive triggering of spikes by graded potentials in the soma as described in PTNRN10.

The input-output flowchart for NEURN30 is shown in Fig. 14.7. Again there are two input functions of time; SCN, representing current applied by a stimulating electrode in the soma, and SCDN, representing current applied by a stimulating electrode at some specific place in the dendritic tree. The neuron model contains NST dendritic stems, each of which has NRG regions. Therefore, the entire neuron has NST × NRG + 1 regions, when one includes the soma. The state variables are as we have seen in PTNRN10 and PTNRN20. The soma has state variables E, TH, and GK, representing membrane potential, threshold, and potassium conductance. Each dendritic compartment has four state variables, V, GCA, CA, and GKD, representing, respectively, the membrane potential, the conductance to calcium, the calcium concentration, and the conductance to potassium. In addition, the signal observed in an intracellular electrode placed in the soma or in any of an arbitrary number of dendritic regions is represented by the functions P_i. The parameters SC, SCSTRT, and SCSTP represent the amplitude and on/off times of the stimulating current applied in the soma; the parameters SD, SDSTRT, SDSTP, IS, and IR represent, respectively, the amplitude, on/off times of current applied at some place in the dendrite, and identities of the stem and region in which the electrode is placed. IFLG1 again determines whether the instantaneous values of all the state variables will be printed out. NDSPY represents the number displayed and signifies the number of dendritic compartments for which potentials will be displayed by P_i functions. ST and RG are the stem and region pairs for each one of these. NST is the number of stems and NRG the number of regions per stem. These might be taken as three and four, for eample. STEP again is the integration step size, typically taken to be 1.0. ECA is the calcium equilibrium potential, typically taken to be about 50 mV. EK is the potassium equilibrium potential, taken to be about − 10 mV. LTSTP is the time of the simulation run, say 60 msec.

All the other parameters are the same as those given for PTNRN10 and PTNRN20 with the exception of CDS and VRT, which are related to the propagation of signals throughout the dendritic tree. VRT is the voltage retention parameter and varies between zero and one. If VRT is 1.0, then potentials in each region are imagined to be totally retained in their own regions and there is no conduction or interaction among compartments. If

PARAMETERS

```
NST, NRG, STEP, ECA, EK
C, TTH, B, TGK, TMEM, THO, CDS, VRT
THD, D, TGC, A, TCA, CAO, BD, TGKD
LTSTOP
```

INPUT FUNCTIONS

STATE AND OUTPUT FUNCTIONS

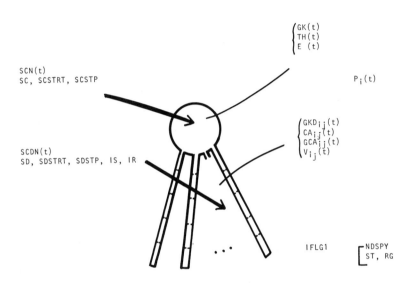

```
SCN(t)
SC, SCSTRT, SCSTP
```

$$\begin{cases} GK(t) \\ TH(t) \\ E\ (t) \end{cases}$$

$P_i(t)$

```
SCDN(t)
SD, SDSTRT, SDSTP, IS, IR
```

$$\begin{cases} GKD_{ij}(t) \\ CA_{ij}(t) \\ GCA_{ij}(t) \\ V_{ij}(t) \end{cases}$$

IFLG1

```
┌NDSPY
└ST, RG
```

Fig. 14.7. Program NEURN30. This program simulates the responses of a model neuron with simplified dendritic tree containing active calcium-related conductances to step currents applied either at the soma or in the dendritic regions.

VRT is 0, there is maximum space averaging or conducting of potentials among regions. Intermediate values give intermediate quantities and speeds of conduction. DCS represents the conduction from the dendrite to the soma. Essentially, this number estimates the ratio of the area of the initial dendritic compartment to that of the soma compartment. If this number is zero there is no effective conduction from dendrite to soma and the soma acts like an isolated point. If the number is very large, the dendritic signals overwhelm anything else going on in the soma and completely determine the potential in the soma. Again, intermediate values correspond to intermediate balances of these effects.

Example of Behavior from NEURN30   Figure 14.8 represents potentials seen in various parts of a dendritic tree of NEURN30 in response to a pulse

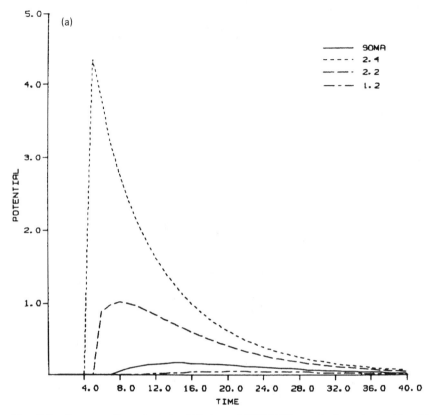

Fig. 14.8. Responses throughout dendritic tree of NEURN30 to pulse input in dendrite. (a) Control case is VRT = 0.1, CDS = 0.2; (b) VRT = 0.2, CDS = 0.2; (c) VRT = 0.1, CDS = 0.1.

input current applied in the dendritic tree. In particular, a stimulating electrode is imagined to be placed in the fourth region of one of three dendritic stems, each of which has five regions. The potential in response to a pulse at this electrode is recorded at the site of stimulation (2.4), at the second region in the same dendrite (2.2), at the soma, and at the second region in one of the other dendrites (1.2). The three figures show the spatiotemporal decay of the potential away from the region of stimulation. Moreover, a comparison of Figs. 14.8a and 14.8b shows that increasing VRT from 0.1 to 0.2 decreases the effective conduction away from the stimulating electrode. Comparison of Fig. 14.8c with either of the other two shows that the effective rise times and time shapes for these potentials can be varied by adjusting CDS.

Program NEURN30, then, allows one to simulate dynamic processes including accommodation, active calcium-related conductances, and more or

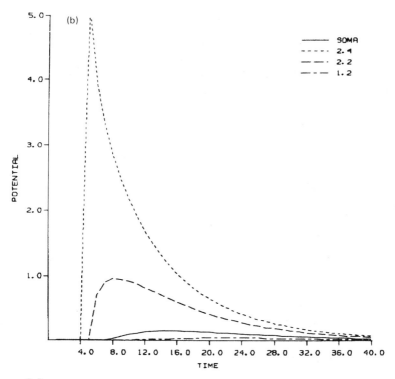

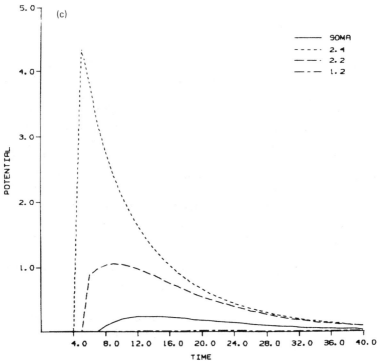

less realistic propagation in dendritic trees with a neuron model that uses integration step sizes as large as 1 msec. Probably the best way to apply this model in studying a particular network or neuron type is to use one of the more elaborate dendritic tree models from Chapter 21 to simulate the neuron realistically and then to adjust the parameters DCS and VRT in NEURN30 to match the essential propagation parameters of that more elaborate model. One can then use the model of the model (so to speak) in network simulation programs or very long simulation runs.

Additional examples of output from NEURN30 and the other models of this chapter are presented in Chapter 15, which includes the bombardment of these simplified neurons with synaptic input.

B. Program Listings and Equations

This section discusses the listings of the FORTRAN computer programs used to simulate the models described in this chapter. It also presents the equation systems underlying these models and includes a more explicit and quantitative definition of the parameters involved than was possible in the preceding qualitative section.

1. State Variable Point Model for Repetitive Firing in Neurons (PTNRN10)

The essential structural feature of program PTNRN10, as can be seen in Appendix 2, is a do-loop over the time variable, L, as contained in the do-1000 and 1000-continue statements of the program. Since the primary state variables of the model E, GK, and TH are defined as shown below by first-order differential equations, one can predict their value at any point in time simply from knowing the input at that time and their values at the previous time. It is not necessary to keep in the memory of the computer any other past history of any of the state variables. Thus the essence of this program is exceedingly simple; one sets initial values for these state variables in the do-loop, which continually updates them step by step in conjunction with the input arriving at any particular step. The portions of the program before the do-1000 loop set various parameter values and initialize the variables. The portion of the program after the do-loop writes out the activity variables and contains the formats. This do-loop approach of updating state variables is the essence of all the neuronal simulation presented in this book.

The specific differential equations governing the behavior of PTNRN10 are shown here as equation system 14.1.

$$\frac{d\mathrm{E}}{dt} = \frac{-\mathrm{E} + \{\mathrm{SCN} + \mathrm{GK} * (\mathrm{EK} - \mathrm{E})\}}{\mathrm{TMEM}}$$

$$\frac{d\mathrm{TH}}{dt} = \frac{-(\mathrm{TH} - \mathrm{TH0}) + \mathrm{c} * \mathrm{E}}{\mathrm{TTH}}$$

$$\mathrm{S} = \begin{cases} 0 & \text{if} \quad \mathrm{E} < \mathrm{TH} \\ 1 & \text{if} \quad \mathrm{E} \geqslant \mathrm{TH} \end{cases} \tag{14.1}$$

$$\frac{d\mathrm{GK}}{dt} = \frac{-\mathrm{GK} + \mathrm{B} * \mathrm{S}}{\mathrm{TGK}}, \quad \mathrm{PS} = \mathrm{ES} + \mathrm{S} * (50 - \mathrm{ES})$$

As indicated in Section A, there is one main input function, SCN, and four output state variables, E, TH, GK, and S. The equation for the transmembrane potential E reflects the standard equivalent circuit model for current fluxes across neuronal membrane (see Sections C,2 and C,3 below). In PTNRN10 fluctuations in membrane potential are brought about only by changes in the applied current SCN and by increases in the neuronal potassium conductance GK, which tends to drive the potential toward the potassium equilibrium potential EK. Accommodation occurs in the model by virtue of the differential equation for TH. In this model, if C is positive, then a positive value of E brings about a positive value of the derivative of TH and therefore an increase in TH. Increases in the potassium conductance GK are brought about by nonzero values of the spiking variable S. This equation system is embodied in the listing for program PTNRN10 through an exponentially based numerical integration scheme, which is discussed in Section C,4.

To use program PTNRN10 to simulate repetitive firing in response to step current inputs, one simply changes the various parameter values as indicated by the listing. For example, to change any of the main cell parameters, C, TTH, B, TGK, THO, or TMEM, one simply changes the numerical value according to its place in the data bank shown at the end of the program. As it stands, the program is set up to simulate the case where C is 0,5, TTH is 30, B is 5, TGK is 3, TH0 is 10, and TMEM is 5. It simulates the response to a step current of amplitude SC = 5 with 0 slope starting at time SCSTRT = 1 and ending at time SCSTP = 30. The entire simulation run lasts for a time LTSTP = 60. The formats for these numbers must be consistent with those indicated by the format statements associated with the appropriate READ statements. For example, C, TTH, B, TGK, TH0, and TMEM are determined by the format 5010.

Definitions and reasonable values for the ranges of the parameters in PTNRN10 and in all programs of this book are collected in Appendix 1.

2. Active Calcium-Related Conductances in a "Two-Point" Model Neuron (PRNRN20)

The FORTRAN listing for PTNRN20, in Appendix 2, shows that this program is only slightly more complicated than that for PTNRN10. Again, the essence of this program is the do-1000 do-loop that updates the state variables, this time, however, for each of the somatic and dendritic compartments. Moreover, the fundamental state variable updating is slightly confounded by the fact that the two points in the model interchange current in proportion to their potential differences and therefore tend to oscillate unnaturally in numerical integration schemes using too large a step size. Therefore in this one program (alone of the simplified neuron models) an extra do-loop, the do-1010 loop, allows more fine-grained integration for the two state variables ESN, the somatic potential, and EDN, the dendritic potential. The other state variables are not subject to this oscillation and may therefore be updated realistically with the larger 1-msec step size, so they are outside the 1010 loop. Looking at the total do-1000 then, one initially sets the current values for both the soma, SCS, and the dendrite, SCD, appropriate at the particular time, then evaluates the total conductance in the soma, GSN, and in the dendrites, GDN, which may vary from time to time depending on the conductance state variables in these regions. Then, on the basis of these, one computes the decay factors associated with the fine-grained step size. The 1010 loop, updating the somatic and dendritic potentials, is then effected. After this the remaining state variables THS, GKS, S, GCA, CA, and GKD are updated. If desired, these instantaneous values of all state variables are printed out both on paper and on a data file. Again, the portion of the program before the do-1000 loop simply reads in various parameter values and writes them out and initializes variables. The section of the code after the 1000 loop simply writes out the time history of potentials in the soma and the dendrites and contains the formats. The data bank containing the current values of the parameters called for in the READ statements is contained at the end of the FORTRAN listing. The differential equations determining the updating of state variables used in PTNRN20 are contained in equation system 14.2.

Soma:

$$\frac{d\text{ES}}{dt} = \frac{-\text{ES} + \{\text{SCS} + \text{GDS} * (\text{ED} - \text{ES}) + \text{GKS} * (\text{EK} - \text{ES})}{\text{TS}}$$

$$\frac{d\text{THS}}{dt} = \frac{-(\text{THS} - \text{TH0}) + \text{C} * \text{ES}}{\text{TTH}}$$

$$\text{S} = \begin{cases} 0 & \text{if } \text{ES} < \text{THS} \\ 1 & \text{if } \text{ES} \geqslant \text{THS} \end{cases}$$

$$\frac{d\text{GKS}}{dt} = \frac{-\text{GKS} + \text{B} * \text{S}}{\tau\text{GK}}, \qquad \text{PS} = \text{ES} + \text{S} * (50 - \text{ES})$$

Dendrite:

$$\frac{d\text{ED}}{dt} = \frac{-\text{ED} + \{\text{SCD} + \text{GSD} * (\text{PS} - \text{ED}) + \text{GCA} * (\text{ECA} - \text{ED}) + \text{GKD} * (\text{EK} - \text{ED})\}}{\text{TD}}$$

$$\frac{d\text{GCA}}{dt} = \begin{cases} \dfrac{-\text{GCA}}{\tau\text{GC}} & \text{if} \quad \text{ED} < \text{THD} \\[2mm] \dfrac{-\text{GCA} + \text{D} * (\text{ED} - \text{THD})}{\tau\text{CA}} & \text{if} \quad \text{ED} \geqslant \text{THD} \end{cases} \qquad (14.2)$$

$$\frac{d\text{CA}}{dt} = \frac{-\text{CA} + \text{A} * \text{GCA}}{\tau\text{CA}}$$

$$\frac{d\text{GKD}}{dt} = \begin{cases} \dfrac{-\text{GKD}}{\tau\text{GKD}} & \text{if} \quad \text{CA} < \text{CA0} \\[2mm] \dfrac{-\text{GKD} + \text{BD}}{\tau\text{GKD}} & \text{if} \quad \text{CD} \geqslant \text{CA0} \end{cases}$$

The equation system for the somatic region is precisely the same as in PTNRN10, except that a term allowing for current exchange between ED and ES in proportion to the parameter GDS is included in the equation for ES. Equation system 14.2 also shows the four differential equations for the four state variables applicable to the dendritic region. The behavior of this equation system is shown qualitatively in Fig. 14.5. Here one can see more precisely what the various parameters determine. The new parameters introduced in equation system 14.2 are identified in Appendix 1, which also contains reasonable best guesses for their values and ranges. To use PTNRN20 to simulate various features of bursty activity associated with active calcium-related conductances, one may simply adjust various parameters of the system in the data bank at the end of the program. For example, to completely eliminate active calcium-related conductance effects one may set THD to a very high value, say 100 or more, or set D = 0. To vary the time scale of these effects, one may adjust the time constants, TGC, TCA, or TGKD.

3. Model for Repetitive Firing in Neurons with Compartmentalized Dendritic Trees (NEURN30)

As can be seen in the attached listing for program NEURN30, the essential feature in the programming is still the updating of state variables by the do-1000 loop over the time variable L. In this program there is no interchange of current between adjacent compartments in porportion to voltage differences; therefore, there are no artificial oscillations in potential and no finer-grained temporal integration is necessary. The situation is complicated a little, though, because there are four state variables for the soma compartment and then four state variables for each of the dendritic compartments. There are NST × NSD dendritic compartments. The logic within the do-1000 loop is as follows: first one updates the values for the somatic current SCN and the dendritic current SCDN. Then one updates the somatic potential E. Then one computes the values of the potentials in all regions in all stems as if the regions were isolated compartments. Then one averages the potentials for each region with its neighboring regions. After all the potentials have been determined, the other dendritic state variables GCA, CA, and GKD are determined. Finally, the soma potential is averaged with its neighboring regions and the other soma state variables TH, GK, and S are updated. If IFLG1 is not zero, all the state variables at this point in time are written out. Again the section of the code before the do-1000 loop merely reads and writes the various parameters and initializes variables. The code after the 1000 loop simply writes out the potentials as functions of time for any regions that have been specified and prescribes the formats. The data block corresponding to the parameters to be read in is contained at the end of the program.

The computational logic used in NEURN30 and the specific equations used in the averaging procedure to approximate propagation are contained in equation system 14.3.

$$E_i = \frac{(E_{i-1} + E_i + E_{i+1})}{3} * (1 - VRT) + VRT * E_i \quad (i = 2, NRG - 1)$$

$$E_{NRG} = \left(\frac{E_{NRG} + E_{NRG-1}}{2} \right) * (1 - VRT) + VRT * E_{NRG}$$

$$E_1 = \left(\frac{E_2 + E_1 + E_s/CDS}{2 + 1/CDS} \right) * (1 - VRT) + VRT * E_1 \quad (14.3)$$

$$ES = \left(\frac{ES + CDS * (E_1^a + E_1^b + \cdots + E_1^{NST})}{1 + CDS * NST} \right)$$

$$* (1 - VRT) + VRT * ES$$

It can be seen, for example, that the voltage retention variable VRT, when set equal to 1, implies maximum voltage retention in the sense that all the potential values remain unchanged by the averaging process. If VRT is zero, on the other hand, the influence of neighboring regions on the potential in a given region is maximum. If CDS, the conduction from dendrite to soma, is zero, the somatic potential remains an isolated variable uninfluenced by events in the dendrites. As CDS becomes increasingly large, the relative significance of dendritic events for the soma potential becomes increasingly large correspondingly. VRT and CDS are the only two new parameters introduced in NEURN30. Estimates of their range are contained in Appendix 1.

One can use program NEURN30 to simulate, for example, the ways in which activity might propagate through dendritic trees in response to various pulse or step currents applied at different places in the dendritic tree as a function of the systemic parameters of the neuron. These can be varied again by changing the appropriate numerical values in the data block at the end of the program in accordance with the READ statements at the beginning of the listing.

C. Theoretical Background

This section presents discussions of various themes and points relevant to the formulation and elaboration of the models discussed above.

1. Observations on Modeling the Single Neuron

Most central vertebrate neurons partition functionally into three regions: dendritic, somatic, and axonal. The dendritic region exhibits graded input activity, primarily postsynaptic potentials; the soma corresponds to the triggering of action potentials by graded activity; and the axonal regions correspond to the production of all-or-none action potentials. This functional partitioning follows primarily from the fact that these nerve cells are largely functionally polarized in the sense of information flow from input to output regions, that activity must funnel through the soma to get to the axons from the dendrites, and that virtually all excitatory synaptic input is effected on dendritic regions. Primitive invertebrates such as coelenterates are not functionally polarized; activity propagates in either direction along processes of such cells. In more advanced invertebrates which are functionally polarized, such as *Aplysia*, some synaptic input is effected on axonal regions and is not necessarily then funneled through a soma. In some vertebrate sensory neurons the axon arises immediately in the input end of the cell and the soma is attached off the side of the axon near its output end. Despite these exceptions, the functional partitioning described above is by far the rule in the vertebrate central nervous system.

Moreover, the functional partitioning has its counterparts in morphological and biochemical partitionings. The overall structural features of the dendritic, somatic, and axonal regions are generally clearly distinct. The axonal regions generally house active molecular channels for increased sodium and potassium conductances during action potentials as described by Hodgkin and Huxley. These channels are generally distributed over the entire axonal region of neurons (excluding, of course, their myelinated regions) up to the axon's junction with the soma. The soma and dendritic regions generally do not contain these specific molecular channels. On the other hand, it appears that at least many cells, particularly in more recently evolved regions of the central nervous system (e.g., granule and pyramidal cells in the hippocampus, pyramidal cells in the cerebral cortex, and some larger thalamic neurons), house molecular channels for gating increased conductances to calcium and potassium to produce so-called calcium spikes in dendritic regions. The triggering of action potentials in the soma occurs where the flow of current from the dendritic regions first encounters the Hodgkin-Huxley sodium-potassium channels near the initial segment of the axon.

The activity in all three of these main functional regions of neurons and indeed virtually all basic electrical phenomena in single neurons are modeled by the same "equivalent circuit model." That is, virtually all serious "bottom-up" modeling of neuroelectric phenomena which purports to approximate realistic descriptions of the actions of the actual physiological processes is based on this equivalent circuit model. This model will be described in Section C,2.

Functional neuron models are usually "broken down" or "pieced together" as follows:

(a) Typically the analysis of action potentials is separated from the analysis of integrative properties in the rest of the neuron. Many interesting questions associated with action potential propagation have been and can continue to be explored with quantitative models. Examples are the conduction speed and dependence of conduction speed on various geometric and membrane parameters, the influence of myelin on speed and reliability of propagation, and the effects of various diseases affecting myelin or other physiological or structural properties on speed and reliability of propagation. One particularly interesting question concerns the reliability with which action potentials generated in the soma of a neuron successfully penetrate to all of its synaptic output terminals in a finely branched axonal termination field. Recently, some theories of plasticity have indicated that the relative likelihood or probability of such successful conduction to a given axonal terminal depends to some extent on the recent firing history of the neuron. Thus a given axon terminal may be more or less likely to be activated for repetitive firing than for single-spike generation. Such questions can be investigated quantitatively with simulation models, the beginning outlines of which might be as indicated in the programs of Chapter 23. Aside from these questions, though, in most neuron modeling the axonal region is considered primarily as a delay line with respect to the flow of information. That is, action potentials are generated at the soma and show up a little later at the axon terminals to be projected as inputs into downstream neurons. Generally in neuron modeling the axon regions are not thought to modulate information processing to any large degree.

(b) In modeling neuronal soma regions it is generally thought that one is dealing with the ongoing generation of spike trains, each essentially by Hodgkin-Huxley mechanisms of sodium-potassium conductance increases. The main problem here is that the Hodgkin-Huxley quantitative model—their equation system—is not accurate for repetitive firing, not even in the giant axon of the squid, where their experimental work was done, let alone any other neuron. Therefore modelers generally use some other quantitative rules for generating action potentials, usually based on a threshold rule either fixed or variable. PTNRN10, for example, is essentially an extension of a model for accommodation determined by variable threshold presented by A. V. Hill in 1936. We extended that model to include adaptive and refractory events following the firing of potentials, consistent with similar models presented by Guftason and Kernell. It seems to us that PTNRN10 is a satisfactory, accurate, and flexible quantitative representation of these events at somatic regions.

(c) The simplest whole-neuron model is the so-called point neuron, wherein one represents purportedly some approximation to the overall input-output information processing dynamics of a neuron by an equivalent circuit localized at one point. In such models one is generally assuming (1) that the axon regions serve as a delay line only, (2) that dendritic regions serve only to channel input into the soma, and (3) that one can get a reasonable representation of the entire cell by the triggering processes at the soma only.

(d) More elaborate functional models include either soma plus passive dendrites or soma plus active dendrites. Rall, Jack, Redman, and others have been largely responsible for exploring properties of passive dendritic regions. Very few individuals have coupled such detailed studies of dendritic regions with spike-generating mechanisms of a somatic region. Traub recently modeled neurons that have both active calcium-related conductances in dendritic regions and somatic triggering sections.

(e) This book approaches the analysis and piecing together of single-neuron models as follows: Chapter 21 contains rather elaborate and presumably accurate models of events in dendritic trees, including both passive and active calcium-related events. Chapter 22 presents quantitative models for generation and propagation of action potentials according to the

Hodgkin-Huxley equations in isolated patches, single and bifurcating dendrites, and arbitrarily branching axonal terminal fields. Finally, Chapters 14 and 15 present three simplified but computer-efficient and relatively realistic models for functional input-output flow in point neurons with compartmentalized dendritic trees.

2. Equivalent Circuit Model for a Patch of Neuronal Membrane

This equivalent circuit model is the basis for all realistic bottom-up modeling of neuronal electrophysiology. It is important to realize that the circuit model when drawn always has reference to some particular area of neuronal surface. For example, as indicated in Fig. 14.9a, it might have reference to increments of area δ chosen over a patch of soma membrane or perhaps an annular region taken from an axon or dendritic region. It may have reference to the entire surface area of the soma, of the dendritic tree, or of the entire neuron. The accuracy of its use depends on the assumption that the potential across the circuit (or the area) is sensibly constant over the entire area being represented. Thus for most neurons it is reasonable to represent the entire soma with one equivalent circuit since the soma is generally isopotential throughout because of the large cross-sectional area for conduction inside the cell. On the other hand, in smaller-diameter processes such as dendrites and axons, where the potential can vary relatively dramatically from one longitudinal position to the next, for the circuit model to be a close approximation to the actual potentials it is necessary that the areas represented by the circuit be correspondingly small.

In terms of the "stratification-of-variable" approach discussed in Chapter 1, the circuit model purports to predict fluctuations of transmembrane potential in terms of modulations of membrane conductance. Specifically, the basic equivalent circuit model of neuron modeling is shown in Figure 14.9b. One node of the circuit, say the top node, represents the electrical potential within the cell; the opposite node, say the node at the bottom of the circuit, represents the potential outside the cell. The various branches in the circuit represent paths through the membrane by which current can be interchanged from the inside to the outside of the cell. For example, the capacitance C represents physically long, electrically polarized lipoprotein molecules in the membrane. These molecules tend to rotate in an electric field, the positive end of the molecule moving away from the positive electric pole and the negative end being attracted toward it. The motion of charge at the ends of these molecules corresponds to an electrical current, and when an applied electric field has displaced such a molecule, there has been a transfer of electrical energy into the realigned position of these molecules. This is a classic capacitive effect in electromagnetic theory, hence the representation of it by the capacitance in the circuit. The essential

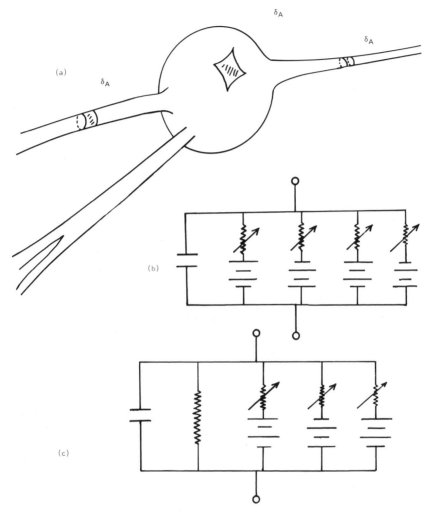

Fig. 14.9. Equivalent circuit model for neuronal signals. (a) Schematic of neuron showing area increments δA; (b) basic equivalent circuit of neuron modeling; (c) circuit of (b) as used in this work.

quantitative feature here is the current through the capacitance. The current associated with the rotating lipoprotein molecules is proportional to the time derivative of the applied electric field. All the other branches in the circuit represent paths by which a specific ionic type may move through the membrane. They are molecular channels or pores for a given ionic type. Such ionic movement is represented as being influenced by a conductance, which measures simply the permeability or ease with which such an ion moves through the membrane, and an equilibrium potential associated with each ionic species. The equilibrium potential is the result of a complex balancing of the different ionic types across the neuronal membrane and represents physically the potential toward which the membrane would go if, in the absence of other changes, the permeability to that particular ionic species were increased without bound. Thus, an increase of sodium conductivity tends to drive the potential toward the sodium equilibrium potential, an increase of potassium permeability tends to drive the potential to the potassium equilibrium potential, and so on. Typically, in the course of ongoing neuronal activity at a given membrane patch, these conductances may be modulated either by synaptic inputs or by intrinsic active changes such as the Hodgkin-Huxley sodium-potassium changes or the active calcium-related calcium-potassium conductance changes. Thus it is reasonable to replace the circuit of Fig. 14.9b by that shown in 14.9c, which simply breaks up the various ionic branches into a resting branch, which is assumed to remain unchanged during all operation, action potential-related conductances, and synaptic input conductances. The synaptic input conductance changes typically involve differential modulations of several ionic conductances. In the circuit model for a given synapse these are all combined into one conductance change, G_i, associated with a given equilibrium potential E_q. Again, it is easy to show there is no loss in generality with such a repackaging of the conductance changes. As shown below, this circuit model can be used to make precise quantitative predictions of the influence of various conductance modulations or overall current flows through the membrane on the transmembrane potential.

The limitations of the circuit model can best be studied theoretically by considering the biophysics of ionic flows and the various molecular gating mechanisms that determine them. Although such investigation is beyond the scope of this book, and indeed much of the molecular biology of membrane-ionic flow is only now being worked out, one can state that as long as (1) the events of interest are greater than about 10^{-9} sec (i.e., greater than 10^{-6} msec), (2) the chemical composition of the surrounding fluid, in particular ionic concentrations, is not changed sufficiently to dramatically alter the equilibrium potentials associated with the different ion paths , and (3) one represents the time courses and magnitudes of the hypothesized conductance changes accurately, then the circuit models of Fig. 14.9 are highly reliable models of the underlying physiological processes.

3. Basic Circuit Equations and Normalization of Terms

The quantitative usefulness of the circuit model is attained by writing an equation which expresses a current balance across the nerve cell membrane. In particular, one states simply that the total amount of current traversing this area of membrane, SC, must be equal to the sum of all the currents that go through the individual branches in the model. This is done for the circuit in Fig. 14.9b as Eq. (14.4a).

$$SC = C * A * \frac{dE}{dt} + \sum g_i * A * (E - E_i) + \sum G_i * A * (E - E_i) \quad (14.4a)$$

$$C * \frac{dE}{dt} = - \sum g_i * E + \sum g_i * E_i + \sum G_i * (E_i - E) + \frac{SC}{A} \quad (14.4b)$$

$$C * \frac{dV}{dt} = - G * V + \sum G_i * (V_i - V) + \frac{SC}{A} \quad (14.4c)$$

$$G \equiv \sum g_i, \quad E_r \equiv \frac{\sum g_i * E_i}{\sum g_i}, \quad V \equiv E - E_r, \quad V_i \equiv E_i - E_r \quad (14.4d)$$

$$\frac{dV}{dt} = \frac{- V + \sum (G_i/G) * (V_i - V) + SC/G * A}{C/G} \quad (14.4e)$$

$$\frac{dV}{dt} = \frac{- V + \sum G_i' * (V_i - V) + SC'}{\tau} \quad (14.4f)$$

$$\tau \equiv C/G, \quad G_i' \equiv G_i/G, \quad SC' \equiv SC/G * A \quad (14.4g)$$

The first term on the right of this equation represents the current through the capacitance, the sum terms with lowercase g's represent currents driven through the resting conductances, and the terms with capital G's represent currents driven through the active conductances (both synaptic and endogenous). To convert this equation to a form common in engineering systems analysis, one simply puts the derivative term on the left of the equation as shown Eq. (14.4b). This form of the equation indicates that changes in E (symbolized by the derivative of E term) are brought about by the various individual terms on the right. For example, the sum $g \times E$ term represents leakage of potential out across the membrane through resting conductances. The SC and G terms represent driving forces applied either by extrinsic experimentation or synaptic input or by endogenous membrane modulations of active conductances.

One can then take some simple mathematical steps to slightly reorganize the equation into more simplified forms. For example, one can define a resting state in which SC, the G's, and the time derivative terms are all zero. In this case Eq. (14.4b) tells us that the potential must attain the value $\sum_i g_i E_i / \sum_i g_i$. This value physically must be the resting potential, and we will denote it by E_r. One can then define a new potential, V, as the potential measured relative to the resting potential, that is, $V = E - E_r$. One can also define $G = \Sigma_i\, g_i$. The equation in these new variables reduces to Eq. (2.4c).

Finally, one can divide through by C and divide the top and bottom of the right-hand side by G. This results in Eq. (14.4e). One then notes that the ratio C/G is equal to the membrane time constant, which is represented by T. Moreover, one can define normalized stimulating currents SC' and normalized conductances G_i. Then Eq. (14.4e) is rewritten as Eq. (14.4f). Equation (14.4f) is the functional operative membrane equation that is usually used in quantitative modeling. Notice that no physiological assumptions have been made in this mathematical manipulation beyond those made in the equivalent circuit drawing. One has effectively only agreed to talk about potentials relative to resting level and to measure applied currents and active conductances relative to resting conductance. For example, in a model based on Eq. (14.4f), if one considers that a given synaptic input system effects at a given instant an active conductance change G_i equal to 2, it means that the input conductance change is 2 times the resting conductance of that patch of membrane. In terms of the circuit diagram, Eq. (14.4f) has effectively moved the equilibrium potential from the resting conductances and replaced all the resting conductances with one branch. This reduced equivalent circuit model is shown in Fig. 14.9c.

Note that the terms C, G_i, and G used here refer to membrane properties per unit area. The total capacitance for a given area of size A is $C \times A$ and the total conductance is $\Sigma_i(g_i + G_i) \times A$. The value C is about 1 microfarad per square centimeter for virtually all neurons in all animals because it represents properties of lipoprotein molecules which are rather unvarying throughout the animal kingdom. On the other hand, the typical resting conductance per unit area, G, might vary considerably from some invertebrates to mammals. This represents structural variations in nerve cell membrane, particularly with regard to the size and number of molecular channels of the various ionic types in different animal species. Invertebrates such as *Aplysia* have smaller conductances and therefore larger time constants (about 200 msec) and are therefore more sluggish in their response to changing neuronal events than, say, mammals, which have relatively larger conductances and smaller time constants (about 5 or 10 msec) and can respond a bit more quickly.

4. Comment on Simulation Methods

The simulations in this book are based on the exponential integration scheme discussed in Subsection 5. The desirable feature of this approach is that it produces relatively realistic behavior with relatively large integration step sizes and thereby achieves high computer efficiency.

The Euler method of integration avoids computing exponentials but requires smaller integration step sizes to achieve comparable accuracy. This alternative might be competitive with the approach used here under some conditions. Until shown otherwise, however, I like the more realistic projection of the exponential rule.

More accurate numerical integation schemes such as the Runge-Kutta method require smaller step sizes and exceeding more computation and produce mathematical accuracy at a price inappropriate for the modeling focuses of most of the book. Some persons might prefer these approaches for the core conductor processes of Chapters 21 and 22.

The SPICE packaged software for simulating electrical circuits is awkward in that it requires one to think explicitly in electrical circuit terms rather than physiological terms and involves considerable electrical engineering jargon. It probably is less efficient than the exponential approach used here. It may be competitive for the core conduction processes of Chapters 21 and 22.

The biggest improvement in simulation techniques will come with parallel-processing machines and languages. These are not yet generally available. When they are, the programs and methods of this book can readily be adapted to them.

5. Rationale for Temporal Integration Scheme

All temporal integration of continuous neuronal membrane-related state variables in this book is effected by the numerical integration method described in this section. This method has the advantages that (1) one can obtain relatively realistic behavior at relatively large integration step sizes (which means that one can simulate larger numbers of compartments, cells, networks, and systems for larger times with the same computer power), and (2) it can be shown to converge in the limit (as integration step size goes to zero) to the continuous mathematical solution. The method has the disadvantage that it is not particularly well adapted to computationally efficient exploration of precise functional forms over shorter time intervals. This disadvantage does not concern us unduly because the focus of interest of this book is on the overall integrative behavior of larger neuronal systems over larger time periods. Moreover, the loss in precision is relatively small: the simulated behavior patterns are realistic and in virtually all cases can be "tuned" as desired by adjusting various parameters in the programs.

All the state variable equation systems used in this book are formulated in terms of first-order equations, all of which can be cast in the generic form shown in Eq. (14.5a).

$$\frac{dE(t)}{dt} = -A(t)E(t) + B(t) \qquad (14.5a)$$

$$E_{i+\Delta} = E_i e^{-A\Delta} + \frac{B}{A}(1 - e^{-A\Delta}) \tag{14.5b}$$

$$\frac{dE}{dt} = \frac{-E + GK(EK - E) + \sum G_i(EQ_i - E) + SCN}{\tau MEM} \tag{14.5c}$$

$$E_{i+1} = E_i e^{-(1 + GK + \sum G_i)\Delta/\tau MEM} + \left(\frac{GK * EK + \sum G_i * EQ_i + SCN}{1 + GK + \sum G_i}\right)$$

$$(1 - e^{-(1 + GK + \sum G_i)\Delta/\tau MEM}) \tag{14.5d}$$

$$\frac{dE}{dt} = \frac{-E + GK(EK - E) + \sum G_i(EQ_i - E) + SCN}{\tau MEM}$$

$$\frac{dTH}{dt} = \frac{-(TH - TH0) + C * E}{TTH}$$

$$S = \begin{cases} 0 & \text{if } E < TH \\ 1 & \text{if } E \geqslant TH \end{cases} \tag{14.5e}$$

$$\frac{dGK}{dt} = \frac{-GK + B * S}{\tau GK}$$

$$GK_{i+\Delta} = GK_i e^{-\Delta/\tau GK} + B * S_i(1 - e^{-\Delta/\tau GK})$$

$$E_{i+\Delta} = E_i e^{-(1 + GK + \sum G_i)\Delta/\tau MEM} + \left(\frac{GK * EK + \sum G_i * EQ_i + SCN}{1 + GK + \sum G_i}\right)$$

$$(1 - e^{-(1 + GK + \sum G_i)\Delta/\tau MEM})$$

$$\tag{14.5f}$$

$$TH_{i+\Delta} = TH_i e^{-\Delta/TTH} + (C * E + TH0)(1 - e^{-\Delta/TTH})$$

$$S_{i+\Delta} = \begin{cases} 0 & \text{if } E_{i+\Delta} < TH_{i+\Delta} \\ 1 & \text{if } E_{i+\Delta} \geqslant TH_{i+\Delta} \end{cases}$$

This can be done without loss of generality since (1) the basic equivalent circuit equation is itself a first-order differential equation, as we saw in Section C,2, and (2) any higher-order equation can be reduced to multiple first-order equations. If the functions A and B in Eq. (14.5a) are constant over any given time interval, then the solution to that equation is as shown in Eq.

(14.5b). That is, the solution tends to decay away from its initial value, E_i and towards its target equilibrium value B/A with an effective time constant of $1/A$. In the programs of this book we use equations of the form (14.5b) as the computed solution to the continuous Eq. (14.5a). This approach may be rationalized in several ways: for example, by saying that one is examining a model system where the inputs are stepwise constant (i.e., that input synaptic conductances last one time step and are constant through such time steps) and (where state variable equations depend on other state variables) that the computed values for state variables represent appropriately computed "averages" of these variables over the integration step size. Or one might state simply that the quantitative model consists fundamentally of discrete updated equations of the form (14.5b) with no particular reference to the continuous forms (14.5a) at all; after all, most of the state variable equations are in the final analysis approximations in themselves. In most cases it is not clear that the differential forms provide better or more desirable representations than the discrete form. Both the discrete and continuous forms contain parameters which, although usually directly translatable to some physical foundation, nonetheless are usually quantified in ways to optimize the observed behavior of the model. Finally, the method using equations of the form (14.5b) does converge to the continuous solution as the step size goes to zero, and in all programs of this book the time integration step size is an adjustable parameter that can easily be varied by the user. The primary contribution of this book is the overall collection of programs and their internal structure: one can easily, in any particular program, substitute an alternative numerical integration scheme by altering a few lines of FORTRAN code. (We believe, however, that this would in most cases be undesirable; it might be appropriate for some applications of the programs in Chapters 21 and 22, which require fine-grained temporal integration because of direct interchange of current between local compartments and therefore tendencies to instabilities at larger step sizes.)

A specific example of this integration method applied to a state variable as used in the programs of this book is illustrated in Eqs. (14.5c) and (14.5d), which show the behavior of the transmembrane potential in program PTNRN11 as driven by applied stimulating current SCN, input synaptic bombardment G_i, and the refractory potassium conductance GK. The synaptic conductance and the potassium conductance are included in the A of Eq. (14.5a) and therefore determine the effective time constant of change of E. These terms also contribute to the B term of Eq. (14.5a). The assumptions implicit in this use are that the stimulating current and the synaptic conductance changes are sensibly constant over the integration step size and that the influence of GK may be approximated by using its latest updated value. The last assumption contains the view that we are considering the numerical values produced by all these state variables collectively to represent some sort of appropriate averages all of which converge in the limit to the exact values defined by the continuous equations. Finally, to fully illustrate both the power and the limitations of the method, Eqs. (14.5e) and (14.5f) contain, respectively, the continuous and discrete forms of the entire coupled equation systems used in updating state variables in PTNRN11. For example, at first glance it might seem mathematically more accurate to update TH before E at a given time step so that the value of TH at $i+$ is influenced by E_i and not E_i+. We have chosen throughout this bok to update TH at $i+$ according to E_i, which slightly augments the accommodative effect by somewhat exaggerating the instantaneous response of TH to E. On the other hand, GK should begin to rise only after an action potential is initiated. Therefore updating GK at $i+$ is made dependent on S_i and is computed prior to updating E at $i+$ so as to effect an influence of S_i on E_i. Again, this example is representative of the approach used to coupled state variables throughout the book.

6. Peculiarities of the Programs

The programs are written in common FORTRAN 77 and FORTRAN IV. They should be adaptable with a minimum of alteration to virtually any computing system that has these compilers. Nonetheless, there are a few idiosyncrasies that should be pointed out. For completeness and to facilitate identification of the data blocks referred to by the READ statements, all program listings here contain the CONTROL statements outside the program which are necessary for their successful execution on the University of Colorado (CU) Computing System. These vary from one computer system to another and will generally not be relevant to users outside CU. The form of the data block in reference to the FORMAT statements of the program will remain constant. Moreover, presumably every user will have to go through a FORTRAN compiler of some sort (flagged for us by MNF(Y) or M77) and some sort of EXECUTION statement (indicated for us by LGO. Programs such as PTNRN20, which write on external data files in the computer system at hand, will have to be modified slightly by users to adapt to their own system of using external files. In the CU system external files are identified in the PROGRAM statement. For example, in PTNRN20, NROUT is an output file and is identified with tape 7. Writing into this data file internal to the program is effected by statements of the form WRITE (7,.....). External to the program, the NRO15 after the LGO statement maps specific data file NRO15 onto the NROUT generic file in the program and the data show up externally in NRO15. Adapting these conventions to one's particular computing system can be accomplished when one knows the system with a minimum of difficulty, involving usually at most four or five lines of code.

As in all conventional FORTRAN programming, the READ statements in the program, unless they have reference to some external data file (which is generally not the case here), refer to the data blocks at the end of the program. The numerical data must be sequential, must correspond precisely in order to the variables called for in the READ statements, and must conform precisely to the formats indicated by the READ statements. There are, of course, many ways to input and output data in FORTRAN; these can be altered easily if one is capable and so desires. Our format system is generally consistent throughout the entire structure of programs and is very functional and easy to use once one becomes accustomed to it. I advise users to forget about the formats, leave them as they are, and spend their time instead on using the programs to produce simulations of interest by varying parameters and perhaps more meaty portions of code if that is appropriate.

Throughout the programs, constants are identified in the parameter statements of the program if they are necessary to dimension various arrays (e.g., see program NEURN30, where the dendritic state variables are arrays having different values for each of the NST × NRG compartments) or if they are numbers that are much less likely to be variable than some of the others—for example, the equilibrium potentials for calcium and potassium as shown in PTNRN20 or the integration step size of 1 msec. Throughout the system of programs, variables and arrays are subscripted with parameters whose ranges are set in the parameters statement of the program. This allows users to require and occupy only the amount of space in computer memory that they are actually going to be making use of. The FORTRAN coding is relatively straightforward. We have tried to be computer efficient with certain obvious approaches, such as computing exponentials related to time rates of decay—the decay constants DCTH and DGK in program PTNRN10 and others—outside do-loops where they are used wherever possible. The coding could probably be made more efficient by a computer programming expert, but as it stands it is reasonably efficient and has the advantage of being very flexible and easily modifiable by anyone conversant with fundamental FORTRAN.

One peculiarity of the programs is the writing of two or more lines of FORTRAN code on a single line of space by the expedient of initializing a new FORTRAN line with a dollar sign. Many computing systems do not allow this device. On systems where this is a not possible, every time you encounter a dollar sign simply delete the dollar sign and write the subsequent code on a new line.

Finally, most of the programs write out only the main input and output functions of the program, whereas some of them have options signaled by IFLGs to print out more detailed information—for example, instantaneous values of all state variables. Moreover, none of the program listings contains instructions for graphic display output, which are peculiar to various computer systems. It is assumed that the users of this program system will have access to the basic skill required to write or plot out on their computer systems any variables that the program has computed. Again, this kind of programming should require very few lines of code.

7. Neurophysiological Basis of PTNRN10

From the computational point of view, PTNRN10 is useful because it provides a realistic model of ongoing spike generation at trigger sections of nerve cells with large integration step sizes and therefore with computational efficiency. From the physiological point of view, PTNRN10 is useful because it provides a believable and quantitatively accurate model of the mechanisms and processes thought to be responsible for fundamental properties of repetitive firing in the triggering sections of neurons, including basic processes of accommodation and adaptation. In essence, the model represents an elaboration and combination of modeling studies presented for accommodation by A. Hill in 1936 and Bradley and Somjen in the 1960s and for adaptation by Guftason and Kernell in the 1960s. The rationale and properties of PTNRN10 were described in detail by MacGregor and Oliver in 1974. That description will be only summarized here.

The main problem of the model of Hodgkin and Huxley for spike generation was that it was not accurate for repetitive firing of multiple action potentials in any cell. Moreover, the model as described by Hodgkin and Huxley did not lend itself readily to a satisfying or comprehensive study of the range of accommodative behaviors in generating single action potentials in neurons. Building on the work of the investigators cited above, the model of PTNRN10 replaces the spike-initiating sodium conductance of the Hodgkin-Huxley model by a variable threshold and represents the accommodative behavior as determined by the two parameters C and TTH. Moreover, PTNRN10 replaces the precise description of potassium conductance changes on the downward side of the spike by triggering at each model action potential an increment in GK which subsequently decays exponentially. The main physiological features of the model include the following (1) The model behavior matches the data of Bradley and Somjin's study of intensity-latency properties of spinal motoneurons very well; in particular, the model describes the whole gamut of the behavior observed in neurons by varying the model parameters C and TTH and matches quantitatively the intensity-latency curve of any single cell. (2) The model produces phasic and tonic cells, including on and off transients, in a predictable manner by varying the appropriate parameters. (3) The model matches data on so-called adaptive behavior as studied quantitatively in steadily firing mononeurons by Guftason and Kernell. (4) The model provides a believable basis for the generation of trains of action potentials reflecting ongoing interaction of excitatory and inhibitory synaptic bombardment and current inflow at the triggering section of neurons.

The model is limited primarily in that it represents only a triggering section of a neuron. More accurate representation of neuronal events should generally take into account interactions in dendritic trees and their funneling into the triggering section. This is not an intrinsic limitation of PTNRN10, however, because that equation system can be coupled with various models for dendritic interaction, as is done for example in NEURN30, NEURN31, and in the dendritic series of Chapter 21. Some persons might say that PTNRN10 fails because, on the one hand, it is not sufficiently simple to be available to a wide number of non-mathematically oriented experimental users while, on the other hand, it ignores the precise shape of the action potential and is not very well adapted to simulating precise forms of input conductance changes and the like; that is, it is not sufficiently detailed to be satisfying to most neurophysiological users. Quite to the contrary, I believe that PTNRN10 represents precisely the balance that is needed between fastidious attention to detail and computational simplicity for effective and realistic simulation of ongoing activity in neuronal triggering sections, cells, networks, and systems.

8. Neurophysiological Basis of PTNRN20

The mechanisms, properties, and extent of dendritic calcium-related active conductances are only now being unraveled. A good review of this area is presented by Krill and Schwindt (1984). The model of PTNRN20 [Eqs. (14.2); see also Fig. 14.5] although providing what we believe to be a reasonable and believable quantitative representation of the basic phenomena involved, is nonetheless simply an early quantitative description. Many uncertainties and gaps in knowledge remain to be filled in before a satisfactory model can be completed. The model as formulated in Eqs. (14.2) and embodied in PTNRN20 has the advantage of being cast as a system of state variables each described by a first-order equation and therefore readily adaptable to the numerical integration scheme described in Section C,3. That is, one can provide a reasonable computational description of the interactions of these state variables through time with a temporal integration scheme that uses an integration step size of 1 msec and is therefore computationally quite efficient. Moreover, the large number of parameters in the model allow one to vary the influences of various features readily. In particular, within the overall structure of all the simulation programs presented in this book which make use of Eqs. (14.2) for calcium-related conductances (including PTNRN20, 21, NEURN30, 31, DENDR51, 52, and JNCT70, SYSTM70, and LNSYSTM70), slight changes in the differential equation system corresponding to the precise model used can be effected quite readily, involving very few lines of code in any one of the programs.

9. Neurophysiological Basis of NEURN30

Program NEURN30 includes the state variable model of PTNRN10 for the triggering section and the model for calcium-related conductances of PTNRN20. The discussions of Sections C,6 and C,7 are directly relevant to this program. The new ingredient in NEURN30 involves the admittedly heuristic "neighbor averaging" method of approximating propagation in dendritic trees. Although one can use numerical analysis techniques to show that a similar kind of averaging can be an exact solution to diffusion equations of the type that govern passive propagation in dendrites, when the step sizes in compartments are sufficiently small, that is not the approach or philosophy governing the structure of NEURN30. To the contrary, the idea here is that one will be able to represent large blocks of dendritic regions including bifurcations with one compartment in the model and thereby represent entire dendritic trees with relatively small numbers of compartments so that the overall computation is efficient. Nonetheless, the approach does have the attribute that the observed propagation exhibits the basic properties of those in passive dendritic trees. The model allows one to allocate different synaptic input systems to different parts of the dendritic tree and thereby realistically simulate the influence of spatial location in information processing in single neurons. Moreover, because of the inclusion of the active calcium-related conductance models in each dendritic compartment, one can simulate the propagation of calcium waves through dendritic regions. Since a given signal can propagate at most through one adjacent compartment at each integration time step, one must use increasingly small step sizes as the number of compartments is increased in a given dendritic stem in order to keep the simulation realistic. For example, if one stimulates in the fourth region of a given dendritic stem, the influence of that stimulation will begin to take effect at the soma with a latency of three integration step sizes. If the integration step size is 1 msec, this begins to push the upper limit of conduction time required in a typical dendritic stem. One can keep the integration step size as 1 msec and simulate dendritic stems having three or four regions each. Such a model gives a considerable increase in flexibility over point neurons in representing the influence of spatial location of inputs. Or one can simulate neurons with dendrites having 10 regions or more by correspondingly reducing the integration step size to ¼ msec or less. This is still a large integration step size compared to those required in many numerical integration techniques. This balance between number of regions and step size is only the upper limit on conduction speed in NEURN30. Effective conduction can be further slowed down by increasing VRT upward toward one.

To use NEURN30 with high accuracy it is recommended that one make some extraneous theoretical (or experimental) evaluation of the desired propagation speed and decays in the neuron being simulated. This can be done either with conventional cable theory or with the more elaborate dendritic tree models presented in Chapter 22. The various parameters (in particular VRT and if appropriate integration step size) can then be adjusted to make model NEURN30 match those theoretical predictions. Once this is done, one has a computationally efficient simulation model for investigating the ongoing dynamics of a cell in the context of its larger system and/or over longer time periods.

D. Exercises

1. Use PTNRN10 to obtain a plot of output firing rate versus amplitude of steady current for a nonaccommodating neuron. How do B and TGK influence this curve?

2. Adjust the accommodative parameters C and TH to produce an on-response.

3. Produce a quantitative version of Fig. 14.3a using PTNRN10.

4. Use a slope current to produce an intensity-latency curve like that shown in Fig. 14.3d from PTNRN10.

5. Stimulate PTNRN20 with a step current in the dendritic compartment to produce a bursty output from active conductances like that shown in Fig. 14.6b. Set the parameter D equal to 0 and repeat the simulation to show that the pattern now reduces to that shown in Fig. 14.6a.

6. Explore the effect of the parameters D, A, and BD on the character of bursty activity produced in PTNRN20. What is the central influence of each of these parameters?

7. How do THD and CA0 influence bursty behavior in PTNRN20?

8. With reference to your results from Exercise 6, how do the time constants TGC, TCA, and TGKD influence the dynamics of activity in PTNRN20?

9. Adjust the active calcium-related parameters in PTNRN20 to produce complex multiple spikes lasting about 7 msec with deep hyperpolarization between complex spikes lasting about 25 msec.

10. Stimulate NEURN30 with a pulse current in a peripheral dendritic region and observe the passsive propagation of activity into several other regions in the neuron, including the somatic region and regions in other dendrites. In this example set the parameter D equal to 0, thereby deactivating the calcium-related conductances. Also adjust the resting threshhold and current amplitude so that no action potentials ensue.

11. Repeat Exercise 10 but now adjust TH0 and SD so that an action potential is generated in the soma. Observe the propagation of the action potential back to the dendrites.

12. Stimulate NEURN30 with a pulse or step current in a peripheral dendritic compartment so as to produce a wave of excitation generated by the calcium-related conductances which propagates to the soma and produces a burst of spikes there.

13. Stimulate NEURN30 with a pulse current of a given amplitude in a peripheral compartment. Repeat the experiment several times with various values of VRT and CDS to explore the influence of these parameters on passive propagation in this model.

14. Add a few lines of FORTRAN code to program PTNRN10 to represent stimulation of the point neuron by a second stimulating electrode with the same descriptive parameters that PTNRN10 uses for its stimulating electrode. With your new program, attempt to produce an off-response as follows. Set C at about 0.8 and TTH at about 25. Set TH0 at 10 and TMEM at about 5. Turn on an excitatory stimulating electrode at time $t = 1$ with amplitude SC equal to 17. Maintain this stimulation indefinitely. Apply strong inhibition with amplitude SC equal to about -40 in the second stimulating electrode for the time period from 100 to 200. Do you see an off-response in the time interval right after 200? Adjust parameters to enhance this response.

15. Add a few lines of FORTRAN code to program PTNRN20 so that the stimulating electrode applied to the soma superimposes a sinusoidally varying current of variable amplitude and frequency on its step current. Use this program to produce the following experiment. First, stimulate with only a step current in the soma. Adjust the parameters to produce rhythmic complex spikes as illustrated in Fig. 14.6b. Now shut off the step current and stimulate instead with the sinusoidal current. At each of several input frequencies (covering a range to include the output frequency observed in the response to the step) determine the amplitude of sinusoidal current necessary to produce output action potentials in the neuron. Make a plot of these threshhold currents versus frequency.

16. Add a few lines of FORTRAN code in program NEURN30 so that you can stimulate with sinusoidal current in either the soma or any dendritic compartment. Study the passive propagation of the sinusoidally varying signals as a function of frequency for several different values of VRT.

15

Synaptic Bombardment in
Model Neurons

This chapter presents computer programs (see Appendix 2) to simulate synaptic bombardment of the simplified model neurons discussed in Chapter 14. The essential ingredient is the production of simulated postsynaptic conductance changes corresponding to the timing of action potentials in presynaptic input fibers. Program PTNRN11 simulates activation of a model neuron by an input fiber system with exponential distributions of input intervals. Programs PTNRN21 and NEURN31 simulate activation of model neurons wherein the occurrence or nonoccurrence of input events in single input fibers at each time step is determined by Bernoulli trials. Programs PTNRN12, PTNRN13, and PTNRN14 introduce packaged function statements which allow one to create Gaussian, gamma, or binomial distributions and which are easily generalizable for the creation of other distributions. These three programs use these distributions for the intervals between input events. Program PTNRN14 also represents the number of synaptic vesicles transferred at a single input event with a binomial distribution. Program PTNRN15 simulates short-term plasticity at input synapses. NEURN32 includes options for the Gaussian, gamma, or binomial distribution of inter-input-event intervals with easy generalizability to other distributions; the binomial model for synaptic vesicle transfer; and short-term synaptic plasticity. Programs PTNRN11, PTNRN12, PTNRN13, PTNRN14, and PTNRN15 are based on the state variable model for repetitive firing in a point neuron; PTNRN10. Program PTNRN21 is based on the two-point model neuron with active, calcium-related conductances, PTNRN20. Programs NEURN31 and NEURN32 are based on the neuron model with compartmentalized dendritic trees, NEURN30.

261

A. Using the Programs: Input–Output Charts and Illustrative Examples

1. State Variable Point Model for Repetitive Firing in Neurons (PTNRN11)

Figure 15.1 shows the input-output flowchart for PTNRN11. The variables and parameters are identical to those for PTNRN10, except that parameters determining the time course of input synaptic conductance changes G_i are now included. To specify this synaptic bombardment, one first specifies the number of synaptic types, SNTP. A single synaptic type may be thought to represent a single particular transmitter-chemoreceptor combination. For each synaptic type, then, one specifies the time constant of decay of the postsynaptic conductance change associated with that type, TG; the equilibrium potential of the synaptic type, EQ; the mean period of input firings of synapses of this type on the cell, PER; the amplitude of

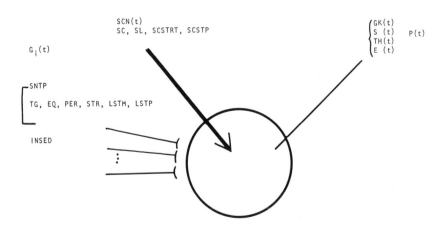

Fig. 15.1.   Program PTNRN11. This program simulates the responses of a four-state-variable model of a single point neuron to interacting synaptic input.

unitary postsynaptic conductance changes of this synaptic type, STR; the starting time LSTM and stopping time LSTP of activity; and INSED, which is a seed determining the set of choices made by a random number generator used in producing the temporal pattern for this synaptic type. The set of parameter values for each synaptic type in the program may be thought to determine the activity for the entire set of input synapses of that type on the point neuron model. These synaptic input channels act by producing a conductance change G toward a particular equilibrium potential EQ. They might also be used to represent current flowing into the point neuron (e.g., current flow into the soma from dendritic regions) by taking EQ very large relative to resting threshhold (say 500) and the corresponding conductance change very small (say 0.01) so that the effective current (which is the product of G × EQ) is the desired level of stimulating current (e.g., 0.01 × 500 = 5). From the string of parameters for a given synaptic type, program PTNRN11 produces a time series of postsynaptic conductance changes, each of amplitude STR and having an exponential distribution of interspike intervals between input events with mean period equal to PER.

Figure 15.2a shows an example of a unitary inhibitory postsynaptic potential (IPSP) (TG = 5 msec, EQ = −10 mV) in PTNRN11. Figure 15.2b shows a representative sampling of interacting excitatory postsynaptic potentials (EPSPs) and IPSPs in PTNRN11. Figure 15.2c shows the generation of a single action potential by a single EPSP preceded by an IPSP in PTNRN11. Figures 15.3a to 15.3d show examples of ongoing repetitive firing produced by PTNRN11 in response to various combinations of excitatory and inhibitory synaptic bombardment. Synaptic conductance changes of various amplitudes and rates and time constants of decay are utilized, as may be seen by observation of the graded generator potentials.

Figure 15.4 shows the production of an off-response by PTNRN11, using the mechanisms of accommodation discussed in Chapter 14 for PTNRN10. In particular, in this simulation, PTNRN11 was bombarded with ongoing excitatory input until time 150. During this time the threshhold accommodated and rose higher than at least 15 mV, as can be seen by the absence of firing during this time period. At time 150 a very strong hyperpolarizing synaptic input was applied, during which time the threshhold decreased. At cessation of the inhibition at time 300, the cell was hypersensitive to the still-maintained excitatory input and fired a burst of spikes in the time interval of 310 to 340. During this time the threshhold rose again to its accommodated state, above 15 mV, so that no more firing was observed.

PTNRN11 allows one to simulate the interaction of an arbitrary number of different synaptic types, with individually specifiable firing rates and synaptic strengths, on the state variable model discussed in Chapter 14.

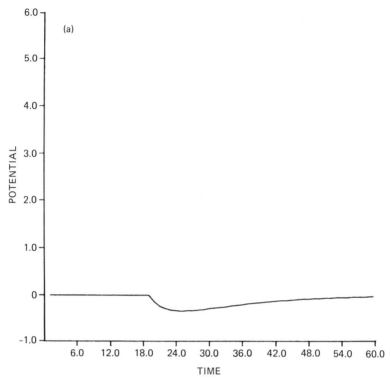

Fig. 15.2. PSP interactions in PTNRN11. (a) Unitary IPSP; (b) interacting EPSPs and IPSPs; (c) single action potential generated by a single EPSP following an IPSP.

2. Stimulating Various Interval and Amplitude Distributions (PTNRN12, PRNRN13, and PTNRN14)

PTNRN11 simulates the bombardment of a simple model neuron with the intervals between input events distributed exponentially with a mean rate which can be fixed by the user. Programs PTNRN12, PTNRN13, and PTNRN14, on the other hand, allow one to specify alternative stochastic distributions for the input. The input-output flowchart for these programs is shown in Fig. 15.5. The only differences between this input-output chart and that shown in Fig. 15.1 for PTNRN11 reside in the parameters describing the input functions. The parameters TG, EQ, PER, SDV, LSTM, LSTP, and INSED are common to all three programs. All of these parameters except SDV have been introduced for PTNRN11 in Fig. 15.1 and have the same meaning here as they did there. PER is the mean period between input events, and SDV is the standard deviation of the distribution

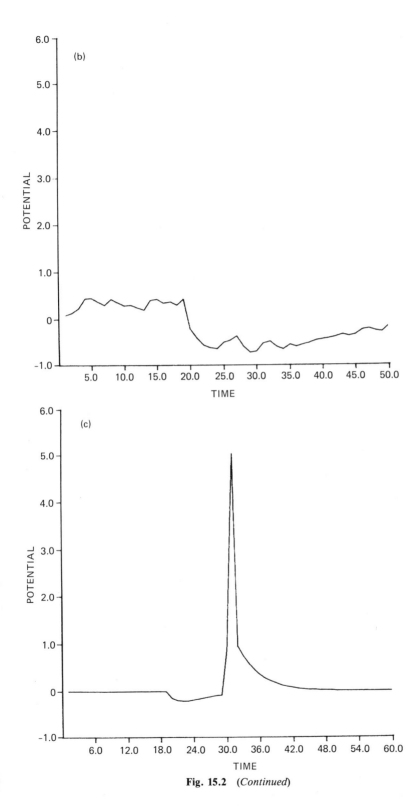

Fig. 15.2 (*Continued*)

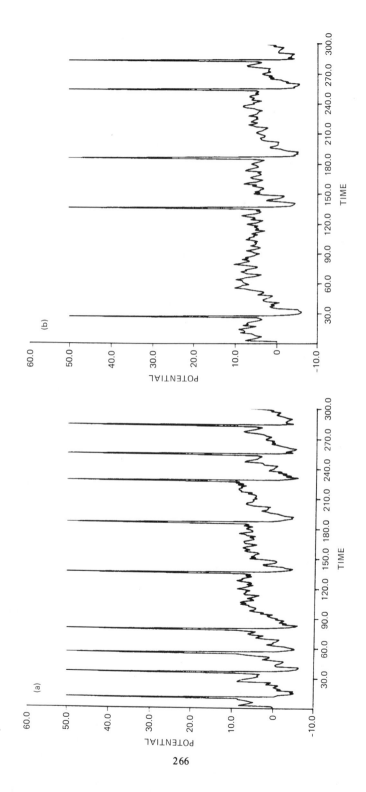

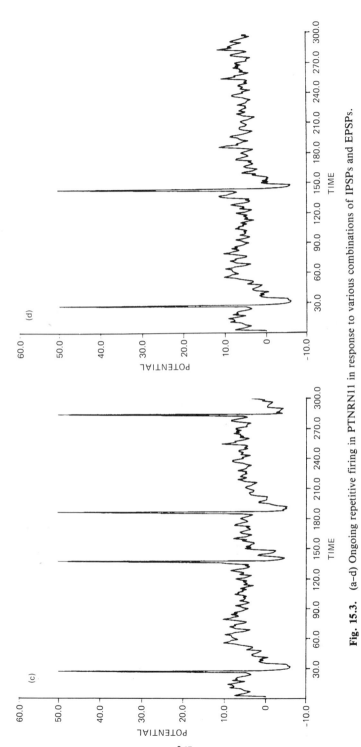

Fig. 15.3. (a–d) Ongoing repetitive firing in PTNRN11 in response to various combinations of IPSPs and EPSPs.

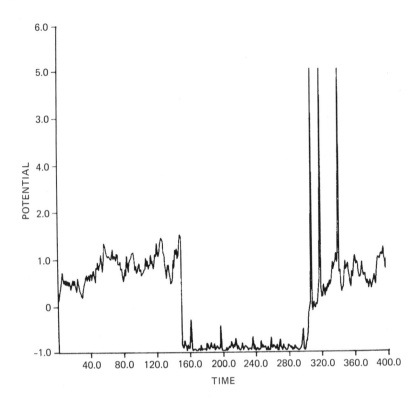

Fig. 15.4.   Postinhibitory rebound ("off response") in PTNRN11.

of inter-input-event intervals. Program PTNRN12 simulates bombardment of the state variable point neuron with an input system with a Gaussian distribution of inter-input-event intervals. To use this program, one specifies, in addition to the input parameters above, the parameter STR, which is the strength of a single input pulse, as discussed for PTNRN11. Figure 15.6a shows a representative result obtained from PTNRN12. Program PTNRN13 incorporates a binomial model for vesicle release. To operate program PTNRN14, the user must specify the parameters P and NM, which represent respectively the probability that a given vesicle will be released by a given action potential and the number of vesicles available for release. The number of vesicles actually released by the model in response to a single input event is chosen from a binomial distribution characterized by these two parameters. The resulting strength of the postsynaptic event is then equal to the number of vesicles released times the strength factor, STRF, which the user specifies in the input parameter list. Figure 15.6b shows an example of activity obtained from PTNRN13. Notice the variable heights of individual PSPs.

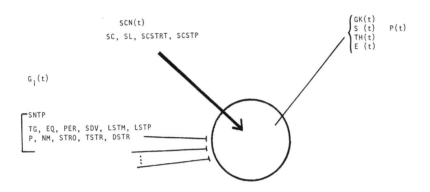

PARAMETERS

STEP, EK,
C, TTH, B, TGK, THO, TMEM
LTSTOP

INPUT FUNCTIONS   STATE AND OUTPUT FUNCTIONS

SCN(t)
SC, SL, SCSTRT, SCSTP

GK(t)
S (t)   P(t)
TH(t)
E (t)

$G_i(t)$

SNTP
TG, EQ, PER, SDV, LSTM, LSTP
P, NM, STRO, TSTR, DSTR

INSED

Fig. 15.5. Programs PTNRN12, PTNRN13, PTNRN14, and PTNRN15. These programs simulate the responses of a three-state-variable model of a single point neuron to interacting synaptic input. (These programs variously include a binomial model for vesicle transfer, short-term plasticity, Gaussian or gamma interval distributions, and a subroutine package easily generalizable to other stochastic distributions.)

Program PTNRN14 incorporates a gamma distribution for the intervals between input events. In using this program, PER and SDV are not the mean and standard deviation of the distribution, but represent rather the parameters A and B, which characterize the gamma distribution, as discussed in Section C,2. PTNRN14 also includes the binomial model for vesicle release. (Notice that one may, if desired, remove the variability from the transmitter release simply by setting both P and NM equal to 1.) Figure 15.6c shows a representative case obtained from PTNRN14 utilizing a gamma distribution of input intervals.

It should be noted at this point that program PTNRN14 is so written as to be very easily generalizable to any stochastic input distribution a user may wish to specify. This may be seen more clearly in Section B.

3. Simulating Short-Term Synaptic Plasticity (PTNRN15)

Figure 15.5 shows the input/output flowchart for PTNRN15, a version of the state variable point neuron which includes a representation of short-term synaptic plasticity. The input and output functions and parameters are

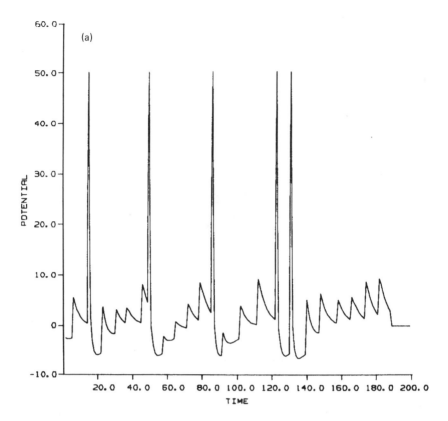

Fig. 15.6. Variable synaptic bombardment in (a) PTNRN12, (b) PTNRN13, and (c) PTNRN14.

the same as in PTNRN10 and PTNRN11, with the addition of two new parameters, TSTR and DSTR. DSTR is an increment in postsynaptic conductance amplitude experienced by the synapse every time it fires (independent of its effect on the postsynaptic cell). This number can be positive, corresponding to facilitation, or negative, corresponding to habituation. TSTR is the time constant of recovery of the synaptic strength to its initial and resting level, STR0. This model is very simple: every time the synapse fires, its strength is either increased or decreased and at any instant in time tends exponentially toward its resting value.

Figure 15.7 shows the response produced by PTNRN15 to synaptic bombardment in a single input channel which facilitates. One can see in Fig. 15.7a that when PSPs are closer together in time, the second PSP tends to be larger than when they are farther apart. This relationship is shown quantitatively in Fig. 3.7b, where the amplitude of the second PSP is plotted

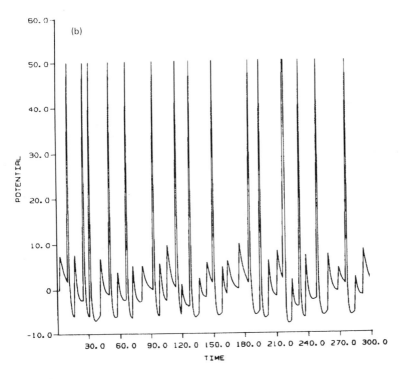

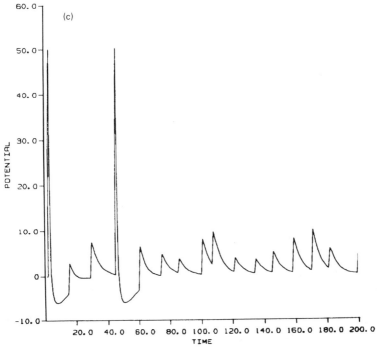

Fig. 15.6 (*Continued*)

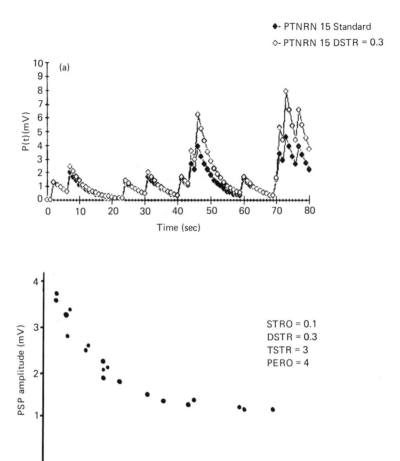

Fig. 15.7. Short-term synaptic facilitation in PTNRN15. (a) Comparison/facilitating synapse to that of nonfacilitating synapse. (b) Time course of decay of facilitation.

against the time interval between adjacent PSPs. The quantitative features of this behavior (percentage rise of PSP amplitude for short intervals and time constant of decay) can be varied simply by adjusting the parameters DSTR and TSTR. As indicated in Chapter 4, more elaborate models for short-term synaptic plasticity are being developed by McNaughton.

4. Active Calcium-Related Conductances in a "Two-Point" Model Neuron (PTNRN21)

Figure 15.8 shows the input-output flowchart for PTNRN21. This program allows one to specify parameters which determine the time course of postsynaptic conductance changes for each of an arbitrary number of synaptic types at both the soma and the dendritic points of the model. Again, the functions G represent the postsynaptic conductance changes; SNTP represents the time constant of decay of the postsynaptic conductance change and EQ represents the equilibrium potential of that synaptic type. PS represents the probability of triggering a postsynaptic conductance change at each time step, and STRS represents the amplitude of the resulting postsynaptic conductance change. PD represents the probability of triggering a postsynaptic conductance change in the dendritic compartment at each integration time step, and STRD represents the strength of the postsynaptic conductance change in the dendritic compartment. Again,

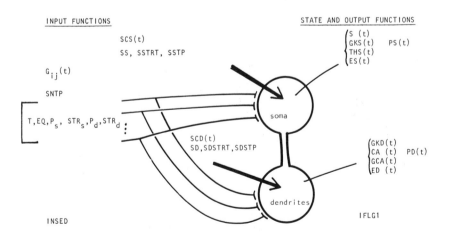

Fig. 15.8. Program PTNRN21. This program simulates the responses of a two-point model of a single neuron with active calcium-related conductances to interacting synaptic input.

INSED is a seed that determines the sequence of choices of a random number generator which determines the precise temporal pattern of firing in each synaptic input channel. Again in this model the parameters representing a single synaptic type represent the total activity in the entire set of synaptic terminals of this particular type imagined to fall on the individual compartments of the model.

PTNRN21 allows one to simulate the interaction of an arbitrary number of different synaptic input types, each of variable firing rate and synaptic strength, on the simplified point neuron model, which combines the state variable representation of repetitive firing at the soma with the active calcium-related dendritic conductances. A representative example showing production of some simulated complex spikes from calcium waves in response to interacting excitatory and inhibitory synaptic bombardment is shown in Fig. 15.9.

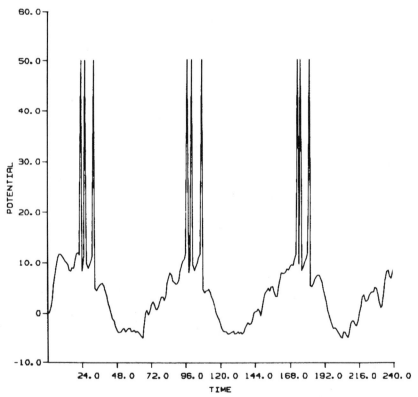

Fig. 15.9. Production of complex spikes by active conductances in response to synaptic bombardment in PTNRN21.

5. Model for Repetitive Firing in Neurons with Compartmentalized Dendritic Trees (NEURN31)

Figure 15.10 shows the input-output flowchart for NEURN31. The state variables and parameters are identical to those for NEURN30 except that one can now specify parameters producing simulated input synaptic conductance changes GS and G. GS represents the time course of post-synaptic conductance changes on the soma, and G represents the time course of postsynaptic conductance changes in the denderitic tree. Again, one first specifies the number of distinct synaptic types denoted by SNTP. Then for each synaptic type he specifies EQ, the equilibrium potential; TG, the time constant of decay of the postsynaptic conductance change; and NTG, the number of targets associated with that synaptic type. The number of targets is the number of compartments in the model neuron activated by this particular synaptic type. Then, for each of the targets, one specifies the target stem, ITS; target region, ITR; strength of the postsynaptic conduc-tance change, STR; probability P of triggering a postsynaptic conductance

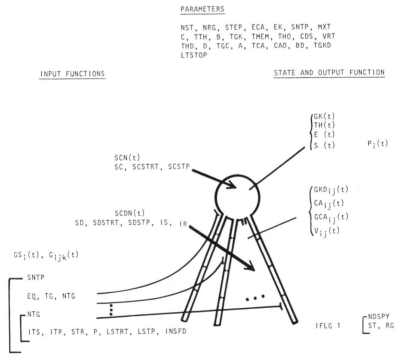

Fig. 15.10. Program NEURN31. This program simulates the responses of a model neuron with simplified dendritic tree containing active calcium-related conductances to interacting synaptic input.

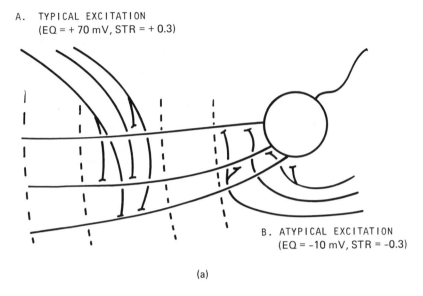

(a)

Fig. 15.11. (a) Excitation of hippocampal pyramidal cells by septal input; (b) simulation of atypical excitation.

change at this target at a given time step; starting time LSTRT and stopping time LSTP of activity at this target; and an input seed, INSED, which determines the random number choices for the particular firing pattern at this target for this synaptic type. The line of parameters for each target for each synaptic type determines postsynaptic conductance changes for the total input activity associated with all synaptic terminals of this particular type falling on a given targeted compartment.

Figure 15.11 shows representative ongoing activity in NEURN31 in response to interacting synaptic bombardment. Cole and Nicoll have theorized that input to hippocampal pyramidal cells from the septum effects an atypical excitation by decreasing conductance to potassium on dendritic stems close to the soma. Figure 15.11b illustrates a simulation of this effect by showing a control level of output activity corresponding to fixed conventional excitatory synaptic bombardment on the periphery of the tree (system A) and an increased level of output activity when the synaptic system B representing the atypical potassium conductance decrease is added in compartments contiguous to the soma. System B is turned on at time 250 in Fig. 15.11b.

6. Variable Stochastic Bombardment of Simplified Model Neuron with Dendritic Tree (NEURN32)

Program NEURN32 incorporates all the features discussed so far in this chapter. It includes the state variable model for repetitive firing at the

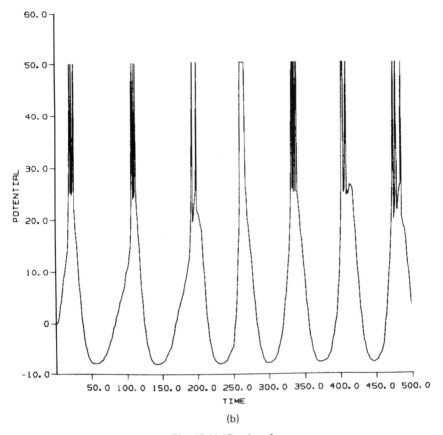

Fig. 15.11 (*Continued*)

soma, the active calcium-related conductances in each of the dendritic compartments, a binomial model for vesicle release at single input events,
variable stochastic distributions for inter-input-event intervals, and short-
term synaptic plasticity. Its input-output flowchart is shown in Fig. 15.12.
The various parameter and function labels have the interpretation discussed
in conjunction with the earlier programs of this chapter.

B. Program Listings and Discussion

1. Exponential and Binomial Distributions of Input Intervals and Simple Deterministic Patterns (PTNRN11, PTNRN21, and NEURN31)

The FORTRAN listings for programs PTNRN11, PTNRN21, and
NEURN31 in Appendix 2 show the coding by which the parameter listings

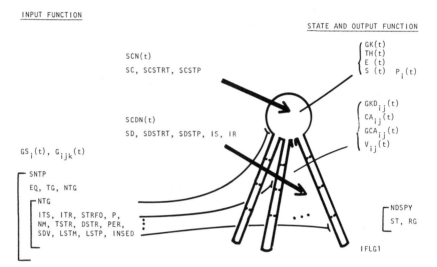

PARAMETERS

NST, NRG, STEP, ECA, EK, SNTP, MXT
C, TTH, B, TGK, TMEM, THO, CDS, VRT
THD, D TGC, A, TCA, CAO, BD, TGKD
LTSTOP

INPUT FUNCTION

STATE AND OUTPUT FUNCTION

SCN(t)

SC, SCSTRT, SCSTP

$\begin{cases} GK(t) \\ TH(t) \\ E(t) \\ S(t) \quad P_i(t) \end{cases}$

SCDN(t)

SD, SDSTRT, SDSTP, IS, IR

$\begin{cases} GKD_{ij}(t) \\ CA_{ij}(t) \\ GCA_{ij}(t) \\ V_{ij}(t) \end{cases}$

$GS_i(t), G_{ijk}(t)$

SNTP

EQ, TG, NTG

NTG

ITS, ITR, STRFO, P,
NM, TSTR, DSTR, PER,
SDV, LSTM, LSTP, INSED

NDSPY

ST, RG

IFLG1

Fig. 15.12. Program NEURN32. This program simulates the responses of a model neuron with simplified dendritic tree containing active calcium-related conductances to interacting synaptic input.

discussed in Section A produce the desired postsynaptic conductance changes in the simulated models. The overall structures of these programs are identical to those for the corresponding models PTNRN10, PTNRN20, and NEURN30 discussed in Chapter 14 except for the addition of the synaptic bombardment functions. In program PTNRN11, for example, the do-910 loop updates the postsynaptic conductance changes for each synaptic type G_i on the point neuron. At each time step, each of these conductance values decays toward zero in accordance with its time constant of decay. The variable LSTM carries, for each synaptic type, the time of the next input event to occur. When the time L is equal to this LSTM value, the postsynaptic conductance associated with that synaptic type is increased by amount STR, and LSTM for the synaptic type is updated. The particular rule for this updating used in PTNRN11 in the line just above the 910 CONTINUE statement corresponds to the rule which produces an exponential distribution of input intervals with period PER. For example, to produce pacemaker inputs of constant period, one would replace that line with the statement LSTM = L + PER. Other rules for deterministic firing patterns could be substituted as well. Usually such alterations would take only one or two lines of substitute code.

The random number system used in these listings includes three functions. The function RANF (0.) produces random numbers between 0 and 1 with a uniform distribution. The function RANSET produces, for a given setting of its argument, a particular sequence of choices made by subsequent calls to RANF. Thus by ensuring that a particular sequence of calls by RANF is preceded by the same setting of RANSET, one can repeatedly produce the same sequence of random numbers. This feature is very useful in reproducing identical input firing patterns in different simulation cases, testing possible errors by repeated runs, etc. The function RANGET resets the value of its argument so that the next time RANSET of that argument is called, the subsequent RANF calls continue sequentially from where they left off prior to the RANGET call, rather than beginning the sequence anew.

Program PTNRN21 uses a slightly different and somewhat simpler approach to produce input intervals. In PTNRN21, at each time interval for each synaptic type, a random number choice is compared against the corresponding probability of firing to determine whether firing occurs at that instant. This is essentially a series of Bernoulli trials for each synaptic type and produces a corresponding binomial interval distribution. One could easily alter PTNRN21 to give input intervals as produced in PTNRN11.

Program NEURN31 uses the same Bernoulli trial approach to generating input temporal structure as PTNRN21, as may be seen by the code in the do-700 loop. This program interprets the specifying of the target stem ITS to zero as implying that the input is to the soma. Again, it would be easy to modify the code in the 700 loop to produce alternative rules for input intervals.

Programs PTNRN11, PTNRN21, and NEURN31, then, specify exponential or Bernoulli trial stochastic input functions simply by adjusting parameter values, allow one to modify widely the temporal structure of input synaptic bombardment of these simplified model neurons by simply altering a few lines of FORTRAN code, and illustrate how one may specify temporal structure for input trains in other contexts as well.

2. Utilizing More General Stochastic Distributions: Gaussian, Gamma, and Binomial Distributions; Generalizability (PTNRN12, PTNRN13, and PTNRN14)

PTNRN12, PTNRN13, and PTNRN14 introduce and gradually expand a set of two function subprograms, STOCH and PROB, which allow one in principle to numerically create random variables with any desired density function and to easily call these to represent any variable in the program. PTNRN12 illustrates the approach with a Gaussian distribution of arbitrary mean and standard deviation for the distribution of intervals between input events. PTNRN13 extends this to represent a binomial distribution for the transfer of synaptic vesicles at one input synapse. Two parameters, P, representing the probability of successful transfer of a given vesicle, and NM, representing the number of vesicles available for release, characterize this distribution. PTNRN14 adds a gamma distribution for input intervals and illustrates thereby how the approach can be easily generalized to other distribuitions.

As may be seen in Appendix 2, PTNRN12 differs from PTNRN11 simply in the statement used to update LSTM in the line two lines above the 910 continue statement. In PTNRN11, LSTM is updated directly from the rule, representing the distribution function for the exponential distribution (see Section C,1). In PTNRN12, on the other hand, LSTM is updated according to the random variables assigned to the function STOCH. The value of STOCH represents the specific value of the interval until next firing in this input channel. This value at any particular call is determined by the two function programs STOCH and PROB shown at the bottom of the listing of PTNRN12. The function subprogram STOCH numerically integrates a given density function defined by PROB from a lower limit, X0, up to a point where the accumulated integral equals a random number, R, chosen from a uniform distribution between 0 and 1. As discussed in Section C,1 for well-behaved probability densities, PROB, this procedure repeated sufficiently often produces random variables, STOCH, with the density distribution PROB. The user must specify three parameters in order to use STOCH successfully: the lower limit of integration X0, the numerical integration step size STEP, and the accuracy criterion CRIT within which the accumulated integral must match the random number R. The limit X0 should be chosen sufficiently small or negative so that the function PROB is sensibly 0 to the left of X0. STEP and CRIT should both be chosen sufficiently small to produce whatever numerical accuracy is acceptable to the user in a given situation. Program PTNRN12 utilizes a Gaussian distribution in the subfunction PROB. The mean of this distribution is A and the standard deviation is B. PTNRN12 calls these programs only once, in the generation of inter-input-event intervals. Clearly, it is an easy matter to substitute alternative lines of code within the function PROB to define alternative probability densities. It is also easy to call STOCH at different points within the parent program for different variables. This flexibility is illustrated in program PTNRN13.

PTNRN13 makes use of two stochastic density functions, the Gaussian distribution for input intervals, as used in PTNRN12, and a binomial distribution applied here to input synaptic strength and intended to represent successful transfer of synaptic vesicles. Notice in the coding in PTNRN13 that the subfunctions STOCH and PROB have been flagged with an index K. If K is equal to 1, these programs produce a value from a binomial distribution. If K is equal to 2, they produce a value from a Gaussian distribution. In subprogram STOCH, the values X0, STEP, and CRIT must be set for each allowable value of K. For example, the values of X0, STEP, and CRIT subscripted with 1's correspond to the binomial distribution and those subscripted with 2's correspond to the Gaussian distribution. Since the binomial distribution is discrete rather than continuous, as is the Gaussian, it is necessary that STEP(1) be equal to 1. In the parent program the function STOCH is called twice, as may be seen in the 910 do-loop for the input synaptic bombardment. First, STOCH is called for the binomial distribution (using the first subscript equal to 1) in fixing the value for the postsynaptic conductance change G associated with a given input event. Second, STOCH is called for a Gaussian distribution (using first subscript equal to 2) in defining the interval until next firing in this input channel.

Program PTNRN13 illustrates how the functions STOCH and PROB may be generalized to produce arbitrary input distributions. This procedure is carried one step further in PTNRN14, which introduces a gamma distribution as a third member distribution in these subprograms. As may be seen in the listing for PTNRN14, the options K = 1 and 2 correspond again to binomial and Gaussian distributions. The option K = 3 now defines the gamma distribution. It is necessary to add the three integration parameters X0, STEP, and CRIT for the new function. The functions STOCH and PROB in PTNRN14 are modified slightly from their form in PTNRN12 and PTNRN13 in that numerical evaluation of some of the constants associated with the individual probability density functions, PROB, are performed within STOCH prior to the numerical integration of the density functions. This simply saves the computer from recalculating these quantities at every step of the integration. PTNRN14 calls STOCH for a binomial distribution in determining postsynaptic conductance G for a given input event, as may be seen in the 910 loop. It also calls STOCH for a gamma distribution of inter-input-event intervals in updating LSTM, as may also be seen in the 910 loop. PTNRN14 does not make use of the Gaussian distribution anywhere in the parent program. One could, however, modify any variable in PTNRN14 to be a Gaussian-distributed random variable simply by calling STOCH with first index 2 in the appropriate place in the parent program.

STOCH and PROB can be generalized readily to represent other probability density functions. It is necessary only that the prescribed density functions be normalized (that is, that their total area, when integrated, is equal to 1) and that the three parameters X0, STEP, and CRIT be defined for each new function to ensure acceptable accuracy of numerical integration. For example, the value of STEP must be 1 for all discrete density functions. The mathematical equations for the probability densities used in these programs are discussed in Sections C,1 and C,2.

3. Simulating Short-Term Synaptic Plasticity (PTNRN15)

Program PTNRN15 includes a simple model of short-term synaptic plasticity at individual input synapses. To utilize this model, the user specifies three parameters: STR0, which represents the baseline resting level of synaptic strength for that particular input channel; DSTR, which represents the increment of synaptic strength, either positive or negative, experienced by the synapse whenever it fires; and TSTR, which represents the time constant with which the synaptic strength returns to its base or resting level from any deviant value. The main ingredients in this model are utilized within the do-910 loop in PTNRN15. As may be seen in the fourth line above the 910 continue statement, the main ingredient is that when a synapse fires, its strength applies not to the postsynaptic effect from this current firing—this has been determined by the value STR—but rather to future events at this input channel. As may be seen in the line of code immediately below the do-910 statement, at each integration time step STR is updated toward STR0 from its current value STR. The incremental term DS represents an exponential decay with time constant TSTR, as may be seen in its definition in the do-310 loop in this section for initializing variables. With this simple model, then, one may simulate individual synapses which facilitate (DSTR positive) or depress (DSTR negative) with repetitive firing and which return to normal states with a given adjustable time constant.

4. Stochastic Synaptic Bombardment of a Model Neuron with Compartmentalized Dendritic Tree (NEURN32)

Program NEURN32 incorporates the options for utilizing random variables of various density functions as determined by the subfunctions STOCH and PROB and for simulating short-term plasticity, which have been discussed for programs PTNRN12 through PTNRN15 within the general program for simulating a neuron with compartmentalized dendritic trees, presented above as NEURN30 and NEURN31. Program NEURN32 is the most general simplified single-neuron model in this book. It can produce all features simply by prescribing input parameters appropriately or by simple modifications of existing code. The other programs, on the other hand, are often more efficient in particular specific applications and illustrate certain programming approaches more clearly.

C. Theoretical Background

1. Generating Various Stochastic Distributions from the Uniformly Distributed Random Variable, RANF(0.)

Two related functions, the probability density function f and the probability distribution function F, quantitatively define the characteristics of a random variable X^*, as shown in Fig. 15.13. The probability density function gives the probability that X^* will be within a region dx centered on x as equal to $f(x)\,dx$. The probability distribution function gives the probability that X^* will be less than or equal to x as $F(x)$. The distribution function at x is equal to the integral of the density function from ∞ to x. To be well behaved, F must go monotonically from 0 to 1; correspondingly, the total area under f must be equal to 1. The illustration of these concepts in Fig. 15.13 is for continuous functions, but the statements are true for discrete functions also.

The important rule in the present context is that if one chooses X^* repeatedly, according to Eq. (15.1) and as indicated in Fig. 15.13, where F is the distribution for a particular probabiliby density and R is a random

$$x^* = F^{-1}(R) \tag{15.1}$$

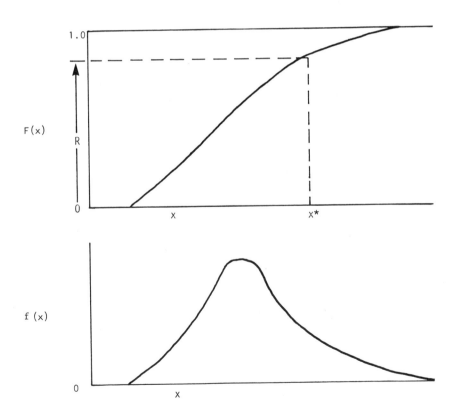

Fig. 15.13. Continuous probability density and distribution functions.

variable uniformly distributed in 0 and 1, then X^* will have a density function f corresponding to the distribution function F used in Eq. (15.1). This rule is illustrated in Fig. 15.13 by the couplet R and X^*. Most probability distribution functions are such that it is not possible to obtain closed functional forms for their inverses, as required by Eq. (15.1).

In these cases, nonetheless, the rule can be approximated by numerical integration. This is exactly what the subfunction STOCH does in the programs of this chapter. One example where Eq. (15.1) can be inverted in closed form is that corresponding to the exponential distribution. Equations (15.2)

$$f(x) = \frac{1}{\beta} e^{-x/\beta}, \qquad x > 0 \tag{15.2a}$$

$$F(x) = \int_0^x f(y)\, dy = 1\, e^{-x/\beta} \tag{15.2b}$$

$$x^* = -\beta \ln(1 - R) \tag{15.2c}$$

show the probability density function and the corresponding distribution function for the exponential distribution. To apply rule (15.1) to this distribution, one simply sets $F(X^*)$ equal to R, where R is a random variable uniformly distributed between 0 and 1, and solves for X^*. The corresponding solution is shown in Eq. (15.2c). Since the distribution of $1 - R$ is the same as the distribution of R for a uniformly distributed variable between 0 and 1, Eq. (15.2c) provides the justification for the closed form used in updating LSTM in the 910 loop in program PRNRN11.

2.   The Exponential, Gaussian, and Binomial Distributions

The exponential density function and its corresponding mean μ and standard deviation σ are given by

$$f(x) = (1/\beta)\, e^{-x/\beta}, \qquad x > 0$$
$$\mu = \beta \tag{15.3}$$
$$\sigma = \beta$$
$$x^* \approx -\beta \ln(R)$$

The probability density function for a Gaussian distributon with mean μ and standard deviation σ is given by

$$f(x) = \frac{1}{\sqrt{2\pi}\, \sigma} \exp -\frac{1}{2} \left(\frac{x - \mu}{\sigma} \right)^2, \qquad -\infty < x < \infty \tag{15.4a}$$

$$x^* \approx \sqrt{-2\ln(R_1)}\, \cos(2\pi R_2) * \sigma + \mu \tag{15.4b}$$

An approximation sometimes used to generate quasi-Gaussian distributions is given by Eq. (15.4b). where R_1 and R_2 are two independently chosen, uniformly distributed random variables.

The probability density function for the gamma distribution in terms of the two parameters α and β, along with its mean and standard deviation, μ and σ, respectively, is given in by

$$f(x) = \frac{1}{\beta^{\alpha}\Gamma(\alpha)}x^{\alpha-1}e^{-x/\beta}, \qquad x > 0$$

$$\Gamma(\alpha) = \int_0^{\infty} y^{\alpha-1}e^{-1}\,dy \tag{15.5}$$

$$\mu = \alpha\beta, \qquad \sigma = \sqrt{\alpha}\,\beta$$

The intervening function, Γ, is the so-called gamma function.

The binomial probability density function for n trials, each with probability of success P, and the corresponding mean μ and standard deviation α are given by

$$f(x) = \frac{x!}{n!(n-x)!}P^x(1-P)^{n-x}, \qquad x = 0, 1, 2, \ldots, n \tag{15.6}$$

$$\mu = np, \qquad \sigma = \sqrt{np(1-P)}$$

This function allows only discrete values of x.

These and other related probability density functions are discussed in more detail in a number of introductory books on probability and statistics.

3.   The Binomial Model for Vesicle Transfer

The basic concept of the binomial model for successful transfer of synaptic vesicles is illustrated in Fig. 15.14. The central idea is that a certain number (n) of synaptic vesicles are available for possible release at a particular point in time. When a given action potential, indicated in Fig. 15.14 by AP, arrives at the presynaptic terminal, each of these vesicles has a certain probability, p, of being successfully released. It is further supposed that the release of any one vesicle is independent of the release of any of the others. This situation conforms to the classical Bernoulli trial situation in probability theory and gives rise to the binomial distribution described quantitatively by Eq. (15.6).

Since the recognition in the early 1960s that synaptic transmission, at least in spinal cord motoneurons, is a stochastic phenomenon and the

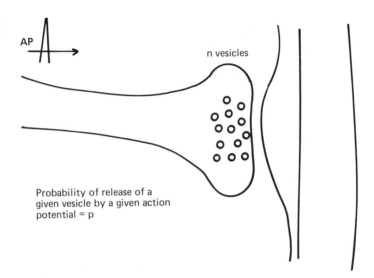

Fig. 15.14.   Binomial model of vesicle release.

application of the binomial model to that transfer, it has been generally pictured that any one single presynaptic bouton could release a variable number of individual vesicles. However, it has been argued by Donald Faber and co-workers, on the basis of studies of brain stem neurons in goldfish, that individual boutons may release at most one vesicle at a given firing and that the binomial model describes, for a single sending neuron, the number of individual boutons that will be activated by a given action potential rather than the number of vesicles that will be required to clarify the extent of validity of the suggestion and its implications for dynamic organization in neuronal networks.

4. Current Experimental Studies of Short-Term Plasticity

Mendel has described short-term synaptic facilitation and depression in spinal motoneurons which look, at first glance at least, like those produced by the simple model for short-term plasticity included in the programs of this chapter. McNaughton discovered that short-term plasticity in granule cells of the hippocampus is considerably more complicated.

D. Exercises

1.   Stimulate PTNRN11 with a single stream of incoming excitatory PSPs at a given mean period PER. Observe the influence of changing PER and the synaptic strength STR on the resulting generator potential.

2. Simulate the response in PTNRN11 to the interaction of a single excitatory input stream and a single inhibitory input stream. Observe the influence of varying the periods and amplitudes of the incoming PSPs.

3. In PTNRN11 take LSTOP equal to 300, EQ equal to 70, and STR equal to 0.3. Take TH0 equal to 10. Make a plot of output firing rate versus input period PER. Repeat this experiment with program PTNRN12, which utilizes a Gaussian distribution for input intervals. Obtain several such curves for PTNRN12, each for a different value of the standard deviation SDV. Finally, repeat this experiment with program PTNRN14, utilizing a gamma distribution of input intervals.

4. Using simulation runs of about 300 msec duration in program PTNRN14, produce curves of output firing rate versus input firing rate for a gamma distribution of input PSPs of constant amplitude. Repeat this experiment, now using a binomial distribution of PSP amplitude wherein the mean value of amplitude is equal to that used in your first case. Repeat the second experiment for several values of NM, the number of vesicles available for release.

5. Use PTNRN15 with a gamma distribution having a relatively large number of small intervals. Prescribe also for the input channel to exhibit short-term potentiation. For a simulation run of about 500 msec or more, produce a curve of output firing rate versus input firing rate. Reproduce this curve for several values of potentiation DSTR down to and including the case of no potentiation, DSTR equal to zero. Do you think that this mechanism can substantially influence the output activity of the receiving cell in ongoing repetitive input firing?

6. Stimulate program PTNRN21 with a single excitatory synaptic stream applied to the dendritic compartment. Adjust the active calcium-related parameters so that this input stream produces from time to time waves of excitation which propagate to the soma and produce complex spikes there.

7. To your stimulation from Exercise 6, add a stream of synaptic inhibition in the soma compartment. Explore the effectiveness of this synaptic inhibition in countering the active calcium-related waves.

8. Bombard program NEURN31 with excitatory synaptic PSPs over the peripheral regions of its dendritic tree. Adjust parameters so that output firing occurs from time to time but at a relatively low rate. Now add an unconventional synaptic excitation with EQ equal to −10 and STR equal to −0.25 to activate only those compartments of the dendritic stems adjacent to the soma. (This simulates the hypothesized action of the septal inputs on parametal cells in the hippocampus.) By activating the second unconventional excitatory system you should be able to produce increased rates of output activity and increased bursting within individual complex spikes. Can you do this?

9. Change one line of code in program PTNRN11 so as to produce deterministic pacemaker input activity.

10. Change one line of code in PTNRN11 so as to produce stochastic input activity varying in a sinusoidal manner with period PER.

11.   Add the functions **STOCH** and **PROB** to program **PTNRN15**. Use these to produce short-term plasticity wherein the increment of PSP amplitude effected at any input firing is a random variable with a gamma distribution.

12.   Stimulate program NEURN32 with excitatory synaptic bombardment in a single input channel distributed over the peripheral regions of its dendritic tree. With no short-term plasticity, adjust parameters so that active calcium conductances are activated from time to time and produce output complex spikes of the soma at a relatively low rate. Now adjust the input channel so that a significant degree of short-term synaptic potentiation occurs at the input synapses. Does this significantly alter the ongoing dynamic response of the neuron?

16

Small Pools of Neurons

This chapter addresses the problem of simulating small homogeneous populations of neurons. The primary purpose of the chapter is to introduce some simple programming techniques for dealing with populations of cells which are used throughout the remainder of the book. (These include nested do-loops for updating the state variables of multiple interrelated neurons; a subroutine, SVUPDT, for updating the state variables of individual cells; an "activity in transit" matrix and corresponding subroutine, TRNSMT, for projecting output synaptic activity from the soma of one cell to the receiving portions of other cells.) The chapter also discusses programs located in Appendix 2 and programming techniques to begin investigations of the influence of extracellular field coupling on the coordinated activity of local neuronal populations and two illustrative examples of the influence of such coupling.

A. Using the Programs: Input–Output Charts and Illustrative Examples

Figure 16.1 shows the input-output flowchart for program POOL10. POOL10 simulates the activity of three pacemaker neurons which may interact synaptically with unit time delay. As indicated in Fig. 16.1, the main input parameters for this model are the values of A and E_0 for each of the three cells. E_0 represents the initial membrane potential for each cell and A represents the intensity of endogenously generated current in the cell, which in turn determines the cell's spontaneous pacemaker firing rate. Each neuron is represented by the state variable model neuron PRNRN10 of Chapter 14. The corresponding state variables E, TH, S, and GK are then

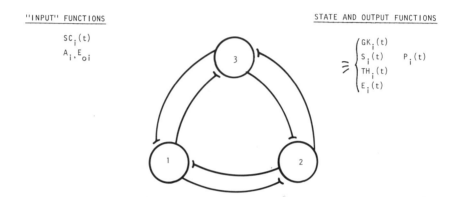

PARAMETERS

STEP

C, TTH, B, TGK, THO, TMEM

A12, A23, A31, A13, A32, A21

LTSTOP

"INPUT" FUNCTIONS

$SC_i(t)$

A_i, E_{oi}

STATE AND OUTPUT FUNCTIONS

$\cong \begin{cases} GK_i(t) \\ S_i(t) \quad P_i(t) \\ TH_i(t) \\ E_i(t) \end{cases}$

Fig. 16.1. Program POOL10. This program simulates the behavior of three pacemaker neurons which interact synaptically with unit time delay.

produced for each of the three neurons, as is the simulated intracellular recording P. The six new parameters introduced in this program are the coupling coefficients A12, A23, A31, A13, A32, and A21, which represent the strength of the synaptic connections between the cells. For example A12 is the strength of the synaptic connection from cell 1 to cell 2. The other parameters are as defined in Chapter 14. With POOL10, then, one can simulate the influence of simple synaptic coupling on the coordination of activity of three spontaneously active pacemaker neurons of various frequencies and phasings.

With program POOL11 as illustrated in Fig. 16.2, one can simulate the influence of synaptic coupling on the coordination of an arbitrary number of spontaneously active pacemakers of various frequencies and phases. Moreover, POOL11 allows one to specify individual and variable conduction times for the different synaptic interconnections. Again, as may be seen in Fig. 16.2, one specifies the parameters A and E_0, which determine the firing rates and phasings of the individual pacemakers. The main output functions are the four state variables and the simulated intracellular recordings P. In this program one first specifies for each of the cells the number of targets (NT) and then for each of these targets the identity of the target (IDT), the conduction time of that synaptic connection (NCT), and the strength of that synaptic connection (STR). The parameter MCPT1 is the maximum conduction time plus one. The

PARAMETERS

MCTP1, NTMX, STEP
C, TTH, B, TGK, THO, TMEM

```
┌  NCLS
│  ┌ NT
│  │ ┌ IDT, NCT, STR
│  │ │
└  └ └
```

"INPUT" FUNCTIONS   STATE AND OUTPUT FUNCTIONS

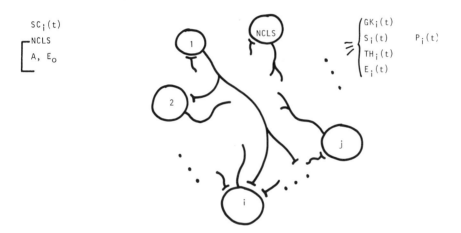

$SC_i(t)$

```
┌ NCLS
│
│  A, E_o
└
```

$$\approx \begin{cases} GK_i(t) \\ S_i(t) \qquad P_i(t) \\ TH_i(t) \\ E_i(t) \end{cases}$$

Fig. 16.2.   Program POOL11. This program simulates a pool of pacemaker neurons which interact synaptically.

parameter NTMX is the maximum number of targets used by any of the cells in the population.

Program POOL12, whose input-output chart is shown in Fig. 16.3, extends program POOL11 to include the possibility of extracellular field coupling between specified members of the neuronal pool. In particular, as seen by the flowchart in Fig. 16.3, the input and output functions and parameters for POOL12 are identical to those for POOL11, except that one reads in for each cell first the number of cells coupled with (NC) and then for each cell coupled with the identity of the coupled cell (IDC) and the amplitude of the coupling (AC). One also specifies the new parameter NCMX, which is the maximum number of cells coupled to by any given member of the pool. With program POOL12, then, an arbitrary number of pacemaker neurons of arbitrary

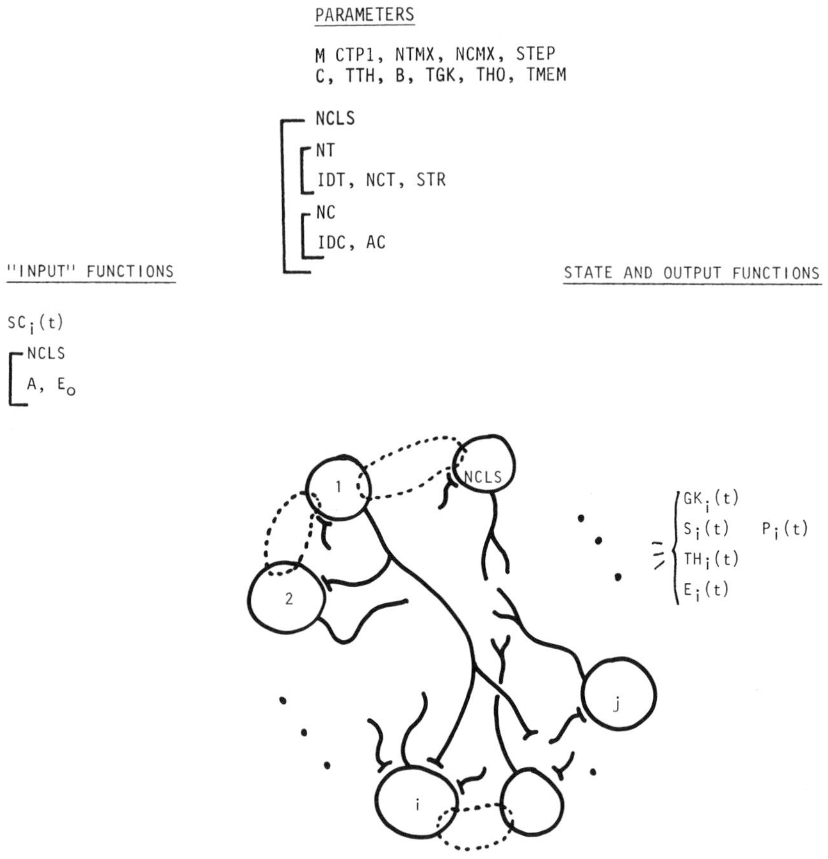

PARAMETERS

M CTP1, NTMX, NCMX, STEP
C, TTH, B, TGK, THO, TMEM

 — NCLS
  ┌ NT
  └ IDT, NCT, STR
  ┌ NC
  └ IDC, AC

"INPUT" FUNCTIONS   STATE AND OUTPUT FUNCTIONS

$SC_i(t)$
┌ NCLS
└ A, E_o

$\left\{\begin{array}{l} GK_i(t) \\ S_i(t) \quad P_i(t) \\ TH_i(t) \\ E_i(t) \end{array}\right.$

Fig. 16.3. Program POOL12. This program simulates a pool of pacemaker neurons which interact synaptically and/or via extracellular field coupling.

frequency and phasing can be simulated. One can explore the influence of both synaptic coupling of various structures, time delays, and strengths and the influence of extracellular field coupling of various structures and strengths, on the coordination of these spontaneous rhythms.

Program POOL20, whose input/output chart is shown in Fig. 16.4, is designed for a special purpose and is somewhat less flexible than POOL12 or POOL11. However, it can easily be used to simulate much larger numbers of cells than either POOL11 or POOL12. Specifically, POOL20 simulates the coordination and possible synchronization of spontaneous pacemaker neurons arranged in a two-dimensional array by the mechanism of extracellur field coupling of each cell with its four nearest neighbors. Again, the main input functions are the pacemaker-determining current A and the initial phasing E_0.

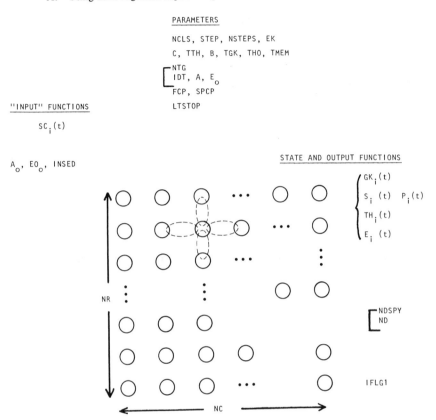

PARAMETERS

NCLS, STEP, NSTEPS, EK

C, TTH, B, TGK, THO, TMEM

$$\left[\begin{array}{l} \text{NTG} \\ \text{IDT, A, E}_0 \\ \text{FCP, SPCP} \end{array}\right.$$

"INPUT" FUNCTIONS   LTSTOP

$SC_i(t)$

A_0, EO_0, INSED

STATE AND OUTPUT FUNCTIONS

$$\left\{\begin{array}{l} GK_i(t) \\ S_i(t) \quad P_i(t) \\ TH_i(t) \\ E_i(t) \end{array}\right.$$

NR

$$\left[\begin{array}{l} \text{NDSPY} \\ \text{ND} \end{array}\right.$$

IFLG1

NC

Fig. 16.4. Program POOL20. This program simulates a two-dimensional pool of neurons which interact by extracellular field coupling.

In this program one specifies mean values for these parameters and the program specifies specific values for individual cells from exponential distributions around these means. Moreover, one specifies an input seed INSED for the random number sequences which determine these choices. Again, the main output functions are the four state variables for each cell and the simulated intracellular recording P. In this program the user may specify the number of cells for which an explicit output function is desired, NDSPY, and then specify this many names of displayed cells ND. IFLG1 is a flag which determines whether or not a two-dimensional spatial display of the activity at each time unit will be produced by the program. NCLS is the number of cells in the population, and NSTEPS is a parameter which determines the number of integration steps per step. This is discussed further in conjunction with the listings below. The parameter NTG is the number of cells in the population for which one specifies individual pacemaker-generating currents A and initial phasing E_0. For each such target one specifies first its name IDT and then the pacemaker current A and the initial value E_0. The parameter FCP is a coupling coefficient which determines the magnitude of the field coupling of each cell with its four nearest neighbors. The parameter SPCP is a spike-coupling field coefficient which determines the amplitude of coupling of the spikes in a given cell on its four nearest neighbors. With program POOL20, one may investigate the influence of various degrees of field coupling between adjacent cells on the coordination of otherwise independent pacemakers. In particular, one may investigate how these various parameters might influence the propensity to develop coordinated waves of synchronization passing through the population. One may try to trigger such waves by initiating coordinated activity in some particular locale through the input target cells represented by the ITD.

EXAMPLE

Sychronization of pacemaker by field coupling.   Figure 16.5 shows an example taken from program POOL12 where three independent pacemaker neurons are brought into synchronized firing by extracellular field coupling

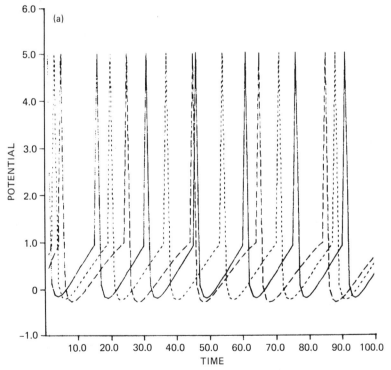

Fig. 16.5.   Entrainment of independent pacemakers by extracellular field coupling. (a) Cells firing independently; (b) sychronized firing with coupling of membrane potentials; (c) expanded view of sychronized firing.

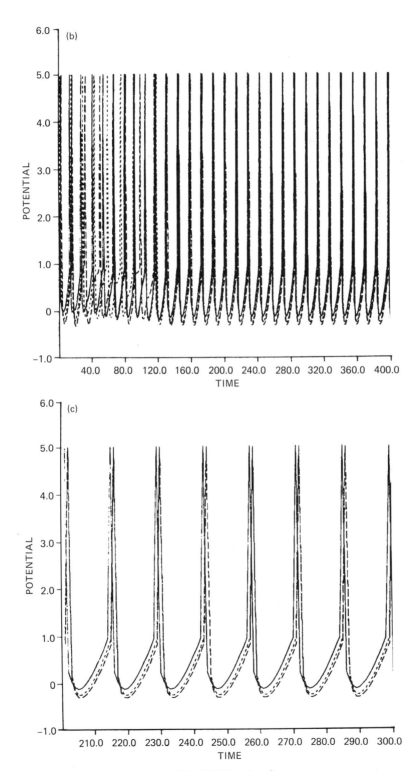

Fig. 16.5 (*Continued*)

among their membrane potentials. Figure 16.5a shows the cells firing independently at different rates (periods of 15, 17.5, and 20 msec, respectively) in the absence of any coupling. Figure 16.5b shows how the same cells come into synchronized firing over a period of about 150 msec when coupling of their membrane potentials is introduced. Figure 16.5c shows an expanded view of this coupling over the time interval from 200 to 300 msec.

B. Program Listings and Discussion (POOL10, POOL11, POOL12, POOL20)

The listing for program POOL10 shows that it has the same essential structure as the programs for the simple neuron models discussed in Chapter 14. In particular, there are sections that read and write the input parameters, sections that initialize variables, a do-1000 loop over time, a section that writes out activity variables, and a section with formats. The essential new addition in this program is the nested do-1500 loop within the do-1000 time loop. In the 1500 loop are the rules which determine the state variables for the individual cells. These rules are identical to those used in the state variable model in Chapter 14, but now they are subscripted with the index I, which refers to different particular cells. This program is quite limited in that it is explicitly restricted to three neurons. The synaptic influence of one cell on another is effected through the input current function SC for the receiving cell. This influence is triggered by the value of the output spiking variable S from the sending cells at the value the spiking variable has from the last time unit. Thus the program is further limited in that all conduction times are simply the step size of the integration scheme nominally taken as 1 msec. Moreover the so-called synaptic effects are mediated simply by a stimulating current so that possible nonlinear mechanisms associated with synaptic conductance are not included. This last limitation is easily corrected and indeed is corrected in most of the programs in later sections of the book.

The listing for POOL11, in Appendix 2, shows that this program also has the same basic structure as that for the single neurons discussed in Chapter 14. There are sections for reading and writing the input parameters, initializing variables, a do-1000 loop over time for updating state variables, a section for writing out activity variables, and a section for the formats. Again as in POOL10 there is a nested do-1500 loop embedded within the do-1000 time loop to update variables for individual cells. In this program, however, the state variables are not explicitly written within the 1500 do-loop. Rather there is a call SVUPDT statement which calls a subroutine to update the state variables for the individual cells. This subroutine in turn contains the basic rules for updating state variables. Again notice that in the 1500 loop if a cell has fired there is a call to another subroutine, TRANSMIT. If the cell has fired it is instructed by the subroutine to transmit synaptic activity to all the target cells to which it projects. The matrix SCIT is an "activity-in-transit" matrix. The first of the two subscripts of this matrix identifies the cell toward which activity is in transit. The second subscript signals the time at which that activity is due to arrive at the target cell. When the second subscript has the value one, the activity is due to arrive at this iteration. In the last subsection in the do-1000 loop the activity in transit matrix SCIT is readjusted by sliding all entries in this matrix one step to the left in time. Thus, if a synaptic input from cell J is due to arrive at cell I in, say, three time units from now, then when cell J fires it puts that increment of activity into the (J, 4) slot of the SCIT matrix. Three time units from now the do-1000 loop will have moved that bit of information over to the SIT of the (J, 1) slot and it will be used then in activating cell J. With this activity-in-transit matrix one can keep track of all the activity due to arrive at a given cell I at the various times from now. It is limited by the fact that all the activity in one slot in the matrix is assumed to be of the same synaptic type. For example, if one sending cell has a synaptic equilibrium potential to a given receiver different from that of another sending cell to that receiver, those pieces of information must be put into different activity-in-transit matrices. Such situations are considered in later chapters. In program POOL11 all synaptic connections are assumed to be effected simply by modifying the total input current SCIT into the given receiving cells. There can be positive or negative increments to this current representing approximations to excitatory and inhibitory effects. Again notice that placing the rules for updating state variables into a subroutine has the advantage that if one wants to modify this program to represent slightly different neurons—for example, simple neurons with dendritic trees as represented by program NEURN30—one simply replaces the rules in the subroutine with the rules for the new model neuron. In such a case, one must also make sure that the parameters transferred in the common statements are now representative of the new model neuron and that the parent program reads in the parameters appropriate to the new model neuron. Otherwise the parent program is left unchanged. In Chapter 23 we will return to this capability of expanding virtually any of the programs in this book to include more complicated models for single neurons. In the meantime, for clarity, all programs involving pools, networks, and systems used in Chapters 16 through 21 will make use of only the state variable model for single neurons discussed above as PTNRN10.

The listing for POOL12, in Appendix 2, is essentially identical to that for POOL11 with the exception that the synaptic input current for individual cells now includes components proportional to the membrane potential in cells with which the given cell is specified to be coupled. This is effected in program POOL12 the do-1200 loop.

Program POOL20, as may be seen from its listing in Appendix 2 is a little more specialized than POOL11 and POOL12. It also has basic sections to read and write input parameters, to initialize variables, a do-1000 loop over time, a section to write out activity variables, and a section to specify the formats. The intricacies of state variable updating are somewhat different in this program than in previous ones, however. In the first place, POOL20 does not have subroutines for state variable updating and for synaptic transmission. Moreover, it splits out the updating of the cell potentials, represented here by both E and V, from the updating of the other state variables TH, GK, and S. The reason for this is that numerical integration of a coupled membrane potential sometimes tends to become unstable at the larger integration step sizes used throughout most of this book. The do-loop over 1200 allows one to specify a more fine-grained integration of the transmembrane potentials to avoid this instability while retaining the computational efficiency associated with larger integration step sizes for the other state variables, which are not subject to this instability. In the do-1200 loop one first updates the membrane potentials for all the cells at each substep in the 1220 loop. Then in the 1240 loop one allows these to interact according to the FCP and SPCP field and spike coupling, respectively. The coding in the 1240 loop identifies the four nearest neighbors under the assumption that the total number of cells are arranged in a two-dimensional rectangle with NC columns and NR rows. The updating of the other state variables TH, GK, and S in the do-1300 loop includes identification of the x and y coordinates of each individual cell, labeled there as JX and JY. The matrix ALINE(JX, JY) is simply a device used to produce the two-dimensional display of asterisks associated with the firing in individual cells. CHAR(1) is a blank assigned to a place when a cell does not fire and CHAR(2) is an asterisk assigned to a place when a cell does fire.

C. Theoretical Background

1. Coupling by Extracellular Field Potentials in Neurons

The possibilty of coupling of neuronal activity by extracellular field currents and potentials in neuronal populations and its possible significance have excited the imagination of neuroscientists for a long time. Some of the earliest studies included observations on so-called ephaptic transmission between individual axons in a single nerve trunk. It is generally thought today that significant coupling of neuronal activity by extracellular field currents probably occurs in at least two cases: (1) in regions of highly dense dendritic neuropile where neighboring neuronal processes (dendrites or axon terminals) are physically very close together and the ability of current to flow away through extracellular fluid is limited because of the high density of other processes, and (2) in local neuronal populations whose cells are firing with a high degree of synchrony—in particular when the ionic current loops from neighboring cells are aligned so as to sum to a global macroscopically detectable event such as might be expected when various prominent rhythms are recorded with macroelectrodes.

The programs presented in this chapter represent first steps to realistic simulation of these phenomena. The next step would be to introduce a model neuron with dendrites, for example, NEURN30, into the program for POOL20 and to effect a field coupling between particular specified regions of the dendrites of adjacent fields. Traub has also modeled these effects, as discussed in Chapter 4.

D. Exercises

1. Use program POOL11 to simulate the synaptic interaction of a dozen pacemakers firing at different rates and phases. Interconnect these synaptically in a closed loop of six groups of two wherein each neuron connects synaptically to each of the two cells in the subsequent link and receives input from each of the two cells in the previous link. Perform simulations of approximately 100 or 200 msec duration at several different levels of synaptic strength, starting from zero synaptic strength. What is the nature of the resulting activity? At higher levels of synaptic strength do action potentials propagate around the loop? Take the conduction times to be 2 and the refractory period time constant TGK to be 3.

2. Repeat Exercise 1, but now arrange the 12 cells sequentially in a single closed loop of 12 steps.

3. Repeat Exercise 1, using POOL12 and substituting field coupling for synaptic coupling between neighboring cells.

4. Program POOL20 was constructed to test the following hypothesis: "If a group of neurons in ongoing activity at reasonable to high rates are structured so that a significant degree of coupling occurs among neighboring cells by extracellular currents, then there should be a tendency for neighboring cells to become synchronized in their firing. Further, if a cluster of neighboring cells becomes synchronized, there should be an even stronger tendency for the extracellular currents associated with their combined activity to entrain yet other cells in their local region into their pattern. This process should perhaps grow until there is a wave of synchronized activity moving through the network." Use POOL20 to investigate this hypothesis by stimulating 6 to 10 cells to fire in perfect synchrony at a reasonable rate within a field of ongoing random background activity. Use a network with 400 cells or more.

5. Notice that Program POOL12 represents extracellular field coupling by a term proportional to the potential in the sending cell, whereas POOL20 represents this coupling by a term porportional to the potential difference between the sending and receiving cells. Which of these do you think is most realistic? Alter the code of POOL12 to represent coupling based on potential differences. What are the main differences in behavior between the two rules?

17

Junctions of Neuronal Populations

This chapter addresses the computational problems associated with activating a neuronal population by activity in a multifiber input system. Such a junction is typically called a convergent-divergent junction: convergent in the sense that any individual receiving cell receives input from a number of input fibers, divergent in the sense that any single input fiber projects to a number of receiving cells. It is the fundamental and generic mode of junction between neuronal populations in the nervous system. The chapter also considers the case of activation of a single neuronal population by an arbitrary number of independent multifiber input systems, each of which might exhibit a distinct biochemical transmitter–receptor combination and therefore unique synaptic physiology. Finally, the chapter shows how to represent spatial organization of neuronal input and receiving populations with particular reference to rectangular two-dimensional arrays.

A. Using the Programs: Input–Output Charts and Illustrative Examples

Figure 17.1 shows the input–output flowchart for programs JNCTN10 and JNCTN11. JNCTN11 differs from JNCTN10 only in that a graphic display is added. The main input functions in JNCTN10 and JNCTN11 are effective synaptic currents flowing into the cell bodies of point neurons, represented here by SC. The parameters one specifies to determine these input currents are NT, the number of synaptic terminals per input fiber; STR0, the mean synaptic strength of the input fibers; PER0, the mean firing period of the individual input fibers; and LSTM0, the mean time of first firing of the input fibers. Internal to the program, then, each input fiber is assigned a unique value

301

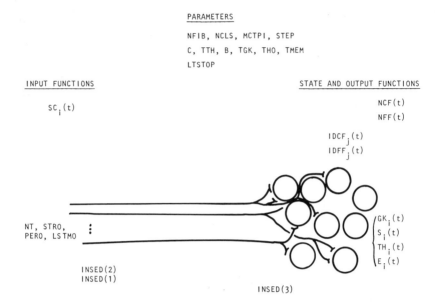

PARAMETERS

NFIB, NCLS, MCTPI, STEP
C, TTH, B, TGK, THO, TMEM
LTSTOP

INPUT FUNCTIONS   STATE AND OUTPUT FUNCTIONS

$SC_i(t)$ $NCF(t)$
                                                            $NFF(t)$

                                                            $IDCF_j(t)$
                                                            $IDFF_j(t)$

NT, STRO,
PERO, LSTMO $GK_i(t)$
                                                            $S_i(t)$
                                                            $TH_i(t)$
                                                            $E_i(t)$

INSED(2)
INSED(1)
                        INSED(3)

Fig. 17.1. Programs JNCTN10 and JNCTN11. These programs simulate the activation of a pool of neurons by an input fiber system.

for strength, period, and first firing time. These three values are chosen from the mean values read in according to exponential distributions. Three random number seeds, INSED(1), INSED(2), and INSED(3), determine the particular distributions obtained for these values, the particular firing pattern in the fibers, and the specific anatomical connections made by the input fiber system to the cells.

The main state and output functions for the population are NCF and NFF, the number of cells and number of fibers firing at each time, respectively; IDCF and IDFF, the identities of the individual cells and fibers which fire at a given time; and the set of four state variables E, TH, S, and GK for each of the cells in the population. The systemic parameters which one must specify to operate the program include NFIB, the number of fibers in the input population; NCLS, the number of cells in the receiving population; MCTP1, the maximum conduction time plus one; and STEP, the integration step size, typically taken as 1 msec. In addition, one specifies the parameters C, TTH, B, TGK, TH0, and TMEM to characterize the cells in the receiving population, as discussed in Chapter 14. In these programs the specified values for these parameters apply to all cells in the given population. LTSTOP is the duration of the simulation in milliseconds.

With programs JNCTN10 and JNCTN11, then, one can simulate and study the ongoing dynamic activity in populations of neurons of different sizes and different characteristic parameter values in response to ongoing synaptic bombardment by arbitrary numbers of fibers with arbitrary numbers of terminals, rates of activity, and strengths of synaptic activation.

Figure 17.2 shows the input–output flowchart for program JNCTN12. Program JNCTN12 also produces ongoing dynamic activity in a pool of neurons of variable character in response to synaptic bombardment. In this case, however, one may simulate activation of the population by an arbitrary number of distinct input fiber populations. One first specifies how many distinct input fiber populations to simulate, NFPOPS, then specifies for each of these fiber populations the following list of parameters: EQ, the synaptic equilibrium potential associated with that fiber population; T, the time constant of decay of postsynaptic conductance changes associated with that population; N, the number of fibers in that population; P, the probability of firing in a given fiber in that population at a given instant in time; INSTRT and INSTP, the starting and stopping times of activity in that fiber population; NCT, the

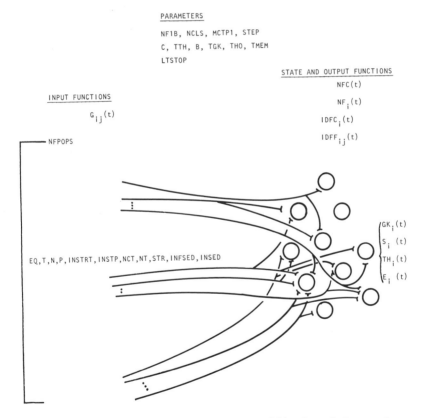

Fig. 17.2. Program JNCTN12. This program simulates the activities of a pool of neurons by an arbitrary number of input fiber systems.

conduction time associated with input in that fiber system; NT, the number of terminals per fiber in that population; STR, the strength of the synapses (measured as the magnitude of the postsynaptic conductance change); and two random number seeds, INFSED associated with firing patterns and INSED which determines the anatomical connections made by the fibers in that system. This string of parameters for a given input fiber system determines the ongoing time course of synaptic conductance changes on the receiving cells in the population. The other systemic parameters and state and output functions are the same in JNCTN12 as those discussed for JNCTN10 and JNCTN11. With program JNCTN12, then, one may simulate the results of interactions of input fiber systems with different chemical transmitters on a single receiving population.

Figure 17.3 shows the input–output flowchart for programs JNCTN20 and JNCTN21. JNCTN21 differs from JNCTN20 only by the inclusion of a two-dimensional graphic display. Programs JNCTN20 and JNCTN21 are precisely analogous to programs JNCTN10 and JNCTN11, except that both the input fiber system and the receiving cell population are spatially

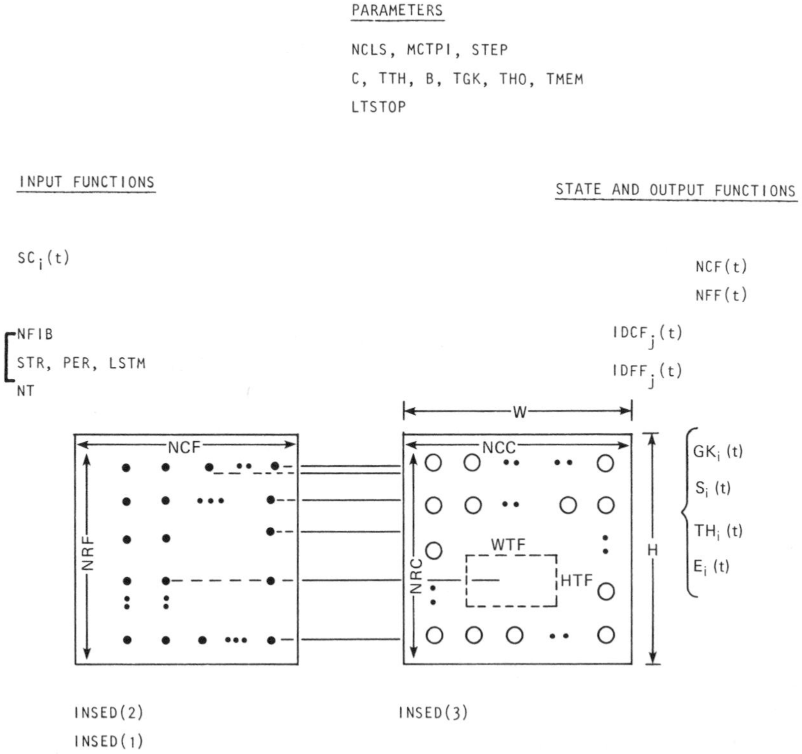
PARAMETERS
NCLS, MCTPI, STEP
C, TTH, B, TGK, THO, TMEM
LTSTOP
INPUT FUNCTIONS
STATE AND OUTPUT FUNCTIONS
SC_i(t)
NCF(t)
NFF(t)
NFIB
STR, PER, LSTM
NT
IDCF_j(t)
IDFF_j(t)
W
NCF
NCC
NRF
NRC
WTF
HTF
H
GK_i (t)
S_i (t)
TH_i (t)
E_i (t)
INSED(2)
INSED(1)
INSED(3)

Fig. 17.3. Programs JNCTN20 and JNCTN21. These programs simulate the activation of a two dimensionally arranged system of input fibers.

arranged in two-dimensional arrays. The state and output functions, systemic parameters, input driving functions, and parameters determining them are the same. In JNCTN20 and 21, however, additional parameters NCF, NRF, NCC, and NRC determine, respectively, the number of columns and number of rows in the fiber and cell populations, respectively. The number of fibers NFIB must equal the product of NCF and NRF, and the number of cells NCLS must equal the product of NCC and NRC. W and H are parameters which define the overall width and height of the space in which the fibers and cells are imagined to exist, respectively. NCF and NCC must be less than or equal to W, and NRF and NRC must be less than or equal to H. In these programs a given input fiber projects all of its output terminals to fall on receiving cells within a rectangular field of width WTF and height HTF. This field is centered on the receiving population around a point which corresponds to the position of the sending fiber.

With programs JNCTN20 and JNCTN21, then, one can simulate the characteristics of ongoing convergent–divergent transfer in a rectangularly organized two-dimensional system of a single receiving population activated by a single input fiber system.

EXAMPLES

Moderately diffuse convergent–divergent transfer. Figure 17.4 illustrates ongoing input/output activity at a moderately convergent–divergent junction. Figure 17.4a shows the total number of input fibers and total number of cells firing as a function of time and Fig. 17.4b shows the generator potential and output spike trains in a representative cell of the population.

Highly diffuse convergent–divergent junctions as synchronization-sensitive filters. The salient dynamic characteristic of highly diffusely connected convergent–divergent junctions is that they are selectively sensitive in responding to synchronized clusters of activity in the input fiber population. This phenomenon is illustrated in Fig. 17.5. The horizontal axis on the charts in Fig. 17.5 is time and the vertical axes are cell number. An asterisk at a particular point in such a display indicates that a particular cell fired an action potential at this time. Random activity is indicated by a more or less uniform distribution of the asterisks over the plane; synchronized clusters in the input or output are represented by vertical lines of astrisks. The example in Fig. 17.5 shows that highly diffuse convergent–divergent junctions tend (1) to respond either with synchronized clusters as an output or not respond at all and (2) to be particularly sensitive to synchronization and synchronized clusters in the input system. In the extreme case, such a network is responding as an organic whole with the unitary output signal and largely giving up its ability to signal

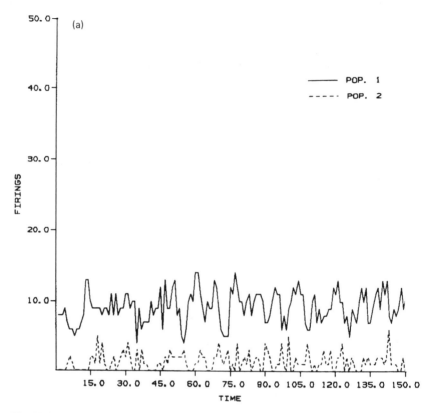

Fig. 17.4. Input-output across a convergent-divergent junction. (a) Total number of input fibers and total number of cells firing; (b) generator potential and output spike trains in a representative cell.

diversified information with more variablility among its individual constituent neurons.

Spatial inversion and integration in moderately diffuse convergent-divergent junctions. Consider now convergent-divergent monosynaptic transfer in two-dimensionally organized populations as exemplified in JNCTN20 and JNCTN21. If such systems are highly discrete so that a given sending fiber projects to a very small number of localized receiving cells, then the system is largely a set of parallel independent channels. At the other extreme, if such systems are highly diffuse so that one fiber projects to cells throughout the entire receiving population then such systems are little different from pools with very little evidence of spatial organization. At intermediate degrees of diffuseness of connection,

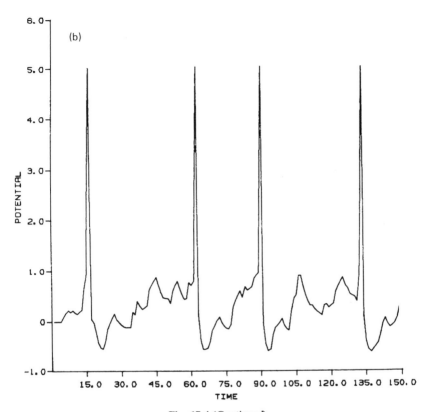

Fig. 17.4 (*Continued*)

the primary ingredient is that given receiving cells tend to integrate information over a certain spatial area of the input. If one pictures ongoing activity as more or less sporadically distributed over the two-dimensional array, then output activity will be engendered in cells which happen to be in neighborhoods of relatively high levels of activity. A special case of this is the so-called spatial inversion effect, which is illustrated in Fig. 17.6. In this phenomenon the peaks of activity in the output system correspond not to regions of peak activity in the input system but rather to positions strategically localized between peaks in the input distribution so as to take advantage of the input from several such peaks. In such cases it appears as if the receiving population is doing a spatial inversion of the input system somewhat as a photograph does a spatial inversion of the negative in the photographic process.

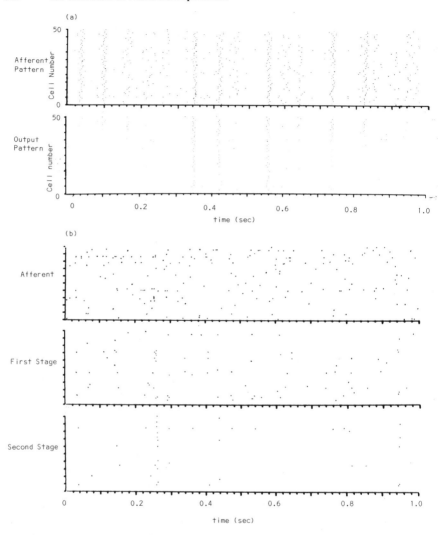

Fig. 17.5. Diffuse convergent–divergent junctions. (a) Diffuse junction as a syncronization-sensitive filter; (b) genesis of clusters in two-stage diffuse transfer.

B. Program Listings and Discussion (JNCTN10, JNCTN11, JNCTN12, JNCTN20, JNCTN21)

The listing for program JNCTN10, in Appendix 2, shows that these programs also exhibit basic sections for reading and writing the input parameters, for initializing variables, a do-1000 loop for updating state variables at each time step, and finally sections for the formats. The main operative section, the do-1000 loop that updates states variables, has two internal loops: do-2000 over the number of input fibers and do-3000 over the individual cells in the receiving population. The do-3000 loop updates the state variables for individual cells in accordance with the rules discussed in Chapter 14 and is driven by the synaptic currents SC(I, 1). These synaptic currents in turn are defined in the do-2000 loop corresponding to activity in the individual input fibers. Specifically, as done in the programs in the chapter on synaptic bombardment, each input fiber carries a flag LSTM which indicates its next time of firing. If the cell fires, (1) it is recorded as a firer, (2) its time of next firing is updated, in this case by an exponential interval distribution, and (3) it projects increments of activity into the activity-in-transit matrix SC in accordance with the specific cellular connections it makes. In this case these connections are chosen at random, uniformly distributed among the entire class of receiving cells. It is easy to interject in the do-2010 loop any specific rule for connectivity one would like to insert. The populations simulated by programs JNCTN10, 11, and 12 are pool exhibiting no spatial structure, so the corresponding rules for interconnections from the input fibers are completely random. The same specific connections are made by a specific input fiber every time that cell fires. This is ensured by the input random number seed INSED(3). It is important to emphasize that the general character and versatility of the do-2000 loop for generating activity in the input fiber system are independent of the specific rules used there for either the interval distribution or the distribution about the terminals. For example, the rule for the exponential interval distribution in updating LSTM can be varied to generate according to the techniques presented in Chapter 15 to produce firing patterns with any desired interval distribution, using the one-line rules for pacemakers or Gaussian distributions or the call to the STOCH subroutine system discussed in Chapter 15. The rule used for generating the particular identities of receiving cells, NREC, can be varied from the random distribution used here to any deterministic or stochastic distribution one chooses simply by writing the desired rule in place of the rule used for NREC in this program, making use if necessary of the STOCH subroutine system.

The listing for JNCTN11 shows that this program is essentially identical to JNCTN10. In JNCTN11, however, the updating of state variables in the do-3000 loop is effected by a call to a subroutine SVUPDT. This offers the flexibility of easily substituting different dynamic rules for the particular cells if desired. Moreover, the distribution of activity from single firing fibers into the activity-in-transit matrix SC is effected by a subroutine TRNSMT. This has the advantage of isolating the specific rule for detailed connectivity in the junction and also, as will be seen below, serves to isolate the code which determines spatial organization at neuronal junctions. JNCTN11 also adds a simple graphic display contained in the variable

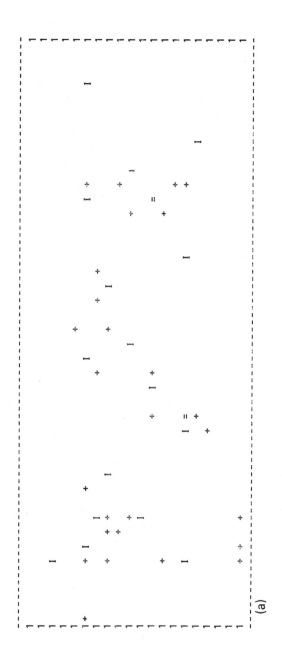

(a)

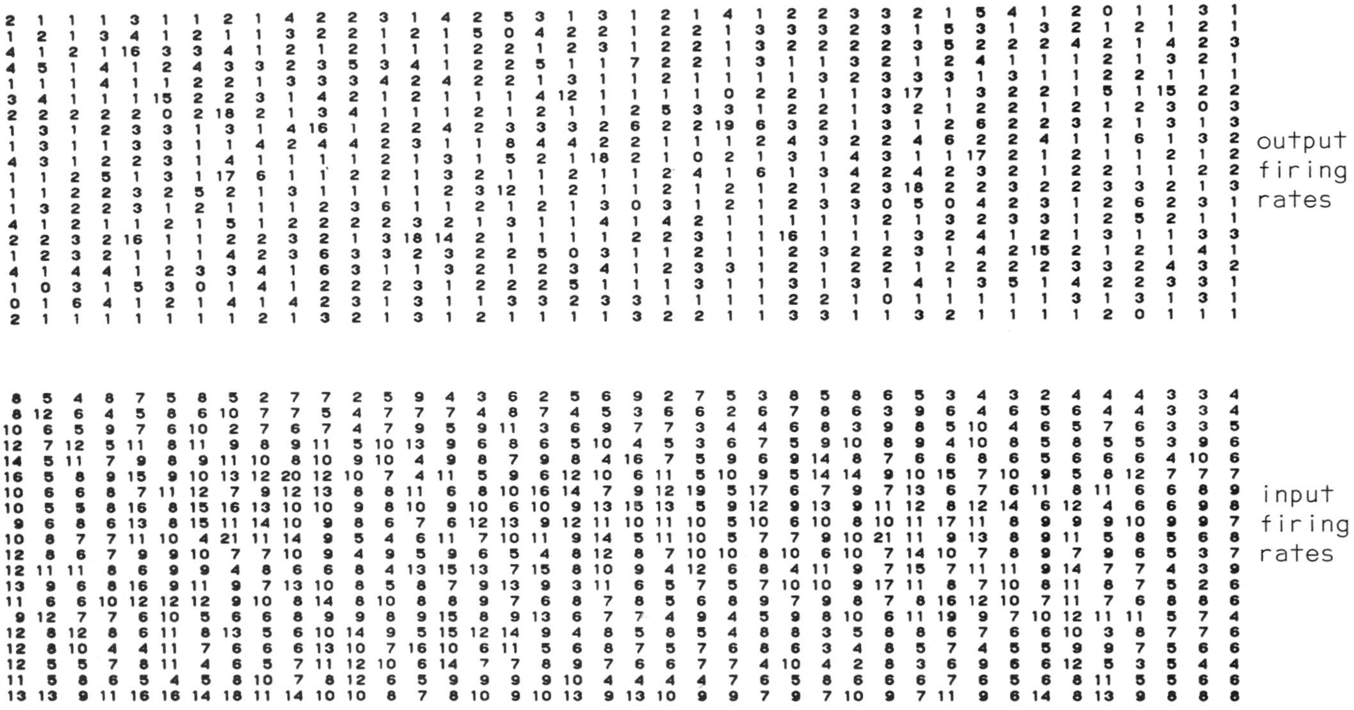

output
firing
rates

input
firing
rates

(b)

Fig. 17.6. Spatial inversion at a convergent–divergent junction. (a) Input (I) and output (+); (b) firing rates.

ALINE. Positions in ALINE correspond to the numbers of specific fibers and specific cells. Firing of a particular cell is flagged by inserting an asterisk (CHAR(2)) in its particular position. Program JNCTN12 differs from JNCTN10 and 11 in two essential features, as may be seen in its listing. First, the do-2000 loop is generalized to update activity in a arbitrary number of input fiber populations rather than a single fiber population. Second, as may be seen in the TRNSMT subroutine, activity from the separate fiber systems projected to receiving cells is stored in separate activity-in-transit matrices G corresponding to the different sending input systems. As may be sen in the SVUPDT subroutine, on arrival at receiving cells, these distinct input activity arrays have different effects on the receiving cells corresponding to the synaptic equilibrium potentials they are associated with, EQ. Moreover, as may be seen in the 4010 do-loop in the parent program, the decay rates in time of these different synaptic input conductances, DCS(K), may vary from one fiber system to another. As may be seen in the do-2000 loop, program JNCTN12 simply compares a random number to a fixed probability for firing in a given cell at each time to determine whether that fiber will fire. Again, if one wishes to use a different firing pattern in the input cells, it is easy to recode that line in correspondence with the techniques introduced in Chapter 15.

The essential technique for converting aspatial pools such as those used in JNCTN10, 11, and 12 into spatially organized populations is illustrated in the listing for program JNCTN20 (Appendix 2). This is simply to convert the cell number or fiber number into a corresponding position in space. This can be done easily for any conceivable spatial organization, not only for the two-dimensional arrays specifically encoded here. Moreover, this must be done only when one is projecting input from one cell or fiber to another and therefore can be constrained entirely to the TRNSMT subroutine. Specifically, one can see in the TRNSMT subroutine in program JNCTN20 that the cell number K is converted to an x coordinate (XF) and a y coordinate (YF) in accordance with the number of columns and rows, NCF and NRF, specified for that fiber array. In the do-100 loop in that subroutine, the x and y coordinates of the target cell (XT and YT, respectively) are generated with reference to the x and y coordinates of the fiber (XF and YF) and the width and height of the terminal field, WTF and HTF. From these coordinates of the receiving cell one generates its column ICC and row IRC and from that information the identity of the receiving cell IDC in accordance with the number of columns and rows in the receiving cell population, NCC and NRC. Finally, the activity-in-transit matrix SC for that particular receiving cell IDC is updated with the appropriate conduction time lag. Again, program JNCTN20 uses simply an exciting synaptic current SC. It would be easy to generalize this program to include a number of fiber systems, each with different synaptic effects, as was done in JNCTN12.

Program JNCTN21 is identical to JNCTN20 with the exception that a two-dimensional graphic display is included in the FORTRAN coding. This is done again through the mechanism of the alphanumeric variable ALINE. Now, however, ALINE has two spatial dimensions instead of one, and whenever a cell or fiber fires that cell or fiber number is converted into its x and y coordinates (JX and JY) so as to identify the appropriate place in the ALINE matrix to insert the asterisk. This is seen in the listing for JNCTN21 in the do-2000 and do-3000 loops.

C. Theoretical Background

1. Functional Types of Synaptic Transmission

It has been traditional in theorizing on the operations of brain networks to classify all neurons as either excitatory or inhibitory. Although this greatly simplifies conceptualizing the operations of interacting neuronal populations, most current research indicates that this is an unwarranted and often misleading oversimplification. The Dale principle of chemical transmission, first enunciated in the 1930s, states that a given neuron emits one and only one chemical transmitter at all its output terminals. In the 1960s it was generally held that a given chemical transmitter had for the most part the same influence on conductance on any neuron it was projected to. By virtue of this, Eccles' extension of the Dale principle to the idea that any single nerve cell had the same influence on postsynaptic conductance at all its output terminals was warranted. This, combined with the assumption that a synapse would be either excitatory or inhibitory depending on whether the equilibrium potential of its postsynaptic conductance change was above or below resting potential, rendered the classification of all cells as either excitatory or inhibitory a reasonable proposition. This simplified picture, however, is brought into question by several observations. First, the postsynaptic effect of a given chemical transmitter depends on the molecular receptor molecules on the receiving cells as well as on the transmitter itself. Examples are known where a given cell sends the same transmitter to different receiving cells which in turn have different receiving receptor molecules and respond with different synaptic equilibrium potentials and therefore effect different functional types of synapses. Second, it is an oversimplification to characterize synaptic action as being simply excitation or inhibition without taking into account more precisely the nature of the synaptic equilibrium potentials and conductance changes associated with synaptic action. For example, a synapse could conceivably have a synaptic equilibrium potential equal to the resting potential of a cell and, if activated when the cell was in a resting state, produce no noticeable change in potential in the cell. However, because it changes the conductance of the cell membrane, it alters the membrane time constant and therefore alters the nature of the response of the cell to any other input it might experience. This also would tend to more strongly clamp the potential toward the resting potential than is normally the case. Similarly, synapses with equilibrium potentials of, say, 25 mV would be excitatory, as would those with equilibrium potentials of 70 mV, but the two would exhibit rather different characteristics in the overall driving of a cell. Third, it seems that some cells do transmit different chemical transmitters at distinct terminals. Although this is probably rare and the Dale principle of one cell–one transmitter is the rule, if does point out the dangers of oversimplifying neuronal populations into either "excitatory" or "inhibitory" neurons.

The models encoded in the programs of this book recognize synaptic diversity to the extent of specifying simply a synaptic equilibrium potential and a rate of decay of postsynaptic conductance change uniquely for a given synaptic "type." These can be specified uniquely for each transmitter receptor junction and therefore make none of the questionable assumptions discussed above. Nonetheless, it is equally clear that one could go considerably beyond what we have done to precisely describe the intricacies and variability of individual synaptic action. For example, p and n in the quantal model (see Chapter 16) clearly differ among synaptic types. Our feeling is that such extensions could be done easily enough within the framework of the general structure of programs presented here and that, for the most part, the degree of refinement in question is at a slightly different level of focus than our concern with the overall ongoing integrative dynamics of interconnected populations.

2. Scaling in Model Neuronal Junctions

It is not totally clear how to optimally scale a model representation of a large nerve network where the model is restricted to containing significantly smaller numbers of cells than the network being modeled. It seems that one must choose between verisimilitude in terms of the network connectivity and intercohesion on the one hand and verisimilitude at the level of individual neuron dynamics on the other. Thus, for a given linkage between two populations it seems that the following nondimensional parameters would be important: the ratio of input fibers to receiving cells $OC = M/N$, and the percentage of receiving cells connected to by a given sending fiber, $\gamma = NT/N$ (NT is the number of output terminals per sending fiber). It turns out that σ is also equal to the percentage of input fibers received from by an individual receiving cell, N_i/M, since $N_i \times N = NT \times M$ (N_i is the average number of input synapses per receiving cell); the percentage of input fibers which must fire simultaneously to trigger a spike in a single receiving cell, σ = Th/Ni (Th is the threshold of the cell); and the minimum percentage of input fibers which must fire to possibly produce any output spikes in the receiving population (assuming no one fiber connects to any one cell more than once), β = Th/M. Three of these numbers α, γ, σ, and β are independent since β = $\sigma \times \gamma$. It seems that if one keeps α, γ, and σ the same in the model as they are in the network being modeled, then certain essential features of the interconnectivity of the network are preserved, and that if one allows any one of these values to be significantly different in the model than in the network being modeled, then some essential feature of the interconnectivity is lost. One discovers, however, that if these values are kept constant, the richness in individual neuron dynamics apparent in the network being modeled is largely lost in the model. For example, present best estimates for the linkage from the perforant path fibers to the granule cells on one side of the hippocampus are N = 988,000, M = 165,000, NT = 60,000, Ni = 10,000, and TH = 400. These give α = 1/6, γ = 0.0607, σ = 0.04, and β = 0.0024. If we simulate with a model system with 4000 granule cells, keeping α, γ, and σ fixed, then N′ = 4000, M′ = 667, Nt′ = 243, Ni′ = 40, and Th′ = 1.6. That is, at the level of individual neuron dynamics, each cell has about 40 synapses and may be activated by coincident firing in any two of them. This is considerable less rich in possibilities for physiological interaction than the real granule cell, which has about 10,000 input synapses and requires simultaneous firing of about 400 to be activated. [See M. J. West and A. H. Andersen, *Brain Res. Rev.* **2**, 317–348 (1980), as an illustration of how nature deals with this problem.]

Such constraints represent unavoidable limitations of scaled-down network modeling. One can use such estimations as these to help determine the optimum size of the scaled-down model and to assess its limitations. One strategy is to keep α, γ, and β constant so that the local circuit model represents the global interconnectivity of the local circuit faithfully and to rely on more detailed single-cell models for high verisimilitude of individual neuron dynamics.

3. Precise Interconnections at Neuronal Junctions

A general question in network modeling is how one specifies the precise interconnections between two populations of neurons. Generally, in the mammalian central nervous system one often knows the classes of cells to which a given afferent population projects, the average numbers of sending terminals per sending fiber and received synapses per receiving cell, and the general topographic organization in terms of size of the receptive field of the receiving cells and size of the target field of the sending fiber. However, within these general constraints it is not known exactly which individual connect to which individual cells. Moreover, it is not known—although certain people have very strong opinions on the matter—whether such microscopic specificity of connections is prescribed genetically or occurs in large part by chance (within constraints mentioned above or even whether such microscopic specifity is important in network dynamics or information processing. In our models we prescribe that a given sending fiber make a certain number of output synapses on cells of a certain class within a certain region of space. The individual specific cells chosen within these regions are then determined by random number generation; the same connections are made by a given sending fiber every time that fiber is activated. On the one hand, it would be easy to substitute some regular or logical rule to specify the microscopic connections in place of the random ones; on the other, it is not clear what regular or logical rule to use, and it seems to us that a highly regular rule might be less realistic than a slightly random rule. Our nets are not "random nets": they have too much ordered structure to be so considered; they are rather "stochastic nets"—that is, nets that incorporate a certain amount of variability within an overall pattern of orderliness. Finally, it will be easy to incorporate into our models whatever connectivity rule one might suggest as an alternative based on convincing arguments on data. At some point in the modeling work it may be of interest to compare model output from two cases which differ only in that one is based on a random rule and the other is based on a regular rule. At a stage somewhat advanced from now, such simulations might provide meaningful input into the debate concerning detailed connectivity of nervous tissue.

D. Exercises

1. Use program JNCTN11 to simulate the ongoing input–output activity of a population of 100 neurons bombarded by a single afferent input system firing at a mean rate PER0. For a simulation time of 100 msec, determine the mean number of cells firing at each time. Plot this number versus the input firing rate and repeat for a number of different input firing rates.

2. Use program JNCTN12 to simulate the ongoing input–output activity in a population of about 100 neurons bombarded by a single excitatory input system and a single inhibitory input system. Produce curves of the average number of cells firing at each time versus the input firing rate at several different levels of inhibitory input.

3. Show how the curve of Exercise 1 depends differentially on the parameters NT and STR0.

4. Show how the curves in Exercise 3 depend differentially on the parameters NT and STR for the inhibitory input.

5. Can you produce synchronization sensitivity like that illustrated in Fig. 17.5 using program JNCTN11?

18

Local Networks, Feed-Forward, and Recurrent Connections: Lateral Inhibition and Internally Sustained Activity

This chapter introduces computational techniques for simulating multiple interacting neuronal populations. Specifically, the chapter considers the case of a single local region containing two populations of neurons: a parent population of excitatory cells and a second set of inhibitory neurons. There is a single afferent fiber system which may activate either or both populations. The parent population projects recurrent connections both to itself and to the inhibitory population. The inhibitory cells project to the excitatory cells. Thus, the fundamental inhibitory mechanisms of feed-forward inhibition, recurrent inhibition, and lateral inhibition are represented. In addition, the recurrent excitatory projections from the parent population onto itself permit the maintenance of internally sustained activity once activity is initiated in the population by some extrinsic source. Two programs represent neuronal pools with no consideration of spatial organization. Three programs represent the neuronal populations as orderly rectangular arrays in two-dimensional space.

A. Using the Programs: Input–Output Charts and Illustrative Examples

Figure 18.1 shows the input–output flowchart for programs NTWRK10 and NTWRK11 (see Appendix 2). The main input functions in these programs are the synaptic currents, SC, which are initiated in the the cell populations by activity in the afferent fiber system. These functions are determined by the parameters P, STRIP, and LIPSTP. P represents the probabilty of firing an action potential in a given fiber at a given time step. STRIP is the strength of the input synapses. LIPSTP is the time that the

PARAMETERS

NEX, NIN, MCTPI, STEP
C, TTH, B, TGK, THO, TMEM
(for each cell group)
NT, NCT, STR, INSED
(for jnctns 11, 12, and 21)
LTSTOP

STATE AND OUTPUT FUNCTIONS

NIF(t)
NEF(t)
NFF(t)

INPUT FUNCTIONS

$SC_i(t)$

P, STRIP, LIPSTP

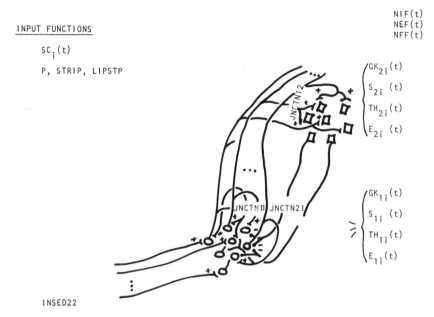

$\begin{cases} GK_{2i}(t) \\ S_{2i}(t) \\ TH_{2i}(t) \\ E_{2i}(t) \end{cases}$

$\begin{cases} GK_{1i}(t) \\ S_{1i}(t) \\ TH_{1i}(t) \\ E_{1i}(t) \end{cases}$

INSED22

Fig. 18.1. Programs NTWRK10 and NTWRK11. These programs simulate the activity of a recurrently connected pool of excitatory and inhibitory neurons.

afferent system is turned off. INSED22 is a random number seed governing the time course of activity in and the specific connections of the afferent fiber system. The output functions for these programs consist of NFF, NEF, and NIF, which represent respectively the numbers of fibers, excitatory cells, and inhibitory cells firing at each time unit, and the state variables E, TH, S, and GK for all the individual cells in both populations. The systemic parameters for the system are the number of excitatory cells NEX, the number of inhibitory cells NIN, the maximum conduction time plus one MCPT1, and STEP, the integration step size. As usual, one specifies the six main parameters C, TTH, B, TGK, TH0, and TMEM for each cell group. Finally, one specifies the number of terminals NT, conduction time NCT, synaptic strength, STR, and input seed INSED for each of the three internal junctions in the system: excitatory to excitatory (11), excitatory to inhibitory (12), and inhibitory to excitatory (21). LTSTOP is, as usual, the duration of the simulation. With these programs one can simulate the ongoing interacting dynamics of these populations in response to afferent bombardment of the parent population. In particular, one can study the influence of recurrent inhibition on through transfer, the balance of recurrent excitation and inhibition in producing internally sustained activity, and questions of this variety. From the user's standpoint, NTWRK10 and NTWRK11 are essentially identical. They differ in internal programming details as discussed in Section B.

Figure 18.2 shows the input–output flowchart for programs NTWRK20, 21, and 22. These programs extend beyond NTWRK10 and 11

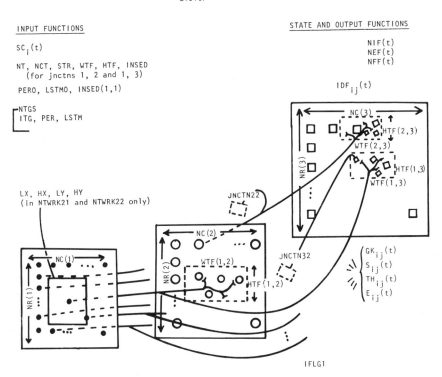

Fig. 18.2. Programs NTWRK21 and NTWRK22. These programs simulate the activity of a recurrently connected and spatially organized network of excitatory and inhibitory neurons activated by a spatially organized system of input fibers.

primarily by introducing spatial organization and feed-forward inhibition from the afferent fiber system through the inhibitory population. Again, the main input functions in these programs are the synaptic currents that project to the cell types from the afferent fiber system SC. To specify the dynamic activity in the afferent system, one specifies the parameters PER0, the mean period between spikes in the individual input fibers, and LSTM, the mean first firing time of the afferent input fibers. Internal to the program, the individual fibers are assigned particular periods and first firing times based on choices from exponential distributions around these means. In these programs the subscript 1 refers to the afferent system, 2 to the excitatory population, and 3 to the inhibitory population. INSED11 governs the choice of firing patterns in the input afferent system. For both output junctions from the afferent systems, one specifies the number of synaptic terminals per fiber NT, mean conduction time NCT, strength of synapse STR, width of the terminal field of the output synaptic terminals WTF, height of the terminal field HTF, and a random number seed INSED which governs the particular connectivity made at that junction. By specifying the parameters PER0 and LSTM0 as indicated above, one prescribes a mean rate of exponentially distributed activity in the individual fibers of the input system. One may define the input activity more precisely by two means. First, one may specify the number of target fibers, NTGS, for which particular activity is to be prescribed. Then, for each of those targets, one prescribes the identity of that target fiber ITG, the period of activity in that afferent fiber PER, and the first firing time in that fiber LSTM. Second, one may constrain the activity in the input system to a rectangular region bounded by low x, high x, low y, and high y (LX, HX, LY, and HY respectively). In NTWRK21 activity outside this region is shut off so that stimulation is effective only from within this rectangular region. In NTWRK22 the activity within this rectangular region is elevated above that outside the region by a factor SNRATIO, a signal-to-noise ratio.

The main state and output functions for these programs are NFF, NEF, and NIF, the numbers of fibers excitatory cells, and inhibitory cells firing at each time; the identities of the individual firers in all three populations IDF; and the four state variables E, TH, S, and GK for all cells in both populations. The systemic parameters are the number of fibers, NFIB, number of cells NCLS, number of excitatory cells NEX, number of inhibitory cells NIN, maximum conduction time plus 1, NCPT1, integration step size, STEP, and width and height of the field within which the fibers and cells exist, W and H, respectively. The sum of NEX and NIN must equal NCLS. Again, one specifies the main parameters C, TTH, B, TGK, TH0, and TMEM for each cell type. Then for each of the internal junctions [excitatory to excitatory (22), excitatory to inhibitory (23), and inhibitory to excitatory (32)] one specifies the number of synaptic terminals per fiber NT, mean conduction time NCT, synaptic strength STR, width and height of the terminal field of fibers from an individual sender, WTF and HTF, and an input seed INSED governing the particular connections made at that junction. LTSTOP again is the duration of the simulation.

For each of the cell populations and the fiber population one must specify the number of columns NC and the number of rows NR. The product of NC and NR must equal the number of members of each of these three populations.

These programs are particularly well adapted to study of the influence of lateral inhibition in systems which might be conceived to be essentially two-dimensional in physical structure. Application to information processing in parts of the visual system comes immediately to mind. Moreover, one can study interesting questions related to properties of internally sustained activity in two-dimensionally arranged networks which may generate concepts of interest with regard to information processing in more central networks such as the hippocampus or cerebral cortex.

EXAMPLES

Internally sustained activity in neuron pools.   Figures 18.3 and 18.4 show two examples of internally sustained activity in pools of cells as simulated by NTWRK11. In parts a of both figures, the solid line shows the number of excitatory cells firing as a function of time. One can notice that whenever the number of excitatory cells tends to increase, there follows immediately a proportionately larger increase in the number of inhibitory cells firing, which in turn tends to bring the number of excitatory cells firing back down to around or below its equilibrium level. The example in Fig. 18.4 differs from that in Fig. 18.3 in that more inhibitory cells are involved, 300 as compared to 50, and the inhibitory cells are much more sensitive to firing in the excitatory population as seen by the large excursions in inhibitory firing shown in Fig. 18.4a. The b and c parts of these figures show unit activity from representative excitatory and inhibitory cells, respectively. The theory for this activity is discussed in Section C,2. The two units in Fig. 18.3 are not synaptically connected, whereas the excitatory unit in Fig. 18.4 is one of the excitatory drivers of the inhibitory unit. Figure 18.4d shows a case where the inhibitory cells are less sensitive. They respond only when significant numbers of excitatory cells fire simultaneously. In this example the inhibition turns down the excitation but does not turn it off. Rhythmic firing patterns result.

Boundary-enhancement produced by lateral inhibition.   Figure 18.5 shows an example of boundary enhancement as determined by feed-forward lateral inhibition from program NTWRK22. In this example a rectangular region of elevated input activity was prescribed in the afferent

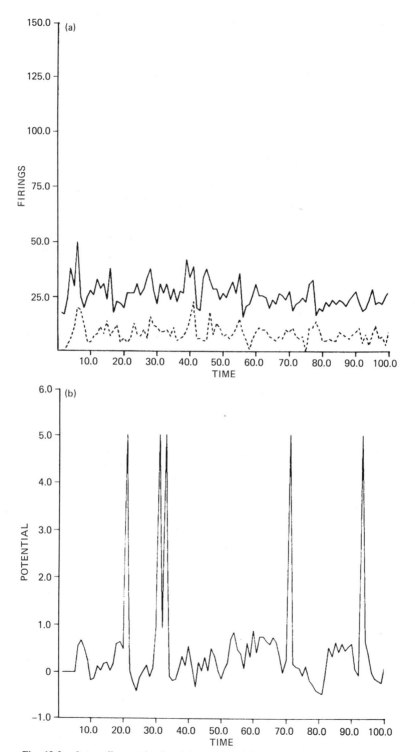

Fig. 18.3. Internally sustained activity produced by NTWRK11. (a) Numbers of (———) excitatory and (---) inhibitory cells firing; (b) unit activity from excitatory cells; (c) unit activity from inhibitory cells.

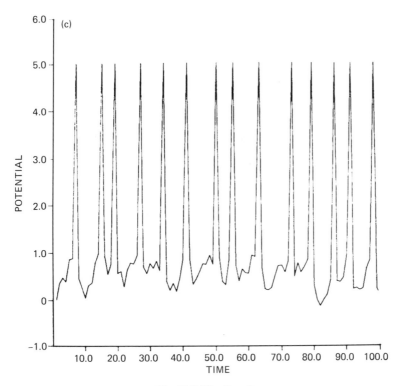

Fig. 18.3 (*Continued*)

fiber system of the program. The widths and heights of the terminal field of the afferent fibers on both the excitatory and inhibitory populations, were all set at 1, constraining the direct excitatory afferent drive to single receiving cells. The terminal fields from the inhibitory to the excitatory population were 3 by 3, effecting a slight spread in the lateral inhibitory effect and thereby producing the boundary-enhancement effect discussed in Section C,3. Figure 18.5 shows the mean level of activity in the four regions of interest. In this example the mean level of activity directly under the stimulus is somewhat higher than that in the extreme periphery away from the stimulus, but the accentuation of spatial position of the boundary is the minimum result of the lateral inhibitory mechanisms. It is possible to adjust parameters so that the mean level of activity under the stimulus is not noticeably different from that in the regions away from the stimulus so that the only result in the receiving population is flagging of the spatial location of the boundary. This is the essence of the concept of "feature extraction" in neuronal populations. It is remarkable that such a simple connectivity device can effect such a striking extraction of a significant feature such as a boundary.

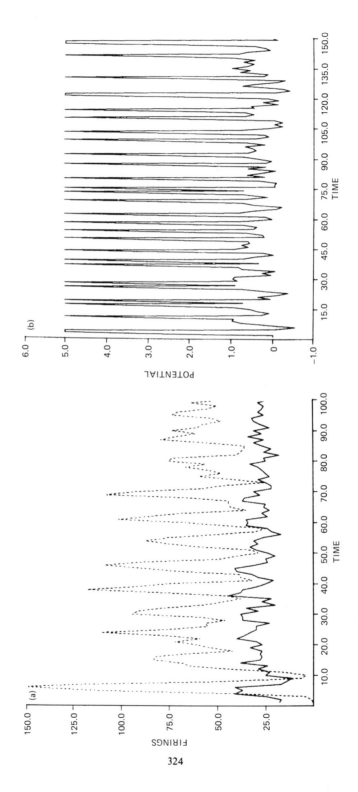

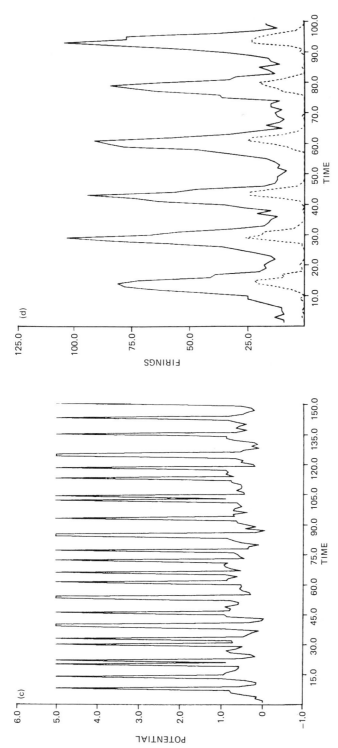

Fig. 18.4. Internally sustained activity with highly sensitive inhibitory cells produced by NETWRK11. Parts (a)–(c) as in Fig. 18.3. (d) Case with less sensitive inhibitory cells.

325

afferent fibers

```
 6  2  1  4  7   3   2   2   7   3   3   2   7   7   5   8  2  3  6  3
14  5  6  9  2   5   2   8   1   6   4   8   2   2   3   5  2  7  6  2
 6  8  4 12  6   9   3   6   5   2   5   2   9   5  11   2  1  1  2 10
 5  1  1  1  4   9   4  11   7   4   5   0   3   5   3   6  2  3  4  6
 3  7  4  5  3   5  10   5   4   3   9   3   3   4   6  16  2  6  4  2
 9  4  3  4  6 169 157  88 123 215  96  38  42   6 131   1  2  1  3  7
13  1  2  1  2  18  99  91  86 130 143   5 114 108 198   1  5  6  5  6
 2  1 13  4  2 152 154  85  34 187 144  29  95  67  23   1  5  4  2  8
 5  8  4  5  4  39 171 114 118 130 122 162 110 158 133   4  2  1  1  3
 4 12  4  6  2 131 169 154 162 114 110 148 136  65 166   3  8  1  1  2
 6  2  2  9  6 133 143  30 105   2 172 119  18  62 117   5  6  3  8  6
 2  4  1  5  6  33  69  30 127 109 139 105   5 116  20   5  4  3  8  3
 5  1  6  2  1 104 177  47  78 150  90 108 130 134 119   2  1  1  1  5
 3  2  7  5 10  82 149 153  92  94  30 130 156  90 226   1  4  7  5  1
 2  6  8  2  5  84  32 160  77 118 130 117 150  39 141   5  1  2  2  3
 9  6  4  6 10   4   7   3   1   7   7   7   8   2   5   6  6  7  5  2
 1  4  6  3  5   4   3  10   6   5   1   4   5   3   2   1  2  7  5  8
 2  7  3  6  5   2   6   5   3   7   6   3   6   1   5   1  4 10  1  5
 4  8  4  1  1   1  12   4   6   2   1   7  11   4   7   6  2  1 10  4
 3  9  1  4  7   3   8   9  12   9   6   5   7   9  11   3  4 16  5  4
```

excitatory cells

```
0  0  0  0  0   0   0   0   0   0   0   0   0   0   0   0  0  0  0  0
1  0  0  0  0   0   0   1   0   0   0   0   0   0   0   1  0  0  0  0
0  0  0  0  0   0   0   0   0   0   0   0   0   0   0   0  0  0  0  1
0  0  0  0  0   1   0   0   0   0   0   0   0   0   0   0  0  0  0  0
0  0  0  0  0   0   0   0   0   0   0   0   0   0   0   0  0  0  0  0
1  0  0  0  0  18  25   7   6   2   6   0   1   0  10   0  0  0  0  1
0  0  0  0  0   0   0   6   2   0  11   0  20   5  14   0  0  0  0  0
0  0  1  0  0   4   8   3   1   3   0   0   2   1   0   0  0  0  0  0
0  1  0  0  0   0   5   0   7   3   2   7   0   1   5   0  0  0  0  0
0  0  0  0  0   7   0   4   7   0   2   0   4   0  15   0  0  0  0  0
0  0  0  1  0   2   0   0   5   0   1   4   0   1  10   0  0  0  0  0
0  0  0  0  0   1   3   0  14   2   3   2   0   2   1   0  0  0  0  0
0  0  0  0  0  11   0   0   0   1   0   4   0   1   7   0  0  0  0  0
0  0  0  0  0   8   9   1   0   4   0   1   6   1  24   0  0  0  0  0
0  0  0  0  0  13   1  27   0  28  11   2  10   0   1   0  0  0  0  0
0  1  0  0  0   0   0   0   0   1   1   0   0   0   0   0  0  0  0  0
0  0  0  0  0   0   0   0   0   0   0   0   0   0   0   0  0  0  0  0
0  0  0  0  0   0   1   0   0   0   0   1   0   0   0   0  0  0  0  0
0  0  0  0  0   0   0   0   0   0   0   0   0   0   0   0  0  0  0  0
0  0  0  0  1   0   1   0   0   0   0   0   0   1   0   0  0  2  0  0
```

inhibitory cells

```
0  0  0  0  0   0   0   0   0   0   0   0   0   0   0   0  0  0  0  0
1  0  0  0  0   0   0   1   0   0   0   0   0   0   0   1  0  0  0  0
0  0  0  0  0   0   0   0   0   0   0   0   0   0   0   0  0  0  0  1
0  0  0  0  0   1   0   0   0   0   0   0   0   0   0   0  0  0  0  0
0  0  0  0  0   0   1   0   0   0   0   0   0   0   0   0  0  0  0  0
1  0  0  0  0  38  54  20  40  53  28   5   6   1  36   0  0  0  0  1
0  0  0  0  0   1  15  24  22  34  49   0  37  25  48   0  0  0  0  0
0  0  1  0  0  35  36  26   6  43  35   3  26   5   2   0  0  0  0  0
0  1  0  0  0   3  43  29  27  35  40  49  32  53  32   0  0  0  0  0
0  0  0  0  0  37  38  37  38  38  27  38  36   5  54   0  0  0  0  0
0  0  0  1  0  38  46   4  25   0  38  39   0  15  39   0  0  0  0  0
0  0  0  0  0   4   8   2  42  33  45  30   0  20   2   0  0  0  0  0
0  0  0  0  0  28  38   0  11  36  22  35  42  23  31   0  0  0  0  0
0  0  0  0  0  17  49  38  19  27   1  36  53  27  56   0  0  0  0  0
0  0  0  0  0  14   5  54  11  39  26  35  49   7  35   0  0  0  0  0
0  1  0  0  0   0   0   0   0   1   1   0   0   0   0   0  0  0  0  0
0  0  0  0  0   0   0   0   0   0   0   0   0   0   0   0  0  0  0  0
0  0  0  0  0   0   1   0   0   0   0   1   0   0   0   0  0  0  0  0
0  0  0  0  0   0   0   0   0   0   0   0   0   0   0   0  0  0  0  0
0  0  0  0  1   0   1   0   0   0   0   0   0   1   0   0  0  2  0  0
```

HERE ARE THE TOTALS FROM EACH REGION: TOTAL FIRINGS , NUMBER OF CELLS

	TOTAL FIRINGS	NUMBER OF CELLS
REGION 1	18	256
REGION 2	2	44
REGION 3	277	36
REGION 4	169	64

HERE ARE THE AVERAGE FIRING RATES FROM EACH REGION

LOW INHIBITION & LOW EXCITATION	0.07
HIGH INHIBITION & LOW EXCITATION	0.05
LOW INHIBITION & HIGH EXCITATION	7.69
HIGH INHIBITION & HIGH EXCITATION	2.64

Fig. 18.5. Boundary enhancement by lateral inhibition.

Recurrently mediated activity in two-dimensional networks (spatially contained or migratory activity). Figures 18.6 and 18.7 show two examples of ongoing recurrent activity in spatially organized networks. Figure 18.6 shows internally sustained activity in the case where the radius of recurrent excitation is smaller than the radius of recurrent inhibition. In this example a brief stimulation initiated activity and then was turned off. It is seen that the network maintains an ongoing pattern of activity and that this activity is spatially constrained to the general region where the stimulus initially was applied.

The example in Fig. 18.7 shows activity in the case where the radius of recurrent excitation is greater than the radius of recurrent inhibition and both recurrent excitation and recurrent inhibition are quite strong. This example shows a very interesting pattern which might be called "migratory activity." It is seen that activity initiated in a given region by some particular input pulse is immediately squelched by the localized recurrent inhibition. However, the lateral recurrent excitation reaching out beyond the region of the recurrent inhibition tends to initiate activity, which in turn spreads away from its origin. Activity tends to move away from its site of origin because of the recurrent inhibition and because of lingering refractoriness at those regions. Thus, there are tentacles or migrations of activity emanating outward through the network from the sources of activity. In these examples the driving activity and the interconnections are somewhat stochastic so the migratory tentacles exhibit some capriciousness of character and direction and eventually generally tend to die out. One can of course adjust parameters to fine-tune such phenomena to amplify or diminish any desired characteristics. Some possible implications of these results are discussed in Section C,4.

B. Program Listings and Discussion

The listing for NTWRK10 shows that this program has the same overall structure as the programs discussed previously. Ther are two new ingredients, both contained in the do-1000 loop for updating state variables. First, there are separate do-loops for updating state variables in each of the two neuronal populations, the excitatory cells and the inhibitory cells, as done in the 1500 and 1700 do-loops, respectively. Second, the activity-in-transit matrices must be extended to take into account the facts that there are two receiving populations and there are both excitatory and inhibitory effects in transit. In NTWRK10 there are three activity-in-transit matrices. SCET and SCIT represent excitatory synaptic currents projected to the excitatory cells and to the inhibitory cells, respectively. In this program excitatory synaptic input to the excitatory cells from the afferent fibers is

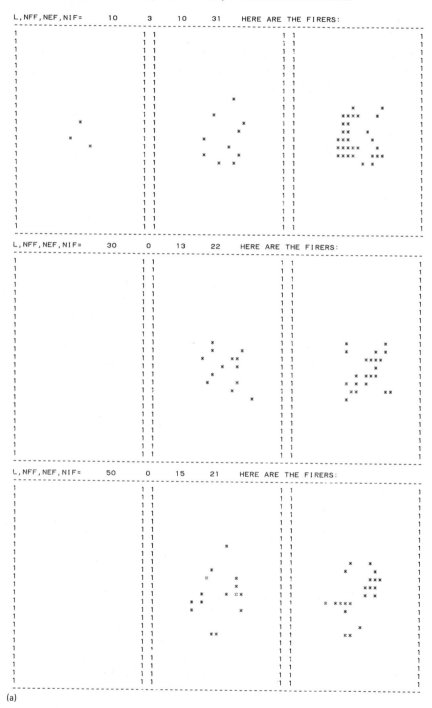

(a)

Fig. 18.6.   Spatially constrained, internally sustained activity ($r_i > r_e$). (a) Instantaneous firing disturbution. (b) Topographic disturbution of activity. (c) Locations of intense stimulation (I) and response (+). (d) Number of cells firing pertime in each population.

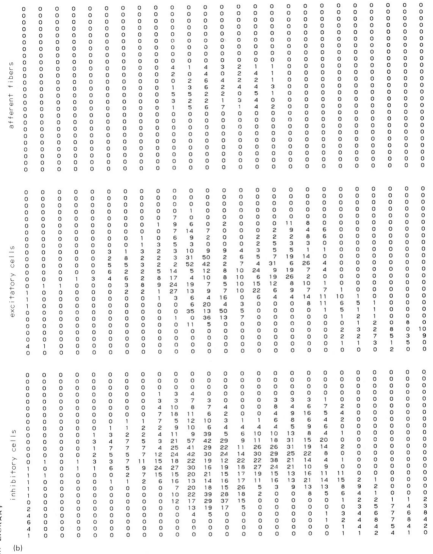

(b)

Fig. 18.6 (*Continued on next page*)

updated at each time unit and effected through the input function SC. The third activity-in-transit matrix, GI, carries the inhibition en route to the excitatory cells from the inhibitory population. GI is updated when inhibitory cells fire. As may be seen in the 1500 loop, GI effects synaptic inhibition by changing membrane conductance toward an equilibrium potential of EI,

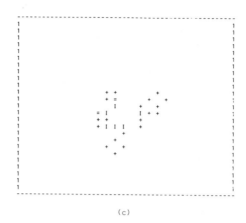

(c)

(d)

Fig.18.6 (*Continued*)

thereby affecting the membrane potentials of the excitatory cells. In this program all these activity-in-transit matrices are shifted to the left at each time unit in the 1910 and 1940 do-loops.

Program NTWRK11 also contains the 1500 and 1700 do-loops for the excitatory and inhibitory cells, respectively. However, updating of state variables and transmitting of information to activity-in-transit matrices are effected in subroutines. In this program excitatory and inhibitory activities in transit are still placed separately in two different activity-in-transit arrays. As in NTWRK10, GI houses inhibitory activity in transit en route to the excitatory population. All excitatory activity in transit, however, is stored in the matrix SCIT. Whether it is delivered to excitatory or inhibitory cells is coded in the first of its subscripts, which identifies the number of the receiving cell.

As may be seen in the listing the overall structure of the program NTWRK20 is very similar to that of program NTWRK11. The main diferences are that (1) the interconnections defined in subroutine TRNSMT are modified to recognize two-dimensional orientation, as discussed in Chapter 17, (2) coordinates for cells (JX) and (JY) are defined whenever a cell or fiber fires in order to flag its position in the ALINE display, and (3) the various parameters related to the two-dimensionality of the system (W, H, NC, NR, WTF, and HTF) are defined and used in the program. Program NTWRK21 adds code between the 105 and 108 CONTINUE statements to define the rectangular region for constraining input stimulation as discussed above. Program NTWRK22 slightly modifies the use of that box in terms of the signal-to-noise ratio and moreover, as may be seen in subroutine TRNSMT, effects a "wraparound" rule for migrating activity which comes to the edge of the rectangular region of the population. In this rule, if activity goes off, say, to the right side of the population, it exerts its influence on the left edge of the same population, etc. This artificial device allows more extended study of migratory patterns and roaming in internally sustained activity patterns.

C. Theoretical Background

1. Significance of Models

The occurrence of a parent population with a second local inhibitory population of cells interconnected as in the models in this chapter is ubiquitous throughout the nervous system. This basic structural arrangement must certainly perform functional operations of fundamental and widespread significance in the nervous system. The computer simulation models presented here can be used to explore and study quantitative characteristics of recurrent connections and lateral inhibition in conjunction with microelectrode recordings of units and small groups of units in such strategic locations as the hippocampus or the spinal cord. Moreover, the models can be used as they stand to illustrate concepts of functional operations. The examples of boundary enhancement and migratory activity briefly described above illustrate this capability. These simulation models, however, represent only the first step in realistic modeling of these phenomena. For example, the NTWRK series of programs as they stand are restricted to excitatory and inhibitory synaptic action. They do not incorporate more subtle refinements of synaptic interaction as discussed in Chapter 17. Second, these programs do not utilize neurons with dendritic trees. Third, the programs simulate responses to only one afferent input system. However, the programs as they stand exhibit the fundamental structure within which extensions to include these and other features can easily be made. Programs presented in Chapters 19 and 23 do remedy most of

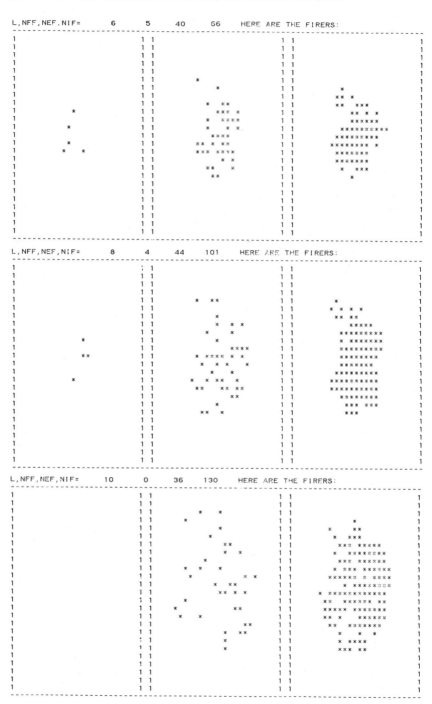

Fig. 18.7. Migratory internally sustained activity ($r_e > r_i$).

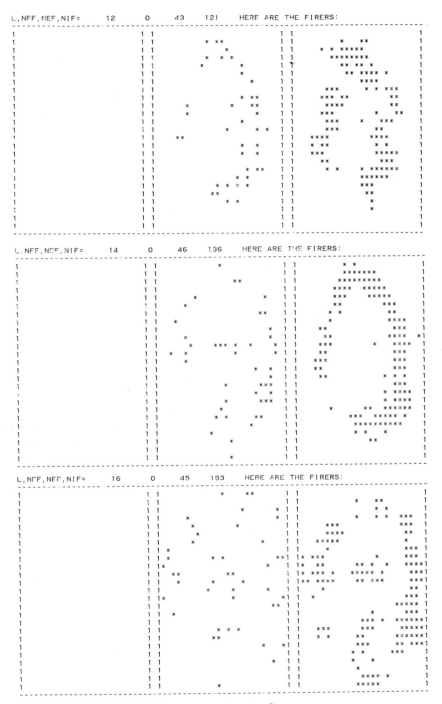

Fig. 18.7 (*Continued*)

these deficiencies and apply directly to the basic structural configuration investigated in this chapter. Therefore there is no real pressure to extend these particular programs of the NTWRK series.

The full range of characteristics and properties of these elemental phenomena associated with lateral and recurrent excitation and inhibition and internally sustained activity are not yet adequately explored and understood. Moreover, their implications are potentially profound for the operations of the nervous system. Indeed, the main guiding theme underlying the production of this book is that the secrets of information coding in neuronal networks have remained refractory to experimental investigation and obscure to our understanding in large part because of the absence of effective tools for studying the network context of neuronal activity. The computer programs and methodology of this book are dedicated to helping provide just such a methodological base for the detailed elucidation of information processing in the networks of the brain.

2. Theory of Internally Sustained Activity in Recurrently Connected Pools

A simple but elegant theory that captures the essential features of internally sustained activity in recurrently connected neuronal populations has been presented by Harth and Anninos. The quantitative essence of this theory is contained in equation system (18.1) and illustrated in Fig. 18.8. One pictures first that if N_i cells in the excitatory population fire at time i and if there are n_{ei} terminals from a given excitatory cell to the inhibitory population, there will be a total of $N_i \times n_{ei}$ excitatory postsynaptic potentials (EPSPs) incident on the inhibitory population at time $i + 1$. In the theory, the likelihood that a given receiving cell will be impinged on by any one of these particular EPSPs is equal to $1/M$, where M is the number of inhibitory cells. Moreover, one assumes that the placement of these incoming EPSPs is independent so that one has essentially a Bernoulli trial situation, where success means for a particular cell that it captures an EPSP and failure means that it does not. The mean number of EPSPs incident on a given inhibitory cell then is given by Eq. (18.1a):

$$\mu_M = \frac{N_i \cdot n_{ei}}{M} \tag{18.1a}$$

$$P_M(x) = \frac{\mu_M^x}{x!} e^{-\mu_M} \tag{18.1b}$$

$$\overline{M}_i = M * P(x \geq \text{Th}_M) = M * \sum_{x=\text{Th}_M}^{N_i \cdot n_{ee}} P_M(x) = M * \left\{ 1 - \sum_{x=0}^{\text{Th}_M - 1} P_M(x) \right\} \tag{18.1c}$$

$$\mu_e = \frac{N_i * n_{ee}}{N}, \qquad \mu_i = \frac{\overline{M}_i * n_{ie}}{N} \tag{18.1d}$$

$$P(x,y) = P(x)\,P(y) = \frac{\mu_e^x e^{-\mu_e}}{x!} \; \frac{\mu_i^y e^{-\mu_i}}{y!} \tag{18.1e}$$

$$\overline{N}_{i+1} = (N - N_i) * P(x - ay \geq \mathrm{Th}_N)$$

$$= (N - N_i) \sum_{x=\mathrm{Th}_N}^{N_i \cdot h_{ee}} \sum_{y=0}^{\min[\mathrm{Th}_N - x,\, M_i * n_{ie}]} P(x, y) \tag{18.1f}$$

Moreover, one assumes that M is large so that $1/M$ is small and the binomial distribution that corresponds to the Bernoulli trial situation can be replaced by the Poisson distribution. That is, the probability that a given inhibitory cell receives x EPSPs is given by Eq. (18.1b). Then one supposes that the likely number of inhibitory cells that fire at time i will be equal to the total number of inhibitory cells M times the probability that a given inhibitory cell fires, which in turn is equal to the probability that the number of EPSPs it receives [x in Eq. (18.1b)] is greater than or equal to the threshhold for the inhibitory cells. Therefore one can write Eq. (18.1c) for the likely number of inhibitory cells to fire at time i. Now consider the PSPs incident on the excitatory population. First there will be EPSPs directed back on these cells from the recurrent excitatory connection of the cells themselves, and second there will be inhibitory PSPs incident on these cells from the connections projected to the cells from the inhibitory cells. Using the sames arguments as above, the mean numbers of EPSPs and of IPSPs incident on a given single excitatory cell are given in Eq. (18.1d). Here, n_{ee} and n_{ie} are the numbers of connections to the excitatory cells from the excitatory and inhibitory populations, respectively. One again assumes that the number of cells, in this case N, is large so that the probability of success, $1/N$, is very small and the binomial distributions can again be replaced by Poisson distributions. One further assumes that the placements of individual EPSPs and IPSPs are independent of each other so that the probability that a given cell receives x EPSPs and y IPSPs is equal to the product of the probability of receiving x EPSPs times the probability of receiving y IPSPs. Specifically, the probability that a given excitatory cell receives exactly x EPSPs and y IPSPs is given in Eq. (18.1e). One then assumes that the expected number of excitatory cells to fire at time $i + 1$ is equal to the probability that any one cell will fire times the number of excitatory cells assumed to be available for firing. The latter quantity is taken to be equal to $N - N_i$, which embodies the assumption that all cells

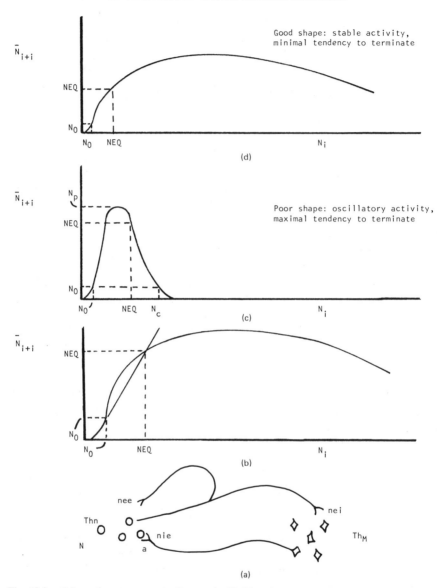

Fig. 18.8. Schematic one-stage transfer curves, N, N_{i+i}, for networks with combined recurrent excitation and recurrent inhibition: (a) combined recurrent excitation and recurrent inhibitions; (b) typical transfer curve; (c) poor transfer curve; (d) good transfer curve.

which fired at the last time will be refractory at this time and unavailable for firing. The probability that a given cell will fire is simply equal to the probability that $x - a \cdot y$ is greater than or equal to the threshold for these cells. Here a is the relative strength of IPSPs compared to EPSPs. Therefore the expected number of excitatory cells firing at $i + 1$ is given by Eq. (18.1f). Thus Eqs. (18.1c) and (18.1f) give precise quantitative estimates of the number of inhibitory and excitatory cells likely to fire as a result of recurrent connections in the network. More important, they give the influence of the various parameters in the network on the dynamics of its activity.

The generic response curve determined by Eqs. (18.1c) and (18.1f) is shown in Fig. 18.8b. The horizontal axes in Fig. 18.8 are the numbers of cells firing at time i. The vertical axes are the expected numbers of cells to fire at the next time, $i + 1$. Figure 18.8 also shows dashed lines where the number of cells firing at $i + 1$ is equal to the number of cells firing at i. Where the predicted transfer curve crosses this line are potential points of equilibrium. Notice that N_0, the lower equilibrium point, is an unstable equilibrium point: if the number of cells firing at time i is less than N_0 it is expected that the number of cells firing at $i + 1$ will be less than this value; by implication, activity, if initailly below N_0, dies out. On the other hand, if activity is initially greater than N_0 but close to N_0, the transfer curve predicts that activity at $i + 1$ will be greater than it was at i. By implication, at time $i + 2$ it will be higher than at $i + 1$ and so on. Without the recurrent inhibition the transfer curve projects up monotonically without the downward phase shown on the right side of Fig. 18.8. The only candidate equilibrium position is N_0 and it is definitely an unstable one. The system essentially goes into seizure activity. This has been called the ignition phenomenon. N_0 is a critical level of firing that a network must maintain in order to stay on. With recurrent inhibition, however, the potential capabilities are a little more subtle. The higher equilibrium point NEQ is a stable equilibrium point. If activity initially is a little greater than NEQ, the transfer curve, being below the dashed line, implies that the activity at $i + 1$ will be less than that at i, therefore driving it back toward NEQ. If the activity is initially slightly below NEQ, the transfer curve is above the dashed line and the activity at $i + 1$ will be slightly greater, therefore projecting it again back toward NEQ.

The above discussion shows that the upper crossover point NEQ is an equilibrium point for internally sustained activity in the network. That is, the recurrent excitation and recurrent inhibition are balanced so as to produce ongoing activity at the level of NEQ each time. Moreover, the shape of the transfer curve around the region of NEQ is critical in determining the nature of that internally sustained activity. For example, if the curve crosses the dashed line at NEQ with a very sharp downward slope to the right, the activity will tend to oscillate. The sharper the slope, the wilder the oscillation. If the peak value of the transfer curve N_p is greater than N_c, which is determined as defined in Fig. 18.8c by N_0, there will be a sharp tendency for this network to turn itself off. On the other hand, if the slope around NEQ is a gentle positive slope as shown in Fig. 18.8d, the internally sustained activity around NEQ is very stable. Specifically, if in this case initial activity is slightly greater than NEQ at time i, then the transfer curve predicts that one will obtain activity at $i + 1$ which is a little less than what it was but still greater than NEQ. By implication, at $i + 2$ the activity will still be diminished but still greater than NEQ. In other words, deviations from NEQ will result in activity which gradually approaches NEQ from the same side of the initial deviation. Fluctuations around NEQ will be unlikely to occur.

The implications of this analysis are theoretically very useful. For example, one may design model networks with parameters to produce various shapes of the engineering transfer curve so as to embody and produce different dynamic types of behavior. Specifically, a network designed to exhibit stable internally sustained activity should generally incorporate a transfer curve with a gentle positive slope across the equilibrium point as shown in Fig. 18.8d. A network designed to exhibit wildly oscillating activity should have parameters chosen to produce a transfer curve as shown in Fig. 18.8c. By the same token, if one encounters in nature recurrently connected populations which differ in their parameter values so as to produce differently shaped transfer curves, it is reasonable to speculate that nature may have made these choices in structural design in part because of the ensuing dynamical propensities associated with them.

The question of how the main parameters in the system influence the shape of the transfer curve is of some significance. In brief, these influences are as follows: first, increasing the strength of the recurrent excitatory drive as embodied primarily in the parameter n_{ei} and $1/Th_N$ tends to drive the transfer curve upward at any level of NT. In designing a population it is essential to have these parameters sufficiently large to ensure at least the possibility of ongoing activity. Second, increasing the drive of the excitatory population onto the inhibitory population, as embodied in the parameters n_{ei} and $1/Th_M$, tends to drive the transfer curve downward at low values of N_i. These two parameters essentially determine when the inhibitory population comes on relative to when the excitatory population comes one. Third, increasing the value of the synaptic strength from the inhibitory cells to the excitatory cells tends to drive the transfer curve downward but only at high values of N_i. At lower values of N_i the critical factor determining the effectiveness of the inhibition is whether or not inhibition falls on a cell. It is only at very high levels of activity when interaction of inhibition and excitation is experienced on large numbers of cells that increasing the synaptic strength of inhibition per se is effected in further driving down the curve. Fourth, increasing the number of inhibitory terminals falling on the excitatory population, as embodied in the parameters n_{ie} and M, tends to drive down the transfer curve more or less uniformly over the entire domain of N_i. These parameters tend to increase the spread of inhibition throughout the excitatory population and effectively amount to a uniform increase in the threshhold required for excitation of the excitatory cells. With these observations it is not difficult to adjust parameters in networks with the structural arrangements discussed in programs of this chapter in a trial-and-error fashion to produce transfer curves and therefore internally sustained activity of any desired character.

3. Theory of Lateral Inhibition

The essential theory of the operations of lateral inhibition as we now know it were worked out by Ratliffe and Hartline some 20 years ago. Undoubtedly the functional implications of lateral inhibition in both its feed forward and recurrent forms with respect to its capacities for sculpturing information processing in neuronal populations are beyond what we have yet been able to grasp. There are implications for sculpturing of information in both spatial and temporal dimensions and moreover within spatiotemporal interactions. The essence of feature extraction by lateral inhibition can be illustrated nicely by the spatial phenomenon of boundary enhancement, as illustrated in Fig. 18.9. Imagine for simplicity a linear array of receiving neurons that include both exitatory and inhibitory cells, which are for the sake of simplicity imagined to be equally bombarded by the input at any given spatial location. The essential ingredient is that the inhibitory cells project inhibition back onto the excitatory cells over a certain lateral dimension. In Fig. 18.9 each inhibitory cell is taken to project to three adjacent-excitatory cells. Suppose, as is illustrated in Fig. 18.9, there is a boundary in the intensity of stimulation incident on the population. One-half of the field experiences high-intensity input, the other half of the field experiences more low-intensity input. It is easy to see that in terms of the interactions of excitatory and inhibitory influences the excitatory population may be partitioned into four regions: (1) cells away from the boundary under the high-intensity stimulation exhibit high-intensity excitation and high-intensity inhibition; (2) cells far away from the boundary in the low-intensity region exhibit low-level excitation and low-level inhibition. However, for cells under the high-intensity stimulation but very close to the boundary, there is high incident direct excitation but only medium level inhibition because some of the inhibitory input to these cells are excited by the low-level intensity input. Moreover for cells in the low-level input region, but close to the boundary, there is a low-level incident excitation but a medium-level inhibition because some of the inhibitory cells acting on these cells are in the

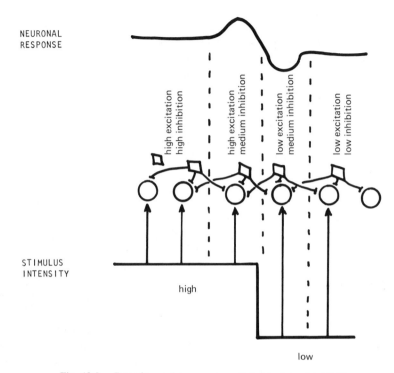

Fig. 18.9. Boundary enhancement mediated by lateral inhibition.

high-level-input region. Qualitatively, the expected level of output for these four regions can be projected as is shown in the curve in the top of Fig. 18.9. There is a high level of activity in the high-intensity region right near its boundary and a low level of activity in the low-intensity region right near its boundary. If one adjusts the parameters associated with the relative strengths of recurrent excitation and inhibition, it is possible to make the levels of activity in the central regions of the high and low intensity about the same. In this case only the boundary is flagged in the information that projects outward from the neuronal populations. It is easy to see that the same kind of reasoning applies to two-dimensional arrays.

4. Recurrently Mediated Activity in Spatially Organized Networks

It is tempting to think of internally sustained activity in neuronal populations as embodying the tendency of the nervous system to take over control from or at least go beyond the influence of the afferent input to which it is subject. For example, one might in particular imagine internally sustained activity to represent some form of internal representation or memory traces of events. The example of spatially constrained, internally sustained activity discussed in Section A could be conceived, for example, to represent a very primitive sort of memory system in a two-dimensional network. One could imagine that particular memory traces were stored in particular locations in the two-dimensional network. Activity in particular regions would correspond to recall of particular memories. Input pulses that would tend to trigger internally sustained activity in particular regions would be reminders of some sort.

Particularly interesting is the influence of the relative sizes of the fields of recurrent excitation and inhibition on activity patterns and the corresponding concept that these might be in part subject to certain variable influences within the nervous system so that on occasion recurrent inhibition might tend to dominate where on other occasions it might be lessened to some degree. The example, shown in Section A of migratory tentacles emenating from a focus of activity when recurrent excitation outreaches recurrent inhibition, for example, has some interesting implications. Suppose, for example, there were two focii of activity in a two-dimensional net, active more or less simultaneously such that the tentacles from the two different focii exhibited some partial overlap. Where the tentacles would cross there would be the likelihood of another focus of activity and again creation of yet a third tentacle system. It is tempting to think of this situation as a possible embodiment of associations of memory traces. The individual tentacles would represent some sort of "features" associated with the individual memory traces; the interactions of the tentacles would represent some association between memory traces. Moreover it is interesting to imagine that the operation of such a system would be heavily dependent on the state of the system. For example, if factors were such that the tentacles tended not to occur on one occasion, this would correspond to a cognitive system that was somewhat highly rigid and constrained, whereas a state of the system which resulted in more ubiquitous crossing of migratory tentacles could correspond to a cognitive system exhibiting widely ubiquitous tentacles and very widespread interacting of tentacles could correspond to a completely undisciplined cognitive system or a breaking down of a cognitive system. Such concepts are entirely speculative, but nonetheless illustrate the level of intricacy of operational ideas which emerge from neuronal network simulation modeling and to illustrate that such speculative insights go considerably beyond what electrophysiological experimentation has as yet been able to provide.

D. Exercises

1. Use program NTWRK11 to attain internally sustained activity. Simulate a population of 300 excitatory cells and 100 inhibitory cells. Adjust the number

of terminals and synaptic strengths at the three internal junctions to obtain this activity. Apply external stimulation in the input fiber population for a period of ten time units. Strive to maintain the ongoing firing rate at about 25 to 40 cells per time unit in the excitatory population.

2. Readjust parameters in your simulation from 1 so as to produce widely oscillating internally sustained activity. This can be obtained by applying a rather strong inhibition from the inhibitory population to the excitatory population, but adjusting other parameters so that the inhibitory cells do not begin to fire until substantial numbers of excitatory cells are firing.

3. Now readjust your parameters from simulations in 1 and 2 so that the number of cells firing at each time unit remains very constant. How constant can you make this? Keep the mean level at around 30 per time unit.

4. Plot the transfer curve of number of excitatory cells firing at time $i + 1$ versus number of excitatory cells firing at time i for your simulations in Exercise 2 and in Exercise 3.

5. Use program NTWRK22 to produce boundary enhancement by lateral inhibition in response to a rectangular region of synaptic input bombardment.

6. Experiment with various other patterns of incident synaptic bombardment to explore further porperties of later inhibition. Can you identify any such effects as candidates for "feature extraction?"

7. Use program NTWRK22 to produce internally sustained activity that is constrained to a particular region in space.

8. Use program NTWRK22 to simulate recurrently mediated activity which tends to migrate across the network.

9. Can you produce internally sustained migratory activity using program NTWRK22?

10. Modify program NTWRK11 so that the input fibers directly activate the inhibitory cells as well as the excitatory cells and thereby produce feedforward inhibition.

11. Further modify NTWRK11 so that the inhibitory cells project inhibitory input to themselves as well as to the excitatory cells. Adjust the parameters in your simulation in #5 so that the response levels away from the boundary, both in the central regions under the stimulation and in the peripheral regions away from the stimulation, are about equal. Thus, only the boundary stands out in response to the stimulation pattern.

12. Repeat Exercise 6 using the parameter adjustments you made in Exercise 12.

13. For some particulary interesting information sculpturing feature you discovered in Exercise 6 or 12, repeat, adjusting only the signal to noise ratio (parameter SNRATIO) so as to explore the tenacity of your phenomenon in the face of background noise. Can you produce by this devise a "stochastic" system wherein the information processing operation is operative on some occasions and not on others?

14. Modify program NTWRK22 so that the synapses from the input fiber population upon the excitatory cell population facilitate according to the rules used in Chapter 15. Explore the significance of this facilitation on various input–output features of this system.

19

Systems of Interacting Neuronal Populations

This chapter presents computer programs to simulate arbitrary configurations of multiple interacting neuronal populations. The user may specify arbitrary numbers of cells and their determining parameters for each population. The populations may be coded to interact synaptically in any desired configuration. The physiological and anatomical parameters for each junction may be specified. The user may define an arbitrary number of external input fiber populations to impinge on the neuronal populations in any desired configuration. For these input systems one may specify the parameters determining their physiological activity and anatomical connections independently. In some of the programs the synaptic systems at each junction are individually adjustable. In the other programs, all synaptic junctions are either excitatory or inhibitory. Some of the programs incorporate two-dimensional rectangular spatial organization, individually adjustable in each of the neuronal populations and input fiber systems. The other programs represent aspatial pools and fiber systems. The dynamics of all single neurons in the programs of this chapter are determined by the state variable point neuron model presented as PTNRN10 in Chapter 14. However, since updating of cell state variables is done in a subroutine, it is a simple matter to substitute more complicated model single neurons. Indeed, the programs of this chapter are extended to represent neurons with simplified dendritic trees in the programs of Chapter 23.

A. Using the Programs: Input–Output Charts and Illustrative Examples

Figure 19.1 shows the input–output flowchart for program SYSTM10. The main input functions in this program are synaptic input currents SCIT

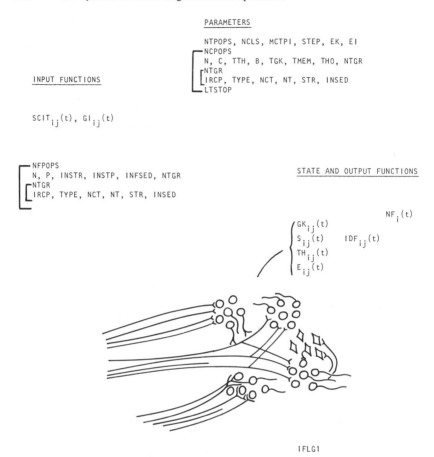

PARAMETERS

NTPOPS, NCLS, MCTPI, STEP, EK, EI
NCPOPS
N, C, TTH, B, TGK, TMEM, THO, NTGR
NTGR
IRCP, TYPE, NCT, NT, STR, INSED
LTSTOP

INPUT FUNCTIONS

$SCIT_{ij}(t)$, $GI_{ij}(t)$

NFPOPS
N, P, INSTR, INSTP, INFSED, NTGR
NTGR
IRCP, TYPE, NCT, NT, STR, INSED

STATE AND OUTPUT FUNCTIONS

$NF_i(t)$

$GK_{ij}(t)$
$S_{ij}(t)$ $IDF_{ij}(t)$
$TH_{ij}(t)$
$E_{ij}(t)$

IFLG1

Fig. 19.1. Program SYSTM10. This program simulates the activity of an arbitrary system of interconnected neuron pools activated by an arbitrary number of input fiber systems. All connections are excitatory or inhibitory.

and input inhibitory synaptic conductance changes GI. These functions of time for each neuron in each population are determined according to the input activity distributed over an arbitrary number of input fiber populations NFPOPS. For each of these fiber populations, the user prescribes the string of parameters N, P, INSTR, INSTP, INFSED, and NTGR. These represent, respectively, the number of fibers in the population, the probability that a particular fiber will fire at a given point in time, the starting and stopping times for activity in that fiber system, a random number seed governing timing of action potentials in that fiber system, and the number of neuronal populations projected to or targeted by that particular fiber population. For each population, one reads the following list of parameters for each of its output junctions: IRCP, TYPE, NCT, NT, STR, and INSED. These represent, respectively, the identity of the receiving population, type of the synaptic junction (1 if excitatory, 2 if inhibitory), conduction time, number of terminals per fiber, strength of the output synapses, and a random number seed governing the specific anatomical connections made by each fiber in that system.

The main state and output functions for SYSTM10 are NF, the number of firers in each population and each fiber system at each instant in time, and IDF, the corresponding identities of firers for each population and fiber system at each instant in time. In addition, the four state variables E, TH, S, and GK are available for each cell in each population. If IFLG1 is greater than zero, values of these state variables are printed out for all cells at each time. If IFLG1 is equal to zero, this information is repressed. Parameters for the system include the number of total populations NTPOPS, number of fiber populations NFPOPS, number of cell populations NCPOPS, maximum number of cells in any cell population NCLS, maximum conduction time plus 1 MCTP1, temporal integration step size STEP, potassium equilibrium potential EK, typically taken as -10, and equilibrium potential of the inhibitory synapses EI, normally taken as -10 mV. For each of the cell populations one reads in first the string of parameters N, C, TTH, B, TGK, TMEM, TH0, and NTGR. N is the number of cells in the population and NTGR the number of neuronal populations to which this population sends output fibers. The other parameters are the familiar parameters for the state variable model as discussed in Chapter 14. Then for each of the cell populations targeted for output from this population, one reads in the parameters IRCP, TYPE, NCT, NT, STR, and INSED. The parameters for these junctions have precisely the same interpretation as for the junctions of the input fiber systems discussed immediately above. Finally, LTSTOP is the duration of the simulation run.

With program SYSTM10, then, one may identify and define an arbitrary number of neuronal populations, each with desired individual characteristics; prescribe interconnections for those populations in any desired fashion; bombard the populations with any desired patterns of input in an arbitrary number of input fiber populations incident on the populations in any desired fashion; and finally observe the resulting ongoing dynamic activity in terms of overall measures such as the total numbers of cells or fibers firing in the populations as functions of time or, more probingly, the identities of specific neurons or fibers firing at each instant in time or, even more probingly, the ongoing patterns of values of state variables in any constituent cells as functions of time.

Figure 19.2 shows the input–output flowchart for program SYSTM11. Program SYSTM11 is essentially identical to program SYSTM10, with the

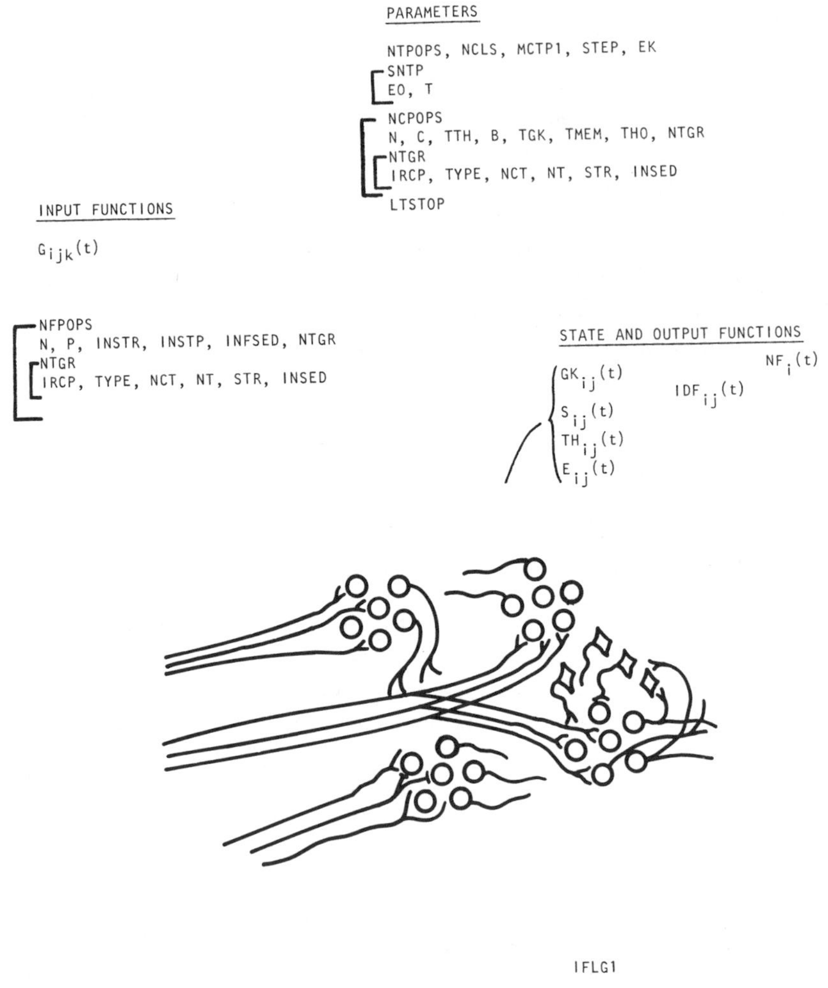

PARAMETERS

NTPOPS, NCLS, MCTP1, STEP, EK
┌SNTP
└ EO, T
┌ NCPOPS
│ N, C, TTH, B, TGK, TMEM, THO, NTGR
│ ┌NTGR
└ └IRCP, TYPE, NCT, NT, STR, INSED
└LTSTOP

INPUT FUNCTIONS

$G_{ijk}(t)$

┌NFPOPS
│ N, P, INSTR, INSTP, INFSED, NTGR
│┌NTGR
└└IRCP, TYPE, NCT, NT, STR, INSED

STATE AND OUTPUT FUNCTIONS

$$\begin{cases} GK_{ij}(t) \\ S_{ij}(t) \\ TH_{ij}(t) \\ E_{ij}(t) \end{cases} \quad IDF_{ij}(t) \quad NF_i(t)$$

IFLG1

Fig. 19.2. Program SYSTM11. This program simulates the activity of an arbitrary number of neuron pools activated by an arbitrary number of input fiber systems.

exception that in SYSTM11 all the individual synaptic junctions are individually adjustable. Specifically, one specifies a parameter SNTP, the number of synaptic types; then for each synaptic type one defines a synaptic equilibrium potential EQ and a time constant T which represents the time constant of decay of postsynaptic conductance changes associated with that synaptic type. Then the variable TYPE prescribed in each synaptic junction may take on any value from 1 up to SNTP and prescribes for that junction the particular values of EQ and T for the corresponding synaptic type.

Notice that in this arrangement different output junctions from a single sending population may be assigned distinct synaptic types, reflecting the concept that the character of a synaptic junction depends on both the presynaptic and postsynaptic elements involved. In all other features, program SYSTM11 is identical to program SYSTM10.

Figure 19.3 shows the input–output flowchart for programs SYSTM20 and SYSTM21. These programs are essentially identical to program SYSTM10 except that now rectangular two-dimensional spatial organization is represented in all neuronal populations and all input fiber systems. The new parameters introduced include W and H, the maximum width and

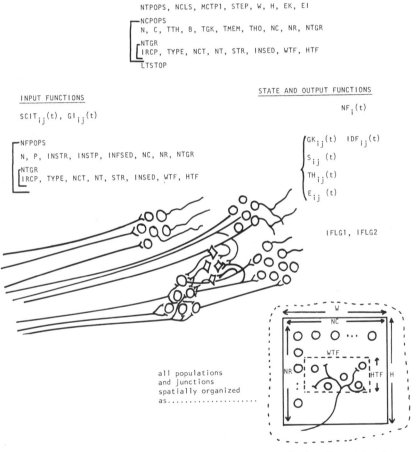

Fig. 19.3. Programs SYSTM20 and SYSTM21. These programs simulate the activity of an arbitrary system of interconnected spatially organized neural networks activated by an arbitrary number of spatially organized input fiber systems. All connections are excitatory or inhibitory.

maximum height of the rectangular arrays, respectively. NC and NR are the number of columns and number of rows which must be prescribed for each cell and fiber population. WTF and HTF are the widths and heights of the terminal fields, which must be prescribed for every synaptic junction, respectively. These new parameters are indicated in the sketch in the lower right corner of Fig. 19.3. Program SYSTM21 includes a simple two-dimensional graphic display. If the parameter IFLG2 is greater than zero, this graphic display is printed at each time step. If IFLG2 is equal to zero, this printing is repressed. Both SYSTM20 and SYSTM21 identify all synaptic junctions as being either excitatory or inhibitory. If they are excitatory they are effected through a synaptic current SCIT; if inhibitory, they are effected through an inhibitory synaptic conductance change GI. In this respect these programs are the same as SYSTM10. In terms of the other systemic parameters and the overall input functions and state and output functions, these programs are identical to the aspatial programs SYSTM10 and SYSTM11.

Figure 19.4 shows the input–output flowchart for program SYSTM22. This program differs from programs SYSTM20 and SYSTM21 only in that all synaptic junctions are now individually adjustable. Again, one specifies the number of synaptic types by SNTP and the characteristics of individual synaptic types by the synaptic equilibrium potential EQ and the time constant of postsynaptic conductance change T. The value of TYPE, which is read in for each junction, may be again assigned any value from 1 to SNTP, and this associates the particular EQ, T couple with that particular synaptic junction. Program SYSTM22 includes the simple two-dimensional graphic display that is included in SYSTM21. This display may be activated or repressed according to the value assigned to IFLG2, as in SYSTM21. In all other respects, program SYSTM22 is equivalent to the other programs of this chapter.

EXAMPLE

Serially coupled oscillators.   Figure 19.5 illustrates an example of 360 neurons in six populations simulated with SYSTM11. The large rhythmic responses (at about 10, 60, 100, and 140) are driven by external bombardment in the common afferent system B, whereas the faster and lower-amplitude oscillations reflect properties of the local circuits.

B. Program Listings and Discussion

As can be seen in the listing in the Appendix 2 for SYSTM10, SYSTM11, SYSTM20, SYSTM21, and SYSTM22, the FORTRAN program

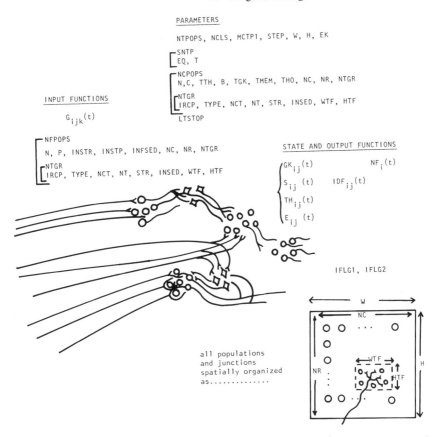

PARAMETERS

NTPOPS, NCLS, MCTP1, STEP, W, H, EK

$\begin{bmatrix} \text{SNTP} \\ \text{EQ, T} \end{bmatrix}$

$\begin{bmatrix} \text{NCPOPS} \\ \text{N,C, TTH, B, TGK, TMEM, THO, NC, NR, NTGR} \end{bmatrix}$

INPUT FUNCTIONS

$G_{ijk}(t)$

$\begin{bmatrix} \text{NTGR} \\ \text{IRCP, TYPE, NCT, NT, STR, INSED, WTF, HTF} \end{bmatrix}$

LTSTOP

$\begin{bmatrix} \text{NFPOPS} \\ \text{N, P, INSTR, INSTP, INFSED, NC, NR, NTGR} \end{bmatrix}$

$\begin{bmatrix} \text{NTGR} \\ \text{IRCP, TYPE, NCT, NT, STR, INSED, WTF, HTF} \end{bmatrix}$

STATE AND OUTPUT FUNCTIONS

$\begin{cases} GK_{ij}(t) & NF_i(t) \\ S_{ij}(t) & IDF_{ij}(t) \\ TH_{ij}(t) \\ E_{ij}(t) \end{cases}$

IFLG1, IFLG2

all populations
and junctions
spatially organized
as..............

W

NC

NR

WTF

HTF

H

Fig. 19.4. Program SYSTM22. This program simulates the activity of an arbitrary system of interconnected spatially organized neural networks activated by an arbitrary number of spatially organized input fiber systems. All synaptic systems are individually adjustable.

SYSTM10 contains the same overall structure as preceding programs discussed in this book. There are sections to read and write input parameters, to initialize variables, a big do-loop to update state variables, and a section prescribing formats. State variables must now have two subscripts corresponding to the cell or fiber population and the individual cell or fiber number. State variables are updated in a SVUPDT subroutine. Synaptic activity is interjected into activity-in-transit matrices SCIT and GI have three subscripts corresponding, respectively, to the cell population targeted, the number of the individual receiving cell, and the time until arrival. The symbols IS and IR in subroutine TRNSMT correspond to the identities of the sending and receiving populations. This program is not complicated in its structure but is remarkably versatile in terms of the variety of neuronal configurations it can simulate.

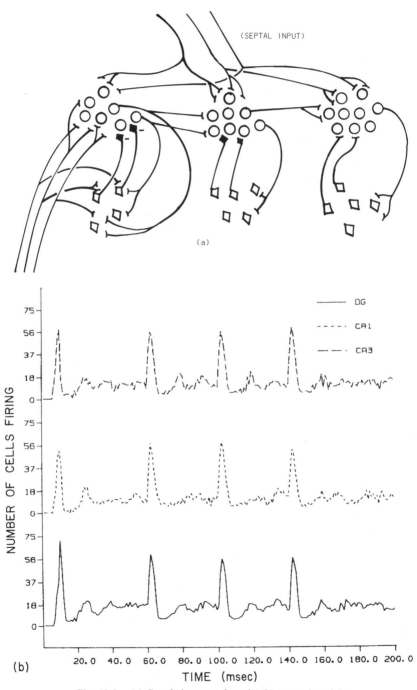

Fig. 19.5. (a) Coupled neuronal pools; (b) network activity.

The main new ingredient in program SYSTM11, as can be seen in the listing, is that the activity-in-transit matrices are now all represented by a single four-dimensional array G. The first subscript corresponds to the synaptic type, the second subscript to the identity of the receiving population, the third subscript to the identity of the individual receiving cell, and the last subscript to the time until arrival of the signal. As can be seen in subroutine TRANSMT in this listing, G replaces the activity-in-transit matrices SCIT and GI used in SYSTM10. As can be seen in the 10 do-loop in subroutine SVUPDT, then, the activity-in-transit matrix G determines synaptic current SYN for each cell in each population at each time according to the appropriate synaptic equilibrium potentials EQ and corresponding contributions GS to the total conductance and therefore to the effective time constant and decay rate DCE for each cell and each population at each time. As usual, as shown in the 1910 do-loop in the parent program, the activity-in-transit matrices are shifted to the left at each time unit.

Program SYSTM20 extends SYSTM10 to include spatial representation. As seen in the listing for SYSTM20, this is effected entirely by first reading in new parameters W, H, NC, NR, WTF, and HTF and then interchanging back and forth in subroutine TRANSMT between cell and fiber numbers and corresponding spatial locations. A cell number K, for example, is converted to x and y coordinates (XS) and (YS) in terms of the number of columns and number of rows NC and NR prescribed for the population of which K is a member. Although the rules for (XS) and (YS) prescribed here represent a two-dimensional rectangular array, it would be simple to code any other spatial organization simply by replacing the rules for those two spatial coordinates and perhaps adding a third spatial coordinate if desired. The x and y coordinates of cells targeted by particular senders (XR) and (YR) are computed relative to the x and y coordinates of the sender (XS) and (YS). These are then converted into the column and row numbers of the receiver, ICR and IRR, respectively, and finally into NREC, which is the number of the receiving cell. This code is contained in the 100 do-loop within subroutine TRANSMT. Again, it would be easy to change the particular rules in this subroutine for different spatial configurations. Except for this interplay between cell number and spatial location contained within subroutine TRANSMT and the prescribing of a few parameters associated with spatial structure, the listing for program SYSTM20 is essentially identical with that for SYSTM10. It is also, in essence, a relatively simple program structurally.

Program SYSTM21 extends program SYSTM20 to include a simple two-dimensional graphic display. This display produces at each point in time for each cell population a two-dimensional array with asterisks at the locations of cells that fire at that point in time. As shown in the listing for program SYSTM21, this is effected in the do-2000 and 3000 loops for cells and fibers, respectively. For each cell or fiber that fires, one computes its x

and y coordinates (JX) and (JY) and puts an asterisk in the corresponding position in the array ALINE. There arrays are then written out in the 2700 and 3700 do-loops if IFLG2 is greater than zero. In all other respects, program SYSTM21 is identical to program SYSTM20.

Both programs SYSTM21 and SYSTM20 incorporate the activity-in-transit matrices SCIT and GI corresponding to the prescription that all junctions are either excitatory or inhibitory. Program SYSTM22, as shown in the listing, generalizes the synaptic junctions to be individually adjustable in the same fashion that program SYSTM11 does. Again, program SYSTM22 replaces SCIT and GI with the four-subscripted matrix G. Synaptic types are defined by their equilibrium potential EQ and time constant postsynaptic conductance decay T. The number of synaptic types SNTP is unlimited. Program SYSTM22 then includes the two-dimensional rectangular spatial organization contained in SYSTM20 and SYSTM21 and the two-dimensional graphic display contained in SYSTM21.

C. Theoretical Background

1. Dealing with Parameters in Large-Scale Computer Models

A main question concerns how such a modeling effort interacts with the large number of parameters it encompasses. A typical model of a neuronal system with, say, four or five neuron populations might contain about 85 parameters. A frequent criticism is that models with many free parameters are suspect because they can be adjusted to match virtually anything. In the first place, this criticism, although often repeated, is simply not true. The mechanisms of a model determine its trends, while its parameters determine largely specific point values. One usually cannot satisfactorily match a model with the wrong mechanisms to both the point values and the overall trends of data no matter how the parameters are adjusted. Moreover, and more to the point for such large and complicated systems, the model system and real system produce a very wide range of behavior in many different contextual situations. For the model to be satisfactory, it has to mimic the observed behavior in trend and point values for all the different situations. If the model has the wrong mechanisms, and if the parameters have been chosen to force the model to match as best it can data obtained under certain conditions, it is most unlikely that it will happen to match behavior under rather different conditions. In the second place, the parameters of many "bottom-up" models are not "free parameters." They are tied explicitly to observable anatomical or physiological quantities. Target values can be specified for them with reasonable confidence. Moreover, most of them can be evaluated rather accurately by specific prescribed experiments. In the third place, one of the main values of a modeling system is to show how significant various parameters are in influencing various features of activity. In fact, there is much variability in anatomical and physiological values associated with individual members of a given class (e.g., different individual granule cells). Models can predict, for example, whether the variations observed in dendritic branching and size adequately account for observed variations in various passive electrotonic parameters within a given class of cells. Moreover, it seems that certain classes of cells and networks may differ from each other primarily by making use of different parameter values within a similar overall framework. For example, the relative strength of recurrent inhibition in recurrently connected networks may in one case completely "squelch" activity producing a phasic or pulsing output signal in one population but in another simply modulate the mean level of output activity. A model system, by illustrating the different kinds of behavior determined by different parameter values, can indicate how different classes of systems may have specialized from a common overall configuration.

2. *Vistas of Simulating Interacting Neuronal Populations*

It is in simulating systems of interacting neuronal populations—in using the programs of the SYSTM series and their extentions and elaborations—that the full value of the quantitative bottom-up approach of engineering modeling of nervous systems should begin to produce its most valuable fruits. It is the primary location where theoretical and modeling studies can penetrate significantly beyond the outer limits of experimental approaches in producing candidate embodiments of the intricacies of brain activity. This is an area where techniques to deal with complexity are most needed. It is an area whose cultivation will require a great deal of effort, but which will be rewarded by glimpses of some of the most fascinating of nature's secrets. The programs developed in this chapter are limited primarily in that they are restricted to point neurons. They can nonetheless be used to advantage in exploring fundamental questions about many of the infinitely varied problems and questions associated with research on different brain networks. Moreover, they can readily be extended to include progressively more realistic and more precisely defined features for specific systems. For example, in Chapter 23 the main programs of this section are extended to included representation of neurons with simplified dendritic trees. The programs of Chapter 20 extend the programs of the SYSTM series discussed here to include the embedding of permanent memory traces according to the Hebbian rule of increased synaptic efficiency at successful synapses.

3. Place of Computer Modeling in Neuroscience

Another fundamental question concerns the way modeling efforts might be used to advance understanding in neuroscience. One point of view is that models cannot be reliably constructed, and therefore presumably not used meaningfully, unless one has complete information concerning all their parameters and inputs. The alternative point of view, which I espouse, is that models are valuable methodological tools which can be used to great advantage in interaction with experimentation in pushing forward at whatever level of ignorance exists on a given system. Constructing a model forces one to spell out precisely the quantities that are unknown. It can help to define experiments that must be performed to clarify issues or fix parameter values. Good models, imaginatively used, can help portray the universe of possibilities beyond established knowledge and therefore help lead questioning and experimentation to significant steps. If one can recognize the truth within the simple summary "theory guides while experiment decides," then it seems accurate to claim that for exceedingly complicated systems like the nervous system, large-scale computer modeling systems help elaborate the interfaces of the guiding and deciding processes.

Another point of view is that models represent a sort of analog wherein one might be asserting that a given neuronal system behaves in part "like" some other removed and simpler physical system—for example, a network with recurrent inhibition is in part like an electrical oscillator. Such thinking sometimes generates valuable physical insights. However, this is not the point of view of this book. The material of this book is purported to provide reasonably accurate mathematical descriptions of the salient physiological mechanisms which govern the organization of electrical activity in neuronal systems. The value of these computer models is that they represent the dynamic behavior predicted to result from these mechanisms operating within the real neuronal systems they simulate.

D. Exercises

1. Use program SYSTM11 to simulate the arrangement illustrated in Fig. 19.5 of the hippocampus as three coupled oscillators. First, with serial connections between the three excitatory populations disengaged (i.e., number of terminals and synaptic strengths set equal to 0), cause each of the three regions to fire rhythmically but out of phase with each other. Second, bring the three regions into phase through the serial excitatory connections between the regions. Third, again disengaging the serial excitatory connections, bring the three regions into synchrony by externally applied input from the septal input system (to successfully do part 3, you must slightly modify the code to produce rhythmic firing in the septal input).

2. Extend your simulation from Exercise 1 so that the serial connections include feed-forward inhibition at both stages. Can you adjust the relative strengths of the excitatory and inhibitory serial connections so that the three regions fire asynchronously even though serial connections are effected? If so, can you adjust the temporal structure of the input systems without adjusting internal parameters so that low-level, fast activity rather than synchronous activity obtains in each of the three regions? If so, can you discern any possible "information processing" operations in the dynamic patterns you observe? Now bombard this system with rhythmic input from the septal input system. Does this produce synchronized rhythmic activity?

3. Use program SYSTM22 to represent a crude model of information processing in the visual system as follows. Simulate six populations of neurons, an excitatory and an inhibitory population at each of three stages in a longitudinally organized system. Bombard both the excitatory and inhibitory populations of the first stage with extrinsically applied input patterns. Let the excitatory cells in each stage project to the inhibitory cells of that stage only. Let the inhibitory cells of each stage project only to the excitatory cells of that stage. Let the excitatory cells of the first and second stages project to both the excitatory and inhibitory cells of the second and third stages, respectively. Now, building on the exercises in Chapter 18, adjust parameters in the first stage to produce boundary enhancement for rectangular stimuli. Adjust parameters in the second and third stages to produce similar effects. Now experiment with various spatial patterns in the extrinsic input system. Do you observe operations that might be related to "feature extraction" in the second and third stages of this system? Manipulate input patterns and intrinsic parameters in a search for such operations.

4. Use program SYSTM11 to simulate dynamic properties of the recurrent inhibition mediated by Renshaw cells in the spinal cord. Go to the original literature to discover appropriate anatomy and physiology for the system. If necessary, modify program SYSTM11 to include such things as short-term synaptic plasticity.

5. Use program SYSTM22 to set up a simplified model of the circuitry of the cerebellum. If necessary, extend the program to include a third spatial dimension.

6. Use program SYSTM11 to represent a simple model of information processing in several stages of the auditory system.

7. Extend your model from Exercise 5 to include a two-dimensional spatial representation by using program SYSTM22. Can you represent a topographic distribution of frequency sensitivity in the cortical region of this model?

20

Embedding of Memory Traces in Neuronal Junctions and Systems by Hebbian Learning

This chapter presents computer programs (located in Appendix 2) to simulate the embedding of memory traces according to the rules of Hebbian learning in several of the structural neural network configurations considered in previous chapters. Specifically, several programs (JNLRNO0, JNLRNOA, and JNLRNOB) simulate Hebbian learning at the junction of a single afferent fiber system with a single receiving neuronal population of cells. These programs illustrate the basic computational approach taken here to simulate Hebbian learning. The second series of programs (LRSYS20, LRSYS2A, LRSYS2B, LRSYS30, LRSYS31, LRSYS32, and LDSYS33) simulate Hebbian learning in arbitrary systems of interacting neuronal populations. Programs JNLRNO0, JNLROA, JNLRNOB, and LRSYS30 are aspatial; the others represent neuronal systems, which are rectangularly organized in two-dimensional space. In LRSYS20, LRSYS2A, and LRSYS2B, all junctions are either excitatory or inhibitory. In programs LRSYS30, LRSYS31, LRSYS32, and LDSYS33, individual synaptic junctions are individually adjustable. All programs of this chapter use neurons represented by the state variable point neuron model of PTNRN10. The programs can easily be extended to incorporate more elaborate model neurons. This is done in Chapter 23.

A. Using the Programs: Input–Output Charts and Illustrative Examples

Figure 20.1 shows the input–output flowchart for programs JNLRNO0, JNLRNOA, and JNLRNOB. From the standpoint of the user, all these programs are identical. They differ only in the way information is stored and processed internally in the programs. The overall flow and structure of

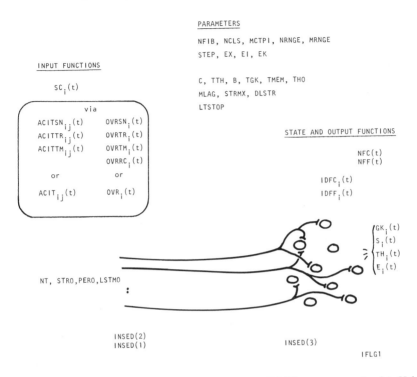

PARAMETERS

NFIB, NCLS, MCTPI, NRNGE, MRNGE

STEP, EX, EI, EK

INPUT FUNCTIONS

$SC_i(t)$

C, TTH, B, TGK, TMEM, THO

MLAG, STRMX, DLSTR

LTSTOP

via

$ACITSN_{ij}(t)$ $OVRSN_i(t)$
$ACITTR_{ij}(t)$ $OVRTR_i(t)$
$ACITTM_{ij}(t)$ $OVRTM_i(t)$
                     $OVRRC_i(t)$

or   or

$ACIT_{ij}(t)$ $OVR_i(t)$

STATE AND OUTPUT FUNCTIONS

NFC(t)
NFF(t)

$IDFC_i(t)$
$IDFF_i(t)$

$\begin{cases} GK_i(t) \\ S_i(t) \\ TH_i(t) \\ E_i(t) \end{cases}$

NT, STRO,PERO,LSTMO

INSED(2)
INSED(1)

INSED(3)

IFLG1

Fig. 20.1.   Programs JNLRNO0, JNLRNOA, and JNLRNOB These programs simulate Hebbian learning at the junction of a single input fiber system with a single pool of receiving neurons.

these programs are very similar to those of the nonlearning junctions discussed in Chapter 17. Specifically, the main input functions are synaptic currents SC, which impinge on the receiving cells as a result of activity in the input fiber system. To determine activity in the input fiber system, one specifies the parameters NT, STR0, PER0, and LSTM0. These represent respectively the number of output terminals per input fiber, the strength of output synapses, the mean period of firing in the afferent system, and the mean first spike time in the fibers of the system. As in the programs in Chapter 17, the program chooses specific periods and first firing times for the individual fibers from exponential distributions around these means. Three input seeds, INSED, govern the particular choice of firing patterns and of anatomical connections made by these fibers. The main output functions are the number of fibers and number of cells firing as functions of time, NFF, and NFC. One may also look at the identities of specific fibers and cells firing, IDFF and IDFC, or the individual state variables, E, TH, S, and GK, for the individual cells. The systemic parameters include the number of fibers NFIB, the number of cells NCLS, the maximum conduction time plus 1

MCTP1, the temporal integration step size STEP, the excitatory and inhibitory equilibrium potentials EX and EI, and the potassium equilibrium potential EK. The cell parameters C, TTH, B, TGK, TMEM, and TH0 are the familiar ones from Chapter 14. LTSTP is the duration of the simulation run. The new parameters related to Hebbian learning include NRNGE, MRNGE, MLAG, STRMX, and DLSTR. NRNGE and MRNGE define the ranges of memory space given to two matrices which store activity in transit, which must be more elaborate if the model system is to know which specific input synapses are to be rewarded on firing. NRNGE is the number of messages contained in hoppers specifically designated for individual cells. Typically its value might be about 20. MRNGE is the maximum number of messages stored in another memory store which contains overflow messages from the NRNGE matrix. MRNGE may vary dramatically depending on the number of cells simulated and the density of activity. Typically its value might be about 2000. Actually, one can trade off NRNGE versus MRNGE to optimize, respectively, computational efficiency or required memory storage. These features are seen more clearly in conjunction with the program listings discussed in Section B. STRMX and DLSTR are parameters which govern the incrementing of synaptic strength when input synapses are successful in triggering action potentials in receiving cells. STRMX is the maximum synaptic strength a given input synapse might ever attain. DLSTR is the increment of strength a synapse experiences on a particular reinforcement, given as the fraction of the distance the strength goes from its present value toward STRMX at each individual reinforcement. MLAG is the time period in milliseconds over which reinforcement might occur for activity in a given fiber; that is, if an output cell fires at time T, then all the input synapses that fired pulses within the time period from T-MLAG until T are considered to have contributed to the success and are therefore increased in strength. With the JNLRN series of programs, then, one may simulate the embedding of memory traces as stored in the strengths of synaptic connections between fibers and cells in a single convergent–divergent neuronal junction for given patterns of activity in the afferent fiber system. At the end of the simulation, the programs write out the synaptic strengths on data files. Moreover, by setting IFLG1, one can direct the programs to read the synaptic strengths from such corresponding data files. This allows one to simulate how such model networks subsequently alter their dynamic characteristics by virtue of the memory traces they have learned in response to some particular afferent fiber input. For example, one might simulate the characteristics of recall of particular memory traces. Progam JNLRN00 is the standard program of the series. JNLRN0A and JNLRN0B both incorporate packing of several pieces of information in individual words in the activity-in-transit matrices so as to conserve required computer memory. JNLRN0B uses a restructured computational algorithm which increases computational efficiency and introduces only a slight theoretical innaccuracy. It is the desired program for exploratory simulations and where computer budgets are in question.

Figure 20.2 shows the input-output flowcharts for programs LRSYS20, LRSYS2A, and LRSYS2B. These programs have the same flexibility and characteristics as programs SYSTM20 and SYSTM21 discussed in Chapter 19. That is, they simulate ongoing dynamic activity in an arbitrary number of interacting neuronal populations and input fiber systems, all of which are independently arrayed in rectangles in two-dimensional space. These programs, however, incorporate the embedding of permanent memory traces by Hebbian learning in the synaptic strengths of their constituent junctions and use slightly different schemes for inputting parameter values. All junctions in these programs are either excitatory or inhibitory. Input activity is

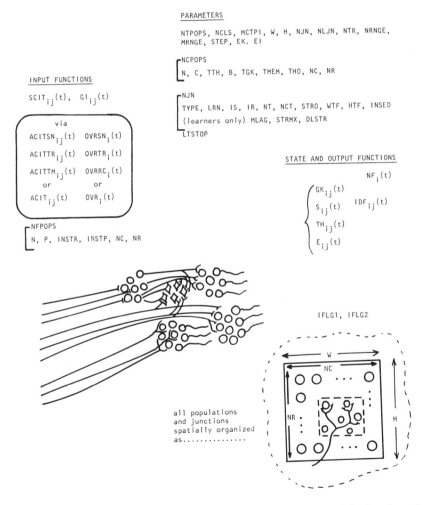

PARAMETERS

NTPOPS, NCLS, MCTPI, W, H, NJN, NLJN, NTR, NRNGE, MRNGE, STEP, EK. EI

┌NCPOPS
N, C, TTH, B, TGK, TMEM, THO, NC, NR

INPUT FUNCTIONS

SCIT$_{ij}$(t), GI$_{ij}$(t)

┌NJN
TYPE, LRN, IS, IR, NT, NCT, STRO, WTF, HTF, INSED
(learners only) MLAG, STRMX, DLSTR
└LTSTOP

via
ACITSN$_{ij}$(t)   OVRSN$_i$(t)
ACITTR$_{ij}$(t)   OVRTR$_i$(t)
ACITTM$_{ij}$(t)   OVRRC$_i$(t)
or   or
ACIT$_{ij}$(t)   OVR$_i$(t)

STATE AND OUTPUT FUNCTIONS

NF$_i$(t)

GK$_{ij}$(t)
                    IDF$_{ij}$(t)
S$_{ij}$(t)

TH$_{ij}$(t)

E$_{ij}$(t)

┌NFPOPS
N, P, INSTR, INSTP, NC, NR

IFLG1, IFLG2

all populations
and junctions
spatially organized
as..............

W
NC
NR
H

Fig. 20.2. Programs LRSYS20 and LRSYS2C. These programs simulate Hebbian learning at the synapses of an arbitrary system of interacting spatially organized neural networks activated by an arbitrary number of input fiber systems. All connections are excitatory or inhibitory.

effected through NFPOPS input fiber systems, for each of which one specifies the number of input fiber, N; the probability of firing in a particular fiber at a given time, P; the starting and stopping times of activity in the fiber population, INSTR and INSTP; a random number seed determining the pattern of activity in the input fibers, INFSED; and the numbers of columns and rows, NC and NR, respectively. The state and output functions include, as usual, the state variables for individual cells in the individual populations, E, TH, S, and GK; the identity of specific cells and fibers firing at each time, IDF; and the numbers of cells and fibers firing in each population at each time, NF. The systemic parameters include the number of total populations, NTPOPS; the maximum number of cells in any population; NCLS; the maximum conduction time plus 1, MCTP1; the maximum width and height of any rectangular two-dimensional array, W and H; the total number of junctions in the system, NJN; the total number of junctions to be designated as Hebbian learning junctions, NLJN; the maximum number of terminals per fiber at any given junction, NTR; the parameters NRNGE and MRNGE, which allocate space to the activity-in-transit matrices as discussed above for programs in the JNLRNR series; the temporal integration size, STEP; the potassium equilibrium potential, EK; and the equilibrium potential of the inhibitory synapses, EI. For each cell population one reads in the number of Cells N, the familiar cell parameters C, TTH, B, TGK, TMEM, and TH0, and the number of columns and rows NC and NR, respectively. One then reads in for each of the junctions the following list of parameters: the TYPE (1 if excitatory, 2 if inhibitory); the value of LRN which signals whether a particular junction is to exhibit properties of Hebbian learning (1 if it learns, 0 if it does not); the names of the sending and receiving populations, IS and IR, respectively (IS ranges from 1 to NTPOPS and IR ranges from 1 to NCPOPS; cell populations are assigned numbers sequentially after the fiber populations); NT, the number of terminals per fiber; the conduction time NCT; the initial synaptic strength STR0; the width and height of the projected fields from individual cells, WTF and HTF, respectively; and input seeds INSED, which determine the specific connections made by specific fibers. Further, for learning junctions only, one reads the parameters MLAG, the time lag over which reinforcement occurs; STRMX, the maximum synaptic strength attainable by a learning synapse; and DLSTR, the increment in strength experienced at a given synapse at a single firing measured in terms of the percentage of the distance between the current value and STRMX incremented at each firing. LTSTP is the duration of the simulation run.

With these programs one can construct an arbitrary configuration of interacting neuronal populations and input fiber systems, each containing individually prescribable cell types and all rectangularly arranged in two-dimensional space, as for the programs discussed in Chapter 19. With these programs one can further specify which of the internal junctions of the networks are to exhibit Hebbian learning. One can use the programs then to simulate the embedding of permanent memory traces in the systems of populations by virtue of the sets of synaptic strengths characteristic of each junction. The programs write out the sets of synaptic strengths at the end of the run and moreover can be directed to read sets of synaptic strengths produced by previous runs or arbitrarily created by the user. They can be thus used to explore the influence on subsequent dynamics of the learning of particular patterns. Specifically, they might be used to explore recall of particular memory traces. LRSYS20, LRSYS2A, and LRSYS2B are identical from the standpoint of the user. They differ internally in how they store and process information. LRSYS20 is the standard for this series. LRSYS2A and LRSYS2B both pack several pieces of information in one word in the activity-in-transit matrices to conserve memory space required by the program. LRSYS2B also separates internally the information used in synaptic transmission from that used in identifying synapses for reinforcement in the process, increasing the efficiency of the computational scheme at only a slight theoretical inaccuracy. All of these programs incorporate a wraparound effect at the edges of the two-dimensional field. This is useful in simulating ongoing internally sustained activity in the population, which is represented by a two-dimensional array that is perhaps somewhat smaller than one might like.

Figure 20.3 shows the input–output flowchart for LRSYS30, LRSYS31, LRSYS32, LDSYS33. LRSYS30 is identical to LRSYS2B, except that individual synaptic junctions are individually adjustable and its neuronal pools are aspatial. Thus the user may prescribe an arbitrary number of synaptic types SNTP, each of which is characterized by its synaptic equilibrium potential EQ and time constant of decay of postsynpatic conductance change T. The parameter type then specified for each junction of the system may have any value from 1 up to SNTP, thereby assigning to that junction the appropriate set of values for EQ and T. These programs also incorporate the packing of several pieces of information into each word in the activity-in-transit matrices to decrease required memory storage and separate the information about individual sending synapses required for synaptic reinforcement from the transmission in order to increase computational efficency. LRSYS30 represents aspatial neuronal populations. LRSYS31 and LRSYS32 represent two-dimensional rectangularly organized populations. LRSYS32 includes a subroutine STIMULS, which allows the user to generate any of a variety of simple stimulation patterns. The parameters associated with STIMULS are NSTM, NSTM1, LSTM, and INTERV. NSTM is the number of cells which are stimulated by STIMULS. NSTM1 is the number of cells which are activated at each call to STIMULS. LSTM is the time of the first call to STIMULS and LSTP is the time when the

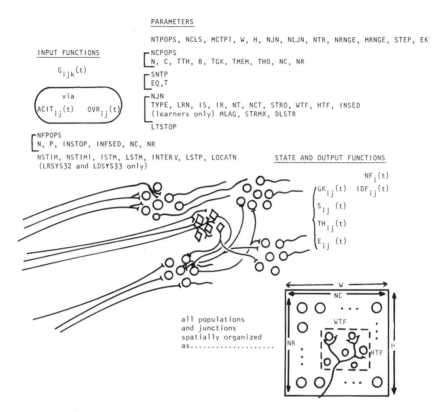

Fig. 20.3. Programs LRSYS30, LRSYS32, and LDSYS33. These programs simulate Hebbian learning at the synapses of an arbitrary system of interacting spatially organized neural networks activated by an arbitrary number of input fiber systems. All synaptic systems are individually adjustable.

external stimulation mediated by STIMULS is shut off. INTERV is the time interval between successive calls to STIMULS. The array LOCATN contains the identities of the specific NSTM cells which are activated by STIMULS, arranged in the sequence desired for their successive activation. With this simple subroutine, the user can generate a wide variety of pulsing or "running" patterns of variable frequency or running speed.

LDSSYS33 is identical to LRSYS32 except that the sections of code which are used to increase synaptic strength at successful synapses are eliminated. Thus, one can use LDSYS33 to explore various "recall" properties in a model neuronal system, some of whose synapses have been increased in strength by previous runs. LDSYS33 recalls the synaptic strengths for junctions targeted as "learning junctions" from a data file produced by earlier runs in LRSYS31 or LRSYS32.

EXAMPLES

The examples described here were produced from earlier programs which were forerunners of the programs presented in this chapter. The first example is illustrated in Fig. 20.4

Multiple, selectively retrievable patterns of internally sustained activity. In brief, this simulation model consisted of five main cell populations in three main regions interconnected so that information flowed in closed loops from entorhinal cortex to dentate gyrus to the subiculum and back to the entorhinal cortex. (The CA3 and CA1 regions, hypothesized to intervene between the dentate gyrus and the subiculum, were short-circuited in this model.) The input fibers activated the cells of the entorhinal cortex. (They might be imagined to project from the association cortex.) In addition to main population cells in the entorhinal cortex, dentate gyrus, and subiculum, there were local inhibitory cells in the enthorhinal cortex and dentate gyrus. There were both feed-forward and recurrent inhibition in these regions mediated by these cells. A total of 2500 cells were used in the simulation, 500 in each population. There was a linear topographic arrangement, wherein the cells of each population were imagined to be arranged sequentially along a line. There were 14 synaptic linkages between cell populations, each of which was convergent–divergent of variable diffusivity and characterized by parameters representing the number of terminals per sending fiber, the spatial range of the target field, and mean individual synaptic strengths. There was potentiation of variable magnitude and decay rate at the entorhinal-to-dentate-gyrus synapses and at the subiculum-to-dentate-gyrus synapses. There was Hebbian enhancement of entorhinal-to-dentate-gyrus synapses. For these simulations there were 200 synapses from each entorhinal cortex cell directed to dentate gyrus cells, giving a total of 10,000 Hebbian synapses in the dentate gyrus. All neurons were instantaneous point neurons. In addition, considerable support software was developed to display and analyze the resulting network activity.

The paradigm studied was as follows: In a "learning phase," particular spatiotemporal input patterns were prescribed for the input fibers into the entorhinal cells. This triggered activity which reverberated through the diffuse linkages around the closed loop. The resulting standing activity pattern reflected both the input pattern and the intrinsic connectivity of the network. Typically, during the learning phase, the input pattern would be activated for some time interval, then terminated so that the system would "free-run" for a further time interval. Throughout this entire learning phase, successful synapses would be increasing in strength according to the Hebbian rule (that is, when a cell fires, all synapses which are active at that time are slightly increased in strength). The simulation of the learning phase was terminated and the set of synaptic strengths, which then reflected the impression made on the network by its response to the particular input, were stored in the computer memory. Subsequent trials with the same network could involve any sequence of further learning phases with the same or different inputs, or "recall phases" wherein any input pattern could be prescribed and the resulting activity pattern observed. Hebbian enhancement did not occur during the recall phases.

Figure 20.4b shows a case where three distinct but overlapping memory traces were embedded in the loop. Each of the three patterns could be activated by virtually any suprathreshhold stimulus which activated their spatial region. Figure 20.4c shows a case where two distinct memory traces share the same region in space. The two memory traces resulted from stimulating cells 1, 4, 7, 10, . . ., 148 and 2, 5, 8, 11, . . ., 149, respectively. Of all the cells participating in the two traces, about one-third appear in both traces, one-third appear only in A, and one-third appear only in B. The response of the network is clearly pattern-sensitive. It responds with pattern A or B, respectively, to the initial A or B stimulus. It responds to

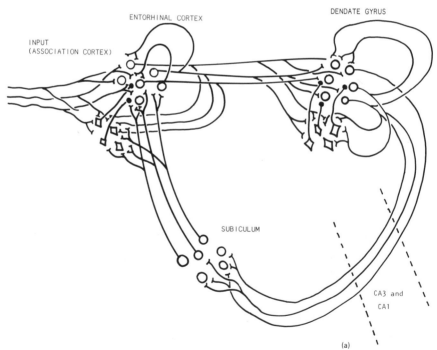

(a)

Fig. 20.4. Embedding of memory traces in a closed-loop model of the hippocampus. (a) Schematic of hippocampus; (b) three distinct but overlapping memory traces embedded in loops; (c) two distinct memory traces sharing same region in space. (*Continued*)

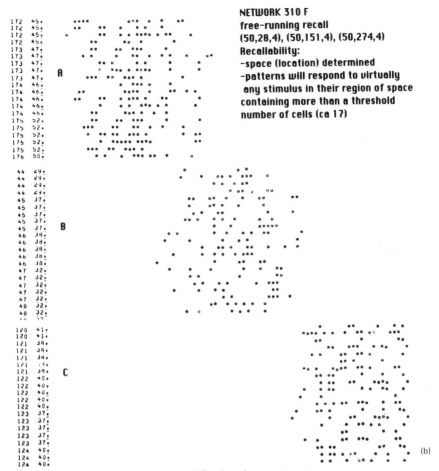

NETWORK 310 F
free-running recall
(50,28,4), (50,151,4), (50,274,4)
Recallability:
-space (location) determined
-patterns will respond to virtually
 any stimulus in their region of space
 containing more than a threshold
 number of cells (ca 17)

Fig. 20.4   (*Continued on next page*)

other stimuli with response A or B in an unpredictable manner. Moreover, associational learning has taken place in the sense that the network responds with the total A (or B) response to any fraction of the initial A (or B) stimulus involving as few as 40 or 50% of its cells. There is a suggestion that the clusters of excitatory cells in patterns A and B are associated with codes of inhibitory cells such that if pattern A becomes dominant, pattern B is held down, and vice versa.

Nature of simulated dynamic memory traces.   A second preliminary example simulation in this context addressed the dynamic nature of memory traces with particular reference to E. Roy John's concept of a "statistical configuration." Initially it did not seem clear to me how a particular dynamic pattern could maintain a unique global coherence but be rather

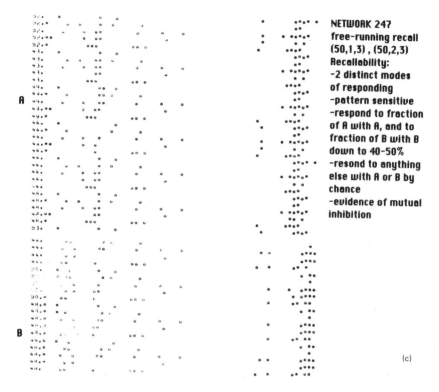

NETWORK 247
free-running recall
(50,1,3), (50,2,3)
Recallability:
-2 distinct modes
of responding
-pattern sensitive
-respond to fraction
of A with A, and to
fraction of B with B
down to 40-50%
-resond to anything
else with A or B by
chance
-evidence of mutual
inhibition

(c)

Fig. 20.4   (*Continued*)

independent—as Roy John suggests—of specific underlying anatomical pathways. However, a preliminary (and somewhat crude) simulation study showed that this could indeed occur. In these examples it is as if the momentary distribution of postsynaptic potentials (PSPs) determined by the momentary distribution of firing cells determines the next distribution of firing cells, so that in this sense physiology (PSPs) rather than anatomy (specific synaptic connections) tends to predominate in leading the evolution of the pattern. The collection of weighted synaptic strengths embedded in the anatomy of the network seems rather to contain and constrain the pattern within some larger limits.

The two examples discussed here are very preliminary. They serve, I hope, to point out the kinds of significant modeling that can be addressed with the programs of this chapter.

B.   Program Listings and Discussion (JNLRN00, JNLRN0A, JNLRN0B, LRSYS20, LRSYS30, LRSYS32, LDSYS33)

The essential programming difficulties presented to the simulator of Hebbian learning in neuronal networks are that (1) present values of individually variable synaptic strengths must somehow be stored in the memory of the computer for every modifiable synapse in the system being simulated, and (2) for every neuron which received modifiable synapses in the system there must be stored a list of messages containing the names of sending cells, the names of sending terminals, and the arrival time for every PSP which has arrived at the cell within the cell's learning time lag. The latter is necessary so that if a given cell fires, it can identify the input synapses which have fired recently and are thereby indicated for increments in strength. In earlier sections of this book where activity-in-transit matrices were used to effect ongoing synaptic activation within neuronal systems it was not necessary to include all this specific detailed information. All that was necessary in those cases was that increments of synaptic activation of particular types were targeted to arrive at particular receiving cells at particular times. The origin of particular increments, or indeed the breakdown of the total input activity of a particular type, was not of interest and therefore not passed along in the matrices. To adequately represent the detailed information about sending cells required to simulate Hebbian learning, the programs of this chapter use two sets of matrices: ACIT and OVR. Use of these matrices can best be illustrated with reference to the listing for program JNLRN00. Note first that the overall structure of JNLRN00 is essentially the same as that of all programs in this book. In particular, the essential dynamic algorithms are contained within the do-1000 loop over time. Notice now that there are three ACIT matrices: ACITSN, which contains the names of sender cells, ACITTR, which contains the name of corresponding sending terminals; and ACITTM, which contains the arrival times. Each potential receiving cell has stored for it in these three matrices a set of messages, numbered 1, 2, and so on up to NRNGE, which correspondingly tell it the name of the sender cell, its sending terminal, and the arriving time of all PSPs en route to it. The ACIT matrices can contain up to NRNGE messages for any particular receiving cell, where NRNGE is a parameter specified by the user. If there are more messages than NRNGE for a particular receiving cell, the extra messages are stored in the overflow matrices: OVRSN, containing the names of sender cells; OVRTR, containing the names of sender terminals; OVRTM, containing the times of arrival; and OVRRC, containing the names of the receiving cells. Thus the overflow matrices are not flagged for particular receiving cells but represent a communal pool of overflow messages which may be directed to any particular receiving cell in the population. At any instant in time there are IVR messages in the overflow matrices. The maximum capacity of the overflow matrices is MRNGE, which is a parameter set by the user.

In the operation of the program, state variables of receiving cells are updated at each time unit in accordance with the input activity that is transmitted to the cells in the ACIT or OVR matrices. This can be seen in the 3000 loop. For a given cell I at a given time, first all the messages in the ACIT matrices en route to this particular cell are scanned, as may be seen in program statement 3570. If a particular message indicates that a particular PSP has arrived either at the present time or within the learning memory of the receiving cell, and therefore is potentially a candidate for reinforcement, the cell and synapse number are stored in temporary arrays ISDR and ISYN. Moreover, if the PSP is incident at the present time, the corresponding increment to the activating synaptic current of the cell SC is effected according to the current strength of that sending synapse. The 3500 loop thus collects all the messages in the ACIT list which may be relevant for the particular receiving cell at the particular time. The 3600 loop is then scanned. All the messages in this list which are directed toward cell I and which may have relevance to the present time are extracted similarly: the names of sending cells and synapses are stored in the temporary matrices ISDR and ISYN and, if appropriate, increments to synaptic stimulating current SC are made in accordance with the strength of the sending terminals. Immediately below the 3600 do-loop, the state variables for cell I are updated. If cell I fires at this particular point in time, the 3700 do-loop is activated, wherein the synaptic strengths of all input terminals which have been active within the memory period of the cell are increased. The rule explicit in statement 3700 drives the synaptic strength of a given synapse from its current value at STR by the fraction DLSTR of the difference between its current value and the maximum strength value STRMX.

The do-2000 loop over the number of input fibers illustrates how information about senders is placed in the ACIT and OVR matrices initially. When a terminal J from a sending fiber I is activated to place a PSP on a receiving cell NREC at a conduction time NCT from now, it checks first to see if the ACIT list for that particular receiving cell has enough room to receive the message. If it does, the corresponding information about I, J, and NCT is placed in the ACIT matrix in accordance with line 2510. If, on the other hand, the ACIT matrices for this particular receiving cell are filled at this point in time, the corresponding information is placed in the OVR matrices, as may be seen in the line of code immediately below the 2500 continue statement. At each integration step through the program, the ACITTM and OVRTM matrices flagging arrival times are reduced by 1, as may be seen in lines 3570 and 4000. When messages in the ACIT matrices have arrival times that outdate the memory interval of the reinforcement process, they are discarded; this is effected in the 3560 do-loop and immediately surrounding lines of code. When the arrival time of messages in the OVR matrices outdates the memory time for reinforcement in the cell, those messages are discarded and the number IVR is downgraded by 1; this is effected in the program in the 3630 do-loop and immediately surrounding lines of code.

By adjusting the parameters NRNGE and MRNGE, one may adjust the proportions of synaptic messages allocated to the ACIT and OVR matrix series, respectively. The ACIT series is computationally more efficient in that a given receiving cell may scan immediately only those messages directed to it. This matrix series, however, may take up a great deal more memory space in the computer than is required at any given instant. If the busiest cell at any particular point in time has NRNGE messages coming toward it, most of the other cells will have considerably less than this number. Since one uses NRNGE spaces for every cell, there is wasted memory space in the central memory of the computer. The matrix OVR is more efficient with respect to memory storage since the only wasted space at any point in time is the difference between MRNGE and the current value of IVR. On the other hand, utilization of the OVR matrices is computationally less efficient since each cell must search all the OVR messages to find those en route to itself. Again, depending on the constraints of one's particular computer system and particular simulation problem, one may allocate these messages in different proportions to these two matrix systems.

In other respects program JNLRN00 is quite similar to the nonlearning junction programs discussed in Chapter 16, JNCTN10 and JNCTN11. JNLRN00 incorporates bombardment by a single afferent fiber system, wherein activity in individual fibers is distributed with exponential interval distributions as indicated in the do-110 do-loop. Again, as discussed in Chapter 15, the specific patterns of activity can readily be varied as desired. Note finally just above the 1500 continue statement that at the end of each successful simulation the sets of synaptic strengths for all the terminals of the fiber system on the receiving population are written out on the hard copy and on data files.

Consider the listing for program JNLRN0A. This program is identical with JNLRN00 except that the information about sending synapses is now data packed in the ACIT and OVR matrices. This is best illustrated for the ACIT matrix by the statements labeled 2510 in both programs. In the earlier program, information about the sending cell I, the sending terminal J, and the conduction time NCT was stored separately in the three matrices ACITSN, ACITTR, and ACITTN. Since each of these variables is adequately represented by three or four digits, one can represent all three of them in a single 14-digit word. Thus the 2510 statement in JNLRN0A shows how all three numbers, J, I, and NCT, can be stored in one 14-digit number.

The statements immediately following the 3570 statement show how the information is recovered from this data packing. Specific information about the arrival time IT, the name of the sending cell ISDR, and the name of the sending synapse ISYS is extracted readily. Similarly, each message in the OVR matrix stores four values corresponding to the name of the sending cell I, the sending synapse J, the conduction time NCT, and the name of the receiving cell NREC, as may be seen by the line of code immediately after the 2500 continue statement. This may be compared with the corresponding line of code immediately below the 2500 continue statement in program JNLRN00, which utilizes four separate words for this information. The specific information about the arrival time IT, names of the sending cell ISDR, and names of the sending synapse ISYN is extracted from OVR as in the code immediately following statement 3640 in JNLRN0A. Functionally, JNLRN0A is equivalent to JNLRN00. Its advantage is that it dramatically reduces the amount of central memory required by the program. To use JNLRN0A on your computer system, it is necessary that your system operate with 64-bit words, that is, words with 14 or more digits. If your computer system uses different-size words, JNLRN0A might serve as a model for similarly data packing on your system.

Computationally, programs JNLRN00 and JNLRN0A are both a little awkward in that each cell sorts through all its incoming messages at each time period. Some of this is necessary because the cell must know what synaptic input is incident on it at a given point in time in order to update its state variables. On the other hand, some of this is unnecessary because, unless the cell fires, it is not interested in the names of sending cells and terminals. To increase computational efficiency, program JNLRN0B separates out the effective synaptic current SC, due to take effect on each particular cell at each time, from the information regarding the antecedents of that activity stored in the ACIT and OVR matrices. Consider the listing for program JNLRN0B. As may be seen in the 2600 do-loop, whenever a given input fiber fires, it both increments the synaptic current matrix SC for a particular receiving cell at a particular arrival time and flags more detailed information about the name of the sending fiber and its terminal in the ACIT or OVR matrices. Then, as may be seen in the 3000 do-loop over the cells, at each time unit state variables are updated immediately and directly according to the value of the input synaptic current SC. If the cell does not fire, the program proceeds to the 3010 and 3000 continue loop statements and then on to the next cell with no consideration of the ACT and OVR matrices. If the cell does fire, then, and only then, the ACIT and OVR matrices are scanned and the information regarding the names of cells and terminals which have contributed to the SC matrix and firing within the memory time period of the cells is extracted. The strengths of the synapses of these specific synaptic terminals are increased according to the rule in the line of code immediately above the 3600 continue statement. Clearly, this separation of information considerably increases the computational efficiency of the program. There is a slight loss of accuracy in this program design however. If there are action potentials in bursts which are closer together than the conduction time for a given sending fiber, and if these succeed in driving a responder to fire, then the strengthening of the synapse due to "success" does not take effect until after the total burst input is effected. Such a distinction makes JNLRN0B behave slightly differently from JNLRN0A or JNLRN00. On the other hand, in most cases such distinctions represent very fine-tuned differences and are perhaps less important than several of the many other assumptions underlying any particular simulation—for example, not including dendritic trees, as in the current models, or purporting to represent a population of cells with hundreds of thousands or millions of neurons with a simulation model of thousands or tens of thousands. Thus I consider JNLRN0B to be the most useful of these three JNLRN models. One problem with JNLRN0A for some users will be the fact that it incorporates data packing in the ACIT and OVR matrices, which assumes that the users have access to 14-digit words. If this is not the case, the users are advised either to employ JNLRN00 as is or, if they wish to obtain the increased computational efficiency of JNLRN0B, to substitute ACITSN, ACITTR, and ACITTN for ACT and OVRSN, OVRTR, OVR-TN, and OVRRC for the OVR matrix in JNLRN0A. This can be done without too much trouble by using JNLRN00 as a guide.

The learning system series of programs is a little more complicated than the junction learning series, primarily because there are multiple cell and fiber populations. Program LRSYS20, for example, simulates an arbitrary number of interacting neuronal populations activated by an arbitrary number of input fiber systems. Junctions may be specified as desired between any of the fiber systems and cell populations. Any of these junctions may be designated to exhibit or not to exhibit Hebbian learning. Consider the subroutine TRNSMT in the listing for program LRSYS20. As may be seen in the line of code immediately above the do-200 statement, if the junction in question is not a learning junction [that is, if LJNN(IS,IR) is equal to zero] then the line of code immediately following the 220 continue statement applies. Activity is transmitted from sender to receiver according to conventional activity-in-transit matrices SCIT and GI. Skipping up the 2000 do-loop over the cell populations in the main program, consider the line of code immediately above the do-2500 statement. If the junction now at the receiving end is not a learning junction, the program is directed to the 2200 continue statement and jumps immediately into state variable updating on the basis only of the SCIT or GI matrices. This logic is the same as that used in the system series described in Chapter 19. Going back in the TRNSMT subroutine in the line just above the do-200 statement, if the junction is a learning junction,then all the code above the do-220 statement is utilized, which means that all the activity in transit is placed in the more elaborate learning matrices—the ACIT and OVR series. Again, back in the do-2000 loop at the receiving end in the main program, if the junction is a learning junction, then all the code from the do-2500 statement down to the 2600 continue statement is activated, wherein the program abstracts the incident synaptic information from the ACIT and OVR matrices. These operations on the ACIT and OVR matrices, both in the sending form in subroutine TRNSMT and in the receiving form in the do-2000 loop, are analogous to those described above for program JNLRN00. The difference is that now these operations are embedded within the overall structure of the system programs from Chapter 19.

LRSYS20, like its counterpart JNLRN00, is the standard program of the LRSYS series. Program LRSYS2A, like JNLRN0A, data-packs information about sending synapses so as to compress the three ACIT matrices into one and the four OVR matrices into one. To use LRSYS2A one must use a computer system with 14-digit words. In other respects LRSYS2A is equivalent to LRSYS20.

LRSYS2B, as may be seen in the listing, separates off the memory and synaptic transmission functions of the activity-in-transit matrices, as does program JNLRN0B. As shown in the do-2000 loop for updating cell state variables, this program, like JNLRN0B, makes use of the detailed information about senders in the ACIT and OVR matrices only if the given receiving cell has fired and then needs to know which input terminals to reinforce. This program then results in considerably greater computational efficiency than either LRSYS20 or LRSYS2A but, like JNLRN0B, incorporates a slight innaccuracy in synaptic reinforcement. Specifically, if spikes cluster in a given sending fiber closer together than the conduction time for that fiber to a particular receiver, and if the first spike in such a burst is successful in triggering a spike in that receiver, then the strength of the synapse at that specific junction will not be increased until after the later spikes in the burst arrive. Again, this is considered a minor inaccuracy compared to many of the other limitations of the simulation models. Program LRSYS2B is the most efficient of the learning series. It is, however, useful only to those employing computing systems with 14-digit words. Those working on other systems are advised either to use LRSYS20 directly or to modify LRSYS2B in accordance with the specific requirements of their system.

Programs LRSYS20, LRSYS2A, and LRSYS2B all utilized junctions which are either excitatory or inhibitory. Program LRSYS30 allows one to specify the synaptic type of each interpopulation junction. As in previous chapters, a synaptic type is defined by its synaptic equilibrium potential EQ and its time constant of decay of postsynaptic conductance change TG. Otherwise, LRSYS30 is very similar to LRSYS2A. That is, the information regarding sending synapses contained in the ACIT and OVR matrices is data packed, and the information processing is arranged so that the ACIT and OVR matrices are called by receiving cells only when those cells fire. Notice that in subroutine TRSNMT for LRSYS30, both at the 220 statement and the two lines immediately below, synaptic activation is effected through the synaptic conductance activity-in-transit matrix for both learning and nonlearning junctions. This activity-in-transit matrix, G, is used in updating the cell state variables called immediately below the do-2900 statement and effected in the subroutine SVUPDT. Program LRSYS30 is the optimal program of the LRSYS series for most users. It simulates ongoing dynamic activity in an arbitrary number of interacting neuronal populations and fiber populations. One may specify internal connectivity at will. One may specify which internal junctions are to exhibit Hebbian learning. One may specify the synaptic types of individual junctions. The program incorporates reasonably efficient algorithms and transmitting information about sending fibers and terminals and is relatively compact in its use of computer memory storage. As it stands, however, it assumes users have computing systems with 14-digit words. If this is not the case, slight modifications must be made in the packing of data into the ACIT and OVR matrices. This should not be difficult for a competent computer programmer.

Program LRSYS31 extends LRSYS30 to represent two-dimensional rectangularly organized neuronal and fiber populations. Program LRSYS32 extends LRSYS31 to include the subroutine STIMULS, which simulates the activation of certain prescribed cells in a population in prescribed temporal patterns by external stimulation by raising the membrane potentials of the target cells above their treshholds.

Program LDSYS33 eliminates the variables and sections of code that deal with the increasing of synaptic strengths at learning junctions. The ACIT and OVR matrices for activity in transit are no longer used. Computation in "recall" situations is thereby made more efficient. The program retains individual synaptic strengths for all terminals at junctions designated as "learning." These values are to be read into LDSYS33 on the ANIN file from ANOUT files produced by LRSYS31 and LRSYS32.

C. Theoretical Background

Our computer simulation studies of memory embedding is conceived within the framework of the following "dynamic correlational" theory of memory in neural networks. Memories are imbedded and represented in neural networks in terms of groups of interrelated unitary memory traces called memory schemata (MS). Memory schemata exhibit long-term plasticity, in that they grow, decay, reorganize, combine with other schemata to form yet higher order schemata, etc. Memory schemata exhibit anatomical foundations called "beds" and dynamic manifestations called "realizations." The bed of a memory schemata consists of a specific ordered sequence of subsets of cells in the network. Equivalently the bed may be thought of as the specific ordered sequences of subsets of synapses that link adjacent subsets of cells in the cell sequence. In either case, the sequence reflects and is determined by the internal recurrent synaptic connectivity of the network and the current level of synaptic strengths as seen relevant to a given initial activation configuration. More precisely, the bed of a memory schematic contains those sets of cells, $\{G_t^j\}$, $t = 0,1, \cdots, L$, which under ideal and optimal conditions and in the absence of external input, would fire at time t given that the set $\{G_o^j\}$ fired at $t = 0$. The number of distinct schemata of this sort is virtually unlimited in large, densely connected networks.

Corresponding to the subscript t, beds exhibit links which may vary in length from only one or several links up to indefinitely large numbers of links. It is assumed that the more interesting schemata have indefinitely long beds. Generally, it is assumed here that beds of short length correspond only to relatively insignificant unitary memory traces; that significant unitary memory traces and significant memory schemata exhibit indefinitely long traces; and that the relative number of cells in the subsets $\{G_t^j\}$ is a measure of the strength or pervasiveness of this schemata.

A "realization" of a memory schemata is a dynamic event that consists of an ordered sequence of firing of subsets of cells $\{[F^i + R]_t\}$ where each set $\{F_t^i\}$ is a subset of the corresponding set $\{G_t^j\}$ in the bed of the schemata, and each $\{R_t\}$ is a subset of extraneous, often randomly determined, cells, for t in t_1, t_z where t_1 and t_z are any numbers within the time range or length of the bed. Realizations exhibit three dimensions of "clarity" in their representation of schemata: (a) The "signal-to-noise" ratio may vary from zero to indefinitely large according to the relative numbers of cells in the R sets. (b) The instantaneous "fidelity" of the realization to the schemata is measured by the percentage of cells of the $\{G_t^j\}$ that are contained in the $\{F_t^i\}$. (c) The "completeness" of the realization is measured by the particular section and range of t values over which the $\{F_t^i\}$ constitute a significant number of the $\{G_t^j\}$.

The unique individuality of a given schemata (including both its corresponding anatomical representation in beds and its dynamical representation in realizations) is signaled only in the relative ordering of the network neurons. Dynamically, this shows up only in the relative timing of firings of the network neurons. Any one neuron may participate in a great many different schemata and, moreover, in more than one position in a given schemata. What signals that the neuron is at a particular time participating in a given schemata is only that that particular neuron tends to fire just after the cells in the $\{G_t^i\}$, just before the cells in the $\{G_{t-1}^i\}$ and simultaneously with the other cells in $\{G_t^i\}$.

It is clear from the above (particularly 5, 6, and 7) that the dynamic realization of a schemata is a stochastic phenomenon. Although beds are assumed to be imprinted according to processes delineated by Hebb, their character is more consonant with the concepts of equipotentiality and mass action of Lashley, the statistical configuration theory of E. Roy John, and the corresponding holograph concepts advocated by Pribrcum, as discussed in Chapter 9, than with Hebb's concepts of cell assemblies and phase sequences. Thus, the indefinitely long sequence $\{G_t^i\}$ makes use of approximately $L \times \bar{N}_t$ cells in the network where L is the length of the bed and N_t is the average number of cells in the sets $\{G_t^i\}$. If, L is about 10^3 and N_t is also about 10^3, then the schemata may make use of approximately 10^6 cells of the network or about one-tenth of 1% or 1% of the cells in the entire network. These cells are assumed to be distributed throughout the network in ways limited only by the dense and diffuse interconnections of the network. Moreover, it is assumed that memory recall may be effected by a dynamic realization of any section of the bed; that the degree of completeness of the recall depends on the length of the section realized. Thus, the bed exhibits the features of mass action, equipotentiality, and holographic distribution. Since the statistics argue that the recurrence of any given set $\{G_t^i\}$ is an extremely unlikely event in large densely connected networks, the beds of schemata are assumed to exhibit indefinitely long sequences and not to recycle as the cell assemblies of Hebb.

Beds are assumed to be formed by the internal reverberations of activity within the network in particular modulated states in response to both particular external input configurations and the ongoing activation of other schemata within the network. Thus, developmentally, one may imagine the network to be a "tabula rasa" which initially collects impressions of the particular external input stream to which it has been subjected. Later on, internal activity correponding to the reactivation of schemata so formed, may interact one with another, or with new input configurations, any of which under different overall states of the entire network to produce various combined, associated higher order, etc. schemata.

A particular realization of a schemata may be triggered by a variety of causes in a mature network including, for example, (a) representation of the external stimulus that originally defined the schemata, say $\{G_o^i\}$; (b) any combination of internal reverberations in the net (corresponding, for example, to the activation of various sets of schemata) that happen to trigger any subset of cells $\{G_t^i\}$ in the schemata; (c) a random scanning served, for example, by a random bombardment of the network by some external input field which happens to trigger any subset $\{G_t^i\}$ of the schemata. Because of the dimensions of clarity exhibited by realizations any of these modes of recall may be fragmentary, masked, or complete or clear to any degree along the various dimensions.

The interactions of realizations of schemata are assumed to mediate the mechanics of cognitive operations. Thus, associations are assumed to reflect the linking up of compatible sequences from the beds of the individual schemata; contradiction is assumed to reflect incompatible overlap of subsets of cells in the beds of the various individual schemata; truth might be assumed to represent the absence of such incompatible overlap of subsets of cells in the individual schemata, that is, the ability of the network to allow the sequences of the individual beds to play off simultaneously without debilitating masking inhibition mediated by recurrent inhibitory synapses in the system. Thus, truth in this sense is a graded quantity and can reach an ideal of 100% clarity. Moreover, because of the assumed plasticity of the network, degrees of incompatibility or contradiction (i.e., the nemesis of truth) can be in part modified by selective adjustment of the individual synaptic strengths involved in defining the original schemata. This process, of course, modifies the original schemata to degrees dependent on the severity of the incompatibility or contradiction between the schemata involved. Transcendence of contradiction, on the other hand, can be obtained by the formation of higher order schemata that may be conceived to make positive use of both those parts of the individual schemata which are incompatible. These concepts can be studied in the mechanics of interactions in model nerve networks.

D.   Exercises

1. Use program LRSYS2B to simulate activation of a single receiving population by a single afferent system. Simulate 1600 such cells and 1600 such fibers, each arranged in 40 by 40 grids. Stimulate the system with a rectangular pattern pulsing at intervals of 4 msec, superimposed on a random, ongoing background pattern. Let the system "learn" this pattern by changing synaptic strengths according to the Hebbian rule for a stimulation period of about 200 msec. Let STRMX be two times STR0. Now use program LDSYS33 to explore recall properties in this system. Is there a pattern of activity in the cells which you can identify as being associated with the particular original input pattern? If so, is the initiation of this particular response pattern especially sensitive to the original input pattern as compared to other input patterns? Is it possible to trigger this response pattern by only some fraction of the initial input pattern?

2. Use program LRSYS32 to simulate the embedding of memory traces in a recurrently connected network as follows: let there be two cell populations, an excitatory population and an inhibitory population, and one afferent fiber system which projects only to the excitatory population. Let the excitatory population project both to itself and to the inhibitory population. Let the inhibitory population project to the excitatory population. Use 900 cells in each population and 900 fibers in the fiber populations. Adjust parameters so that with no learning you produce internally sustained activity as you did in the exercises of

Chapter 18. Now prescribe an input pattern consisting of a closed square, pulsating every 4 msec and localized at some particular region in the rectangular array. Let the system learn in the junction of the excitatory-to-excitatory synapses. Let STRMX be two times STRO. Let the stimulus pulsate from time 0 up until time 100 msec; let the system continue to reverberate internally up to 200 msec, at which time the simulation is stopped. Now use program LDSYS33 to simulate recall of the pattern associated with this learning situation. Can you identify a pattern particularly associated with this stimulus? If so, explore its response properties. Is it more sensitive to patterns similar to the original stimulus? Does it respond to some fraction of the original stimulus?

3. Now, building your results from Exercise 2, apply a second pulsating square located in a different region of the rectangular array. Repeat the learning exercise you did in Exercise 2, this time being careful to read in the initial strength pattern obtained in your work in that exercise. Can you now find two distinct response patterns which are associated with the two distinct stimuli? If so, show that they are selectively retrievable and explore the future of the selectivity. Do the response patterns tend to respond particularly to stimulus patterns similar to those initially applied? Do they respond to fractions of the original stimuli? How does the system respond to unrelated stimuli?

4. Modify program LRSYS32 so that Hebbian reinforcement occurs when the total number of input pulses arriving at the synapses of the learning junction over the last five time units may be estimated to be greater than some critical number NCRIT.

5. Modify program LRSYS32 such that Hebbian reinforcement of synapse occurs whenever the number of input pulses at the learning junction over the last five time units is greater than some critical number NCRIT and the number of input pulses at another input system is less that some critical number NCRIT2.

21

More Accurate Models of Neurons with Dendritic Trees

This chapter presents models providing mathematically accurate descriptions of spatiotemporal integration in dendritic trees. The numerical integrations over space and time are handled as follows. The dendritic tree is compartmentalized into an arbitrary number of compartments, each of which can be made arbitrarily small. Each compartment is assumed to be isopotential. Current is allowed to leak between adjacent compartments in proportion to the potential difference between the compartments and a mathematical model for the resistance between the compartments. The dynamic activity of each compartment is represented by a single equivalent circuit for that compartment. Mathematically, each compartment is characterized by one or more first-order ordinary differential equations in time. Numerical integration of the state variable equations for each compartment is effected by the difference technique introduced for point neurons in Chapter 14. The programs of this chapter, however, require much smaller step sizes of numerical integration because of the continuous graded interchange of current between neighboring compartments. Because of the large number of spatial compartments and numerical integration steps required, these programs require considerably more computation than the programs of the earlier chapters. Factors of several thousand to 25,000 are typical.

Models described in this chapter include representations of both passive electrotonic properties and active calcium-related conductances in dendritic regions and a state variable model for repetitive firing of all-or-none action potentials in the soma. The models described include representations of single dentrites, a single bifurcating dendrite, and dendrtric trees of arbitrary morphology, adjustable by the user. The models include representation of activation by arbitrary numbers of synaptic input systems, each of which effects individually adjustable synaptic action and can be targeted at will within the dendritic tree. Some of the programs include representation of synaptic action by a mathematically equivalent synaptic conductance representing the influence of dendritic spines. Most of the programs include a subroutine which automatically computes the overall passive electrotonic properties of dendritic trees—the total admittance and its partitioning among the dendritic regions, the length constant of various sections, and so forth.

A. Using the Programs: Input–Output Charts and Illustrative Examples

Figure 21.1 shows the input–output flowchart for DENDR01. This program simulates the dynamic response of a neuron with a single passive

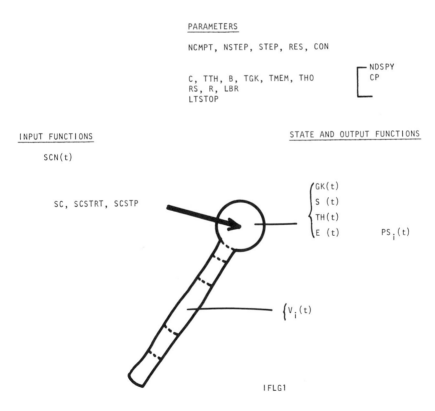

Fig. 21.1. Program DENDR01. This program simulates the activity of a neuron with a single passive dendrite attached to the soma in response to a step current input.

dendrite to a current applied at the cell body. The main input function is the magnitude of the current applied at the soma, SCN. The parameters SC, representing the amplitude of the current, and SCSTRT and SCSTP, representing the starting and stopping times of its application, respectively, determine SCN. The state and output functions for this model include the four state variables E, TH, S, and GK for the soma and the values of potential V_i for each compartment in the dendrite. IFLG1 determines whether the values of these potentials are printed at each time or repressed. The main output functions are $PS_i(t)$. These represent the ongoing potentials that would be recorded with a microelectrode into various regions of the model. The first of these always corresponds to the soma potential. To determine the others, the user specifies first NDSPY, which is the number of regions to be displayed (in addition to the soma), and then NDSPY values for CP, which identify the names of the compartments to be so displayed. The systemic parameters for this model include NCMPT, the number of compartments in the dendrite. Generally the larger the value of NCMPT, the more accurate the simulation. This is discussed more fully in Section C. STEP is the integration step size for the main do-loop in the program and is usually taken to be about 1 msec. NSTEP is the number of numerical integration steps effected on the compartment potentials within a given step; typically it should be of the order of 500–1000 to obtain mathematically accurate results. RES is the resistivity of the intra- and extracellular fluid. CON is the membrane conductance per unit area. The parameters C, TTH, B, TGK, TMEM, and TH0 are the parameters for the state variable model for the soma in micro-meters, R the radius of the dendrite in micro-meters, and LBR the length of the branch in micro-meters. LTSTP is the duration of the simulation. With program DENDR01, one may study quantitatively basic phenomena of passive electrotonic conduction in single dendrites. One may also study the influence of various parameters representing either the physiological system or the computational methodology.

   Figure 21.2 shows the input–output flowchart for program DENDR02. This program extends DENDR01 to include synaptic activation by input fibers. An arbitrary number of synaptic types, SNTP, may be specified, each characterized by its synaptic equilibrium potential EQ. The user then specifies the number of compartments to be targeted by input fibers, NTG. For each of these, the user specifies the identity of the target compartment, ITC; the synaptic type, TYPE; the probability that an impulse will fire at a given time, P; the strength of the synapse, STR; and the starting and stopping times, LSTRT and LSTP. If ITC is equal to zero, the input is directed to the soma. If ITC is a number between 1 and NCMPT, i t is directed to the corresponding compartment. The dendritic compartment adjacent to the soma is number 1; the most distal compartment is number NCMPT. The random number seed INSED determines the particular firing pattern in the input fibers.

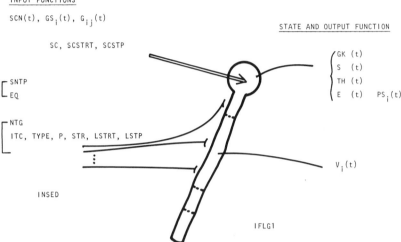

Fig. 21.2. Program DENDR02. This program simulates the ongoing input-output activity of a neuron with a single passive dendrite attached to the soma in response to synaptic input bombardment.

Program DENDR02 introduces synaptic bombardment of dendritic trees and allows users to investigate basic phenomena of electrotonic conduction of synaptic activation in dendrites, including influences of both physiological parameters and parameters relevant to the numerical integration methods.

Figure 21.3 shows the input-output flowchart for program DENTR03. Program DENDR03 introduces the numerical methodology required to simulate bifurcating dendrites. The new parameters introduced are NDN, which is the number of dendrites, in this model equal to three; and BR, which is the branch number of a given compartment. Individual compartments are referred to by their branch number BR and then by the compartment number CP within that branch. The parent branch is number 1; the two distal branches are 2 and 3.

Figure 21.4 shows the input-output flowchart for program DENDR21. Program DENDR21 simulates the ongoing input-output activity of a single neuron with a passive dendritic tree of arbitrary morphology, activated by

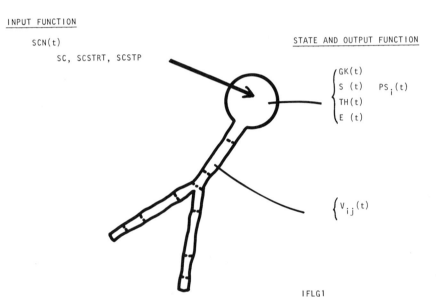

PARAMETERS

NDN, NCMPT, NSTEP, STEP, RES, CON
C, TTH, B, TGK, TMEM, THO
RS, RST, LBR
LTSTOP

$\lceil$ NDSPY
$\lfloor$ BR, CP

INPUT FUNCTION

SCN(t)

SC, SCSTRT, SCSTP

STATE AND OUTPUT FUNCTION

$\begin{cases} GK(t) \\ S\ (t) \quad PS_i(t) \\ TH(t) \\ E\ (t) \end{cases}$

$\{ V_{ij}(t) \}$

IFLG1

Fig. 21.3. Program DENDR03. This program simulates the response of a neuron with a single passive bifurcating dendrite attached to the soma to a step current input.

an arbitrary number of synaptic input systems which are individually adjustable. The main input functions are SCN, which is current applied at a stimulating electrode applied at the soma, GS, which represents synaptic input of different types incident on the individual dendritic compartments. The main state and output functions include the four state variables for the soma compartment, E, TH, S, and GK; the membrane potentials at each dendritic compartment, V_{ij}, and finally the model intracellular records PS_i for the soma and any other desired particular compartments. To use this program, the user specifies the number of synaptic types SNTP and for each of these the synaptic equilibrium potential EQ and the time constant of postsynaptic conductance decay T. Then the user specifies the number of fiber populations NFPOPS and for each of these specifies the synaptic type,

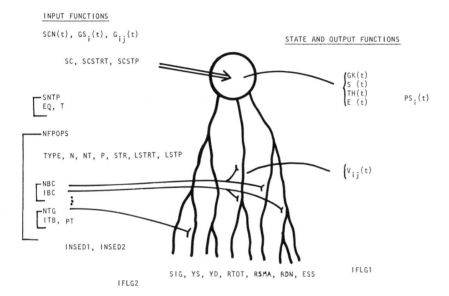

PARAMETERS

NREG, NCMPT, NSTEPS, STEP, NTGS, EK, RES, CON

C, TTH, B, TGK, TMEM, THO

RS, NST, RST, LBRNCH

┌NDN

│   ┌NDSPY
│   └BR, CP

└ J, RG, TPCN1, TPCN2, BTCN1, BTCN2

LTSTOP

INPUT FUNCTIONS

$SCN(t)$, $GS_i(t)$, $G_{ij}(t)$

SC, SCSTRT, SCSTP

STATE AND OUTPUT FUNCTIONS

┌SNTP
└EQ, T

$\begin{cases}GK(t)\\S(t)\\TH(t)\\E(t)\end{cases}$ $PS_i(t)$

┌NFPOPS

TYPE, N, NT, P, STR, LSTRT, LSTP

$\{V_{ij}(t)$

┌NBC
└IBC

┌NTG
└ITB, PT

INSED1, INSED2

SIG, YS, YD, RTOT, RSMA, RDN, ESS   IFLG1

IFLG2

Fig. 21.4. Program DENDR21. This program simulates the ongoing input–output activity of a single neuron with a passive dendritic tree of arbitrary morphology activated by an arbitrary number of synaptic input systems which are individually adjustable. Subroutine TREE computes the total resistance of the dendritic tree.

TYPE; number of fibers in the system, N; number of terminals per individual fiber, NT; proability that an individual fiber in the system will fire at a given time, P; synaptic strength, STR; and starting and stopping times of activity in the fiber system, LSTRT and LTSTOP. Then for each type one specifies the number of branches connected to NBC and for each of these the identity of the branch connected to IBC. The user may have tighter control over the input by specifying a number of targets, NTG, and for each of these the identity of the target branch, ITB, and the probability of firing at this target, PT. INSED1 and INSED2 are random number seeds which govern firing patterns in the input systems and the distribution of terminals in the IBC series. Systemic parameters that the user specifies include NREG, the number of regions in the dendritic tree. This is determined as follows. The overall dendritic tree is composed of a certain number NDN of branches, all of equal length LBRNCH. Each branch in turn is broken into a number NCMPT of compartments. A given branch is said to be in region $i + 1$ if one must traverse i bifurcations in going from the soma to that branch. Thus, the branches emanating directly from the soma are in region 1, as are any other branches connected to these without intervening bifurcations. The two branches determined by the first bifurcation as one proceeds outward from the soma are in region 2, and so on. The number of regions in the dendritic tree sketched in Fig. 21.4 is 4. The reason for defining regions is that at each bifurcation, the radii of the branches are altered according to the three-halves rule (see Chapter 3) and therefore require slightly different computations in the program. In this program, NTGS is the maximum mumber of targets for a given input fiber system. EK is the potassium equilibrium potential in millivolts, NST the number of dendritic stems, and RST the radius of the stems. In Fig. 21.4, NST is three. For each dendritic branch, one reads in first the branch number J, then the region RG, then the names of the other branches this branch connects to, represented by the four numbers TPNC1, TPCN2, BTCN1, and BTCN2. These represent, respectively, the top branches connected to and bottom branches connected to. This can be understood best by looking at Fig. 21.5. For branch 1, for example, both TPCN1 and TPCN2 are zero, BTCN1 is equal to 4, and BTCN2 is equal to 5. For branch 13, TPCN1 is 6, TPCN2 is 14, BTCN1 is 23, and BTCN2 is 24. If a given branch connects to only one other branch at its top end, then TPCN2 is equal to 0. If it connects to only one other branch at its bottom end, then BTCN2 is equal to zero. If a given branch connects to the soma, then both TPCN1 and TPCN2 are equal to zero. If a branch is immediately below a bifurcation and therefore connects to two fibers at its top end, one always identifies the parent fiber as TPCN1 and the parallel fiber as TPCN2, as illustrated above for branch 13. With these rules, one constructs dendritic trees of abritrary morphology.

Program DENDR21 includes a subroutine called TREE, which produces various values of interest related to the electrotonic properties of the dendritc tree. If IFLG2 is equal to 1, the subroutine is activated. If IFLG2 is equal to 0, it is suppressed. SIG represents the ratio of the somatic resistance to the dendritic resistance. YS represents the admittance of the soma, YD the admittance of the dendritic tree. RTOT is the total resistance of the neuron as seen from the soma, RSMA the somatic resistance and RDN the dendritic resistance. ESS is the predicted steady-state potential at the soma, corresponding to a constant current SC applied at the soma.

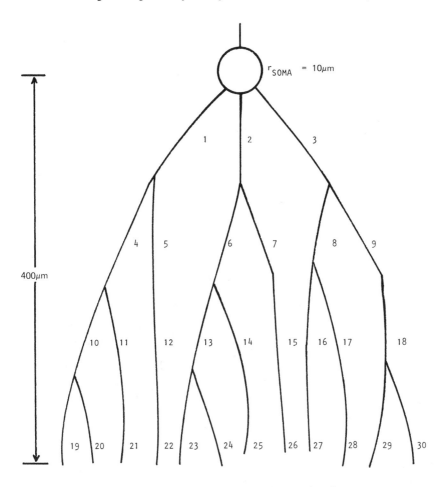

Fig. 21.5. Global morphology of the granule cell.

With program DENDR21, then, the user may simulate various electrotonic properties or ongoing dynamic activity in model neurons of arbitrary morphology with passive dendritic trees, as bombarded by individually adjustable synaptic input systems of various synaptic types and placement.

Figure 21.6 shows the input–output flowchart for program DENDR31. This program extends DENDR21 to include the state variable model for active calcium-related conductances introduced in Chapter 14. In DENDR31, each dendritic compartment is capable of exhibiting these active conductances. Therefore the state variables GCA, CA, and GKD are introduced for each individual dendritic compartment. New parameters include ECA,

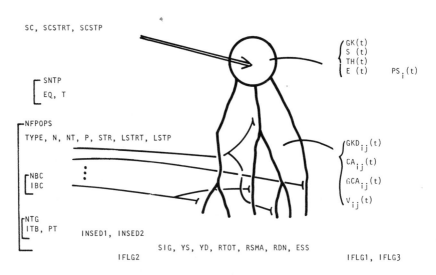

PARAMETERS

NREG, NCMPT, NSTEPS, STEP, NTGS, EK, ECA, RES, CON
C, TTH, B, TGK, TMEM, THO
RS, NST, RST, LBRNCH
THD, D, TGC, A, TCA, CAO, BD, TGKD

INPUT FUNCTIONS

SCN(t), $GS_i(t)$, $G_{ij}(t)$

NDN
J, RG, TPCN1, TPCN2, BTCN1, BTCN2
LTSTOP

NDSPY
BR, CP

STATE AND OUTPUT FUNCTIONS

SC, SCSTRT, SCSTP

SNTP
EQ, T

NFPOPS
TYPE, N, NT, P, STR, LSTRT, LSTP

NBC
IBC

NTG
ITB, PT   INSED1, INSED2

SIG, YS, YD, RTOT, RSMA, RDN, ESS
IFLG2

GK(t)
S (t)
TH(t)
E (t) $PS_i(t)$

$GKD_{ij}(t)$
$CA_{ij}(t)$
$GCA_{ij}(t)$
$V_{ij}(t)$

IFLG1, IFLG3

Fig. 21.6.   Program DENDR31. This program simulates the ongoing input–output activity of a single neuron with a dendritic tree of arbitrary morphology activated by an arbitrary number of synaptic input systems which are individually adjustable. Active calcium-related conductances are included. Subroutine TREE computes the total resistance of the dendritic tree.

the equilibrium potential for calcium, and the string of parameters THD, D, TGC, A, TCA, CA0, BD, and TGKD descriptive of the active calcium-related conductances, as introduced in Chapter 14. Specifically, THD is the voltage threshold in dendrites for activation of the calcium conductance GCA, D the amplitude coefficient driving the magnitude of GCA, TGC the time constant of response of the calcium conductance GCA, A the amplitude coefficient which determines the magnitude of calcium concentration CA in response to a given level of calcium conductance GCA, TCA the time constant of response of the calcium concentration to the calcium conductance GCA, CA0 the threshold the calcium level must attain to trigger the potassium conductance GKD, BD the amplitude constant which drives the potassium conductance GKD for a given level of calcium concentration, and TGKD the time constant of response of the potassium conductance. In all other respects, DENDR31 is equivalent to DENDR21.

With program DENDR31, the user may study various features of dendritic spatiotemporal integration, including the production and propagation of so-called calcium waves and the interactions of various synaptic input systems distributed at will over the dendritic tree. The user may also simulate responses of the model neuron to various step or pulse input current supplied by an electrode at the soma, ongoing input–output dynamic activity for various arrangements of synaptic bombardment, the influence of anatomical or physiological parameters on the dynamics of the system, or the influence of parameters reflecting the numerical integration scheme on the dymanics of the model.

Figure 21.7 shows the input–output flowchart for program DENDR51. This program extends DENDR21 to include both the active calcium-related conductances as described for DENDR31 and the representation of dendritic spines by the mathematically equivalent synaptic conductance change as discussed for DENDR41.

EXAMPLES

Responses to current step inputs in a neuron with a single dendrite and in a neuron with a single bifucating dendrite.   Figure 21.8 shows representative activity from DENDR01 and DENDR03. In both cases, a current pulse of duration 2 msec was applied to the somatic regions of these neurons and ensuing membrane potentials recorded at the soma, 100 μm out in the dendrite, and 200 μm out in the dendrite. Figure 21.8a from DENDR01 represents an unbifurcating dendrite with potentials recorded at the middle and distal compartments of the dendrite. Figure 21.8b from DENDR03 represents a bifurcating dendrite with the dendritic potentials recorded in compartments just proximal to the bifucation and at the most distal compartment, respectively. In this example the bifurcation markedly diminishes the amount of current which effectively propagates to the distal reaches of the tree.

Basic electrotonic properties of the granule cell of the hippocampus.   Figure 21.9 shows representative electrotonic properties of single excitatory postsynaptic potentials (PSPs) induced in a passive model of the granule cell of the hippocampus. Figure 21.9a shows representative unitary PSPs recorded at the soma, initiated by pulses and synapses in various parts of the dendritic tree. Figure 21.9b shows the peak amplitudes of individual unitary PSPs as a function of the input synaptic conductance change G at three regions in the dendritic tree. Finally,

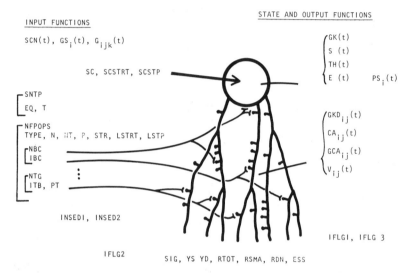

PARAMETERS

NST, NREG, NCMPT, NSTEPS, STEP, NTGS, EK, ECA, RES, CON

C, TTH, B, TGK, TMEM, THO

RS, RST

THD, D, TGC, A, TCA, CAO, BD, TGKD

┌NDSPY
└ BR, CP

┌NDN
│ J, RG, LBR, TPCNI, TPCN2, BTCNI, BTCN2, NS, RN, RSK, LSK
└ LTSTOP

STATE AND OUTPUT FUNCTIONS

INPUT FUNCTIONS

$SCN(t)$, $GS_i(t)$, $G_{ijk}(t)$

SC, SCSTRT, SCSTP

┌SNTP
└ EQ, T

┌NFPOPS
│ TYPE, N, ÑT, P, STR, LSTRT, LSTP
┌NBC
└ IBC

┌NTG
└ ITB, PT

INSEDI, INSED2

IFLG2

$\left\{\begin{array}{l} GK(t) \\ S(t) \\ TH(t) \\ E(t) \quad\quad PS_i(t) \end{array}\right.$

$\left\{\begin{array}{l} GKD_{ij}(t) \\ CA_{ij}(t) \\ GCA_{ij}(t) \\ V_{ij}(t) \end{array}\right.$

IFLGI, IFLG 3

SIG, YS YD, RTOT, RSMA, RDN, ESS

Fig. 21.7. Program DENDR51. This program simulates the ongoing input–output activity of a single neuron with a dendritic tree of arbitrary morphology activated by an arbitrary number of synaptic input systems which are individually adjustable. Active calcium-related conductances are included. Synaptic input on spines is represented by an equivalent synaptic conductance. Subroutine TREE computes the resistance of the dendritic tree.

Fig. 21.9c shows the peak amplitude of the composite PSP recorded at the soma as a function of the number of contributing unitary PSPs incident on the peripheral regions of the dendritic tree. Electrophysiological data from granule cells of the rat hippocampus indicate that unitary PSPs should be of the order of 0.1 mV and that approximately 400 simultaneous PSPs are required to drive the somatic potential to its threshold of about 24 mV.

Predicted increases in a unitary PSP in a granule cell obtained by increasing the spine stalk radius. Figure 21.10 shows the difference in amplitude in a unitary PSP as recorded at the soma corresponding to the case where the radius of the stalk of the dendritic spine is increased by 40% from 0.067 to 0.079 μm. The simulation model predicts that such an increase in dendritic anatomy, which has been observed in carefully controlled studies, should produce an increase of approximately 20% in the physiological response at the soma. This result lends credence to the hypothesis that significant plasticity in neuronal dynamics may be mediated by such morphological changes in dendritic spines.

Production of complex spines and bursting by calcium-related active conductance in the dendrites of granule cells. Figure 21.11 shows the propagation of a calcium wave produced in DENDR51 by afferent synaptic bombardment in the peripheral regions of the dendritic tree. The wave of depolarization propagate through the dendritic tree to the soma, where it produces a burst or "complex spike" of all-or-none action potentials at the soma triggering mechanisms.

B. Program Listings and Discussions (DENDR01, DENDR02, DENDR03, DENDR21, DENDR31, DENDR51)

Program DENDR01, as may be seen in Appendix 2, has the same overall structure as most other programs in this book. In particular, the do-loop for updating state variables at each time step is the heart of the program. The essential ingredient of the programs in this chapter, however, is the do-4000 loop nested within the do-1000 loop, which affects fine-grained temporal integration of the potentials in the soma and all dendritic compartments. Since updating of these membrane potentials involves continuous graded interchanging of currents in proportion to continuous differences in potential in neighboring compartments, instabilities and oscillations in numerical integration occur unless the integration time step is sufficiently small. In the integration scheme used here in the do-4000 loop, the set of potentials U at a given step are used to project values at the next increment temporarily represented by V. Note also in the formation used throughout this chapter that only the membrane potentials are integrated with such fine temporal resolution. The other state variables, for example, GK, TH, and S for the soma, are incremented with the larger temporal step, STEP, usually about 1 msec. This approach is computationally efficient and does not introduce significant errors in overall dynamics, since synaptic gates are typically open for periods of the order of 1 msec and the other state variables GK, TH, and S can be imangined to vary piecewise in STEP intervals without signigicant consequences. Among other things, this allows one to update certain functions which are used in the do-4000 loop only at STEP intervals; this is done for the somatic decay factor ZS, as may

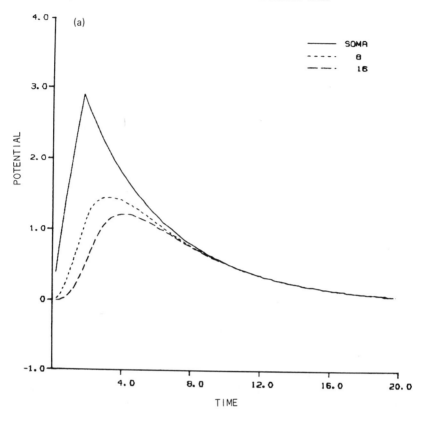

Fig. 21.8. Responses along dendrite to current pulse at soma in (a) DENDR01 and (b) DENDR03.

be seen two lines below the do-1000 statement in the listing for DENDR01. Program DENDR01, then, illustrates the basic computational approach to updating state variables used throughout the programs of this chapter.

Program DENDR02, which incorporates synaptic activation of a single dendrite, is shown in Appendix 2. This program differs from DENDR01 primarily in the do-2000 loop, which activates input synaptic conductance changes, and the lines of code surrounding the 3000 series of statements. Again notice that the do-4000 loop is the essential core of this program. The section of code around the 3000 series of statements is involved in ascertaining the influences of instantaneous input synaptic conductance changes on both the decay factors and driving currents for all the compartments in the neuron. Since the conductances are updated piecewise in steps of STEP, these operations may be done outside the do-4000 loop for computational efficiency, as discussed above.

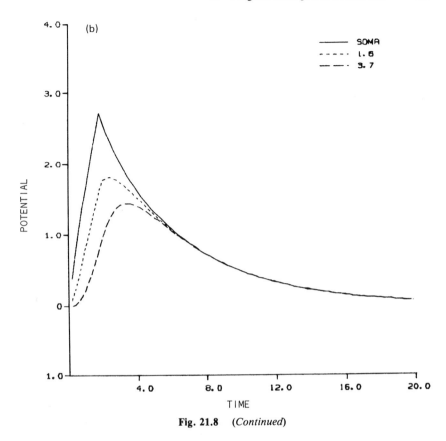

Fig. 21.8   (*Continued*)

The state variables in program DENDR03 have two subscripts since each compartment is now assigned both a branch and a compartment number. Structurally, however, the program is essentially the same as DENDR01. There is a fine-scale temporal integration loop, the do-4000 loop, embedded within the larger do-1000 time loop. The essential new ingredient in DENDR03 is that the compartment contiguous to the bifurcation must incorporate the proper mathematical description of current exchange corresponding to that bifurcation. The mathematical theory for this coupling is presented in Section C. In DENDR03 it is incorporated in the values for RTR, which represents R transfer, and the corresponding intercompartmental conductances GBT and GTU, which are used for interchanging current between the (1, NCMPT), (2, 1), and (3, 1) compartments in the do-4000 loop. The values for RTR, GBT, and GTU are defined in the section that reads and writes the input parameters.

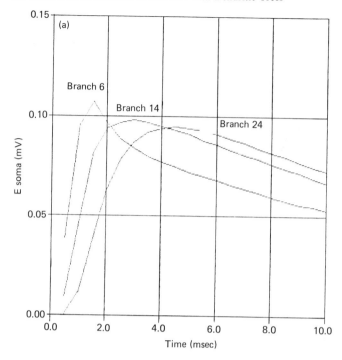

Fig. 21.9. Unitary PSPs in model of grandule cell. (a) Soma response from a single EPSP; (b) response EPSP conductances; (c) response to pulse of EPSPs.

The listing in Appendix 2 shows the program for DENDR21. Although the code is somewhat complicated, the basic structure of the program is essentially identical to that of DENDR01. It includes sections for reading and writing parameters, for initializing variables, a large do-1000 loop for updating state variables, a section for writing out activity variables, another for identifying formats, and finally a subroutine. The heart of the do-1000 loop is the do-4000 loop for fine-grained temporal updating of membrane potentials is all compartments of the neuron. Membrane potentials are updated on the basis of the synaptic current they experience at a given point in time and in accordance with the current they interchange with adjacent compartments. Adjacent compartments are identified according to the TPCN and BTCN arrays read into the program by the user. Conductances between compartments are represented by GDD for compartments within the same branch and by GTRT (G transfer top) and GTRB (G transfer bottom) for compartments at the ends of branches, while are liable to be affected by bifurcations. The rules determining GDD, GTRT, and GTRB are encoded in the section that reads and writes the input parameters. This program also allows the user to prescribe synaptic bombardment by an

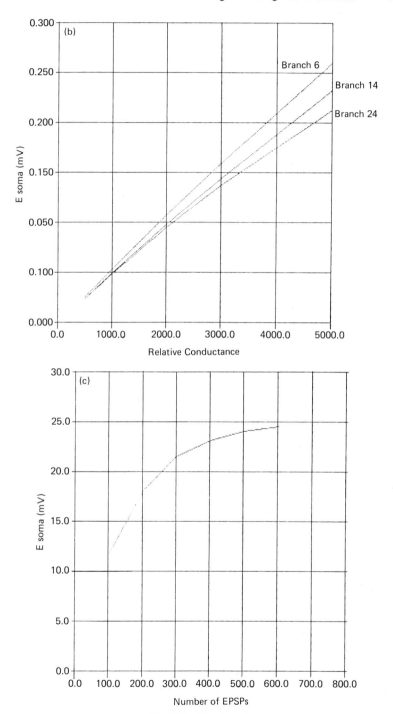

Fig. 21.9 (*Continued*)

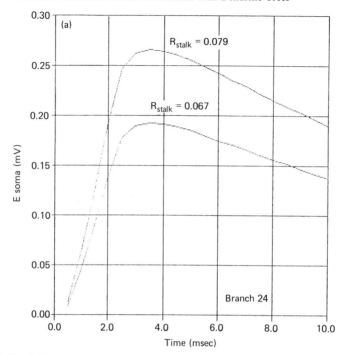

Fig. 21.10. Influence of size of dendritic spine on PSPs in model granule cell. (a) Time course of single EPSPs; (b) response to EPSP conductances; (c) response to a pulse of EPSPs.

arbitrary number of fiber populations, each with an adjustable synaptic type. The generation of firing in input systems used in this listing involves comparing a randon number against a probability of firing for every fiber at every time unit. This code could be generalized easily to represent any of the options presented in Chapter 15 for alternative distributions of interspike intervals, modulations in synaptic amplitude, the binomial model for vesicle transfer, or various forms of short-term plasticity. Such modifications could be restricted entirely to the do-2000 loop. Again, the somatic state variables TH, GK, and S are updated outside the do-4000 loop, as are the synaptic conductance values and corresponding synaptic currents and decay factors for the potentials.

If IFLG2 is greater than zero, program DENDR21 calls for the execution of subroutine TREE. As may be seen in the listing, this subroutine contains algorithms which produce values for the admittance and corresponding resistance of different regions of the dendritic tree as a whole. The theory underlying these computations is presented in Section C. The program also produces values for the radius, area, resistance per unit length, admittance, and length constant for each region in the dendritic tree. It also produces values for DCS and DCB, which represent the fraction of exponential decay for steady potentials between adjacent compartments of a branch in this region and along an entire branch in this region, respectively.

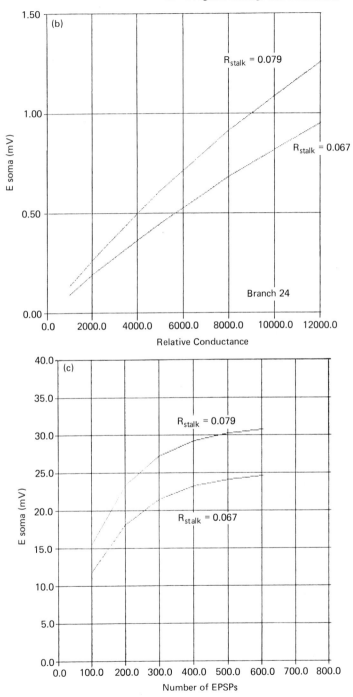

Fig. 10. (*Continued*)

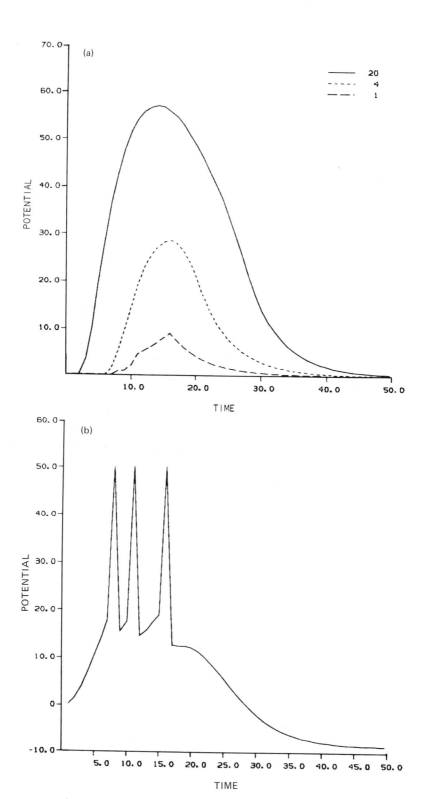

The structure of program DENDR31 differs from that of DENDR21 only in that active calcium-related conductances are included for each dendritic compartment, as may be seen in the listing. The essential new ingredient is the do-4330 loop, which updates the calcium-related state variables for each dendritic compartment. Notice that these variables are outside the fine-grained temporal integration loop do-4000. It is assumed that these variables can be represented by functions stepwise constant over intervals of size STEP without introducing serious error in the overall functioning of the model. The rule governing the dynamics of these variables is the same as that discussed in Chapter 14. The only other new additions in DENDR31 represents reading in of the parameters describing the dynamics of these new state variables and initializing their values in the initializtion section.

Program DENDR41 introduces the representation of dendritic spines by the expedient of an equivalent synaptic conductance placed on the parent spine. The mathematical theory underlying this is discussed in Section C. Its execution in DENDR41 is shown in the 10 lines of code immediately above the write (6, 6050) statement. In particular, two parameters, G1 and G2, are defined for each dendritic branch in terms of the parameters of the dendritic spines assigned to that branch. Functions G1 and G2 are then used to produce the appropriate equivalent synaptic conductance in the 3210 loop, which collects the input synaptic conductance for each compartment at each time step. DENDR41 also allows the user to specify separate lengths LBR for the individual branches in the dendritic tree. It is not clear that this is much of an advantage, however, since one can in principle represent any dendritic tree by the programs DENDR21 and DENDR31 by choosing LBRNCH small enough, and introducing different length sizes for individual branches in DENDR41 may require tighter constraints on the integration parameters NSTEP and NCMPT to attain a given level of accuracy. In other respects, DENDR41 is identical to DENDR21.

The program for DENDR51 extends DENDR21 to include both the active calcium-related conductances for dendritic compartments, as introduced in DENDR31, and the equivalent effective input synaptic conductance to represent dendritic spines, as introduced in DENDR41. In the listing for DENDR51, the inclusion of the calcium-related conductances shows up primarily in the do-4330 loop and in reading the appropriate parameters and initializing the state variables. The representation of dendritic spines is included by identifying parameters G1 and G2 for each dendritic branch in terms of the geometry of the spines prescribed for that branch, as may be seen in the code immediately above the write (6, 6050) statement and in the identification of GEQ in the do-3210 loop.

Programs DENDR31 and DENDR51 represent the internal calcium concentration in a single compartment by a rate equation which allows calcium to leak out at a rate specifiable by the user, independent of the

---

Fig. 21.11. Propagation of calcium-mediated potential wave in DENDR51. (a) Active calcium conductance; (b) neuron potential.

values of the calcium concentration in neighboring compartments. Since the overall model used for these calcium-related conductances is only a first approximation to the actual phenomena involved, although we believe a reasonable first approximation, it is possible that ignoring the diffusion of calcium explicitly through the dendritic branches is not a serious defect in the model compared to its other limitations. On the other hand, it is possible that this state variable model for calcium-related conductances can be fine-tuned or extended to the point where a reasonable approximation to the diffusion of calcium would be desirable. To do this accurately, one would need to include updating of calcium concentrations for each of the dendritic compartments within the fine-scaled do-4000 integration loop. This could be done without too much trouble, but we prefer to leave this in the category of future work, since the need does not seem to be pressing at this time.

C. Theoretical Background

1. Diffusion Equation for Longitudinal Conduction in Dendrites and Axons

In this section we will derive the classical equation for electrotonic conduction in cylindrical neuronal elements. This is the fundamental differential equation which governs longitudinal conduction of transmembrane electrical signals and axons.

Figure 21.12 illustrates a section of a dendrite of axon. Equation (21.1)

$$I_i = i\, 2\pi r\, dx + I_i + \frac{\partial I_i}{\partial x}\, dx + \cdots$$

(21.1)

$$\frac{\partial I_i}{\partial x} = -i\pi r$$

is obtrained by writing a current balance for the control volume of length dx defined in the figure. I_i is the total longitudinal current inside the dendrite as a function of position x. As dx is assumed to be differentially small and I_i is assumed to be continuous, its value at the right-hand side of the control volume, $x + dx$, is expressible in a Taylor series expansion, as indicated in Fig. 21.12. Here i is the current per unit area exiting the control volume through the membrane. Physically, Eq. (21.1) says that since no charge accumulates in the control volume, the net change is current from one side of

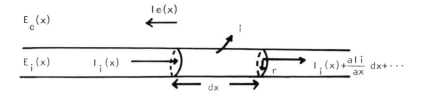

Fig. 21.12.   Section of a dendrite of axon.

the control volume to the other is equal to the amount of current lost across the membrane.

Equations (21.2a) state the assumption that the intra- and extracellular fluids are ohmic conductors. That is, the currents are proportional to the

$$I_i = -\frac{1}{R_i}\frac{\partial E_i}{\partial x}, \qquad I_e = -\frac{1}{R_e}\frac{\partial E_e}{\partial x} \tag{21.2a}$$

$$I_i + I_e = 0, \qquad \frac{\partial I_i}{\partial x} + \frac{\partial I_e}{\partial x} = 0 \tag{21.2b}$$

$$-\frac{1}{R_i}\frac{\partial^2 E_i}{\partial x^2} - \frac{1}{R_e}\frac{\partial^2 E_e}{\partial x^2} = 0, \qquad \frac{\partial^2 E_e}{\partial x^2} = -\frac{R_e}{R_i}\frac{\partial^2 E_i}{\partial x^2} \tag{21.2c}$$

$$E \equiv E_i - E_e, \qquad \frac{\partial^2 E}{\partial x^2} = \frac{\partial^2 E_i}{\partial x^2} - \frac{\partial^2 E_e}{\partial x^2} \tag{21.2d}$$

$$\frac{\partial^2 E}{\partial x^2} = \frac{\partial^2 E_i}{\partial x^2}\left(1 + \frac{R_e}{R_i}\right) \tag{21.2e}$$

$$\frac{\partial^2 E}{\partial x^2} = -R_i\frac{\partial I_i}{\partial x}\left(1 + \frac{R_e}{R_i}\right) = \frac{\partial I_i}{\partial x}(R_i + R_e) \tag{21.2f}$$

gradients of the electric fields. R_i and R_e are the resistances per unit length, respectively, in the internal and external fluid. These in turn are equal to the conductivity of the fluid times the effective cross-sectional area. Equation (21.2b) expresses the Kirchhoffian law that currents must flow in closed loops when there are no sources and sinks of charge. For a neuronal dendrite or axon, the current that is passing longitudinally in one direction within the dendrite must be counterbalanced by an equal amount of total current passing in the other direction external to the dendrite at any instant in time. Equations (21.2c) result from differentiating (21.2a) and substituting them into (21.2b). The transmembrane potential E is defined as the difference between the internal and external potentials, as defined in Eq. (21.2d). Combining Eq. (21.2c) and (21.2d) results in Eq. (21.2e). Substituting (21.2a) into (21.2e) results in Eq. (21.2f). Finally, substitution of Eq. (21.1) into (21.2f) results in Eq. (21.3).

Equation 21.3 is the fundamental conduction equation which relates the transmembrane current at any point in space to the second derivative of the transmembrane potential at that point in terms of the radius of the dendrite and the external and internal resistances per unit length. In effecting the mathematical operations Eq. (21.2) it has been assumed that the resistances per unit length both internal and external to the dendrite do not vary with x. Thus the derivations are not strictly applicable to dendrites or axons which taper or otherwise change radius with coordinate x. They are also not applicable directly to regions where a dense extracellular neuropile changes in density with position x. Generally, however, it is reasonable to model most dendritic and axonal fields by cylinders of constant radius where the radius changes only at bifurcations. In this situation, Eq. (21.3) may be applied to each continuous segment between bifurcations.

$$\frac{1}{2\pi r (R_i + R_e)} \frac{\partial^2 E}{\partial x^2} = i \qquad (21.3)$$

Equation (21.3) then may be extended to the full electrotonic equation by expressing the transmembrane current i in terms of the current–voltage relations for the equivalent circuit model for the membrane as developed in Section C,3 in Chapter 14. This results in Eq. (21.4). Here C is the membrane

$$\frac{1}{2\pi r (R_i + R_e)} \frac{\partial^2 E}{\partial x^2} = C \frac{\partial E}{\partial t} + \Sigma (g_i + G_i)(E - E_i) \qquad (21.4)$$

to ionic species i, an G_i and increment in membrane conductance to ionic species i, which may reflect any neuronal process (eg., synaptic activaton, spike generation, or active conductance changes) that acts by modulating membrane conductances. Equations (21.5a) define the total resting

$$G \equiv \Sigma g_i, \qquad \Sigma g_i E_i \equiv 0$$

$$\frac{1}{2\pi r (R_i + R_e)} \frac{\partial^2 E}{\partial x^2} = C \frac{\partial E}{\partial t} + G * E + \Sigma G_i (E - E_i)$$

$$(21.5a)$$

$$\frac{1}{2\pi r (R_i + R_e) G} \frac{\partial^2 E}{\partial x^2} = \frac{C}{G} \frac{\partial E}{\partial t} + E + \Sigma \frac{G_i}{G} (E - E_i)$$

$$(21.5b)$$

$$\lambda \equiv \frac{1}{\sqrt{2\pi r(R_i + R_e)G}}, \qquad \tau \equiv \frac{C}{G},$$

$$G_i' \equiv \frac{G'}{G} \tag{21.5c}$$

conductance of the membrane G and express the convention used here that potentials are measured relative to the resting potential of the membrane which is taken as zero. Equation (21.4) can then be rewritten as Eq. (21.5b). The length constant λ and the time constant τ are defined as in Eq. (21.5c). Moreover, it is convenient to normalize the active conductance modulations G_i, as indicated in Eq. (21.5c). The standard form for the classical equation of electrotonus which governs the dynamics and longitudinal conduction of electrical signals in circular dendrites and axons then is expressed in Eq. (21.6a). Finally, if one chooses to work with spatial coordinates normalized to the length constant and time units normalized to the time constant τ, Eq. (21.6a) can be expressed as in Eq. (21.6b):

$$\lambda^2 \frac{\partial^2 E}{\partial x^2} = \tau \frac{\partial E}{\partial t} + E + \sum G_i'(E - E_i) \tag{21.6a}$$

$$\frac{\partial^2 E}{\partial x'^2} = \frac{\partial E}{\partial t'} + E + \sum G_i'(E - E_i)$$

$$x' \equiv \frac{x}{\lambda}, \qquad t' \equiv \frac{t}{\tau}, \qquad G_i' \equiv \frac{G_i}{G} \tag{21.6b}$$

2. Calculation of Admittances for Dendrites and Axonal Fields of Arbitary Configuration

This section presents four simple rules which allow one to compute the steady state electrical admittance for dendritic and axonal fields of any arbitrary configuration.

The main rules for this simple calculus are contained in Eqs. (21.7a–d):

$$\overline{Y}_0 = \frac{1}{\lambda R} \tanh\left(\frac{l}{\tau}\right) \tag{21.7a}$$

$$\overline{Y} = \frac{\overline{Y}_0 + \overline{Y}^*}{1 + \overline{Y}_0 \overline{Y}^* (\lambda R)^2} \tag{21.7b}$$

$$\overline{Y} = \overline{Y}_a + \overline{Y}_b \tag{21.7c}$$

$$\overline{Y}_t = \frac{1}{\lambda R} \left\{ \frac{\tanh(l/\lambda) + r/2\lambda}{1 + (r/2\lambda)\tanh(l/\lambda)} \right\} \tag{21.7d}$$

$$R = 1/\sigma \pi r^2 = \rho/\pi r^2, \qquad G\rho = r/2\lambda^2 \tag{21.7e}$$

Equation (21.7a) gives the electrical admittance Y_0 seen looking into a single nonbifurcating segment of dendrite of length l; λ is the length constant of the dendrite and R is the resistance per unit length, equal to the resistivity of the fluid divided by the cross-sectional area. Equation (21.7b) is the admittance seen looking into a single nonbifurcating length of dendrite which has attached at its distal end some admittance Y^*. Equation (21.7c) gives the total admittance at a point when two separate admittances Y_a and Y_b are attached in parallel at the point. Finally, Eq. (21.7d) gives the admittance for a single nonbifurcating section of dendrite of length l whose distal end is closed with neuronal membrane. The situations described by these equations are illustrated in Figure 21.13.

To see how these equations are used to find the total admittance of a dendritic tree as seen from a given position, consider the dendritic tree with five branches shown in Fig. 21.14. First, the admittances for the individual branches are given according to Eqs. (21.7a) and (21.7d), as shown in Eqs. (21.8a):

$$\overline{Y}_i = \frac{1}{\lambda_i R_i} \tanh\left(\frac{l_i}{\lambda_i}\right), \qquad i = 1 \text{ and } 3$$

$$\overline{Y}_i = \frac{1}{\lambda_i R_i} \left\{ \frac{\tanh(l_i/\lambda_i) + r_i/2\lambda_i}{1 + (r_i/2\lambda_i) \tanh(l_i/\lambda_i)} \right\}, \qquad i = 2, 4, \text{ and } 5 \qquad (21.8a)$$

$$\overline{Y}_3^* = \frac{\overline{Y}_3 + (\overline{Y}_4 + \overline{Y}_5)}{1 + \overline{Y}_3(\overline{Y}_4 + \overline{Y}_5) (\lambda_3 R_3)^2} \qquad (21.8b)$$

$$\overline{Y} = \frac{\overline{Y}_1 + (\overline{Y}_2 + \overline{Y}_3^*)}{1 + \overline{Y}_1(\overline{Y}_2 + \overline{Y}_3^*) (\lambda_1 R_1)^2} \qquad (21.8c)$$

Second, an intermediate admittance Y_3, which represents the admittance seen looking out from the first bifurcation into the branch containing Y_3, Y_4, and Y_5, can be calculated according to Eq. (21.8b) by using Eqs. (21.7b) and (21.7c). Finally, the total admittance of the dendritic tree as seen looking out from the soma, Y, is given by Eq.(21.8c), again using rules (21.7b) and (21.7c).

This procedure can be used similarly to get the admittance for any configuration of bifurcating dendritic segments. Subroutine TREE, incorporated in most of the simulation programs of this chapter, automatically performs these calculations for whatever dendritic configuration the user has specified. Moreover, one may extend the methodology of Eqs. (21.7) to time-varying admittances by generalizing Eqs. (21.7a) and (21.7d) as shown in Eqs. (21.9).

$$\overline{Y} = \frac{1}{\lambda R} \left\{ \frac{\sqrt{1 + j\omega\tau}\ \tanh((x/\lambda)\sqrt{j\omega\tau + 1}) + (r/2\lambda)(j\omega\tau + 1)}{1 + (r/2\lambda)\sqrt{1 + j\omega\tau}\ \tanh((x/\lambda)\sqrt{j\omega\tau + 1})} \right\}$$

(21.9a)

$$\overline{Y} = \frac{1}{R_{eq}} + j\omega\, C_{eq}$$

$$R_{eq} = \frac{\lambda R \left[1 + \dfrac{r}{\lambda}\left\{\dfrac{a\sinh(2xa/\lambda) - b\sin(2xb/\lambda)}{\mathrm{cash}(2xa/\lambda) + \cos(2xb/\lambda)}\right\} + \dfrac{r}{2\lambda}\sqrt{1 + (\omega\tau)^2}\cdot\left\{\dfrac{\sinh^2(2xa/\lambda) + \sin^2(2xb/\lambda)}{\xi^2}\right\}\right]}{\dfrac{a\sinh(2xa/\lambda) - b\sin(2xb/\lambda)}{\xi} + \dfrac{r}{2\lambda}\left\{1 + \sqrt{1 + (\omega\tau)^2}\,\dfrac{[\sinh^2(2xa/\lambda) + \sin^2(2xb/\lambda)]}{\xi^2}\right\} + \left(\dfrac{r}{2\lambda}\right)^2}$$

$$\cdot\left\{\frac{[a + \omega\tau b]\sinh(2xa/\lambda) + [a\omega\tau - b]\sin(2xb/\lambda)}{\xi}\right\}$$

(21.9b)

$$\omega C_{eq} = \frac{1}{\lambda R} \frac{\{b\sinh(2xa/\lambda) + a\sin(2xb/\lambda)\} + (r/2\lambda)\omega\tau}{1 + (r/\lambda)[\{a\sinh(2xa/\lambda) - b\sin(2xb/\lambda)\}/\xi] + (r/2\lambda)^2} \frac{+ (r/2\lambda)^2\{[a\omega\tau - b]\sinh(2xa/\lambda) - [b\omega\tau a]\sin(2xb/\lambda)\}}{\sqrt{1 + (\omega\tau)^2}\ [\{\sinh^2(2xa/\lambda) + \sin^2(2xb/\lambda)\}/\xi^2\}}$$

(21.9c)

$$a = \sqrt{\frac{\sqrt{1 + (\omega\tau)^2} + 1}{2}}\,, \qquad b = \sqrt{\frac{\sqrt{1 + (\omega\tau)^2} - 1}{2}}$$

$$\xi = \cosh\frac{2xa}{\lambda} + \cos\frac{2xb}{\lambda}\,, \qquad R = \frac{1}{\sigma\pi r^2} = \frac{\zeta}{\pi r^2}\,,$$

(21.9d)

$$\lambda = \sqrt{\frac{r}{2G\zeta}}$$

In these equations j is the square root of -1 and is the membrane time constant. In these expressions and in those that would be derived from them by the operational calculus described here, the real part would be the reciprocal of the equivalent resistance and the imaginary part would be the

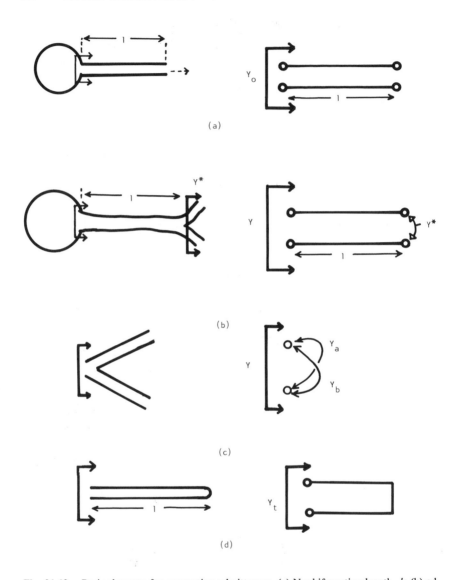

Fig. 21.13. Basic elements for computing admittances. (a) Nonbifurcating length, l; (b) admittance Y^* at the end of nonbifurcating length, l; (c) bifuration; (d) terminal length, l.

product of ω and the equivalent capacitance. Some useful limiting exprerssions for the equivalent resistance and equivalent capacitance, respectively, of a single, unbifurcating, passive dendrite of variable length as x or $\omega\tau$ independently go to 0 or ∞ can be readily obtained from Eqs. (21.9b) and (21.9c).

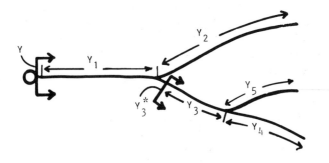

Fig. 21.14.   Illustrative admittance configuration.

In the remainder of this section I will derive the equations given above as Eqs. (21.7). First, consider some basic governing equations presented here as Eqs. (21.10):

$$\frac{\partial^2 E}{\partial x^2} - \lambda^2 E = 0 \tag{21.10a}$$

$$E = A \sinh\left(\frac{x}{\lambda}\right) + B \cosh\left(\frac{x}{\lambda}\right) \tag{21.10b}$$

$$I = -\sigma \frac{\partial E}{\partial x} A_x$$

$$A_x = \pi r^2 \tag{21.10c}$$

$$I = -\frac{\sigma A_x}{\lambda}\left[A \cosh\left(\frac{x}{\lambda}\right) + B \sinh\left(\frac{x}{\lambda}\right)\right] \tag{21.10d}$$

$$I_0 \equiv E_0 \overline{Y} \tag{21.10e}$$

$$\overline{Y} = -\frac{1}{\lambda R}\frac{A}{B} \tag{21.10f}$$

which will be used to derive the remaining results. The steady-state potential in a single unbifurcating dendrite is described by Eq. (21.10a), which may be obtained from Eq. (21.6). Its solution is given by Eq. (21.10b), where A and B are constants of integration to be determined by matching the appropriate boundary conditions. Equation (21.10c) relates the current to the derivative of the potential according to Ohm's law. σ is the conductivity of the fluid and R is the resistance per unit length. Using Eq. (21.10b) in Eq. (21.10c) results in Eq. (21.10d). The admittance Y seen looking into the dendrite at position $x = 0$ is given by the ratio of the current and potential, both evaluated at the position $x = 0$, as indicated by Eq. (21.10e). The latter terms can be evaluated from Eqs. (21.10b) and (21.10d). The resulting expression for the total admittance seen from the position $x = 0$ is given by Eq. (21.10f), where Y is given in terms of the length constant of the dendrite, the resistance per unit length, and the ratio of the parameters A and B, which must be determined for specific cases according to the boundary conditions placed on the dendrite.

Now consider a single finite unbifurcating dendritic segment of length l. For this case the boundary condition is that the current at the far end $x = l$ must be 0. From Eq. (21.10d), this implies the relationship given as Eq. (21.11a):

$$I(l) = 0$$

$$A \cosh\left(\frac{l}{\lambda}\right) + B \sinh\left(\frac{l}{\lambda}\right) = 0 \qquad (21.11a)$$

$$\frac{A}{B} = -\frac{sinh(l/\lambda)}{cosh(l/\lambda)} = -\tanh\left(\frac{l}{\lambda}\right) \qquad (21.11b)$$

$$\overline{Y}_0 = \frac{1}{\lambda R} \tanh\left(\frac{l}{\lambda}\right) \qquad (21.11c)$$

The first fundamental rule, Eq. (21.7a), then is obtained by using Eq. (21.11a) in (21.10f).

Now consider the case where an admittance Y^* is attached to the distal end of a single unbifurcating dendritic segment of length l as illustrated in Fig. 21.14b. In this case the governing boundary condition is that the current at position $x = l$ is equal to the potential at $x = l$ times the admittance Y^*, as indicated in Eq. (21.12a):

$$I_l = E_l \overline{Y}^* \qquad (21.12a)$$

$$-\frac{1}{\lambda R}\left[A \cosh\left(\frac{l}{\lambda}\right) + B \sinh\left(\frac{l}{\lambda}\right)\right] = \left[A \sinh\left(\frac{l}{\lambda}\right) + B \cosh\left(\frac{l}{\lambda}\right)\right]\overline{Y}^*$$

$$(21.12b)$$

Using Eqs. (21.10b) and (21.10d) in Eq. (21.12a) results in Eq. (21.12b). The governing equation (21.7b) then follows directly by algebraic manipulation of Eq. (21.12b).

The constraining condition for the case of two admittances placed in parallel at a given point, as illustrated in Fig. 21.14c, is that the total output current must equal the sum of the currents into the two branches, as given in Eq. (21.13a):

$$I = I_a + I_b \tag{21.13a}$$

$$\overline{EY} = \overline{EY}_a + \overline{EY}_b \tag{21.13b}$$

This current balance can be expressed eqivalently by Eq. (21.13b). Since all the potentials are measured at the same point and therefore represent the same value, Eq. (21.7c) follows immediately from Eq. (21.13b).

The governing constraint for the case where a single unbifurcating dendritic segment of length l is capped at its distal end by a neuronal membrane and therefore represents a terminating section of dendrite, as indicated in Fig. 21.13d, is expressed by Eq. (21.14a):

$$I(l) = G * A_x * E(l) = G * \pi r^2 E(l) \tag{21.14a}$$

$$-\frac{1}{\lambda R}\left[A\cosh\left(\frac{l}{\lambda}\right) + B\sinh\left(\frac{l}{\lambda}\right)\right] = G\pi r^2\left[A\sinh\left(\frac{l}{\lambda}\right) + B\cosh\left(\frac{l}{\lambda}\right)\right]$$

$$\tag{21.14b}$$

$$\lambda = \sqrt{\frac{r}{2G\varsigma}}, \quad R = \frac{1}{\sigma \pi r^2} = \frac{\varsigma}{\pi r^2} \tag{21.14c}$$

Using Eqs. (21.10b) and (21.10d), one may express this constraint as Eq. (21.14b). The final governing equation (21.7d), follows by algebraic manipulations of (21.14b). In this manipulation the interrelations between the membrane conductance per unit are G, the length constant λ, the conductivity σ, the resistivity ς, and the resistance per unit length R contained in Eqs. (21.14c) are used.

3.   Circuit Models and Equations for Compartmentalized Models of Dendritic Trees

The compartmentalized approach to analyzing dendritic potentials consists of breaking up a dendritic tree into a relatively large number of individual compartments, each of which may be considered approximately isopotential, rather than analyzing differentially small regions in terms of derivatives as was done in Section C,1. In this case, the analysis consists of simultaneously manipulating a set of variables—the compartment potentials—instead of integrating continuous differential equations of the form of Eq. (21.6) The compartmental analysis of an arbitrarily general dendritic tree may be constructed in terms of four constituent types of elements: compartments in internal regions, at terminal regions; adjacent to bifurcations; and adjacent to soma–dendritic junctions.

A typical internal compartment is schematized in Fig. 21.15a. Its corresponding equivalent circuit model is shown in Fig. 21.15b. The model represents the electrodynamics of the compartment by supposing that current can leak out of the compartment by Ohmic conduction into adjacent compartments on either side of it as well as into the extracellular fluid through the membrane of the compartment. The governing equation for the potential in this compartment is given as Eq. (21.15a),

$$\frac{dE}{dt/\tau} = -E + \frac{1}{\text{RDD} * \text{G} * \text{AM}} \{(E_l - E) + (E_r - E)\} \qquad (21.15a)$$

$$+ \sum \frac{G_i}{\text{G} * \text{AM}} (E_i - E)$$

$$\text{RDD} = \zeta l/(\pi r^2) * \text{FD} \qquad (21.15b)$$

where E is the transmembrane potential of the compartment and E_l and E_r are the potentials of the adjacent compartments. The quantity RDD

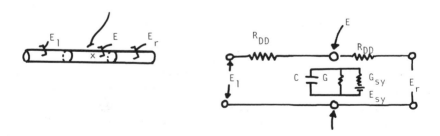

Fig. 21.15.   Equivalent circuit for internal compartment.

represents the electrical resistance between the compartments of each of their neighbors, and, AM is the surface area of membrane of the compartment. The resistance RDD is given by Eq. (21.15b). This expression assumes that the resistance to transporting current from the midpoint of its neighboring is equal to that presented by a medium of resistivity ρ, cross-sectional area A, and length equal to the length of the compartment l. The total resistance for the current loop from the midpoint of the compartment to the midpoint of the adjacent compartment and traversing back again external to the dendrite, as indicated in Fig. 21.15a, must also take into account the resistance of the extracellular medium. This in turn depends of the density and packing of other cells (neurons and glia) in the immediately surrounding area. For most cases this packing is probably not too dense, and the effective volume of extracellular fluid available for conduction is sufficiently large that the extracellular contribution to the resistance can be neglected compared to the intracelluar component. On the other hand, in regions of dense packing of neuropile, one must make some estimate of the relative contribution of extracellular resistance to the total intracompartmental resistance. In our programs this is expressed by a scaling factor FD, as indicated in Eq. (21.15b).

Figure 21.16a illustrates the case of a compartment at the terminal end of a dendrite. Figure 21.16b shows the equivalent circuit for this case. Such a compartment leaks current into the single adjacent compartment, laterally outward across its lateral surface membrane and longitudinally out of the membrane capping the dendrite. The equation governing the potential in the compartment is Eq. (21.16):

$$\frac{dE}{dt/\tau} = -E\left(1 + \frac{AT}{AM}\right) + \frac{1}{RDD * G * AM}(E_l - E_r)$$

$$+ \sum \frac{G_i}{G * AM}(E_i - E) \tag{21.16}$$

The intercompartmental resistance RDD is the same for this case as that given in Eq. (21.15b) for an internal segment. AT is the area of the terminal cap on the terminating dendrite.

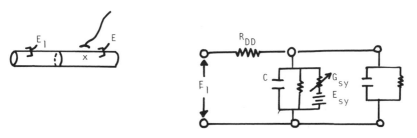

Fig. 21.16. Equivalent circuit for terminal compartment.

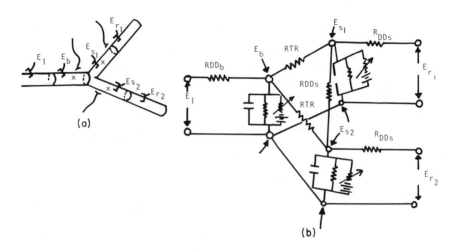

Fig. 21.17. Equivalent circuit for compartments adjacent to a bifurcation.

Figure 21.17a illustrates the structural relationships for compartments adjacent to a bifurcation. The equivalent electrical circuits for the bifurcation are shown in Fig. 21.17b. The governing equations for the transmembrane potentials in the adjacent compartments of the parent branch and the smaller branches are, respectively, Eqs. (21.17a) and (21.17b).

$$\frac{dE}{dt/\tau} = -E + \frac{1}{\text{RTR} * G * \text{AM}} \{(E_{\text{sml}_1} - E) + (E_{\text{sml}_2} - E)\}$$

$$+ \frac{1}{\text{RDD} * G * \text{AM}} (E_l - E) + \sum \frac{1}{G * \text{AM}} (E_i - E)$$

$$\tag{21.17a}$$

$$\frac{dE}{dt/\tau} = -E + \frac{1}{\text{RTR} * G * \text{AM}} (E_{\text{big}} - E) + \frac{1}{\text{RDD} * G * \text{AM}}$$

$$\cdot \{(E_{\text{op}} - E) + (E_r - E)\} + \sum \frac{1}{G * \text{AM}} (E_i - E) \tag{21.17b}$$

$$\text{RTR} = \zeta l/(2\pi) * (1/r_b^2 + 1/r^2) * \text{FD} \tag{21.17c}$$

Each of these compartments exchanges current with three neighboring compartments. In both cases the RDD resistances between compartments in a common dentritic branch are given by the same expression discussed above in Eq. (21.15b). On the other hand, for resistances between compartments that traverse the bifurcation one must take into account the possibility that the geometry is different on either side of the bifurcation. In caluolating the resistance between the smaller branches of the bifurcation, the geometries are assumed to be identical so that the resistance is the same as if these two compartments were in the same dendritic branch. On the other hand, the resistance between a compartment in the parent branch and one of those in the smaller branch is given by Eq. (21.17c). Notice that the governing equations for potentials are based on current balances which in turn depend on the quantity of surface area for the compartment being considered. Therefore the compartmental areas AM referred to in Eq. (21.17a) for the parent branch and (21.17b) for the smaller branches will have different values.

Finally, consider the case sketched in Fig. 21.18a, illustrating several dendritic compartments continuous to a soma. The circuit diagram for this case is shown in Fig. 21.18b. Equations (21.18a) and (21.18b) show the

$$\frac{dE_s}{dt/\tau} = -E_s + \frac{1}{\text{RSD} * G * \text{AS}} \sum_{j=1}^{\text{NST}} (E_{d_j} - E_s)$$

$$+ \sum \frac{G_i}{G * \text{AS}} (E_i - E_s) \tag{21.18a}$$

$$\frac{dE}{dt/\tau} = -E + \frac{1}{\text{RSD} * G * \text{AM}} (E_s - E)$$

$$+ \frac{1}{\text{RBB} * G * \text{AM}} \sum_{j=1}^{\text{NST}-1} (E_{\text{op}_j} - E)$$

$$+ \frac{1}{\text{RDD} * G * \text{AM}} (E_r - E) + \sum \frac{G_i}{G * \text{AM}} (E_i - E) \tag{21.18b}$$

$$\text{RSD} = \zeta/\pi * \left(\frac{l/2}{r^2} + \frac{r_s}{(r_s/\sqrt{2})^2} \right) * \text{FSD} \tag{21.18c}$$

$$\text{RBB} = \zeta/\pi * \left(\frac{l}{r^2} + \frac{r_s}{(r_s/\sqrt{2})^2} \right) * \text{FSD} \tag{21.18d}$$

$$\text{AM} = 2\pi r l, \quad \text{AT} = \pi r^2, \quad \text{AS} = 4\pi r_s^2 - \text{NST} * \pi r^2, \quad \tau \equiv C/G \tag{21.18e}$$

governing equations for the potentials in the soma and in the dendritic compartments, respectively. Current from the soma leaks out across the somatic

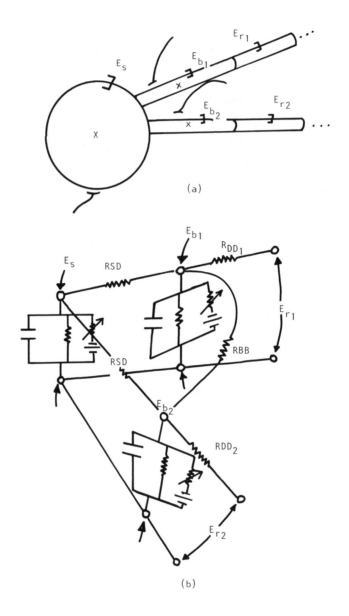

Fig. 21.18. Equivalent circuit for compartments adjacent to the soma.

membrane and into the adjacent dendritic compartments in proportion to the potential drops into these compartments. The corresponding resistance for this current leakage is RSD and its magnitude is given in Eq. (21.18c).

The dendritic compartments in turn leak current to neighboring compartments of the neighboring dendrites. The resistances for these three paths are respectively, RDD, which is the same as that given in Eq. (21.15b), RSD, given in Fig. (21.18b), and RBB, given in Eq. (21.18d). The expressions for RSD and RBB approximate the contribution of the soma region to these resistances by the quantity FSD. A slightly more exact expression is used for RSD in some of the programs in this chapter. The actual contribution to RBB would be expected to vary somewhat depending on the specific morphology of the cell under consideration.

With these generic equations for interactions between the variously arranged compartments, one may piece together the ingredients required to simulate arbitrarily general dendritic configurations, as done in the computer programs of this chapter.

4. Compartmentalized Circuit Model for Dendritic Spines; Equivalent Effective Synapse for a Dendritic Spine

Figure 21.19 illustrates a single dendritic spine and the equivalent circuit model used here to represent it. The spine is represented electrically by two compartments, the first representing its head, which is taken as a sphere, and the second representing its stalk, taken as a circular cylinder. The diferential equations governing the dynamics of the spine and the compartment to which it is attached are Eqs. (21.19):

$$C * A_h \frac{dE_h}{dt} = -G * E_h + G_{sy} * (E_{sy} - E_h) + \frac{E_s - E_h}{R_{hs}}$$

$$C * A_s \frac{dE_s}{dt} = -G * A * E_s + \frac{E_h - E_s}{R_{hs}} + \frac{E_d - E_s}{R_{sd}} \qquad (21.19)$$

$$C * A_d \frac{dE_d}{dt} = -G * A_d * E_d + \sum_{j=1}^{N_{sp}} \frac{E_{s_j} - E_d}{R_{sd_j}} + \frac{E_l - E}{R_{dd}} + \frac{E_r - E}{R_{dd}}$$

In these expressions the subscripts d, s, and h refer, respectively, to the dendrite, stalk, and head. The intercompartmental resistances are R_{hs} between the head and the stalk, R_{sd} between the stalk and the dendrite, and R_{dd} between the dendritic compartment and its neighboring compartments in the dendrite. The quantitative expressions for these intercompartmental resistances are given in Eqs. (21.20):

$$R_{hs} = (\zeta/\pi) [2/r_h + l_s/2r_s^2)] * FSP$$

$$R_{sd} = (\zeta/2\pi) [l_s/r_s^2 + (l_d + 4r_s)/2r_d^2] * FSP \qquad (21.20)$$

$$R_{dd} = (\zeta l_d/\pi r_d^2) * FD$$

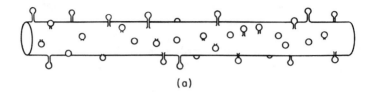

(a)

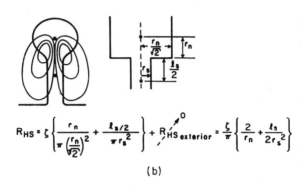

(b)

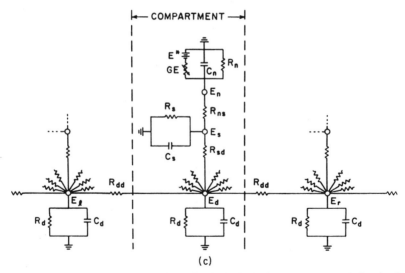

(c)

Fig. 21.19. (a) Schematic representation of a single dendritic compartment; (b) longitudinal resistance between spine head and spine stalk subcompartments; (c) equivalent circuit for a compartment and its relationship to its neighboring compartments (each compartment contains about 52 branches representing individual spines radiating out from the E_d point; the circuit for one such branch is shown in entirety).

They incorporate the same physical ideas discussed for intercompartmental resistances in section C-3 immediately above. Equations (21.19) can easily be manipulated into the form shown in Eq. (21.21a),

$$\frac{dE_h}{d(t/\tau)} = -E_h + G'_{sy} * (E_{sy} - E_h) + a * (E_s - E_h)$$

$$\frac{dE_s}{d(t/\tau)} = -E_s + b * (E_h - E_s) + c * (E_d - E_s) \qquad (21.21a)$$

$$\frac{dE_d}{d(t/\tau)} = -E_d + \sum_{j=1}^{N_{sp}} d_j * (E_{s_j} - E_d) + e_l (E_l - E_d) + e_r - E_d)$$

$$a = \frac{1}{A_h * G * R_{hs}}, \qquad b = \frac{1}{A_s * G * R_{hs}}$$

$$c = \frac{1}{A_s * G * R_{sd}}, \qquad d = \frac{1}{A_d * G * R_{sd}} \qquad (21.21b)$$

$$e = \frac{1}{A_d * G * R_{dd}}, \qquad \tau \equiv C/G, \qquad G'_{sy} \equiv \frac{1}{G * A_h}$$

where the nondimensional parameters a, b, c, d, and e, shown in Eqs. (21.21b), have been introduced. The value of this manipulation is that the numerical values for a, b, c, d, and e for representative sizes of dendritic spines and representative sizes of dendritic compartmentalization can be shown to be exceedingly large compared to 1. For example, the numbers given in Table 21.1 show that the values of a, b, and c which govern the dynamic properties of potentials in the spines are of the order of several hundred thousand. This suggest that the time derivative terms in the first two of Eqs. (21.21) may be neglected compared to other terms in the equation system. This in turn implies that Eqs. (21.21) may be rewritten as shown in Eqs. (21.22a):

$$0 = G'_{sy} * (E_{sy} - E_h) + a * (E_s - E_h)$$

$$0 = b * (E_h - E_s) + c * (E_d - E_s) \qquad (21.22a)$$

$$\frac{dE_d}{d(t/\tau)} = -E_d + d * (E_s - E_d) + e_l * (E_l - E_d) + e_r * (E_r - E_d)$$

Table 21.1(a)

Numerical Values of Model Granule Cell Compartment

	Region 2	Region 3	Region 4
Number of branches	11	12	4
Number spines/region	4074	4444	1481
Radius spine head (mic)	.1450	.1660	.1780
Radius spine stalk (mic)	.0530	.0645	.0670
Length spine stalk (mic)	.3750	.3750	.3750
Radius dendrite (mic)	.7937	.6804	.6140
Area spine head (sq mic)	.2554	.3332	.3840
Area spine stalk (sq mic)	.1249	.1520	.1579
Area stalk x-sect (sq mic)	.0088	.0131	.0141
Area spine total (sq mic)	.3803	.4852	.5419
Number spines/comp	52.91	52.91	52.91
Area dend prop (sq mic)	99.27	84.81	76.41
Area dend eff (sq mic)	119.4	110.5	105.1
Area dend effl (sq mic)	119.0	110.0	104.5
Resist head-stalk (ohms)	.000018(**12)	.000013(**12)	.000012(**12)
Resist stalk-dend (ohms)	.000017(**12)	.000012(**12)	.000012(**12)
Resist dend-dend (ohms)	.000007(**12)	.000009(**12)	.000012(**12)
Resist mem head (ohms)	4542(**12)	3481(**12)	3020(**12)
Resist mem stalk (ohms)	9289(**12)	7633(**12)	7348(**12)
Resist mem spine (ohms)	3050(**12)	2391(**12)	2141(**12)
Resist mem dend pr (ohms)	11.69(**12)	13.68(**12)	15.18(**12)
Resist mem spines tot (ohms)	57.66(**12)	45.19(**12)	40.46(**12)
Resist mem dend pr (ohms)	9.716(**12)	10.50(**12)	11.04(**12)
Res mem/length (ohm/mic)	.1943(**12)	.2100(**12)	.2208(**12)
Long res/length (ohm/mic)	.354(**6)	.481(**6)	.591(**6)
Length constant (mic)	741.2	660.5	611.2
$1-\exp(-l/lc)$	.0266	.0298	.03219
a	253,100	253,500	255,745
b	517,600	599,700	622,200
c	558,200	613,100	599,300
d	585.7	847.1	904.9
e	1,378	1096	938.7
ac/(b + c)	131,300	138,300	124,500
db/(b + c)	281.8	418.9	460.9

Table 21.1(b)

Limiting (Steady-State) Values for an Active Spine Synapse
on a Model Granule Cell Compartment

	Region 1	Region 2	Region 3
E_{stalk}/E_{head}	0.892	0.931	0.942
E_{dend}/E_{head}	0.791	0.863	0.882

Table 21.1(c)

Steady-State Potentials in Passive Spines

E_{head}/E_{dend}	ca .999991 (all regions)
E_{stalk}/E_{dend}	ca .999995 (all regions)

$$E_s - E_d = \left(\frac{b}{b + c}\right)\left[\frac{G_{sy}}{G_{sy} + ac/(b + c)}\right](E_{sy} - E_h) \qquad (21.22b)$$

Physically, this mathmatical assumption is equivalent to stating that the current flows across the neuronal membrane in the spine regions are insignificant compared to the current flow between the spinal regions and the dendrite and may be neglected in comparison with these other currents. By algebraic manipulation of Eqs. (21.22a) one can obtain the intermediate equation shown in (21.22b). This expression in turn can be used in conjuction with the last of Eqs. (21.22a) to produce the equivalent overall governing equation system given as Eqs. (21.23):

$$\frac{dE_d}{dt} = -E_d + G_{eff} * (E_{sx} - E_d) + e_l(E_l - E_d) + e_r(E_r - E_d)$$

$$(21.23a)$$

$$G_{eff} = \left(\frac{db}{b + c}\right)\left[\frac{G_{sy}}{G_{sy} + ac/(b + c)}\right] \qquad (21.23b)$$

These expression states that the influence of a synaptic conductance modulation of the head of a spine may be represented by an equivalent synaptic conductance change placed directly on the surface of the parent dendritic compartment. Equation (21.23a) is the governing equation for the potential in the parent dendritic compartment in terms of the equivalent synaptic conductance change G_{eff}; Eq. (21.23b) expresses the equivalent synaptic conductance G_{eff} in terms of the actual conductance G experienced at the head of the spine and the various parameters a through e which characterize the spinal system. Recall that both G and G_{eff} are normalized to the resting membrane conductance.

It is of interest to compute the relative steady-state variables of potentials in the dendritic head and spine for active and passive synapses, respectively. Tables 21.1b and 21.1c, respectively, represent such estimates, which are obtained from Eqs. (21.24) and (21.25):

$$E_{s_{ss}} = \frac{1 + (m + 1)/d}{1 + (m + 1)(b + c)/db}$$

$$\frac{E_{d_{ss}}}{E_{h_{ss}}} = \frac{1}{1 + (m + 1)(b + c)/db} \qquad (21.24)$$

$$E_{d_{ss}} = \frac{G_{eff}}{G_{eff} + (m + 1)}E_{sy}$$

$$m = \frac{2\pi r_d^2}{A_d G \zeta \lambda} = \frac{I_0/(A_d G)}{E_d}$$

$$E_{s_{ss}} = \left[\frac{c}{1 + c + b/(1 + a)}\right] E_d$$

$$E_{h_{ss}} = \left(\frac{a}{1 + a}\right) \left[\frac{c}{1 + c\, b/(1 + a)}\right] E_d \tag{21.25}$$

These equations in turn are readily derivable from the governing equation system (21.21). The estimates shown in Table 21.1b indicate that under an active synapse at the head of the spine, the potential in the stalk might be approximately 90% of that at the head, whereas the potential in the parent dendritic branch might be approximately 80% of that in the head. On the other hand, potentials in the stalk and head regions of inactive spines are insensibly different from the potentials in the parent dendritic compartments. The numerical values used in constructing these estimates are based on compartmentalization of the granule cells of the rat hippocampus. The values assumed are shown in Table 21.1a.

Equation (21.23b) for an effective equivalent synaptic conductance for a dendritic spine is used in programs DENDR41 and DENDR51. In addition, those programs incorporate the required changes in membrane surface area to correctly account for total membrane resistance and capacitance and effective longitudinal resistance necessitated by the imposition of a number of spines of a particular geometry in a given dendritic compartment.

5. Convergence Properties of the DENDR Series of Computer Programs

This series of computer programs tends to converge most accurately to theoretically predicted steady-state values when the integration parameters NCMPT and NSTEPS are adjusted so that the terms in the exponents which govern the numerical integration scheme used here, as discussed in Section C,4 in Chapter 14, are approximately the same for all the different potentials in the system. Errors are introduced when any two of these values are markedly different and particularly when values for interfacing compartments are markedly different. Moreover, these numbers should have values which do not strain the accuracy of numerical evaluation of expotential functions within the FORTRAN system.

Specifically, Fig. 21.20 maps the steady-state error obtained in applying program DENDR21 to a case of a single unbifurcating dendrite which is attached to a soma where a step current is applied. In this case $\Delta 1$ and $\Delta 2$ represent the arguments of the exponents applicable to the numerical integration of the somatic region and the adjacent dendritic compartment, respectively. The error is smallest when these values are about equal to each

BASED ON 46 DATA POINTS:

$$r_d \quad \text{in} \quad 0.7\text{-}1.0$$
$$r_s \quad \text{in} \quad 3.0\text{-}10.0$$
$$\text{NCMPTS} \quad \text{in} \quad 34\text{-}43$$
$$\text{NSTEPS} \quad \quad 1000$$

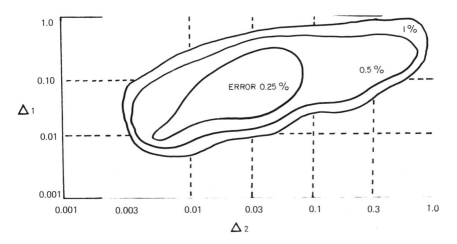

Fig. 21.20.   Steady-state convergence properties of DENDR21.

other and to about 0.01. The error increases in surrounding regions even when these two values are about equal because some of the other exponents in the system become largely disparate. The expressions for $\Delta 1$ and $\Delta 2$ are given in Eqs. (21.26):

$$\Delta_1 = \left[1 + \frac{1}{(AS/4\pi\lambda r)(l/\lambda + 4r^2/\lambda r_s)} \right] \left(\frac{\Delta}{\text{TMEM} * \text{NSTEPS}} \right)$$

$$\Delta_2 = \left\{ 1 + \frac{1}{(l/\lambda)^2} + \frac{1}{\frac{1}{2}[(l/\lambda)^2 + (l/\lambda) \, 4r^2/\lambda r_s^2]} \right\} \left(\frac{\Delta}{\text{TMEM} * \text{NSTEPS}} \right)$$

$$(21.26)$$

The primary free parameters one has to control are NCMPT and NSTEPS. For this case, highly accurate numerical convergence is obtained for NCMPT about 35 and NSTEPS about 1000.

It is more difficult to guarantee numerical convergence for more elaborate dendritic trees because of the larger numbers of possible disparate

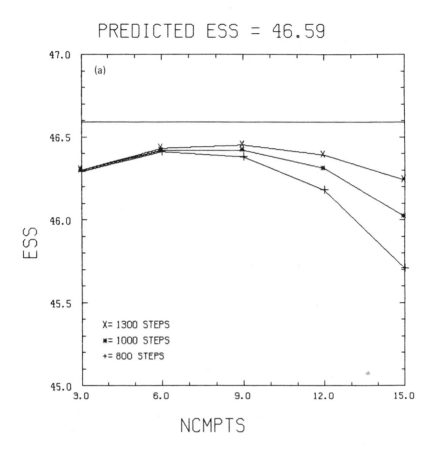

Fig. 21.21. Steady-state error in model of granule cell: (a) assymptotic steady-state potential as a function of NCMPTS; (b) potential as a function of time.

expotential terms encountered. It would be possible and perhaps desirable to refine these programs internally so that the exponential terms were guaranteed to be approximately equal. This could be achieved by varying the number of compartments in the different regions in the dendritic tree to counterbalance the influence of changing dendritic radius in these regions. We leave this task in the realm of "future work." With the programs as they stand, one can obtain realistic and representative behavior and moreover in most cases, with a little juggling, can obtain relatively good numerical accuracy. For example, Fig. 21.21 shows the steady-state error obtained from program DENDR21 applied to the 30-branched dendritic tree model of the hippocampal granule cell discussed in Section A. The figure shows that the error can be reduced to less than 1% by judicious choices of NCMPT and NSTEPS.

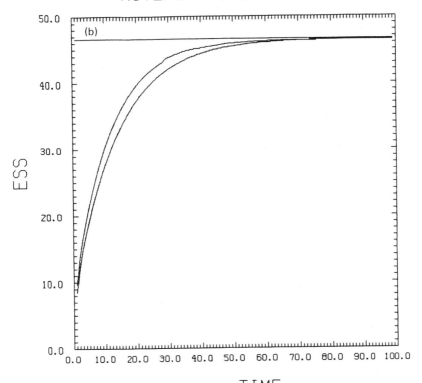

Fig. 21.21 (*Continued*)

We have not systematically explored the temporal convergence properties of the computer programs of this series. However, the results we have
obtained so far lead us to believe that the programs do produce temporally
accurate behavior. Moreover, in accordance with the discussion of
numerical accuracy given in Chapter 14, our focus of interest is on the
overall, ongoing, information processing features of neurons as they are
subjected to various interacting input strains and their mutual interactions
in large populations and systems. We consider these models sufficiently accurate for meaningful simulation in such contexts. Such refinements and
improvements as may be desired by researchers with other points of view
may be effected by slight internal adjustments of the exponents governing
the numerical integration of by sustaining alternative numerical integration
schemes with the overall structure of the programs.

D. Exercises

1. Use program DENDR01 to examine the time course of the response to a step current, as stimulated and recorded in the soma of a neuron with a very long dendrite. Take LRB equal to 3500 and NCMPT equal to 100. Adjust the othetr parameters at reasonable values. Take the threshold to be very large, say 100; take SC to be 25. Plot the time course of the response for each of LRB = 3500, 2000, 350, and 0. Take NCMPT = 100, 50, and 10, respectively, for these four cases.

2. Use program DENDR02 to study the passive electrotonic decay of a single PSP initiated at a dendritic compartment some distance from the soma.

3. Use DENDR02 to illustrate the effectiveness of synaptic inhibition in compartments contiguous to the soma on synaptic excitation applied in more distal regions of the dendrite.

4. Use program DENDR21 to simulate the dendritic tree illustrated in Fig. 21.5. Project one excitatory system to branches 10, 11, and 12 and a second excitatory system to branches 13, 14, and 15. Project and inhibitory input system to branches 6 and 7. Perform various comparative simulations to illustrate the effectiveness of the inhibitory system in blocking excitatory input.

5. Use program DENDR31 to simulate the granule cell illustrated in Fig. 21.5. Project and excitatory synaptic input system to activate branches 10–18. Activate another synaptic input system to target branches 19–30. Now project an atypical excitatory system (EQ = -10, STR = -0.1) to activate regions 1, 2, and 3. For the first runs leave the atypical excitory system quiescent. Carefully adjust the parameters related to the active calcium conductances (THD, D, . . .) so that the excitatory input triggers "calcium waves" which propagate up to the soma and there produce bursts or complex spikes. Obtain a control level of this type of activity wherein several bursts or complex spikes occur over a reasonable time period. Now apply the atypical excitatory system and observe an increase in the number of bursts and the number of spikes per burst.

6. Use program DENDR41 to illustrate the change in the potential observed at the soma in response to a single unitary excitatory PSP applied in a dendritic region as a function of the diameter of the stalk of the spine on which the synapse is placed. Explore the influence of various critical parameters on this change.

7. Sometimes one may want to simulate "atypical" synaptic excitation wherein conductance toward a negative equilibrium potential is decreased. For example, this is the hypothisized action of septal input on granule and parametal cells in the hippocampus. In simulating this phenomenon with the models of this chapter, there is danger that under conditions of high synapic input of this type the programs might exceed the upper limit of conductance change allowed to blocking conductance to go negative, which is physically meaningless. Adjust program DENDR51 so that the total negative conductance change at such input system can never exceed a certain negative limit, say 0.3. (Hint: this can be done wth one line of code at a strategic point in the program.)

8. Use program DENDR51 to simulate the spiny cell of the substantia nigra as described in the research of Dr. Philip Groves (cf., Chapter 6).

9. Use program DENDR21 to obtain and intensity–duration curve for accommodation as discussed in Chapter 14 for a neuron with the dendritic tree illustrated in Fig. 21.5. You will have to extend program DENDR21 to simulate RAM input currents of slope SL.

10. Modify program DENDR51 to include short-term plasticity in its incident synapses as modeled in Chapter 15.

11. Modify program DENDR51 to approximate the influence of recurrent inhibition as follows: insert an activity-in-transit matrix, G_i, as introduced in earlier chapters. Now project output signals from the model neuron when it fires into the activity-in-transit matrix such that synaptic inhibition is effected on the soma of the model neuron with a time lag of about 5 time units.

12. Use the model of Exercise 11 together with two distinct streams of synaptic excitation in the dendritic tree and atypical synaptic excitation on branches 1, 2, and 3 to simulate ongoing input–output activity in this neuron. Study the influence on the ongoing activity of the various parameters relating to the recurrent inhibition, short-term plasticity, spine geometry, active calcium conductance-related changes, and accommodative properties.

22

Propagation of Action Potentials in Axons and Axon Termination Fields

This chapter presents computer programs to simulate the generation and propagation of action potentials according to the equation system presented by Hodgkin and Huxley. This equation system produces fine-grained descriptions of the time course of the membrane potential, primarily as determined by excursions in the membrane conductance to sodium and potassium. The behavior of these latter functions is governed in turn by the time courses of three state functions, *n, m,* and *h.* Since the essential focus of this equation system is on the detailed time course of the action potential and the events immediately leading up to it, the computer programs in this chapter necessarily involve considerably more temporal integration steps than the programs in the previous chapters. The chapter presents programs which describe the generation of single action potentials in a point neuron, the generation and propagation of action potentials down a single unbifurcating axon, the propagation of action potentials down a myelinated axon, the propagation of action potentials into an axon that bifurcates, and the propagation of action potentials into an axon termination field of arbitrary structural configuration. The chapter also presents a program which modifies the Hodgkin–Huxley model to simulate contraction of the Purkinje fibers of heart muscle.

A. Using the Programs: Input–Output Charts and Illustrative Examples

Figure 22.1 shows the input–output flowchart for program HHNRN01. This program represents a point neuron driven by a current from an experimentally applied electrode which generates action potentials according

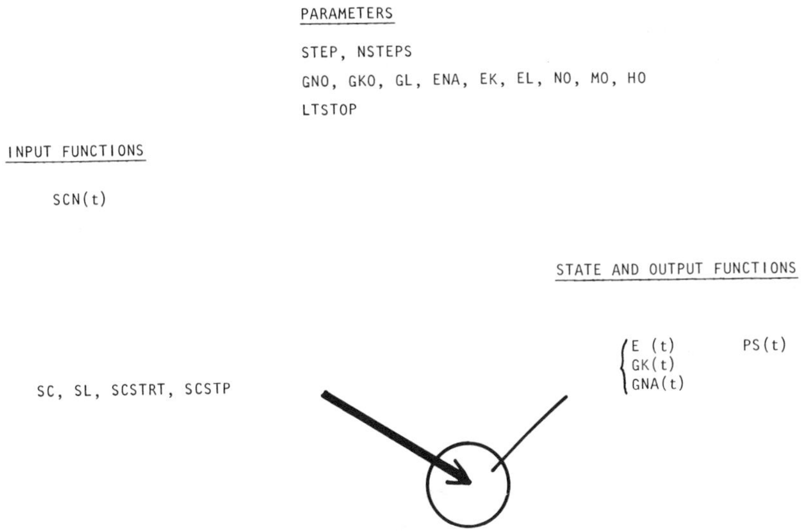

PARAMETERS

STEP, NSTEPS

GNO, GKO, GL, ENA, EK, EL, NO, MO, HO

LTSTOP

INPUT FUNCTIONS

SCN(t)

STATE AND OUTPUT FUNCTIONS

$\begin{cases} E\ (t) \\ GK(t) \\ GNA(t) \end{cases}$ PS(t)

SC, SL, SCSTRT, SCSTP

Fig. 22.1. Program HHNRN01. This program simulates the Hodgkin–Huxley model of the response in a neuron patch to a stimulating electrode.

to the Hodgkin–Huxley equation system. The parameters which determine the input stimulating current are SC, SL, SCSTRT, and SCSTP. These determine, respectively, the amplitudes of step and ramp currents and the starting and stopping times for input stimulation. The state and output functions for this model are the transmembrane potential and the sodium and potassium conductances GNA and GK. The systemic parameters for use of this model include STEP and NSTEPS. STEP is the main cycle step size in the program, typically taken as 1 msec. The state and output functions are outputted at every step size STEP. NSTEPS defines the number of temporal integration steps in each unit of STEP; typically this number should be 50 or higher to obtain acceptable accuracy in numerical integration. The next line of systemic parameters refers to the Hodgkin–Huxley model directly. GN0 and GK0 represent the resting values for sodium conductance and potassium conductance, respectively. GN0 is taken as 120 and GK0 as 36. GL is the resting value of the leakage conductance, taken as 0.3. ENA, EK, and EL are the equilibrium potentials for sodium, potassium, and leakage, respectively. Their values are specified respectively as -115, 12, -10.6. N0, M0, and H0 are the resting levels of the state functions which determine the time courses of GNA and GK. Their values are 0.31767691, 0.05293249, and 0.59612075. LTSTOP is the duration of the simulation.

Figure 22.2 shows the input–output flowchart for program AXON01. This program simulates the generation of action potentials in the soma by

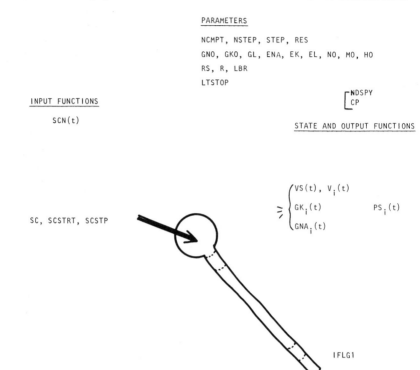

PARAMETERS

NCMPT, NSTEP, STEP, RES

GNO, GKO, GL, ENA, EK, EL, NO, MO, HO

RS, R, LBR

LTSTOP

INPUT FUNCTIONS

SCN(t)

NDSPY
CP

STATE AND OUTPUT FUNCTIONS

SC, SCSTRT, SCSTP

$\begin{cases} VS(t), V_i(t) \\ GK_i(t) \qquad PS_i(t) \\ GNA_i(t) \end{cases}$

IFLG1

Fig. 22.2. Program AXON01. This program simulates the propagation of action potentials along a single nonbifurcating axon according to the Hodgkin–Huxley model.

applied stimulating current from an experimental electrode and subsequent propagation of those action potentials down a finite axon. The new parameters introduced in AXON01 include NCMPT, the number of compartments used to represent the finite axon; RES, the resistivity of the intracellular fluid; RS the radius of the soma; R the radius of the axon; and LBR, the length of the axon branch. Program AXON01 generates values for the membrane potential, potassium conductance, and sodium conductance for the soma and each compartment of the axon. The user specifies how many compartments the potentials are to be displayed for, NDSPY, and for each of these the identity of the compartment, CP. The resulting set of potentials is stored in PS. If IFKG1 is equal to 1, the potentials in all compartments are printed out in each step in time. If IFLG1 is equal to 0, this printing is suppressed.

Figure 22.3 shows the input–output flowchart for program AXON02. This program is identical to AXON01 except that it simulates propagation down myelinated axons. The new parameter introduced in AXON02 is

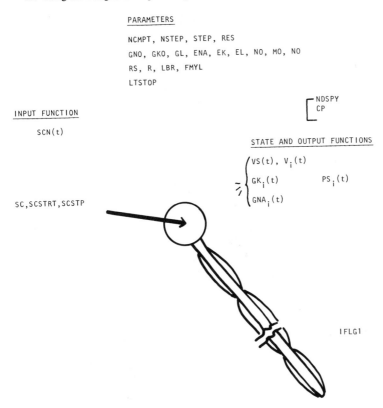

PARAMETERS

NCMPT, NSTEP, STEP, RES

GNO, GKO, GL, ENA, EK, EL, NO, MO, NO

RS, R, LBR, FMYL

LTSTOP

NDSPY
CP

INPUT FUNCTION

SCN(t)

STATE AND OUTPUT FUNCTIONS

VS(t), V_i(t)

GK_i(t) PS_i(t)

GNA_i(t)

SC,SCSTRT,SCSTP

IFLGl

Fig. 22.3. Program AXON02. This program simulates the propagation of action potentials along a myelinated axon with equally spaced nodes of Ranvier according to the Hodgkin–Huxley model.

FMYL, which represents the fraction of membrane surface in the axon which is myelinated.

Figure 22.4 shows the input–output flowchart for program AXON03, which simulates propogation of action potentials into a axon that bifurcates. The new parameter introduced in AXON03 is NAX, the number of axons. In this program, NAX is three. To identify the compartments for which potentials are to be displayed, one must identify both the branch number and a compartment number. There are NCMPT compartments in each branch. The branch emanating from the soma is number 1; the other two branches are numbers 2 and 3. Compartments in a given branch are numbered such that the compartment closest to the soma is number 1 and the compartment most distal from the soma is number NCMPT. Otherwise, the parameters in AXON03 are identical to those in the earlier programs. LBR refers to the length of individual branches in the tree.

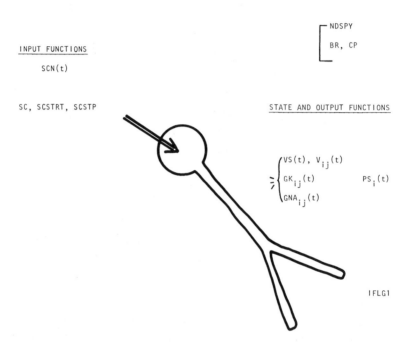

PARAMETERS

NAX, NCMPT, NSTEP, STEP, RES
GNO, GKO, GL, ENA, EK, EL, NO, MO, HO
RS, RST, LBR
LTSTOP

NDSPY

BR, CP

INPUT FUNCTIONS

SCN(t)

SC, SCSTRT, SCSTP

STATE AND OUTPUT FUNCTIONS

$$\gtrsim \begin{cases} VS(t), \ V_{ij}(t) \\ GK_{ij}(t) \qquad\qquad PS_i(t) \\ GNA_{ij}(t) \end{cases}$$

IFLG1

Fig. 22.4. Program AXON03. This program simulates the propagation of action potentials along an axon with a single bifurcation according to the Hodgkin–Huxley model.

Figure 22.5 shows the input–output flowchart for program AXON04. Program AXON04 simulates the generations of action potentials in a soma and their subsequent propagation into an axon terimination field of arbitrary structural configuration. The methodology and parameters used to define the structure of the axonal field are the same as those used to define the dendritic fields in Chapter 21. In fact, as seen in Section B, the structure of AXON04 is essentially identical to that of program DENDR21. The parameter NREG, number of regions, has the same interpretation as discussed above for the dendritic programs of Chapter 21. Specificallly, a branch is said to be in region i if one must traverse $i - 1$ bifurcations in progressing from this branch to the soma. For each axon branch, one reads in the number of branch J, radius of the branch R, length of the branch LBR, and top and bottom connections TPCN 1 and 2 and BTCN 1 and 2. Please

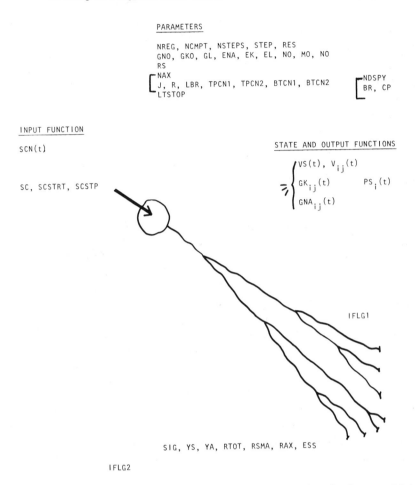

PARAMETERS

NREG, NCMPT, NSTEPS, STEP, RES
GNO, GKO, GL, ENA, EK, EL, NO, MO, NO
RS
NAX
J, R, LBR, TPCN1, TPCN2, BTCN1, BTCN2
LTSTOP

NDSPY
BR, CP

INPUT FUNCTION

SCN(t)

STATE AND OUTPUT FUNCTIONS

$VS(t)$, $V_{ij}(t)$
$GK_{ij}(t)$ $PS_i(t)$
$GNA_{ij}(t)$

SC, SCSTRT, SCSTP

IFLG1

SIG, YS, YA, RTOT, RSMA, RAX, ESS

IFLG2

Fig. 22.5. Program AXON04. This program simulates the propagation of action potentials into axon termination fields of arbitrary bifurcation patterns according to the Hodgkin–Huxley model.

refer to Chapter 21 for explanations of how to define these parameters in creating specific termination configurations. Program AXON04, like the dendrite series of programs, contains subroutine TREE, which computes various electrotonic parameters of interest regarding the axon termination field. These parameters have the same interpretation as they do in Chapter 21. Other parameters in AXON04 have the same meaning as in earlier programs in this chapter.

With the programs of this chapter one may stimulate the initiation of single action potentials in cell bodies and their subsequent propagation into axonal fields of various configurations. One may study the influence of various physiological and anatomical parameters on such features of propagation as conduction speed and reliability of successful propagation through bifurcations. Although the models can be driven to produce propagation of bursts of multiple individual action potentials, these results are of questionable validity because the Hodgkin–Huxley model itself does not accurately describe repetitive firing, but is accurate only for single isolated action potentials.

Figure 22.6 shows the input–output flowchart for program HEART01. This program simulates contractions of Purkinje fibers of the heart according to a model presented by McAllister and Noble. Please refer to their original publications for detailed information (see citation in Chapter 6).

EXAMPLES

Propagation of an action potential along myelinated and unmyelinated axons. Figure 22.7 shows the propagation of a single action potential down a single unbifurcating axon. Figure 22.7a shows propagation simulated with program AXON01 for an unmyelinated axon. Propagation from compartment 1 to compartment 8 requires approximately 3.5 msec. Figure 22.7b, simulated with program AXON02, represents propagation in an axon whose surface is 90% myelinated but which is otherwise structurally and physiologically equivalent to the one simulated in 22.7a. In this case, conduction from compartment 1 to compartment 8 took approximately 1 msec.

Propagation of action potentials into axonal bifurcating field. Figure 22.8 illustrates propagation of a single action potential into the branches of a bifurcating axon, as simulated by program AXON03. The curve shows the arrival of the action potential in the soma, compartments 4 and 8 of branch 1, compartment 5 of branch 2, and compartment 8 of branch 3.

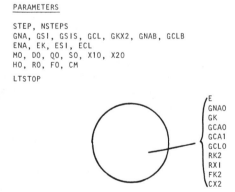

```
PARAMETERS

STEP, NSTEPS
GNA, GSI, GSIS, GCL, GKX2, GNAB, GCLB
ENA, EK, ESI, ECL
MO, DO, QO, SO, X10, X20
HO, RO, FO, CM

LTSTOP
```

```
E
GNAO
GK
GCAO
GCA1
GCLO
RK2
RXI
FK2
CX2
```

Fig. 22.6. Program HEART01. This program simulates the McAllister–Noble model of the Purkinje fiber.

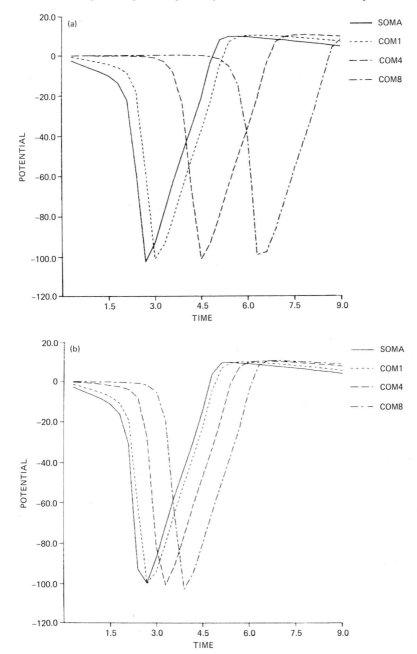

Fig. 22.7. Influence of myelin on conduction of action potential: simulations with (a) AXON01 and (b) AXON02.

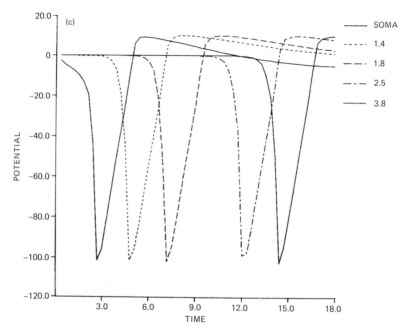

Fig. 22.8. Propagation of action potential into bifurcating branches.

Figure 22.9 illustrates a contraction of a Purkinje fiber of heart muscle as simulated by HEART01.

B. Program Listings and Discussion (HHNRN01, AXON01, AXON02, AXON03, AXON04, HEART01)

Consider the listing in Appendix 2 for program HHNRN01. The essential structure of this program is the same as that for the simple model neurons introduced in Chapter 14. The heart of the program in the do-1000 loop for updating state variables. Since the state variables in the Hodgkin–Huxley model are defined by first-order differential equations, as are all the state variables in our simple model neurons presented in Chapter 14, the same numerical method of updating the state variables is used here as was used in the earlier chapters. At each iteration of the do-900 loop, one updates the values of the A's and B's (corresponding the Hodgkin and Huxley α's and β's, respectively) in accordance with the present value of the membrane potential E. One then uses these values to update the values for the functions n, m, and h, represented here by FN, FM, and FH. On the basis of these values, one updates the values of the sodium and potassium

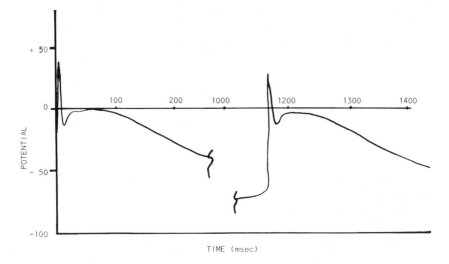

Fig. 22.9. Purkinje fiber action potential simulated with HEART01.

conductances GN and GK. These are then used in conjunction with the applied stimulating current at this time, SCN, to update the membrane potential E.

Notice that only state variables directly relevant to the fine-grained temporal integration of the Hodgkin–Huxley state variables are contained in the do-900 loop. Other state variables are adjusted as piecewise constant over the larger step size STEP.

The program in Appendex 2 for AXON01 differs from HHNRN01 in that Hodgkin–Huxley equations must be simulateld for each compartment of the finite axon. Again the overall structural configuration of the program is the same. The heart of this program is the fine-grained temporal integration in the do-4000 loop contained within the larger-scale do-1000 loop. The logic of the Hodgkin–Huxley computation has been relegated to a subroutine HHEQN. The equations for the membrane potentials in these compartments must now contain terms which reflect current interchange between adjacent compartments in proportion to the potential difference between those compartments and the intercompartmental conductances. The latter are represented by GDS, GSD, and GDD, representing conductances from axon to soma, from soma to axon, and from axon to axon. Logic embedded in the 4000 loop in this program is as follows: one first calls the Hodgkin–Huxley subroutine for all compartments in the model to determine the updated values of the sodium and potassium conductances and therefore the total transmembrane current driven by these conductances in each compartment. One then updates the membrane potentials in all compartments of the model on the basis of these active action currents and the currents interchanged between compartments because of differences in potentials. This process is reiterated at whatever step size is required to make the technique converge to the correct value. Typically, for these multicompartmental programs several hundred to a thousand intergration steps are needed per millisecond of model time.

The primary effect of myelinating an axon between equally spaced nodes of Ranvier is to decrease the effective length of segments LSEG, across which transmembrane current may flow, and to alter the intercompartmental resistances RSD and RDD. One may see in the listing for program AXON02 how this is done in the lines of code between the RES = 700,000 and write (6, 6000) statements. In all other respects program AXON02 is identical to program AXON01. This theory is elaborated in Section C.

Program AXON03 simulates the propagation of an axon potential into a bifurcating axon. Its basic structure is very similar to that of program DENDR03, discussed in Chapter 21. Essential features are that potentials must be given with two subscripts, indicating both branch number and compartment, and that intercompartmental conductances must be expressed accurately for compartments adjacent to bifurcations and take cognizance of the particular branches in which they reside. In program AXON03 this is effected as discussed for program DENDR03. AXON03 has a fine-grained time intergration loop, do-4000. As discussed for AXON02, the Hodgkin–Huxley subroutine is first called for each compartment at each time step to update the potassium and sodium conductances and determine the driving current engendered by these for each compartment. Second, all potentials in the model are updated in accordance with these driving currents and current interchange between adjacent compartments. To attain satisfactory accuracy, this procedure must be effected several hundred to a thousand times per model millisecond.

Progam AXON04 simulates the propagation of action potentials into bifurcating axonal termination fields of arbitrary structural configuration. As may be seen in the program listing, the detailed structure of program AXON04 is essentially identical to that of program DENDR21, discussed in Chapter 21. The essential features of defining structural configurations according to the TPCN and BTCN arrays and allowing current interchange between compartments in accordance with carefully defined values for intercompartmental conductances, particularly taking into account interchange at bifurcations, are included in AXON04 in the same manner as they are in DENDR21. Moreover, AXON04, like the earlier programs of this chapter, adds the element of calling the Hodgkin–Huxley subroutine to determine membrane current activated by the GNA and GK functions at each intergration time step. This, as usual, is effected in and around the do-4005 loop.

It would not be difficult to include possible myelination in programs AXON03 and AXON04. The appropriate lines of code in AXON02 illustrate how this could be done. However, it seems that the axonal termination fields to which AXON04 and AXON03 apply are generally unmyelinated. Therefore this does not appear to be a pressing need.

A number of researchers have presented modifications of the Hodgkin–Huxley equation system to describe various active neuroelectric or muscular events. All of these can easily be simulated by making the appropriate modifications in HHNRN01. This is illustrated in the listing for program HEART01, which adapts a model presented by McAllister and Noble for contraction of the Purkinje fibers of the heart. Please see the reference in Chapter 6 for further details on this model. Moreover, any of this class of extensions of the Hodgkin–Huxley equation can be further extended in simulations to include spatial conduction by appropriate modifications of the AXON series of programs.

C. Theoretical Background

1. The Hodgkin–Huxley Equation System

The Hodgkin–Huxley equation system, as presented in 1952, is shown in Chapter 2 as Eq. (22.1). Essentially, this equation system describes the driving of the transmembrane potential E by active sodium and potassium conductances GNA and GK. In the Hodgkin–Huxley model, these conductances are driven by three intermediate state functions, n, m, and h, which in turn are driven by first-order rate equations according to the parameters and as may be seen in Eq. (2.2). The α's and β's in turn depend on the potential E, as also indicated in Eq. (2.2). Hodgkin and Huxley did an extensive series of voltage clamp experiments, wherein they measured empirically the sodium and potassium currents and conductances and membrane potentials. On the basis of these data, they empirically determined the functional equations for the α's and β's given in Eq. (2.2) to make the entire equation system consistent with the data. Although speculation concerning possible molecular mechanisms underlying these conductance changes was part of the creative process Hodgkin and Huxley generated, nonetheless the particular functional forms for n, m, and h are somewhat arbitrary and have no valid physical basis. Hodgkin and Huxley speculated that n, m, and h might represent concentrations of different kinds of gating molecules. For example, $1 - n$ would represent the concentration of n on one side of the membrane and n would represent the concentration on the other side. α and β would thus be diffusion constants governing the diffusion of these molecules across the membrane in either direction. Since the total potassium conductance, for example, is GK $\times$ n^4, one might speculate that the confluence of four molecules of type n was necessary for the opening of a gate to a potassium ion, since it could be argued that the probability that four would coalesce could be proportional to the fourth power of the relative density. To open a pore for a sodium molecule might require the presence of three m molecules and one h molecule. These speculations were without solid basis, in fact, but nonetheless served to excite the imagination and drive the research forward. The central lasting contribution of Hodgkin and Huxley was the qualitative definition of the action potential's dependence on sodium and potassium conductances and the precise quantitative empirical description of those conductance changes. The fact that the specific equation system they used is of dubious validity and does not describe any known underlying physical processes does not subtract from those fundamental contributions. Nonetheless, potential modelers should be aware of the shortcomings of the equation system, as discussed in Chapter 2. Most notably, the equation system accurately describes only the generation of single action potentials and does not adequately describe repetitive firing of multiple action potentials.

2. Modeling Conduction in Myelinated Axons

Figure 22.10 illustrates the simple modifications required to simulate the longitudinal flows of currents in myelinated axons. Essentially the myelin forces the current to proceed longitudinally through the axon and to traverse the membrane only at the so-called nodes of Ranvier interspersed between the sections of myelin. For simplicity, consider the axon to be composed of repeated equal segments of length LSEG, each of which is covered by myelin in a length LMYL, equal to LSEG $\times$ FMYL, and is unmyelinated in a length LSEG', equal to LSEG $\times$ (1 – FMYL). Figure 22.10 shows the essential current loop one must describe. There are two essential features. First, current traverses the membrane through an effective area of 2 R* LSEG'.

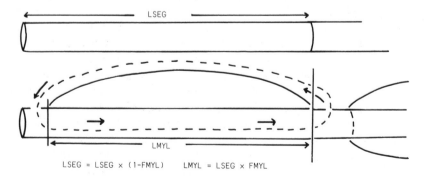

LSEG = LSEG × (1-FMYL)   LMYL = LSEG × FMYL

Fig. 22.10. Influences of myelin on electrical conduction in axon.

Second, current traverses longitudinally through a resistance equal to the total length between the centers of adjacent nodes of Ranvier, given by LSEG. These are the essential features introduced into the coding of AXON02 between the lines RES = 700,000 and write (6, 6000).

D. Exercises

1. Use program HHNRN01 to study the nature of Hodgkin–Huxley action potentials initiated by step currents. Consider the artificial case of repetive firing and plot the frequency of output pulses versus the amplitude of steady-state current.

2. Use program AXON02 to quantify the enhancement of propagation speed by myelination in axons. In particular, for a given LBR, plot the conduction time required from one axon to another as a function of the parameter FMYL. Now repeat this plotting for several increasing values of LBR. Are there certain LBR, FMYL combinations which result in no transmission?

3. Use program AXON03 to study the safety factor in propagation of a Hodgkin–Huxley action potential across an axonal bifurcation. Are there critical ranges of geometric parameters (for example, RST)?

4. Use program AXON04 to simulate the propagation of Hodgkin–Huxley action potentials into progressively finer axonal termination fields. If you use the three-halves rule for reduction of axonal diameter at bifurcations and reasonable values for the radius of the parent axon, do you encounter ranges where propagation is problematic?

5. Modify program AXON04 to include myelination of axonal branches as done in program AXON02.

6. Kalvin, Schwindt, and Connors attempted to extend the Hodgkin–Huxley equation system to describe repetive firing (see reference in Chapter 2). Modify program HHNRN01 to simulate this model. How do repetitive firing properties of this model compare to those in Exercise 1?

7. Modify program AXON04 to include the Connor and Stevens model (see Chapter 2). Reassess the safety of action potential propagation across bifurcations in light of this modification.

8. Frankenhauser extended the Hodgkin–Huxley model to simulate frog motoneurons. (See Chapter 2). Modify program HHNEURM1 to simulate this model.

9. Over the years, Noble has performed a number of extensions of the Hodgkin–Huxley equation system to simulate contractions of various parts of the heart muscle. Program HEART1 simulates McAllister and Nobel's recent model of the Purkinje fibers of the heart. Simulate normal pulsing of Purkinje fibers with HEART1.

10. Can you produce high-intensity oscillations (fibrillations) by adjusting parameters in HEART1?

23

Systems of Neuronal Populations with Dendritic Trees

The programs in this chapter are generalizations of programs from Chapters 17, 19, and 20 to include representation of neurons with dendritic trees. Specifically, the programs simulate a single neuronal population bombarded by an arbitrary number of input fiber systems of variable synaptic types, systems of arbitrary numbers of interacting neuronal populations, and systems of arbitrary numbers of interacting neuronal populations wherein the individual synaptic junctions may be specified to embed memory traces by Hebbian learning. Aspatial versions of all three types of programs are included. Programs including two-dimensional rectangular organization are included for the last two types. In all the programs, parameters determining the geometry and dynamic properties of the dendritic trees may be specified for the individual regions in the program.

In all programs in this chapter, which are located in Apprendix 2, the individual neurons are represented dynamically by the simplified model for a compartmentalized neuron introduced as NEURN30 and NEURN31 in Chapters 14 and 15. The technique used is simply to replace the subroutine used for updating state variables of individual cells, SVUPDT, in the JNCTN and SYSTM programs of Chapters 17, 19, and 20 from the rules representing the state variable point neuron model introduced as PTNRN10 and PTNRN11 in Chapters 14 and 15 to the corresponding rules for the neuron model with dendritic trees, NEURN30 and NEURN31.

A. Using the Programs: Input–Output Charts

Figures 23.1 through 23.5 show the input–output flowcharts for the seven programs of this chapter. As may be seen by comparing the flowcharts,

PARAMETERS

NFIB, NCLS, MCTP1, STEP, NST, NRG, NRG1

C, TTH, B, TGK, THO, TMEM, CDS, VRT

THD, D, TGC, A, TCA, CAO, BD, TGKD

LTSTOP

INPUT FUNCTIONS

$$G_{ijk}(t)$$

STATE AND OUTPUT FUNCTIONS

NFC(t)
$NF_i(t)$
$IDFC_i(t)$
$IDFF_{ij}(t)$

NFPOPS

EQ,T,N,P,INSTRT,INSTP,NCT,NT_j,STR_j,INFSED,INSED

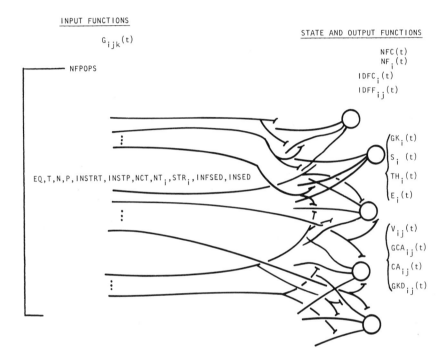

$GK_i(t)$
$S_i(t)$
$TH_i(t)$
$E_i(t)$

$V_{ij}(t)$
$GCA_{ij}(t)$
$CA_{ij}(t)$
$GKD_{ij}(t)$

Fig. 23.1.   Program JNCTN70. This program simulates the activation of a pool of neurons by an arbitrary number of input fiber systems.

these programs represent direct extensions, respectively, of program JNCTN12 from Chapter 17, programs SYSTM11 and SYSTM22 from Chapter 19, and programs LRSYS30 and LRSYS31 from Chapter 20. In each case, the input functions, parameters, and state and output functions are the same for the program in this chapter as for its antecedent program with the following exceptions. First, all individual neurons have dendritic trees. Second, the input parameter list must be extended to include NSTMX, NRGMX, and NRGP1M, which represent, respectively, the maximum number of dendritic stems on any neuron type in the simulation, the maximum number of dendritic regions in any neuron type in the population,

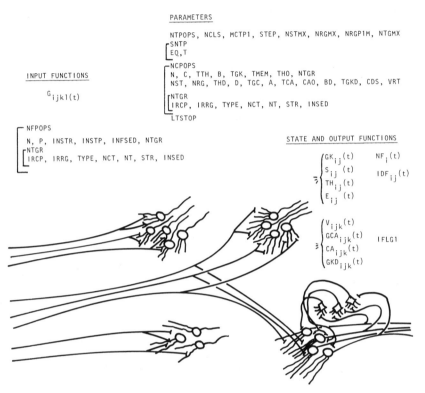

Fig. 23.2. Program SYSTM70. This program simulates an arbitrary number of interactive neuronal populations with dendritic trees activated by an arbitrary number of individually adjustable input fiber systems.

and the maximum number of regions of any neuron type in the population plus 1. Third, for each cell population, parameters corresponding to both the structure of the dendritic tree and the model for active calcium-related conductances, as introduced in Chapters 14 and 15, must be specified. As introduced in those earlier chapters, these include the number of stems and number of regions in the dendritic tree, NST and NRG; the dendritic-to-soma conductance, GDS, and the voltage retention parameter, VRT; and the parameters for calcium-related conductances, THD, D, TGC, A, TCA, CA0, BD, and TGKD, as defined in Chapter 14. Fourth, every system of outgoing axons must be targeted to hit a particular receiving region, IRRG, as well as a given receiving population, IRCP. In these programs, if IRRG is equal to 0 the corresponding synaptic terminals fall at any place in the dendritic tree or on the soma. If IRRG is equal to the number of regions in the receiving populations plus 1, the input falls on the soma of the receiving

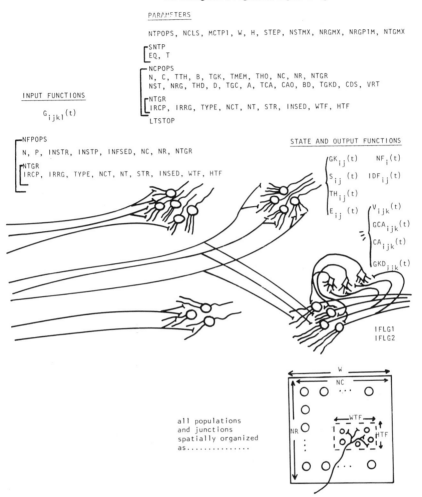

PARAMETERS

NTPOPS, NCLS, MCTP1, W, H, STEP, NSTMX, NRGMX, NRGP1M, NTGMX

SNTP
EQ, T

NCPOPS
N, C, TTH, B, TGK, TMEM, THO, NC, NR, NTGR
NST, NRG, THD, D, TGC, A, TCA, CAO, BD, TGKD, CDS, VRT

NTGR
IRCP, IRRG, TYPE, NCT, NT, STR, INSED, WTF, HTF

LTSTOP

INPUT FUNCTIONS

$G_{ijkl}(t)$

NFPOPS
N, P, INSTR, INSTP, INFSED, NC, NR, NTGR

NTGR
IRCP, IRRG, TYPE, NCT, NT, STR, INSED, WTF, HTF

STATE AND OUTPUT FUNCTIONS

$GK_{ij}(t)$ $NF_i(t)$

$S_{ij}(t)$ $IDF_{ij}(t)$

$TH_{ij}(t)$

$E_{ij}(t)$ $V_{ijk}(t)$

$GCA_{ijk}(t)$

$CA_{ijk}(t)$

$GKD_{ijk}(t)$

IFLG1
IFLG2

all populations
and junctions
spatially organized
as . . . .

Fig. 23.3. Program SYSTM71. This program simulates an arbitrary number of interactive neuronal populations with dendritic trees activated by an arbitrary number of individually adjusted input fiber systems. All populations are spatially organized in two dimensions.

population. Fifth, the state variables corresponding to the active calcium-related conductance model, V, GCA, CA, and GKD, are produced as state and output functions for each compartment in every cell of every population in the system.

With these few additions, the programs of this chapter are used in the same way as their antecedents in Chapters 17, 19, and 20. The parameters used here have the same meaning as defined in these earlier chapters. The reader is encouraged to review the discussions in those chapters and the summary listing of parameters in Appendix 1 to clear up any uncertainties.

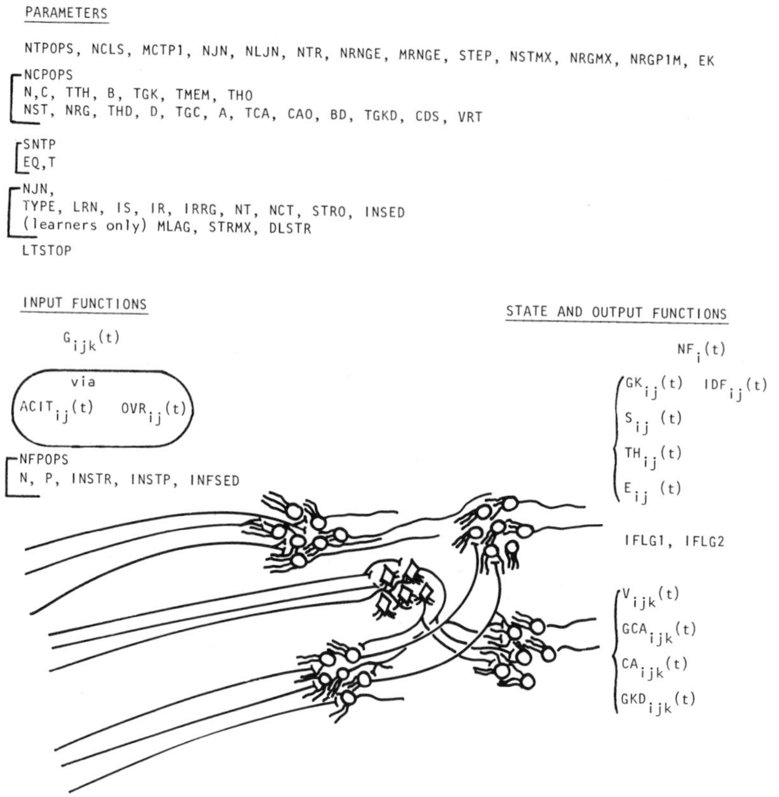

PARAMETERS

NTPOPS, NCLS, MCTP1, NJN, NLJN, NTR, NRNGE, MRNGE, STEP, NSTMX, NRGMX, NRGP1M, EK

$\begin{bmatrix} \text{NCPOPS} \\ \text{N,C, TTH, B, TGK, TMEM, THO} \\ \text{NST, NRG, THD, D, TGC, A, TCA, CAO, BD, TGKD, CDS, VRT} \end{bmatrix}$

$\begin{bmatrix} \text{SNTP} \\ \text{EQ,T} \end{bmatrix}$

$\begin{bmatrix} \text{NJN,} \\ \text{TYPE, LRN, IS, IR, IRRG, NT, NCT, STRO, INSED} \\ \text{(learners only) MLAG, STRMX, DLSTR} \end{bmatrix}$

LTSTOP

INPUT FUNCTIONS   STATE AND OUTPUT FUNCTIONS

$G_{ijk}(t)$

$NF_i(t)$

via

$ACIT_{ij}(t)$ $OVR_{ij}(t)$

$\begin{cases} GK_{ij}(t) & IDF_{ij}(t) \\ S_{ij}(t) \\ TH_{ij}(t) \\ E_{ij}(t) \end{cases}$

$\begin{bmatrix} \text{NFPOPS} \\ \text{N, P, INSTR, INSTP, INFSED} \end{bmatrix}$

IFLG1, IFLG2

$\begin{cases} V_{ijk}(t) \\ GCA_{ijk}(t) \\ CA_{ijk}(t) \\ GKD_{ijk}(t) \end{cases}$

Fig. 23.4. Program LRSYS70. This program simulates Hebbian learning at the synapses of an arbitrary system of interacting neuronal populations with dendritic trees activatged by an arbitrary number of input fiber systems. All synaptic systems are individually adjustable.

B.   Program Listings and Discussion (JNCTN70, SYSTM70, SYSTM71, LRSYS70, LRSYS72, LDSYS73)

The program listings in Appendix 2 for JNCTN70, SYSTM70, SYSTM71, LRSYS70, LRSYS71, LRSYS72, and LDSYS73 constitute the simulation models introduced in this chapter. As may readily be seen in these listings, these programs are but slight modifications of their respective antecedents, JNCTN12 in Chapter 17, SYSTM11 and SYSTM22 in Chapter 19, and LRSYS30, LRSYS31, LRSYS32, and LDSYS33 in Chapter 20. In all cases, the primary difference is a change in subroutine SVUPDT to include the dynamic rules for the compartmentalized neuron with dendrites, NEURN30 and NEURN31 introduced in Chapters 14 and 15. This entails changing the common statements and the read and write statements in the parent program to include expanded parameters and state variables. Also,

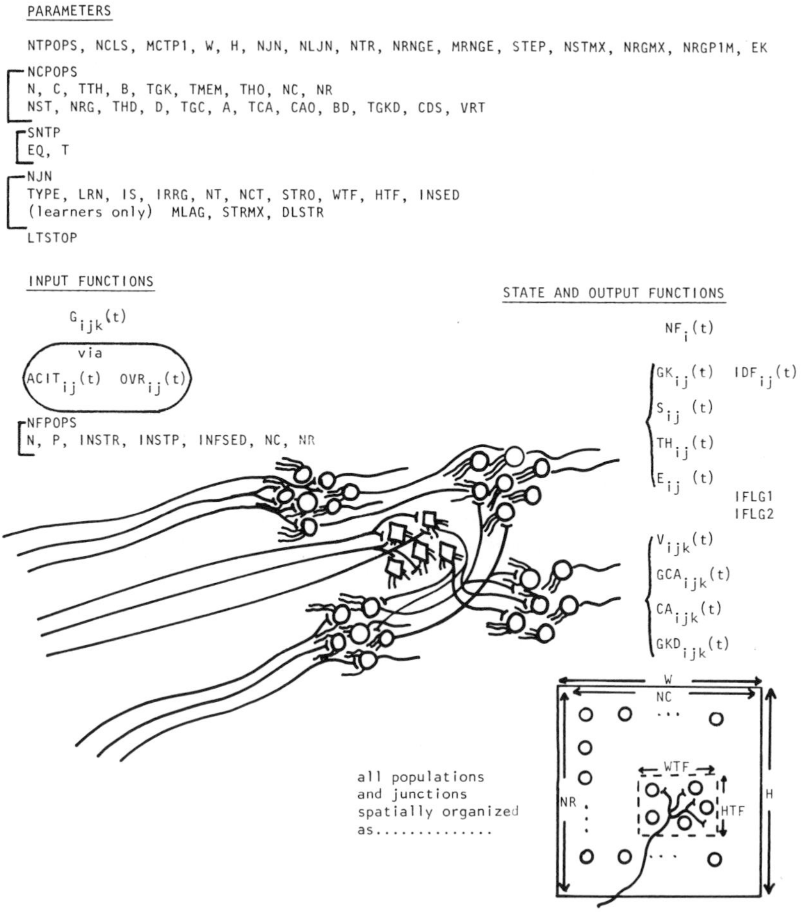

PARAMETERS

NTPOPS, NCLS, MCTP1, W, H, NJN, NLJN, NTR, NRNGE, MRNGE, STEP, NSTMX, NRGMX, NRGP1M, EK

```
┌─NCPOPS
│ N, C, TTH, B, TGK, TMEM, THO, NC, NR
└ NST, NRG, THD, D, TGC, A, TCA, CAO, BD, TGKD, CDS, VRT
  ┌─SNTP
  └ EQ, T
┌─NJN
│ TYPE, LRN, IS, IRRG, NT, NCT, STRO, WTF, HTF, INSED
│ (learners only)  MLAG, STRMX, DLSTR
└ LTSTOP
```

INPUT FUNCTIONS

$G_{ijk}(t)$

via

$ACIT_{ij}(t)$ $OVR_{ij}(t)$

```
┌─NFPOPS
└ N, P, INSTR, INSTP, INFSED, NC, NR
```

STATE AND OUTPUT FUNCTIONS

$NF_i(t)$

$GK_{ij}(t)$ $IDF_{ij}(t)$

$S_{ij}(t)$

$TH_{ij}(t)$

$E_{ij}(t)$ $IFLG1$ $IFLG2$

$V_{ijk}(t)$

$GCA_{ijk}(t)$

$CA_{ijk}(t)$

$GKD_{ijk}(t)$

all populations
and junctions
spatially organized
as.............

W
NC
NR
WTF
HTF
H

Fig. 23.5. Programs LRSYS72 and LDSYS73. These programs simulate Hebbian learning at the synapses of an arbitrary system of interacting spatially organized neural networks with dendritic trees activated by an arbitrary number of input fiber systems. All synaptic systems are individually adjustable.

in all programs the subroutine TRNSMT is modified to distribute synaptic input appropriately over the dendritic tree of the receiving neuron. It is necessary to expand the basic activity-in-transit matrix G to discriminate input activity targeted to the different dendritic regions. Finally, it is necessary to initialize the variables and decay rates associated with the calcium-related conductances for the dendritic regions in the section for initializing varibles.

The programs of this chapter are capable of a very wide range of highly articulated realistic simulation of dynamic activity in neuronal systems.

Moreover, users may readily extend or modify these programs to include either more specialized or more elaborate representations of neuronal physiology or dendritic tree morphology. Such things can be done readily enough by modifying or substituting appropriate lines of code within the subroutines of the programs, as done in this chapter to the programs from Chapters 17, 19, and 20.

C. Theoretical Background

1. Significance of the Programs

The programs of this chapter represent the culmination of the methodology developed in this book to this point in time. They have capabilities and potential usefulness far beyond what has been explored so far. It is our hope that the next decade will see a great elaboration of the possible dynamical properties of neuronal systems produced from models such as these and that this book may help play a part in that elaboration.

As we argued in an earlier book on modeling, the area of neural networks represents a critical chasm or hiatus in theoretical neuroscience. It is a level at which the psychology of mental activity promises to meet the biology of brain activity. Its overall governing principles of operation, which probably can be conceived as in some way representing "emergent" phenomena from underlying single unit neurophysiology, have only begun to be hinted at. It is an area so complex that special tools such as the computer simulation programs presented in this book are essential in our exploration of it. Model systems realized on computer programs like the ones developed in this chapter can be used to elaborate the properties and specific predictions of particular hypotheses or theories; to generate theories and hypotheses; to explore the influence of various parameters on systemic behavior; to serve as a context within which one may imagine different particular systems to represent specialization toward specific parameter values or structural or physiological conditions; to predict the outcome of particular experiments for a given system, such as applying certain stimulation, injecting certain drugs, or performing certain surgical operations.

The programs of this chapter appear to be particularly useful for exploring the functional significance of the structure of dendritic trees in various neuronal regions. For example, it seems a general rule that particular systems of inputs into neuronal populations are often targeted to specific regions of the dendritic tress. The possible functional significance of interactions of different input systems differentially placed within dendritic trees can be explored with computer models such as those in this chapter.

D. Exercises

1. Use program JNCTN70 to simulate the highly convergent–divergent bombardment of a single population of neurons by a single afferent system. Effect the synaptic input on the peripheral regions of the dendritic trees. Try to obtain synchronization sensitivity as discussed in Chapter 17. Do the dendritic trees enhance or diminish the sensitivity to synchronization?

2. Repeat Exercises 1 and 2 of Chapter 19 which simulates the hippocampal theta rythm in a model with six neuronal populations. Now use program SYSTM70 and include dendritic trees on all the neurons. Effect excitation of the parent populations by entophinal inputs on the peripheral regions of the dendritic trees. Effect excitation projected outward by the excitatory cells on the dendritic trees of receiving cells. Project synapses from the inhibitory cells onto the cell bodies and more proximal dendritic regions of receiving cells. Project the septal input onto the proximal dendritic trees of the excitatory cells. What differences or increased degrees of freedom are provided to the dynamics by the dendritic trees?

3. Repeat the exercises of Chapter 19 in simulating feature extraction in the visual system. This time use program SYSTM71 to include neurons with dendritic trees.

4. Use program LRSYS72 to repeat the exercises of Chapter 20, this time using neurons with dendritic trees. What degrees of freedom do the dendritic trees introduce regarding the embedding of permanent memory traces in these systems?

24

The Place of Computer Modeling in Neuroscience

A guiding idea in collecting the programs for this book has been that, because neuronal systems are so complicated in structure and physiology, it should be valuable to have computational simulation models to help predict the varieties of dynamic activity in a variety of situations. This observation becomes obvious when one considers the extremely rapid increase in power and availability of computers and the daily use of simulation models in other areas of science dealing with complex sytems—weather prediction, cosmology, astrophysics, turbulent fluid mechanics, etc. With respect to the nervous system specifically, such capabilities can be especially useful in a variety of contexts including:

(a) *Instructional use*: As computational capabilities become increasingly common and entire generations of young students and researchers grow up throughly imbued with basic computer literacy, it seems inevitable that every serious senior/graduate level neuroscience course and even introductory biopsychology courses should include as standard fare, realistic computer simulation of fundamental neurophysiological mechanisms and concepts.

(b) *Clinical use*: Hospitals and clinics should before long incorporate realistic simulation models of various neuronal systems in clinical contexts with technicians trained to explore the possible influences of various prescribed treatments. Representation of individual cases by specifying several to several dozen idiosyncratic parameters is standard in such computer simulation.

(c) *Use in experimental neuroscience research*: Every neurophysiological research laboratory should include significant computational capability. Increasingly, in coming decades, such capabilities should be expanded to incorporate realistic computer simulation modeling as part of the standard repertoire of experimental investigation.

(d) *Use in "top-down" brain/mind research*: Students, scholars, and researchers in the areas of cognitive psychology and artificial intelligence and in more nebulous areas on the outer edges of psychology, biology, computer science, systems engineerings, and medicine should find these programs useful in generating and studying concepts regarding systemic and global properties of nervous systems and their possible "emergent" properties.

(e) *Use in neural modeling research*: The programs presented here can serve as a reference base from which subsequent efforts to expand, refine, or improve engineering and computer modeling of neuronal activity might move forward. For example, researchers in computer science or computer engineering may find these approaches suggestive or helpful in attempts to design, for example, committed parallel hardware systems or hybrid systems to simulate brain activity.

In any application context, many users should be able to employ the programs appropriate for their purpose exactly as they are presented here. Most of the programs are quite general within their various domains; many neuronal situations can be described simply by adjusting parameter values and stimulus conditions within the existing structure of the programs.

Many other applications, however, particularly in clinical and experimental research contexts, might require minor modifications or extensions of particular code to represent some focus of interest with adequate accuracy. Experimental and clinical researchers should find it easy to have competent computer programmers modify and/or extend the packaged programs to represent ever more accurate and refined representations of the particular systems they are researching. Such programmers can also connect the detailed output of these simulation programs to various graphic desplay and data analysis capabilitites in individual laboratories.

In a smaller set of applications, some workers may wish to structure a simulation in a manner different from that presented in this book. In such cases the programs and equations presented here might serve a useful purpose as reference bases or as "jumping-off" points.

There are, of course, limits to the use and value of computer simulation. It is in itself no substitute for either experimental investigation or conceptual or even mathematical theorizing. Moreover, intelligent use of computer simulation in generating reliable understanding requires intelligence, experience, and persistence, just as does competent use of the microelectrode and other experimental tools. In many ways, the use of a computer simulation program is similar to the use of an experimental preparation. One provides a context and some sort of stimulation and observes the resulting behavior; one adjusts the context or stimulation and again observes the response. The purpose of the studies is to find the underlying principles governing the behavior from which outcomes of subsequent experiments can be predicted. The computer model has a number of advantages: (1) many experiments can be performed with much greater ease on the computer than on animals, (2) parameters can easily be varied through extremely wide ranges, (3) the computer model can provide a much more detailed picture of system activity than is possible in an animal, (4) the computer model can spell out explicitly the implications of some specific hypothesized rule of operation or malfunction, (5) the computer model can embody a theory supposed to be applicable to an entire class of which a particular animal may represent only one specialized instance, and (6) explicating the implications of theories by computer simulation exposes many vagaries and idiosyncracies of theories which are not otherwise immediately clear. My opinion is that, on balance, computer simulation is best seen as a working tool which serves to mediate between theory and experiment in searching out the fundamental principles and properties of operation. If, ideally, "theory guides and experiment decides," then large-scale simulation helps elaborate the interfaces of the guiding and deciding processes.

This book is intended to be more broad than deep. That is, I see it as containing a methodology that should be useful in enhancing and deepening understanding in a variety of applications, to degrees beyond that described by any of the illustrative applications included here.

Appendix 1

Symbols Used in FORTRAN Programs

A	Sensitivity to calcium concentration (2)
A(i, j)	Coupling coefficient between cells i and j (0–1)
ACT	Activation current in Hodgkin–Huxley model
ACIT	Activity in transit, data-packed matrix
ACITSN	Number of fiber system
ACITTR	Terminal number
ACITTM	Delay in learning action
AD	Admittance of the ith branch
AH, AM, AN	Alpha for h, m, n, respectively, in Hodgkin–Huxley model
AE	Parameter for dendritic compartmentalization representing the quantity A defined in Eq. (21.21b)
AM	Membrane area of a segment in the dendrites or axon
AS	Area of soma
ASK	Area of stock
B	Sensitivity to potassium conductance (4)
BD	Dendritic potassium sensitivity (4000)
BE	Parameter for dendritic compartmentalization representing the quantity B defined in Eq. (21.21b)
BH, BM, BN	Beta for h, m, n, respectively, in Hodgkin–Huxley model
BTCN	Number of bottom connection for dendritic segment
C	Threshold sensitivity (0–1)
CA(t)	Calcium concentration
CAD	Calcium concentration threshold for potassium conductance activation (20 mV)
CDN	Compartment number for displayed potential
CDS	Conductance proximal dendrite to soma

CON	Resting membrane conductance
CRIT	Critical value for probability distribution
D	Sensitivity to potassium conductance
DCA	Decay constant for calcium action
DCB	Decay of potential in a branch
DCG	Decay constant for potassium action
DCGE	Decay constant for excitatory conductance
DCGI	Decay constant for inhibitory conductance
DCS	Decay of potential in a compartment
DCTH	Decay constant for threshold
DDN	Dendrite number for displayed potential
DE	Parameter for dendritic compartmentalization representing the quantity D defined in Eq. (21.21b)
DG	Decay constant for synaptic action
DGC	Decay constant for calcium action
DGK	Decay constant for potassium action
DGKD	Decay constant for dendritic potassium action
DL	Step size, delta (1 msec)
DLSTR	Percent change in terminal strength on successful transmission
DTH	Decay constant for threshold
E(t)	Cell membrane potential
EO(i)	Initial potential
ECA	Resting calcium potential
ED	Prior dendritic potential
EDN	Updated dendritic potential
EEQ	Equilibrium potential for synaptic type
EI	Inhibitory resting potential (-10 mV)
EK	Resting potassium potential (-10)
EQ	Equilibrium potential for synaptic type
ES	Prior soma potential
ESS	Updated soma potential
EX	Excitatory resting potential (70 mV)
FCP	Field coupling (0–1.0: complete coupling)
FD	Adjustment to dendritic intracompartmental resistances to account for extracellular resistance (1.0: negligible)
FH, FM, FN	Gamma for $h, m, n,$ respectively in Hodgkin–Huxley model
FMYL	Fraction of myelinated membrane (0.9)
FS	Adjustment to soma intracompartmental resistances to account for extracellular resistance (1.0: negligible)
G1, G2	Combination of AE, BE, CE, DE defined in dendritic programs

G	Normalized conductance into cell i at time t
G(i, t)	Normalized conductance change for cell i at time t
GBB	Normalized conductance corresponding to RBB normalized
GBT	Normalized conductance of distal (bottom) dendritic segment
GCA	Calcium conductance, normalized
GDD	Conductance from compartment to compartment, normalized
GDS	Conductance from proximal compartment to soma, normalized
GE	Excitatory conductance
*gi	Inhibitory conductance
GK	Potassium conductance
GK0	Initial potassium conductance (30–35 mV)
GKD	Dendritic potassium conductance
GKS	Soma potassium conductance
GL	Conductance leakage (0.3)
GLK	Normalized leakage conductance out the peripheral tips of axon or dendrite
GN	Normalized sodium conductance
GN0	Initial sodium conductance (120 mV)
G0	Normalized resting conductance of dendritic compartment
G0S	Normalized resting conductance of soma compartment
GS	Normalized conductance to soma
GSD	Conductance leakage from soma to proximal dendritic compartment
GSM	Active conductance at soma
GTOT	Total conductance into cell i
GTP	Normalized conductance of proximal (top) dendritic segment
GTS	Total conductance of soma, normalized
H	Height of cell population
H0	Resting value of Hodgin–Huxley variable H (0.59612)
HTF	Height of terminal field
HX	High x value—to define a noisy box in neuron programs
HY	High y value—to define a noisy box in neuron programs
IBC	Identity of compartment on which synaptic activity falls
ICC	Temporary column for identifying receiving cell
ICN	Identity of compartment
IDB	Identity of branch—for display purposes
IDC	Identity of receiving cell's column

IDC	Identity of compartment—for display purposes
IDF	Identity of firing cell
IDFC	Identity of firing cell
IDF	Identity of firing fiber
IDN	Identity of branch
IDR	Identity of receiving cell's row
IDT	Cell identity—for POOL programs
INSED	Input seed for random number generator
INSTOP	Stop time for activation of input fiber system
INSTRT	Start time for activation of input fiber system
INTERV	Teaching interval
IR	Identity of receiving cell
IRC	Temporary row to identify receiving cell
IRCP	Identity of receiving cell population
IRG	Identity of region
IRRG	Identity of receiving region
IS	Identity of sending cell
IS	Identity of stem
ISTM	Numnber of cells to be taught each period
ITB	Identity of specific target branch
ITG	Target identity—cell number
ITR	Identity of target region
ITS	Identity of stem
J	Number of dendritic segment
JX	x coordinate for spatially organized network, used for graphic display
JY	y coordinate for spatially organized network, used for graphic display
KRG	Identity of receiving region
LBR	Length of axonal branch (1000 μm)
LBRNCH	Length of dendritic branch
LC	Length constant of dendritic branch
LMYL	Length of myelinated segment
LOCATN	Number of each cell to be taught
LRN	Identification for learning junctions
LSK	Length of dendritic stock (0.375 μm)
LSEG	Length of segment
LSTIM	Time for next stimulus to occur
LSTM0	Initial time cell was stimulated
LSTP	Last iteration for teaching
LSTP	Stop time for synaptic input
LSTRT	Start time for synaptic input
LTSTOP	Last iteration for simulation

LX	Low x value—used to define a noisy box in the neuron programs
LY	Low y value—used to define a noisy box in neuron programs
MARKER	Location of last cell to be stimulated
MCTP1	Maximum conduction time plus one
MLAG	Maxium lag time for learning action
M0	Resting value of Hodgkin–Huxley variable M (0.0529)
MRNGE	Storage parameter—maximum allowable number of "overflow" input messages
N	Number of cells
NBC	Number of compartment on which synaptic input falls
NC	Number of columns
NCC	Number of columns—cell population
NCF	Number of columns—fiber population
NCF(t)	Number of cells that have fired
NCLS	Number of cells
NCMPT	Number of compartments
NCPOPS	Total number of cell populations
NCT	Conductance time
ND	Number of cell to be displayed
NDSPY	Number of compartments to be displayed
NEIP	Number of excitatory inputs
NEX	Number of excitatory cells
NFC	Number of cells that have fired at time t
NFF	Number of fibers that have fired at time t
NFIB	Number of fibers
NFPOPS	Total number of fiber populations
NIIP	Number of inhibitory input fibers
NIN	Number of inhibitory cells
NIP	Maximum number of inputs
NJN	Total number of junctions
NLJN	Total number of learning junctions
NM	Name of dendritic segment being operated on in subroutine tree
N0	Resting value of Hodgkin–Huxley variable N (0.3177)
NR	Number of rows
NRC	Number of rows—cell population
NREC	Number of receiving cell
NRF	Number of rows—fiber system
NRG	Number of regions in dendritic field
NRNGE	Storage parameter—maximum allowable number of input messages

NST	Number of stems in dendritic field
NSTEPS	Number of steps to be taken per iteration
NSTIM	Number of cells to be taught
NSTIM1	First cell to be taught
NT	Number of terminals
NTG	Number of target cells
NTGS	Number of target cells to be recorded from
NYMX	Maximum number of terminals
NTPOPS	Total number of populations—cells and fibers
NYR	Maximum number of terminals
NUP	Total number of inputs—excitatory and inhibitory
OVR	Overflow data-packed matrix for activity in transit
OVRSN	Number of fiber system
OVRTR	Terminal number
OVTRM	Delay in learning action
OVRRC	Data for receiving cell
P	Average period for input firing
P(t)	Cell potential
PD	Potential in dendritic compartment
PER0	Initial period for input firing
PER	Average period for input firing
PS	Potential in soma
PT	Probability of firing for synaptic input on specific target brand
R	Radius of dendritic stem contiguous to soma
RBB	Resistance between top compartment of separate dendritic stems contiguous to soma
RDD	Resistance compartment to compartment
RDN	Radius of dendrite
REG	Equivalent resistance
RES	Axon resistivity
RG	Region within a dendritic tree
RG	Region to be displayed
RH	Radius of spine head (0.145–0.175 μm)
RL	Resistivity of intracellular fluid
RS	Radius of soma (5 μm)
RSA	Resistance soma to axon
RSK	Radius of stock (0.5–0.07 μm)
RSMA	Resistivity of soma
RST	Radius of the dendrite contiguous to soma (1.25 μm)
RTOT	Total resistivity
RTP1, RTP2	Resistances to proximal (top) compartments
S(t)	Spiking variable—to indicate whether a cell has fired

SC	Magnitude of step current injected into a point neuron
SC	Synaptic action present at cell i
SC(i, t)	Step current injected into cell i
SCD(t)	Total synaptic current into dendrites at time t
SCIT	Synaptic current in transit
SCS(t)	Total synaptic current into soma at time t
SD	Magnitude of step current into dendrites
SDP	Sum of potentials in compartments adjacent to soma
SDSTP	Stop time of step current into dendrites
SDSTRT	Start time of step current into dendrites
SDT	Temporary activation factor
SDV	Standard deviation
SIG	Ratio of dendritic and somatic admittance
*sl	Slope of injected ramp current
SPCP	Spike coupling
SNT	Synaptic type
SNTP	Synaptic type
SS	Magnitude of step current into soma
SSTP	Stop time of step current into soma
SSTRT	Start time of step current into soma
ST	Stem to be displayed
STOCH	x position along probability distribution
STRO	Initial strength of synaptic action
STR	Conductance change per successful transmission at synapse
SYN	Total synaptic action from all inputs on cell i
SYNS	Temporary membrane current
T	Time constant for synaptic action (0–2 msec)
TCA	Time constant for calcium concentration decay (5 msec)
TD	Dendritic membrane time constant (10–20 msec)
TEP	Time constant for excitatory synaptic action (0–2 msec)
TG	Time constant for synaptic action (0–2 msec)
TGC	Time constant for calcium conductance (5 msec)
TGK	Refractory time constant (3–10 msec)
TGKD	Dendritic refractory time constant (50 msec)
TH(t)	Cell threshold
TH0	Initial threshold (10–20 mV)
THD(t)	Dendritic compartment threshold
THS(t)	Soma threshold
TIP	Time constant for inhibitory synaptic action
TMEM	Membrane time constant (5–11 msec)
TPCN	Number of top connection for dendritic segment
TTH	Time constant for accomodation (20–25 msec)

TYPE	Synaptic type
U	Potential at time $t-1$
US	Prior voltage potential
USED	Flag used in subroutine tree to indicate whether dendritic segment has been considered yet
V(i, j, t)	Potential in any compartment i, j at time t
VRT	Voltage ratio between dendritic compartments
VS	Somatic potential
W	Width of cell population
WTF	Width of terminal field
X	Temporary parameter in computing intercompartmental resistances
XF	x coordinate of firing cell
XT	x coordinate of receiving cell
YD	Admittance of dendritic tree
YF	Y coordinate of firing cell
YN	Admittance of some portion of dendritic tree (temporary)
YS	Soma admittance
Yt	Y coordinate of receiving cell
Z	Temporary integration factor
ZS	Temporary integration factor

Appendix 2

Computer Programs

```
/JOB
A224.
FAKE.
PDFV.
MNF(Y).
LGO.
/EOR
      PROGRAM PTNRN10 (INPUT,OUTPUT,TAPE6=OUTPUT)
*
****      THIS PROGRAM SIMULATES THE RESPONSES OF A THREE-STATE-
*         VARIABLE MODEL OF A SINGLE POINT NEURON TO APPLIED STEP
*         AND RAMP INPUT CURRENTS.
****
      INTEGER SCSTRT,SCSTP
      PARAMETER STEP=1.,EK=-10.
      DIMENSION P(500)
*
****         THIS SECTION READS AND WRITES THE INPUT PARAMETERS
*
      READ 5010, C,TTH,B,TGK,THØ,TMEM
      READ 5020, SC,SL,SCSTRT,SCSTP
      READ 5030, LTSTOP
*
      WRITE(6,6000)
      WRITE(6,6010) C,TTH,B,TGK,THØ,TMEM
      WRITE(6,6020) SC,SL,SCSTRT,SCSTP
      WRITE(6,6030) LTSTOP
*
****         THIS SECTION INITIALIZES VARIABLES
*
      E=Ø. $TH=THØ $S=Ø. $GK=Ø.
      DCTH=EXP(-STEP/TTH) $DGK=EXP(-STEP/TGK)
*
****         THIS SECTION UPDATES STATE VARIABLES AT EACH TIME STEP
*
      DO 1000 L=1,LTSTOP
*
      SCN=Ø. $IF(L.GE.SCSTRT.AND.L.LT.SCSTP) SCN=SC+SL*FLOAT(L-SCSTRT)
*
      GK=GK*DGK+B*S*(1.-DGK) $GTOT=1.+GK $DCE=EXP(-GTOT*STEP/TMEM)
      E=E*DCE+(SCN+GK*EK)*(1.-DCE)/GTOT
      TH=THØ+(TH-THØ)*DCTH+C*E*(1.-DCTH)
      S=Ø. $IF(E.GE.TH) S=1. $P(L)=E+S*(50.-E)
 1000 CONTINUE
*
****         THIS SECTION WRITES OUT ACTIVITY VARIABLES
*
      WRITE(6,6050) (P(L),L=1,LTSTOP)
*
****         THESE ARE THE FORMATS
*
 5010 FORMAT(12F6.2)
 5020 FORMAT(2F6.2,2I6)
 5030 FORMAT(I6)
 6000 FORMAT(1H1)
 6010 FORMAT(5X,*C,TTH,B,TGK,THØ,TMEM=*,6F7.2)
 6020 FORMAT(5X,*SC,SL,SCSTRT,SCSTP=*,2F7.2,2I7)
 6030 FORMAT(5X,*LSTSTOP=*,I7)
 6050 FORMAT(/,(2X,20F6.2))
      STOP
      END
/EOR
  .5  30.    5.    3.   10.    5.
 20.   Ø.         1   100
  100
/EOR
/EOI
```

```
/JOB
A224,TL=100.
FAKE.
PDFV.
MNF(Y).
LGO,NRØ15.
SAVE,NRØ15.
EXIT.
SAVE,NRØ15.
/EOR
      PROGRAM PTNRN2Ø(NROUT,INPUT,OUTPUT,TAPE6=OUTPUT,TAPE7=NROUT)
*
****      THIS PROGRAM SIMULATES THE REPONSES OF A TWO-POINT MODEL
*         OF A SINGLE NEURON WITH ACTIVE CALCIUM-RELATED CONDUCTANCES
****      TO STEP CURRENTS APPLIED IN EITHER OF THE TWO REGIONS.
*
      INTEGER SSTRT,SSTP,SDSTRT,SDSTP
      PARAMETER STEP=1.,NSTEPS=1Ø,EK=-1Ø.,ECA=5Ø.
      DIMENSION PS(5ØØ),PD(5ØØ)
*
****          THIS SECTION READS AND WRITES THE INPUT PARAMETERS
*
      READ 5ØØØ, IFLG1
      READ 5Ø1Ø, TS,GDS,THØ,C,TTH,B,TGK
      READ 5Ø1Ø, TD,GSD,THD,D,TGC,A,TCA,CAØ,BD,TGKD
      READ 5Ø2Ø, SS,SSTRT,SSTP,SD,SDSTRT,SDSTP
      READ 5ØØØ, LTSTOP
*
      WRITE(6,6ØØØ) $WRITE(6,6Ø1Ø) TS,GDS,THØ,C,TTH,B,TGK
      WRITE(6,6Ø2Ø) TD,GSD,THD,D,TGC,A,TCA,CAØ,BD,TGKD
      WRITE(6,6Ø3Ø) SS,SSTRT,SSTP,SD,SDSTRT,SDSTP
      IF(IFLG1.GT.Ø) WRITE(6,6Ø4Ø)
*
****          THIS SECTION INITIALIZES VARIABLES
*
      ES=Ø. $THS=THØ $GKS=Ø. $ESS=Ø.
      ED=Ø. $GCA=Ø. $CA=Ø. $GKD=Ø.
      DTHS=EXP(-STEP/TTH) $DGKS=EXP(-STEP/TGK)
      DGC=EXP(-STEP/TGC) $DCA=EXP(-STEP/TCA) $DGKD=EXP(-STEP/TGKD)
*
****          THIS SECTION UPDATES STATE VARIABLES AT EACH TIME STEP
*
      DO 1ØØØ L=1,LTSTOP
*
      SCS=Ø. $IF(L.GE.SSTRT.AND.L.LT.SSTP) SCS=SS
      SCD=Ø. $IF(L.GE.SDSTRT.AND.L.LT.SDSTP) SCD=SD
*
      GSM=1.+GDS+GKS $Z=EXP(-STEP*GSM/TS/FLOAT(NSTEPS))
      GDN=1.+GSD+GCA+GKD $ZD=EXP(-STEP*GDN/TD/FLOAT(NSTEPS))
*
****              THIS SUB-SECTION ALLOWS MORE FINE-GRAINED INTEGRATION
*
      DO 1Ø1Ø I=1,NSTEPS
      ESN=ES*Z+(GDS*ED+GKS*EK+SCS)*(1.-Z)/GSM
      EDN=ED*ZD+(GSD*ESS+GCA*ECA+GKD*EK+SCD)*(1.-ZD)/GDN
      ES=ESN $ED=EDN
 1Ø1Ø CONTINUE
****
      THS=THS*DTHS+(THØ+C*ES)*(1.-DTHS)
      S=Ø. $IF(ES.GE.THS) S=1.
      GKS=GKS*DGKS+B*S*(1.-DGKS)
      GCA=GCA*DGC $IF(ED.GE.THD) GCA=GCA+D*(ED-THD)*(1.-DGC)
      CA=CA*DCA+A*GCA*(1.-DCA)
      GKD=GKD*DGKD $IF(CA.GE.CAØ) GKD=GKD+BD*(1.-DGKD)
```

```
        PS(L)=ES+(50.-ES)*S  $ESS=PS(L)  $PD(L)=ED
*
        IF(IFLG1.GT.0)  WRITE(6,6110)  L,ES,ED,THS,S,GKS,GCA,CA,GKD
        IF(IFLG1.GT.0)  WRITE(7,7110)  L,ES,ED,THS,S,GKS,GCA,CA,GKD
*
 1000 CONTINUE
*
****          THIS SECTION WRITES OUT ACTIVITY VARIABLES
*
        WRITE(6,6140)  $WRITE(6,6150)  (PS(L),L=1,LTSTOP)
        WRITE(6,6140)  $WRITE(6,6150)  (PD(L),L=1,LTSTOP)
        WRITE(7,6140)  $WRITE(7,7150)  (PS(L),L=1,LTSTOP)
        WRITE(7,6140)  $WRITE(7,7150)  (PD(L),L=1,LTSTOP)
*
****          THESE ARE THE FORMATS
*
        STOP
 5000 FORMAT(12I6)
 5010 FORMAT(12F6.2)
 5020 FORMAT(F6.2,2I6,F6.2,2I6)
 6000 FORMAT(1H1)
 6010 FORMAT(5X,*TS,GDS,TH0,C,TTH,B,TGK=*,7F7.2)
 6020 FORMAT(5X,*TD,GSD,THD,D,TGC,A,TCA,CA0,BD,TGKD=*,10F7.2)
 6030 FORMAT(5X,*SS,SSTRT,SSTP,SD,SDSTRT,SDSTP=*,F7.2,2I7,F7.2,2I7)
 6040 FORMAT(/,5X,*      L        ES        ED        THS        S
       *GKS       GCA        CA        GKD     *)
 6110 FORMAT(5X,I8,8F10.2)
 6140 FORMAT(//)
 6150 FORMAT(5X,20F6.2)
 7110 FORMAT(I8,8F8.2)
 7150 FORMAT(12F6.2)
        END
/EOR
     1
   5.    5.    12.    0.    20.    33.    5.
   5.    5.    16.    2.2    5.    2.    5.    20.    100.    15.
  30.        1   1000   0.        1    1
    240
/EOR
/EOI




/JOB
A224,TL=300.
FAKE.
PDFV.
MNF(Y).
LGO,DN603.
SAVE,DN603.
EXIT.
SAVE,DN603.
/EOR
        PROGRAM NEURN30 (DNOUT,INPUT,OUTPUT,TAPE6=OUTPUT,TAPE7=DNOUT)
*
****          THIS PROGRAM SIMULATES THE RESPONSES OF A MODEL NEURON WITH
*             SIMPLIFIED DENDRITIC TREE CONTAINING ACTIVE CALCIUM-RELATED
*             CONDUCTANCES TO STEP CURRENTS APPLIED EITHER AT THE SOMA OR
****          IN THE DENDRITIC REGIONS.
*
        INTEGER SCSTRT,SCSTP,SDSTRT,SDSTP,ST,RG
        PARAMETER NST=3,NRG=4,STEP=1.,ECA=50.,EK=-10.,NDSPY=1
        DIMENSION V(NST,NRG),GCA(NST,NRG),CA(NST,NRG),GKD(NST,NRG),
       *          GT(NST,NRG),Z(NST,NRG),U(NRG),ST(NDSPY),RG(NDSPY),
       *          PS(3,500)
*
****          THIS SECTION READS AND WRITES THE INPUT PARAMETERS
```

```
*
      READ 5000, IFLG1
      READ 5010, C,TTH,B,TGK,TMEM,THØ,CDS,VRT
      READ 5010, THD,D,TGC,A,TCA,CAØ,BD,TGKD
      READ 5000, ((ST(I),RG(I)),I=1,NDSPY)
      READ 5020, SC,SCSTRT,SCSTP
      READ 5020, SD,SDSTRT,SDSTP,IS,IRG
      READ 5000, LTSTOP
*
      WRITE(6,6000) $WRITE(6,6010) NST,NRG,STEP,IFLG1
      WRITE(6,6020) C,TTH,B,TGK,TMEM,THØ,CDS,VRT
      WRITE(6,6030) THD,D,TGC,A,TCA,CAØ,BD,TGKD
      WRITE(6,6040) ((ST(I),RG(I)),I=1,NDSPY)
      WRITE(6,6050) SC,SCSTRT,SCSTP,SD,SDSTRT,SDSTP,IS,IRG,LTSTOP
      WRITE(6,6000)
*
****       THIS SECTION INITIALIZES VARIABLES
*
      E=Ø. $TH=THØ $GK=Ø. $DO 210 I=1,NST $DO 210 J=1,NRG
      V(I,J)=Ø. $GCA(I,J)=Ø. $CA(I,J)=Ø. $GKD(I,J)=Ø.
  210 CONTINUE
      DTH=EXP(-STEP/TTH) $DGK=EXP(-STEP/TGK)
      DGC=EXP(-STEP/TGC) $DCA=EXP(-STEP/TCA) $DGKD=EXP(-STEP/TGKD)
*
****       THIS SECTION UPDATES STATE VARIABLES AT EACH TIME STEP
*
      DO 1000 L=1,LTSTOP
*
      SCN=Ø. $IF(L.GE.SCSTRT.AND.L.LT.SCSTP) SCN=SC
      SCDN=Ø. $IF(L.GE.SDSTRT.AND.L.LT.SDSTP) SCDN=SD
*
      GTS=1.+GK $ZS=EXP(-GTS*STEP/TMEM)
      E=E*ZS+(SCN+GK*EK)*(1.-ZS)/GTS $DM=Ø.
      DO 980 I=1,NST $DO 920 J=1,NRG
      GT(I,J)=1.+GCA(I,J)+GKD(I,J) $Z(I,J)=EXP(-GT(I,J)*STEP/TMEM)
      SCI=Ø. $IF(I.EQ.IS.AND.J.EQ.IRG) SCI=SCDN
  920 V(I,J)=V(I,J)*Z(I,J)+(SCI+GCA(I,J)*ECA+GKD(I,J)*EK)*(1.-Z(I,J))/
     *                  GT(I,J)
      DM=DM+V(I,1)
      U(1)=(V(I,1)+V(I,2)+E/CDS)/(2.+1./CDS)*(1.-VRT)+VRT*V(I,1)
      DO 930 J=2,NRG-1
  930 U(J)=(V(I,J-1)+V(I,J)+V(I,J+1))/3.*(1.-VRT)+VRT*V(I,J)
      U(NRG)=(V(I,NRG)+V(I,NRG-1))/2.*(1.-VRT)+VRT*V(I,NRG)
      DO 940 J=1,NRG
  940 V(I,J)=U(J)
*
      DO 950 J=1,NRG
      GCA(I,J)=GCA(I,J)*DGC
      IF(V(I,J).GE.THD) GCA(I,J)=GCA(I,J)+D*(V(I,J)-THD)*(1.-DGC)
      CA(I,J)=CA(I,J)*DCA+A*GCA(I,J)*(1.-DCA)
      GKD(I,J)=GKD(I,J)*DGKD
      IF(CA(I,J).GE.CAØ) GKD(I,J)=GKD(I,J)+BD*(1.-DGKD)
  950 CONTINUE
*
  980 CONTINUE
*
      E=(E+CDS*DM)/(1.+CDS*FLOAT(NST))*(1.-VRT)+VRT*E
      TH=TH*DTH+(THØ+C*E)*(1.-DTH) $GK=GK*DGK
      S=Ø. $IF(E.LT.TH) GO TO 995
      S=1. $GK=GK+B*(1.-DGK)
  995 CONTINUE $PS(1,L)=E+S*(50.-E)
      DO 996 I=1,NDSPY
  996 PS(I+1,L)=V(ST(I),RG(I))
      IF(IFLG1.GT.Ø)THEN $WRITE(6,6279)L,E,TH,GK$WRITE(7,7279)L,E,TH,GK
      DO 998 I=1,NST
      WRITE(6,6280) ((V(I,J),GCA(I,J),CA(I,J),GKD(I,J)),J=1,NRG)
  998 WRITE(7,7280) ((V(I,J),GCA(I,J),CA(I,J),GKD(I,J)),J=1,NRG) $ENDIF
```

```
*
 1000 CONTINUE
*
****          THIS SECTION WRITES OUT ACTIVITY VARIABLES
*
      DO 1100 I=1,NDSPY+1 $WRITE(6,6002) $WRITE(7,6002)
      WRITE(6,6280) (PS(I,L),L=1,LTSTOP)
 1100 WRITE(7,7280) (PS(I,L),L=1,LTSTOP)
*
****          THESE ARE THE FORMATS
*
      STOP
 5000 FORMAT(12I6)
 5010 FORMAT(12F6.2)
 5020 FORMAT(F6.2,4I6)
 6000 FORMAT(1H1)
 6002 FORMAT(//)
 6010 FORMAT(5X,*NST,NRG,STEP,IFLG1=*,2I7,F7.2,I7)
 6020 FORMAT(5X,*C,TTH,B,TGK,TMEM,TH0,CDS,VRT=*,8F7.2)
 6030 FORMAT(5X,*THD,D,TGC,A,TCA,CA0,BD,TGKD=*,8F7.2)
 6040 FORMAT(5X,*STS,TGS=*,12I7)
 6050 FORMAT(5X,*SC,SCSTRT,SCSTP,SD,SDSTRT,SDSTP,IS,IRG,LTSTOP=*,
     *         F7.2,2I7,F7.2,5I7)
 6279 FORMAT(5X,I5,3F7.2)
 6280 FORMAT(5X,20F6.2)
 7279 FORMAT(I5,3F7.2)
 7280 FORMAT(16F5.1)
*
      END
/EOR
    1
  0.    25.    20.    10.    11.    24.    3.      .5
 32.     0.    10.     2.    10.    20.    80.    15.
     1     3
  0.        1     0
125.        1  9999       1       3
     30
/EOR
/EOI




/JOB
A224.
FAKE.
PDFV.
MNF(Y).
LGO.
/EOR
      PROGRAM PTNRN11 (INPUT,OUTPUT,TAPE6=OUTPUT)
*
****          THIS PROGRAM SIMULATES THE RESPONSES OF A THREE-STATE-
*             VARIABLE MODEL OF A SINGLE POINT NEURON TO INTERACTING
****          SYNAPTIC INPUT (EXPONENTIAL INTERVAL DISTRIBUTIONS ARE USED).
*
      INTEGER SCSTRT,SCSTP,SNTP
      PARAMETER STEP=1.,EK=-10.,SNTP=2
      DIMENSION TG(SNTP),EQ(SNTP),PER(SNTP),STR(SNTP),LSTM(SNTP),
     *          LSTP(SNTP),INSED(SNTP),G(SNTP),DG(SNTP),P(500)
*
****          THIS SECTION READS AND WRITES THE INPUT PARAMETERS
*
      READ 5010, C,TTH,B,TGK,TH0,TMEM
      READ 5020, SC,SL,SCSTRT,SCSTP
      DO 110 I=1,SNTP
  110 READ 5030, TG(I),EQ(I),PER(I),STR(I),LSTM(I),LSTP(I),INSED(I)
      READ 5000, LTSTOP
```

```
*
      WRITE(6,6000)
      WRITE(6,6010) C,TTH,B,TGK,THØ,TMEM
      WRITE(6,6020) SC,SL,SCSTRT,SCSTP
      WRITE(6,6030) $DO 210 I=1,SNTP
  210 WRITE(6,6035) TG(I),EQ(I),PER(I),STR(I),LSTM(I),LSTP(I),INSED(I)
      WRITE(6,6040) LTSTOP
*
****        THIS SECTION INITIALIZES VARIABLES
*
      E=Ø. $TH=THØ $S=Ø. $GK=Ø.
      DCTH=EXP(-STEP/TTH) $DGK=EXP(-STEP/TGK)
      DO 310 I=1,SNTP $G(I)=Ø.
  310 DG(I)=EXP(-STEP/TG(I))
*
****        THIS SECTION UPDATES STATE VARIABLES AT EACH TIME STEP
*
      DO 1000 L=1,LTSTOP
*
      SCN=Ø. $IF(L.GE.SCSTRT.AND.L.LT.SCSTP) SCN=SC
*
      DO 910 I=1,SNTP $G(I)=G(I)*DG(I)
      IF(L.EQ.LSTM(I).AND.L.LT.LSTP(I)) THEN $G(I)=G(I)+STR(I)
      CALL RANSET(INSED(I))
      LSTM(I)=L+INT(-PER(I)*ALOG(RANF(Ø.)))+1
      CALL RANGET(INSED(I)) $ENDIF
  910 CONTINUE
*
      GS=Ø. $GSI=Ø. $DO 920 I=1,SNTP
      GS=GS+G(I) $GSI=GSI+G(I)*EQ(I)
  920 CONTINUE
*
      GK=GK*DGK+B*S*(1.-DGK) $GTOT=1.+GS+GK $DCE=EXP(-GTOT*STEP/TMEM)
      E=E*DCE+(SCN+GSI+GK*EK)*(1.-DCE)/GTOT
      TH=THØ+(TH-THØ)*DCTH+C*E*(1.-DCTH)
      S=Ø. $IF(E.GE.TH) S=1. $P(L)=E+S*(50.-E)
 1000 CONTINUE
*
****        THIS SECTION WRITES OUT ACTIVITY VARIABLES
*
      WRITE(6,6050) (P(L),L=1,LTSTOP)
*
****        THESE ARE THE FORMATS
*
 5000 FORMAT(12I6)
 5010 FORMAT(12F6.2)
 5020 FORMAT(2F6.2,2I6)
 5030 FORMAT(4F6.2,3I6)
 6000 FORMAT(1H1)
 6010 FORMAT(5X,*C,TTH,B,TGK,THØ,TMEM=*,6F7.2)
 6020 FORMAT(5X,*SC,SL,SCSTRT,SCSTP=*,2F7.2,2I7)
 6030 FORMAT(5X,*T,EQ,PER,STR,LSTM,LSTP,INSED=:*)
 6035 FORMAT(5X,4F7.2,3I7)
 6040 FORMAT(5X,*LTSTOP=*,I7)
 6050 FORMAT(/,(2X,20F6.2))
      STOP
      END
/EOR
   .5   30.    5.    3.   10.    5.
   5.    0.         1        30
   .3   70.    4.    3.         3   999  1227
  10.  -10.   10.   15.        10   999  1936
    60
/EOR
/EOI
```

```
/JOB
A224,TL=200.
FAKE.
PDFV.
MNF(Y).
LGO,NR124A.
SAVE,NR123A.
EXIT.
SAVE,NR123A.
/EOR
      PROGRAM PTNRN12 (NROUT,INPUT,OUTPUT,TAPE6=OUTPUT,TAPE7=NROUT)
*
****      THIS PROGRAM SIMULATES THE RESPONSES OF A THREE-STATE-
*         VARIABLE MODEL OF A SINGLE POINT NEURON TO INTERACTING
*         SYNAPTIC INPUT. (GAUSSIAN INTERVAL DISTRIBUTIONS ARE EFFECTED
****      BY A SUBROUTINE EASILY GENERALIZEABLE TO OTHER DISTRIBUTIONS.)
*
      INTEGER SCSTRT,SCSTP,SNTP
      PARAMETER STEP=1.,EK=-10.,SNTP=2
      DIMENSION TG(SNTP),EQ(SNTP),PER(SNTP),SDV(SNTP),STR(SNTP),
     *          LSTM(SNTP),LSTP(SNTP),G(SNTP),DG(SNTP),P(500),
     *          INSED(SNTP)
*
****      THIS SECTION READS AND WRITES THE INPUT PARAMETERS
*
      READ 5010, C,TTH,B,TGK,THØ,TMEM
      READ 5020, SC,SL,SCSTRT,SCSTP
      DO 110 I=1,SNTP
  110 READ 5030, TG(I),EQ(I),PER(I),SDV(I),STR(I),LSTM(I),LSTP(I),
     *          INSED(I)
      READ 5000, LTSTOP
*
      WRITE(6,6000)
      WRITE(6,6010) C,TTH,B,TGK,THØ,TMEM
      WRITE(6,6020) SC,SL,SCSTRT,SCSTP
      WRITE(6,6030) $DO 210 I=1,SNTP
  210 WRITE(6,6035) TG(I),EQ(I),PER(I),SDV(I),STR(I),LSTM(I),LSTP(I),
     *          INSED(I)
      WRITE(6,6040) LTSTOP
*
****      THIS SECTION INITIALIZES VARIABLES
*
      E=0. $TH=THØ $S=0. $GK=0.
      DCTH=EXP(-STEP/TTH) $DGK=EXP(-STEP/TGK)
      DO 310 I=1,SNTP $G(I)=0.
  310 DG(I)=EXP(-STEP/TG(I))
*
****      THIS SECTION UPDATES STATE VARIABLES AT EACH TIME STEP
*
      DO 1000 L=1,LTSTOP
*
      SCN=0. $IF(L.GE.SCSTRT.AND.L.LT.SCSTP) SCN=SC
*
      DO 910 I=1,SNTP $G(I)=G(I)*DG(I)
      IF(L.EQ.LSTM(I).AND.L.LT.LSTP(I)) THEN $G(I)=G(I)+STR(I)
      CALL RANSET(INSED(I))
      LSTM(I)=L+INT(STOCH(PER(I),SDV(I))-.5)+1
      CALL RANGET(INSED(I)) $ENDIF
  910 CONTINUE
*
      GS=0. $GSI=0. $DO 920 I=1,SNTP
      GS=GS+G(I) $GSI=GSI+G(I)*EQ(I)
  920 CONTINUE
*
      GK=GK*DGK+B*S*(1.-DGK) $GTOT=1.+GS+GK $DCE=EXP(-GTOT*STEP/TMEM)
      E=E*DCE+(SCN+GSI+GK*EK)*(1.-DCE)/GTOT
```

```
      TH=THØ+(TH-THØ)*DCTH+C*E*(1.-DCTH)
      S=Ø. $IF(E.GE.TH) S=1. $P(L)=E+S*(5Ø.-E)
 1ØØØ CONTINUE
*
****          THIS SECTION WRITES OUT ACTIVITY VARIABLES
*
      WRITE(6,6Ø5Ø) (P(L),L=1,LTSTOP)
      WRITE(7,7Ø5Ø) (P(L),L=1,LTSTOP)
*
****          THESE ARE THE FORMATS
*
 5ØØØ FORMAT(12I6)
 5Ø1Ø FORMAT(12F6.2)
 5Ø2Ø FORMAT(2F6.2,2I6)
 5Ø3Ø FORMAT(5F6.2,3I6)
 6ØØØ FORMAT(1H1)
 6Ø1Ø FORMAT(5X,*C,TTH,B,TGK,THØ,TMEM=*,6F7.2)
 6Ø2Ø FORMAT(5X,*SC,SL,SCSTRT,SCSTP=*,2F7.2,2I7)
 6Ø3Ø FORMAT(5X,*T,EQ,PER,SDV,STR,LSTM,LSTP=:*)
 6Ø35 FORMAT(5X,5F7.2,3I7)
 6Ø4Ø FORMAT(5X,*LTSTOP=*,I7)
 6Ø5Ø FORMAT(/,(2X,2ØF6.2))
 7Ø5Ø FORMAT(12F6.2)
      STOP
      END
*
      FUNCTION STOCH(A,B)
      PARAMETER XØ=-8.,STEP=1.,CRIT=.Ø5
*
      STOCH=XØ $F=Ø. $R=RANF(Ø.) $DFF=-1. $D=STEP
    5 F=F+PROB(STOCH+D/2.,A,B)*D $DF=F-R
      IF(ABS(DF).LE.CRIT) GO TO 1ØØ
      IF(DFF*DF.LT.Ø.) D=-D/2. $DFF=DF
      STOCH=STOCH+D $GO TO 5
  1ØØ CONTINUE $IF(STOCH.LT.Ø.) STOCH=Ø.
      RETURN
      END
*
      FUNCTION PROB(X,A,B)
*
      PROB=EXP(-((X-A)/B)**2/2.)/B/SQRT(2.*3.14159)
      RETURN
      END
*
/EOR
   .Ø  3Ø.    5.    3.   1Ø.    5.
   Ø.   Ø.        1    3Ø
   .3  7Ø.    8.    2.    .3     3    999  1227
  1.  -1Ø.   1Ø.    4.    .5   1ØØ    2ØØ  1936
   3ØØ
/EOR
/EOI


/JOB
A224,TL=2ØØ.
FAKE.
PDFV.
MNF(Y).
LGO,NR135A.
SAVE,NR135A.
EXIT.
SAVE,NR135A.
/EOR
      PROGRAM PTNRN13 (NROUT,INPUT,OUTPUT,TAPE6=OUTPUT,TAPE7=NROUT)
*
```

```
****     THIS PROGRAM SIMULATES THE RESPONSES OF A THREE-STATE-
*        VARIABLE MODEL OF A SINGLE POINT NEURON TO INTERACTING
*        SYNAPTIC INPUT. (A BINOMIAL MODEL FOR VESSICLE TRANSFER IS
*        INCLUDED.  A GAUSSIAN INTERVAL DISTRIBUTION IS USED.  THESE ARE
*        EFFECTED BY A SUBROUTINE PACKAGE EASILY GENERALIAZEABLE TO IN-
****     CLUDE OTHER DISTRIBUTIONS AS WELL.)
*
       INTEGER SCSTRT,SCSTP,SNTP
       REAL NM
       PARAMETER STEP=1.,EK=-10.,SNTP=2
       DIMENSION TG(SNTP),EQ(SNTP),PER(SNTP),SDV(SNTP),STRF(SNTP),
      *         LSTM(SNTP),LSTP(SNTP),G(SNTP),DG(SNTP),PS(500),
      *         INSED(SNTP),P(SNTP),NM(SNTP)
*
****          THIS SECTION READS AND WRITES THE INPUT PARAMETERS
*
       READ 5010, C,TTH,B,TGK,THØ,TMEM
       READ 5020, SC,SL,SCSTRT,SCSTP
       DO 110 I=1,SNTP
  110 READ 5030, TG(I),EQ(I),PER(I),SDV(I),STRF(I),LSTM(I),LSTP(I),
      *         INSED(I),P(I),NM(I)
       READ 5000, LTSTOP
*
       WRITE(6,6000)
       WRITE(6,6010) C,TTH,B,TGK,THØ,TMEM
       WRITE(6,6020) SC,SL,SCSTRT,SCSTP
       WRITE(6,6030) $DO 210 I=1,SNTP
  210 WRITE(6,6035) TG(I),EQ(I),PER(I),SDV(I),STRF(I),LSTM(I),LSTP(I),
      *         INSED(I),P(I),NM(I)
       WRITE(6,6040) LTSTOP
*
****          THIS SECTION INITIALIZES VARIABLES
*
       E=Ø. $TH=THØ $S=Ø. $GK=Ø.
       DCTH=EXP(-STEP/TTH) $DGK=EXP(-STEP/TGK)
       DO 310 I=1,SNTP $G(I)=Ø.
  310 DG(I)=EXP(-STEP/TG(I))
*
****          THIS SECTION UPDATES STATE VARIABLES AT EACH TIME STEP
*
       DO 1000 L=1,LTSTOP
*
       SCN=Ø. $IF(L.GE.SCSTRT.AND.L.LT.SCSTP) SCN=SC
*
       DO 910 I=1,SNTP $G(I)=G(I)*DG(I)
       IF(L.EQ.LSTM(I).AND.L.LT.LSTP(I)) THEN
       CALL RANSET(INSED(I))
       G(I)=G(I)+STRF(I)*STOCH(1,P(I),NM(I))
       LSTM(I)=L+INT(STOCH(2,PER(I),SDV(I))-.5)+1
       CALL RANGET(INSED(I)) $ENDIF
  910 CONTINUE
*
       GS=Ø. $GSI=Ø. $DO 920 I=1,SNTP
       GS=GS+G(I) $GSI=GSI+G(I)*EQ(I)
  920 CONTINUE
*
       GK=GK*DGK+B*S*(1.-DGK) $GTOT=1.+GS+GK $DCE=EXP(-GTOT*STEP/TMEM)
       E=E*DCE+(SCN+GSI+GK*EK)*(1.-DCE)/GTOT
       TH=THØ+(TH-THØ)*DCTH+C*E*(1.-DCTH)
       S=Ø. $IF(E.GE.TH) S=1. $PS(L)=E+S*(5Ø.-E)
 1000 CONTINUE
*
****          THIS SECTION WRITES OUT ACTIVITY VARIABLES
*
       WRITE(6,6050) (PS(L),L=1,LTSTOP)
       WRITE(7,7050) (PS(L),L=1,LTSTOP)
*
```

```
****         THESE ARE THE FORMATS
*
 5000 FORMAT(12I6)
 5010 FORMAT(12F6.2)
 5020 FORMAT(2F6.2,2I6)
 5030 FORMAT(5F6.2,3I6,F6.4,F6.0)
 6000 FORMAT(1H1)
 6010 FORMAT(5X,*C,TTH,B,TGK,THO,TMEM=*,6F7.2)
 6020 FORMAT(5X,*SC,SL,SCSTRT,SCSTP=*,2F7.2,2I7)
 6030 FORMAT(5X,*T,EQ,PER,SDV,STR,LSTM,LSTP,P,NM=:*)
 6035 FORMAT(5X,5F7.2,3I7,F7.4,F7.0)
 6040 FORMAT(5X,*LTSTOP=*,I7)
 6050 FORMAT(/,(2X,20F6.2))
 7050 FORMAT(12F6.2)
      STOP
      END
*
      FUNCTION STOCH(K,A,B)
      PARAMETER NFS=2
      DIMENSION XO(NFS),STEP(NFS),CRIT(NFS)
      XO(1)=0.  $STEP(1)=1.  $CRIT(1)=0.
      XO(2)=-8. $STEP(2)=.1 $CRIT(2)=.05
*
      STOCH=XO(K)  $F=0. $R=RANF(0.) $DFF=-1.  $D=STEP(K)
    5 F=F+PROB(K,STOCH+D/2.,A,B)*D  $DF=F-R
      IF(CRIT(K).LE.0..AND.F.GE.R) GO TO 100
      IF(ABS(DF).LE.CRIT(K)) GO TO 100
      IF(DFF*DF.LT.0.) D=-D/2.  $DFF=DF
      STOCH=STOCH+D $GO TO 5
  100 CONTINUE $IF(STOCH.LT.0.) STOCH=0.
      RETURN
      END
*
      FUNCTION PROB(K,X,A,B)
*
      IF(K.EQ.1) THEN $I=INT(X) $J=INT(B) $XR=FLOAT(I) $BR=FLOAT(J)
      FCT=1. $IF(I.EQ.0) GO TO 100
      DO 10 L=1,I $FCT=FCT*BR/XR $BR=BR-1. $XR=XR-1.
   10 CONTINUE
  100 CONTINUE
      PROB=FCT*A**I*(1.-A)**(J-I)  $ENDIF
      IF(K.EQ.2) THEN
      PROB=EXP(-((X-A)/B)**2/2.)/B/SQRT(2.*3.14159) $ENDIF
*
      RETURN
      END
*
/EOR
   .0   30.   20.    5.   10.    5.
   0.    0.         1    30
   .3   70.    8.    2.   .03    3   999  1227 .1        50.
   1.  -10.   10.    5.    .1   999   200  1936 .05       25.
  300
/EOR
/EOI


/JOB
A224,TL=200.
FAKE.
PDFV.
MNF(Y).
LGO,NR137A.
SAVE,NR137A.
EXIT.
SAVE,NR137A.
```

```
/EOR
      PROGRAM PTNRN14 (NROUT,INPUT,OUTPUT,TAPE6=OUTPUT,TAPE7=NROUT)
*
****      THIS PROGRAM SIMULATES THE RESPONSES OF A THREE-STATE-
*         VARIABLE MODEL OF A SINGLE POINT NEURON TO INTERACTING
*         SYNAPTIC INPUT. (A BINOMIAL MODEL FOR VESSICLE TRANSFER IS
*         INCLUDED.  A GAMMA INTERVAL DISTRIBUTION IS USED.  THESE ARE
*         EFFECTED BY A SUBROUTINE PACKAGE EASILY GENERALIZABLE TO IN-
****      CLUDE OTHER DISTRIBUTIONS AS WELL.)
*
      INTEGER SCSTRT,SCSTP,SNTP
      REAL NM
      PARAMETER STEP=1.,EK=-10.,SNTP=2
      DIMENSION TG(SNTP),EQ(SNTP),PER(SNTP),SDV(SNTP),STRF(SNTP),
     *          LSTM(SNTP),LSTP(SNTP),G(SNTP),DG(SNTP),PS(500),
     *          P(SNTP),NM(SNTP),INSED(SNTP)
*
****      THIS SECTION READS AND WRITES THE INPUT PARAMETERS
*
      READ 5010, C,TTH,B,TGK,TH0,TMEM
      READ 5020, SC,SL,SCSTRT,SCSTP
      DO 110 I=1,SNTP
  110 READ 5030, TG(I),EQ(I),PER(I),SDV(I),STRF(I),LSTM(I),LSTP(I),
     *           INSED(I),P(I),NM(I)
      READ 5000, LTSTOP
*
      WRITE(6,6000)
      WRITE(6,6010)  C,TTH,B,TGK,TH0,TMEM
      WRITE(6,6020)  SC,SL,SCSTRT,SCSTP
      WRITE(6,6030)  $DO 210 I=1,SNTP
  210 WRITE(6,6035)  TG(I),EQ(I),PER(I),SDV(I),STRF(I),LSTM(I),LSTP(I),
     *               INSED(I),P(I),NM(I)
      WRITE(6,6040)  LTSTOP
*
****      THIS SECTION INITIALIZES VARIABLES
*
      E=0. $TH=TH0 $S=0. $GK=0.
      DCTH=EXP(-STEP/TTH) $DGK=EXP(-STEP/TGK)
      DO 310 I=1,SNTP $G(I)=0.
  310 DG(I)=EXP(-STEP/TG(I))
*
****      THIS SECTION UPDATES STATE VARIABLES AT EACH TIME STEP
*
      DO 1000 L=1,LTSTOP
*
      SCN=0. $IF(L.GE.SCSTRT.AND.L.LT.SCSTP) SCN=SC
*
      DO 910 I=1,SNTP $G(I)=G(I)*DG(I)
      IF(L.EQ.LSTM(I).AND.L.LT.LSTP(I)) THEN
      CALL RANSET(INSED(I))
      G(I)=G(I)+STRF(I)*STOCH(1,P(I),NM(I))
      LSTM(I)=L+INT(STOCH(3,PER(I),SDV(I))-.5)+1
      CALL RANGET(INSED(I)) $ENDIF
  910 CONTINUE
*
      GS=0. $GSI=0. $DO 920 I=1,SNTP
      GS=GS+G(I) $GSI=GSI+G(I)*EQ(I)
  920 CONTINUE
*
      GK=GK*DGK+B*S*(1.-DGK) $GTOT=1.+GS+GK $DCE=EXP(-GTOT*STEP/TMEM)
      E=E*DCE+(SCN+GSI+GK*EK)*(1.-DCE)/GTOT
      TH=TH0+(TH-TH0)*DCTH+C*E*(1.-DCTH)
      S=0. $IF(E.GE.TH) S=1. $PS(L)=E+S*(50.-E)
 1000 CONTINUE
*
****      THIS SECTION WRITES OUT ACTIVITY VARIABLES
*
```

```
      WRITE(6,6050)  (PS(L),L=1,LTSTOP)
      WRITE(7,7050)  (PS(L),L=1,LTSTOP)
*
****          THESE ARE THE FORMATS
*
 5000 FORMAT(12I6)
 5010 FORMAT(12F6.2)
 5020 FORMAT(2F6.2,2I6)
 5030 FORMAT(5F6.2,3I6,F6.4,F6.0)
 6000 FORMAT(1H1)
 6010 FORMAT(5X,*C,TTH,B,TGK,TH0,TMEM=*,6F7.2)
 6020 FORMAT(5X,*SC,SL,SCSTRT,SCSTP=*,2F7.2,2I7)
 6030 FORMAT(5X,*T,EQ,PER,SDV,STR,LSTM,LSTP,INSED,P,NM=:*)
 6035 FORMAT(5X,5F7.2,3I7,F7.4,F7.0)
 6040 FORMAT(5X,*LTSTOP=*,I7)
 6050 FORMAT(/,(2X,20F6.2))
 7050 FORMAT(12F6.2)
      STOP
      END
*
      FUNCTION STOCH(K,A,B)
      PARAMETER NFS=3
      DIMENSION X0(NFS),STEP(NFS),CRIT(NFS)
      X0(1)=0.  $STEP(1)=1. $CRIT(1)=0.
      X0(2)=-8. $STEP(2)=.1 $CRIT(2)=.05
      X0(3)=0.  $STEP(3)=.1 $CRIT(3)=.05
*
      IF(K.EQ.1) G=1.
      IF(K.EQ.2) G=B*SQRT(2.*3.14159)
      IF(K.EQ.3) THEN $STEP3=.1 $CRIT3=.01
      G=0. $X=STEP3/2.
    3 DG=X**(A-1.)*EXP(-X)*STEP3 $G=G+DG
      IF(DG/G.LT.CRIT3) GO TO 300
      X=X+STEP3 $GO TO 3
  300 G=G*B**A $ENDIF
      CRIT(K)=G*CRIT(K)
*
      STOCH=X0(K) $F=0. $R=G*RANF(0.) $DFF=-1. $D=STEP(K)
    5 F=F+PROB(K,STOCH+D/2.,A,B)*D $DF=F-R
      IF(CRIT(K).LE.0..AND.F.GE.R) GO TO 100
      IF(ABS(DF).LE.CRIT(K)) GO TO 100
      IF(DFF*DF.LT.0.) D=-D/2. $DFF=DF
      STOCH=STOCH+D $GO TO 5
  100 CONTINUE $IF(STOCH.LT.0.) STOCH=0.
      RETURN
      END
*
      FUNCTION PROB(K,X,A,B)
*
      IF(K.EQ.1) THEN $I=INT(X) $J=INT(B) $XR=FLOAT(I) $BR=FLOAT(J)
      FCT=1. $IF(I.EQ.0) GO TO 100
      DO 10 L=1,I $FCT=FCT*BR/XR $BR=BR-1. $XR=XR-1.
   10 CONTINUE
  100 CONTINUE
      PROB=FCT*A**I*(1.-A)**(J-I) $ENDIF
      IF(K.EQ.2) PROB=EXP(-((X-A)/B)**2/2.)
      IF(K.EQ.3) PROB=X**(A-1.)*EXP(-X/B)
*
      RETURN
      END
*
/EOR
   .0  30.   20.    5.   10.    5.
   0.    0.         1    30
   .3  70.    8.    2.    .03    3    999  1227 .1      50.
   1.  -10.  10.    5.    .1    999   200  1936 .05     25.
    300
```

```
/EOR
/EOI


/JOB
A224.
FAKE.
PDFV.
MNF(Y).
LGO,NR122.
SAVE,NR122.
EXIT.
SAVE,NR122.
/EOR
      PROGRAM PTNRN15 (NROUT,INPUT,OUTPUT,TAPE6=OUTPUT,TAPE7=NROUT)
*
****      THIS PROGRAM SIMULATES THE RESPONSES OF A THREE-STATE-
*         VARIABLE MODEL OF A SINGLE POINT NEURON TO INTERACTING
****      SYNAPTIC INPUT. SHORT-TERM SYNAPTIC PLASTICITY IS INCLUDED.
*
      INTEGER SCSTRT,SCSTP,SNTP
      PARAMETER STEP=1.,EK=-10.,SNTP=1
      DIMENSION TG(SNTP),EQ(SNTP),PER(SNTP),STR(SNTP),LSTM(SNTP),
     *          LSTP(SNTP),G(SNTP),DG(SNTP),P(500),STRØ(SNTP),DS(SNTP),
     *          TSTR(SNTP),DSTR(SNTP),INSED(SNTP)
*
****      THIS SECTION READS AND WRITES THE INPUT PARAMETERS
*
      READ 5010, C,TTH,B,TGK,THØ,TMEM
      READ 5020, SC,SL,SCSTRT,SCSTP
      DO 110 I=1,SNTP
  110 READ 5030, TG(I),EQ(I),TSTR(I),DSTR(I),PER(I),STRØ(I),LSTM(I),
     *          LSTP(I),INSED(I)
      READ 5000, LTSTOP
*
      WRITE(6,6000)
      WRITE(6,6010) C,TTH,B,TGK,THØ,TMEM
      WRITE(6,6020) SC,SL,SCSTRT,SCSTP
      WRITE(6,6030) $DO 210 I=1,SNTP
  210 WRITE(6,6035) TG(I),EQ(I),TSTR(I),DSTR(I),PER(I),STRØ(I),LSTM(I),
     *          LSTP(I),INSED(I)
      WRITE(6,6040) LTSTOP
*
****      THIS SECTION INITIALIZES VARIABLES
*
      E=0. $TH=THØ $S=0. $GK=0.
      DCTH=EXP(-STEP/TTH) $DGK=EXP(-STEP/TGK)
      DO 310 I=1,SNTP $G(I)=0.
      STR(I)=STRØ(I) $DS(I)=EXP(-STEP/TSTR(I))
  310 DG(I)=EXP(-STEP/TG(I))
*
****      THIS SECTION UPDATES STATE VARIABLES AT EACH TIME STEP
*
      DO 1000 L=1,LTSTOP
*
      SCN=0. $IF(L.GE.SCSTRT.AND.L.LT.SCSTP) SCN=SC
*
      DO 910 I=1,SNTP $G(I)=G(I)*DG(I)
      STR(I)=STRØ(I)+(STR(I)-STRØ(I))*DS(I)
      IF(L.EQ.LSTM(I).AND.L.LT.LSTP(I)) THEN $G(I)=G(I)+STR(I)
      STR(I)=STR(I)+DSTR(I)
      CALL RANSET(INSED(I))
      LSTM(I)=L+INT(-PER(I)*ALOG(RANF(Ø.)))+1
      CALL RANGET(INSED(I)) $ENDIF
  910 CONTINUE
*
```

```
        GS=0. $GSI=0. $DO 920 I=1,SNTP
        GS=GS+G(I) $GSI=GSI+G(I)*EQ(I)
  920 CONTINUE
*
        GK=GK*DGK+B*S*(1.-DGK) $GTOT=1.+GS+GK $DCE=EXP(-GTOT*STEP/TMEM)
        E=E*DCE+(SCN+GSI+GK*EK)*(1.-DCE)/GTOT
        TH=TH0+(TH-TH0)*DCTH+C*E*(1.-DCTH)
        S=0. $IF(E.GE.TH) S=1. $P(L)=E+S*(50.-E)
 1000 CONTINUE
*
****        THIS SECTION WRITES OUT ACTIVITY VARIABLES
*
        WRITE(6,6050) (P(L),L=1,LTSTOP)
        WRITE(7,7050) (P(L),L=1,LTSTOP)
*
****        THESE ARE THE FORMATS
*
 5000 FORMAT(12I6)
 5010 FORMAT(12F6.2)
 5020 FORMAT(2F6.2,2I6)
 5030 FORMAT(6F6.2,3I6)
 6000 FORMAT(1H1)
 6010 FORMAT(5X,*C,TTH,B,TGK,TH0,TMEM=*,6F7.2)
 6020 FORMAT(5X,*SC,SL,SCSTRT,SCSTP=*,2F7.2,2I7)
 6030 FORMAT(5X,*T,EQ,TSTR,DSTR,PER,STR,LSTM,LSTP=:*)
 6035 FORMAT(5X,6F7.2,3I7)
 6040 FORMAT(5X,*LTSTOP=*,I7)
 6050 FORMAT(/,(2X,20F6.2))
 7050 FORMAT(12F6.2)
        STOP
        END
/EOR
   .5  30.    5.    3.    70.    5.
   0.    0.    999    30
   .3  70.    3.    .1   4.     .1      3    999   1227
    120
/EOR
/EOI



/JOB
A224,TL=100.
FAKE.
PDFV.
MNF(Y).
LGO,NR015.
SAVE,NR015.
EXIT.
SAVE,NR015.
/EOR
        PROGRAM PTNRN21(NROUT,INPUT,OUTPUT,TAPE6=OUTPUT,TAPE7=NROUT)
*
****        THIS PROGRAM SIMULATES THE RESPONSES OF A TWO-POINT MODEL
*           OF A SINGLE NEURON WITH ACTIVE CALCIUM-RELATED CONDUCTANCES
*           TO INTERACTING SYNAPTIC INPUT. (EXPONENTIAL INTERVAL DISTRI-
****        BUTIONS ARE USED.)
*
        INTEGER SSTRT,SSTP,SDSTRT,SDSTP,SNTP
        PARAMETER STEP=1.,NSTEPS=10,EK=-10.,ECA=50.,SNTP=2
        DIMENSION PS(500),PD(500),T(SNTP),EQ(SNTP),P(SNTP,2),STR(SNTP,2),
      *         DG(SNTP),G(SNTP,2),INSED(SNTP)
*
****        THIS SECTION READS AND WRITES THE INPUT PARAMETERS
*
        READ 5000, IFLG1
        READ 5010, TS,GDS,TH0,C,TTH,B,TGK
```

```
      READ 5010, TD,GSD,THD,D,TGC,A,TCA,CAØ,BD,TGKD
      READ 5020, SS,SSTRT,SSTP,SD,SDSTRT,SDSTP
      DO 110 I=1,SNTP
  110 READ 5030, T(I),EQ(I),((P(I,J),STR(I,J)),J=1,2),INSED(I)
      READ 5000, LTSTOP
*
      WRITE(6,6000) $WRITE(6,6010) TS,GDS,THØ,C,TTH,B,TGK
      WRITE(6,6020) TD,GSD,THD,D,TGC,A,TCA,CAØ,BD,TGKD
      WRITE(6,6030) SS,SSTRT,SSTP,SD,SDSTRT,SDSTP
      DO 210 I=1,SNTP
  210 WRITE(6,6035) T(I),EQ(I),((P(I,J),STR(I,J)),J=1,2),INSED(I)
      IF(IFLG1.GT.Ø) WRITE(6,6040)
*
****          THIS SECTION INITIALIZES VARIABLES
*
      ES=Ø. $THS=THØ $GKS=Ø. $ESS=Ø.
      ED=Ø. $GCA=Ø. $CA=Ø. $GKD=Ø.
      DO 310 I=1,SNTP $DO 310 J=1,2
  310 G(I,J)=Ø.
      DTHS=EXP(-STEP/TTH) $DGKS=EXP(-STEP/TGK)
      DGC=EXP(-STEP/TGC) $DCA=EXP(-STEP/TCA) $DGKD=EXP(-STEP/TGKD)
      DO 320 I=1,SNTP
  320 DG(I)=EXP(-STEP/T(I))
*
****          THIS SECTION UPDATES STATE VARIABLES AT EACH TIME STEP
*
      DO 1000 L=1,LTSTOP
*
      SCS=Ø. $IF(L.GE.SSTRT.AND.L.LT.SSTP) SCS=SS
      SCD=Ø. $IF(L.GE.SDSTRT.AND.L.LT.SDSTP) SCD=SD
*
      DO 910 I=1,SNTP $CALL RANSET(INSED(I)) $DO 908 J=1,2
      G(I,J)=G(I,J)*DG(I) $IF(RANF(Ø.).LT.P(I,J)) G(I,J)=G(I,J)+STR(I,J)
  908 CONTINUE $CALL RANGET(INSED(I))
  910 CONTINUE
      GS=Ø. $GD=Ø. $GSI=Ø. $GDI=Ø. $DO 920 I=1,SNTP
      GS=GS+G(I,1) $GD=GD+G(I,2) $GSI=GSI+G(I,1)*EQ(I)
  920 GDI=GDI+G(I,2)*EQ(I)
*
      GSM=1.+GS+GDS+GKS $Z=EXP(-STEP*GSM/TS/FLOAT(NSTEPS))
      GDN=1.+GD+GSD+GCA+GKD $ZD=EXP(-STEP*GDN/TD/FLOAT(NSTEPS))
*
****          THIS SUB-SECTION ALLOWS FOR MORE FINE-GRAINED INTEGRATION
*
      DO 1010 I=1,NSTEPS
      ESN=ES*Z+(GDS*ED+GKS*EK+SCS+GSI)*(1.-Z)/GSM
      EDN=ED*ZD+(GSD*ESS+GCA*ECA+GKD*EK+SCD+GDI)*(1.-ZD)/GDN
      ES=ESN $ED=EDN
 1010 CONTINUE
****
      THS=THS*DTHS+(THØ+C*ES)*(1.-DTHS)
      S=Ø. $IF(ES.GE.THS) S=1.
      GKS=GKS*DGKS+B*S*(1.-DGKS)
      GCA=GCA*DGC $IF(ED.GE.THD) GCA=GCA+D*(ED-THD)*(1.-DGC)
      CA=CA*DCA+A*GCA*(1.-DCA)
      GKD=GKD*DGKD $IF(CA.GE.CAØ) GKD=GKD+BD*(1.-DGKD)
      PS(L)=ES+(5Ø.-ES)*S $ESS=PS(L) $PD(L)=ED
*
      IF(IFLG1.GT.Ø) WRITE(6,6110) L,ES,ED,THS,S,GKS,GCA,CA,GKD
      IF(IFLG1.GT.Ø) WRITE(7,7110) L,ES,ED,THS,S,GKS,GCA,CA,GKD
*
 1000 CONTINUE
*
****          THIS SECTION WRITES OUT ACTIVITY VARIABLES
*
      WRITE(6,6140) $WRITE(6,6150) (PS(L),L=1,LTSTOP)
      WRITE(6,6140) $WRITE(6,6150) (PD(L),L=1,LTSTOP)
```

```
      WRITE(7,6140) $WRITE(7,7150) (PS(L),L=1,LTSTOP)
      WRITE(7,6140) $WRITE(7,7150) (PD(L),L=1,LTSTOP)
*
****          THESE ARE THE FORMATS
*
      STOP
 5000 FORMAT(12I6)
 5010 FORMAT(12F6.2)
 5020 FORMAT(F6.2,2I6,F6.2,2I6)
 5030 FORMAT(6F6.2,I6)
 6000 FORMAT(1H1)
 6010 FORMAT(5X,*TS,GDS,THØ,C,TTH,B,TGK=*,7F7.2)
 6020 FORMAT(5X,*TD,GSD,THD,D,TGC,A,TCA,CAØ,BD,TGKD=*,10F7.2)
 6030 FORMAT(5X,*SS,SSTRT,SSTP,SD,SDSTRT,SDSTP=*,F7.2,2I7,F7.2,2I7)
 6035 FORMAT(5X,*T,EQ,P1,STR1,P2,STR2,INSED=*,6F7.2,I7)
 6040 FORMAT(/,5X,*   L     ES      ED        THS      S
     *GKS    GCA      CA      GKD    *)
 6110 FORMAT(5X,I8,8F10.2)
 6140 FORMAT(//)
 6150 FORMAT(5X,20F6.2)
 7110 FORMAT(I8,8F8.2)
 7150 FORMAT(12F6.2)
      END
/EOR
    1
   5.    5.   12.    Ø.   20.   33.    5.
   5.    5.   16.   2.2    5.    2.    5.   20.  100.   15.
  30.       1  1000   Ø.        1      1
    .5  70.    Ø.    Ø.        .1    .3   1227
   5.  -10.    .05  1.5    Ø.    Ø.   1936
    240
/EOR
/EOI


/JOB
A224,CM=70000,TL=300.
FAKE.
PDFV.
MNF(Y).
LGO,DN6Ø4.
SAVE,DN6Ø4.
EXIT.
SAVE,DN6Ø4.
/EOR
      PROGRAM NEURN31 (DNOUT,INPUT,OUTPUT,TAPE6=OUTPUT,TAPE7=DNOUT) ·
*
****      THIS PROGRAM SIMULATES THE RESPONSES OF A MODEL NEURON WITH
*         SIMPLIFIED DENDRITIC TREE CONTAINING ACTIVE CALCIUM-RELATED
*         CONDUCTANCES TO INTERACTING SYNAPTIC INPUT. (EXPONENTIAL
****      INTERVAL DISTRIBUTIONS ARE USED.)
*
      INTEGER SCSTRT,SCSTP,SDSTRT,SDSTP,ST,RG,SNTP
      PARAMETER NST=3,NRG=4,STEP=1.,ECA=50.,EK=-10.,NDSPY=1,SNTP=2,
     *         MXT=3
      DIMENSION V(NST,NRG),GCA(NST,NRG),CA(NST,NRG),GKD(NST,NRG),
     *          GT(NST,NRG),Z(NST,NRG),U(NRG),ST(NDSPY),RG(NDSPY),
     *          PS(3,500),EQ(SNTP),TG(SNTP),NTG(SNTP),ITS(NST,MXT),
     *          ITR(SNTP,MXT),STR(SNTP,MXT),P(SNTP,MXT),
     *          LSTRT(SNTP,MXT),LSTP(SNTP,MXT),DG(SNTP),GS(SNTP),
     *          G(SNTP,NST,NRG),NF(SNTP,MXT),INSED(SNTP,MXT)
*
****      THIS SECTION READS AND WRITES INPUT PARAMETERS
*
      READ 5000, IFLG1
      READ 5010, C,TTH,B,TGK,TMEM,THØ,CDS,VRT
```

```
      READ 5010, THD,D,TGC,A,TCA,CA0,BD,TGKD
      READ 5000, ((ST(I),RG(I)),I=1,NDSPY)
      READ 5020, SC,SCSTRT,SCSTP
      READ 5020, SD,SDSTRT,SDSTP,IS,IRG
      DO 110 I=1,SNTP
      READ 5030, EQ(I),TG(I),NTG(I)
      DO 110 J=1,NTG(I)
  110 READ 5040, ITS(I,J),ITR(I,J),STR(I,J),P(I,J),LSTRT(I,J),
     *           LSTP(I,J),INSED(I,J)
      READ 5000, LTSTOP
*
      WRITE(6,6000) $WRITE(6,6010) NST,NRG,STEP,IFLG1
      WRITE(6,6020) C,TTH,B,TGK,TMEM,TH0,CDS,VRT
      WRITE(6,6030) THD,D,TGC,A,TCA,CA0,BD,TGKD
      WRITE(6,6040) ((ST(I),RG(I)),I=1,NDSPY)
      WRITE(6,6050) SC,SCSTRT,SCSTP,SD,SDSTRT,SDSTP,IS,IRG,LTSTOP
      WRITE(6,6060) $DO 120 I=1,SNTP
      WRITE(6,6070) EQ(I),TG(I),NTG(I) $DO 120 J=1,NTG(I)
  120 WRITE(6,6075) ITS(I,J),ITR(I,J),STR(I,J),P(I,J),
     *           LSTRT(I,J),LSTP(I,J),INSED(I,J)
      WRITE(6,6000)
*
****        THIS SECTION INITIALIZES VARIABLES
*
      E=0. $TH=TH0 $GK=0. $DO 210 I=1,NST $DO 210 J=1,NRG
      V(I,J)=0. $GCA(I,J)=0. $CA(I,J)=0. $GKD(I,J)=0.
  210 CONTINUE
      DTH=EXP(-STEP/TTH) $DGK=EXP(-STEP/TGK)
      DGC=EXP(-STEP/TGC) $DCA=EXP(-STEP/TCA) $DGKD=EXP(-STEP/TGKD)
      DO 220 K=1,SNTP $GS(K)=0. $DG(K)=EXP(-STEP/TG(K))
      DO 220 I=1,NST $DO 220 J=1,NRG
  220 G(K,I,J)=0.
*
****        TH IS SECTION UPDATES STATE VARIABLES AT EACH TIME STEP
*
      DO 1000 L=1,LTSTOP
*
      SCN=0. $IF(L.GE.SCSTRT.AND.L.LT.SCSTP) SCN=SC
      SCDN=0. $IF(L.GE.SDSTRT.AND.L.LT.SDSTP) SCDN=SD
*
      DO 700 I=1,SNTP $DO 700 J=1,NTG(I)
      NF(I,J)=0 $IF(L.LT.LSTRT(I,J).OR.L.GE.LSTP(I,J)) GO TO 700
      CALL RANSET(INSED(I,J))
      IF(RANF(0.).GT.P(I,J)) GO TO 695 $NF(I,J)=NF(I,J)+1
      IF(ITS(I,J).EQ.0) THEN $GS(I)=GS(I)+STR(I,J) $ELSE
      G(I,ITS(I,J),ITR(I,J))=G(I,ITS(I,J),ITR(I,J))+STR(I,J) $ENDIF
  695 CALL RANGET(INSED(I,J))
  700 CONTINUE
*
      GTS=1.+GK $SY=0. $DO 510 K=1,SNTP $GTS=GTS+GS(K)
  510 SY=SY+GS(K)*EQ(K) $ZS=EXP(-GTS*STEP/TMEM)
      E=E*ZS+(SCN+SY+GK*EK)*(1.-ZS)/GTS $DM=0.
      DO 980 I=1,NST $DO 920 J=1,NRG $GT(I,J)=1.+GCA(I,J)+GKD(I,J)
      SY=0. $DO 910 K=1,SNTP
      GT(I,J)=GT(I,J)+G(K,I,J)
  910 SY=SY+G(K,I,J)*EQ(K) $Z(I,J)=EXP(-GT(I,J)*STEP/TMEM)
      SCI=0. $IF(I.EQ.IS.AND.J.EQ.IRG) SCI=SCDN
  920 V(I,J)=V(I,J)*Z(I,J)+(SCI+SY+GCA(I,J)*ECA+GKD(I,J)*EK)*(1.-Z(I,J))
     *       /GT(I,J)
      DM=DM+V(I,1)
      U(1)=(V(I,1)+V(I,2)+E/CDS)/(2.+1./CDS)*(1.-VRT)+VRT*V(I,1)
      DO 930 J=2,NRG-1
  930 U(J)=(V(I,J-1)+V(I,J)+V(I,J+1))/3.*(1.-VRT)+VRT*V(I,J)
      U(NRG)=(V(I,NRG)+V(I,NRG-1))/2.*(1.-VRT)+VRT*V(I,NRG)
      DO 940 J=1,NRG
  940 V(I,J)=U(J)
*
```

```
      DO 950 J=1,NRG
      GCA(I,J)=GCA(I,J)*DGC
      IF(V(I,J).GE.THD) GCA(I,J)=GCA(I,J)+D*(V(I,J)-THD)*(1.-DGC)
      CA(I,J)=CA(I,J)*DCA+A*GCA(I,J)*(1.-DCA)
      GKD(I,J)=GKD(I,J)*DGKD
      IF(CA(I,J).GE.CAØ) GKD(I,J)=GKD(I,J)+BD*(1.-DGKD)
  950 CONTINUE
*
  980 CONTINUE
*
      E=(E+CDS*DM)/(1.+CDS*FLOAT(NST))*(1.-VRT)+VRT*E
      TH=TH*DTH+(THØ+C*E)*(1.-DTH) $GK=GK*DGK
      S=0. $IF(E.LT.TH) GO TO 995
      S=1. $GK=GK+B*(1.-DGK)
  995 CONTINUE $PS(1,L)=E+S*(50.-E)
      DO 996 I=1,NDSPY
  996 PS(I+1,L)=V(ST(I),RG(I))
      IF(IFLG1.GT.Ø) THEN $DO 997 I=1,SNTP
  997 WRITE(6,6278) (NF(I,J),J=1,NTG(I))
      WRITE(6,6279) L,E,TH,GK $WRITE(7,7279) L,E,TH,GK
      DO 998 I=1,NST
      WRITE(6,6280) ((V(I,J),GCA(I,J),CA(I,J),GKD(I,J)),J=1,NRG)
  998 WRITE(7,7280) ((V(I,J),GCA(I,J),CA(I,J),GKD(I,J)),J=1,NRG) $ENDIF
*
      DO 410 K=1,SNTP $GS(K)=GS(K)*DG(K) $DO 410 I=1,NST $DO 410 J=1,NRG
  410 G(K,I,J)=G(K,I,J)*DG(K)
*
 1000 CONTINUE
*
****       THIS SECTION WRITES OUT ACTIVITY VARIABLES
*
      DO 1100 I=1,NDSPY+1 $WRITE(6,6002) $WRITE(7,6002)
      WRITE(6,6280) (PS(I,L),L=1,LTSTOP)
 1100 WRITE(7,7280) (PS(I,L),L=1,LTSTOP)
*
****       THESE ARE THE FORMATS
*
      STOP
 5000 FORMAT(12I6)
 5010 FORMAT(12F6.2)
 5020 FORMAT(F6.2,4I6)
 5030 FORMAT(2F6.2,I6)
 5040 FORMAT(2I6,2F6.2,3I6)
 6000 FORMAT(1H1)
 6002 FORMAT(//)
 6010 FORMAT(5X,*NST,NRG,STEP,IFLG1=*,2I7,F7.2,I7)
 6020 FORMAT(5X,*C,TTH,B,TGK,TMEM,THØ,CDS,VRT=*,8F7.2)
 6030 FORMAT(5X,*THD,D,TGC,A,TCA,CAØ,BD,TGKD=*,8F7.2)
 6040 FORMAT(5X,*STS,TGS=*,12I7)
 6050 FORMAT(5X,*SC,SCSTRT,SCSTP,SD,SDSTRT,SDSTP,IS,IRG,LTSTOP=*,
      *       F7.2,2I7,F7.2,5I7)
 6060 FORMAT(5X,*EQ,TG,NTG,/,ITB,ITR,STR,P,STRT,STP,INSED:*)
 6070 FORMAT(15X,2F7.2,I7)
 6075 FORMAT(15X,2I7,2F7.2,3I7)
 6278 FORMAT(5X,20I5)
 6279 FORMAT(5X,I5,3F7.2)
 6280 FORMAT(5X,20F6.2)
 7279 FORMAT(I5,3F7.2)
 7280 FORMAT(16F5.1)
*
      END
/EOR
      1
   Ø.    25.   2Ø.   1Ø.   11.   24.    3.    .5
  32.    Ø.    1Ø.    2.   1Ø.   2Ø.   8Ø.   15.
      1     3
   Ø.      1     Ø
```

```
125.      1     Ø     1     3
 70.    2.        3
      1    3   1.       .10     1   9999   1227
      2    3   1.       .10     1   9999   1936
      3    3   1.       .10     1   9999   4800
-10.    8.        1
      2    1   1.       .10     1   9999   3412
     30
/EOR
/EOI



/JOB
A224,CM=70000,TL=300.
FAKE.
PDFV.
MNF(Y).
LGO,NR308.
SAVE,NR308.
EXIT.
SAVE,NR308.
/EOR
      PROGRAM NEURN32 (DNOUT,INPUT,OUTPUT,TAPE6=OUTPUT,TAPE7=DNOUT)
*
****    THIS PROGRAM SIMULATES THE RESPONSES OF A MODEL NEURON WITH
*       SIMPLIFIED DENDRITIC TREE CONTAINING ACTIVE CALCIUM-RELATED
*       CONDUCTANCES TO INTERACTING SYNAPTIC INPUT. (A BINOMIAL MODEL
*       FOR VESSICLE TRANSFER IS INCLUDED.  A GAMMA INTERVAL DISTRIBUTION
*       IS USED.  SHORT-TERM SYNAPTIC PLASTICITY ISINCLUDED.  THE
*       STOCHASTIC DISTRIBUTIONS ARE EFFECTED BY A SUBROUTINE PACKAGE
****    EASILY GENERALIZEABLE TO OTHER DISTRIBUTIONS.)
*
      INTEGER SCSTRT,SCSTP,SDSTRT,SDSTP,ST,RG,SNTP
      REAL NM
      PARAMETER NST=3,NRG=4,STEP=1.,ECA=50.,EK=-10.,NDSPY=1,SNTP=2,
     *          MXT=3
      DIMENSION V(NST,NRG),GCA(NST,NRG),CA(NST,NRG),GKD(NST,NRG),
     *          GT(NST,NRG),Z(NST,NRG),U(NRG),ST(NDSPY),RG(NDSPY),
     *          PS(3,500),EQ(SNTP),TG(SNTP),NTG(SNTP),ITS(NST,MXT),
     *          ITR(SNTP,MXT),STRFØ(SNTP,MXT),STRF(SNTP,MXT),
     *          TSTR(SNTP,MXT),DS(SNTP,MXT),DSTR(SNTP,MXT),
     *          PER(SNTP,MXT),SDV(SNTP,MXT),P(SNTP,MXT),NM(SNTP,MXT),
     *          LSTM(SNTP,MXT),LSTP(SNTP,MXT),DG(SNTP),GS(SNTP),
     *          G(SNTP,NST,NRG),NF(SNTP,MXT),INSED(SNTP,MXT)
*
****         THIS SECTION READS AND WRITES INPUT PARAMETERS
*
      READ 5000, IFLG1
      READ 5010, C,TTH,B,TGK,TMEM,THØ,CDS,VRT
      READ 5010, THD,D,TGC,A,TCA,CAØ,BD,TGKD
      READ 5000, ((ST(I),RG(I)),I=1,NDSPY)
      READ 5020, SC,SCSTRT,SCSTP
      READ 5020, SD,SDSTRT,SDSTP,IS,IRG
      DO 110 I=1,SNTP
      READ 5030, EQ(I),TG(I),NTG(I)
      DO 110 J=1,NTG(I)
  110 READ 5040, ITS(I,J),ITR(I,J),STRFØ(I,J),P(I,J),NM(I,J),
     *           TSTR(I,J),DSTR(I,J),PER(I,J),SDV(I,J),
     *           LSTM(I,J),LSTP(I,J),INSED(I,J)
      READ 5000, LTSTOP
*
      WRITE(6,6000) $WRITE(6,6010) NST,NRG,STEP,IFLG1
      WRITE(6,6020) C,TTH,B,TGK,TMEM,THØ,CDS,VRT
      WRITE(6,6030) THD,D,TGC,A,TCA,CAØ,BD,TGKD
      WRITE(6,6040) ((ST(I),RG(I)),I=1,NDSPY)
      WRITE(6,6050) SC,SCSTRT,SCSTP,SD,SDSTRT,SDSTP,IS,IRG,LTSTOP
```

```
      WRITE(6,6060) $DO 120 I=1,SNTP
      WRITE(6,6070) EQ(I),TG(I),NTG(I) $DO 120 J=1,NTG(I)
  120 WRITE(6,6075) ITS(I,J),ITR(I,J),STRFØ(I,J),P(I,J),NM(I,J),
     *              TSTR(I,J),DSTR(I,J),PER(I,J),SDV(I,J),
     *              LSTM(I,J),LSTP(I,J),INSED(I,J)
      WRITE(6,6000)
*
****       THIS SECTION INITIALIZES VARIABLES
*
      E=0. $TH=THØ $GK=0. $DO 210 I=1,NST $DO 210 J=1,NRG
      V(I,J)=0. $GCA(I,J)=0. $CA(I,J)=0. $GKD(I,J)=0.
  210 CONTINUE
      DTH=EXP(-STEP/TTH) $DGK=EXP(-STEP/TGK)
      DGC=EXP(-STEP/TGC) $DCA=EXP(-STEP/TCA) $DGKD=EXP(-STEP/TGKD)
      DO 220 K=1,SNTP $GS(K)=0. $DG(K)=EXP(-STEP/TG(K))
      DO 225 KK=1,NTG(K)
      STRF(K,KK)=STRFØ(K,KK) $DS(K,KK)=EXP(-STEP/TSTR(K,KK))
  225 CONTINUE
      DO 220 I=1,NST $DO 220 J=1,NRG
  220 G(K,I,J)=0.
*
****       TH IS SECTION UPDATES STATE VARIABLES AT EACH TIME STEP
*
      DO 1000 L=1,LTSTOP
*
      SCN=0. $IF(L.GE.SCSTRT.AND.L.LT.SCSTP) SCN=SC
      SCDN=0. $IF(L.GE.SDSTRT.AND.L.LT.SDSTP) SCDN=SD
*
      DO 700 I=1,SNTP $DO 700 J=1,NTG(I)
      STRF(I,J)=STRFØ(I,J)+(STRF(I,J)-STRFØ(I,J))*DS(I,J)
      NF(I,J)=0 $IF(L.EQ.LSTM(I,J).AND.L.LT.LSTP(I,J)) THEN
      CALL RANSET(INSED(I,J)) $NF(I,J)=NF(I,J)+1
      STR=STRF(I,J)*STOCH(1,P(I,J),NM(I,J))
      STRF(I,J)=STRF(I,J)+DSTR(I,J)
      LSTM(I,J)=L+INT(STOCH(3,PER(I,J),SDV(I,J))-.5)+1
      IF(ITS(I,J).EQ.0) THEN $GS(I)=GS(I)+STR $ELSE
      G(I,ITS(I,J),ITR(I,J))=G(I,ITS(I,J),ITR(I,J))+STR $ENDIF
      CALL RANGET(INSED(I,J)) $ENDIF
  700 CONTINUE
*
      GTS=1.+GK $SY=0. $DO 510 K=1,SNTP $GTS=GTS+GS(K)
  510 SY=SY+GS(K)*EQ(K) $ZS=EXP(-GTS*STEP/TMEM)
      E=E*ZS+(SCN+SY+GK*EK)*(1.-ZS)/GTS $DM=0.
      DO 980 I=1,NST $DO 920 J=1,NRG $GT(I,J)=1.+GCA(I,J)+GKD(I,J)
      SY=0. $DO 910 K=1,SNTP
      GT(I,J)=GT(I,J)+G(K,I,J)
  910 SY=SY+G(K,I,J)*EQ(K) $Z(I,J)=EXP(-GT(I,J)*STEP/TMEM)
      SCI=0. $IF(I.EQ.IS.AND.J.EQ.IRG) SCI=SCDN
  920 V(I,J)=V(I,J)*Z(I,J)+(SCI+SY+GCA(I,J)*ECA+GKD(I,J)*EK)*(1.-Z(I,J))
     *              /GT(I,J)
      DM=DM+V(I,1)
      U(1)=(V(I,1)+V(I,2)+E/CDS)/(2.+1./CDS)*(1.-VRT)+VRT*V(I,1)
      DO 930 J=2,NRG-1
  930 U(J)=(V(I,J-1)+V(I,J)+V(I,J+1))/3.*(1.-VRT)+VRT*V(I,J)
      U(NRG)=(V(I,NRG)+V(I,NRG-1))/2.*(1.-VRT)+VRT*V(I,NRG)
      DO 940 J=1,NRG
  940 V(I,J)=U(J)
*
      DO 950 J=1,NRG
      GCA(I,J)=GCA(I,J)*DGC
      IF(V(I,J).GE.THD) GCA(I,J)=GCA(I,J)+D*(V(I,J)-THD)*(1.-DGC)
      CA(I,J)=CA(I,J)*DCA+A*GCA(I,J)*(1.-DCA)
      GKD(I,J)=GKD(I,J)*DGKD
      IF(CA(I,J).GE.CAØ) GKD(I,J)=GKD(I,J)+BD*(1.-DGKD)
  950 CONTINUE
*
  980 CONTINUE
```

```
*
      E=(E+CDS*DM)/(1.+CDS*FLOAT(NST))*(1.-VRT)+VRT*E
      TH=TH*DTH+(TH0+C*E)*(1.-DTH)  $GK=GK*DGK
      S=0.  $IF(E.LT.TH) GO TO 995
      S=1.  $GK=GK+B*(1.-DGK)
  995 CONTINUE $PS(1,L)=E+S*(50.-E)
      DO 996 I=1,NDSPY
  996 PS(I+1,L)=V(ST(I),RG(I))
      IF(IFLG1.GT.0) THEN $DO 997 I=1,SNTP
  997 WRITE(6,6278) (NF(I,J),J=1,NTG(I))
      WRITE(6,6279) L,E,TH,GK  $WRITE(7,7279) L,E,TH,GK
      DO 998 I=1,NST
      WRITE(6,6280) ((V(I,J),GCA(I,J),CA(I,J),GKD(I,J)),J=1,NRG)
  998 WRITE(7,7280) ((V(I,J),GCA(I,J),CA(I,J),GKD(I,J)),J=1,NRG) $ENDIF
*
      DO 410 K=1,SNTP $GS(K)=GS(K)*DG(K) $DO 410 I=1,NST $DO 410 J=1,NRG
  410 G(K,I,J)=G(K,I,J)*DG(K)
*
 1000 CONTINUE
*
****       THIS SECTION WRITES OUT ACTIVITY VARIABLES
*
      DO 1100 I=1,NDSPY+1 $WRITE(6,6002) $WRITE(7,6002)
      WRITE(6,6280) (PS(I,L),L=1,LTSTOP)
 1100 WRITE(7,7280) (PS(I,L),L=1,LTSTOP)
*
****       THESE ARE THE FORMATS
*
      STOP
 5000 FORMAT(12I6)
 5010 FORMAT(12F6.2)
 5020 FORMAT(F6.2,4I6)
 5030 FORMAT(2F6.2,I6)
 5040 FORMAT(2I6,7F6.2,3I6)
 6000 FORMAT(1H1)
 6002 FORMAT(//)
 6010 FORMAT(5X,*NST,NRG,STEP,IFLG1=*,2I7,F7.2,I7)
 6020 FORMAT(5X,*C,TTH,B,TGK,TMEM,TH0,CDS,VRT=*,8F7.2)
 6030 FORMAT(5X,*THD,D,TGC,A,TCA,CA0,BD,TGKD=*,8F7.2)
 6040 FORMAT(5X,*STS,TGS=*,12I7)
 6050 FORMAT(5X,*SC,SCSTRT,SCSTP,SD,SDSTRT,SDSTP,IS,IRG,LTSTOP=*,
     *       F7.2,2I7,F7.2,5I7)
 6060 FORMAT(5X,*EQ,TG,NTG,/,ITB,ITR,STRF0,P,NM,TSTR,DSTR,PER,SDV,LSTM,
     *       LSTP,INSED:*)
 6070 FORMAT(15X,2F7.2,I7)
 6075 FORMAT(15X,2I7,7F7.2,3I7)
 6278 FORMAT(5X,20I5)
 6279 FORMAT(5X,I5,3F7.2)
 6280 FORMAT(5X,20F6.2)
 7279 FORMAT(I5,3F7.2)
 7280 FORMAT(16F5.1)
*
      END
*
      FUNCTION STOCH(K,A,B)
      PARAMETER NFS=3
      DIMENSION X0(NFS),STEP(NFS),CRIT(NFS)
      X0(1)=0.  $STEP(1)=1.  $CRIT(1)=0.
      X0(2)=-8. $STEP(2)=.1  $CRIT(2)=.05
      X0(3)=0.  $STEP(3)=.1  $CRIT(3)=.05
*
      IF(K.EQ.1)  G=1.
      IF(K.EQ.2)  G=B*SQRT(2.*3.14159)
      IF(K.EQ.3)  THEN $STEP3=.1 $CRIT3=.01
      G=0.  $X=STEP3/2.
    3 DG=X**(A-1.)*EXP(-X)*STEP3 $G=G+DG
      IF(DG/G.LT.CRIT3) GO TO 300
```

```
      X=X+STEP3 $GO TO 3
  300 G=G*B**A $ENDIF
      CRIT(K)=G*CRIT(K)
*
      STOCH=X0(K) $F=0. $R=G*RANF(0.) $DFF=-1. $D=STEP(K)
    5 F=F+PROB(K,STOCH+D/2.,A,B)*D $DF=F-R
      IF(CRIT(K).LE.0..AND.F.GE.R) GO TO 100
      IF(ABS(DF).LE.CRIT(K)) GO TO 100
      IF(DFF*DF.LT.0.) D=-D/2. $DFF=DF
      STOCH=STOCH+D $GO TO 5
  100 CONTINUE $IF(STOCH.LT.0.) STOCH=0.
      RETURN
      END
*
      FUNCTION PROB(K,X,A,B)
*
      IF(K.EQ.1) THEN $I=INT(X) $J=INT(B) $XR=FLOAT(I) $BR=FLOAT(J)
      FCT=1. $IF(I.EQ.0) GO TO 100
      DO 10 L=1,I $FCT=FCT*BR/XR $BR=BR-1. $XR=XR-1.
   10 CONTINUE
  100 CONTINUE
      PROB=FCT*A**I*(1.-A)**(J-I) $ENDIF
      IF(K.EQ.2) PROB=EXP(-((X-A)/B)**2/2.)
      IF(K.EQ.3) PROB=X**(A-1.)*EXP(-X/B)
*
      RETURN
      END
*
/EOR
      1
   0.    25.    20.    10.    11.    24.    3.      .1
  32.     2.2  10.     2.    10.    10.    200.   15.
      1    3
   0.         1      0
 125.         1      0       1       3
  70.    2.         3
      1    3  1.      .2   50.    3.      .1   7.    2.      3   999  1227
      2    3  1.      .2   50.    3.      .1   7.    2.      6   999  1936
      3    3  1.      .2   50.    3.      .1   7.    2.      9   999  2348
 -10.    8.         1
      2    1   .4    .2   25.    1.      .5  10.    3.    100   200  1960
    100
/EOR
/EOI


/JOB
A224,CM=130000.
FAKE.
PDFV.
MNF(Y).
LGO.
/EOR
      PROGRAM POOL10 (INPUT,OUTPUT,TAPE6=OUTPUT)
*
****      THIS PROGRAM SIMULATES THE BEHAVIOR OF THREE PACEMAKER
****      NEURONS WHICH INTERACT SYNAPTICALLY WITH UNIT TIME DELAY.
*
      PARAMETER STEP=1.,EK=-10.
      DIMENSION E(3),TH(3),S(3),GK(3),E0(3),P(3,140),SC(3)
*
****      THIS SECTION READS AND WRITES THE INPUT PARAMETERS
*
      READ 2010, C,TTH,B,TGK,THO,TMEM
      READ 2010, A1,A2,A3,(E0(I),I=1,3)
      READ 2010, A12,A13,A21,A23,A31,A32
```

```
      READ 2020, LTSTOP
*
      WRITE(6,6000)
      WRITE(6,6010) C,TTH,B,TGK,THO,TMEM
      WRITE(6,6012) A1,A2,A3,(E0(I),I=1,3)
      WRITE(6,6014) A12,A13,A21,A23,A31,A32
      WRITE(6,6020) LTSTOP
*
****       THIS SECTION INITIALIZES VARIABLES
*
      DCTH=EXP(-STEP/TTH) $DCG=EXP(-STEP/TGK)
      DO 110 I=1,3
      E(I)=E0(I) $TH(I)=THO $GK(I)=0. $S(I)=0.
  110 CONTINUE
*
****       THIS SECTION UPDATES STATE VARIABLES AT EACH TIME STEP
*
      DO 1000 L=1,LTSTOP
      SC(1)=A1+A21*S(2)+A31*S(3)
      SC(2)=A2+A12*S(1)+A32*S(3)
      SC(3)=A3+A13*S(1)+A23*S(2)
      DO 1500 I=1,3
      GK(I)=GK(I)*DCG+B*(1.-DCG)*S(I) $DCE=EXP(-(1.+GK(I))*STEP/TMEM)
      E(I)=E(I)*DCE+(SC(I)+GK(I)*EK)*(1.-DCE)/(1.+GK(I))
      TH(I)=THO+(TH(I)-THO)*DCTH+C*E(I)*(1.-DCTH)
      S(I)=0. $IF(E(I).GE.TH(I)) S(I)=1.
      P(I,L)=E(I)+S(I)*(50.-E(I))
 1500 CONTINUE
 1000 CONTINUE
*
****       THIS SECTION WRITES OUT ACTIVITY VARIABLES
*
      DO 4010 I=1,3
      WRITE(6,6030)
 4010 WRITE(6,6035) (P(I,L),L=1,LTSTOP)
*
      STOP
*
****       THESE ARE THE FORMATS
*
 2010 FORMAT(12F6.2)
 2020 FORMAT(12I6)
 6000 FORMAT(1H1)
 6010 FORMAT(5X,*C,TTH,B,TGK,THO,TMEM=*,6F7.2)
 6012 FORMAT(5X,*A1,A2,A3,E01,E02,E03=*6F7.2)
 6014 FORMAT(5X,*A12,A13,A21,A23,A31,A32=*6F7.2)
 6020 FORMAT(5X,*LTSTOP=*,I7)
 6030 FORMAT(//)
 6035 FORMAT(5X,20F6.2)
      END
/EOR
   0.   25.   20.    5.   10.    5.
   2.   1.75  1.5    .8    .5    .2
   0.    0.    0.    0.    0.    0.
    100
/EOR
/EOI


/JOB
A224,CM=70000.
FAKE.
PDFV.
MNF(Y).
LGO.
/EOR
```

```
        PROGRAM POOL11 (INPUT,OUTPUT,TAPE6=OUTPUT)
*
****        THIS PROGRAM SIMULATES A POOL OF PACEMAKER NEURONS
*           WHICH INTERACT SYNAPTICALLY WITH ARBITRARY INTER-
****        CONNECTIONS AND TIME DELAYS.
*
        PARAMETER NCLS=3,MCTP1=6,NTMX=2,STEP=1.
        DIMENSION A(NCLS),P(NCLS,140),E0(NCLS)
        COMMON/SV/E(NCLS),TH(NCLS),S(NCLS),GK(NCLS),B,C,THO,TMEM,DCG,
       *        DCTH,SC(NCLS)
        COMMON/TR/NT(NCLS),IDT(NCLS,NTMX),NCT(NCLS,NTMX),STR(NCLS,
       *        NTMX),SCIT(NCLS,MCTP1)
*
****        THIS SECTION READS AND WRITES THE INPUT PARAMETERS
*
        READ 2010, C,TTH,B,TGK,THO,TMEM
        READ 2010, ((A(I),E0(I)),I=1,NCLS)
        DO 120 I=1,NCLS
        READ 2020, NT(I)
  120 READ 2025, ((IDT(I,J),NCT(I,J),STR(I,J)),J=1,NT(I))
        READ 2020, LTSTOP
*
        WRITE(6,6000)
        WRITE(6,6010) C,TTH,B,TGK,THO,TMEM
        WRITE(6,6012) ((A(I),E0(I)),I=1,NCLS)
        DO 210 I=1,NCLS
        WRITE(6,6020) NT(I)
        DO 220 J=1,NT(I)
  220 WRITE(6,6025) IDT(I,J),NCT(I,J),STR(I,J)
  210 CONTINUE
        WRITE(6,6022) LTSTOP
*
****        THIS SECTION INITIALIZES VARIABLES
*
        DCTH=EXP(-STEP/TTH) $DCG=EXP(-STEP/TGK)
        DO 310 I=1,NCLS
        E(I)=E0(I) $TH(I)=THO $GK(I)=0. $S(I)=0.
        DO 305 J=1,MCTP1
  305 SCIT(I,J)=0.
  310 CONTINUE
*
****        THIS SECTION UPDATES STATE VARIABLES AT EACH TIME STEP
*
        DO 1000 L=1,LTSTOP
*
        DO 1500 I=1,NCLS
        SC(I)=A(I)+SCIT(I,1)
        CALL SVUPDT(I)
        IF(E(I).LT.TH(I)) GO TO 1490
        S(I)=1.
        CALL TRNSMT(I)
 1490 P(I,L)=E(I)+S(I)*(50.-E(I))
 1500 CONTINUE
*
        DO 1200 I=1,NCLS
        DO 1210 J=1,MCTP1-1
 1210 SCIT(I,J)=SCIT(I,J+1)
 1200 SCIT(I,MCTP1)=0.
*
 1000 CONTINUE
*
****        THIS SECTION WRITES OUT ACTIVITY VARIABLES
*
        DO 4010 I=1,NCLS
        WRITE(6,6030)
 4010 WRITE(6,6035) (P(I,L),L=1,LTSTOP)
*
```

```
      STOP
****          THESE ARE THE FORMATS
*
 2010 FORMAT(12F6.2)
 2020 FORMAT(12I6)
 2025 FORMAT(2I6,F6.2)
 6000 FORMAT(1H1)
 6010 FORMAT(5X,*C,TTH,B,TGK,THO,TMEM=*,6F7.2)
 6012 FORMAT(5X,*  (A,E0)= *,16F7.2)
 6020 FORMAT(5X,*NT=*,I7)
 6022 FORMAT(5X,*LTSTOP=*,I7)
 6025 FORMAT(10X,*IDT,NCT,STR=*,2I7,F7.2)
 6030 FORMAT(//)
 6035 FORMAT(5X,20F6.2)
      END
*
      SUBROUTINE SVUPDT(K)
*
***   THIS SUBROUTINE UPDATES THE STATE VARIABLES OF AN INDIVIDUAL NEURON
*
      PARAMETER NCLS=3,STEP=1.,EK=-10.
      COMMON/SV/E(NCLS),TH(NCLS),S(NCLS),GK(NCLS),B,C,THO,TMEM,DCG,
     *         DCTH,SC(NCLS)
*
      GK(K)=GK(K)*DCG+B*(1.-DCG)*S(K)  $DCE=EXP(-(1.+GK(K))*STEP/TMEM)
      E(K)=E(K)*DCE+(SC(K)+GK(K)*EK)*(1.-DCE)/(1.+GK(K))
      TH(K)=THO+(TH(K)-THO)*DCTH+C*E(K)*(1.-DCTH)
      S(K)=0.
      RETURN
      END
*
      SUBROUTINE TRNSMT(K)
*
****          THIS SUBROUTINE PROJECTS SYNAPTIC ACTIVATION FROM A SENDING
****          FIBER TO ITS TARGETED RECEIVING CELLS.
*
      PARAMETER NCLS=3,MCTP1=6,NTMX=2
      COMMON/TR/NT(NCLS),IDT(NCLS,NTMX),NCT(NCLS,NTMX),STR(NCLS,
     *         NTMX),SCIT(NCLS,MCTP1)
*
      DO 100 J=1,NT(K)
      SCIT(IDT(K,J),NCT(K,J)+1)=SCIT(IDT(K,J),NCT(K,J)+1)+STR(K,J)
  100 CONTINUE
      RETURN
      END
*
/EOR
   0.   25.   20.    5.   10.    5.
   2.     .8  1.75    .5   1.5    .2
      2
      2    1  0.
      3    1  0.
      2
      1    1  0.
      3    1  0.
      2
      1    1  0.
      2    1  0.
    100
/EOR
/EOI


/JOB
A224,CM=70000.
FAKE.
```

```
PDFV.
MNF(Y).
LGO.
/EOR
      PROGRAM POOL12 (INPUT,OUTPUT,TAPE6=OUTPUT)
*
****      THIS PROGRAM SIMULATES A POOL OF PACEMAKER NEURONS
*         WHICH INTERACT SYNAPTICALLY AND/OR VIA EXTRACELLULAR
****      CURRENT COUPLING.
*
      PARAMETER NCLS=3,MCTP1=6,NTMX=2,NCMX=2,STEP=1.
      DIMENSION A(NCLS),P(NCLS,140),E0(NCLS),NC(NCLS),IDC(NCLS,NCMX),
     *          AC(NCLS,NCMX)
      COMMON/SV/E(NCLS),TH(NCLS),S(NCLS),GK(NCLS),B,C,THO,TMEM,DCG,
     *          DCTH,SC(NCLS)
      COMMON/TR/NT(NCLS),IDT(NCLS,NTMX),NCT(NCLS,NTMX),STR(NCLS,
     *          NTMX),SCIT(NCLS,MCTP1)
*
****      THIS SECTION READS AND WRITES THE INPUT PARAMETERS
*
      READ 2010, C,TTH,B,TGK,THO,TMEM
      READ 2010, ((A(I),E0(I)),I=1,NCLS)
      DO 120 I=1,NCLS
      READ 2020, NT(I),NC(I)
      READ 2025, ((IDT(I,J),NCT(I,J),STR(I,J)),J=1,NT(I))
  120 READ 2015, ((IDC(I,J),AC(I,J)),J=1,NC(I))
      READ 2020, LTSTOP
*
      WRITE(6,6000)
      WRITE(6,6010) C,TTH,B,TGK,THO,TMEM
      WRITE(6,6012) ((A(I),E0(I)),I=1,NCLS)
      DO 210 I=1,NCLS
      WRITE(6,6020) NT(I),NC(I)
      DO 220 J=1,NT(I)
  220 WRITE(6,6025) IDT(I,J),NCT(I,J),STR(I,J)
      WRITE(6,6022) $WRITE(6,6023) ((AC(I,J),IDC(I,J)),J=1,NC(I))
  210 CONTINUE
      WRITE(6,6028) LTSTOP
*
****      THIS SECTION INITIALIZES VARIABLES
*
      DCTH=EXP(-STEP/TTH) $DCG=EXP(-STEP/TGK)
      DO 310 I=1,NCLS
      E(I)=E0(I) $TH(I)=THO $GK(I)=0. $S(I)=0.
      DO 305 J=1,MCTP1
  305 SCIT(I,J)=0.
  310 CONTINUE
*
****      THIS SECTION UPDATES STATE VARIABLES AT EACH TIME STEP
*
      DO 1000 L=1,LTSTOP
*
      DO 1200 I=1,NCLS
      SC(I)=A(I) $DO 1200 J=1,NC(I)
 1200 SC(I)=SC(I)+AC(I,J)*E(IDC(I,J))
*
      DO 1500 I=1,NCLS
      SC(I)=SC(I)+SCIT(I,1)
      CALL SVUPDT(I)
      IF(E(I).LT.TH(I)) GO TO 1490
      S(I)=1.
      CALL TRNSMT(I)
 1490 P(I,L)=E(I)+S(I)*(50.-E(I))
 1500 CONTINUE
*
      DO 1600 I=1,NCLS
      DO 1610 J=1,MCTP1-1
```

```
 1610 SCIT(I,J)=SCIT(I,J+1)
 1600 SCIT(I,MCTP1)=0.
*
 1000 CONTINUE
*
**** THIS SECTION WRITES OUT ACTIVITY VARIABLES
*
      DO 4010 I=1,NCLS
      WRITE(6,6030)
 4010 WRITE(6,6035) (P(I,L),L=1,LTSTOP)
*
      STOP
**** THESE ARE THE FORMATS
*
 2010 FORMAT(12F6.2)
 2015 FORMAT(6(I6,F6.2))
 2020 FORMAT(12I6)
 2025 FORMAT(4(2I6,F6.2))
 6000 FORMAT(1H1)
 6010 FORMAT(5X,*C,TTH,B,TGK,THO,TMEM=*,6F7.2)
 6012 FORMAT(5X,* (A,E0)= *,16F7.2)
 6020 FORMAT(5X,*NT=*,I7)
 6022 FORMAT(5X,*COUPLERS:*)
 6023 FORMAT(7X,10(F6.2,I6))
 6025 FORMAT(10X,*IDT,NCT,STR=*,2I7,F7.2)
 6028 FORMAT(5X,*LTSTOP=*,I7)
 6030 FORMAT(//)
 6035 FORMAT(5X,20F6.2)
      END
*
      SUBROUTINE SVUPDT(K)
*
*** THIS SUBROUTINE UPDATES THE STATE VARIABLES OF AN INDIVIDUAL NEURON
*
      PARAMETER NCLS=3,STEP=1.,EK=-10.
      COMMON/SV/E(NCLS),TH(NCLS),S(NCLS),GK(NCLS),B,C,THO,TMEM,DCG,
     *          DCTH,SC(NCLS)
*
      GK(K)=GK(K)*DCG+B*(1.-DCG)*S(K) $DCE=EXP(-(1.+GK(K))*STEP/TMEM)
      E(K)=E(K)*DCE+(SC(K)+GK(K)*EK)*(1.-DCE)/(1.+GK(K))
      TH(K)=THO+(TH(K)-THO)*DCTH+C*E(K)*(1.-DCTH)
      S(K)=0.
      RETURN
      END
*
      SUBROUTINE TRNSMT(K)
*
**** THIS SUBROUTINE PROJECTS SYNAPTIC ACTIVATION FROM A SENDING
**** FIBER TO ITS TARGETED RECEIVING CELLS.
*
      PARAMETER NCLS=3,MCTP1=6,NTMX=2
      COMMON/TR/NT(NCLS),IDT(NCLS,NTMX),NCT(NCLS,NTMX),STR(NCLS,
     *          NTMX),SCIT(NCLS,MCTP1)
*
      DO 100 J=1,NT(K)
      SCIT(IDT(K,J),NCT(K,J)+1)=SCIT(IDT(K,J),NCT(K,J)+1)+STR(K,J)
  100 CONTINUE
      RETURN
      END
*
/EOR
  0.    25.    20.    5.    10.    5.
  2.     .8    1.75   .5    1.5    .2
     2     2
     2     1   9.         3     1  9.
     2    .25        3    .25
     2     2
```

```
      1     1   9.        3     1   9.
      1    .25       3   .25
      2     2
      1     1   9.        2     1   9.
      1    .25       2   .25
  100
/EOR
/EOI


/JOB
A224,CM=70000,TL=300.
FAKE.
PDFV.
MNF(Y).
LGO,PL006.
SAVE,PL006.
EXIT.
SAVE,PL006.
/EOR
      PROGRAM POOL20 (PLOUT,INPUT,OUTPUT,TAPE6=OUTPUT,TAPE7=PLOUT)
*
****       THIS PROGRAM SIMULATES A TWO-DIMENIONAL POOL OF PACEMAKER
****       NEURONS WHICH INTERACT VIA EXTRACELLULAR CURRENT COUPLING.
*
      PARAMETER NCLS=800,STEP=1.,NSTEPS=1,EK=-10.,NDSPY=2,NTG=4,
     *          NC=40,NR=20
      DIMENSION A(NCLS),E0(NCLS),SC(NCLS),P(NDSPY,500),ND(NDSPY),
     *          E(NCLS),TH(NCLS),S(NCLS),GK(NCLS),DCE(NCLS),V(NCLS),
     *          IDT(NTG),ALINE(NC,NR),CHAR(2)
*
****       THIS SECTION READS AND WRITES THE INPUT PARAMETERS
*
      READ 2040, IFLG1
      READ 2010, C,TTH,B,TGK,THO,TMEM
      READ 2020, A0,E00,INSED
      CALL RANSET(INSED) $DO 120 I=1,NCLS
      A(I)=-A0*ALOG(RANF(0.)) $E0(I)=-E00*ALOG(RANF(0.))
  120 CONTINUE
      READ 2050, ((IDT(I),A(IDT(I)),E0(IDT(I))),I=1,NTG)
      READ 2010, FCP,SPCP
      READ 2040, ((ND(I)),I=1,NDSPY)
      READ 2040, LTSTOP
*
      WRITE(6,6000)
      WRITE(6,6005) NCLS,STEP,NSTEPS,EK,NDSPY,NTG,IFLG1
      WRITE(6,6010) C,TTH,B,TGK,THO,TMEM
      WRITE(6,6020) A0,E00,INSED
      WRITE(6,6050) ((IDT(I),A(IDT(I)),E0(IDT(I))),I=1,NTG)
      WRITE(6,6030) NC,NR,FCP,SPCP
      WRITE(6,6045) (ND(I),I=1,NDSPY)
      WRITE(6,6040) LTSTOP
*
****       THIS SECTION INITIALIZES VARIABLES
*
      DCTH=EXP(-STEP/TTH) $DCG=EXP(-STEP/TGK)
      DO 310 I=1,NCLS
      E(I)=E0(I) $TH(I)=THO $GK(I)=0. $S(I)=0. $SC(I)=0.
      DCE(I)=EXP(-STEP/FLOAT(NSTEPS)/TMEM)
  310 CONTINUE
      CHAR(1)=1H  $CHAR(2)=1H*
*
****       THIS SECTION UPDATES STATE VARIABLES AT EACH TIME STEP
*
      DO 1000 L=1,LTSTOP
*
```

```
****                    THIS SUBSECTION ALLOWS MORE FINE-GRAINED INTEGRATION
*
      DO 1200 J=1,NSTEPS
      DO 1220 I=1,NCLS
      SC(I)=A(I)+SC(I)
 1220 V(I)=E(I)*DCE(I)+(SC(I)+GK(I)*EK)*(1.-DCE(I))/(1.+GK(I))
*
      DO 1240 I=1,NCLS $X=0.
      K=I-1 $IF(MOD(K,NC).GT.0) X=X+FCP*(V(K)-V(I))+SPCP*S(K)
      K=I+1 $IF(MOD(I,NC).GT.0) X=X+FCP*(V(K)-V(I))+SPCP*S(K)
      K=I-NC $IF(K.GT.0) X=X+FCP*(V(K)-V(I))+SPCP*S(K)
      K=I+NC $IF(K.LE.NCLS) X=X+FCP*(V(K)-V(I))+SPCP*S(K)
 1240 SC(I)=X
*
      DO 1260 I=1,NCLS
 1260 E(I)=V(I)
*
 1200 CONTINUE
*
      DO 1300 I=1,NCLS
      TH(I)=THO+(TH(I)-THO)*DCTH+C*E(I)*(1.-DCTH)
      JX=MOD((I-1),NC)+1 $JY=(I-1)/NC+1
      IF(E(I).GE.TH(I)) THEN $S(I)=1. $ALINE(JX,JY)=CHAR(2)
      ELSE $S(I)=0. $ALINE(JX,JY)=CHAR(1) $ENDIF
      GK(I)=GK(I)*DCG+B*(1.-DCG)*S(I)
      DCE(I)=EXP(-(1.+GK(I))*STEP/FLOAT(NSTEPS)/TMEM)
 1300 CONTINUE
*
      DO 1400 I=1,NDSPY
 1400 P(I,L)=E(ND(I))+S(ND(I))*(50.-E(ND(I)))
*
      IF(IFLG1.GT.0) THEN
      WRITE(6,6055) (E(ND(I)),I=1,NDSPY) $DO 4110 JY=1,NR
 4110 WRITE(6,6060) (ALINE(JX,NR+1-JY),JX=1,NC) $ENDIF
*
 1000 CONTINUE
*
****          THIS SECTION WRITES OUT ACTIVITY VARIABLES
*
      DO 4010 I=1,NDSPY
      WRITE(6,6033) $WRITE(7,6033)
      WRITE(7,7055) ((P(ND(I),L)),L=1,LTSTOP)
 4010 WRITE(6,6035) ((P(ND(I),L)),L=1,LTSTOP)
*
      STOP
****          THESE ARE THE FORMATS
*
 2010 FORMAT(12F6.2)
 2020 FORMAT(2F6.2,I6)
 2040 FORMAT(12I6)
 2050 FORMAT(4(I6,2F6.2))
 6000 FORMAT(1H1)
 6005 FORMAT(5X,*NCLS,STEP,NSTEPS,EK,NDSPY,NTG,IFLG1=*,I7,F7.2,I7,F7.2,
     *           3I7)
 6010 FORMAT(5X,*C,TTH,B,TGK,THO,TMEM=*,6F7.2)
 6020 FORMAT(5X,*A0,E00,INSED=*,2F7.2,I7)
 6030 FORMAT(5X,*NC,NR,FCP,SPCP=*,2I7,2F7.2)
 6040 FORMAT(5X,*LTSTOP=*,I7)
 6033 FORMAT(//)
 6035 FORMAT(5X,20F6.2)
 6045 FORMAT(5X,*ND=*,12I7)
 6050 FORMAT(5X,*IDT,A,E0=*,5(I7,2F7.2))
 6055 FORMAT(3X,*POT=*,12(F7.2))
 6060 FORMAT(3X,120A1)
 7055 FORMAT(12F6.2)
      END
/EOR
```

```
        1
    0.   25.    20.     5.    10.     5.
   15.    2.  734692
   380 25.     5.       381 25.     5.      420 25.     5.       421 25.      5.
   .0    .0
    420    500
     20
/EOR
/EOI



/JOB
A224,CM=70000,TL=200.
FAKE.
PDFV.
MNF(Y).
LGO,LTS01.
SAVE,LTS01.
EXIT.
SAVE,LTS01.
/EOR
      PROGRAM JNCTN10 (JNOUT,INPUT,OUTPUT,TAPE7=JNOUT,TAPE6=OUTPUT)
*
****      THIS PROGRAM SIMULATES THE ACTIVATION OF A POOL OF NEURONS
****      BY AN INPUT FIBER SYSTEM.
*
      PARAMETER NFIB=50,NCLS=50,MCTP1=3,STEP=1.,
     *          EX=70.,EI=-10.,EK=-10.
      DIMENSION SC(NCLS,MCTP1),E(NCLS),TH(NCLS),S(NCLS),GK(NCLS),
     *          IDFF(NFIB),IDFC(NCLS),STR(NFIB),
     *          PER(NFIB),LSTM(NFIB),INSED(3)
*
****      THIS SECTION READS AND WRITES THE INPUT PARAMETERS
*
      READ 5010, NT,INSED(2),INSED(3),LTSTOP
      READ 5020, C,TTH,B,TGK,THO,TMEM
      READ 5030, STR0,PER0,LSTM0,INSED(1)
      CALL RANSET(INSED(1))
      DO 110 I=1,NFIB
      STR(I)=-STR0*ALOG(RANF(0.))
      PER(I)=-PER0*ALOG(RANF(0.))
      LSTM(I)=INT(-FLOAT(LSTM0)*ALOG(RANF(0.)))+1
  110 CONTINUE
*
      WRITE(6,6000)
      WRITE(6,6010) NFIB,NCLS,NT,INSED(2),INSED(3),LTSTOP
      WRITE(6,6020) C,TTH,B,TGK,THO,TMEM
      WRITE(6,6030) STR0,PER0,LSTM0,INSED(1)
      WRITE(6,6040)
*
****      THIS SECTION INITIALIZES VARIABLES
*
      DO 310 I=1,NCLS
      E(I)=0. $TH(I)=THO $S(I)=0. $GK(I)=0.
      DO 309 J=1,MCTP1
  309 SC(I,J)=0.
  310 CONTINUE
      DCTH=EXP(-STEP/TTH) $DCG=EXP(-STEP/TGK)
*
****      THIS SECTION UPDATES STATE VARIABLES AT EACH TIME STEP
*
      DO 1000 L=1,LTSTOP
      NFF=0 $NFC=0 $GE=0. $GI=0.
*
      DO 2000 I=1,NFIB
      IF(L.NE.LSTM(I)) GO TO 2000
```

```
      CALL RANSET(INSED(2))
      NFF=NFF+1 $IDFF(NFF)=I
      LSTM(I)=L+INT(-PER(I)*ALOG(RANF(Ø.)))+1
      CALL RANGET(INSED(2)) $CALL RANSET(I*INSED(3))
      DO 2010 J=1,NT
      NREC=INT(RANF(Ø.)*FLOAT(NCLS))+1$NCT=INT(RANF(Ø.)*FLOAT(MCTP1-1))
     *          +2
 2010 SC(NREC,NCT)=SC(NREC,NCT)+STR(I)
 2000 CONTINUE
*
      DO 3000 I=1,NCLS
      GK(I)=GK(I)*DCG+B*(1.-DCG)*S(I)
      GTOT=1.+GE+GI+GK(I) $DCE=EXP(-GTOT*STEP/TMEM)
      E(I)=E(I)*DCE+(SC(I,1)+GE*EX+GI*EI+GK(I)*EK)*(1.-DCE)/GTOT
      TH(I)=THO+(TH(I)-THO)*DCTH+C*E(I)*(1.-DCTH) $S(I)=Ø.
      IF(E(I).LT.TH(I)) GO TO 3000
      S(I)=1. $NFC=NFC+1 $IDFC(NFC)=I
*
 3000 CONTINUE
*
      WRITE(6,6050) L,NFF,NFC,(IDFF(I),I=1,NFF),(IDFC(J),J=1,NFC)
      WRITE(7,7050) L,NFF,NFC,(IDFF(I),I=1,NFF),(IDFC(J),J=1,NFC)
      DO 4010 I=1,NCLS $DO 4020 J=1,MCTP1-1
 4020 SC(I,J)=SC(I,J+1)
 4010 SC(I,MCTP1)=Ø.
*
 1000 CONTINUE
*
      STOP
****          THESE ARE THE FORMATS
*
 5010 FORMAT(12I6)
 5020 FORMAT(12F6.2)
 5030 FORMAT(2F6.2,2I6)
 6000 FORMAT(1H1)
 6010 FORMAT(5X,*NFIB,NCLS,NT,INSEDS,LTSTOP=*,6I6)
 6020 FORMAT(5X,*C,TTH,B,TGK,THO,TMEM=*,6F8.2)
 6030 FORMAT(5X,*STRØ,PERØ,LSTMØ,INSEDØ=*,2F7.2,2I7)
 6040 FORMAT(///)
 6050 FORMAT(5X,30I4)
 7050 FORMAT(30I4)
      END
/EOR
    10  1227  1936    10
   .ØØ 20.   20.    5.   10.     5.
   2.    5.     25  4748
/EOR
/EOI




/JOB
A224,CM=70000,TL=200.
FAKE.
PDFV.
MNF(Y).
LGO,JNØØ1.
SAVE,JNØØ1.
EXIT.
SAVE,JNØØ1.
/EOR
      PROGRAM JNCTN11 (JNOUT,INPUT,OUTPUT,TAPE7=JNOUT,TAPE6=OUTPUT)
*
****          THIS PROGRAM SIMULATES THE ACTIVATION OF A POOL OF NEURONS
****          BY AN INPUT FIBER SYSTEM. A SIMPLE GRAPHIC DISPLAY IS ADDED.
*
      PARAMETER NFIB=50,NCLS=50,MCTP1=2,STEP=1.
```

```
      DIMENSION IDFF(NFIB),IDFC(NCLS),CHAR(2),ALINE(110),
     *           PER(NFIB),LSTM(NFIB)
      COMMON SC(NCLS,MCTP1)
      COMMON/SV/E(NCLS),TH(NCLS),S(NCLS),GK(NCLS),B,C,THO,TMEM,DCG,DCTH
      COMMON/TR/ NT,STR(NFIB),INSED(3)
*
****        THIS SECTION READS AND WRITES THE INPUT PARAMETERS
*
      READ 5010, NT,INSED(2),INSED(3),LTSTOP
      READ 5020, C,TTH,B,TGK,THO,TMEM
      READ 5030, STRØ,PERØ,LSTMØ,INSED(1)
      CALL RANSET(INSED(1))
      DO 110 I=1,NFIB
      STR(I)=-STRØ*ALOG(RANF(Ø.))
      PER(I)=-PERØ*ALOG(RANF(Ø.))
      LSTM(I)=INT(-FLOAT(LSTMØ)*ALOG(RANF(Ø.)))+1
  110 CONTINUE
*
      WRITE(6,6000)
      WRITE(6,6010) NFIB,NCLS,NT,INSED(2),INSED(3),LTSTOP
      WRITE(6,6020) C,TTH,B,TGK,THO,TMEM
      WRITE(6,6030) STRØ,PERØ,LSTMØ,INSED(1)
      WRITE(6,6040)
*
****        THIS SECTION INITIALIZES VARIABLES
*
      DO 310 I=1,NCLS
      E(I)=Ø. $TH(I)=THO $S(I)=Ø. $GK(I)=Ø.
      DO 309 J=1,MCTP1
  309 SC(I,J)=Ø.
  310 CONTINUE
      CHAR(1)=1H  $CHAR(2)=1H*
      DCTH=EXP(-STEP/TTH) $DCG=EXP(-STEP/TGK)
*
****        THIS SECTION UPDATES STATE VARIABLES AT EACH TIME STEP
*
      DO 1000 L=1,LTSTOP
      DO 1010 I=1,110
 1010 ALINE(I)=CHAR(1) $NFF=Ø $NFC=Ø
*
      DO 2000 I=1,NFIB
      IF(L.NE.LSTM(I)) GO TO 2000
      CALL RANSET(INSED(2))
      NFF=NFF+1 $IDFF(NFF)=I $ALINE(I)=CHAR(2)
      LSTM(I)=L+INT(-PER(I)*ALOG(RANF(Ø.)))+1
      CALL RANGET(INSED(2)) $CALL TRNSMT(I)
 2000 CONTINUE
*
      DO 3000 I=1,NCLS
      CALL SVUPDT(I)
      IF(E(I).LT.TH(I)) GO TO 3000
      S(I)=1. $NFC=NFC+1 $IDFC(NFC)=I $ALINE(60+I)=CHAR(2)
 3000 CONTINUE
*
      WRITE(6,6050) L,NFF,(ALINE(I),I=1,50),NFC,(ALINE(I),I=61,110)
      WRITE(7,7010) L,NFF,NFC,(IDFF(I),I=1,NFF),(IDFC(J),J=1,NFC)
      DO 4010 I=1,NCLS $DO 4020 J=1,MCTP1-1
 4020 SC(I,J)=SC(I,J+1)
 4010 SC(I,MCTP1)=Ø.
*
 1000 CONTINUE
*
      STOP
****        THESE ARE THE FORMATS
*
 5010 FORMAT(12I6)
 5020 FORMAT(12F6.2)
```

```
 5030 FORMAT(2F6.2,2I6)
 6000 FORMAT(1H1)
 6010 FORMAT(5X,*NFIB,NCLS,NT,INSEDS,LTSTOP=*,6I6)
 6020 FORMAT(5X,*C,TTH,B,TGK,THO,TMEM=*,6F8.2)
 6030 FORMAT(5X,*STR0,PER0,LSTM0,INSED0=*,2F7.2,2I7)
 6040 FORMAT(///)
 6050 FORMAT(2X,2I4,2X,50A1,2X,I4,50A1)
 7010 FORMAT(30I4)
      END
*
*
      SUBROUTINE SVUPDT(K)
*
***   THIS SUBROUTINE UPDATES THE STATE VARIABLES OF AN INDIVIDUAL NEURON
*
      PARAMETER NCLS=50,MCTP1=2,STEP=1.,EK=-10.
      COMMON SC(NCLS,MCTP1)
      COMMON/SV/E(NCLS),TH(NCLS),S(NCLS),GK(NCLS),B,C,THO,TMEM,DCG,DCTH
*
      GK(K)=GK(K)*DCG+B*(1.-DCG)*S(K)  $DCE=EXP(-(1.+GK(K))*STEP/TMEM)
      E(K)=E(K)*DCE+(SC(K,1)+GK(K)*EK)*(1.-DCE)/(1.+GK(K))
      TH(K)=THO+(TH(K)-THO)*DCTH+C*E(K)*(1.-DCTH)
      S(K)=0.
      RETURN
      END
*
      SUBROUTINE TRNSMT(K)
      PARAMETER NFIB=50,NCLS=50,MCTP1=2
      COMMON SC(NCLS,MCTP1)
      COMMON/TR/ NT,STR(NFIB),INSED(3)
*
****      THIS SUBROUTINE PROJECTS SYNAPTIC ACTIVATION FROM A SENDING
****      FIBER TO ITS TARGETED RECEIVING CELLS.
*
      CALL RANSET (K*INSED(3))
      DO 10 J=1,NT
      NREC=INT(RANF(0.)*FLOAT(NCLS))+1
      NCT=INT(RANF(0.)*FLOAT(MCTP1-1))+2
   10 SC(NREC,NCT)=SC(NREC,NCT)+STR(K)
*
      RETURN
      END
*
/EOR
    50  1227  1936    200
   .00 20.   50.      .5 10.       .1
   .1   5.       25  4748
/EOR
/EOI


/JOB
A224,CM=70000,TL=200.
FAKE.
PDFV.
MNF(Y).
LGO,JN001.
SAVE,JN001.
EXIT.
SAVE,JN001.
/EOR
      PROGRAM JNCTN12 (JNOUT,INPUT,OUTPUT,TAPE7=JNOUT,TAPE6=OUTPUT)
*
****      THIS PROGRAM SIMULATES THE ACTIVATION OF A POOL OF NEURONS
*         BY AN ARBITRARY NUMBER OF INPUT FIBER SYSTEMS.  A SIMPLE
****      GRAPHIC DISPLAY IS INCLUDED.
```

```
*
      PARAMETER NFPOPS=2,NFIB=50,NCLS=50,MCTP1=2,STEP=1.
      DIMENSION T(NFPOPS),N(NFPOPS),P(NFPOPS),
     *         INSTRT(NFPOPS),INSTP(NFPOPS),INFSED(NFPOPS),
     *         NF(NFPOPS),IDFF(NFPOPS,NFIB),IDFC(NCLS),CHAR(2),
     *         ALINE(110),PER(NFPOPS),LSTM(NFPOPS),DCS(NFPOPS)
      COMMON G(NFPOPS,NCLS,MCTP1)
      COMMON/SV/E(NCLS),TH(NCLS),S(NCLS),GK(NCLS),B,C,THO,TMEM,DCG,DCTH
     *         ,EQ(NFPOPS)
      COMMON/TR/ NCT(NFPOPS),NT(NFPOPS),STR(NFPOPS),INSED(NFPOPS)
*
****      THIS SECTION READS AND WRITES THE INPUT PARAMETERS
*
      READ 5020, C,TTH,B,TGK,THO,TMEM
      DO 110 I=1,NFPOPS
  110 READ 5030, EQ(I),T(I),N(I),P(I),INSTRT(I),INSTP(I),INFSED(I),
     *         NCT(I),NT(I),STR(I),INSED(I)
      READ 5010, LTSTOP
*
      WRITE(6,6000)
      WRITE(6,6010) NFPOPS,NCLS,LTSTOP
      WRITE(6,6020) C,TTH,B,TGK,THO,TMEM
      DO 210 I=1,NFPOPS
  210 WRITE(6,6030) I,EQ(I),T(I),N(I),P(I),INSTRT(I),INSTP(I),INFSED(I),
     *         NCT(I),NT(I),STR(I),INSED(I)
      WRITE(6,6040)
*
****      THIS SECTION INITIALIZES VARIABLES
*
      DO 310 I=1,NCLS
      E(I)=0. $TH(I)=1. $S(I)=0. $GK(I)=0.
      DO 309 K=1,NFPOPS $DO 309 J=1,MCTP1
  309 G(K,I,J)=0.
  310 CONTINUE
      CHAR(1)=1H  $CHAR(2)=1H*
      DCTH=EXP(-STEP/TTH) $DCG=EXP(-STEP/TGK)
      DO 320 I=1,NFPOPS
  320 DCS(I)=EXP(-STEP/T(I))
*
****      THIS SECTION UPDATES STATE VARIABLES AT EACH TIME STEP
*
      DO 1000 L=1,LTSTOP
      NFC=0 $DO 1010 I=1,110
 1010 ALINE(I)=CHAR(1)
*
      DO 2000 I=1,NFPOPS $NF(I)=0
      IF(L.LT.INSTRT(I).OR.L.GE.INSTP(I)) GO TO 2000
      CALL RANSET(INFSED(I)) $DO 1900 J=1,N(I)
      IF(RANF(0.).GT.P(I)) GO TO 1900
      CALL RANGET(INFSED(I))
      NF(I)=NF(I)+1 $IDFF(I,NF(I))=J $CALL TRNSMT(I,J)
      IF(I.EQ.1) ALINE(J)=CHAR(2)
      CALL RANSET(INFSED(I))
 1900 CONTINUE $CALL RANGET(INFSED(I))
 2000 CONTINUE
*
      DO 3000 I=1,NCLS
      CALL SVUPDT(I)
      IF(E(I).LT.TH(I)) GO TO 3000
      S(I)=1. $NFC=NFC+1 $IDFC(NFC)=I $ALINE(60+I)=CHAR(2)
 3000 CONTINUE
*
      WRITE(6,6050) L,NF(1),(ALINE(I),I=1,50),NFC,(ALINE(I),I=61,110)
      WRITE(7,7010) L,(NF(I),I=1,NFPOPS),NFC,((IDFF(I,J),J=1,NF(I)),
     *         I=1,NFPOPS),(IDFC(J),J=1,NFC)
*
      DO 4010 K=1,NFPOPS $DO 4010 I=1,NCLS
```

```
      G(K,I,1)=G(K,I,1)*DCS(K)+G(K,I,2) $DO 4020 J=2,MCTP1-1
      G(K,I,J)=G(K,I,J+1)
 4020 CONTINUE $G(K,I,MCTP1)=0.
 4010 CONTINUE
*
 1000 CONTINUE
*
      STOP
****        THESE ARE THE FORMATS
*
 5010 FORMAT(12I6)
 5020 FORMAT(12F6.2)
 5030 FORMAT (2F6.2,I6,F6.2,5I6,F6.2,I6)
 6000 FORMAT(1H1)
 6010 FORMAT(5X,*NFPOPS,NCLS,LTSTOP=*,3I6)
 6020 FORMAT(5X,*C,TTH,B,TGK,THO,TMEM=*,6F8.2)
 6030 FORMAT(5X,*I,EQ,T,N,P,INSTRT,INSTP,INFSED,NCT,NT,STR,INSED=:*,
      *        I6,2F6.2,I6,F6.2,5I6,F6.2,I6)
 6040 FORMAT(///)
 6050 FORMAT(2X,2I4,2X,50A1,2X,I4,50A1)
 7010 FORMAT(30I4)
      END
*
*
      SUBROUTINE SVUPDT(K)
*
***  THIS SUBROUTINE UPDATES THE STATE VARIABLES OF AN INDIVIDUAL NEURON
*
      PARAMETER NFPOPS=2,NCLS=50,MCTP1=2,STEP=1.,EK=-10
      COMMON G(NFPOPS,NCLS,MCTP1)
      COMMON/SV/E(NCLS),TH(NCLS),S(NCLS),GK(NCLS),B,C,THO,TMEM,DCG,DCTH
      *          ,EQ(NFPOPS)
*
      GS=0. $SYN=0. $DO 10 I=1,NFPOPS
      GS=GS+G(I,K,1) $SYN=SYN+G(I,K,1)*EQ(I)
   10 CONTINUE
      GK(K)=GK(K)*DCG+B*(1.-DCG)*S(K) $GTOT=1.+GK(K)+GS
      DCE=EXP(-GTOT*STEP/TMEM)
      E(K)=E(K)*DCE+(SYN+GK(K)*EK)*(1.-DCE)/GTOT
      TH(K)=THO+(TH(K)-THO)*DCTH+C*E(K)*(1.-DCTH)
      S(K)=0.
      RETURN
      END
*
      SUBROUTINE TRNSMT(K,M)
      PARAMETER NFPOPS=2,NCLS=50,MCTP1=2
      COMMON G(NFPOPS,NCLS,MCTP1)
      COMMON/TR/ NCT(NFPOPS),NT(NFPOPS),STR(NFPOPS),INSED(NFPOPS)
*
****        THIS SUBROUTINE PROJECTS SYNAPTIC ACTIVATION FROM A SENDING
****        FIBER TO ITS TARGETED RECEIVING CELLS.
*
      CALL RANSET(M*INSED(K))
      DO 10 J=1,NT(K)
      NREC=INT(RANF(0.)*FLOAT(NCLS))+1
      KCT=INT(RANF(0.)*FLOAT(NCT(K)))+2
   10 G(K,NREC,KCT)=G(K,NREC,KCT)+STR(K)
      CALL RANGET(M*INSED(K))
*
      RETURN
      END
*
/EOR
   .00 20.    50.     .5 10.      .1
 70.   1.     50   .1      0   999 1227    1    5   .5    4748
-10.   5.     50   .1     15    50 1936    1    4  2.    1010
    60
```

```
/EOR
/EOI


/JOB
A224,CM=70000,TL=50.
FAKE.
PDFV.
MNF(Y).
LGO,JN002.
SAVE,JN002.
EXIT.
SAVE,JN002.
/EOR
      PROGRAM JNCTN20 (JNOUT,INPUT,OUTPUT,TAPE7=JNOUT,TAPE6=OUTPUT)
*
****      THIS PROGRAM SIMULATES THE ACTIVATION OF A TWO-DIMENSIONALLY
*         ARRANGED POPULATION OF NEURONS BY A TWO-DIMENSIONALLY ARRANGED
****      SYSTEM OF INPUT FIBERS.
*
      INTEGER W $REAL HTF
      PARAMETER NFIB=20,NCLS=30,MCTP1=6,W=6,H=5,STEP=1.,
     *          EX=70.,EI=-10.,EK=-10.
      DIMENSION E(NCLS),TH(NCLS),S(NCLS),GK(NCLS),
     *          IDFF(NFIB),IDFC(NCLS),CHAR(2),ALINE(70),
     *          PER(NFIB),LSTM(NFIB)
      COMMON/TR/ NT,STR(NFIB),NCF,NRF,NCC,NRC,WTF,HTF,INSED(2),
     *           SC(NCLS,MCTP1)
*
****      THIS SECTION READS AND WRITES THE INPUT PARAMETERS
*
      READ 5010, NT,INSED(1),INSED(2),LTSTOP
      READ 5020, C,TTH,B,TGK,THO,TMEM
      READ 5030, STR0,PER0,LSTM0,INSED(1)
      CALL RANSET(INSED(1))
      DO 110 I=1,NFIB
      STR(I)=-STR0*ALOG(RANF(0.))
      PER(I)=-PER0*ALOG(RANF(0.))
      LSTM(I)=INT(-FLOAT(LSTM0)*ALOG(RANF(0.)))+1
  110 CONTINUE
      READ 5035, NCF,NRF,NCC,NRC,WTF,HTF
*
      WRITE(6,6000)
      WRITE(6,6010) NFIB,NCLS,NT,INSED(1),INSED(2),LTSTOP
      WRITE(6,6020) C,TTH,B,TGK,THO,TMEM
      WRITE(6,6030) STR0,PER0,LSTM0,INSED(1)
      WRITE(6,6035) NCF,NRF,NCC,NRC,WTF,HTF
      WRITE(6,6040)
*
****      THIS SECTION INITIALIZES VARIABLES
*
      DO 310 I=1,NCLS
      E(I)=0. $TH(I)=THO $S(I)=0. $GK(I)=0.
      DO 309 J=1,MCTP1
  309 SC(I,J)=0.
  310 CONTINUE
      CHAR(1)=1H  $CHAR(2)=1H*
      DCTH=EXP(-STEP/TTH) $DCG=EXP(-STEP/TGK)
*
****      THIS SECTION UPDATES STATE VARIABLES AT EACH TIME STEP
*
      DO 1000 L=1,LTSTOP
      DO 1010 I=1,70
 1010 ALINE(I)=CHAR(1) $NFF=0 $NFC=0 $GE=0. $GI=0.
*
      DO 2000 I=1,NFIB
```

```
      IF(L.NE.LSTM(I)) GO TO 2000
      CALL RANSET(INSED(1))
      NFF=NFF+1 $IDFF(NFF)=I $ALINE(I)=CHAR(2)
      LSTM(I)=L+INT(-PER(I)*ALOG(RANF(0.)))+1
      CALL RANGET(INSED(1)) $CALL TRNSMT(I)
 2000 CONTINUE
*
      DO 3000 I=1,NCLS
      GK(I)=GK(I)*DCG+B*(1.-DCG)*S(I)
      GTOT=1.+GE+GI+GK(I) $DCE=EXP(-GTOT*STEP/TMEM)
      E(I)=E(I)*DCE+(SC(I,1)+GE*EX+GI*EI+GK(I)*EK)*(1.-DCE)/GTOT
      TH(I)=THO+(TH(I)-THO)*DCTH+C*E(I)*(1.-DCTH) $S(I)=0.
      IF(E(I).LT.TH(I)) GO TO 3000
      S(I)=1. $NFC=NFC+1 $IDFC(NFC)=I $ALINE(40+I)=CHAR(2)
 3000 CONTINUE
*
      WRITE(6,6050) L,NFF,NFC,(ALINE(I),I=1,70)
      WRITE(7,7010) L,NFF,NFC,(IDFF(I),I=1,NFF),(IDFC(J),J=1,NFC)
      DO 4010 I=1,NCLS $DO 4020 J=1,MCTP1-1
 4020 SC(I,J)=SC(I,J+1)
 4010 SC(I,MCTP1)=0.
*
 1000 CONTINUE
*
      STOP
****          THESE ARE THE FORMATS
*
 5010 FORMAT(12I6)
 5020 FORMAT(12F6.2)
 5030 FORMAT(2F6.2,2I6)
 5035 FORMAT(4I6,2F6.2)
 6000 FORMAT(1H1)
 6010 FORMAT(5X,*NFIB,NCLS,NT,INSEDS,LTSTOP=*,6I6)
 6020 FORMAT(5X,*C,TTH,B,TGK,THO,TMEM=*,6F8.2)
 6030 FORMAT(5X,*STR0,PER0,LSTM0,INSED0=*,2F7.2,2I7)
 6035 FORMAT(5X,*NCF,NRF,NCC,NRC,WTF,HTF=*,4I7,2F7.2)
 6040 FORMAT(///)
 6050 FORMAT(2X,3I4,2X,70A1)
 7010 FORMAT(30I4)
      END
*
      SUBROUTINE TRNSMT(K)
      INTEGER W $REAL HTF
      PARAMETER NFIB=20,NCLS=30,MCTP1=6,W=6,H=5
      COMMON/TR/ NT,STR(NFIB),NCF,NRF,NCC,NRC,WTF,HTF,INSED(2),
     *           SC(NCLS,MCTP1)
*
****      THIS SUBROUTINE PROJECTS SYNAPTIC ACTIVATION FROM A SENDING
****      FIBER TO ITS TARGETED RECEIVING CELLS.
*
      XF=FLOAT(MOD(K,NCF))*FLOAT(W)/FLOAT(NCF)
      IF(XF.EQ.0.) XF=FLOAT(W)
      YF=FLOAT(K/NCF+1)*FLOAT(H)/FLOAT(NRF)
      CALL RANSET(K*INSED(2))
      DO 100 J=1,NT
      XT=XF+(RANF(0.)-.5)*WTF $YT=YF+(RANF(0.)-.5)*HTF
      ICC=INT(FLOAT(NCC)*(XT-.5)/FLOAT(W))+1
      IRC=INT(FLOAT(NRC)*(YT-.5)/FLOAT(H))+1
      IF(ICC.LE.0.OR.ICC.GT.NCC) GO TO 80
      IF(IRC.LE.0.OR.IRC.GT.NRC) GO TO 80
      IDC=NCC*(IRC-1)+ICC
      NCT=INT(RANF(0.)*MCTP1)+1
      SC(IDC,NCT)=SC(IDC,NCT)+STR(K)
   80 CONTINUE
  100 CONTINUE
*
      RETURN
```

```
      END
/EOR
    10   1227   1936     20
   .00 20.    20.     5.   10.    5.
  1.00  5.       1
  1.00  5.       2
  1.00  5.       3
  1.00  5.       4
  1.00  5.       5
  1.00  6.       6
  1.00  6.       7
  1.00  7.       8
  1.00  7.       9
  1.00  4.      10
  1.00  4.       1
  1.00  8.       2
  1.00  4.       3
  1.00  5.       4
  1.00  6.       5
  1.00  7.       6
  1.00  5.       7
  1.00  6.       8
  1.00  4.       9
  1.00  9.      10
     5      4      6     5 3.    2.
/EOR
/EOI


/JOB
A224,CM=75000,TL=1000.
FAKE.
PDFV.
MNF(Y,PL=10000).
SETLIM,PR.
LGO,JN007.
SAVE,JN007.
EXIT.
SAVE,JN007.
/EOR
      PROGRAM JNCTN21 (JNOUT,INPUT,OUTPUT,TAPE7=JNOUT,TAPE6=OUTPUT)
*
****      THIS PROGRAM SIMULATES THE ACTIVATION OF A TWO-DIMENSIONALLY
*         ARRANGED POPULATION OF NEURONS BY A TWO-DIMENSIONALLY ARRANGED
****      SYSTEM OF INPUT FIBERS.  A SIMPLE GRAPHIC DISPLAY IS ADDED.
*
      INTEGER W $REAL HTF
      PARAMETER NFIB=800,NCLS=800,MCTP1=6,W=40,H=20,IFLG1=0,NTGS=4,
     *      STEP=1.,EX=70.,EI=-10.,EK=-10.
********** ADJUST 6060 AND 6070 FORMATS TO WA1, AND WI2, ETC ********
      DIMENSION E(NCLS),TH(NCLS),S(NCLS),GK(NCLS),
     *      IDFF(NFIB),IDFC(NCLS),CHAR(2),ALINE(2,W,H),LINE(2,W,H),
     *      PER(NFIB),LSTM(NFIB),NFF(140),NFC(140),ITG(NTGS)
      COMMON/TR/ NT,STR(NFIB),NCF,NRF,NCC,NRC,WTF,HTF,INSED(3),
     *           SC(NCLS,MCTP1)
*
****      THIS SECTION READS AND WRITES THE INPUT PARAMETERS
*
      READ 5020, C,TTH,B,TGK,THO,TMEM
      READ 5035, NCF,NRF,NCC,NRC,WTF,HTF
      READ 5010, NT,INSED(2),INSED(3),LTSTOP
      READ 5030, STR0,PER0,LSTM0,INSED(1)
      CALL RANSET(INSED(1))
      DO 110 I=1,NFIB
      STR(I)=-STR0*ALOG(RANF(0.))
      PER(I)=-PER0*ALOG(RANF(0.))
```

```
      LSTM(I)=INT(-FLOAT(LSTM0)*ALOG(RANF(0.)))+1
  110 CONTINUE
      DO 105 I=1,NTGS
  105 READ 5040, ITG(I),STR(ITG(I)),PER(ITG(I)),LSTM(ITG(I))
*
      WRITE(6,6000)
      WRITE(6,6010) NFIB,NCLS,NT,INSED(2),INSED(3),LTSTOP
      WRITE(6,6020) C,TTH,B,TGK,THO,TMEM
      WRITE(6,6030) STR0,PER0,LSTM0,INSED(1)
      WRITE(6,6035) NCF,NRF,NCC,NRC,WTF,HTF
      WRITE(6,6040)
*
****        THIS SECTION INITIALIZES VARIABLES
*
      DO 310 I=1,NCLS
      E(I)=0. $TH(I)=THO $S(I)=0. $GK(I)=0.
      DO 309 J=1,MCTP1
  309 SC(I,J)=0.
  310 CONTINUE
      CHAR(1)=1H $CHAR(2)=1H*
      DO 320 I=1,2 $DO 320 J=1,W $DO 320 K=1,H
  320 LINE(I,J,K)=0
      DCTH=EXP(-STEP/TTH) $DCG=EXP(-STEP/TGK)
*
****        THIS SECTION UPDATES STATE VARIABLES AT EACH TIME STEP
*
      DO 1000 L=1,LTSTOP
      DO 1010 I=1,2 $DO 1010 J=1,W $DO 1010 K=1,H
 1010 ALINE(I,J,K)=CHAR(1) $NFF(L)=0 $NFC(L)=0 $GE=0. $GI=0.
*
      DO 2000 I=1,NFIB
      IF(L.NE.LSTM(I)) GO TO 2000
      JX=INT(FLOAT(MOD((I-1),NCF)+1)*FLOAT(W)/FLOAT(NCF))
      JY=INT(FLOAT((I-1)/NCF+1)*FLOAT(H)/FLOAT(NRF))
      CALL RANSET(INSED(2))
      NFF(L)=NFF(L)+1 $IDFF(NFF(L))=I $ALINE(1,JX,JY)=CHAR(2)
      LINE(1,JX,JY)=LINE(1,JX,JY)+1
      LSTM(I)=L+INT(-PER(I)*ALOG(RANF(0.)))+1
      CALL RANGET(INSED(2)) $CALL TRNSMT(I)
 2000 CONTINUE
*
      DO 3000 I=1,NCLS
      GK(I)=GK(I)*DCG+B*(1.-DCG)*S(I)
      GTOT=1.+GE+GI+GK(I) $DCE=EXP(-GTOT*STEP/TMEM)
      E(I)=E(I)*DCE+(SC(I,1)+GE*EX+GI*EI+GK(I)*EK)*(1.-DCE)/GTOT
      TH(I)=THO+(TH(I)-THO)*DCTH+C*E(I)*(1.-DCTH) $S(I)=0.
      IF(E(I).LT.TH(I)) GO TO 3000
      JX=INT(FLOAT(MOD((I-1),NCC)+1)*FLOAT(W)/FLOAT(NCC))
      JY=INT(FLOAT((I-1)/NCC+1)*FLOAT(H)/FLOAT(NRC))
      S(I)=1. $NFC(L)=NFC(L)+1 $IDFC(NFC(L))=I $ALINE(2,JX,JY)=CHAR(2)
      LINE(2,JX,JY)=LINE(2,JX,JY)+1
 3000 CONTINUE
*
      IF(IFLG1.EQ.0) GO TO 4120
      IF(MOD(L,2).EQ.1) WRITE (6,6000)
      WRITE(6,6050) L,NFF(L),NFC(L)
      WRITE(6,6052) (IDFF(I),I=1,NFF(L))
      WRITE(6,6052) (IDFC(I),I=1,NFC(L))
      DO 4110 JY=1,H
 4110 WRITE(6,6060)  (ALINE(1,JX,H+1-JY),JX=1,W),(ALINE(2,JX,H+1-JY),JX=
     *               1,W)
 4120 CONTINUE
      WRITE(7,7010) L,NFF(L),NFC(L),(IDFF(I),I=1,NFF(L)),(IDFC(J)
     *              ,J=1,NFC(L))
      DO 4010 I=1,NCLS $DO 4020 J=1,MCTP1-1
 4020 SC(I,J)=SC(I,J+1)
 4010 SC(I,MCTP1)=0.
```

```
*
 1000 CONTINUE
*
****          THIS SECTION WRITES OUT ACTIVITY VARIABLES
*
      WRITE(6,6000)
      WRITE(6,6052)  (NFF(L),L=1,LTSTOP)
      WRITE(6,6052)  (NFC(L),L=1,LTSTOP)
      WRITE(6,6040) $DO 4210 JY=1,H
 4210 WRITE(6,6070)  (LINE(1,JX,H+1-JY),JX=1,W)
      WRITE(6,6040) $DO 4220 JY=1,H
 4220 WRITE(6,6070)  (LINE(2,JX,H+1-JY),JX=1,W)
*
      STOP
****          THESE ARE THE FORMATS
*
 5010 FORMAT(12I6)
 5020 FORMAT(12F6.2)
 5030 FORMAT(2F6.2,2I6)
 5035 FORMAT(4I6,2F6.2)
 5040 FORMAT(I6,2F6.2,I6)
 6000 FORMAT(1H1)
 6010 FORMAT(5X,*NFIB,NCLS,NT,INSEDS,LTSTOP=*,6I6)
 6020 FORMAT(5X,*C,TTH,B,TGK,THO,TMEM=*,6F8.2)
 6030 FORMAT(5X,*STR0,PER0,LSTM0,INSED0=*,2F7.2,2I7)
 6035 FORMAT(5X,*NCF,NRF,NCC,NRC,WTF,HTF=*,4I7,2F7.2)
 6040 FORMAT(///)
 6050 FORMAT(////,2X,*L,NFF=*,3I7,4X,*HERE ARE THE FIBERS AND CELLS:*)
 6052 FORMAT(3X,20I5)
 6060 FORMAT(3X,40A1,30X,40A1)
 6070 FORMAT(3X,40I3)
 7010 FORMAT(20I4)
      END
*
      SUBROUTINE TRNSMT(K)
      INTEGER W $REAL HTF
      PARAMETER NFIB=800,NCLS=800,MCTP1=6,W=40,H=20
      COMMON/TR/ NT,STR(NFIB),NCF,NRF,NCC,NRC,WTF,HTF,INSED(3),
     *           SC(NCLS,MCTP1)
*
****          THIS SUBROUTINE PROJECTS SYNAPTIC ACTIVATION FROM A SENDING
****          FIBER TO ITS TARGETED RECEIVING CELLS.
*
      XF=FLOAT(MOD(K,NCF))*FLOAT(W)/FLOAT(NCF)
      YF=FLOAT(K/NCF+1)*FLOAT(H)/FLOAT(NRF)
      CALL RANSET(K*INSED(3))
      DO 100 J=1,NT
      XT=XF+(RANF(0.)-.5)*WTF $YT=YF+(RANF(0.)-.5)*HTF
      ICC=INT(FLOAT(NCC)*(XT-.5)/FLOAT(W))+1
      IRC=INT(FLOAT(NRC)*(YT-.5)/FLOAT(H))+1
      IF(ICC.LE.0.OR.ICC.GT.NCC) GO TO 80
      IF(IRC.LE.0.OR.IRC.GT.NRC) GO TO 80
      IDC=NCC*(IRC-1)+ICC
      NCT=INT(RANF(0.)*FLOAT(MCTP1-1))+2
      SC(IDC,NCT)=SC(IDC,NCT)+STR(K)
   80 CONTINUE
  100 CONTINUE
*
      RETURN
      END
/EOR
   .00 20.    20.     5.    10.     5.
    40    20     40     20 13.    13.
    20  1227   1936     60
  1.  800.     800   4748
   298  1.    10.      3
   303  1.    10.      6
```

```
   498 1.   10.       1
   503 1.   10.       4
/EOR
/EOI


/JOB
A224,CM=60000.
FAKE.
PDFV.
MNF(Y).
LGO,SST003.
SAVE,SST003.
EXIT.
SAVE,SST003.
/EOR
      PROGRAM NTWRK10 (NTOUT,INPUT,OUTPUT,TAPE7=NTOUT,TAPE6=OUTPUT)
*
****       THIS PROGRAM SIMULATES THE ACTIVITY OF A RECURRENTLY-CONNECTED
****       POOL OF EXCITATORY AND INHIBITORY NEURONS.
*
      PARAMETER NEX=300,NIN=50,MCTP1=6,STEP=1.,EI=-10.,EK=-10.
      DIMENSION E1(NEX),TH1(NEX),S1(NEX),GK1(NEX),E2(NIN),
     *          TH2(NIN),S2(NIN),GK2(NIN),SC(NEX),SCET(NEX,MCTP1),
     *          GI(NEX,MCTP1),SCIT(NIN,MCTP1),NEF(140),NIF(140)
*
****       THIS SECTION READS AND WRITES THE INPUT PARAMETERS
*
      READ 2010, C1,TTH1,B1,TGK1,THO1,TMEM1
      READ 2010, C2,TTH2,B2,TGK2,THO2,TMEM2
      READ 2020, NT11,NT12,NT21
      READ 2020, NCT11,NCT12,NCT21
      READ 2010, STR11,STR12,STR21
      READ 2020 ,INSED11,INSED12,INSED21,INSED22
      READ 2030, P,STRIP,LIPSTP
      READ 2020, LTSTOP
*
      WRITE(6,6000)
      WRITE(6,6010) C1,TTH1,B1,TGK1,THO1,TMEM1
      WRITE(6,6015) C2,TTH2,B2,TGK2,THO2,TMEM2
      WRITE(6,6020) NT11,NT12,NT21
      WRITE(6,6022) NCT11,NCT12,NCT21
      WRITE(6,6024) STR11,STR12,STR21
      WRITE(6,6030) P,STRIP,LIPSTP
      WRITE(6,6035) INSED11,INSED12,INSED21,INSED22
      WRITE(6,6040) LTSTOP
*
****       THIS SECTION INITIALIZES VARIABLES
*
      DCTH1=EXP(-STEP/TTH1) $DCTH2=EXP(-STEP/TTH2)
      DCG1=EXP(-STEP/TGK1)  $DCG2= EXP(-STEP/TGK2)
      DO 110 I=1,NEX
      E1(I)=0. $TH1(I)=THO1 $S1(I)=0. $GK1(I)=0.
      DO 115 J=1,MCTP1
      SCET(I,J)=0. $GI(I,J)=0.
  115 CONTINUE
  110 CONTINUE
      DO 120 I=1,NIN
      E2(I)=0. $TH2(I)=THO2 $S2(I)=0. $GK2(I)=0.
      DO 125 J=1,MCTP1
      SCIT(I,J)=0.
  125 CONTINUE
  120 CONTINUE
*
****       THIS SECTION UPDATES STATE VARIABLES AT EACH TIME STEP
*
```

```
      DO 1000 L=1,LTSTOP
      NEF(L)=0 $NIF(L)=0
*
****              UPDATING EXCITATORY CELLS
*
      CALL RANSET(INSED22)
      DO 1500 I=1,NEX
      SC(I)=SCET(I,1)
      IF(L.LE.LIPSTP.AND.RANF(0.).LE.P) SC(I)=SC(I)+STRIP
      GK1(I)=GK1(I)*DCG1+B1*(1.-DCG1)*S1(I) $GTOT=1.+GK1(I)+GI(I,1)
      DCE=EXP(-GTOT*STEP/TMEM1)
      E1(I)=E1(I)*DCE+(SC(I)+GK1(I)*EK+GI(I,1)*EI)*(1.-DCE)/GTOT
      TH1(I)=THO1+(TH1(I)-THO1)*DCTH1+C1*E1(I)*(1.-DCTH1)
      S1(I)=0. $IF(E1(I).LT.TH1(I)) GO TO 1490
      CALL RANGET(INSED22)
      S1(I)=1. $NEF(L)=NEF(L)+1 $CALL RANSET(I*INSED11)
      DO 1410 J=1,NT11
      NREC=INT(RANF(0.)*FLOAT(NEX))+1
      NCT=INT(RANF(0.)*FLOAT(NCT11))+2
      SCET(NREC,NCT)=SCET(NREC,NCT)+STR11
 1410 CONTINUE
      CALL RANSET(I*INSED12)
      DO 1420 J=1,NT12
      NREC=INT(RANF(0.)*FLOAT(NIN))+1
      NCT=INT(RANF(0.)*FLOAT(NCT12))+2
      SCIT(NREC,NCT)=SCIT(NREC,NCT)+STR12
 1420 CONTINUE $CALL RANSET(INSED22)
 1490 CONTINUE
 1500 CONTINUE $CALL RANGET(INSED22)
*
****              UPDATING INHIBITORY CELLS
*
      DO 1700 I=1,NIN
      GK2(I)=GK2(I)*DCG2+B2*(1.-DCG2)*S2(I) $GTOT=1.+GK2(I)
      DCE=EXP(-GTOT*STEP/TMEM2)
      E2(I)=E2(I)*DCE+(SCIT(I,1)+GK2(I)*EK)*(1.-DCE)/GTOT
      TH2(I)=THO2+(TH2(I)-THO2)*DCTH2+C2*E2(I)*(1.-DCTH2)
      S2(I)=0. $IF(E2(I).LT.TH2(I)) GO TO 1690
      S2(I)=1. $NIF(L)=NIF(L)+1 $CALL RANSET(I*INSED21)
      DO 1610 J=1,NT21
      NREC=INT(RANF(0.)*FLOAT(NEX))+1
      NCT=INT(RANF(0.)*FLOAT(NCT21))+2
      GI(NREC,NCT)=GI(NREC,NCT)+STR21
 1610 CONTINUE
 1690 CONTINUE
 1700 CONTINUE
*
      DO 1910 I=1,NEX $DO 1920 J=1,MCTP1-1
      SCET(I,J)=SCET(I,J+1) $GI(I,J)=GI(I,J+1)
 1920 CONTINUE $SCET(I,MCTP1)=0. $GI(I,MCTP1)=0.
 1910 CONTINUE
      DO 1940 I=1,NIN $DO 1950 J=1,MCTP1-1
      SCIT(I,J)=SCIT(I,J+1)
 1950 CONTINUE $GI(I,MCTP1)=0.
 1940 CONTINUE
*
 1000 CONTINUE
*
****         THIS SECTION WRITES OUT ACTIVITY VARIABLES
*
      WRITE(6,6105) $WRITE(6,6110) (NEF(L),L=1,LTSTOP)
      WRITE(6,6115) $WRITE(6,6110) (NIF(L),L=1,LTSTOP)
      WRITE(7,7110) (NEF(L),L=1,LTSTOP)
      WRITE(7,7105) $WRITE(7,7110) (NIF(L),L=1,LTSTOP)
*
      STOP
****         THESE ARE THE FORMATS
```

```
*
 2010 FORMAT(12F6.2)
 2020 FORMAT(12I6)
 2030 FORMAT(2F6.2,I6)
 6000 FORMAT(1H1)
 6010 FORMAT(5X,*EXC:C,TTH,B,TGK,THO,TMEM=*,6F7.2)
 6015 FORMAT(5X,*INH:C,TTH,B,TGK,THO,TMEM=*,6F7.2)
 6020 FORMAT(5X,*NT11,NT12,NT21=*,3I6)
 6022 FORMAT(5X,*NCT11,NCT12,NCT21=*,3I6)
 6024 FORMAT(5X,*STR11,STR12,STR21=*,3F7.2)
 6030 FORMAT(5X,*P,STRIP,LIPSTP=*,2F7.2,I7)
 6035 FORMAT(5X,*INSEEDS=*,4I7)
 6040 FORMAT(5X,*LTSTOP=*,I7)
 6105 FORMAT(/,10X,*HERE ARE THE OUTPUTS:*)
 6110 FORMAT(5X,20I6)
 6115 FORMAT(/)
 7105 FORMAT(/)
 7110 FORMAT(20I4)
      END
/EOR
   0.    25.    20.    5.    10.    5.
   0.    25.    10.    4.    10.    10.
     20      4     10
      1      1      1
   1.    1.    1.
   5629   7143   3517   2243
    .05 10.        10
     30
/EOR
/EOI



/JOB
A224,CM=60000,TL=700.
FAKE.
PDFV.
MNF(Y).
LGO,SST081.
SAVE,SST081.
EXIT.
SAVE,SST081.
/EOR
      PROGRAM NTWRK11 (NTOUT,INPUT,OUTPUT,TAPE7=NTOUT,TAPE6=OUTPUT)
*
****      THIS PROGRAM SIMULATES THE ACTIVITY OF A RECURRENTLY-CONNECTED
****      POOL OF EXCITATORY AND INHIBITORY NEURONS.
*
      PARAMETER NCLS=600,NEX=300,NIN=300,MCTP1=6,STEP=1.
      DIMENSION NEF(140),NIF(140),T(140),TTH(2),TGK(2)
      COMMON SCIT(NCLS,MCTP1),GI(NCLS,MCTP1)
      COMMON/SV/ E(NCLS),TH(NCLS),S(NCLS),GK(NCLS),B(2),C(2),
     *           DCG(2),DCTH(2),THO(2),TMEM(2)
      COMMON/TR/ NT(2,2),NCT(2,2),STR(2,2),NCL(2),INSED(2,2)
*
****      THIS SECTION READS AND WRITES THE INPUT PARAMETERS
*
      READ 2010, C(1),TTH(1),B(1),TGK(1),THO(1),TMEM(1)
      READ 2010, C(2),TTH(2),B(2),TGK(2),THO(2),TMEM(2)
      READ 2020, NT(1,1),NT(1,2),NT(2,1)
      READ 2020, NCT(1,1),NCT(1,2),NCT(2,1)
      READ 2010, STR(1,1),STR(1,2),STR(2,1)
      READ 2020, INSED(1,1),INSED(1,2),INSED(2,1),INSED(2,2)
      READ 2030, P,STRIP,LIPSTP
      READ 2020, LTSTOP
*
      WRITE(6,6000)
```

```
        WRITE(6,6010)  C(1),TTH(1),B(1),TGK(1),THO(1),TMEM(1)
        WRITE(6,6015)  C(2),TTH(2),B(2),TGK(2),THO(2),TMEM(2)
        WRITE(6,6020)  NT(1,1),NT(1,2),NT(2,1)
        WRITE(6,6022)  NCT(1,1),NCT(1,2),NCT(2,1)
        WRITE(6,6024)  STR(1,1),STR(1,2),STR(2,1)
        WRITE(6,6030)  P,STRIP,LIPSTP
        WRITE(6,6035)  INSED(1,1),INSED(1,2),INSED(2,1),INSED(2,2)
        WRITE(6,6040)  LTSTOP
*
****          THIS SECTION INITIALIZES VARIABLES
*
        NCL(1)=NEX $NCL(2)=NIN
        DO 102 I=1,2
        DCTH(I)=EXP(-STEP/TTH(I)) $DCG(I)=EXP(-STEP/TGK(I))
  102 CONTINUE
        DO 110 I=1,NCLS
        E(I)=0. $S(I)=0. $GK(I)=0.
        IF(I.LE.NEX) THEN $TH(I)=THO(1)
        ELSE $TH(I)=THO(2) $ENDIF
        DO 115 J=1,MCTP1
        SCIT(I,J)=0. $GI(I,J)=0.
  115 CONTINUE
  110 CONTINUE
*
****          THIS SECTION UPDATES STATE VARIABLES AT EACH TIME STEP
*
        DO 1000 L=1,LTSTOP
        NEF(L)=0 $NIF(L)=0
*
****                    UPDATING THE EXCITATORY CELLS
*
        CALL RANSET(INSED(2,2))
        DO 1500 I=1,NEX
        IF(L.LE.LIPSTP.AND.RANF(0.).LE.P) SCIT(I,1)=SCIT(I,1)+STRIP
        CALL SVUPDT(1,I)
        IF(E(I).LT.TH(I)) GO TO 1490
        S(I)=1. $NEF(L)=NEF(L)+1 $CALL RANGET(INSED(2,2))
        CALL TRNSMT(1,1,I)
        CALL TRNSMT(1,2,I) $CALL RANSET(INSED(2,2))
 1490 CONTINUE
 1500 CONTINUE $CALL RANGET(INSED(2,2))
*
****                    UPDATING THE INHIBITORY CELLS
*
        DO 1700 I=NEX+1,NCLS
        CALL SVUPDT(2,I)
        S(I)=0. $IF(E(I).LT.TH(I)) GO TO 1690
        S(I)=1. $NIF(L)=NIF(L)+1
        CALL TRNSMT(2,1,I-NEX)
 1690 CONTINUE
 1700 CONTINUE
*
        DO 1910 I=1,NCLS $DO 1920 J=1,MCTP1-1
        SCIT(I,J)=SCIT(I,J+1) $GI(I,J)=GI(I,J+1)
 1920 CONTINUE $SCIT(I,MCTP1)=0. $GI(I,MCTP1)=0.
 1910 CONTINUE
        T(L)=L
*
 1000 CONTINUE
*
****          THIS SECTION WRITES OUT ACTIVITY VARIABLES
*
        WRITE(6,6105) $WRITE(6,6110) (NEF(L),L=1,LTSTOP)
        WRITE(6,6115) $WRITE(6,6110) (NIF(L),L=1,LTSTOP)
        WRITE(7,7110) (NEF(L),L=1,LTSTOP)
        WRITE(7,7105) $WRITE(7,7110) (NIF(L),L=1,LTSTOP)
*
```

```
      STOP
****          THESE ARE THE FORMATS
*
 2010 FORMAT(12F6.2)
 2020 FORMAT(12I6)
 2030 FORMAT(2F6.2,I6)
 6000 FORMAT(1H1)
 6010 FORMAT(5X,*EXC:C,TTH,B,TGK,THO,TMEM=*,6F7.2)
 6015 FORMAT(5X,*INH:C,TTH,B,TGK,THO,TMEM=*,6F7.2)
 6020 FORMAT(5X,*NT11,NT12,NT21=*,3I6)
 6022 FORMAT(5X,*NCT11,NCT12,NCT21=*,3I6)
 6024 FORMAT(5X,*STR11,STR12,STR21=*,3F7.2)
 6030 FORMAT(5X,*P,STRIP,LIPSTP=*,2F7.2,I7)
 6035 FORMAT(5X,*INSEDS=*,4I7)
 6040 FORMAT(5X,*LTSTOP=*,I7)
 6105 FORMAT(/,10X,*HERE ARE THE OUTPUTS:*)
 6110 FORMAT(5X,20I6)
 6115 FORMAT(/)
 7105 FORMAT(/)
 7110 FORMAT(20I4)
      END
*
      SUBROUTINE SVUPDT(IS,K)
*
***   THIS SUBROUTINE UPDATES THE STATE VARIABLES OF AN INDIVIDUAL NEURON
*
      PARAMETER NCLS=600,NEX=300,MCTP1=6,STEP=1.,EI=-10.,EK=-10.
      COMMON SCIT(NCLS,MCTP1),GI(NCLS,MCTP1)
      COMMON/SV/ E(NCLS),TH(NCLS),S(NCLS),GK(NCLS),B(2),C(2),
     *           DCG(2),DCTH(2),THO(2),TMEM(2)
*
      GK(K)=GK(K)*DCG(IS)+B(IS)*(1.-DCG(IS))*S(K)$GTOT=1.+GK(K)+GI(K,1)
      DCE=EXP(-GTOT*STEP/TMEM(IS))
      E(K)=E(K)*DCE+(SCIT(K,1)+GI(K,1)*EI+GK(K)*EK)*(1.-DCE)/GTOT
      TH(K)=THO(IS)+(TH(K)-THO(IS))*DCTH(IS)+C(IS)*E(K)*(1.-DCTH(IS))
      S(K)=0.
      RETURN
      END
*
      SUBROUTINE TRNSMT(IS,IR,K)
****      THIS SUBROUTINE PROJECTS SYNAPTIC ACTIVATION FROM A SENDING
****      FIBER TO ITS TARGETED RECEIVING CELLS.
*
      PARAMETER NCLS=600,NEX=300,MCTP1=6
      COMMON SCIT(NCLS,MCTP1),GI(NCLS,MCTP1)
      COMMON/TR/NT(2,2),NCT(2,2),STR(2,2),NCL(2),INSED(2,2)
*
      CALL RANSET(K*INSED(IS,IR))
      DO 100 J=1,NT(IS,IR)
      NREC=INT(RANF(0.)*FLOAT(NCL(IR)))+1
      IF(IR.EQ.2) NREC=NREC+NEX
      KCT=INT(RANF(0.)*FLOAT(NCT(IS,IR)))+2
      IF(IS.EQ.1) SCIT(NREC,KCT)=SCIT(NREC,KCT)+STR(IS,IR)
      IF(IS.EQ.2) GI(NREC,KCT)=GI(NREC,KCT)+STR(IS,IR)
  100 CONTINUE
      RETURN
      END
*
/EOR
   0.    25.   10.    2.    10.    5.
   0.    25.   10.    2.    10.    5.
     12     26     12
      4      2      2
   2.     1.     .35
  5629   7143   3517   2243
   .06 10.        4
   100
```

```
/EOR
/EOI


/JOB
A224,CM=117000,TL=600.
FAKE.
PDFV.
MNF(Y).
LGO,JN216.
SAVE,JN216.
EXIT.
SAVE,JN216.
/EOR
      PROGRAM NTWRK21 (NTOUT,INPUT,OUTPUT,TAPE7=NTOUT,TAPE6=OUTPUT)
*
****  THIS PROGRAM SIMULATES THE ACTIVITY OF A RECURRENTLY-CONNECTED
*     AND SPATIALLY-ORGANIZED NETWOK OF EXCITATORY AND INHIBITORY
****  NEURONS ACTIVATED BY A SPATIALLY-ORGANIZED INPUT SYSTEM OF FIBERS.
*
      INTEGER W,H,HX,HY
      PARAMETER NFIB=800,NCLS=1600,NEX=800,NIN=800,MCTP1=5,W=40,H=20,
     *          IFLG1=0,NTGS=0,STEP=1.
*
**********    ADJUST 6060 & 6070 FORMATS TO WA1, WI3 ETC    **********
*
      DIMENSION NFF(300),NEF(300),NIF(300),TTH(2),TGK(2),
     *          ALINE(3,W,H),LINE(3,W,H),CHAR(2),PER(NFIB),LSTM(NFIB),
     *          IDF(3,NEX),ITG(NFIB)
      COMMON SCIT(NCLS,MCTP1),GI(NCLS,MCTP1)
      COMMON/SV/ E(NCLS),TH(NCLS),S(NCLS),GK(NCLS),B(2),C(2),
     *           DCG(2),DCTH(2),THO(2),TMEM(2)
      COMMON/TR/ NT(3,3),NCT(3,3),STR(3,3),NCL(3),INSED(3,3),
     *           WTF(3,3),HTF(3,3),NC(3),NR(3)
*
****       THIS SECTION READS AND WRITES THE INPUT PARAMETERS
*
      READ 2010, C(1),TTH(1),B(1),TGK(1),THO(1),TMEM(1)
      READ 2010, C(2),TTH(2),B(2),TGK(2),THO(2),TMEM(2)
      READ 2020, ((NC(I),NR(I)),I=1,3)
      READ 2020, NT(1,2),NT(1,3),NT(2,2),NT(2,3),NT(3,2)
      READ 2020, NCT(1,2),NCT(1,3),NCT(2,2),NCT(2,3),NCT(3,2)
      READ 2010, STR(1,2),STR(1,3),STR(2,2),STR(2,3),STR(3,2)
      READ 2010, WTF(1,2),WTF(1,3),WTF(2,2),WTF(2,3),WTF(3,2)
      READ 2010, HTF(1,2),HTF(1,3),HTF(2,2),HTF(2,3),HTF(3,2)
      READ 2020, INSED(1,2),INSED(1,3),INSED(2,2),INSED(2,3),
     *           INSED(3,2)
      READ 2030, PER0,LSTM0,INSED(1,1),INSTOP
      CALL RANSET(INSED(1,1))
      DO 105 I=1,NFIB
      PER(I)=-PER0*ALOG(RANF(0.))
      LSTM(I)=INT(-FLOAT(LSTM0)*ALOG(RANF(0.)))+1
  105 CONTINUE
      READ 2020, LX,HX,LY,HY
      IF(LX.LE.1.AND.HX.GE.W.AND.LY.LE.1.AND.HY.GE.H) GO TO 108
      DO 109 I=1,NFIB
      JX=INT(FLOAT(MOD((I-1),NC(1))+1)*FLOAT(W)/FLOAT(NC(1)))
      JY=INT(FLOAT((I-1)/NC(1)+1)*FLOAT(H)/FLOAT(NR(1)))
      IF(JX.LT.LX.OR.JX.GT.HX.OR.JY.LT.LY.OR.JY.GT.HY) LSTM(I)=1000000
  109 CONTINUE
  108 CONTINUE
      DO 106 I=1,NTGS
  106 READ 2040, ITG(I),PER(ITG(I)),LSTM(ITG(I))
      READ 2020, LTSTOP
*
      WRITE(6,6000)
```

```
      WRITE(6,6010) C(1),TTH(1),B(1),TGK(1),THO(1),TMEM(1)
      WRITE(6,6015) C(2),TTH(2),B(2),TGK(2),THO(2),TMEM(2)
      WRITE(6,6020) NT(1,2),NT(1,3),NT(2,2),NT(2,3),NT(3,2)
      WRITE(6,6022) NCT(1,2),NCT(1,3),NCT(2,2),NCT(2,3),NCT(3,2)
      WRITE(6,6024) STR(1,2),STR(1,3),STR(2,2),STR(2,3),STR(3,2)
      WRITE(6,6026) WTF(1,2),WTF(1,3),WTF(2,2),WTF(2,3),WTF(3,2)
      WRITE(6,6028) HTF(1,2),HTF(1,3),HTF(2,2),HTF(2,3),HTF(3,2)
      WRITE(6,6035) INSED(1,2),INSED(1,3),INSED(2,2),INSED(2,3),
     *              INSED(3,2)
      WRITE(6,6030) PER0,LSTM0,INSED(1,1),INSTOP
      WRITE(6,6042) LX,HX,LY,HY
      WRITE(6,6047)
      DO 107 I=1,NTGS
  107 WRITE(6,6048) ITG(I),PER(ITG(I)),LSTM(ITG(I))
      WRITE(6,6040) LTSTOP
*
****          THIS SECTION INITIALIZES VARIABLES
*
      NCL(1)=NFIB $NCL(2)=NEX $NCL(3)=NIN
      DO 102 I=1,2
      DCTH(I)=EXP(-STEP/TTH(I)) $DCG(I)=EXP(-STEP/TGK(I))
  102 CONTINUE
      DO 110 I=1,NCLS
      E(I)=0. $S(I)=0. $GK(I)=0.
      IF(I.LE.NEX) THEN $TH(I)=THO(1)
      ELSE $TH(I)=THO(2) $ENDIF
      DO 115 J=1,MCTP1
      SCIT(I,J)=0. $GI(I,J)=0.
  115 CONTINUE
  110 CONTINUE
      CHAR(1)=1H  $CHAR(2)=1H*
      DO 120 I=1,3 $DO 120 J=1,W $DO 120 K=1,H
  120 LINE(I,J,K)=0
*
****          THIS SECTION UPDATES STATE VARIABLES AT EACH TIME STEP
*
      DO 1000 L=1,LTSTOP
      NFF(L)=0 $NEF(L)=0 $NIF(L)=0
      DO 1010 I=1,3 $DO 1010 J=1,W $DO 1010 K=1,H
 1010 ALINE(I,J,K)=CHAR(1)
*
****          UPDATING THE INPUT FIBERS
*
      IF(L.GE.INSTOP) GO TO 1350
      DO 1300 I=1,NFIB
      IF(L.NE.LSTM(I)) GO TO 1300
      JX=INT(FLOAT(MOD((I-1),NC(1))+1)*FLOAT(W)/FLOAT(NC(1)))
      JY=INT(FLOAT((I-1)/NC(1)+1)*FLOAT(H)/FLOAT(NR(1)))
      NFF(L)=NFF(L)+1 $IDF(1,NFF(L))=I $ALINE(1,JX,JY)=CHAR(2)
      LINE(1,JX,JY)=LINE(1,JX,JY)+1
      CALL RANSET(INSED(3,3))
      LSTM(I)=L+INT(-PER(I)*ALOG(RANF(0.)))+1
      CALL RANGET(INSED(3,3))
      CALL TRNSMT(1,2,I)
      CALL TRNSMT(1,3,I)
 1300 CONTINUE
 1350 CONTINUE
*
****          UPDATING THE EXCITATORY CELLS
*
      DO 1500 I=1,NEX
      CALL SVUPDT(1,I)
      IF(E(I).LT.TH(I)) GO TO 1490
      JX=INT(FLOAT(MOD((I-1),NC(2))+1)*FLOAT(W)/FLOAT(NC(2)))
      JY=INT(FLOAT((I-1)/NC(2)+1)*FLOAT(H)/FLOAT(NR(2)))
      S(I)=1. $NEF(L)=NEF(L)+1 $IDF(2,NEF(L))=I
      ALINE(2,JX,JY)=CHAR(2) $LINE(2,JX,JY)=LINE(2,JX,JY)+1
```

```
      CALL TRNSMT(2,2,I)
      CALL TRNSMT(2,3,I)
 1490 CONTINUE
 1500 CONTINUE
*
****                UPDATING THE INHIBITORY CELLS
*
      DO 1700 I=NEX+1,NCLS
      CALL SVUPDT(2,I)
      S(I)=0. $IF(E(I).LT.TH(I)) GO TO 1690
      JX=INT(FLOAT(MOD((I-NEX-1),NC(3))+1)*FLOAT(W)/FLOAT(NC(3)))
      JY=INT(FLOAT((I-NEX-1)/NC(3)+1)*FLOAT(H)/FLOAT(NR(3)))
      S(I)=1. $NIF(L)=NIF(L)+1 $IDF(3,NIF(L))=I-NEX
      ALINE(3,JX,JY)=CHAR(2) $LINE(3,JX,JY)=LINE(3,JX,JY)+1
      CALL TRNSMT(3,2,I-NEX)
 1690 CONTINUE
 1700 CONTINUE
*
      IF(IFLG1.EQ.0) GO TO 4120
      IF(MOD(L,2).EQ.1) WRITE(6,6000)
      WRITE(6,6050) L,NFF(L),NEF(L),NIF(L)
      WRITE(6,6052) (IDF(1,I),I=1,NFF(L))
      WRITE(6,6052) (IDF(2,I),I=1,NEF(L))
      WRITE(6,6052) (IDF(3,I),I=1,NIF(L))
      DO 4110 JY=1,H
 4110 WRITE(6,6060) (ALINE(1,JX,H+1-JY),JX=1,W),(ALINE(2,JX,H+1-JY),
     *              JX=1,W),(ALINE(3,JX,H+1-JY),JX=1,W)
 4120 CONTINUE
      WRITE(7,7100) L,NFF(L),NEF(L),NIF(L)
      DO 1910 I=1,NCLS $DO 1920 J=1,MCTP1-1
      SCIT(I,J)=SCIT(I,J+1) $GI(I,J)=GI(I,J+1)
 1920 CONTINUE $SCIT(I,MCTP1)=0. $GI(I,MCTP1)=0.
 1910 CONTINUE
*
 1000 CONTINUE
*
****        THIS SECTION WRITES OUT ACTIVITY VARIABLES
*
      WRITE(6,6000)
      WRITE(6,6105) $WRITE(6,6110) (NFF(L),L=1,LTSTOP)
      WRITE(6,6115) $WRITE(6,6110) (NEF(L),L=1,LTSTOP)
      WRITE(6,6115) $WRITE(6,6110) (NIF(L),L=1,LTSTOP)
      WRITE(6,6045) $DO 4210 JY=1,H
 4210 WRITE(6,6070) (LINE(1,JX,H+1-JY),JX=1,W)
      WRITE(6,6045) $DO 4220 JY=1,H
 4220 WRITE(6,6070) (LINE(2,JX,H+1-JY),JX=1,W)
      WRITE(6,6045) $DO 4230 JY=1,H
 4230 WRITE(6,6070) (LINE(3,JX,H+1-JY),JX=1,W)
      WRITE(7,7100) (NFF(L),L=1,LTSTOP)
      WRITE(7,7105) $WRITE(7,7100) (NEF(L),L=1,LTSTOP)
      WRITE(7,7105) $WRITE(7,7100) (NIF(L),L=1,LTSTOP)
*
      STOP
****        THESE ARE THE FORMATS
*
 2010 FORMAT(12F6.2)
 2020 FORMAT(12I6)
 2030 FORMAT(F6.2,3I6)
 2040 FORMAT(I6,F6.2,I6)
 6000 FORMAT(1H1)
 6010 FORMAT(5X,*EXC:C,TTH,B,TGK,THO,TMEM=*,6F7.2)
 6015 FORMAT(5X,*INH:C,TTH,B,TGK,THO,TMEM=*,6F7.2)
 6020 FORMAT(5X,*NT12,NT13,NT22,NT23,NT32=*,5I7)
 6022 FORMAT(5X,*NCT12,NCT13,NCT22,NCT3,NCT32=*,5I7)
 6024 FORMAT(5X,*STR12,STR13,STR22,STR23,STR32=*,5F7.2)
 6026 FORMAT(5X,*WTFS=*,5F7.2)
 6028 FORMAT(5X,*HTFS=*,5F7.2)
```

```
6030 FORMAT(5X,*PER0,LSTM0,INSED11,INSTOP=*,F7.2,3I7)
6035 FORMAT(5X,*INSEDS=*,5I7)
6040 FORMAT(5X,*LTSTOP=*,I7)
6042 FORMAT(5X,*LX,HX,LY,HY=*,4I7)
6045 FORMAT(///)
6047 FORMAT(5X,*SELECTED INPUTS; ITG,PER,LSTM:*)
6048 FORMAT(5X,I7,F7.2,I7)
6050 FORMAT(///,2X,*L,NFF,NEF,NIF=*,4I7,4X,*HERE ARE THE FIRERS:*)
6052 FORMAT(3X,20I5)
6060 FORMAT(2X,40A1,3X,40A1,3X,40A1)
6070 FORMAT(3X,40I3)
6105 FORMAT(/,10X,*HERE ARE THE OUTPUTS:*)
6110 FORMAT(5X,20I6)
6115 FORMAT(/)
7100 FORMAT(20I4)
7105 FORMAT(/)
     END
*
     SUBROUTINE SVUPDT(IS,K)
*
*** THIS SUBROUTINE UPDATES THE STATE VARIABLES OF AN INDIVIDUAL NEURON,
*
     PARAMETER NCLS=1600,NEX=800,MCTP1=5,STEP=1.,EI=-10.,EK=-10.
     COMMON SCIT(NCLS,MCTP1),GI(NCLS,MCTP1)
     COMMON/SV/ E(NCLS),TH(NCLS),S(NCLS),GK(NCLS),B(2),C(2),
    *            DCG(2),DCTH(2),THO(2),TMEM(2)
*
     GK(K)=GK(K)*DCG(IS)+B(IS)*(1.-DCG(IS))*S(K)$GTOT=1.+GK(K)+GI(K,1)
     DCE=EXP(-GTOT*STEP/TMEM(IS))
     E(K)=E(K)*DCE+(SCIT(K,1)+GI(K,1)*EI+GK(K)*EK)*(1.-DCE)/GTOT
     TH(K)=THO(IS)+(TH(K)-THO(IS))*DCTH(IS)+C(IS)*E(K)*(1.-DCTH(IS))
     S(K)=0.
     RETURN
     END
*
     SUBROUTINE TRNSMT(IS,IR,K)
**** THIS SUBROUTINE PROJECTS SYNAPTIC ACTIVATION FROM A SENDING
**** FIBER TO ITS TARGETED RECEIVING CELLS.
*
     INTEGER W,H
     PARAMETER NCLS=1600,NEX=800,MCTP1=5,W=40,H=20
     COMMON SCIT(NCLS,MCTP1),GI(NCLS,MCTP1)
     COMMON/TR/NT(3,3),NCT(3,3),STR(3,3),NCL(3),INSED(3,3),
    *            WTF(3,3),HTF(3,3),NC(3),NR(3)
*
     XS=FLOAT(MOD(K,NC(IS)))*FLOAT(W)/FLOAT(NC(IS))
     IF(XS.EQ.0.) XS=FLOAT(W)
     YS=FLOAT(K/NC(IS)+1)*FLOAT(H)/FLOAT(NR(IS))
     CALL RANSET(K*INSED(IS,IR))
     DO 100 J=1,NT(IS,IR)
     XR=XS+(RANF(0.)-.5)*WTF(IS,IR) $YR=YS+(RANF(0.)-.5)*HTF(IS,IR)
     ICR=INT(FLOAT(NC(IR))*(XR-.5)/FLOAT(W))+1
     IRR=INT(FLOAT(NR(IR))*(YR-.5)/FLOAT(H))+1
     IF(ICR.LE.0.OR.ICR.GT.NC(IR)) GO TO 80
     IF(IRR.LE.0.OR.IRR.GT.NR(IR)) GO TO 80
     NREC=NC(IR)*(IRR-1)+ICR
     IF(IR.EQ.3) NREC=NREC+NEX
     KCT=INT(RANF(0.)*FLOAT(NCT(IS,IR)))+2
     IF(IS.NE.3) SCIT(NREC,KCT)=SCIT(NREC,KCT)+STR(IS,IR)
     IF(IS.EQ.3) GI(NREC,KCT)=GI(NREC,KCT)+STR(IS,IR)
  80 CONTINUE
 100 CONTINUE
*
     RETURN
     END
*
/EOR
```

```
Ø.    25.    1Ø.    2.    1Ø.    5.
Ø.    25.    1Ø.    2.    1Ø.    5.
   4Ø     2Ø     4Ø     2Ø     4Ø     2Ø
   25     25     12     18      6
    1      1      4      2      2
   2.     2.     2.     1.    .35
   5.     5.     9.     9.     9.
   5.     5.     5.     5.     5.
 2157   7539   1426   6927   5559
  45.     2Ø   1227      5
   16     24      8     12
  1ØØ
/EOR
/EOI



/JOB
A224,CM=117ØØØ,TL=6ØØ.
FAKE.
PDFV.
MNF(Y,PL=1ØØØØ).
SETLIM,PR.
LGO,JN221.
SAVE,JN221.
EXIT.
SAVE,JN221.
/EOR
      PROGRAM NTWRK22 (NTOUT,INPUT,OUTPUT,TAPE7=NTOUT,TAPE6=OUTPUT)
*
*** THIS PROGRAM SIMULATES THE ACTIVITY OF A RECURRENTLY-CONNECTED
*   AND SPATIALLY-ORGANIZED NETWORK OF EXCITATORY AND INHIBITORY
*   NEURONS ACTIVATED BY A SPATIALLY-ORGAMNIZED SYSTEM OF INPUT FIBERS.
*** INPUT IS EASILY LOCALIZABLE.  NETWORK ACTIVITY "WRAPS AROUND" EDGES.
*
      INTEGER W,H,HX,HY
      PARAMETER NFIB=8ØØ,NCLS=16ØØ,NEX=8ØØ,NIN=8ØØ,MCTP1=5,W=4Ø,H=2Ø,
     *         IFLG1=Ø,NTGS=Ø,STEP=1.
*
**********    ADJUST 6Ø6Ø & 6Ø7Ø FORMATS TO WA1, WI3 ETC    **********
*
      DIMENSION NFF(3ØØ),NEF(3ØØ),NIF(3ØØ),TTH(2),TGK(2),
     *         ALINE(3,W,H),LINE(3,W,H),CHAR(2),PER(NFIB),LSTM(NFIB),
     *         IDF(3,NEX),ITG(NFIB)
      COMMON SCIT(NCLS,MCTP1),GI(NCLS,MCTP1)
      COMMON/SV/ E(NCLS),TH(NCLS),S(NCLS),GK(NCLS),B(2),C(2),
     *         DCG(2),DCTH(2),THO(2),TMEM(2)
      COMMON/TR/ NT(3,3),NCT(3,3),STR(3,3),NCL(3),INSED(3,3),
     *         WTF(3,3),HTF(3,3),NC(3),NR(3)
*
****         THIS SECTION READS AND WRITES THE INPUT PARAMETERS
*
      READ 2Ø1Ø, C(1),TTH(1),B(1),TGK(1),THO(1),TMEM(1)
      READ 2Ø1Ø, C(2),TTH(2),B(2),TGK(2),THO(2),TMEM(2)
      READ 2Ø2Ø, ((NC(I),NR(I)),I=1,3)
      READ 2Ø2Ø, NT(1,2),NT(1,3),NT(2,2),NT(2,3),NT(3,2)
      READ 2Ø2Ø, NCT(1,2),NCT(1,3),NCT(2,2),NCT(2,3),NCT(3,2)
      READ 2Ø1Ø, STR(1,2),STR(1,3),STR(2,2),STR(2,3),STR(3,2)
      READ 2Ø1Ø, WTF(1,2),WTF(1,3),WTF(2,2),WTF(2,3),WTF(3,2)
      READ 2Ø1Ø, HTF(1,2),HTF(1,3),HTF(2,2),HTF(2,3),HTF(3,2)
      READ 2Ø2Ø, INSED(1,2),INSED(1,3),INSED(2,2),INSED(2,3),
     *         INSED(3,2)
      READ 2Ø3Ø, PERØ,LSTMØ,INSED(1,1),INSTOP
      CALL RANSET(INSED(1,1))
      DO 1Ø5 I=1,NFIB
      PER(I)=-PERØ*ALOG(RANF(Ø.))
      LSTM(I)=INT(-FLOAT(LSTMØ)*ALOG(RANF(Ø.)))+1
```

```
  105 CONTINUE
      READ 2025, LX,HX,LY,HY,SNRATIO
      IF(LX.LE.1.AND.HX.GE.W.AND.LY.LE.1.AND.HY.GE.H) GO TO 108
      DO 109 I=1,NFIB
      JX=INT(FLOAT(MOD((I-1),NC(1))+1)*FLOAT(W)/FLOAT(NC(1)))
      JY=INT(FLOAT((I-1)/NC(1)+1)*FLOAT(H)/FLOAT(NR(1)))
      IF(JX.LT.LX.OR.JX.GT.HX.OR.JY.LT.LY.OR.JY.GT.HY) GO TO 109
      PER(I)=PER(I)/SNRATIO
  109 CONTINUE
  108 CONTINUE
      DO 106 I=1,NTGS
  106 READ 2040, ITG(I),PER(ITG(I)),LSTM(ITG(I))
      READ 2020, LTSTOP
*
      WRITE(6,6000)
      WRITE(6,6010) C(1),TTH(1),B(1),TGK(1),THO(1),TMEM(1)
      WRITE(6,6015) C(2),TTH(2),B(2),TGK(2),THO(2),TMEM(2)
      WRITE(6,6020) NT(1,2),NT(1,3),NT(2,2),NT(2,3),NT(3,2)
      WRITE(6,6022) NCT(1,2),NCT(1,3),NCT(2,2),NCT(2,3),NCT(3,2)
      WRITE(6,6024) STR(1,2),STR(1,3),STR(2,2),STR(2,3),STR(3,2)
      WRITE(6,6026) WTF(1,2),WTF(1,3),WTF(2,2),WTF(2,3),WTF(3,2)
      WRITE(6,6028) HTF(1,2),HTF(1,3),HTF(2,2),HTF(2,3),HTF(3,2)
      WRITE(6,6035) INSED(1,2),INSED(1,3),INSED(2,2),INSED(2,3),
     *              INSED(3,2)
      WRITE(6,6030) PER0,LSTM0,INSED(1,1),INSTOP
      WRITE(6,6042) LX,HX,LY,HY,SNRATIO
      WRITE(6,6047)
      DO 107 I=1,NTGS
  107 WRITE(6,6048) ITG(I),PER(ITG(I)),LSTM(ITG(I))
      WRITE(6,6040) LTSTOP
*
****        THIS SECTION INITIALIZES VARIABLES
*
      NCL(1)=NFIB $NCL(2)=NEX $NCL(3)=NIN
      DO 102 I=1,2
      DCTH(I)=EXP(-STEP/TTH(I)) $DCG(I)=EXP(-STEP/TGK(I))
  102 CONTINUE
      DO 110 I=1,NCLS
      E(I)=0. $S(I)=0. $GK(I)=0.
      IF(I.LE.NEX) THEN $TH(I)=THO(1)
      ELSE $TH(I)=THO(2) $ENDIF
      DO 115 J=1,MCTP1
      SCIT(I,J)=0. $GI(I,J)=0.
  115 CONTINUE
  110 CONTINUE
      CHAR(1)=1H  $CHAR(2)=1H*
      DO 120 I=1,3 $DO 120 J=1,W $DO 120 K=1,H
  120 LINE(I,J,K)=0
*
****        THIS SECTION UPDATES STATE VARIABLES AT EACH TIME STEP
*
      DO 1000 L=1,LTSTOP
      NFF(L)=0 $NEF(L)=0 $NIF(L)=0
      DO 1010 I=1,3 $DO 1010 J=1,W $DO 1010 K=1,H
 1010 ALINE(I,J,K)=CHAR(1)
*
****                  UPDATING THE INPUT FIBERS
*
      IF(L.GE.INSTOP) GO TO 1350
      DO 1300 I=1,NFIB
      IF(L.NE.LSTM(I)) GO TO 1300
      JX=INT(FLOAT(MOD((I-1),NC(1))+1)*FLOAT(W)/FLOAT(NC(1)))
      JY=INT(FLOAT((I-1)/NC(1)+1)*FLOAT(H)/FLOAT(NR(1)))
      NFF(L)=NFF(L)+1 $IDF(1,NFF(L))=I $ALINE(1,JX,JY)=CHAR(2)
      LINE(1,JX,JY)=LINE(1,JX,JY)+1
      CALL RANSET(INSED(3,3))
      LSTM(I)=L+INT(-PER(I)*ALOG(RANF(0.)))+1
```

```
      CALL RANGET(INSED(3,3))
      CALL TRNSMT(1,2,I)
      CALL TRNSMT(1,3,I)
 1300 CONTINUE
 1350 CONTINUE
*
****                    UPDATING THE EXCITATORY CELLS
*
      DO 1500 I=1,NEX
      CALL SVUPDT(1,I)
      IF(E(I).LT.TH(I)) GO TO 1490
      JX=INT(FLOAT(MOD((I-1),NC(2))+1)*FLOAT(W)/FLOAT(NC(2)))
      JY=INT(FLOAT((I-1)/NC(2)+1)*FLOAT(H)/FLOAT(NR(2)))
      S(I)=1. $NEF(L)=NEF(L)+1 $IDF(2,NEF(L))=I
      ALINE(2,JX,JY)=CHAR(2) $LINE(2,JX,JY)=LINE(2,JX,JY)+1
      CALL TRNSMT(2,2,I)
      CALL TRNSMT(2,3,I)
 1490 CONTINUE
 1500 CONTINUE
*
****                    UPDATING THE INHIBITORY CELLS
*
      DO 1700 I=NEX+1,NCLS
      CALL SVUPDT(2,I)
      S(I)=0. $IF(E(I).LT.TH(I)) GO TO 1690
      JX=INT(FLOAT(MOD((I-NEX-1),NC(3))+1)*FLOAT(W)/FLOAT(NC(3)))
      JY=INT(FLOAT((I-NEX-1)/NC(3)+1)*FLOAT(H)/FLOAT(NR(3)))
      S(I)=1. $NIF(L)=NIF(L)+1 $IDF(3,NIF(L))=I-NEX
      ALINE(3,JX,JY)=CHAR(2) $LINE(3,JX,JY)=LINE(3,JX,JY)+1
      CALL TRNSMT(3,2,I-NEX)
 1690 CONTINUE
 1700 CONTINUE
*
      IF(IFLG1.EQ.0) GO TO 4120
      IF(MOD(L,2).EQ.1) WRITE(6,6000)
      WRITE(6,6050) L,NFF(L),NEF(L),NIF(L)
      WRITE(6,6052) (IDF(1,I),I=1,NFF(L))
      WRITE(6,6052) (IDF(2,I),I=1,NEF(L))
      WRITE(6,6052) (IDF(3,I),I=1,NIF(L))
      DO 4110 JY=1,H
 4110 WRITE(6,6060) (ALINE(1,JX,H+1-JY),JX=1,W),(ALINE(2,JX,H+1-JY),
     *               JX=1,W),(ALINE(3,JX,H+1-JY),JX=1,W)
 4120 CONTINUE
      WRITE(7,7100) L,NFF(L),NEF(L),NIF(L)
      DO 1910 I=1,NCLS $DO 1920 J=1,MCTP1-1
      SCIT(I,J)=SCIT(I,J+1) $GI(I,J)=GI(I,J+1)
 1920 CONTINUE $SCIT(I,MCTP1)=0. $GI(I,MCTP1)=0.
 1910 CONTINUE
*
 1000 CONTINUE
*
****          THIS SECTION WRITES OUT ACTIVITY VARIABLES
*
      WRITE(6,6000)
      WRITE(6,6105) $WRITE(6,6110) (NFF(L),L=1,LTSTOP)
      WRITE(6,6115) $WRITE(6,6110) (NEF(L),L=1,LTSTOP)
      WRITE(6,6115) $WRITE(6,6110) (NIF(L),L=1,LTSTOP)
      WRITE(6,6045) $DO 4210 JY=1,H
 4210 WRITE(6,6070) (LINE(1,JX,H+1-JY),JX=1,W)
      WRITE(6,6045) $DO 4220 JY=1,H
 4220 WRITE(6,6070) (LINE(2,JX,H+1-JY),JX=1,W)
      WRITE(6,6045) $DO 4230 JY=1,H
 4230 WRITE(6,6070) (LINE(3,JX,H+1-JY),JX=1,W)
      WRITE(7,7100) (NFF(L),L=1,LTSTOP)
      WRITE(7,7105) $WRITE(7,7100) (NEF(L),L=1,LTSTOP)
      WRITE(7,7105) $WRITE(7,7100) (NIF(L),L=1,LTSTOP)
*
```

```
      STOP
****          THESE ARE THE FORMATS
*
 2010 FORMAT(12F6.2)
 2020 FORMAT(12I6)
 2025 FORMAT(4I6,F6.2)
 2030 FORMAT(F6.2,3I6)
 2040 FORMAT(I6,F6.2,I6)
 6000 FORMAT(1H1)
 6010 FORMAT(5X,*EXC:C,TTH,B,TGK,THO,TMEM=*,6F7.2)
 6015 FORMAT(5X,*INH:C,TTH,B,TGK,THO,TMEM=*,6F7.2)
 6020 FORMAT(5X,*NT12,NT13,NT22,NT23,NT32=*,5I7)
 6022 FORMAT(5X,*NCT12,NCT13,NCT22,NCT3,NCT32=*,5I7)
 6024 FORMAT(5X,*STR12,STR13,STR22,STR23,STR32=*,5F7.2)
 6026 FORMAT(5X,*WTFS=*,5F7.2)
 6028 FORMAT(5X,*HTFS=*,5F7.2)
 6030 FORMAT(5X,*PER0,LSTM0,INSED11,INSTOP=*,F7.2,3I7)
 6035 FORMAT(5X,*INSEDS=*,5I7)
 6040 FORMAT(5X,*LTSTOP=*,I7)
 6042 FORMAT(5X,*LX,HX,LY,HY,SNRATIO=*,4I7,F7.2)
 6045 FORMAT(///)
 6047 FORMAT(5X,*SELECTED INPUTS; ITG,PER,LSTM:*)
 6048 FORMAT(5X,I7,F7.2,I7)
 6050 FORMAT(///,2X,*L,NFF,NEF,NIF=*,4I7,4X,*HERE ARE THE FIRERS:*)
 6052 FORMAT(3X,20I5)
 6060 FORMAT(2X,40A1,3X,40A1,3X,40A1)
 6070 FORMAT(3X,40I3)
 6105 FORMAT(/,10X,*HERE ARE THE OUTPUTS:*)
 6110 FORMAT(5X,20I6)
 6115 FORMAT(/)
 7100 FORMAT(20I4)
 7105 FORMAT(/)
      END
*
      SUBROUTINE SVUPDT(IS,K)
*
***   THIS SUBROUTINE UPDATES THE STATE VARIABLES OF AN INDIVIDUAL NEURON
*
      PARAMETER NCLS=1600,NEX=800,MCTP1=5,STEP=1.,EI=-10.,EK=-10.
      COMMON SCIT(NCLS,MCTP1),GI(NCLS,MCTP1)
      COMMON/SV/ E(NCLS),TH(NCLS),S(NCLS),GK(NCLS),B(2),C(2),
     *           DCG(2),DCTH(2),THO(2),TMEM(2)
*
      GK(K)=GK(K)*DCG(IS)+B(IS)*(1.-DCG(IS))*S(K)$GTOT=1.+GK(K)+GI(K,1)
      DCE=EXP(-GTOT*STEP/TMEM(IS))
      E(K)=E(K)*DCE+(SCIT(K,1)+GI(K,1)*EI+GK(K)*EK)*(1.-DCE)/GTOT
      TH(K)=THO(IS)+(TH(K)-THO(IS))*DCTH(IS)+C(IS)*E(K)*(1.-DCTH(IS))
      S(K)=0.
      RETURN
      END
*
      SUBROUTINE TRNSMT(IS,IR,K)
*
****      THIS SUBROUTINE PROJECTS SYNAPTIC ACTIVATION FROM A SENDING
****      FIBER TO ITS TARGETED RECEIVING CELLS.
*
      INTEGER W,H
      PARAMETER NCLS=1600,NEX=800,MCTP1=5,W=40,H=20
      COMMON SCIT(NCLS,MCTP1),GI(NCLS,MCTP1)
      COMMON/TR/NT(3,3),NCT(3,3),STR(3,3),NCL(3),INSED(3,3),
     *          WTF(3,3),HTF(3,3),NC(3),NR(3)
*
      XS=FLOAT(MOD(K,NC(IS)))*FLOAT(W)/FLOAT(NC(IS))
      IF(XS.EQ.0.) XS=FLOAT(W)
      YS=FLOAT(K/NC(IS)+1)*FLOAT(H)/FLOAT(NR(IS))
      CALL RANSET(K*INSED(IS,IR))
      DO 100 J=1,NT(IS,IR)
```

```
      XR=XS+(RANF(Ø.)-.5)*WTF(IS,IR) $YR=YS+(RANF(Ø.)-.5)*HTF(IS,IR)
      ICR=INT(FLOAT(NC(IR))*(XR-.5)/FLOAT(W))+1
      IRR=INT(FLOAT(NR(IR))*(YR-.5)/FLOAT(H))+1
      IF(ICR.LE.Ø) ICR=NC(IR)+ICR
      IF(ICR.GT.NC(IR)) ICR=ICR-NC(IR)
      IF(IRR.LE.Ø) IRR=NR(IR)+IRR
      IF(IRR.GT.NR(IR)) IRR=IRR-NR(IR)
      NREC=NC(IR)*(IRR-1)+ICR
      IF(IR.EQ.3) NREC=NREC+NEX
      KCT=INT(RANF(Ø.)*FLOAT(NCT(IS,IR)))+2
      IF(IS.NE.3) SCIT(NREC,KCT)=SCIT(NREC,KCT)+STR(IS,IR)
      IF(IS.EQ.3) GI(NREC,KCT)=GI(NREC,KCT)+STR(IS,IR)
  100 CONTINUE
*
      RETURN
      END
*
/EOR
   Ø.    25.    1Ø.    2.    1Ø.    5.
   Ø.    25.    1Ø.    2.    1Ø.    5.
     40      20      40      20      40      20
     25      25       8      12       8
      1       1       4       2       2
   2.    2.    2.    1.    .35
   5.    5.    9.    9.    15.
   5.    5.    5.    5.    9.
   2157   7539   1426   6927   5559
  45.       20   1227      5
     16      24       8      12    8.
    100
/EOR
/EOI



/JOB
A224,CM=6ØØØØ,TL=100.
FAKE.
PDFV.
MNF(Y).
LGO,SYSØØA.
SAVE,SYSØØA.
EXIT.
SAVE,SYSØØA.
/EOR
      PROGRAM SYSTM1Ø (SYSTM,INPUT,OUTPUT,TAPE6=OUTPUT,TAPE7=SYSTM)
*
****      THIS PROGRAM SIMULATES THE ACTIVITY OF AN ARBITRARY SYSTEM OF
*         INTERCONNECTED NEURON POOLS ACTIVATED BY AN ARBITRARY NUMBER
*         OF INPUT FIBER SYSTEMS.  ALL CONNECECTIONS ARE EXCITATORY
****      OR INHIBITORY.
*
      INTEGER TYPE
      PARAMETER NTPOPS=5,NFPOPS=2,NCPOPS=3,NCLS=5Ø,MCTP1=2,
     *          STEP=1.,NTGMX=2
      DIMENSION P(NFPOPS),INSTR(NFPOPS),INSTP(NFPOPS),INFSED(NFPOPS),
     *          NF(NTPOPS),IDF(NCLS),TTH(NCPOPS),TGK(NCPOPS),
     *          NTGR(NTPOPS)
      COMMON SCIT(NCPOPS,NCLS,MCTP1),GI(NCPOPS,NCLS,MCTP1)
      COMMON/SV/E(NCPOPS,NCLS),TH(NCPOPS,NCLS),S(NCPOPS,NCLS),
     *          GK(NCPOPS,NCLS),B(NCPOPS),C(NCPOPS),DCG(NCPOPS),
     *          DCTH(NCPOPS),TMEM(NCPOPS),THO(NCPOPS)
      COMMON/TR/NCT(NTPOPS,NTGMX),NT(NTPOPS,NTGMX),STR(NTPOPS,
     *          NTGMX),INSED(NTPOPS,NTGMX),N(NTPOPS),TYPE(NTPOPS,
     *          NTGMX),IRCP(NTPOPS,NTGMX)
*
****          THIS SECTION READS AND WRITES THE INPUT PARAMETERS
```

```
*
      READ 5110, IFLG1,IFLG2
      DO 110 I=1,NTPOPS $IF(I.LE.NCPOPS) THEN
      READ 5130, N(I),C(I),TTH(I),B(I),TGK(I),TMEM(I),THO(I),
     *           NTGR(I) $ELSE $II=I-NCPOPS
      READ 5140, N(I),P(II),INSTR(II),INSTP(II),INFSED(II),
     *           NTGR(I) $ENDIF
      DO 120 J=1,NTGR(I)
  120 READ 5150, IRCP(I,J),TYPE(I,J),NCT(I,J),NT(I,J),
     *           STR(I,J),INSED(I,J)
  110 CONTINUE
      READ 5110, LTSTOP
*
      WRITE(6,6000) $WRITE(6,6005) NTPOPS,NFPOPS,NCPOPS,NCLS,MCTP1,
     *           NTGMX,IFLG1,IFLG2
      WRITE(6,6112) $WRITE(6,6116) $WRITE(6,6118)
      DO 210 I=1,NTPOPS $IF(I.LE.NCPOPS) THEN
      WRITE(6,6130) N(I),C(I),TTH(I),B(I),TGK(I),TMEM(I),THO(I),
     *           NTGR(I) $ELSE $II=I-NCPOPS
      WRITE(6,6140) N(I),P(II),INSTR(II),INSTP(II),INFSED(II),
     *           NTGR(I) $ENDIF
      DO 220 J=1,NTGR(I)
  220 WRITE(6,6150) IRCP(I,J),TYPE(I,J),NCT(I,J),NT(I,J),
     *           STR(I,J),INSED(I,J)
  210 CONTINUE
      WRITE(6,6160) LTSTOP
*
****        THIS SECTION INITIALIZES VARIABLES
*
      DO 310 I=1,NCPOPS
      DCTH(I)=EXP(-STEP/TTH(I)) $DCG(I)=EXP(-STEP/TGK(I))
      DO 310 J=1,NCLS
      E(I,J)=0. $TH(I,J)=THO(I) $S(I,J)=0. $GK(I,J)=0.
      DO 310 K=1,MCTP1
      SCIT(I,J,K)=0. $GI(I,J,K)=0.
  310 CONTINUE
*
****        THIS SECTION UPDATES STATE VARIABLES AT EACH TIME STEP
*
      DO 1000 L=1,LTSTOP
*
*
****             UPDATING THE CELLS
*
      DO 2000 I=1,NCPOPS
      NF(I)=0 $IF(IFLG1.GT.0) WRITE(7,7060) L,I
      DO 2900 J=1,N(I)
      CALL SVUPDT(I,J)
      IF(E(I,J).LT.TH(I,J)) GO TO 2850
      S(I,J)=1. $NF(I)=NF(I)+1 $IDF(NF(I))=J
      DO 2800 K=1,NTGR(I)
      IF(NT(I,K).GT.0) CALL TRNSMT(I,K,J)
 2800 CONTINUE
 2850 CONTINUE
 2900 CONTINUE $IF(IFLG1.GT.0) WRITE(7,7060) (IDF(J),J=1,NF(I))
 2000 CONTINUE
*
****             UPDATING THE INPUT FIBERS
*
      DO 3000 I=1,NFPOPS
      II=I+NCPOPS $NF(II)=0 $IF(IFLG1.GT.0) WRITE(7,7060) L,II
      IF(L.LT.INSTR(I).OR.L.GE.INSTP(I)) GO TO 3000
      CALL RANSET(INFSED(I))
      DO 3900 J=1,N(II)
      IF(RANF(0.).GT.P(I)) GO TO 3850 $CALL RANGET(INFSED(I))
      NF(II)=NF(II)+1 $IDF(NF(II))=J
      DO 3800 K=1,NTGR(II)
```

```
      IF(NT(II,K).GT.0) CALL TRNSMT(II,K,J)
 3800 CONTINUE $CALL RANSET(INFSED(I))
 3850 CONTINUE
 3900 CONTINUE $IF(IFLG1.GT.0) WRITE(7,7060) (IDF(J),J=1,NF(II))
      CALL RANGET(INFSED(I))
 3000 CONTINUE
*
      WRITE(6,6510) L,(NF(I),I=1,NTPOPS)
      WRITE(7,7510) L,(NF(I),I=1,NTPOPS)
*
      DO 1910 I=1,NCPOPS $DO 1910 J=1,NCLS $DO 1920 K=1,MCTP1-1
      SCIT(I,J,K)=SCIT(I,J,K+1) $GI(I,J,K)=GI(I,J,K+1)
 1920 CONTINUE $SCIT(I,J,MCTP1)=0. $GI(I,J,MCTP1)=0.
 1910 CONTINUE
*
 1000 CONTINUE
*
      STOP
*
**** 	     THESE ARE THE FORMATS
*
 5110 FORMAT(12I6)
 5130 FORMAT(I6,6F6.2,I6)
 5140 FORMAT(I6,F6.4,6I6)
 5150 FORMAT(4I6,F6.2,I6)
 6000 FORMAT(1H1)
 6005 FORMAT(5X,*NTPOPS,NFPOPS,NCPOPS,NCLS,MCTP1,NTGMX,IFLG1,*
     *,*IFLG2=:*,8I7)
 6112 FORMAT(//,5X,*HERE ARE, FOR EACH POPULATION:,*,/,10X,*N,C,TTH,B,TG
     *K,TMEM,THO,NTGR*,/,20X,*OR*)
 6116 FORMAT(10X,*N,P,INSTR,INSTP,INFSED,NTGR*,/,20X,*AND*)
 6118 FORMAT(10X,*IRCP,TYPE,NCT,NT,STR,INSED..FOR EACH TG*)
 6130 FORMAT(/,5X,I7,6F7.2,I7)
 6140 FORMAT(/,5X,I7,F7.4,6I7)
 6150 FORMAT(4I7,F7.2,I7)
 6160 FORMAT(5X,*LTSTOP=*,I7)
 6510 FORMAT(5X,15I7)
 7510 FORMAT(15I5)
 7060 FORMAT(20I4)
      END
*
      SUBROUTINE SVUPDT(I,K)
*
*** THIS SUBROUTINE UPDATES THE STATE VARIABLES OF AN INDIVIDUAL NEURON
*
      PARAMETER NCPOPS=3,NCLS=50,MCTP1=2,STEP=1.,EK=-10.,EI=-10.
      COMMON SCIT(NCPOPS,NCLS,MCTP1),GI(NCPOPS,NCLS,MCTP1)
      COMMON/SV/E(NCPOPS,NCLS),TH(NCPOPS,NCLS),S(NCPOPS,NCLS),
     *        GK(NCPOPS,NCLS),B(NCPOPS),C(NCPOPS),DCG(NCPOPS),
     *        DCTH(NCPOPS),TMEM(NCPOPS),THO(NCPOPS)
*
      GK(I,K)=GK(I,K)*DCG(I)+B(I)*(1.-DCG(I))*S(I,K)
      GTOT=1.+GK(I,K)+GI(I,K,1) $DCE=EXP(-GTOT*STEP/TMEM(I))
      E(I,K)=E(I,K)*DCE+(SCIT(I,K,1)+GI(I,K,1)*EI+GK(I,K)*EK)*(1.-DCE)/
     *        GTOT
      TH(I,K)=THO(I)+(TH(I,K)-THO(I))*DCTH(I)+C(I)*E(I,K)*(1.-DCTH(I))
      S(I,K)=0.
*
      RETURN
      END
*
      SUBROUTINE TRNSMT(IS,L,K)
*
**** 	    THIS SUBROUTINE PROJECTS SYNAPTIC ACTIVATION FROM A SENDING
**** 	    FIBER TO ITS TARGETED RECEIVING CELLS.
*
      INTEGER TYPE
```

```
      PARAMETER NTPOPS=5,NCPOPS=3,NCLS=50,MCTP1=2,NTGMX=2
      COMMON SCIT(NCPOPS,NCLS,MCTP1),GI(NCPOPS,NCLS,MCTP1)
      COMMON/TR/NCT(NTPOPS,NTGMX),NT(NTPOPS,NTGMX),STR(NTPOPS,
     *          NTGMX),INSED(NTPOPS,NTGMX),N(NTPOPS),TYPE(NTPOPS,
     *          NTGMX),IRCP(NTPOPS,NTGMX)
*
      IR=IRCP(IS,L)
      CALL RANSET(K*INSED(IS,L))
      DO 100 J=1,NT(IS,L)
      NREC=INT(RANF(0.)*FLOAT(N(IR)))+1
      KCT=INT(RANF(0.)*FLOAT(NCT(IS,L)))+2
      IF(TYPE(IS,L).EQ.1) SCIT(IR,NREC,KCT)=SCIT(IR,NREC,KCT)+
     *                    STR(IS,L)
      IF(TYPE(IS,L).EQ.2) GI(IR,NREC,KCT)=GI(IR,NREC,KCT)+
     *                    STR(IS,L)
  100 CONTINUE
*
      RETURN
      END
/EOR
    0      0
   50  0.   25.   20.    5.    5.   10.         2
    3     1    1    12 10.    1227
    2     1    1    12 10.    1936
   50  .5   25.   20.    5.    5.   10.         0
   50  0.   25.   20.    5.    5.   10.         1
    1     2    1    10 10.    3421
   50 .1         1   999  4165     1
    1     1    1    12 10.    8917
   50 .1        15    50  3315     1
    3     1    1    12 10.    6373
   20
/EOR
/EOI




/JOB
A224,CM=60000,TL=100.
FAKE.
PDFV.
M77.
LGO,SYS00B.
SAVE,SYS00B.
EXIT.
SAVE,SYS00B.
/EOR
      PROGRAM SYSTM11 (SYSTM,INPUT,OUTPUT,TAPE6=OUTPUT,TAPE7=SYSTM)
*
***   THIS PROGRAM SIMULATES THE ACTIVITY OF AN ARBITRARY NUMBER
***   OF NEURON POOLS ACTIVATED BY AN ARBITRARY NUMBER OF INPUT
***   FIBER SYSTEMS.
*
      INTEGER TYPE,SNTP
      PARAMETER NTPOPS=5,NFPOPS=2,NCPOPS=3,NCLS=50,MCTP1=2,
     *          SNTP=2,STEP=1.,NTGMX=2
      DIMENSION P(NFPOPS),INSTR(NFPOPS),INSTP(NFPOPS),INFSED(NFPOPS),
     *          IDF(NCLS),TTH(NCPOPS),TGK(NCPOPS),T(SNTP),DCS(SNTP),
     *          NF(NTPOPS),NTGR(NTPOPS)
      COMMON G(SNTP,NCPOPS,NCLS,MCTP1)
      COMMON/SV/E(NCPOPS,NCLS),TH(NCPOPS,NCLS),S(NCPOPS,NCLS),
     *          GK(NCPOPS,NCLS),B(NCPOPS),C(NCPOPS),DCG(NCPOPS),
     *          DCTH(NCPOPS),TMEM(NCPOPS),THO(NCPOPS),EQ(SNTP)
      COMMON/TR/NCT(NTPOPS,NTGMX),NT(NTPOPS,NTGMX),STR(NTPOPS,
     *          NTGMX),INSED(NTPOPS,NTGMX),N(NTPOPS),TYPE(NTPOPS,
     *          NTGMX),IRCP(NTPOPS,NTGMX)
*
```

```
****          THIS SECTION READS AND WRITES THE INPUT PARAMETERS
*
      READ 5110, IFLG1,IFLG2
      READ 5120, ((EQ(I),T(I)),I=1,SNTP)
      DO 110 I=1,NTPOPS $IF(I.LE.NCPOPS) THEN
      READ 5130, N(I),C(I),TTH(I),B(I),TGK(I),TMEM(I),THO(I),
     *          NTGR(I) $ELSE $II=I-NCPOPS
      READ 5140, N(I),P(II),INSTR(II),INSTP(II),INFSED(II),
     *          NTGR(I) $ENDIF
      DO 120 J=1,NTGR(I)
  120 READ 5150, IRCP(I,J),TYPE(I,J),NCT(I,J),NT(I,J),
     *          STR(I,J),INSED(I,J)
  110 CONTINUE
      READ 5110, LTSTOP
*
      WRITE(6,6000) $WRITE(6,6005) NTPOPS,NFPOPS,NCPOPS,NCLS,MCTP1,
     *          NTGMX,IFLG1,IFLG2
      WRITE(6,6110) ((EQ(I),T(I)),I=1,SNTP)
      WRITE(6,6112) $WRITE(6,6116) $WRITE(6,6118)
      DO 210 I=1,NTPOPS $IF(I.LE.NCPOPS) THEN
      WRITE(6,6130) N(I),C(I),TTH(I),B(I),TGK(I),TMEM(I),THO(I),
     *          NTGR(I) $ELSE $II=I-NCPOPS
      WRITE(6,6140) N(I),P(II),INSTR(II),INSTP(II),INFSED(II),
     *          NTGR(I) $ENDIF
      DO 220 J=1,NTGR(I)
  220 WRITE(6,6150) IRCP(I,J),TYPE(I,J),NCT(I,J),NT(I,J),
     *          STR(I,J),INSED(I,J)
  210 CONTINUE
      WRITE(6,6160) LTSTOP
*
****          THIS SECTION INITIALIZES VARIABLES
*
      DO 310 I=1,NCPOPS
      DCTH(I)=EXP(-STEP/TTH(I)) $DCG(I)=EXP(-STEP/TGK(I))
      DO 310 J=1,NCLS
      E(I,J)=0. $TH(I,J)=THO(I) $S(I,J)=0. $GK(I,J)=0.
      DO 310 K=1,MCTP1 $DO 310 L=1,SNTP
      G(L,I,J,K)=0.
  310 CONTINUE
      DO 320 I=1,SNTP
  320 DCS(I)=EXP(-STEP/T(I))
*
****          THIS SECTION UPDATES STATE VARIABLES AT EACH TIME STEP
*
      DO 1000 L=1,LTSTOP
*
*
****              UPDATING THE CELLS
*
      DO 2000 I=1,NCPOPS
      NF(I)=0 $IF(IFLG1.GT.0) WRITE(7,7060) L,I
      DO 2900 J=1,N(I)
      CALL SVUPDT(I,J)
      IF(E(I,J).LT.TH(I,J)) GO TO 2850
      S(I,J)=1. $NF(I)=NF(I)+1 $IDF(NF(I))=J
      DO 2800 K=1,NTGR(I)
      IF(NT(I,K).GT.0) CALL TRNSMT(I,K,J)
 2800 CONTINUE
 2850 CONTINUE
 2900 CONTINUE $IF(IFLG1.GT.0) WRITE(7,7060) (IDF(J),J=1,NF(I))
 2000 CONTINUE
*
****              UPDATING THE INPUT FIBERS
*
      DO 3000 I=1,NFPOPS
      II=I+NCPOPS $NF(II)=0 $IF(IFLG1.GT.0) WRITE(7,7060) L,II
      IF(L.LT.INSTR(I).OR.L.GE.INSTP(I)) GO TO 3000
```

```
      CALL RANSET(INFSED(I))
      DO 3900 J=1,N(II)
      IF(RANF(0.).GT.P(I)) GO TO 3850 $CALL RANGET(INFSED(I))
      NF(II)=NF(II)+1 $IDF(NF(II))=J
      DO 3800 K=1,NTGR(II)
      IF(NT(II,K).GT.0) CALL TRNSMT(II,K,J)
 3800 CONTINUE $CALL RANSET(INFSED(I))
 3850 CONTINUE
 3900 CONTINUE $IF(IFLG1.GT.0) WRITE(7,7060) (IDF(J),J=1,NF(II))
      CALL RANGET(INFSED(I))
 3000 CONTINUE
*
      WRITE(6,6510) L,(NF(I),I=1,NTPOPS)
      WRITE(7,7510) L,(NF(I),I=1,NTPOPS)
*
      DO 1910 I=1,NCPOPS $DO 1910 J=1,NCLS $DO 1910 M=1,SNTP
      G(M,I,J,1)=G(M,I,J,1)*DCS(M)+G(M,I,J,2) $DO 1920 K=2,MCTP1-1
      G(M,I,J,K)=G(M,I,J,K+1)
 1920 CONTINUE $G(M,I,J,MCTP1)=0.
 1910 CONTINUE
*
 1000 CONTINUE
*
      STOP
*
****          THESE ARE THE FORMATS
*
 5110 FORMAT(12I6)
 5120 FORMAT(12F6.2)
 5130 FORMAT(I6,6F6.2,I6)
 5140 FORMAT(I6,F6.4,4I6)
 5150 FORMAT(4I6,F6.2,I6)
 6000 FORMAT(1H1)
 6005 FORMAT(5X,*NTPOPS,NFPOPS,NCPOPS,NCLS,MCTP1,NTGMX,IFLG1,*
     *,*IFLG2=:*,8I7)
 6110 FORMAT(5X,*EQ,T=*,12F7.2)
 6112 FORMAT(//,5X,*HERE ARE, FOR EACH POPULATION:,*,/,10X,*N,C,TTH,B,TG
     *K,TMEM,THO,NTGR*,/,20X,*OR*)
 6116 FORMAT(10X,*N,P,INSTR,INSTP,INFSED,NTGR*,/,20X,*AND*)
 6118 FORMAT(10X,*IRCP,TYPE,NCT,NT,STR,INSED..FOR EACH TG*)
 6130 FORMAT(/,5X,I7,6F7.2,I7)
 6140 FORMAT(/,5X,I7,F7.4,4I7)
 6150 FORMAT(4I7,F7.2,I7)
 6160 FORMAT(5X,*LTSTOP=*,I7)
 6510 FORMAT(5X,15I7)
 7510 FORMAT(15I5)
 7060 FORMAT(20I4)
      END
*
      SUBROUTINE SVUPDT(I,K)
*
***  THIS SUBROUTINE UPDATES THE STATE VARIABLES OF AN INDIVIDUAL NEURON
*
      INTEGER SNTP
      PARAMETER NCPOPS=3,NCLS=50,MCTP1=2,SNTP=2,EK=-10,STEP=1.
      COMMON G(SNTP,NCPOPS,NCLS,MCTP1)
      COMMON/SV/E(NCPOPS,NCLS),TH(NCPOPS,NCLS),S(NCPOPS,NCLS),
     *          GK(NCPOPS,NCLS),B(NCPOPS),C(NCPOPS),DCG(NCPOPS),
     *          DCTH(NCPOPS),TMEM(NCPOPS),THO(NCPOPS),EQ(SNTP)
*
      GS=0. $SYN=0. $DO 10 J=1,SNTP
      GS=GS+G(J,I,K,1) $SYN=SYN+G(J,I,K,1)*EQ(J)
   10 CONTINUE
      GK(I,K)=GK(I,K)*DCG(I)+B(I)*(1.-DCG(I))*S(I,K)
      GTOT=1.+GK(I,K)+GS $DCE=EXP(-GTOT*STEP/TMEM(I))
      E(I,K)=E(I,K)*DCE+(SYN+GK(I,K)*EK)*(1.-DCE)/GTOT
      TH(I,K)=THO(I)+(TH(I,K)-THO(I))*DCTH(I)+C(I)*E(I,K)*(1.-DCTH(I))
```

```
      S(I,K)=0.
*
      RETURN
      END
*
      SUBROUTINE TRNSMT(IS,L,K)
*
****      THIS SUBROUTINE PROJECTS SYNAPTIC ACTIVATION FROM A SENDING
****      FIBER TO ITS TARGETED RECEIVING CELLS.
*
      INTEGER TYPE,SNTP
      PARAMETER NTPOPS=5,NCPOPS=3,NCLS=50,MCTP1=2,SNTP=2,NTGMX=2
      COMMON G(SNTP,NCPOPS,NCLS,MCTP1)
      COMMON/TR/NCT(NTPOPS,NTGMX),NT(NTPOPS,NTGMX),STR(NTPOPS,
     *          NTGMX),INSED(NTPOPS,NTGMX),N(NTPOPS),TYPE(NTPOPS,
     *          NTGMX),IRCP(NTPOPS,NTGMX)
*
      IR=IRCP(IS,L)
      CALL RANSET(K*INSED(IS,L))
      DO 100 J=1,NT(IS,L)
      NREC=INT(RANF(0.)*FLOAT(N(IR)))+1
      KCT=INT(RANF(0.)*FLOAT(NCT(IS,L)))+2
      G(TYPE(IS,L),IR,NREC,KCT)=G(TYPE(IS,L),IR,NREC,KCT)+STR(IS,L)
  100 CONTINUE
*
      RETURN
      END
/EOR
      0       0
 70.    1.  -10.    5.
    50  0.   25.   20.    5.    5.   10.            2
     3    1     1     8    .5  1227
     2    1     1     8    .5  1936
    50   .5   25.   20.    5.    5.   10.            0
    50  0.   25.   20.    5.    5.   10.            1
     1    2     1    10  1.    3421
    50  .1          1   999  4165      1
     1    1     1    10    .5  8917
    50  .1         15    50  3315      1
     3    1     1    10    .5  6373
    20
/EOR
/EOI




/JOB
A224,CM=70000,TL=200.
FAKE.
PDFV.
MNF(Y).
LGO,SYS00C.
SAVE,SYS00C.
EXIT.
SAVE,SYS00C.
/EOR
      PROGRAM SYSTM20 (SYSTM,INPUT,OUTPUT,TAPE6=OUTPUT,TAPE7=SYSTM)
*
****      THIS PROGRAM SIMULATES THE ACTIVITY OF AN ARBITRARY SYSTEM OF
*         INTERCONNECTED SPATIALLY-ORGANIZED NEURAL NETWORKS ACTIVATED BY
*         AN ARBITRARY NUMBER OF SPATIALLY-ORGANIZED INPUT FIBER SYSTEMS.
****      ALL CONNECTIONS ARE EXCITATORY OR INHIBITORY.
*
      INTEGER TYPE,W,H
      PARAMETER NTPOPS=5,NFPOPS=2,NCPOPS=3,NCLS=200,MCTP1=2,
     *          W=20,H=10,STEP=1.,NTGMX=2
      DIMENSION P(NFPOPS),INSTR(NFPOPS),INSTP(NFPOPS),INFSED(NFPOPS),
```

```
     *          IDF(NCLS),TTH(NCPOPS),TGK(NCPOPS),NF(NTPOPS),
     *          NTGR(NTPOPS)
      COMMON SCIT(NCPOPS,NCLS,MCTP1),GI(NCPOPS,NCLS,MCTP1)
      COMMON/SV/E(NCPOPS,NCLS),TH(NCPOPS,NCLS),S(NCPOPS,NCLS),
     *          GK(NCPOPS,NCLS),B(NCPOPS,NCLS),C(NCPOPS,NCLS),DCG(NCPOPS),
     *          DCTH(NCPOPS),TMEM(NCPOPS),THO(NCPOPS)
      COMMON/TR/NCT(NTPOPS,NTGMX),NT(NTPOPS,NTGMX),STR(NTPOPS,
     *          NTGMX),INSED(NTPOPS,NTGMX),N(NTPOPS),TYPE(NTPOPS,
     *          NTGMX),IRCP(NTPOPS,NTGMX),NC(NTPOPS),NR(NTPOPS),
     *          WTF(NTPOPS,NTGMX),HTF(NTPOPS,NTGMX)
*
****          THIS SECTION READS AND WRITES THE INPUT PARAMETERS
*
      READ 5110, IFLG1,IFLG2
      DO 110 I=1,NTPOPS $IF(I.LE.NCPOPS) THEN
      READ 5130, N(I),C(I),TTH(I),B(I),TGK(I),TMEM(I),THO(I),
     *          NC(I),NR(I),NTGR(I) $ELSE $II=I-NCPOPS
      READ 5140, N(I),P(II),INSTR(II),INSTP(II),INFSED(II),
     *          NC(I),NR(I),NTGR(I) $ENDIF
      DO 120 J=1,NTGR(I)
  120 READ 5150, IRCP(I,J),TYPE(I,J),NCT(I,J),NT(I,J),
     *          STR(I,J),INSED(I,J),WTF(I,J),HTF(I,J)
  110 CONTINUE
      READ 5110, LTSTOP
*
      WRITE(6,6000) $WRITE(6,6005) NTPOPS,NFPOPS,NCPOPS,NCLS,MCTP1,
     *          NTGMX,IFLG1,IFLG2
      WRITE(6,6112) $WRITE(6,6116) $WRITE(6,6118)
      DO 210 I=1,NTPOPS $IF(I.LE.NCPOPS) THEN
      WRITE(6,6130) N(I),C(I),TTH(I),B(I),TGK(I),TMEM(I),THO(I),
     *          NC(I),NR(I),NTGR(I) $ELSE $II=I-NCPOPS
      WRITE(6,6140) N(I),P(II),INSTR(II),INSTP(II),INFSED(II),
     *          NC(I),NR(I),NTGR(I) $ENDIF
      DO 220 J=1,NTGR(I)
  220 WRITE(6,6150) IRCP(I,J),TYPE(I,J),NCT(I,J),NT(I,J),
     *          STR(I,J),INSED(I,J),WTF(I,J),HTF(I,J)
  210 CONTINUE
      WRITE(6,6160) LTSTOP
*
****          THIS SECTION INITIALIZES VARIABLES
*
      DO 310 I=1,NCPOPS
      DCTH(I)=EXP(-STEP/TTH(I)) $DCG(I)=EXP(-STEP/TGK(I))
      DO 310 J=1,NCLS
      E(I,J)=0. $TH(I,J)=THO(I) $S(I,J)=0. $GK(I,J)=0.
      DO 310 K=1,MCTP1
      SCIT(I,J,K)=0. $GI(I,J,K)=0.
  310 CONTINUE
*
****          THIS SECTION UPDATES STATE VARIABLES AT EACH TIME STEP
*
      DO 1000 L=1,LTSTOP
*
*
****                    UPDATING THE CELLS
*
      DO 2000 I=1,NCPOPS
      NF(I)=0 $IF(IFLG1.GT.0) WRITE(7,7060) L,I
      DO 2900 J=1,N(I)
      CALL SVUPDT(I,J)
      IF(E(I,J).LT.TH(I,J)) GO TO 2850
      S(I,J)=1. $NF(I)=NF(I)+1 $IDF(NF(I))=J
      DO 2800 K=1,NTGR(I)
      IF(NT(I,K).GT.0) CALL TRNSMT(I,K,J)
 2800 CONTINUE
 2850 CONTINUE
 2900 CONTINUE $IF(IFLG1.GT.0) WRITE(7,7060) (IDF(J),J=1,NF(I))
```

```
 2000 CONTINUE
*
****               UPDATING THE INPUT FIBERS
*
      DO 3000 I=1,NFPOPS
      II=I+NCPOPS $NF(II)=0 $IF(IFLG1.GT.0) WRITE(7,7060) L,II
      IF(L.LT.INSTR(I).OR.L.GE.INSTP(I)) GO TO 3000
      CALL RANSET(INFSED(I))
      DO 3900 J=1,N(II)
      IF(RANF(0.).GT.P(I)) GO TO 3850 $CALL RANGET(INFSED(I))
      NF(II)=NF(II)+1 $IDF(NF(II))=J
      DO 3800 K=1,NTGR(II)
      IF(NT(II,K).GT.0) CALL TRNSMT(II,K,J)
 3800 CONTINUE $CALL RANSET(INFSED(I))
 3850 CONTINUE
 3900 CONTINUE $IF(IFLG1.GT.0) WRITE(7,7060) (IDF(J),J=1,NF(II))
      CALL RANGET(INFSED(I))
 3000 CONTINUE
*
      WRITE(6,6510) L,(NF(I),I=1,NTPOPS)
      WRITE(7,7510) L,(NF(I),I=1,NTPOPS)
*
      DO 1910 I=1,NCPOPS $DO 1910 J=1,NCLS $DO 1920 K=1,MCTP1-1
      SCIT(I,J,K)=SCIT(I,J,K+1) $GI(I,J,K)=GI(I,J,K+1)
 1920 CONTINUE $SCIT(I,J,MCTP1)=0. $GI(I,J,MCTP1)=0.
 1910 CONTINUE
*
 1000 CONTINUE
*
      STOP
*
****          THESE ARE THE FORMATS
*
 5110 FORMAT(12I6)
 5130 FORMAT(I6,6F6.2,3I6)
 5140 FORMAT(I6,F6.4,6I6)
 5150 FORMAT(4I6,F6.2,I6,2F6.2)
 6000 FORMAT(1H1)
 6005 FORMAT(5X,*NTPOPS,NFPOPS,NCPOPS,NCLS,MCTP1,NTGMX,IFLG1,*
     * ,*IFLG2=:*,8I7)
 6112 FORMAT(//,5X,*HERE ARE, FOR EACH POPULATION:,*,/,10X,*N,C,TTH,B,TG
     *K,TMEM,THO,NC,NR,NTGR*,/,20X,*OR*)
 6116 FORMAT(10X,*N,P,INSTR,INSTP,INFSED,NC,NR,NTGR*,/,20X,*AND*)
 6118 FORMAT(10X,*IRCP,TYPE,NCT,NT,STR,INSED,WTF,HTF..FOR EACH TG*)
 6130 FORMAT(/,5X,I7,6F7.2,3I7)
 6140 FORMAT(/,5X,I7,F7.4,6I7)
 6150 FORMAT(4I7,F7.2,I7,2F7.2)
 6160 FORMAT(5X,*LTSTOP=*,I7)
 6510 FORMAT(5X,15I7)
 7510 FORMAT(15I5)
 7060 FORMAT(20I4)
      END
*
      SUBROUTINE SVUPDT(I,K)
*
***   THIS SUBROUTINE UPDATES THE STATE VARIABLES OF AN INDIVIDUAL NEURON
*
      PARAMETER NCPOPS=3,NCLS=200,MCTP1=2,STEP=1.,EK=-10.,EI=-10.
      COMMON SCIT(NCPOPS,NCLS,MCTP1),GI(NCPOPS,NCLS,MCTP1)
      COMMON/SV/E(NCPOPS,NCLS),TH(NCPOPS,NCLS),S(NCPOPS,NCLS),
     *          GK(NCPOPS,NCLS),B(NCPOPS),C(NCPOPS),DCG(NCPOPS),
     *          DCTH(NCPOPS),TMEM(NCPOPS),THO(NCPOPS)
*
      GK(I,K)=GK(I,K)*DCG(I)+B(I)*(1.-DCG(I))*S(I,K)
      GTOT=1.+GK(I,K)+GI(I,K,1) $DCE=EXP(-GTOT*STEP/TMEM(I))
      E(I,K)=E(I,K)*DCE+(SCIT(I,K,1)+GI(I,K,1)*EI+GK(I,K)*EK)*(1.-DCE)/
     *       GTOT
```

```
      TH(I,K)=THO(I)+(TH(I,K)-THO(I))*DCTH(I)+C(I)*E(I,K)*(1.-DCTH(I))
      S(I,K)=0.
*
      RETURN
      END
*
      SUBROUTINE TRNSMT(IS,L,K)
*
****      THIS SUBROUTINE PROJECTS SYNAPTIC ACTIVATION FROM A SENDING
****      FIBER TO ITS TARGETED RECEIVING CELLS.
*
      INTEGER TYPE,W,H
      PARAMETER NTPOPS=5,NCPOPS=3,NCLS=200,MCTP1=2,W=20,H=10,NTGMX=2
      COMMON SCIT(NCPOPS,NCLS,MCTP1),GI(NCPOPS,NCLS,MCTP1)
      COMMON/TR/NCT(NTPOPS,NTGMX),NT(NTPOPS,NTGMX),STR(NTPOPS,
     *          NTGMX),INSED(NTPOPS,NTGMX),N(NTPOPS),TYPE(NTPOPS,
     *          NTGMX),IRCP(NTPOPS,NTGMX),NC(NTPOPS),NR(NTPOPS),
     *          WTF(NTPOPS,NTGMX),HTF(NTPOPS,NTGMX)
*
      IR=IRCP(IS,L)
      XS=FLOAT(MOD(K,NC(IS)))*FLOAT(W)/FLOAT(NC(IS))
      IF(XS.EQ.0.) XS=FLOAT(W)
      YS=FLOAT(K/NC(IS)+1)*FLOAT(H)/FLOAT(NR(IS))
      CALL RANSET(K*INSED(IS,L))
      DO 100 J=1,NT(IS,L)
      XR=XS+(RANF(0.)-.5)*WTF(IS,L) $YR=YS+(RANF(0.)-.5)*HTF(IS,L)
      ICR=INT(FLOAT(NC(IR))*(XR-.5)/FLOAT(W))+1
      IRR=INT(FLOAT(NR(IR))*(YR-.5)/FLOAT(H))+1
      IF(ICR.LE.0) ICR=NC(IR)+ICR
      IF(ICR.GT.NC(IR)) ICR=ICR-NC(IR)
      IF(IRR.LE.0) IRR=NR(IR)+IRR
      IF(IRR.GT.NR(IR)) IRR=IRR-NR(IR)
      NREC=NC(IR)*(IRR-1)+ICR
      KCT=INT(RANF(0.)*FLOAT(NCT(IS,L)))+2
      IF(TYPE(IS,L).EQ.1) SCIT(IR,NREC,KCT)=SCIT(IR,NREC,KCT)+
     *                    STR(IS,L)
      IF(TYPE(IS,L).EQ.2) GI(IR,NREC,KCT)=GI(IR,NREC,KCT)+
     *                    STR(IS,L)
  100 CONTINUE
*
      RETURN
      END
/EOR
    0    0
   50  0.   25.   20.    5.    5.   10.       10      5     2
    3     1     1    12 10.    1227  9.    5.
    2     1     1    12 10.    1936  9.    5.
   50  .5   25.   20.    5.    5.   10.       10      5     0
   50  0.   25.   20.    5.    5.   10.       10      5     1
    1     2     1    12 12.    3421 20.   10.
   50  .1     1   999  4165    10     5     1
    1     1     1    12 12.    8917 20.   10.
   50  .1    15    50  3315    10     5     1
    3     1     1    12 12.    6373 20.   10.
   20
/EOR
/EOI


/JOB
A224,CM=70000,TL=200.
FAKE.
PDFV.
MNF(Y,PL=20000).
SETLIM,PR.
LGO,SYS00D.
```

```
SAVE,SYSØØD.
EXIT.
SAVE,SYSØØD.
/EOR
      PROGRAM SYSTM21 (SYSTM,INPUT,OUTPUT,TAPE6=OUTPUT,TAPE7=SYSTM)
*
****     THIS PROGRAM SIMULATES THE ACTIVITY OF AN ARBITRARY SYSTEM OF
*        INTERCONNECTED SPATIALLY-ORGANIZED NEURAL NETWORKS ACTIVATED BY
*        AN ARBITRARY NUMBER OF SPATIALLY-ORGANIZED INPUT FIBER SYSTEMS.
****     ALL CONNECTIONS ARE EXCITATORY OR INHIBITORY.
*
      INTEGER TYPE,W,H
      PARAMETER NTPOPS=5,NFPOPS=2,NCPOPS=3,NCLS=200,MCTP1=2,W=20,H=10,
     *        STEP=1.,NTGMX=2
      DIMENSION P(NFPOPS),INSTR(NFPOPS),INSTP(NFPOPS),INFSED(NFPOPS),
     *        IDF(NCLS),TTH(NCPOPS),TGK(NCPOPS),CHAR(2),ALINE(W,H),
     *        LINE(NTPOPS,W,H),NF(NTPOPS),NTGR(NTPOPS)
      COMMON SCIT(NCPOPS,NCLS,MCTP1),GI(NCPOPS,NCLS,MCTP1)
      COMMON/SV/E(NCPOPS,NCLS),TH(NCPOPS,NCLS),S(NCPOPS,NCLS),
     *        GK(NCPOPS,NCLS),B(NCPOPS),C(NCPOPS),DCG(NCPOPS),
     *        DCTH(NCPOPS),TMEM(NCPOPS),THO(NCPOPS)
      COMMON/TR/NCT(NTPOPS,NTGMX),NT(NTPOPS,NTGMX),STR(NTPOPS,
     *        NTGMX),INSED(NTPOPS,NTGMX),N(NTPOPS),TYPE(NTPOPS,
     *        NTGMX),IRCP(NTPOPS,NTGMX),NC(NTPOPS),NR(NTPOPS),
     *        WTF(NTPOPS,NTGMX),HTF(NTPOPS,NTGMX)
*
****          THIS SECTION READS AND WRITES THE INPUT PARAMETERS
*
      READ 5110, IFLG1,IFLG2
      DO 110 I=1,NTPOPS $IF(I.LE.NCPOPS) THEN
      READ 5130, N(I),C(I),TTH(I),B(I),TGK(I),TMEM(I),THO(I),
     *        NC(I),NR(I),NTGR(I) $ELSE $II=I-NCPOPS
      READ 5140, N(I),P(II),INSTR(II),INSTP(II),INFSED(II),
     *        NC(I),NR(I),NTGR(I) $ENDIF
      DO 120 J=1,NTGR(I)
120   READ 5150, IRCP(I,J),TYPE(I,J),NCT(I,J),NT(I,J),
     *        STR(I,J),INSED(I,J),WTF(I,J),HTF(I,J)
110   CONTINUE
      READ 5110, LTSTOP
*
      WRITE(6,6000) $WRITE(6,6005) NTPOPS,NFPOPS,NCPOPS,NCLS,MCTP1,
     *        NTGMX,IFLG1,IFLG2
      WRITE(6,6112) $WRITE(6,6116) $WRITE(6,6118)
      DO 210 I=1,NTPOPS $IF(I.LE.NCPOPS) THEN
      WRITE(6,6130) N(I),C(I),TTH(I),B(I),TGK(I),TMEM(I),THO(I),
     *        NC(I),NR(I),NTGR(I) $ELSE $II=I-NCPOPS
      WRITE(6,6140) N(I),P(II),INSTR(II),INSTP(II),INFSED(II),
     *        NC(I),NR(I),NTGR(I) $ENDIF
      DO 220 J=1,NTGR(I)
220   WRITE(6,6150) IRCP(I,J),TYPE(I,J),NCT(I,J),NT(I,J),
     *        STR(I,J),INSED(I,J),WTF(I,J),HTF(I,J)
210   CONTINUE
      WRITE(6,6160) LTSTOP
*
****          THIS SECTION INITIALIZES VARIABLES
*
      CHAR(1)=1H  $CHAR(2)=1H*
      DO 310 I=1,NCPOPS
      DCTH(I)=EXP(-STEP/TTH(I)) $DCG(I)=EXP(-STEP/TGK(I))
      DO 310 J=1,NCLS
      E(I,J)=0. $TH(I,J)=THO(I) $S(I,J)=0. $GK(I,J)=0.
      DO 310 K=1,MCTP1
      SCIT(I,J,K)=0. $GI(I,J,K)=0.
310   CONTINUE
      DO 320 I=1,NTPOPS $DO 320 J=1,W $DO 320 K=1,H
320   LINE(I,J,K)=0
*
```

```
****          THIS SECTION UPDATES STATE VARIABLES AT EACH TIME STEP
*
      DO 1000 L=1,LTSTOP
*
*
****                      UPDATING THE CELLS
*
      DO 2000 I=1,NCPOPS
      NF(I)=0 $IF(IFLG1.GT.0) WRITE(7,7060) L,I
      DO 2010 J=1,W $DO 2010 K=1,H
 2010 ALINE(J,K)=CHAR(1)
      DO 2900 J=1,N(I)
      CALL SVUPDT(I,J)
      IF(E(I,J).LT.TH(I,J)) GO TO 2850
      S(I,J)=1. $NF(I)=NF(I)+1 $IDF(NF(I))=J
      JX=INT(FLOAT(MOD((J-1),NC(I))+1)*FLOAT(W)/FLOAT(NC(I)))
      JY=INT(FLOAT((J-1)/NC(I)+1)*FLOAT(H)/FLOAT(NR(I)))
      ALINE(JX,JY)=CHAR(2)
      LINE(I,JX,JY)=LINE(I,JX,JY)+1
      DO 2800 K=1,NTGR(I)
      IF(NT(I,K).GT.0) CALL TRNSMT(I,K,J)
 2800 CONTINUE
 2850 CONTINUE
 2900 CONTINUE $IF(IFLG1.GT.0) WRITE(7,7060) (IDF(J),J=1,NF(I))
      IF(IFLG2.GT.0) THEN $WRITE(6,6122) $DO 2700 JY=1,H
 2700 WRITE(6,6060) (ALINE(JX,H+1-JY),JX=1,W) $ENDIF
 2000 CONTINUE
*
****                  UPDATING THE INPUT FIBERS
*
      DO 3000 I=1,NFPOPS
      II=I+NCPOPS $NF(II)=0 $IF(IFLG1.GT.0) WRITE(7,7060) L,II
      IF(L.LT.INSTR(I).OR.L.GE.INSTP(I)) GO TO 3000
      IF(IFLG2.GT.0) THEN $DO 3010 J=1,W $DO 3010 K=1,H
 3010 ALINE(J,K)=CHAR(1) $ENDIF
      CALL RANSET(INFSED(I))
      DO 3900 J=1,N(II)
      IF(RANF(0.).GT.P(I)) GO TO 3850 $CALL RANGET(INFSED(I))
      NF(II)=NF(II)+1 $IDF(NF(II))=J
      JX=INT(FLOAT(MOD((J-1),NC(I))+1)*FLOAT(W)/FLOAT(NC(I)))
      JY=INT(FLOAT((J-1)/NC(I)+1)*FLOAT(H)/FLOAT(NR(I)))
      ALINE(JX,JY)=CHAR(2)
      LINE(II,JX,JY)=LINE(II,JX,JY)+1
      DO 3800 K=1,NTGR(II)
      IF(NT(II,K).GT.0) CALL TRNSMT(II,K,J)
 3800 CONTINUE $CALL RANSET(INFSED(I))
 3850 CONTINUE
 3900 CONTINUE $IF(IFLG1.GT.0) WRITE(7,7060) (IDF(J),J=1,NF(II))
      IF(IFLG2.GT.0) THEN $WRITE(6,6122) $DO 3700 JY=1,H
 3700 WRITE(6,6060) (ALINE(JX,H+1-JY),JX=1,W) $ENDIF
      CALL RANGET(INFSED(I))
 3000 CONTINUE
*
      WRITE(6,6510) L,(NF(I),I=1,NTPOPS)
      WRITE(7,7510) L,(NF(I),I=1,NTPOPS)
*
      DO 1910 I=1,NCPOPS $DO 1910 J=1,NCLS $DO 1920 K=1,MCTP1-1
      SCIT(I,J,K)=SCIT(I,J,K+1) $GI(I,J,K)=GI(I,J,K+1)
 1920 CONTINUE $SCIT(I,J,MCTP1)=0. $GI(I,J,MCTP1)=0.
 1910 CONTINUE
*
 1000 CONTINUE
*
****          THIS SECTION WRITES OUT ACTIVITY VARIABLES
*
      DO 4700 I=1,NTPOPS $WRITE(6,6122) $DO 4710 JY=1,H
 4710 WRITE(6,6070) (LINE(I,JX,H+1-JY),JX=1,W)
```

```
 4700 CONTINUE
*
      STOP
*
****          THESE ARE THE FORMATS
*
 5110 FORMAT(12I6)
 5130 FORMAT(I6,6F6.2,3I6)
 5140 FORMAT(I6,F6.4,6I6)
 5150 FORMAT(4I6,F6.2,I6,2F6.2)
 6000 FORMAT(1H1)
 6005 FORMAT(5X,*NTPOPS,NFPOPS,NCPOPS,NCLS,MCTP1,NTGMX,IFLG1,*
     *,*IFLG2=:*,8I7)
 6112 FORMAT(//,5X,*HERE ARE, FOR EACH POPULATION:,*,/,10X,*N,C,TTH,B,TG
     *K,TMEM,THO,NC,NR,NTGR*,/,20X,*OR*)
 6116 FORMAT(10X,*N,P,INSTR,INSTP,INFSED,NC,NR,NTGR*,/,20X,*AND*)
 6118 FORMAT(10X,*IRCP,TYPE,NCT,NT,STR,INSED,WTF,HTF..FOR EACH TG*)
 6122 FORMAT(//)
 6130 FORMAT(/,5X,I7,6F7.2,3I7)
 6140 FORMAT(/,5X,I7,F7.4,6I7)
 6150 FORMAT(4I7,F7.2,I7,2F7.2)
 6180 FORMAT(5X,*LTSTOP=*,I7)
 6060 FORMAT(5X,80A1)
 6070 FORMAT(2X,40I3)
 6510 FORMAT(5X,15I7)
 7510 FORMAT(15I5)
 7060 FORMAT(20I4)
      END
*
      SUBROUTINE SVUPDT(I,K)
*
***  THIS SUBROUTINE UPDATES THE STATE VARIABLES OF AN INDIVIDUAL NEURON
*
      PARAMETER NCPOPS=3,NCLS=200,MCTP1=2,STEP=1.,EK=-10.,EI=-10.
      COMMON SCIT(NCPOPS,NCLS,MCTP1),GI(NCPOPS,NCLS,MCTP1)
      COMMON/SV/E(NCPOPS,NCLS),TH(NCPOPS,NCLS),S(NCPOPS,NCLS),
     *         GK(NCPOPS,NCLS),B(NCPOPS),C(NCPOPS),DCG(NCPOPS),
     *         DCTH(NCPOPS),TMEM(NCPOPS),THO(NCPOPS)
*
      GK(I,K)=GK(I,K)*DCG(I)+B(I)*(1.-DCG(I))*S(I,K)
      GTOT=1.+GK(I,K)+GI(I,K,1)  $DCE=EXP(-GTOT*STEP/TMEM(I))
      E(I,K)=E(I,K)*DCE+(SCIT(I,K,1)+GI(I,K,1)*EI+GK(I,K)*EK)*(1.-DCE)/
     *      GTOT
      TH(I,K)=THO(I)+(TH(I,K)-THO(I))*DCTH(I)+C(I)*E(I,K)*(1.-DCTH(I))
      S(I,K)=0.
*
      RETURN
      END
*
      SUBROUTINE TRNSMT(IS,L,K)
*
****      THIS SUBROUTINE PROJECTS SYNAPTIC ACTIVATION FROM A SENDING
****      FIBER TO ITS TARGETED RECEIVING CELLS.
*
      INTEGER TYPE,W,H
      PARAMETER NTPOPS=5,NCPOPS=3,NCLS=200,MCTP1=2,W=20,H=10,NTGMX=2
      COMMON SCIT(NCPOPS,NCLS,MCTP1),GI(NCPOPS,NCLS,MCTP1)
      COMMON/TR/NCT(NTPOPS,NTGMX),NT(NTPOPS,NTGMX),STR(NTPOPS,
     *         NTGMX),INSED(NTPOPS,NTGMX),N(NTPOPS),TYPE(NTPOPS,
     *         NTGMX),IRCP(NTPOPS,NTGMX),NC(NTPOPS),NR(NTPOPS),
     *         WTF(NTPOPS,NTGMX),HTF(NTPOPS,NTGMX)
*
      IR=IRCP(IS,L)
      XS=FLOAT(MOD(K,NC(IS)))*FLOAT(W)/FLOAT(NC(IS))
      IF(XS.EQ.0.) XS=FLOAT(W)
      YS=FLOAT(K/NC(IS)+1)*FLOAT(H)/FLOAT(NR(IS))
      CALL RANSET(K*INSED(IS,L))
```

```
      DO 100 J=1,NT(IS,L)
      XR=XS+(RANF(0.)-.5)*WTF(IS,L) $YR=YS+(RANF(0.)-.5)*HTF(IS,L)
      ICR=INT(FLOAT(NC(IR))*(XR-.5)/FLOAT(W))+1
      IRR=INT(FLOAT(NR(IR))*(YR-.5)/FLOAT(H))+1
      IF(ICR.LE.0) ICR=NC(IR)+ICR
      IF(ICR.GT.NC(IR)) ICR=ICR-NC(IR)
      IF(IRR.LE.0) IRR=NR(IR)+IRR
      IF(IRR.GT.NR(IR)) IRR=IRR-NR(IR)
      NREC=NC(IR)*(IRR-1)+ICR
      KCT=INT(RANF(0.)*FLOAT(NCT(IS,L)))+2
      IF(TYPE(IS,L).EQ.1) SCIT(IR,NREC,KCT)=SCIT(IR,NREC,KCT)+
     *                    STR(IS,L)
      IF(TYPE(IS,L).EQ.2) GI(IR,NREC,KCT)=GI(IR,NREC,KCT)+
     *                    STR(IS,L)
  100 CONTINUE
*
      RETURN
      END
/EOR
    0    0
   50  0.   25.   20.    5.    5.  10.      10    5    2
    3    1    1    8  5.   1227  9.    5.
    2    1    1    8  5.   1936  9.    5.
   50  .5  25.   20.    5.    5.  10.      10    5    0
   50  0.   25.   20.    5.    5.  10.      10    5    1
    1    2    1   10 10.   3421 20.   10.
   50  .1       1  999  4165   10      5    1
    1    1    1   10 10.   8917 20.   10.
   50  .1      15   50  3315   10      5    1
    3    1    1   10 10.   6373 20.   10.
   20
/EOR
/EOI


/JOB
A224,CM=70000,TL=200.
FAKE.
PDFV.
M77(PL=20000).
SETLIM,PR.
LGO,SYS00E.
SAVE,SYS00E.
EXIT.
SAVE,SYS00E.
/EOR
      PROGRAM SYSTM22 (SYSTM,INPUT,OUTPUT,TAPE6=OUTPUT,TAPE7=SYSTM)
*
****     THIS PROGRAM SIMULATES THE ACTIVITY OF AN ARBITRARY SYSTEM OF
*        INTERCONNECTED SPATIALLY-ORGANIZED NEURAL NETWORKS ACTIVATED BY
*        AN ARBITRARY NUMBER OF SPATIALLY-ORGANIZED INPUT FIBER SYSTEMS.
****     ALL SYNAPTIC SYSTEMS ARE INDIVIDUALLY ADJUSTABLE.
*
      INTEGER TYPE,W,H,SNTP
      PARAMETER NTPOPS=5,NFPOPS=2,NCPOPS=3,NCLS=200,MCTP1=2,W=20,H=10,
     *       STEP=1.,SNTP=2,NTGMX=2
      DIMENSION P(NFPOPS),INSTR(NFPOPS),INSTP(NFPOPS),INFSED(NFPOPS),
     *          IDF(NCLS),TTH(NCPOPS),TGK(NCPOPS),CHAR(2),ALINE(W,H),
     *          LINE(NTPOPS,W,H),T(SNTP),DCS(SNTP),NF(NTPOPS),
     *          NTGR(NTPOPS)
      COMMON G(SNTP,NCPOPS,NCLS,MCTP1)
      COMMON/SV/E(NCPOPS,NCLS),TH(NCPOPS,NCLS),S(NCPOPS,NCLS),
     *          GK(NCPOPS,NCLS),B(NCPOPS),C(NCPOPS),DCG(NCPOPS),
     *          DCTH(NCPOPS),TMEM(NCPOPS),THO(NCPOPS),EQ(SNTP)
      COMMON/TR/NCT(NTPOPS,NTGMX),NT(NTPOPS,NTGMX),STR(NTPOPS,
     *          NTGMX),INSED(NTPOPS,NTGMX),N(NTPOPS),TYPE(NTPOPS,
```

```
      *              NTGMX),IRCP(NTPOPS,NTGMX),NC(NTPOPS),NR(NTPOPS),
      *              WTF(NTPOPS,NTGMX),HTF(NTPOPS,NTGMX)
*
****          THIS SECTION READS AND WRITES THE INPUT PARAMETERS
*
      READ 5110, IFLG1,IFLG2
      READ 5120, ((EQ(I),T(I)),I=1,SNTP)
      DO 110 I=1,NTPOPS $IF(I.LE.NCPOPS) THEN
      READ 5130, N(I),C(I),TTH(I),B(I),TGK(I),TMEM(I),THO(I),
      *          NC(I),NR(I),NTGR(I) $ELSE $II=I-NCPOPS
      READ 5140, N(I),P(II),INSTR(II),INSTP(II),INFSED(II),
      *          NC(I),NR(I),NTGR(I) $ENDIF
      DO 120 J=1,NTGR(I)
 120  READ 5150, IRCP(I,J),TYPE(I,J),NCT(I,J),NT(I,J),
      *          STR(I,J),INSED(I,J),WTF(I,J),HTF(I,J)
 110  CONTINUE
      READ 5110, LTSTOP
*
      WRITE(6,6000) $WRITE(6,6005) NTPOPS,NFPOPS,NCPOPS,NCLS,MCTP1,
      *          NTGMX,IFLG1,IFLG2
      WRITE(6,6110) ((EQ(I),T(I)),I=1,SNTP)
      WRITE(6,6112) $WRITE(6,6116) $WRITE(6,6118)
      DO 210 I=1,NTPOPS $IF(I.LE.NCPOPS) THEN
      WRITE(6,6130) N(I),C(I),TTH(I),B(I),TGK(I),TMEM(I),THO(I),
      *          NC(I),NR(I),NTGR(I) $ELSE $II=I-NCPOPS
      WRITE(6,6140) N(I),P(II),INSTR(II),INSTP(II),INFSED(II),
      *          NC(I),NR(I),NTGR(I) $ENDIF
      DO 220 J=1,NTGR(I)
 220  WRITE(6,6150) IRCP(I,J),TYPE(I,J),NCT(I,J),NT(I,J),
      *          STR(I,J),INSED(I,J),WTF(I,J),HTF(I,J)
 210  CONTINUE
      WRITE(6,6160) LTSTOP
*
****          THIS SECTION INITIALIZES VARIABLES
*
      CHAR(1)=1H  $CHAR(2)=1H*
      DO 310 I=1,NCPOPS
      DCTH(I)=EXP(-STEP/TTH(I)) $DCG(I)=EXP(-STEP/TGK(I))
      DO 310 J=1,NCLS
      E(I,J)=0. $TH(I,J)=THO(I) $S(I,J)=0. $GK(I,J)=0.
      DO 310 K=1,MCTP1 $DO 310 M=1,SNTP
      G(M,I,J,K)=0.
 310  CONTINUE
      DO 320 I=1,NTPOPS $DO 320 J=1,W $DO 320 K=1,H
 320  LINE(I,J,K)=0
      DO 330 I=1,SNTP
 330  DCS(I)=EXP(-STEP/T(I))
*
****          THIS SECTION UPDATES STATE VARIABLES AT EACH TIME STEP
*
      DO 1000 L=1,LTSTOP
*
*
****                UPDATING THE CELLS
*
      DO 2000 I=1,NCPOPS
      NF(I)=0 $IF(IFLG1.GT.0) WRITE(7,7060) L,I
      DO 2010 J=1,W $DO 2010 K=1,H
 2010 ALINE(J,K)=CHAR(1)
      DO 2900 J=1,N(I)
      CALL SVUPDT(I,J)
      IF(E(I,J).LT.TH(I,J)) GO TO 2850
      S(I,J)=1. $NF(I)=NF(I)+1 $IDF(NF(I))=J
      JX=INT(FLOAT(MOD((J-1),NC(I))+1)*FLOAT(W)/FLOAT(NC(I)))
      JY=INT(FLOAT((J-1)/NC(I)+1)*FLOAT(H)/FLOAT(NR(I)))
      ALINE(JX,JY)=CHAR(2)
      LINE(I,JX,JY)=LINE(I,JX,JY)+1
```

```
      DO 2800 K=1,NTGR(I)
      IF(NT(I,K).GT.0) CALL TRNSMT(I,K,J)
 2800 CONTINUE
 2850 CONTINUE
 2900 CONTINUE $IF(IFLG1.GT.0) WRITE(7,7060) (IDF(J),J=1,NF(I))
      IF(IFLG2.GT.0) THEN $WRITE(6,6122) $DO 2700 JY=1,H
 2700 WRITE(6,6060) (ALINE(JX,H+1-JY),JX=1,W) $ENDIF
 2000 CONTINUE
*
****                UPDATING THE INPUT FIBERS
*
      DO 3000 I=1,NFPOPS
      II=I+NCPOPS $NF(II)=0 $IF(IFLG1.GT.0) WRITE(7,7060) L,II
      IF(L.LT.INSTR(I).OR.L.GE.INSTP(I)) GO TO 3000
      IF(IFLG2.GT.0) THEN $DO 3010 J=1,W $DO 3010 K=1,H
 3010 ALINE(J,K)=CHAR(1) $ENDIF
      CALL RANSET(INFSED(I))
      DO 3900 J=1,N(II)
      IF(RANF(0.).GT.P(I)) GO TO 3850 $CALL RANGET(INFSED(I))
      NF(II)=NF(II)+1 $IDF(NF(II))=J
      JX=INT(FLOAT(MOD((J-1),NC(I))+1)*FLOAT(W)/FLOAT(NC(I)))
      JY=INT(FLOAT((J-1)/NC(I)+1)*FLOAT(H)/FLOAT(NR(I)))
      ALINE(JX,JY)=CHAR(2)
      LINE(II,JX,JY)=LINE(II,JX,JY)+1
      DO 3800 K=1,NTGR(II)
      IF(NT(II,K).GT.0) CALL TRNSMT(II,K,J)
 3800 CONTINUE $CALL RANSET(INFSED(I))
 3850 CONTINUE
 3900 CONTINUE $IF(IFLG1.GT.0) WRITE(7,7060) (IDF(J),J=1,NF(II))
      IF(IFLG2.GT.0) THEN $WRITE(6,6122) $DO 3700 JY=1,H
 3700 WRITE(6,6060) (ALINE(JX,H+1-JY),JX=1,W) $ENDIF
      CALL RANGET(INFSED(I))
 3000 CONTINUE
*
      WRITE(6,6510) L,(NF(I),I=1,NTPOPS)
      WRITE(7,7510) L,(NF(I),I=1,NTPOPS)
*
      DO 1910 I=1,NCPOPS $DO 1910 J=1,NCLS $DO 1910 M=1,SNTP
      G(M,I,J,1)=G(M,I,J,1)*DCS(M)+G(M,I,J,2) $DO 1920 K=2,MCTP1-1
 1920 G(M,I,J,K)=G(M,I,J,K+1) $G(M,I,J,MCTP1)=0.
 1910 CONTINUE
*
 1000 CONTINUE
*
****        THIS SECTION WRITES OUT ACTIVITY VARIABLES
*
      DO 4700 I=1,NTPOPS $WRITE(6,6122) $DO 4710 JY=1,H
 4710 WRITE(6,6070) (LINE(I,JX,H+1-JY),JX=1,W)
 4700 CONTINUE
*
      STOP
*
****        THESE ARE THE FORMATS
*
 5110 FORMAT(12I6)
 5120 FORMAT(12F6.2)
 5130 FORMAT(I6,6F6.2,3I6)
 5140 FORMAT(I6,F6.4,6I6)
 5150 FORMAT(4I6,F6.2,I6,2F6.2)
 6000 FORMAT(1H1)
 6005 FORMAT(5X,*NTPOPS,NFPOPS,NCPOPS,NCLS,MCTP1,NTGMX,IFLG1,*
     *,*IFLG2=:*,8I7)
 6110 FORMAT(5X,*EQ,T=*,12F7.2)
 6112 FORMAT(//,5X,*HERE ARE, FOR EACH POPULATION:,*,/,10X,*N,C,TTH,B,TG
     *K,TMEM,THO,NC,NR,NTGR*,/,20X,*OR*)
 6116 FORMAT(10X,*N,P,INSTR,INSTP,INFSED,NC,NR,NTGR*,/,20X,*AND*)
 6118 FORMAT(10X,*IRCP,TYPE,NCT,NT,STR,INSED,WTF,HTF..FOR EACH TG*)
```

```
6122 FORMAT(//)
6130 FORMAT(/,5X,I7,6F7.2,3I7)
6140 FORMAT(/,5X,I7,F7.4,6I7)
6150 FORMAT(4I7,F7.2,I7,2F7.2)
6160 FORMAT(5X,*LTSTOP=*,I7)
6060 FORMAT(5X,80A1)
6070 FORMAT(2X,40I3)
6510 FORMAT(5X,15I7)
7510 FORMAT(15I5)
7060 FORMAT(20I4)
     END
*
     SUBROUTINE SVUPDT(I,K)
*
*** THIS SUBROUTINE UPDATES THE STATE VARIABLES OF AN INDIVIDUAL NEURON
*
     INTEGER SNTP
     PARAMETER NCPOPS=3,NCLS=200,MCTP1=2,STEP=1.,EK=-10.,SNTP=2
     COMMON G(SNTP,NCPOPS,NCLS,MCTP1)
     COMMON/SV/E(NCPOPS,NCLS),TH(NCPOPS,NCLS),S(NCPOPS,NCLS),
    *         GK(NCPOPS,NCLS),B(NCPOPS),C(NCPOPS),DCG(NCPOPS),
    *         DCTH(NCPOPS),TMEM(NCPOPS),THO(NCPOPS),EQ(SNTP)
*
     GS=0. $SYN=0. $DO 10 J=1,SNTP
     GS=GS+G(J,I,K,1) $SYN=SYN+G(J,I,K,1)*EQ(J)
  10 CONTINUE
     GK(I,K)=GK(I,K)*DCG(I)+B(I)*(1.-DCG(I))*S(I,K)
     GTOT=1.+GK(I,K)+GS $DCE=EXP(-GTOT*STEP/TMEM(I))
     E(I,K)=E(I,K)*DCE+(SYN+GK(I,K)*EK)*(1.-DCE)/
    *        GTOT
     TH(I,K)=THO(I)+(TH(I,K)-THO(I))*DCTH(I)+C(I)*E(I,K)*(1.-DCTH(I))
     S(I,K)=0.
*
     RETURN
     END
*
     SUBROUTINE TRNSMT(IS,L,K)
*
**** THIS SUBROUTINE PROJECTS SYNAPTIC ACTIVATION FROM A SENDING
**** FIBER TO ITS TARGETED RECEIVING CELLS.
*
     INTEGER TYPE,W,H,SNTP
     PARAMETER NTPOPS=5,NCPOPS=3,NCLS=200,MCTP1=2,W=20,H=10,SNTP=2,
    *        NTGMX=2
     COMMON G(SNTP,NCPOPS,NCLS,MCTP1)
     COMMON/TR/NCT(NTPOPS,NTGMX),NT(NTPOPS,NTGMX),STR(NTPOPS,
    *         NTGMX),INSED(NTPOPS,NTGMX),N(NTPOPS),TYPE(NTPOPS,
    *         NTGMX),IRCP(NTPOPS,NTGMX),NC(NTPOPS),NR(NTPOPS),
    *         WTF(NTPOPS,NTGMX),HTF(NTPOPS,NTGMX)
*
     IR=IRCP(IS,L)
     XS=FLOAT(MOD(K,NC(IS)))*FLOAT(W)/FLOAT(NC(IS))
     IF(XS.EQ.0.) XS=FLOAT(W)
     YS=FLOAT(K/NC(IS)+1)*FLOAT(H)/FLOAT(NR(IS))
     CALL RANSET(INSED(IS,L)+K*N(IS))
     DO 100 J=1,NT(IS,L)
     XR=XS+(RANF(0.)-.5)*WTF(IS,L) $YR=YS+(RANF(0.)-.5)*HTF(IS,L)
     ICR=INT(FLOAT(NC(IR))*(XR-.5)/FLOAT(W))+1
     IRR=INT(FLOAT(NR(IR))*(YR-.5)/FLOAT(H))+1
     IF(ICR.LE.0) ICR=NC(IR)+ICR
     IF(ICR.GT.NC(IR)) ICR=ICR-NC(IR)
     IF(IRR.LE.0) IRR=NR(IR)+IRR
     IF(IRR.GT.NR(IR)) IRR=IRR-NR(IR)
     NREC=NC(IR)*(IRR-1)+ICR
     KCT=INT(RANF(0.)*FLOAT(NCT(IS,L)))+2
     I=TYPE(IS,L)
     G(I,IR,NREC,KCT)=G(I,IR,NREC,KCT)+STR(IS,L)
```

```
    100 CONTINUE
*
      RETURN
      END
/EOR
       0       0
   70.    1.    -10.    5.
     50  0.    25.    20.    5.    5.   10.       10    5    2
      3    1    1     8   .5   1227  9.    5.
      2    1    1     8   .5   1936  9.    5.
     50  .5   25.    20.    5.    5.   10.       10    5    0
     50  0.    25.    20.    5.    5.   10.       10    5    1
      1    2    1    10   1.   3421 20.   10.
     50  .1         1   999  4165   10       5    1
      1    1    1    10   .5   8917 20.   10.
     50  .1        15    50  3315   10       5    1
      3    1    1    10   .5   6373 20.   10.
     20
/EOR
/EOI


/JOB
A224,CM=70000,TL=200.
FAKE.
PDFV.
MNF(Y).
GET,STR000.
LGO,STR000,LN003,STR003.
SAVE,LN003,STR003.
EXIT.
SAVE,LN003,STR003.
/EOR
      PROGRAM JNLRN00 (STRIN,DYN,STROUT,INPUT,OUTPUT,TAPE3=STRIN,TAPE6=
     *            OUTPUT,TAPE7=DYN,TAPE8=STROUT)
*
****        THIS PROGRAM SIMULATES HEBBIAN LEARNING AT THE JUNCTION OF AN
****        INPUT FIBER SYSTEM AND A SINGLE POOL OF RECEIVING NEURONS.
*
      INTEGER ACITSN,ACITTR,ACITTM,OVRSN,OVRTR,OVRTM,OVRRC
      PARAMETER NT=10,NFIB=50,NCLS=50,MCTP1=3,NRNGE=10,MRNGE=1000,
     *        STEP=1.,EX=70.,EI=-10.,EK=-10.
      DIMENSION E(NCLS),TH(NCLS),S(NCLS),GK(NCLS),
     *        IDFF(NFIB),IDFC(NCLS),STR(NFIB,NT),
     *        PER(NFIB),LSTM(NFIB),INSED(3),
     *        ACITSN(NCLS,NRNGE),ACITTR(NCLS,NRNGE),ACITTM(NCLS,NRNGE),
     *        OVRSN(MRNGE),OVRTR(MRNGE),OVRTM(MRNGE),OVRRC(MRNGE),
     *        ISDR(100),ISYN(100)
*
****        THIS SECTION READS AND WRITES THE INPUT PARAMETERS
*
      READ 5010, IFLG1,INSED(2),INSED(3),LTSTOP
      READ 5020, C,TTH,B,TGK,TMEM,TH0
      READ 5030, STR0,PER0,LSTM0,INSED(1)
      CALL RANSET(INSED(1))
      DO 110 I=1,NFIB
      IF(IFLG1.EQ.0) THEN $DO 109 J=1,NT
  109 STR(I,J)=STR0 $ENDIF
      PER(I)=-PER0*ALOG(RANF(0.))
      LSTM(I)=INT(-FLOAT(LSTM0)*ALOG(RANF(0.)))+1
  110 CONTINUE
      IF(IFLG1.GT.0) READ(3,5900) ((STR(I,J),J=1,NT),I=1,NFIB)
      READ 5035, MLAG,STRMX,DLSTR
*
      WRITE(6,6000)
      WRITE(6,6010) NFIB,NCLS,NT,INSED(2),INSED(3),LTSTOP
```

```
      WRITE(6,6020) C,TTH,B,TGK,TMEM,THØ
      WRITE(6,6030) STRØ,PERØ,LSTMØ,INSED(1)
      WRITE(6,6035) MLAG,STRMX,DLSTR
      WRITE(6,6040)
*
****          THIS SECTION INITIALIZES VARIABLES
*
      DO 310 I=1,NCLS
      E(I)=0. $TH(I)=THØ $S(I)=0. $GK(I)=0.
      DO 310 J=1,NRNGE
      ACITSN(I,J)=Ø $ACITTR(I,J)=Ø $ACITTM(I,J)=Ø
  310 CONTINUE
      IVR=0 $DCTH=EXP(-STEP/TTH) $DCG=EXP(-STEP/TGK)
      DO 320 J=1,MRNGE
      OVRSN(J)=Ø $OVRTR(J)=Ø $OVRTM(J)=Ø $OVRRC(J)=Ø
  320 CONTINUE
*
****          THIS SECTION UPDATES STATE VARIABLES AT EACH TIME STEP
*
      DO 1000 L=1,LTSTOP
      NFF=Ø $NFC=Ø $GE=0. $GI=0.
*
****              UPDATING THE INPUT FIBERS
*
      DO 2000 I=1,NFIB
      IF(L.NE.LSTM(I)) GO TO 2000
      CALL RANSET(INSED(2))
      NFF=NFF+1 $IDFF(NFF)=I
      LSTM(I)=L+INT(-PER(I)*ALOG(RANF(Ø.)))+1
      CALL RANGET(INSED(2)) $CALL RANSET(I*INSED(3))
      DO 2600 J=1,NT
      NREC=INT(RANF(Ø.)*FLOAT(NCLS))+1
      NCT=INT(RANF(Ø.)*FLOAT(MCTP1-1))+1
      DO 2500 JR=1,NRNGE
      IF(ACITSN(NREC,JR).EQ.Ø) GO TO 2510
 2500 CONTINUE $IVR=IVR+1 $IF(IVR.GT.(MRNGE-1)) GO TO 1500
      OVRSN(IVR)=I $OVRTR(IVR)=J $OVRTM(IVR)=NCT+1+MLAG $OVRRC(IVR)=NREC
      GO TO 2600
 2510 ACITSN(NREC,JR)=I $ACITTR(NREC,JR)=J $ACITTM(NREC,JR)=NCT+2+MLAG
 2600 CONTINUE
 2000 CONTINUE
*
*
****              UPDATING THE CELLS
*
      DO 3000 I=1,NCLS
      JCR=Ø $SC=0.
      DO 3500 J=1,NRNGE
 3570 ACITTM(I,J)=ACITTM(I,J)-1 $IF(ACITSN(I,J).EQ.Ø) GO TO 3599
      IF(ACITTM(I,J).GT.MLAG+1) GO TO 3500
      JCR=JCR+1 $ISDR(JCR)=ACITSN(I,J) $ISYN(JCR)=ACITTR(I,J)
      IF(ACITTM(I,J).GT.MLAG) SC=SC+STR(ISDR(JCR),ISYN(JCR))
      IF(ACITTM(I,J).GT.1) GO TO 3500 $JUP=NRNGE-1
      DO 3560 J1=J,JUP
      ACITSN(I,J1)=ACITSN(I,J1+1) $ACITTR(I,J1)=ACITTR(I,J1+1)
 3560 ACITTM(I,J1)=ACITTM(I,J1+1)
      ACITSN(I,NRNGE)=Ø $ACITTR(I,NRNGE)=Ø $ACITTM(I,NRNGE)=Ø
      GO TO 3570
 3500 CONTINUE
 3599 CONTINUE
      JUP=IVR
      DO 3600 J=1,JUP
 3640 IF(I.NE.OVRRC(J)) GO TO 3600
      IF(OVRTM(J).GT.MLAG+1) GO TO 3600
      JCR=JCR+1 $ISDR(JCR)=OVRSN(J) $ISYN(JCR)=OVRTR(J)
      IF(OVRTM(J).GT.MLAG) SC=SC+STR(ISDR(JCR),ISYN(JCR))
      IF(OVRTM(J).GT.1) GO TO 3600
```

```
      IVR=IVR-1 $DO 3630 J1=J,IVR
      OVRSN(J1)=OVRSN(J1+1)$OVRTR(J1)=OVRTR(J1+1)$OVRTM(J1)=OVRTM(J1+1)
 3630 OVRRC(J1)=OVRRC(J1+1)
      OVRSN(IVR+1)=0 $OVRTR(IVR+1)=0 $OVRTM(IVR+1)=0 $OVRRC(IVR+1)=0
      GO TO 3640
 3600 CONTINUE
*
      GK(I)=GK(I)*DCG+B*(1.-DCG)*S(I)  $GTOT=1.+GE+GI+GK(I)
      DCE=EXP(-GTOT*STEP/TMEM)
      E(I)=E(I)*DCE+(SC+GE*EX+GI*EI+GK(I)*EK)*(1.-DCE)/GTOT
      TH(I)=TH0+(TH(I)-TH0)*DCTH+C*E(I)*(1.-DCTH) $S(I)=0.
      IF(E(I).LT.TH(I)) GO TO 3000
      S(I)=1. $NFC=NFC+1 $IDFC(NFC)=I
*
      DO 3700 J=1,JCR $K=ISDR(J) $K1=ISYN(J)
 3700 STR(K,K1)=STR(K,K1)+DLSTR*(STRMX-STR(K,K1))
*
 3000 CONTINUE
*
      WRITE(6,6050) L,NFF,NFC,(IDFF(I),I=1,NFF),(IDFC(J),J=1,NFC)
      WRITE(7,7050) L,NFF,NFC,(IDFF(I),I=1,NFF),(IDFC(J),J=1,NFC)
      DO 4000 J=1,IVR
 4000 OVRTM(J)=OVRTM(J)-1
*
 1000 CONTINUE
*
****          THIS SECTION WRITES OUT ACTIVITY VARIABLES
*
      WRITE(6,6040)
      WRITE(6,6900) ((STR(I,J),J=1,NT),I=1,NFIB)
      WRITE(8,5900) ((STR(I,J),J=1,NT),I=1,NFIB)
 1500 CONTINUE
      STOP
****          THESE ARE THE FORMATS
*
 5010 FORMAT(12I6)
 5020 FORMAT(12F6.2)
 5030 FORMAT(2F6.2,2I6)
 5035 FORMAT(I6,2F6.2)
 5900 FORMAT(F6.3)
 6000 FORMAT(1H1)
 6010 FORMAT(5X,*NFIB,NCLS,NT,INSEDS,LTSTOP=*,6I6)
 6020 FORMAT(5X,*C,TTH,B,TGK,TMEM,TH0=*,6F8.2)
 6030 FORMAT(5X,*STR0,PER0,LSTM0,INSED0=*,2F7.2,2I7)
 6035 FORMAT(5X,*MLAG,STRMX,DLSTR=*,I7,2F7.2)
 6040 FORMAT(///)
 6050 FORMAT(5X,30I4)
 6900 FORMAT(5X,20F6.3)
 7050 FORMAT(30I4)
      END
/EOR
    0  1227  1936   200
  .00 20.    20.    5.    5.   10.
  2.   20.     25  4748
    3  4.    .25
/EOR
/EOI


/JOB
A224,CM=70000,TL=200.
FAKE.
PDFV.
MNF(Y).
GET,STR000.
LGO,STR000,LN005,STR005.
```

```
SAVE,LN005,STR005.
EXIT.
SAVE,LN005,STR005.
/EOR
      PROGRAM JNLRN0A (STRIN,DYN,STROUT,INPUT,OUTPUT,TAPE3=STRIN,TAPE6=
     *           OUTPUT,TAPE7=DYN,TAPE8=STROUT)
*
*** THIS PROGRAM SIMULATES HEBBIAN LEARNING AT THE JUNCTION OF A
*   SINGLE INPUT FIBER SYSTEM WITH A SINGLE POOL OF RECEIVING NEURONS.
*** THE SYNAPTIC ACTIVITY IN TRANSIT IS DATA-PACKED.
*
      INTEGER ACIT,OVR
      PARAMETER NT=10,NFIB=50,NCLS=50,MCTP1=3,NRNGE=10,MRNGE=1000,
     *          STEP=1.,EX=70.,EI=-10.,EK=-10.
      DIMENSION E(NCLS),TH(NCLS),S(NCLS),GK(NCLS),
     *          IDFF(NFIB),IDFC(NCLS),STR(NFIB,NT),
     *          PER(NFIB),LSTM(NFIB),INSED(3),
     *          ACIT(NCLS,NRNGE),OVR(MRNGE),ISDR(200),ISYN(200)
*
****       THIS SECTION READS AND WRITES THE INPUT PARAMETERS
*
      READ 5010, IFLG1,INSED(2),INSED(3),LTSTOP
      READ 5020, C,TTH,B,TGK,TMEM,TH0
      READ 5030, STR0,PER0,LSTM0,INSED(1)
      CALL RANSET(INSED(1))
      DO 110 I=1,NFIB
      IF(IFLG1.EQ.0) THEN $DO 109 J=1,NT
  109 STR(I,J)=STR0 $ENDIF
      PER(I)=-PER0*ALOG(RANF(0.))
      LSTM(I)=INT(-FLOAT(LSTM0)*ALOG(RANF(0.)))+1
  110 CONTINUE
      IF(IFLG1.GT.0) READ(3,5900) ((STR(I,J),J=1,NT),I=1,NFIB)
      READ 5035, MLAG,STRMX,DLSTR
*
      WRITE(6,6000)
      WRITE(6,6010) NFIB,NCLS,NT,INSED(2),INSED(3),LTSTOP
      WRITE(6,6020) C,TTH,B,TGK,TMEM,TH0
      WRITE(6,6030) STR0,PER0,LSTM0,INSED(1)
      WRITE(6,6035) MLAG,STRMX,DLSTR
      WRITE(6,6040)
*
****       THIS SECTION INITIALIZES VARIABLES
*
      DO 310 I=1,NCLS
      E(I)=0. $TH(I)=TH0 $S(I)=0. $GK(I)=0.
      DO 310 J=1,NRNGE
  310 ACIT(I,J)=0
      IVR=0 $DCTH=EXP(-STEP/TTH) $DCG=EXP(-STEP/TGK)
      DO 320 J=1,MRNGE
  320 OVR(J)=0
      LG0=10000000*MLAG $LG1=10000000*(MLAG+1)
*
****       THIS SECTION UPDATES STATE VARIABLES AT EACH TIME STEP
*
      DO 1000 L=1,LTSTOP
      NFF=0 $NFC=0 $GE=0. $GI=0.
*
****             UPDATING THE INPUT FIBERS
*
      DO 2000 I=1,NFIB
      IF(L.NE.LSTM(I)) GO TO 2000
      CALL RANSET(INSED(2))
      NFF=NFF+1 $IDFF(NFF)=I
      LSTM(I)=L+INT(-PER(I)*ALOG(RANF(0.)))+1
      CALL RANGET(INSED(2)) $CALL RANSET(I*INSED(3))
      DO 2600 J=1,NT
      NREC=INT(RANF(0.)*FLOAT(NCLS))+1
```

```
      NCT=INT(RANF(0.)*FLOAT(MCTP1-1))+1
      DO 2500 JR=1,NRNGE
      IF(ACIT(NREC,JR).EQ.0) GO TO 2510
2500  CONTINUE $IVR=IVR+1 $IF(IVR.GT.(MRNGE-1)) GO TO 1500
      OVR(IVR)=J+1000*I+10000000*(NCT+MLAG)+10000000000*NREC $GO TO 2600
2510  ACIT(NREC,JR)=J+1000*I+10000000*(NCT+1+MLAG)
2600  CONTINUE
2000  CONTINUE
*
*
****                  UPDATING THE CELLS
*
      DO 3000 I=1,NCLS
      JCR=0 $SC=0.
      DO 3500 J=1,NRNGE
3570  IF(ACIT(I,J).EQ.0) GO TO 3599 $IT=ACIT(I,J)-10000000
      IF(IT.GT.LG1) GO TO 3500 $ID=MOD(IT,10000000)
      JCR=JCR+1 $ISDR(JCR)=ID/1000 $ISYN(JCR)=ID-(ID/1000)*1000
      IF(IT.GT.LG0) SC=SC+STR(ISDR(JCR),ISYN(JCR))
      IF(IT.GT.10000000) GO TO 3500 $JUP=NRNGE-1
      DO 3560 J1=J,JUP
3560  ACIT(I,J1)=ACIT(I,J1+1) $ACIT(I,NRNGE)=0 $GO TO 3570
3500  ACIT(I,J)=IT
3599  CONTINUE
      JUP=IVR
      DO 3600 J=1,JUP
3640  IF(OVR(J)/10000000000.NE.I) GO TO 3600
      IT=OVR(J)-10000000000*I $IF(IT.GT.LG1) GO TO 3600
      ID=MOD(IT,10000000) $JCR=JCR+1
      ISDR(JCR)=ID/1000 $ISYN(JCR)=ID-(ID/1000)*1000
      IF(IT.GT.LG0) SC=SC+STR(ISDR(JCR),ISYN(JCR))
      IF(IT.GT.10000000) GO TO 3600
      IVR=IVR-1 $DO 3630 J1=J,IVR
3630  OVR(J1)=OVR(J1+1) $OVR(IVR+1)=0 $GO TO 3640
3600  CONTINUE
*
      GK(I)=GK(I)*DCG+B*(1.-DCG)*S(I)
      GTOT=1.+GE+GI+GK(I) $DCE=EXP(-GTOT*STEP/TMEM)
      E(I)=E(I)*DCE+(SC+GE*EX+GI*EI+GK(I)*EK)*(1.-DCE)/GTOT
      TH(I)=TH0+(TH(I)-TH0)*DCTH+C*E(I)*(1.-DCTH) $S(I)=0.
      IF(E(I).LT.TH(I)) GO TO 3000
      S(I)=1. $NFC=NFC+1 $IDFC(NFC)=I
*
      DO 3700 J=1,JCR $K=ISDR(J) $K1=ISYN(J)
3700  STR(K,K1)=STR(K,K1)+DLSTR*(STRMX-STR(K,K1))
*
3000  CONTINUE
*
      WRITE(6,6050) L,NFF,NFC,(IDFF(I),I=1,NFF),(IDFC(J),J=1,NFC)
      WRITE(7,7050) L,NFF,NFC,(IDFF(I),I=1,NFF),(IDFC(J),J=1,NFC)
      DO 4000 J=1,IVR
4000  OVR(J)=OVR(J)-10000000
*
1000  CONTINUE
*
****        THIS SECTION WRITES OUT ACTIVITY VARIABLES
*
      WRITE(6,6040)
      WRITE(6,6900) ((STR(I,J),J=1,NT),I=1,NFIB)
      WRITE(8,5900) ((STR(I,J),J=1,NT),I=1,NFIB)
1500  CONTINUE
      STOP
****        THESE ARE THE FORMATS
*
5010  FORMAT(12I6)
5020  FORMAT(12F6.2)
5030  FORMAT(2F6.2,2I6)
```

```
5035 FORMAT(I6,2F6.2)
5900 FORMAT(F6.3)
6000 FORMAT(1H1)
6010 FORMAT(5X,*NFIB,NCLS,NT,INSEDS,LTSTOP=*,6I6)
6020 FORMAT(5X,*C,TTH,B,TGK,TMEM,TH0=*,6F8.2)
6030 FORMAT(5X,*STR0,PER0,LSTM0,INSED0=*,2F7.2,2I7)
6035 FORMAT(5X,*MLAG,STRMX,DLSTR=*,I7,2F7.2)
6040 FORMAT(///)
6050 FORMAT(5X,30I4)
6900 FORMAT(5X,20F6.3)
7050 FORMAT(30I4)
     END
/EOR
     0  1227  1936    31
  .00 20.    20.     5.     5.    10.
   2.  20.       25   4748
     3  4.       .25
/EOR
/EOI


/JOB
A224,CM=70000,TL=200.
FAKE.
PDFV.
MNF(Y).
GET,STR000.
LGO,STR000,LN004,STR004.
SAVE,LN004,STR004.
EXIT.
SAVE,LN004,STR004.
/EOR
     PROGRAM JNLRN0B (STRIN,DYN,STROUT,INPUT,OUTPUT,TAPE3=STRIN,TAPE6=
    *                 OUTPUT,TAPE7=DYN,TAPE8=STROUT)
*
*** THIS PROGRAM SIMULATES HEBBIAN LEARNING AT THE JUNCTION OF A
*   SINGLE INPUT FIBER SYSTEM AND A SINGLE POOL OF RECEIVING NEURONS.
*   SYNAPTIC ACTIVITY IN TRANSIT IS DATA-PACKED. SEPARTATE MEMORY AND
*** TRANSMISSION LINKS.
*
     INTEGER ACIT,OVR
     PARAMETER NT=10,NFIB=50,NCLS=50,MCTP1=3,NRNGE=10,MRNGE=1000,
    *          STEP=1.,EX=70.,EI=-10.,EK=-10.
     DIMENSION E(NCLS),TH(NCLS),S(NCLS),GK(NCLS),SC(NCLS,MCTP1),
    *          IDFF(NFIB),IDFC(NCLS),STR(NFIB,NT),
    *          PER(NFIB),LSTM(NFIB),INSED(3),
    *          ACIT(NCLS,NRNGE),OVR(MRNGE)
*
****     THIS SECTION READS AND WRITES THE INPUT PARAMETERS
*
     READ 5010, IFLG1,INSED(2),INSED(3),LTSTOP
     READ 5020, C,TTH,B,TGK,TMEM,TH0
     READ 5030, STR0,PER0,LSTM0,INSED(1)
     CALL RANSET(INSED(1))
     DO 110 I=1,NFIB
     IF(IFLG1.EQ.0) THEN $DO 109 J=1,NT
 109 STR(I,J)=STR0 $ENDIF
     PER(I)=-PER0*ALOG(RANF(0.))
     LSTM(I)=INT(-FLOAT(LSTM0)*ALOG(RANF(0.)))+1
 110 CONTINUE
     IF(IFLG1.GT.0) READ(3,5900) ((STR(I,J),J=1,NT),I=1,NFIB)
     READ 5035, MLAG,STRMX,DLSTR
*
     WRITE(6,6000)
     WRITE(6,6010) NFIB,NCLS,NT,INSED(2),INSED(3),LTSTOP
     WRITE(6,6020) C,TTH,B,TGK,TMEM,TH0
```

```
      WRITE(6,6030) STRØ,PERØ,LSTMØ,INSED(1)
      WRITE(6,6035) MLAG,STRMX,DLSTR
      WRITE(6,6040)
*
****          THIS SECTION INITIALIZES VARIABLES
*
      DO 310 I=1,NCLS
      E(I)=Ø. $TH(I)=THØ $S(I)=Ø. $GK(I)=Ø.
      DO 308 J=1,MCTP1
  308 SC(I,J)=Ø.
      DO 310 J=1,NRNGE
  310 ACIT(I,J)=Ø
      IVR=Ø $DCTH=EXP(-STEP/TTH) $DCG=EXP(-STEP/TGK)
      DO 320 J=1,MRNGE
  320 OVR(J)=Ø
      LG1=1ØØØØØØØ*(MLAG+1)
*
****          THIS SECTION UPDATES STATE VARIABLES AT EACH TIME STEP
*
      DO 1000 L=1,LTSTOP
      NFF=Ø $NFC=Ø $GE=Ø. $GI=Ø.
*
****                UPDATING THE INPUT FIBERS
*
      DO 2000 I=1,NFIB
      IF(L.NE.LSTM(I)) GO TO 2000
      CALL RANSET(INSED(2))
      NFF=NFF+1 $IDFF(NFF)=I
      LSTM(I)=L+INT(-PER(I)*ALOG(RANF(Ø.)))+1
      CALL RANGET(INSED(2)) $CALL RANSET(I*INSED(3))
      DO 2600 J=1,NT
      NREC=INT(RANF(Ø.)*FLOAT(NCLS))+1
      NCT=INT(RANF(Ø.)*FLOAT(MCTP1-1))+1
      SC(NREC,NCT+1)=SC(NREC,NCT+1)+STR(I,J)
      DO 2500 JR=1,NRNGE
      IF(ACIT(NREC,JR).EQ.Ø) GO TO 2510
 2500 CONTINUE $IVR=IVR+1 $IF(IVR.GT.(MRNGE-1)) GO TO 1500
      OVR(IVR)=J+1000*I+10000000*(NCT+MLAG)+10000000000*NREC $GO TO 2600
 2510 ACIT(NREC,JR)=J+1000*I+10000000*(NCT+MLAG)
 2600 CONTINUE
 2000 CONTINUE
*
*
****                UPDATING THE CELLS
*
      DO 3000 I=1,NCLS
      GK(I)=GK(I)*DCG+B*(1.-DCG)*S(I) $GTOT=1.+GE+GI+GK(I)
      DCE=EXP(-GTOT*STEP/TMEM)
      E(I)=E(I)*DCE+(SC(I,1)+GE*EX+GI*EI+GK(I)*EK)*(1.-DCE)/GTOT
      TH(I)=THØ+(TH(I)-THØ)*DCTH+C*E(I)*(1.-DCTH) $S(I)=Ø.
      IF(E(I).LT.TH(I)) GO TO 3010
      S(I)=1. $NFC=NFC+1 $IDFC(NFC)=I
      DO 3500 J=1,NRNGE $IF(ACIT(I,J).EQ.Ø) GO TO 3599
      IF(ACIT(I,J).GT.LG1) GO TO 3500
      ID=MOD(ACIT(I,J),10000000) $IS=ID/1000 $JS=ID-IS*1000
      STR(IS,JS)=STR(IS,JS)+DLSTR*(STRMX-STR(IS,JS))
 3500 CONTINUE
 3599 CONTINUE $JUP=IVR $DO 3600 J=1,JUP
      IF(OVR(J)/10000000000.NE.I) GO TO 3600
      IT=OVR(J)-10000000000*I $IF(IT.GT.LG1) GO TO 3600
      ID=MOD(IT,10000000) $IS=ID/1000 $JS=ID-IS*1000
      STR(IS,JS)=STR(IS,JS)+DLSTR*(STRMX-STR(IS,JS))
 3600 CONTINUE
 3010 CONTINUE
*
 3000 CONTINUE
*
```

```
      WRITE(6,6050) L,NFF,NFC,(IDFF(I),I=1,NFF),(IDFC(J),J=1,NFC)
      WRITE(7,7050) L,NFF,NFC,(IDFF(I),I=1,NFF),(IDFC(J),J=1,NFC)
      JUP=IVR $DO 4000 J=1,JUP
 4010 IF(OVR(J).EQ.0) GO TO 4019
      IF(MOD(OVR(J),10000000000).GT.10000000) GO TO 4000 $IVR=IVR-1
      DO 4560 J1=J,IVR
 4560 OVR(J1)=OVR(J1+1) $OVR(IVR+1)=0 $GO TO 4010
 4000 OVR(J)=OVR(J)-10000000
 4019 CONTINUE
      DO 4300 I=1,NCLS $DO 4700 J=1,NRNGE
 4710 IF(ACIT(I,J).EQ.0) GO TO 4799
      IF(ACIT(I,J).GT.10000000) GO TO 4700 $DO 4360 J1=J,NRNGE-1
 4360 ACIT(I,J1)=ACIT(I,J1+1) $ACIT(I,NRNGE)=0 $GO TO 4710
 4700 ACIT(I,J)=ACIT(I,J)-10000000
 4799 CONTINUE $DO 4400 J=1,MCTP1-1
 4400 SC(I,J)=SC(I,J+1) $SC(I,MCTP1)=0.
 4300 CONTINUE
*
 1000 CONTINUE
*
**** 	 THIS SECTION WRITES OUT ACTIVITY VARIABLES
*
      WRITE(6,6040)
      WRITE(6,6900) ((STR(I,J),J=1,NT),I=1,NFIB)
      WRITE(8,5900) ((STR(I,J),J=1,NT),I=1,NFIB)
 1500 CONTINUE
      STOP
**** 	 THESE ARE THE FORMATS
*
 5010 FORMAT(12I6)
 5020 FORMAT(12F6.2)
 5030 FORMAT(2F6.2,2I6)
 5035 FORMAT(I6,2F6.2)
 5900 FORMAT(F6.3)
 6000 FORMAT(1H1)
 6010 FORMAT(5X,*NFIB,NCLS,NT,INSEDS,LTSTOP=*,6I6)
 6020 FORMAT(5X,*C,TTH,B,TGK,TMEM,TH0=*,6F8.2)
 6030 FORMAT(5X,*STR0,PER0,LSTM0,INSED0=*,2F7.2,2I7)
 6035 FORMAT(5X,*MLAG,STRMX,DLSTR=*,I7,2F7.2)
 6040 FORMAT(///)
 6050 FORMAT(5X,30I4)
 6900 FORMAT(5X,20F6.3)
 7050 FORMAT(30I4)
      END
/EOR
     0  1227  1936    31
  .00 20.   20.    5.    5.   10.
   2.    20.      25  4748
     3  4.     .25
/EOR
/EOI


/JOB
A224,CM=130000,TL=1000.
FAKE.
PDFV.
MNF(Y).
GET,STR000.
LGO,STR000,LN008,STR008.
SAVE,LN008,STR008.
EXIT.
SAVE,LN008,STR008.
/EOR
      PROGRAM LRSYS20 (ANIN,LRSYS,ANOUT,INPUT,OUTPUT,
     *          TAPE3=ANIN,TAPE6=OUTPUT,TAPE7=LRSYS,TAPE8=ANOUT)
```

```
*
*** THIS PROGRAM SIMULATES HEBBIAN LEARNING AT THE SYNAPSES OF AN
*   ARBITRARY SYSTEM OF INTERACTING SPATIALLY-ORGANIZED NEURAL NETWORKS
*   ACTIVATED BY AN ARBITRARY NUMBER OF INPUT FIBER SYSTEMS.  ALL
*** CONNECTIONS ARE EXCITATORY OR INHIBITORY. STANDARD.
*
      INTEGER TYPE,W,H,ACITSN,ACITTR,ACITTM,OVRRC,OVRSN,OVRTR,OVRTM
      PARAMETER NTPOPS=10,NFPOPS=2,NCPOPS=8,NCLS=200,MCTP1=5,W=20,H=10,
     *          NJN=22,NLJN=1,NTR=10,NRNGE=10,MRNGE=1000,STEP=1.
      DIMENSION P(NFPOPS),INSTP(NFPOPS),INFSED(NFPOPS),NF(NTPOPS),
     *          IDF(NCLS),TTH(NCPOPS),TGK(NCPOPS),IFLG(NJN),
     *          STRMX(NLJN),DLSTR(NLJN),LRN(NJN),IS(NJN),IR(NJN),
     *          LG0(NLJN),LG1(NLJN),IJL(100),ISDR(100),ISYN(100),
     *          INSTR(NFPOPS)
      COMMON SCIT(NCPOPS,NCLS,MCTP1),GI(NCPOPS,NCLS,MCTP1)
      COMMON/SV/E(NCPOPS,NCLS),TH(NCPOPS,NCLS),S(NCPOPS,NCLS),
     *          GK(NCPOPS,NCLS),B(NCPOPS),C(NCPOPS),DCG(NCPOPS),
     *          DCTH(NCPOPS),TMEM(NCPOPS),TH0(NCPOPS)
      COMMON/TR/TYPE(NJN),NT(NJN),NCT(NJN),STR0(NJN),WTF(NJN),HTF(NJN),
     *          INSED(NJN),LJNN(NTPOPS,NCPOPS),MLAG(NLJN),
     *          ACITSN(NLJN,NCLS,NRNGE),ACITTR(NLJN,NCLS,NRNGE),
     *          ACITTM(NLJN,NCLS,NRNGE),OVRSN(NLJN,MRNGE),
     *          OVRTR(NLJN,MRNGE),OVRTM(NLJN,MRNGE),
     *          OVRRC(NLJN,MRNGE),IVR(NLJN),
     *          STR(NLJN,NCLS,NTR),N(NTPOPS),NC(NTPOPS),NR(NTPOPS),
     *          JNN(NTPOPS,NCPOPS),L
*
****      THIS SECTION READS AND WRITES THE INPUT PARAMETERS
*
      READ 5100, IFLG1,IFLG2
      DO 110 I=1,NCPOPS
110   READ 5110, N(I),C(I),TTH(I),B(I),TGK(I),TMEM(I),TH0(I),
     *          NC(I),NR(I)
      DO 120 I=1,NFPOPS $II=NCPOPS+I
120   READ 5120, N(II),P(I),INSTR(I),INSTP(I),INFSED(I),NC(II),NR(II)
      DO 128 I=1,NCPOPS $IFLG(I)=0 $DO 128 J=1,NTPOPS
      JNN(J,I)=0 $LJNN(J,I)=0
128   CONTINUE $J=0 $DO 130 I=1,NJN
      READ 5130, TYPE(I),LRN(I),IS(I),IR(I),NT(I),NCT(I),STR0(I),
     *          WTF(I),HTF(I),INSED(I)
      JNN(IS(I),IR(I))=I
      IF(LRN(I).GT.0) THEN $J=J+1 $LJNN(IS(I),IR(I))=J $IFLG(IR(I))=1
      READ 5140, MLAG(J),STRMX(J),DLSTR(J) $ENDIF
130   CONTINUE
      READ 5100, LTSTOP
*
      WRITE(6,6000) $WRITE(6,6100) NTPOPS,NFPOPS,NCPOPS,NCLS,MCTP1,W,H,
      WRITE(6,6102) NJN,NLJN,NTR,NRNGE,MRNGE,IFLG1
      WRITE(6,6110) $DO 111 I=1,NCPOPS
111   WRITE(6,6112) N(I),C(I),TTH(I),B(I),TGK(I),TMEM(I),TH0(I),
     *          NC(I),NR(I)
      WRITE(6,6120) $DO 121 I=1,NFPOPS $II=NCPOPS+I
121   WRITE(6,6122) N(II),P(I),INSTR(I),INSTP(I),INFSED(I),
     *          NC(II),NR(II)
      WRITE(6,6130) $J=0 $DO 131 I=1,NJN
      WRITE(6,6132) TYPE(I),LRN(I),IS(I),IR(I),NT(I),NCT(I),
     *          STR0(I),WTF(I),HTF(I),INSED(I)
      IF(LRN(I).GT.0) THEN $J=J+1
      WRITE(6,6140) MLAG(J),STRMX(J),DLSTR(J) $ENDIF
131   CONTINUE
      WRITE(6,6200) LTSTOP $WRITE(6,6000)
*
****      THIS SECTION INITIALIZES VARIABLES
*
      DO 310 I=1,NCPOPS
      DCTH(I)=EXP(-STEP/TTH(I)) $DCG(I)=EXP(-STEP/TGK(I))
      DO 310 J=1,NCLS
```

```
      E(I,J)=0. $TH(I,J)=TH0(I) $S(I,J)=0. $GK(I,J)=0.
      DO 310 K=1,MCTP1
      SCIT(I,J,K)=0. $GI(I,J,K)=0.
  310 CONTINUE
      DO 330 I=1,NLJN $IVR(I)=0 $LG1(I)=MLAG(I)+1
      LG0(I)=MLAG(I)
      DO 332 J=1,NCLS $DO 332 K=1,NRNGE
      ACITSN(I,J,K)=0 $ACITTR(I,J,K)=0 $ACITTM(I,J,K)=0
  332 CONTINUE $DO 334 K=1,MRNGE
      OVRSN(I,K)=0 $OVRTR(I,K)=0 $OVRTM(I,K)=0 $OVRRC(I,K)=0
  334 CONTINUE
  330 CONTINUE
      IF(IFLG2.EQ.0) THEN $DO 320 I=1,NJN $IF(LRN(I).EQ.0) GO TO 320
      II=LJNN(IS(I),IR(I)) $DO 320 J=1,NCLS $DO 320 K=1,NTR
      STR(II,J,K)=STR0(I)
  320 CONTINUE $ELSE
      READ(3,8900) (((STR(I,J,K),K=1,NTR),J=1,NCLS),I=1,NLJN) $ENDIF
*
****      THIS SECTION UPDATES STATE VARIABLES AT EACH TIME STEP
*
      DO 1000 L=1,LTSTOP
*
*
****          UPDATING THE CELLS
*
      DO 2000 I=1,NCPOPS
      NF(I)=0
      DO 2900 J=1,N(I)
      IF(IFLG(I).EQ.0) GO TO 2199 $JCR=0 $DO 2200 LS=1,NTPOPS
      IF(LJNN(LS,I).EQ.0) GO TO 2200 $LJ=LJNN(LS,I)
      DO 2500 K=1,NRNGE
 2570 IF(ACITSN(LJ,J,K).EQ.0)GO TO 2599 $ACITTM(LJ,J,K)=ACITTM(LJ,J,K)-1
      IT=ACITTM(LJ,J,K) $IF(IT.GE.LG1(LJ)) GO TO 2500 $JCR=JCR+1
      IJL(JCR)=LJ $ISDR(JCR)=ACITSN(LJ,J,K) $ISYN(JCR)=ACITTR(LJ,J,K)
      IF(IT.GE.LG0(LJ)) SCIT(I,J,1)=SCIT(I,J,1)+STR(LJ,ISDR(JCR),
     *                                               ISYN(JCR))
      IF(IT.GE.1) GO TO 2500 $JUP=NRNGE-1
      DO 2560 K1=K,JUP
      ACITSN(LJ,J,K1)=ACITSN(LJ,J,K1+1) $ACITTR(LJ,J,K1)=ACITTR(LJ,J,
     *        K1+1)
 2560 ACITTM(LJ,J,K1)=ACITTM(LJ,J,K1+1) $ACITSN(LJ,J,NRNGE)=0
      ACITTR(LJ,J,NRNGE)=0 $ACITTM(LJ,J,NRNGE)=0 $GO TO 2570
 2500 CONTINUE
 2599 CONTINUE
      JUP=IVR(LJ)
      DO 2600 K=1,JUP
 2640 IF(OVRRC(LJ,K).NE.I) GO TO 2600
      IT=OVRTM(LJ,K) $IF(IT.GE.LG1(LJ)) GO TO 2600
      JCR=JCR+1 $IJL(JCR)=LJ $ISDR(JCR)=OVRSN(LJ,K) $ISYN(JCR)=
     *                                               OVRTR(LJ,K)
      IF(IT.GE.LG0(LJ)) SCIT(I,J,1)=SCIT(I,J,1)+STR(LJ,ISDR(JCR),
     *                                               ISYN(JCR))
      IF(IT.GE.1) GO TO 2600
      IVR(LJ)=IVR(LJ)-1 $DO 2630 K1=K,IVR(LJ)
      OVRSN(LJ,K1)=OVRSN(LJ,K1+1) $OVRTR(LJ,K1)=OVRTR(LJ,K1+1)
      OVRTM(LJ,K1)=OVRTM(LJ,K1+1) $OVRRC(LJ,K1)=OVRRC(LJ,K1+1)
 2630 CONTINUE $OVRSN(LJ,IVR(LJ)+1)=0 $OVRTR(LJ,IVR(LJ)+1)=0
      OVRTM(LJ,IVR(LJ)+1)=0 $OVRRC(LJ,IVR(LJ)+1)=0 $GO TO 2640
 2600 CONTINUE
 2200 CONTINUE
 2199 CONTINUE
      CALL SVUPDT(I,J)
      IF(E(I,J).LT.TH(I,J)) GO TO 2850
      S(I,J)=1. $NF(I)=NF(I)+1 $IDF(NF(I))=J
      DO 2700 K=1,JCR $LJ=IJL(K) $JC=ISDR(K) $JT=ISYN(K)
 2700 STR(LJ,JC,JT)=STR(LJ,JC,JT)+DLSTR(LJ)*
     *                          (STRMX(LJ)-STR(LJ,JC,JT))
```

```
      DO 2800 K=1,NCPOPS
      IF(JNN(I,K).GT.0) CALL TRNSMT(I,K,J)
2800 CONTINUE
2850 CONTINUE
2900 CONTINUE $IF(IFLG1.GT.0) WRITE(7,7060) (L,I,NF(I),
    *                (IDF(J),J=1,NF(I)))
2000 CONTINUE
*
****                 UPDATING THE INPUT FIBERS
*
      DO 3000 I=1,NFPOPS
      II=I+NCPOPS $NF(II)=0
      IF(L.LT.INSTR(I).OR.L.GE.INSTP(I)) GO TO 3000
      CALL RANSET(INFSED(I))
      DO 3900 J=1,N(II)
      IF(RANF(0.).GT.P(I)) GO TO 3850 $CALL RANGET(INFSED(I))
      NF(II)=NF(II)+1 $IDF(NF(II))=J
      DO 3800 K=1,NCPOPS
      IF(JNN(II,K).GT.0) CALL TRNSMT(II,K,J)
3800 CONTINUE $CALL RANSET(INFSED(I))
3850 CONTINUE
3900 CONTINUE $IF(IFLG1.GT.0) WRITE(7,7060) (L,II,NF(II),
    *                (IDF(J),J=1,NF(II)))
      CALL RANGET(INFSED(I))
3000 CONTINUE
*
      WRITE(6,6510) L,(NF(I),I=1,NTPOPS)
      WRITE(7,7510) L,(NF(I),I=1,NTPOPS)
*
      DO 1910 I=1,NCPOPS $DO 1910 J=1,NCLS $DO 1920 K=1,MCTP1-1
      SCIT(I,J,K)=SCIT(I,J,K+1) $GI(I,J,K)=GI(I,J,K+1)
1920 CONTINUE $SCIT(I,J,MCTP1)=0. $GI(I,J,MCTP1)=0.
1910 CONTINUE
      DO 1700 LJ=1,NLJN $DO 1700 J=1,IVR(LJ)
1700 OVRTM(LJ,J)=OVRTM(LJ,J)-1
*
1000 CONTINUE
*
****        THIS SECTION WRITES OUT ACTIVITY VARIABLES
*
      DO 4100 JS=1,NTPOPS $DO 4100 JR=1,NCPOPS
      IF(LJNN(JS,JR).EQ.0) GO TO 4100 $WRITE(6,6040)
      LJ=LJNN(JS,JR)
      WRITE(6,6900) ((STR(LJ,I,K),K=1,NTR)=1,NCLS)
      WRITE(8,8900) ((STR(LJ,I,K),K=1,NTR)=1,NCLS)
4100 CONTINUE
*
      STOP
****        THESE ARE THE FORMATS
*
5100 FORMAT(12I6)
5110 FORMAT(I6,6F6.2,2I6)
5120 FORMAT(I6,F6.2,5I6)
5130 FORMAT(6I6,3F6.2,I6)
5140 FORMAT(I6,2F6.2)
6000 FORMAT(1H1)
6040 FORMAT(////)
6100 FORMAT(5X,*NTPOPS,NCPOPS,NFPOPS,NCLS,MCTP1,W,H=*,7I7)
6102 FORMAT(5X,*NJN,NLJN,NTR,NRNGE,MRNGE,IFLG1=*,6I7)
6110 FORMAT(5X,*CELLS: N,C,TTH,B,TGK,TMEM,TH0,NC,NR=*)
6112 FORMAT(9X,I7,6F7.2,2I7)
6120 FORMAT(///,5X,*FIBERS: N,P,INSTR,INSTP,INFSED,NC,NR=*,//)
6122 FORMAT(5X,I7,F7.2,5I7)
6130 FORMAT(///,5X,*JNCTNS: TYPE,LRN,IS,IR,NT,NCT,STR0,WTF,HTF,*
    *        ,*INSED=*,/10X,*MLAG,STRMX,DLSTR=*,//)
6132 FORMAT(5X,6I7,3F7.2,I7)
6140 FORMAT(10X,I7,2F7.2)
```

```
6200 FORMAT(5X,*LTSTOP=*,I7)
6510 FORMAT(5X,15I7)
6900 FORMAT(5X,20F6.3)
7510 FORMAT(15I5)
7060 FORMAT(20I4)
8900 FORMAT(F6.3)
     END
*
     SUBROUTINE SVUPDT(I,K)
*
*** THIS SUBROUTINE UPDATES THE STATE VARIABLES OF AN INDIVIDUAL NEURON
*
     PARAMETER NCPOPS=8,NCLS=200,MCTP1=5,STEP=1.,EK=-10.,EI=-10.
     COMMON SCIT(NCPOPS,NCLS,MCTP1),GI(NCPOPS,NCLS,MCTP1)
     COMMON/SV/E(NCPOPS,NCLS),TH(NCPOPS,NCLS),S(NCPOPS,NCLS),
    *          GK(NCPOPS,NCLS),B(NCPOPS),C(NCPOPS),DCG(NCPOPS),
    *          DCTH(NCPOPS),TMEM(NCPOPS),TH0(NCPOPS)
*
     GK(I,K)=GK(I,K)*DCG(I)+B(I)*(1.-DCG(I))*S(I,K)
     GTOT=1.+GK(I,K)+GI(I,K,1) $DCE=EXP(-GTOT*STEP/TMEM(I))
     E(I,K)=E(I,K)*DCE+(SCIT(I,K,1)+GI(I,K,1)*EI+GK(I,K)*EK)*(1.-DCE)/
    *     GTOT
     TH(I,K)=TH0(I)+(TH(I,K)-TH0(I))*DCTH(I)+C(I)*E(I,K)*(1.-DCTH(I))
     S(I,K)=0.
*
     RETURN
     END
*
     SUBROUTINE TRNSMT(IS,IR,K)
*
****    THIS SUBROUTINE PROJECTS SYNAPTIC ACTIVATION FROM A SENDING
****    FIBER TO ITS TARGETED RECEIVING CELLS.
*
     INTEGER TYPE,W,H,ACITSN,ACITTR,ACITTM,OVRRC,OVRSN,OVRTR,OVRTM
     PARAMETER NTPOPS=10,NCPOPS=8,NCLS=200,MCTP1=5,W=20,H=10,
    *          NJN=22,NLJN=1,NTR=10,NRNGE=10,MRNGE=1000
     COMMON SCIT(NCPOPS,NCLS,MCTP1),GI(NCPOPS,NCLS,MCTP1)
     COMMON/TR/TYPE(NJN),NT(NJN),NCT(NJN),STR0(NJN),WTF(NJN),HTF(NJN),
    *          INSED(NJN),LJNN(NTPOPS,NCPOPS),MLAG(NLJN),
    *          ACITSN(NLJN,NCLS,NRNGE),ACITTR(NLJN,NCLS,NRNGE),
    *          ACITTM(NLJN,NCLS,NRNGE),OVRSN(NLJN,MRNGE),
    *          OVRTR(NLJN,MRNGE),OVRTM(NLJN,MRNGE),
    *          OVRRC(NLJN,MRNGE),IVR(NLJN),
    *          STR(NLJN,NCLS,NTR),N(NTPOPS),NC(NTPOPS),NR(NTPOPS),
    *          JNN(NTPOPS,NCPOPS),L
*
     JN=JNN(IS,IR)
     XS=FLOAT(MOD(K,NC(IS)))*FLOAT(W)/FLOAT(NC(IS))
     IF(XS.EQ.0.) XS=FLOAT(W)
     YS=FLOAT(K/NC(IS)+1)*FLOAT(H)/FLOAT(NR(IS))
     CALL RANSET(K*INSED(JN))
     DO 100 J=1,NT(JN)
     XR=XS+(RANF(0.)-.5)*WTF(JN) $YR=YS+(RANF(0.)-.5)*HTF(JN)
     ICR=INT(FLOAT(NC(IR))*(XR-.5)/FLOAT(W))+1
     IRR=INT(FLOAT(NR(IR))*(YR-.5)/FLOAT(H))+1
     IF(ICR.LE.0) ICR=NC(IR)+ICR
     IF(ICR.GT.NC(IR)) ICR=ICR-NC(IR)
     IF(IRR.LE.0) IRR=NR(IR)+IRR
     IF(IRR.GT.NR(IR)) IRR=IRR-NR(IR)
     NREC=NC(IR)*(IRR-1)+ICR
     KCT=INT(RANF(0.)*FLOAT(NCT(JN)))+1
     IF(LJNN(IS,IR).GT.0) THEN $I=LJNN(IS,IR)
     DO 200 JR=1,NRNGE
     IF(ACITSN(I,NREC,JR).EQ.0) GO TO 210
 200 CONTINUE $IVR(I)=IVR(I)+1 $IF(IVR(I).GT.(MRNGE-1)) L=99999
     OVRRC(I,IVR(I))=NREC $OVRTM(I,IVR(I))=KCT+MLAG(I)
     OVRSN(I,IVR(I))=K $OVRTR(I,IVR(I))=J $GO TO 220
```

```
  210 ACITSN(I,NREC,JR)=K $ACITTR(I,NREC,JR)=J $ACITTM(I,NREC,JR)=
     *                                              KCT+MLAG(I)
  220 CONTINUE $ELSE
      IF(TYPE(JN).EQ.1) SCIT(IR,NREC,KCT+1)=SCIT(IR,NREC,KCT+1)+
     *                            STRØ(JN)
      IF(TYPE(JN).EQ.2) GI(IR,NREC,KCT+1)=GI(IR,NREC,KCT+1)+
     *                            STRØ(JN) $ENDIF
  100 CONTINUE
*
      RETURN
      END
/EOR
     Ø      Ø
    200  Ø.    25.    20.     5.     5.    10.        20     10
    200  Ø.    25.    10.     3.     5.    10.        20     10
    200  Ø.    25.    20.     5.     5.    10.        20     10
    200  Ø.    25.    10.     3.     5.    10.        20     10
    200  Ø.    25.    20.     5.     5.    10.        20     10
    200  Ø.    25.    10.     3.     5.    10.        20     10
    200  Ø.    25.    20.     5.     5.    10.        20     10
    200  Ø.    25.    10.     3.     5.    10.        20     10
    200  .10     1     50   7935    20     10
    200  .ØØ     1      Ø   8642    20     10
      1    Ø     1      2      8     2  14.     5.     4.    1516
      1    Ø     3      4      8     2  14.     5.     4.    3423
      1    Ø     5      6      8     2  14.     5.     4.    6238
      1    Ø     7      8      8     2  14.     5.     4.       Ø
      2    Ø     2      1     10     2  10.     7.     5.    2127
      2    Ø     4      3     10     2  10.     7.     5.    4447
      2    Ø     6      5     10     2  10.     7.     5.    5663
      2    Ø     8      7     10     2  10.     7.     5.    8486
      1    Ø     1      3     10     4  14.     5.     4.    1028
      1    Ø     3      5     10     4  14.     5.     4.    3822
      1    Ø     5      7     10     4  14.     5.     4.    6224
      1    Ø     7      1     10     4  14.     5.     4.    7156
      1    Ø     1      4      3     4  14.     5.     4.    1602
      1    Ø     3      6      3     4  14.     5.     4.    3020
      1    Ø     5      8      3     4  14.     5.     4.    6208
      1    Ø     7      2      3     4  14.     5.     4.    7082
      1    1     9      1     10     4  14.     5.     4.    9549
      3   20.          .25
      1    Ø     9      2      3     4  14.     5.     4.    9411
      1    Ø    10      2      1     4  14.    10.     8.    1039
      1    Ø    10      4      1     4  14.    10.     8.    6290
      1    Ø    10      6      1     4  14.    10.     8.    8756
      1    Ø    10      8      1     4  14.    10.     8.    3111
     20
/EOR
/EOI


/JOB
A224,CM=130000,TL=1000.
FAKE.
PDFV.
M77.
GET,STRØØØ.
LGO,STRØØØ,LNØØ7,STRØØ7.
SAVE,LNØØ7,STRØØ7.
EXIT.
SAVE,LNØØ7,STRØØ7.
/EOR
      PROGRAM LRSYS30 (ANIN,LRSYS,ANOUT,INPUT,OUTPUT,
     *          TAPE3=ANIN,TAPE6=OUTPUT,TAPE7=LRSYS,TAPE8=ANOUT)
*
****      THIS PROGRAM SIMULATES HEBBIAN LEARNING AT THE SYNAPSES OF AN
```

```
*           ARBITRARY SYSTEM OF INTERACTING NEURONAL POPULATIONS
*           ACTIVATED BY AN ARBITRARY NUMBER OF INPUT FIBER SYSTEMS.  ALL
*           SYNAPTIC SYSTEMS ARE INDIVIDUALLY ADJUSTABLE. SEPARATE TRANS-
****        MISSION AND MEMORY LINKS. ACIT,OVR PACKED.
*
      INTEGER TYPE,ACIT,OVR,SNTP
      PARAMETER NTPOPS=10,NFPOPS=2,NCPOPS=8,NCLS=200,MCTP1=5,
     *          NJN=22,NLJN=1,NTR=10,NRNGE=10,MRNGE=1000,STEP=1.,SNTP=2
      DIMENSION P(NFPOPS),INSTP(NFPOPS),INFSED(NFPOPS),NF(NTPOPS),
     *          IDF(NCLS),TTH(NCPOPS),TGK(NCPOPS),IFLG(NJN),LG1(NLJN),
     *          STRMX(NLJN),DLSTR(NLJN),LRN(NJN),IS(NJN),IR(NJN),
     *          T(SNTP),DCS(SNTP),INSTR(NFPOPS)
      COMMON G(SNTP,NCPOPS,NCLS,MCTP1)
      COMMON/SV/E(NCPOPS,NCLS),TH(NCPOPS,NCLS),S(NCPOPS,NCLS),
     *          GK(NCPOPS,NCLS),B(NCPOPS),C(NCPOPS),DCG(NCPOPS),
     *          DCTH(NCPOPS),TMEM(NCPOPS),THØ(NCPOPS),EQ(SNTP)
      COMMON/TR/TYPE(NJN),NT(NJN),NCT(NJN),STRØ(NJN),
     *          INSED(NJN),LJNN(NTPOPS,NCPOPS),MLAG(NLJN),
     *          ACIT(NLJN,NCLS,NRNGE),OVR(NLJN,MRNGE),IVR(NLJN),
     *          STR(NLJN,NCLS,NTR),N(NTPOPS),
     *          JNN(NTPOPS,NCPOPS),L
*
****        THIS SECTION READS AND WRITES THE INPUT PARAMETERS
*
      READ 5100, IFLG1,IFLG2
      READ 5105, ((EQ(I),T(I)),I=1,SNTP)
      DO 110 I=1,NCPOPS
  110 READ 5110, N(I),C(I),TTH(I),B(I),TGK(I),TMEM(I),THØ(I)
      DO 120 I=1,NFPOPS $II=NCPOPS+I
  120 READ 5120, N(II),P(I),INSTR(I),INSTP(I),INFSED(I)
      DO 128 I=1,NCPOPS $IFLG(I)=Ø $DO 128 J=1,NTPOPS
      JNN(J,I)=Ø $LJNN(J,I)=Ø
  128 CONTINUE $J=Ø $DO 130 I=1,NJN
      READ 5130, TYPE(I),LRN(I),IS(I),IR(I),NT(I),NCT(I),STRØ(I),
     *           INSED(I)
      JNN(IS(I),IR(I))=I
      IF(LRN(I).GT.Ø) THEN $J=J+1 $LJNN(IS(I),IR(I))=J $IFLG(IR(I))=1
      READ 5140, MLAG(J),STRMX(J),DLSTR(J) $ENDIF
  130 CONTINUE
      READ 5100, LTSTOP
*
      WRITE(6,6000) $WRITE(6,6100) NTPOPS,NFPOPS,NCPOPS,NCLS,MCTP1
      WRITE(6,6102) NJN,NLJN,NTR,NRNGE,MRNGE,IFLG1
      WRITE(6,6105) ((EQ(I),T(I)),I=1,SNTP)
      WRITE(6,6110) $DO 111 I=1,NCPOPS
  111 WRITE(6,6112) N(I),C(I),TTH(I),B(I),TGK(I),TMEM(I),THØ(I)
      WRITE(6,6120) $DO 121 I=1,NFPOPS $II=NCPOPS+I
  121 WRITE(6,6122) N(II),P(I),INSTR(I),INSTP(I),INFSED(I)
      WRITE(6,6130) $J=Ø $DO 131 I=1,NJN
      WRITE(6,6132) TYPE(I),LRN(I),IS(I),IR(I),NT(I),NCT(I),
     *              STRØ(I),INSED(I)
      IF(LRN(I).GT.Ø) THEN $J=J+1
      WRITE(6,6140) MLAG(J),STRMX(J),DLSTR(J) $ENDIF
  131 CONTINUE
      WRITE(6,6200) LTSTOP $WRITE(6,6000)
*
****        THIS SECTION INITIALIZES VARIABLES
*
      DO 310 I=1,NCPOPS
      DCTH(I)=EXP(-STEP/TTH(I)) $DCG(I)=EXP(-STEP/TGK(I))
      DO 310 J=1,NCLS
      E(I,J)=Ø. $TH(I,J)=THØ(I) $S(I,J)=Ø. $GK(I,J)=Ø.
      DO 310 K=1,MCTP1 $DO 310 M=1,SNTP
      G(M,I,J,K)=Ø.
  310 CONTINUE
      DO 330 I=1,NLJN $IVR(I)=Ø $LG1(I)=10000000*(MLAG(I)+1)
      DO 332 J=1,NCLS $DO 332 K=1,NRNGE
```

```
  332 ACIT(I,J,K)=0 $DO 334 K=1,MRNGE
  334 OVR(I,K)=0
  330 CONTINUE
      DO 340 I=1,SNTP
  340 DCS(I)=EXP(-STEP/T(I))
      IF(IFLG2.EQ.0) THEN $DO 320 I=1,NJN $IF(LRN(I).EQ.0) GO TO 320
      II=LJNN(IS(I),IR(I)) $DO 320 J=1,NCLS $DO 320 K=1,NTR
      STR(II,J,K)=STR0(I)
  320 CONTINUE $ELSE
      READ(3,8900) (((STR(I,J,K),K=1,NTR),J=1,NCLS),I=1,NLJN) $ENDIF
*
****       THIS SECTION UPDATES STATE VARIABLES AT EACH TIME STEP
*
      DO 1000 L=1,LTSTOP
*
*
****                    UPDATING THE CELLS
*
      DO 2000 I=1,NCPOPS
      NF(I)=0
      DO 2900 J=1,N(I)
      CALL SVUPDT(I,J)
      IF(E(I,J).LT.TH(I,J)) GO TO 2850
      S(I,J)=1. $NF(I)=NF(I)+1 $IDF(NF(I))=J
*
      IF(IFLG(I).EQ.0) GO TO 2699 $DO 2700 LS=1,NTPOPS
      IF(LJNN(LS,I).EQ.0) GO TO 2700 $LJ=LJNN(LS,I)
      DO 2500 K=1,NRNGE $IF(ACIT(LJ,J,K).EQ.0) GO TO 2499
      IF(ACIT(LJ,J,K).GT.LG1(LJ)) GO TO 2500
      ID=MOD(ACIT(LJ,J,K),10000000) $JC=ID/1000 $JT=ID-JC*1000
      STR(LJ,JC,JT)=STR(LJ,JC,JT)+DLSTR(LJ)*
     *                    (STRMX(LJ)-STR(LJ,JC,JT))
 2500 CONTINUE
 2499 CONTINUE $JUP=IVR(LJ) $DO 2600 K=1,JUP
      IF(OVR(LJ,K)/10000000000.NE.J) GO TO 2600
      IF(MOD(OVR(LJ,K),10000000000).GT.LG1(LJ)) GO TO 2600
      ID=MOD(OVR(LJ,K),10000000) $JC=ID/1000 $JT=ID-JC*1000
      STR(LJ,JC,JT)=STR(LJ,JC,JT)+DLSTR(LJ)*
     *                    (STRMX(LJ)-STR(LJ,JC,JT))
 2600 CONTINUE
 2700 CONTINUE
*
 2699 CONTINUE
      DO 2800 K=1,NCPOPS
      IF(JNN(I,K).GT.0) CALL TRNSMT(I,K,J)
 2800 CONTINUE
 2850 CONTINUE
 2900 CONTINUE $IF(IFLG1.GT.0) WRITE(7,7060) (L,I,NF(I),
     *              (IDF(J),J=1,NF(I)))
 2000 CONTINUE
*
****                  UPDATING THE INPUT FIBERS
*
      DO 3000 I=1,NFPOPS
      II=I+NCPOPS $NF(II)=0
      IF(L.LT.INSTR(I).OR.L.GE.INSTP(I)) GO TO 3000
      CALL RANSET(INFSED(I))
      DO 3900 J=1,N(II)
      IF(RANF(0.).GT.P(I)) GO TO 3850 $CALL RANGET(INFSED(I))
      NF(II)=NF(II)+1 $IDF(NF(II))=J
      DO 3800 K=1,NCPOPS
      IF(JNN(II,K).GT.0) CALL TRNSMT(II,K,J)
 3800 CONTINUE $CALL RANSET(INFSED(I))
 3850 CONTINUE
 3900 CONTINUE $IF(IFLG1.GT.0) WRITE(7,7060) (L,II,NF(II),
     *              (IDF(J),J=1,NF(II)))
      CALL RANGET(INFSED(I))
```

```
 3000 CONTINUE
*
      WRITE(6,6510) L,(NF(I),I=1,NTPOPS)
      WRITE(7,7510) L,(NF(I),I=1,NTPOPS)
*
      DO 1910 I=1,NCPOPS $DO 1910 J=1,NCLS $DO 1910 M=1,SNTP
      G(M,I,J,1)=G(M,I,J,1)*DCS(M)+G(M,I,J,2) $DO 1920 K=2,MCTP1-1
 1920 G(M,I,J,K)=G(M,I,J,K+1) $G(M,I,J,MCTP1)=0.
 1910 CONTINUE
      DO 1600 LJ=1,NLJN $JUP=IVR(LJ) $DO 1700 J=1,JUP
 1710 IF(OVR(LJ,J).EQ.0) GO TO 1699
      IF(MOD(OVR(LJ,J),10000000000).GT.10000000) GO TO 1700
      IVR(LJ)=IVR(LJ)-1 $DO 1760 J1=J,IVR(LJ)
 1760 OVR(LJ,J1)=OVR(LJ,J1+1) $OVR(LJ,IVR(LJ)+1)=0 $GO TO 1710
 1700 OVR(LJ,J)=OVR(LJ,J)-10000000
 1699 CONTINUE
      DO 1800 I=1,NCLS $DO 1850 J=1,NRNGE
 1810 IF(ACIT(LJ,I,J).EQ.0) GO TO  1849
      IF(ACIT(LJ,I,J).GT.10000000) GO TO 1850 $DO 1860 J1=1,NRNGE-1
 1860 ACIT(LJ,I,J1)=ACIT(LJ,I,J1+1) $ACIT(LJ,I,NRNGE)=0 $GO TO 1810
 1850 ACIT(LJ,I,J)=ACIT(LJ,I,J)-10000000
 1849 CONTINUE
 1800 CONTINUE
 1600 CONTINUE
*
 1000 CONTINUE
*
**** 	  THIS SECTION WRITES OUT ACTIVITY VARIABLES
*
      DO 4100 JS=1,NTPOPS $DO 4100 JR=1,NCPOPS
      IF(LJNN(JS,JR).EQ.0) GO TO 4100 $WRITE(6,6040)
      LJ=LJNN(JS,JR)
      WRITE(6,6900) ((STR(LJ,I,K),K=1,NTR),I=1,NCLS)
      WRITE(8,8900) ((STR(LJ,I,K),K=1,NTR),I=1,NCLS)
 4100 CONTINUE
*
      STOP
**** 	  THESE ARE THE FORMATS
*
 5100 FORMAT(12I6)
 5105 FORMAT(12F6.2)
 5110 FORMAT(I6,6F6.2)
 5120 FORMAT(I6,F6.2,3I6)
 5130 FORMAT(6I6,F6.2,I6)
 5140 FORMAT(I6,2F6.2)
 6000 FORMAT(1H1)
 6040 FORMAT(///)
 6100 FORMAT(5X,*NTPOPS,NCPOPS,NFPOPS,NCLS,MCTP1=*,5I7)
 6102 FORMAT(5X,*NJN,NLJN,NTR,NRNGE,MRNGE,IFLG1=*,6I7)
 6105 FORMAT(5X,*EQ,T=*,12F7.2)
 6110 FORMAT(5X,*CELLS: N,C,TTH,B,TGK,TMEM,THO=*)
 6112 FORMAT(9X,I7,6F7.2)
 6120 FORMAT(///,5X,*FIBERS: N,P,INSTR,INSTP,INFSED=*,//)
 6122 FORMAT(5X,I7,F7.2,3I7)
 6130 FORMAT(///,5X,*JNCTNS: TYPE,LRN,IS,IR,NT,NCT,STRØ,*
      *        ,*INSED=*,/10X,*MLAG,STRMX,DLSTR=*,//)
 6132 FORMAT(5X,6I7,F7.2,I7)
 6140 FORMAT(10X,I7,2F7.2)
 6200 FORMAT(5X,*LTSTOP=*,I7)
 6510 FORMAT(5X,15I7)
 6900 FORMAT(5X,20F6.3)
 7510 FORMAT(15I5)
 7060 FORMAT(20I4)
 8900 FORMAT(F6.3)
      END
*
      SUBROUTINE SVUPDT(I,K)
```

```
*
*** THIS SUBROUTINE UPDATES THE STATE VARIABLES OF AN INDIVIDUAL NEURON
*
      INTEGER SNTP
      PARAMETER NCPOPS=8,NCLS=200,MCTP1=5,STEP=1.,EK=-10.,SNTP=2
      COMMON G(SNTP,NCPOPS,NCLS,MCTP1)
      COMMON/SV/E(NCPOPS,NCLS),TH(NCPOPS,NCLS),S(NCPOPS,NCLS),
     *          GK(NCPOPS,NCLS),B(NCPOPS),C(NCPOPS),DCG(NCPOPS),
     *          DCTH(NCPOPS),TMEM(NCPOPS),THØ(NCPOPS),EQ(SNTP)
*
      GS=Ø. $SYN=Ø. $DO 10 J=1,SNTP
      GS=GS+G(J,I,K,1) $SYN=SYN+G(J,I,K,1)*EQ(J)
   10 CONTINUE
      GK(I,K)=GK(I,K)*DCG(I)+B(I)*(1.-DCG(I))*S(I,K)
      GTOT=1.+GK(I,K)+GS $DCE=EXP(-GTOT*STEP/TMEM(I))
      E(I,K)=E(I,K)*DCE+(SYN+GK(I,K)*EK)*(1.-DCE)/
     *       GTOT
      TH(I,K)=THØ(I)+(TH(I,K)-THØ(I))*DCTH(I)+C(I)*E(I,K)*(1.-DCTH(I))
      S(I,K)=Ø.
*
      RETURN
      END
*
      SUBROUTINE TRNSMT(IS,IR,K)
*
****      THIS SUBROUTINE PROJECTS SYNAPTIC ACTIVATION FROM A SENDING
****      FIBER TO ITS TARGETED RECEIVING CELLS.
*
      INTEGER TYPE,ACIT,OVR,SNTP
      PARAMETER NTPOPS=10,NCPOPS=8,NCLS=200,MCTP1=5,
     *          NJN=22,NLJN=1,NTR=10,NRNGE=10,MRNGE=1000,SNTP=2
      COMMON G(SNTP,NCPOPS,NCLS,MCTP1)
      COMMON/TR/TYPE(NJN),NT(NJN),NCT(NJN),STRØ(NJN),
     *          INSED(NJN),LJNN(NTPOPS,NCPOPS),MLAG(NLJN),
     *          ACIT(NLJN,NCLS,NRNGE),OVR(NLJN,MRNGE),IVR(NLJN),
     *          STR(NLJN,NCLS,NTR),N(NTPOPS),
     *          JNN(NTPOPS,NCPOPS),L
*
      JN=JNN(IS,IR)
      CALL RANSET(K*INSED(JN))
      DO 100 J=1,NT(JN)
      NREC=INT(RANF(Ø.)*FLOAT(N(IR)))+1
      KCT=INT(RANF(Ø.)*FLOAT(NCT(JN)))+1
      IF(LJNN(IS,IR).GT.Ø) THEN $I=LJNN(IS,IR)
      DO 200 JR=1,NRNGE
      IF(ACIT(I,NREC,JR).EQ.Ø) GO TO 210
  200 CONTINUE $IVR(I)=IVR(I)+1 $IF(IVR(I).GT.(MRNGE-1)) L=99999
      OVR(I,IVR(I))=J+1000*K+10000000*(KCT+MLAG(I))+10000000000*NREC
      GO TO 220
  210 ACIT(I,NREC,JR)=J+1000*K+10000000*(KCT+MLAG(I))
  220 G(TYPE(JN),IR,NREC,KCT+1)=G(TYPE(JN),IR,NREC,KCT+1)+STR(I,K,J)
      ELSE
      G(TYPE(JN),IR,NREC,KCT+1)=G(TYPE(JN),IR,NREC,KCT+1)+STRØ(JN)
      ENDIF
  100 CONTINUE
*
      RETURN
      END
/EOR      Ø      Ø
        Ø      Ø
  70.00   3.00-10.00   2.00
     200   0.00 25.00 20.00   5.00   5.00 10.00
     200   0.00 25.00 20.00   5.00   5.00 10.00
     200   0.00 25.00 20.00   5.00   5.00 10.00
     200   0.00 25.00 20.00   5.00   5.00 10.00
     200   0.00 25.00 20.00   5.00   5.00 10.00
     200   0.00 25.00 20.00   5.00   5.00 10.00
```

```
200   0.00 25.00 20.00   5.00   5.00 10.00
200   0.00 25.00 20.00   5.00   5.00 10.00
200   .10    1    50   3118
200   .10    1     0   4938
  1    0    1    2    8    2   .2   8765
  1    0    3    4    8    2   .2   1164
  1    0    5    6    8    2   .2   6238
  1    0    7    8    8    2   .2   1109
  2    0    2    1   10    2  1.00  2127
  2    0    4    3   10    2  1.00  4447
  2    0    6    5   10    2  1.00  5197
  2    0    8    7   10    2  1.00  6120
  1    0    1    3   10    4   .2   1028
  1    0    3    5   10    4   .2   4862
  1    0    5    7   10    4   .2   1310
  1    0    7    1   10    4   .2   7156
  1    0    1    4    3    4   .2   3853
  1    0    3    6    3    4   .2   6163
  1    0    5    8    3    4   .2   4401
  1    0    7    2    3    4   .2   7082
  1    1    9    1   10    4   .2   9562
  3   2.        .25
  1    0    9    2    3    4   .2   9411
  1    0   10    2    1    4   .2   1039
  1    0   10    4    1    4   .2   6221
  1    0   10    6    1    4   .2   8756
  1    0   10    8    1    4   .2   3111
 20
/EOR
/EOI


/JOB
A224,CM=130000,TL=1000.
FAKE.
PDFV.
M77.
GET,STR000.
LGO,STR000,LN001,STR001.
SAVE,LN001,STR001.
EXIT.
SAVE,LN001,STR001.
/EOR
      PROGRAM LRSYS32 (ANIN,LRSYS,ANOUT,INPUT,OUTPUT,
     *            TAPE3=ANIN,TAPE6=OUTPUT,TAPE7=LRSYS,TAPE8=ANOUT)
*
****   THIS PROGRAM SIMULATES HEBBIAN LEARNING AT THE SYNAPSES OF AN
*      ARBITRARY SYSTEM OF INTERACTING SPATIALLY-ORGANIZED NEURAL NETS
*      ACTIVATED BY AN ARBITRARY NUMBER OF INPUT FIBER SYSTEMS.  ALL
*      SYNAPTIC SYSTEMS ARE INDIVIDUALLY ADJUSTABLE. SEPARATE TRANS-
****   MISSION AND MEMORY LINKS. ACIT,OVR PACKED.
*
      INTEGER TYPE,W,H,ACIT,OVR,SNTP
      PARAMETER NTPOPS=10,NFPOPS=2,NCPOPS=8,NCLS=200,MCTP1=5,W=20,H=10,
     *        NJN=22,NLJN=1,NTR=10,NRNGE=10,MRNGE=1000,STEP=1.,SNTP=2,
     *        NSTIM=2
      DIMENSION P(NFPOPS),INSTP(NFPOPS),INFSED(NFPOPS),NF(NTPOPS),
     *        IDF(NCLS),TTH(NCPOPS),TGK(NCPOPS),IFLG(NJN),LG1(NLJN),
     *        STRMX(NLJN),DLSTR(NLJN),LRN(NJN),IS(NJN),IR(NJN),
     *        T(SNTP),DCS(SNTP),INSTR(NFPOPS)
      COMMON G(SNTP,NCPOPS,NCLS,MCTP1)
      COMMON/SV/E(NCPOPS,NCLS),TH(NCPOPS,NCLS),S(NCPOPS,NCLS),
     *        GK(NCPOPS,NCLS),B(NCPOPS),C(NCPOPS),DCG(NCPOPS),
     *        DCTH(NCPOPS),TMEM(NCPOPS),TH0(NCPOPS),EQ(SNTP)
      COMMON/STIM/LOCATN(NSTIM)
      COMMON/TR/TYPE(NJN),NT(NJN),NCT(NJN),STR0(NJN),WTF(NJN),HTF(NJN),
```

```
     *          INSED(NJN),LJNN(NTPOPS,NCPOPS),MLAG(NLJN),
     *          ACIT(NLJN,NCLS,NRNGE),OVR(NLJN,MRNGE),IVR(NLJN),
     *          STR(NLJN,NCLS,NTR),N(NTPOPS),NC(NTPOPS),NR(NTPOPS),
     *          JNN(NTPOPS,NCPOPS),L
*
****        THIS SECTION READS AND WRITES THE INPUT PARAMETERS
*
       READ 5100, IFLG1,IFLG2
       READ 5105, ((EQ(I),T(I)),I=1,SNTP)
       DO 110 I=1,NCPOPS
 110   READ 5110, N(I),C(I),TTH(I),B(I),TGK(I),TMEM(I),TH0(I),
     *          NC(I),NR(I)
       DO 120 I=1,NFPOPS $II=NCPOPS+I
 120   READ 5120, N(II),P(I),INSTR(I),INSTP(I),INFSED(I),NC(II),NR(II)
       DO 128 I=1,NCPOPS $IFLG(I)=0 $DO 128 J=1,NTPOPS
       JNN(J,I)=0 $LJNN(J,I)=0
 128   CONTINUE $J=0 $DO 130 I=1,NJN
       READ 5130, TYPE(I),LRN(I),IS(I),IR(I),NT(I),NCT(I),STR0(I),
     *          WTF(I),HTF(I),INSED(I)
       JNN(IS(I),IR(I))=I
       IF(LRN(I).GT.0) THEN $J=J+1 $LJNN(IS(I),IR(I))=J $IFLG(IR(I))=1
       READ 5140, MLAG(J),STRMX(J),DLSTR(J) $ENDIF
 130   CONTINUE
       READ 5100, ISTM,LSTIM,INTERV,LSTP,NSTIM1
       READ 5100, (LOCATN(I),I=1,NSTIM)
       READ 5100, LTSTOP
*
       WRITE(6,6000) $WRITE(6,6100) NTPOPS,NFPOPS,NCPOPS,NCLS,MCTP1,W,H,
       WRITE(6,6102) NJN,NLJN,NTR,NRNGE,MRNGE,IFLG1
       WRITE(6,6105) ((EQ(I),T(I)),I=1,SNTP)
       WRITE(6,6110) $DO 111 I=1,NCPOPS
 111   WRITE(6,6112) N(I),C(I),TTH(I),B(I),TGK(I),TMEM(I),TH0(I),
     *          NC(I),NR(I)
       WRITE(6,6120) $DO 121 I=1,NFPOPS $II=NCPOPS+I
 121   WRITE(6,6122) N(II),P(I),INSTR(I),INSTP(I),INFSED(I),
     *          NC(II),NR(II)
       WRITE(6,6130) $J=0 $DO 131 I=1,NJN
       WRITE(6,6132) TYPE(I),LRN(I),IS(I),IR(I),NT(I),NCT(I),
     *          STR0(I),WTF(I),HTF(I),INSED(I)
       IF(LRN(I).GT.0) THEN $J=J+1
       WRITE(6,6140) MLAG(J),STRMX(J),DLSTR(J) $ENDIF
 131   CONTINUE
       WRITE(6,6150) ISTM,LSTIM,INTERV,LSTP,NSTIM,NSTIM1
       WRITE(6,6154) $WRITE(6,6155) (LOCATN(I),I=1,NSTIM)
       WRITE(6,6200) LTSTOP $WRITE(6,6000)
*
****        THIS SECTION INITIALIZES VARIABLES
*
       MARKER=0
       DO 310 I=1,NCPOPS
       DCTH(I)=EXP(-STEP/TTH(I)) $DCG(I)=EXP(-STEP/TGK(I))
       DO 310 J=1,NCLS
       E(I,J)=0. $TH(I,J)=TH0(I) $S(I,J)=0. $GK(I,J)=0.
       DO 310 K=1,MCTP1 $DO 310 M=1,SNTP
       G(M,I,J,K)=0.
 310   CONTINUE
       DO 330 I=1,NLJN $IVR(I)=0 $LG1(I)=10000000*(MLAG(I)+1)
       DO 332 J=1,NCLS $DO 332 K=1,NRNGE
 332   ACIT(I,J,K)=0 $DO 334 K=1,MRNGE
 334   OVR(I,K)=0
 330   CONTINUE
       DO 340 I=1,SNTP
 340   DCS(I)=EXP(-STEP/T(I))
       IF(IFLG2.EQ.0) THEN $DO 320 I=1,NJN $IF(LRN(I).EQ.0) GO TO 320
       II=LJNN(IS(I),IR(I)) $DO 320 J=1,NCLS $DO 320 K=1,NTR
       STR(II,J,K)=STR0(I)
 320   CONTINUE $ELSE
```

```
      READ(3,8900) (((STR(I,J,K),K=1,NTR),J=1,NCLS),I=1,NLJN) $ENDIF
*
****          THIS SECTION UPDATES STATE VARIABLES AT EACH TIME STEP
*
      DO 1000 L=1,LTSTOP
*
      IF(L.EQ.LSTIM.AND.L.LE.LSTP) CALL STIMULS(ISTM,LSTIM,INTERV,
     *                                          NSTIM1,MARKER)
*
****                    UPDATING THE CELLS
*
      DO 2000 I=1,NCPOPS
      NF(I)=0
      DO 2900 J=1,N(I)
      CALL SVUPDT(I,J)
      IF(E(I,J).LT.TH(I,J)) GO TO 2850
      S(I,J)=1. $NF(I)=NF(I)+1 $IDF(NF(I))=J
*
      IF(IFLG(I).EQ.0) GO TO 2699 $DO 2700 LS=1,NTPOPS
      IF(LJNN(LS,I).EQ.0) GO TO 2700 $LJ=LJNN(LS,I)
      DO 2500 K=1,NRNGE $IF(ACIT(LJ,J,K).EQ.0) GO TO 2499
      IF(ACIT(LJ,J,K).GT.LG1(LJ)) GO TO 2500
      ID=MOD(ACIT(LJ,J,K),10000000) $JC=ID/1000 $JT=ID-JC*1000
      STR(LJ,JC,JT)=STR(LJ,JC,JT)+DLSTR(LJ)*
     *                          (STRMX(LJ)-STR(LJ,JC,JT))
2500  CONTINUE
2499  CONTINUE $JUP=IVR(LJ) $DO 2600 K=1,JUP
      IF(OVR(LJ,K)/10000000000.NE.J) GO TO 2600
      IF(MOD(OVR(LJ,K),10000000000).GT.LG1(LJ)) GO TO 2600
      ID=MOD(OVR(LJ,K),10000000) $JC=ID/1000 $JT=ID-JC*1000
      STR(LJ,JC,JT)=STR(LJ,JC,JT)+DLSTR(LJ)*
     *                          (STRMX(LJ)-STR(LJ,JC,JT))
2600  CONTINUE
2700  CONTINUE
*
2699  CONTINUE
      DO 2800 K=1,NCPOPS
      IF(JNN(I,K).GT.0) CALL TRNSMT(I,K,J)
2800  CONTINUE
2850  CONTINUE
2900  CONTINUE $IF(IFLG1.GT.0) WRITE(7,7060) (L,I,NF(I),
     *              (IDF(J),J=1,NF(I)))
2000  CONTINUE
*
****                    UPDATING THE INPUT FIBERS
*
      DO 3000 I=1,NFPOPS
      II=I+NCPOPS $NF(II)=0
      IF(L.LT.INSTR(I).OR.L.GE.INSTP(I)) GO TO 3000
      CALL RANSET(INFSED(I))
      DO 3900 J=1,N(II)
      IF(RANF(0.).GT.P(I)) GO TO 3850 $CALL RANGET(INFSED(I))
      NF(II)=NF(II)+1 $IDF(NF(II))=J
      DO 3800 K=1,NCPOPS
      IF(JNN(II,K).GT.0) CALL TRNSMT(II,K,J)
3800  CONTINUE $CALL RANSET(INFSED(I))
3850  CONTINUE
3900  CONTINUE $IF(IFLG1.GT.0) WRITE(7,7060) (L,II,NF(II),
     *              (IDF(J),J=1,NF(II)))
      CALL RANGET(INFSED(I))
3000  CONTINUE
*
      WRITE(6,6510) L,(NF(I),I=1,NTPOPS)
      WRITE(7,7510) L,(NF(I),I=1,NTPOPS)
*
      DO 1910 I=1,NCPOPS $DO 1910 J=1,NCLS $DO 1910 M=1,SNTP
      G(M,I,J,1)=G(M,I,J,1)*DCS(M)+G(M,I,J,2) $DO 1920 K=2,MCTP1-1
```

```
 1920 G(M,I,J,K)=G(M,I,J,K+1) $G(M,I,J,MCTP1)=0.
 1910 CONTINUE
      DO 1600 LJ=1,NLJN $JUP=IVR(LJ) $DO 1700 J=1,JUP
 1710 IF(OVR(LJ,J).EQ.0) GO TO 1699
      IF(MOD(OVR(LJ,J),10000000000).GT.10000000) GO TO 1700
      IVR(LJ)=IVR(LJ)-1 $DO 1760 J1=J,IVR(LJ)
 1760 OVR(LJ,J1)=OVR(LJ,J1+1) $OVR(LJ,IVR(LJ)+1)=0 $GO TO 1710
 1700 OVR(LJ,J)=OVR(LJ,J)-10000000
 1699 CONTINUE
      DO 1800 I=1,NCLS $DO 1850 J=1,NRNGE
 1810 IF(ACIT(LJ,I,J).EQ.0) GO TO  1849
      IF(ACIT(LJ,I,J).GT.10000000) GO TO 1850 $DO 1860 J1=1,NRNGE-1
 1860 ACIT(LJ,I,J1)=ACIT(LJ,I,J1+1) $ACIT(LJ,I,NRNGE)=0 $GO TO 1810
 1850 ACIT(LJ,I,J)=ACIT(LJ,I,J)-10000000
 1849 CONTINUE
 1800 CONTINUE
 1600 CONTINUE
*
 1000 CONTINUE
*
**** 	     THIS SECTION WRITES OUT ACTIVITY VARIABLES
*
      DO 4100 JS=1,NTPOPS $DO 4100 JR=1,NCPOPS
      IF(LJNN(JS,JR).EQ.0) GO TO 4100 $WRITE(6,6040)
      LJ=LJNN(JS,JR)
      WRITE(6,6900) ((STR(LJ,I,K),K=1,NTR),I=1,NCLS)
      WRITE(8,8900) ((STR(LJ,I,K),K=1,NTR),I=1,NCLS)
 4100 CONTINUE
*
      STOP
****	     THESE ARE THE FORMATS
*
 5100 FORMAT(12I6)
 5105 FORMAT(12F6.2)
 5110 FORMAT(I6,6F6.2,2I6)
 5120 FORMAT(I6,F6.2,5I6)
 5130 FORMAT(6I6,3F6.2,I6)
 5140 FORMAT(I6,2F6.2)
 6000 FORMAT(1H1)
 6040 FORMAT(///)
 6100 FORMAT(5X,*NTPOPS,NCPOPS,NFPOPS,NCLS,MCTP1,W,H=*,7I7)
 6102 FORMAT(5X,*NJN,NLJN,NTR,NRNGE,MRNGE,IFLG1=*,6I7)
 6105 FORMAT(5X,*EQ,T=*,12F7.2)
 6110 FORMAT(5X,*CELLS: N,C,TTH,B,TGK,TMEM,THO,NC,NR=*)
 6112 FORMAT(9X,I7,6F7.2,2I7)
 6120 FORMAT(///,5X,*FIBERS: N,P,INSTR,INSTP,INFSED,NC,NR=*,//)
 6122 FORMAT(5X,I7,F7.2,5I7)
 6130 FORMAT(///,5X,*JNCTNS: TYPE,LRN,IS,IR,NT,NCT,STR0,WTF,HTF,*
     *          ,*INSED=*,/10X,*MLAG,STRMX,DLSTR=*,//)
 6132 FORMAT(5X,6I7,3F7.2,I7)
 6140 FORMAT(10X,I7,2F7.2)
 6150 FORMAT(5X,*ISTM,LSTIM,INTERV,LSTP,NSTIM,NSTIM1=*,6I7)
 6154 FORMAT(5X,*LOCATNS:*)
 6155 FORMAT(10X,20I6)
 6200 FORMAT(5X,*LTSTOP=*,I7)
 6510 FORMAT(5X,15I7)
 6900 FORMAT(5X,20F6.3)
 7510 FORMAT(15I5)
 7060 FORMAT(20I4)
 8900 FORMAT(F6.3)
      END
*
      SUBROUTINE SVUPDT(I,K)
*
***  THIS SUBROUTINE UPDATES THE STATE VARIABLES OF AN INDIVIDUAL NEURON
*
      INTEGER SNTP
```

```
      PARAMETER NCPOPS=8,NCLS=200,MCTP1=5,STEP=1.,EK=-10.,SNTP=2
      COMMON G(SNTP,NCPOPS,NCLS,MCTP1)
      COMMON/SV/E(NCPOPS,NCLS),TH(NCPOPS,NCLS),S(NCPOPS,NCLS),
     *          GK(NCPOPS,NCLS),B(NCPOPS),C(NCPOPS),DCG(NCPOPS),
     *          DCTH(NCPOPS),TMEM(NCPOPS),THØ(NCPOPS),EQ(SNTP)
*
      GS=Ø. $SYN=Ø. $DO 10 J=1,SNTP
      GS=GS+G(J,I,K,1) $SYN=SYN+G(J,I,K,1)*EQ(J)
   10 CONTINUE
      GK(I,K)=GK(I,K)*DCG(I)+B(I)*(1.-DCG(I))*S(I,K)
      GTOT=1.+GK(I,K)+GS $DCE=EXP(-GTOT*STEP/TMEM(I))
      E(I,K)=E(I,K)*DCE+(SYN+GK(I,K)*EK)*(1.-DCE)/
     *          GTOT
      TH(I,K)=THØ(I)+(TH(I,K)-THØ(I))*DCTH(I)+C(I)*E(I,K)*(1.-DCTH(I))
      S(I,K)=Ø.
*
      RETURN
      END
*
      SUBROUTINE TRNSMT(IS,IR,K)
*
****      THIS SUBROUTINE PROJECTS SYNAPTIC ACTIVATION FROM A SENDING
****      FIBER TO ITS TARGETED RECEIVING CELLS.
*
      INTEGER TYPE,W,H,ACIT,OVR,SNTP
      PARAMETER NTPOPS=10,NCPOPS=8,NCLS=200,MCTP1=5,W=20,H=10,
     *          NJN=22,NLJN=1,NTR=10,NRNGE=10,MRNGE=1000,SNTP=2
      COMMON G(SNTP,NCPOPS,NCLS,MCTP1)
      COMMON/TR/TYPE(NJN),NT(NJN),NCT(NJN),STRØ(NJN),WTF(NJN),HTF(NJN),
     *          INSED(NJN),LJNN(NTPOPS,NCPOPS),MLAG(NLJN),
     *          ACIT(NLJN,NCLS,NRNGE),OVR(NLJN,MRNGE),IVR(NLJN),
     *          STR(NLJN,NCLS,NTR),N(NTPOPS),NC(NTPOPS),NR(NTPOPS),
     *          JNN(NTPOPS,NCPOPS),L
*
      JN=JNN(IS,IR)
      XS=FLOAT(MOD(K,NC(IS)))*FLOAT(W)/FLOAT(NC(IS))
      IF(XS.EQ.Ø.) XS=FLOAT(W)
      YS=FLOAT(K/NC(IS)+1)*FLOAT(H)/FLOAT(NR(IS))
      CALL RANSET(K*INSED(JN))
      DO 100 J=1,NT(JN)
      XR=XS+(RANF(Ø.)-.5)*WTF(JN) $YR=YS+(RANF(Ø.)-.5)*HTF(JN)
      ICR=INT(FLOAT(NC(IR))*(XR-.5)/FLOAT(W))+1
      IRR=INT(FLOAT(NR(IR))*(YR-.5)/FLOAT(H))+1
      IF(ICR.LE.Ø) ICR=NC(IR)+ICR
      IF(ICR.GT.NC(IR)) ICR=ICR-NC(IR)
      IF(IRR.LE.Ø) IRR=NR(IR)+IRR
      IF(IRR.GT.NR(IR)) IRR=IRR-NR(IR)
      NREC=NC(IR)*(IRR-1)+ICR
      KCT=INT(RANF(Ø.)*FLOAT(NCT(JN)))+1
      IF(LJNN(IS,IR).GT.Ø) THEN $I=LJNN(IS,IR)
      DO 200 JR=1,NRNGE
      IF(ACIT(I,NREC,JR).EQ.Ø) GO TO 210
  200 CONTINUE $IVR(I)=IVR(I)+1 $IF(IVR(I).GT.(MRNGE-1)) L=99999
      OVR(I,IVR(I))=J+1000*K+10000000*(KCT+MLAG(I))+10000000000*NREC
      GO TO 220
  210 ACIT(I,NREC,JR)=J+1000*K+10000000*(KCT+MLAG(I))
  220 G(TYPE(JN),IR,NREC,KCT+1)=G(TYPE(JN),IR,NREC,KCT+1)+STR(I,K,J)
      ELSE
      G(TYPE(JN),IR,NREC,KCT+1)=G(TYPE(JN),IR,NREC,KCT+1)+STRØ(JN)
      ENDIF
  100 CONTINUE
*
      RETURN
      END
*
      SUBROUTINE STIMULS(IR,LSTIM,INTERV,NSTIM1,MARKER)
*
```

```
****        THIS SUBROUTINE
        PARAMETER NSTIM=2,NCLS=200,NCPOPS=8
        COMMON/SV/E(NCPOPS,NCLS),TH(NCPOPS,NCLS)
        COMMON/STIM/LOCATN(NSTIM)
*
        DO 10 I=1,NSTIM1
        J=I+MARKER $E(IR,LOCATN(J))=TH(IR,LOCATN(J))+3.
    10 CONTINUE
        MARKER=MARKER+NSTIM1 $IF(MARKER.GE.NSTIM) MARKER=0
        LSTIM=LSTIM+INTERV
        RETURN
        END
*
/EOR
        0        0
    70.    1.    -10.      3.
        200  0.    25.    20.     5.     5.    10.         20      10
        200  0.    25.    10.     3.     5.    10.         20      10
        200  0.    25.    20.     5.     5.    10.         20      10
        200  0.    25.    10.     3.     5.    10.         20      10
        200  0.    25.    20.     5.     5.    10.         20      10
        200  0.    25.    10.     3.     5.    10.         20      10
        200  0.    25.    20.     5.     5.    10.         20      10
        200  0.    25.    10.     3.     5.    10.         20      10
        200    .10      1    50   7935     20      10
        200    .00      1     0   8642     20      10
          1     0      1     2      8      2     .2    5.     4.      1516
          1     0      3     4      8      2     .2    5.     4.      3423
          1     0      5     6      8      2     .2    5.     4.      6238
          1     0      7     8      8      2    0.    5.     4.         0
          2     0      2     1     10      2    1.    7.     5.      2127
          2     0      4     3     10      2    1.    7.     5.      4447
          2     0      6     5     10      2    1.    7.     5.      5663
          2     0      8     7     10      2    1.    7.     5.      8486
          1     0      1     3     10      4     .2    5.     4.      1028
          1     0      3     5     10      4     .2    5.     4.      3822
          1     0      5     7     10      4     .2    5.     4.      6224
          1     0      7     1     10      4     .2    5.     4.      7156
          1     0      1     4      3      4     .2    5.     4.      1602
          1     0      3     6      3      4     .2    5.     4.      3020
          1     0      5     8      3      4     .2    5.     4.      6208
          1     0      7     2      3      4     .2    5.     4.      7082
          1     1      9     1     10      4     .2    5.     4.      9549
          3    2.           .25
          1     0      9     2      3      4     .2    5.     4.      9411
          1     0     10     2      1      4     .2   10.     8.      1039
          1     0     10     4      1      4     .2   10.     8.      6290
          1     0     10     6      1      4     .2   10.     8.      8756
          1     0     10     8      1      4     .2   10.     8.      3111
          1     1      1     2      2      5
          9    10
         20
/EOR
/EOI


/JOB
A224,CM=130000,TL=1000.
FAKE.
PDFV.
M77.
GET,STR000.
LGO,STR000,LN007.
SAVE,LN007.
EXIT.
SAVE,LN007.
```

```
/EOR
      PROGRAM LDSYS33 (ANIN,LRSYS,INPUT,OUTPUT,
     *          TAPE3=ANIN,TAPE6=OUTPUT,TAPE7=LRSYS)
*
****      THIS PROGRAM SIMULATES HEBBIAN LEARNING AT THE SYNAPSES OF AN
*         ARBITRARY SYSTEM OF INTERACTING SPATIALLY-ORGANIZED NEURAL NETS
*         ACTIVATED BY AN ARBITRARY NUMBER OF INPUT FIBER SYSTEMS.  ALL
*         SYNAPTIC SYSTEMS ARE INDIVIDUALLY ADJUSTABLE. SEPARATE TRANS-
****      MISSION AND MEMORY LINKS. ACIT,OVR PACKED.
*
      INTEGER TYPE,W,H,SNTP
      PARAMETER NTPOPS=10,NFPOPS=2,NCPOPS=8,NCLS=200,MCTP1=5,W=20,H=10,
     *          NJN=22,NLJN=1,NTR=10,STEP=1.,SNTP=2,
     *          NSTIM=2
      DIMENSION P(NFPOPS),INSTP(NFPOPS),INFSED(NFPOPS),NF(NTPOPS),
     *          IDF(NCLS),TTH(NCPOPS),TGK(NCPOPS),IFLG(NJN),
     *          LRN(NJN),IS(NJN),IR(NJN),
     *          T(SNTP),DCS(SNTP),INSTR(NFPOPS)
      COMMON G(SNTP,NCPOPS,NCLS,MCTP1)
      COMMON/SV/E(NCPOPS,NCLS),TH(NCPOPS,NCLS),S(NCPOPS,NCLS),
     *          GK(NCPOPS,NCLS),B(NCPOPS),C(NCPOPS),DCG(NCPOPS),
     *          DCTH(NCPOPS),TMEM(NCPOPS),TH0(NCPOPS),EQ(SNTP)
      COMMON/STIM/LOCATN(NSTIM)
      COMMON/TR/TYPE(NJN),NT(NJN),NCT(NJN),STR0(NJN),WTF(NJN),HTF(NJN),
     *          INSED(NJN),LJNN(NTPOPS,NCPOPS),
     *          STR(NLJN,NCLS,NTR),N(NTPOPS),NC(NTPOPS),NR(NTPOPS),
     *          JNN(NTPOPS,NCPOPS),L
*
****          THIS SECTION READS AND WRITES THE INPUT PARAMETERS
*
      READ 5100, IFLG1,IFLG2,IFLG3
      READ 5105, ((EQ(I),T(I)),I=1,SNTP)
      DO 110 I=1,NCPOPS
  110 READ 5110, N(I),C(I),TTH(I),B(I),TGK(I),TMEM(I),TH0(I),
     *          NC(I),NR(I)
      DO 120 I=1,NFPOPS $II=NCPOPS+I
  120 READ 5120, N(II),P(I),INSTR(I),INSTP(I),INFSED(I),NC(II),NR(II)
      DO 128 I=1,NCPOPS $IFLG(I)=0 $DO 128 J=1,NTPOPS
      JNN(J,I)=0 $LJNN(J,I)=0
  128 CONTINUE $J=0 $DO 130 I=1,NJN
     .READ 5130, TYPE(I),LRN(I),IS(I),IR(I),NT(I),NCT(I),STR0(I),
     *          WTF(I),HTF(I),INSED(I)
      JNN(IS(I),IR(I))=I
      IF(LRN(I).GT.0) THEN J=J+1 $LJNN(IS(I),IR(I))=J $IFLG(IR(I))=1 $ENDIF
  130 CONTINUE
      READ 5100, ISTM,LSTIM,INTERV,LSTP,NSTIM1
      READ 5100, (LOCATN(I),I=1,NSTIM)
      READ 5100, LTSTOP
*
      WRITE(6,6000) $WRITE(6,6100) NTPOPS,NFPOPS,NCPOPS,NCLS,MCTP1,W,H,
      WRITE(6,6102) NJN,NLJN,NTR,IFLG1,IFLG2,IFLG3
      WRITE(6,6105) ((EQ(I),T(I)),I=1,SNTP)
      WRITE(6,6110) $DO 111 I=1,NCPOPS
  111 WRITE(6,6112) N(I),C(I),TTH(I),B(I),TGK(I),TMEM(I),TH0(I),
     *          NC(I),NR(I)
      WRITE(6,6120) $DO 121 I=1,NFPOPS $II=NCPOPS+I
  121 WRITE(6,6122) N(II),P(I),INSTR(I),INSTP(I),INFSED(I),
     *          NC(II),NR(II)
      WRITE(6,6130) $J=0 $DO 131 I=1,NJN
      WRITE(6,6132) TYPE(I),LRN(I),IS(I),IR(I),NT(I),NCT(I),
     *          STR0(I),WTF(I),HTF(I),INSED(I)
  131 CONTINUE
      WRITE(6,6150) ISTM,LSTIM,INTERV,LSTP,NSTIM,NSTIM1
      WRITE(6,6154) $WRITE(6,6155) (LOCATN(I),I=1,NSTIM)
      WRITE(6,6200) LTSTOP $WRITE(6,6000)
*
****          THIS SECTION INITIALIZES VARIABLES
```

```
*
      MARKER=0
      DO 310 I=1,NCPOPS
      DCTH(I)=EXP(-STEP/TTH(I)) $DCG(I)=EXP(-STEP/TGK(I))
      DO 310 J=1,NCLS
      E(I,J)=0. $TH(I,J)=TH0(I) $S(I,J)=0. $GK(I,J)=0.
      DO 310 K=1,MCTP1 $DO 310 M=1,SNTP
      G(M,I,J,K)=0.
  310 CONTINUE
      DO 340 I=1,SNTP
  340 DCS(I)=EXP(-STEP/T(I))
      IF(IFLG2.EQ.0) THEN $DO 320 I=1,NJN $IF(LRN(I).EQ.0) GO TO 320
      II=LJNN(IS(I),IR(I)) $DO 320 J=1,NCLS $DO 320 K=1,NTR
      STR(II,J,K)=STR0(I)
  320 CONTINUE $ELSE
      READ(3,8900) (((STR(I,J,K),K=1,NTR),J=1,NCLS),I=1,NLJN) $ENDIF
*
****      THIS SECTION UPDATES STATE VARIABLES AT EACH TIME STEP
*
      DO 1000 L=1,LTSTOP
*
      IF(L.EQ.LSTIM.AND.L.LE.LSTP) CALL STIMULS(ISTM,LSTIM,INTERV,
     *                                      NSTIM1,MARKER)
*
****              UPDATING THE CELLS
*
      DO 2000 I=1,NCPOPS
      NF(I)=0
      DO 2900 J=1,N(I)
      CALL SVUPDT(I,J)
      IF(E(I,J).LT.TH(I,J)) GO TO 2850
      S(I,J)=1. $NF(I)=NF(I)+1 $IDF(NF(I))=J
*
      DO 2800 K=1,NCPOPS
      IF(JNN(I,K).GT.0) CALL TRNSMT(I,K,J)
 2800 CONTINUE
 2850 CONTINUE
 2900 CONTINUE $IF(IFLG1.GT.0) WRITE(7,7060) (L,I,NF(I),
     *          (IDF(J),J=1,NF(I)))
 2000 CONTINUE
*
****              UPDATING THE INPUT FIBERS
*
      DO 3000 I=1,NFPOPS
      II=I+NCPOPS $NF(II)=0
      IF(L.LT.INSTR(I).OR.L.GE.INSTP(I)) GO TO 3000
      CALL RANSET(INFSED(I))
      DO 3900 J=1,N(II)
      IF(RANF(0.).GT.P(I)) GO TO 3850 $CALL RANGET(INFSED(I))
      NF(II)=NF(II)+1 $IDF(NF(II))=J
      DO 3800 K=1,NCPOPS
      IF(JNN(II,K).GT.0) CALL TRNSMT(II,K,J)
 3800 CONTINUE $CALL RANSET(INFSED(I))
 3850 CONTINUE
 3900 CONTINUE $IF(IFLG1.GT.0) WRITE(7,7060) (L,II,NF(II),
     *          (IDF(J),J=1,NF(II)))
      CALL RANGET(INFSED(I))
 3000 CONTINUE
*
      WRITE(6,6510) L,(NF(I),I=1,NTPOPS)
      WRITE(7,7510) L,(NF(I),I=1,NTPOPS)
*
      DO 1910 I=1,NCPOPS $DO 1910 J=1,NCLS $DO 1910 M=1,SNTP
      G(M,I,J,1)=G(M,I,J,1)*DCS(M)+G(M,I,J,2) $DO 1920 K=2,MCTP1-1
 1920 G(M,I,J,K)=G(M,I,J,K+1) $G(M,I,J,MCTP1)=0.
 1910 CONTINUE
*
```

```
 1000 CONTINUE
*
****         THIS SECTION WRITES OUT ACTIVITY VARIABLES
*
      IF(IFLG3.GT.0) THEN $DO 4100 JS=1,NTPOPS $DO 4100 JR=1,NCPOPS
      IF(LJNN(JS,JR).EQ.0) GO TO 4100 $WRITE(6,6040)
      LJ=LJNN(JS,JR) $JN=JNN(JS,JR)
      WRITE(6,6900) ((STR(LJ,I,K),K=1,NT(JN)),I=1,N(JS))
 4100 CONTINUE $ENDIF
*
      STOP
****         THESE ARE THE FORMATS
*
 5100 FORMAT(12I6)
 5105 FORMAT(12F6.2)
 5110 FORMAT(I6,6F6.2,2I6)
 5120 FORMAT(I6,F6.2,5I6)
 5130 FORMAT(6I6,3F6.2,I6)
 6000 FORMAT(1H1)
 6040 FORMAT(///)
 6100 FORMAT(5X,*NTPOPS,NCPOPS,NFPOPS,NCLS,MCTP1,W,H=*,7I7)
 6102 FORMAT(5X,*NJN,NLJN,NTR,IFLG1,IFLG2,IFLG3=*,6I7)
 6105 FORMAT(5X,*EQ,T=*,12F7.2)
 6110 FORMAT(5X,*CELLS: N,C,TTH,B,TGK,TMEM,THO,NC,NR=*)
 6112 FORMAT(9X,I7,6F7.2,2I7)
 6120 FORMAT(///,5X,*FIBERS: N,P,INSTR,INSTP,INFSED,NC,NR=*,//)
 6122 FORMAT(5X,I7,F7.2,5I7)
 6130 FORMAT(///,5X,*JNCTNS: TYPE,LRN,IS,IR,NT,NCT,STR0,WTF,HTF,*
     *          ,*INSED=*,//)
 6132 FORMAT(5X,6I7,3F7.2,I7)
 6150 FORMAT(5X,*ISTM,LSTIM,INTERV,LSTP,NSTIM,NSTIM1=*,6I7)
 6154 FORMAT(5X,*LOCATNS:*)
 6155 FORMAT(10X,20I6)
 6200 FORMAT(5X,*LTSTOP=*,I7)
 6510 FORMAT(5X,15I7)
 6900 FORMAT(5X,20F6.3)
 7510 FORMAT(15I5)
 7060 FORMAT(20I4)
 8900 FORMAT(F6.3)
      END
*
      SUBROUTINE SVUPDT(I,K)
*
***  THIS SUBROUTINE UPDATES THE STATE VARIABLES OF AN INDIVIDUAL NEURON
*
      INTEGER SNTP
      PARAMETER NCPOPS=8,NCLS=200,MCTP1=5,STEP=1.,EK=-10.,SNTP=2
      COMMON G(SNTP,NCPOPS,NCLS,MCTP1)
      COMMON/SV/E(NCPOPS,NCLS),TH(NCPOPS,NCLS),S(NCPOPS,NCLS),
     *          GK(NCPOPS,NCLS),B(NCPOPS),C(NCPOPS),DCG(NCPOPS),
     *          DCTH(NCPOPS),TMEM(NCPOPS),TH0(NCPOPS),EQ(SNTP)
*
      GS=0. $SYN=0. $DO 10 J=1,SNTP
      GS=GS+G(J,I,K,1) $SYN=SYN+G(J,I,K,1)*EQ(J)
   10 CONTINUE
      GK(I,K)=GK(I,K)*DCG(I)+B(I)*(1.-DCG(I))*S(I,K)
      GTOT=1.+GK(I,K)+GS $DCE=EXP(-GTOT*STEP/TMEM(I))
      E(I,K)=E(I,K)*DCE+(SYN+GK(I,K)*EK)*(1.-DCE)/
     *          GTOT
      TH(I,K)=TH0(I)+(TH(I,K)-TH0(I))*DCTH(I)+C(I)*E(I,K)*(1.-DCTH(I))
      S(I,K)=0.
*
      RETURN
      END
*
      SUBROUTINE TRNSMT(IS,IR,K)
*
```

```
****      THIS SUBROUTINE PROJECTS SYNAPTIC ACTIVATION FROM A SENDING
****      FIBER TO ITS TARGETED RECEIVING CELLS.
*
      INTEGER TYPE,W,H,SNTP
      PARAMETER NTPOPS=10,NCPOPS=8,NCLS=200,MCTP1=5,W=20,H=10,
     *          NJN=22,NLJN=1,NTR=10,SNTP=2
      COMMON G(SNTP,NCPOPS,NCLS,MCTP1)
      COMMON/TR/TYPE(NJN),NT(NJN),NCT(NJN),STRØ(NJN),WTF(NJN),HTF(NJN),
     *      INSED(NJN),LJNN(NTPOPS,NCPOPS),
     *      STR(NLJN,NCLS,NTR),N(NTPOPS),NC(NTPOPS),NR(NTPOPS),
     *      JNN(NTPOPS,NCPOPS),L
*
      JN=JNN(IS,IR)
      XS=FLOAT(MOD(K,NC(IS)))*FLOAT(W)/FLOAT(NC(IS))
      IF(XS.EQ.Ø.) XS=FLOAT(W)
      YS=FLOAT(K/NC(IS)+1)*FLOAT(H)/FLOAT(NR(IS))
      CALL RANSET(K*INSED(JN))
      DO 100 J=1,NT(JN)
      XR=XS+(RANF(Ø.)-.5)*WTF(JN) $YR=YS+(RANF(Ø.)-.5)*HTF(JN)
      ICR=INT(FLOAT(NC(IR))*(XR-.5)/FLOAT(W))+1
      IRR=INT(FLOAT(NR(IR))*(YR-.5)/FLOAT(H))+1
      IF(ICR.LE.Ø) ICR=NC(IR)+ICR
      IF(ICR.GT.NC(IR)) ICR=ICR-NC(IR)
      IF(IRR.LE.Ø) IRR=NR(IR)+IRR
      IF(IRR.GT.NR(IR)) IRR=IRR-NR(IR)
      NREC=NC(IR)*(IRR-1)+ICR
      KCT=INT(RANF(Ø.)*FLOAT(NCT(JN)))+1
      IF(LJNN(IS,IR).GT.Ø) THEN $X=STR(LJNN(IS,IR),K,J)
      ELSE $X=STRØ(JN) $ENDIF
      G(TYPE(JN),IR,NREC,KCT+1)=G(TYPE(JN),IR,NREC,KCT+1)+X
  100 CONTINUE
*
      RETURN
      END
*
      SUBROUTINE STIMULS(IR,LSTIM,INTERV,NSTIM1,MARKER)
*
****      THIS SUBROUTINE
      PARAMETER NSTIM=2,NCLS=200,NCPOPS=8
      COMMON/SV/E(NCPOPS,NCLS),TH(NCPOPS,NCLS)
      COMMON/STIM/LOCATN(NSTIM)
*
      DO 10 I=1,NSTIM1
      J=I+MARKER $E(IR,LOCATN(J))=TH(IR,LOCATN(J))+3.
   10 CONTINUE
      MARKER=MARKER+NSTIM1 $IF(MARKER.GE.NSTIM) MARKER=0
      LSTIM=LSTIM+INTERV
      RETURN
      END
*
/EOR
      Ø        Ø        Ø
 70.      1.   -10.     3.
    200  Ø.    25.    20.     5.     5.    10.        20    10
    200  Ø.    25.    10.     3.     5.    10.        20    10
    200  Ø.    25.    20.     5.     5.    10.        20    10
    200  Ø.    25.    10.     3.     5.    10.        20    10
    200  Ø.    25.    20.     5.     5.    10.        20    10
    200  Ø.    25.    10.     3.     5.    10.        20    10
    200  Ø.    25.    20.     5.     5.    10.        20    10
    200  Ø.    25.    10.     3.     5.    10.        20    10
    200  .10        1    50  7935      20    10
    200  .00        1     Ø  8642      20    10
      1    Ø        1     2     8      2    .2   5.     4.      1516
      1    Ø        3     4     8      2    .2   5.     4.      3423
      1    Ø        5     6     8      2    .2   5.     4.      6238
      1    Ø        7     8     8      2   Ø.    5.     4.         Ø
```

```
2    Ø    2    1    10    2    1.    7.    5.    2127
2    Ø    4    3    10    2    1.    7.    5.    4447
2    Ø    6    5    10    2    1.    7.    5.    5663
2    Ø    8    7    10    2    1.    7.    5.    8486
1    Ø    1    3    10    4    .2    5.    4.    1028
1    Ø    3    5    10    4    .2    5.    4.    3822
1    Ø    5    7    10    4    .2    5.    4.    6224
1    Ø    7    1    10    4    .2    5.    4.    7156
1    Ø    1    4    3     4    .2    5.    4.    1602
1    Ø    3    6    3     4    .2    5.    4.    3020
1    Ø    5    8    3     4    .2    5.    4.    6208
1    Ø    7    2    3     4    .2    5.    4.    7082
1    1    9    1    10    4    .2    5.    4.    9549
1    Ø    9    2    3     4    .2    5.    4.    9411
1    Ø    10   2    1     4    .2    10.   8.    1039
1    Ø    10   4    1     4    .2    10.   8.    6290
1    Ø    10   6    1     4    .2    10.   8.    8756
1    Ø    10   8    1     4    .2    10.   8.    3111
1    1    1    2    2     5
9    10
20
/EOR
/EOI


/JOB
A224,TL=200.
FAKE.
PDFV.
MNF(Y).
LGO,DNØ1A.
SAVE,DNØ1A.
EXIT.
SAVE,DNØ1A.
/EOR
      PROGRAM DENDRØ1(DNOUT,INPUT,OUTPUT,TAPE6=OUTPUT,TAPE7=DNOUT)
*
****      THIS PROGRAM SIMULATES THE ACTIVITY OF A NEURON WITH A SINGLE
*         PASSIVE DENDRITE ATTACHED TO THE SOMA IN RESPONSE TO A STEP
****      CURRENT INPUT.
*
      INTEGER SCSTRT,SCSTP
      REAL LBR,LSEG
      PARAMETER NCMPT=10, NSTEP=1000, STEP=1.,NDSPY=2,FD=1.,FS=1.
      DIMENSION GØ(NCMPT),Z(NCMPT),V(NCMPT),U(NCMPT),IDC(10),
     *          PS(10,500)
*
****      THIS SECTION READS AND WRITES THE INPUT PARAMETERS
*
      READ 5000, IFLG1
      READ 5010, C,TTH,B,TGK,TMEM,THØ
      READ 5010, RS,R,LBR
      READ 5020, SC,SCSTRT,SCSTP
      READ 5000, (IDC(J),J=1,NDSPY)
      READ 5000, LTSTOP
      RES=700000. $CON=1./(TMEM*10.**11)
*
      AS=4.*3.14159*RS**2 $LSEG=LBR/FLOAT(NCMPT)
      AM=2.*3.14159*R*LSEG $X=SQRT(RS**2-R**2)
      RSD=RES*(LSEG/R**2+ALOG((RS+X)/(RS-X))/RS)/2./3.14159*FS
      RDD=RES*LSEG/3.14159/R**2*FD
      GDS=1./(RSD*CON*AS) $GSD=1./(RSD*CON*AM)
      GDD=1./(RDD*CON*AM) $GLK=3.14159*R**2/AM
*
      WRITE(6,6000) $WRITE(6,6005) NCMPT,NSTEP,STEP,NDSPY,RES,CON,FD,FS
      WRITE(6,6010) C,TTH,B,TGK,TMEM,THØ
```

```
      WRITE(6,6020) RS,R,LBR,LSEG,AS,AM
      WRITE(6,6025) RSD,RDD,GDS,GSD,GDD,GLK
      WRITE(6,6030) SC,SCSTRT,SCSTP,LTSTOP
*
****          THIS SECTION INITIALIZES VARIABLES
*
      GØS=1.+GDS
      GØ(1)=1.+GSD+GDD
      DO 110 I=2,NCMPT-1
  110 GØ(I)=1.+2.*GDD
      GØ(NCMPT)=1.+GLK+GDD
*
      DO 210 I=1,NCMPT
  210 Z(I)=EXP(-GØ(I)*STEP/TMEM/FLOAT(NSTEP))
*
      US=0. $VS=0. $GK=0. $TH=THØ
      DGK=EXP(-STEP/TGK) $DTH=EXP(-STEP/TTH)
      DO 310 I=1,NCMPT $U(I)=0.
  310 V(I)=0.
*
****          THIS SECTION UPDATES STATE VARIABLES AT EACH TIME STEP
*
      DO 1000 L=1,LTSTOP
      SCN=0. $IF(L.GE.SCSTRT.AND.L.LT.SCSTP) SCN=SC
      ZS=EXP(-(GØS+GK)*STEP/TMEM/FLOAT(NSTEP))
*
*
****                THIS SUB-SECTION PERFORMS FINE-GRAINED INTEGRATION
****                OF COMPARTMENT POTENTIALS.
*
      DO 4000 IS=1,NSTEP
      VS=US*ZS+(GDS*U(1)+SCN)*(1.-ZS)/GØS
      V(1)=U(1)*Z(1)+(GSD*US+GDD*U(2))*(1.-Z(1))/GØ(1)
      DO 4010 J=2,NCMPT-1
 4010 V(J)=U(J)*Z(J)+GDD*(U(J-1)+U(J+1))*(1.-Z(J))/GØ(J)
      V(NCMPT)=U(NCMPT)*Z(NCMPT)+GDD*U(NCMPT-1)*(1.-Z(NCMPT))/GØ(NCMPT)
      DO 4020 J=1,NCMPT
 4020 U(J)=V(J) $US=VS
 4000 CONTINUE
*
      TH=TH*DTH+(THØ+C*VS)*(1.-DTH) $GK=GK*DGK
      S=0. $IF(VS.LT.TH) GO TO 4300
      S=1. $GK=GK+B
 4300 CONTINUE $PS(1,L)=VS+S*(50.-VS)
      IF(IFLG1.GT.0) THEN
      WRITE(6,6280) VS,(V(J),J=1,NCMPT) $ENDIF
      DO 4500 I=1,NDSPY
 4500 PS(I+1,L)=V(IDC(I))
*
 1000 CONTINUE
*
****          THIS SECTION WRITES OUT ACTIVITY VARIABLES
*
      DO 1600 I=1,NDSPY+1 $WRITE(6,6002) $WRITE(7,6002)
      WRITE(6,6280) (PS(I,L),L=1,LTSTOP)
 1600 WRITE(7,7280) (PS(I,L),L=1,LTSTOP)
*
      STOP
****          THESE ARE THE FORMATS
*
 5000 FORMAT(12I6)
 5010 FORMAT(12F6.2)
 5020 FORMAT(F6.2,2I6)
 6000 FORMAT(1H1)
 6002 FORMAT(/)
 6005 FORMAT(5X,*NCMPT,NSTEP,STEP,NDSPY,RES,CON,FD,FS=*,2I7,F7.2,I7,
      *     2F10.2,2F7.2)
```

```
 6010 FORMAT(5X,*C,TTH,B,TGK,TMEM,THØ=*,6F7.2)
 6020 FORMAT(5X,*RS,R,LBR,LSEG,AS,AM=*,6F7.2)
 6025 FORMAT(5X,*RSD,RDD,GDS,GSD,GDD,GLK=*,6E12.2)
 6030 FORMAT(5X,*SC,SCSTRT,SCSTP,LTSTOP=*,F7.2,3I7)
 6280 FORMAT(5X,20F6.2)
 7280 FORMAT(13F6.2)
      END
/EOR
      Ø
  Ø.   20.    4.    5.    5.   24.
  5.    1.  100.
 10.       1  9999
      3     8
     35
/EOR
/EOI




/JOB
A224,TL=200.
FAKE.
PDFV.
MNF(Y).
LGO,DNØ2A.
SAVE,DNØ2A.
EXIT.
SAVE,DNØ2A.
/EOR
      PROGRAM DENDRØ2(DNOUT,INPUT,OUTPUT,TAPE6=OUTPUT,TAPE7=DNOUT)
*
****      THIS PROGRAM SIMULATES THE ONGOING INPUT-OUTPUT ACTIVITY OF
*         A NEURON WITH A SINGLE PASSIVE DENDRITE ATTACHED TO THE SOMA
****      IN RESPONSE TO SYNAPTIC INPUT BOMBARDMENT.
*
      INTEGER SCSTRT,SCSTP,SNTP,TYPE
      REAL LBR,LSEG
      PARAMETER NCMPT=10,NSTEP=1000,STEP=1.,NDSPY=2,SNTP=1,NTG=1,FD=1.,
     *         FS=1.
      DIMENSION GØ(NCMPT),Z(NCMPT),V(NCMPT),U(NCMPT),IDC(10),
     *          PS(10,500),GS(SNTP),G(SNTP,NCMPT),SYN(NCMPT),
     *          ITC(NTG),TYPE(NTG),P(NTG),STR(NTG),LSTRT(NTG),LSTP(NTG),
     *          INSED(NTG),IDF(NCMPT),EQ(SNTP),GT(NCMPT)
*
****        THIS SECTION READS AND WRITES THE INPUT PARAMETERS
*
      READ 5000, IFLG1
      READ 5010, C,TTH,B,TGK,TMEM,THØ
      READ 5010, RS,R,LBR
      DO 102 I=1,NTG
  102 READ 5015, ITC(I),TYPE(I),P(I),STR(I),LSTRT(I),LSTP(I),INSED(I)
      READ 5010, (EQ(K),K=1,SNTP)
      READ 5020, SC,SCSTRT,SCSTP
      READ 5000, (IDC(J),J=1,NDSPY)
      READ 5000, LTSTOP
      RES=700000. $CON=1./(TMEM*10.**11)
*
      AS=4.*3.14159*RS**2 $LSEG=LBR/FLOAT(NCMPT)
      AM=2.*3.14159*R*LSEG $X=SQRT(RS**2-R**2)
      RSD=RES*(LSEG/R**2+ALOG((RS+X)/(RS-X))/RS)/2./3.14159*FS
      RDD=RES*LSEG/3.14159/R**2*FD
      GDS=1./(RSD*CON*AS) $GSD=1./(RSD*CON*AM)
      GDD=1./(RDD*CON*AM) $GLK=3.14159*R**2/AM
*
      WRITE(6,6000) $WRITE(6,6005) NCMPT,NSTEP,STEP,NDSPY,RES,CON,FD,FS
      WRITE(6,6010) C,TTH,B,TGK,TMEM,THØ
      WRITE(6,6020) RS,R,LBR,LSEG,AS,AM
```

```
      WRITE(6,6025) RSD,RDD,GDS,GSD,GDD,GLK
      DO 104 I=1,NTG
 104  WRITE(6,6027) ITC(I),TYPE(I),P(I),STR(I),LSTRT(I),LSTP(I),INSED(I)
      WRITE(6,6028) (EQ(K),K=1,SNTP)
      WRITE(6,6030) SC,SCSTRT,SCSTP,LTSTOP
*
****          THIS SECTION INITIALIZES VARIABLES
*
      G0S=1.+GDS
      G0(1)=1.+GSD+GDD
      DO 110 I=2,NCMPT-1
 110  G0(I)=1.+2.*GDD
      G0(NCMPT)=1.+GLK+GDD
*
      US=0. $VS=0. $GK=0. $TH=TH0
      DGK=EXP(-STEP/TGK) $DTH=EXP(-STEP/TTH)
      DO 310 I=1,NCMPT $U(I)=0. $DO 305 K=1,SNTP
 305  G(K,I)=0.
 310  V(I)=0. $DO 320 K=1,SNTP
 320  GS(K)=0.
*
****          THIS SECTION UPDATES STATE VARIABLES AT EACH TIME STEP
*
      DO 1000 L=1,LTSTOP
      SCN=0. $IF(L.GE.SCSTRT.AND.L.LT.SCSTP) SCN=SC
      NF=0
*
      CALL RANSET(INSED(I))
      DO 2000 I=1,NTG
      IF(RANF(0.).GT.P(I)) GO TO 2000
      NF=NF+1 $IDF(NF)=I
      IF(ITC(I).EQ.0) GS(TYPE(I))=GS(TYPE(I))+STR(I)
      IF(ITC(I).GT.0) G(TYPE(I),ITC(I))=G(TYPE(I),ITC(I))+STR(I)
 2000 CONTINUE $CALL RANGET(INSED(I))
*
      GSM=0. $SYNS=0. $DO 3000 K=1,SNTP
      SYNS=SYNS+GS(K)*EQ(K)
 3000 GSM=GSM+GS(K) $GTS=G0S+GSM $ZS=EXP(-GTS*STEP/TMEM/FLOAT(NSTEP))
      DO 3200 J=1,NCMPT $SYNG=0. $SYN(J)=0. $DO 3210 K=1,SNTP
      SYN(J)=SYN(J)+G(K,J)*EQ(K)
 3210 SYNG=SYNG+G(K,J) $GT(J)=G0(J)+SYNG
 3200 Z(J)=EXP(-GT(J)*STEP/TMEM/FLOAT(NSTEP))
*
*
****                THIS SUB-SECTION PERFORMS FINE-GRAINED INTEGRATION
****                OF COMPARTMENT POTENTIALS.
*
      DO 4000 IS=1,NSTEP
      VS=US*ZS+(GDS*U(1)+SYNS+SCN)*(1.-ZS)/GTS
      V(1)=U(1)*Z(1)+(GSD*US+GDD*U(2)+SYN(1))*(1.-Z(1))/GT(1)
      DO 4010 J=2,NCMPT-1
 4010 V(J)=U(J)*Z(J)+(GDD*(U(J-1)+U(J+1))+SYN(J))*(1.-Z(J))/GT(J)
      V(NCMPT)=U(NCMPT)*Z(NCMPT)+(GDD*U(NCMPT-1)+SYN(NCMPT))*
     *         (1.-Z(NCMPT))/GT(NCMPT)
      DO 4020 J=1,NCMPT
 4020 U(J)=V(J) $US=VS
 4000 CONTINUE
*
      TH=TH*DTH+(TH0+C*VS)*(1.-DTH) $GK=GK*DGK
      S=0. $IF(VS.LT.TH) GO TO 4300
      S=1. $GK=GK+B
 4300 CONTINUE $PS(1,L)=VS+S*(50.-VS)
      IF(IFLG1.GT.0) THEN
      WRITE(6,6280) VS,(V(J),J=1,NCMPT) $ENDIF
      DO 4500 I=1,NDSPY
 4500 PS(I+1,L)=V(IDC(I))
*
```

```
 1000 CONTINUE
*
****          THIS SECTION WRITES OUT ACTIVITY VARIABLES
*
      DO 1600 I=1,NDSPY+1 $WRITE(6,6002) $WRITE(7,6002)
      WRITE(6,6280) (PS(I,L),L=1,LTSTOP)
 1600 WRITE(7,7280) (PS(I,L),L=1,LTSTOP)
*
      STOP
****          THESE ARE THE FORMATS
*
 5000 FORMAT(12I6)
 5010 FORMAT(12F6.2)
 5015 FORMAT(2I6,2F6.2,3I6)
 5020 FORMAT(F6.2,2I6)
 6000 FORMAT(1H1)
 6002 FORMAT(/)
 6005 FORMAT(5X,*NCMPT,NSTEP,STEP,NDSPY,RES,CON,FD,FS=*,2I7,F7.2,I7,
     *          2F10.2,2F7.2)
 6010 FORMAT(5X,*C,TTH,B,TGK,TMEM,TH0=*,6F7.2)
 6020 FORMAT(5X,*RS,R,LBR,LSEG,AS,AM=*,6F7.2)
 6025 FORMAT(5X,*RSD,RDD,GDS,GSD,GDD,GLK=*,6E12.2)
 6027 FORMAT(5X,*ITC,SNT,P,STR,LSTRT,LSTP,INSED=*,2I7,2F7.2,3I7)
 6028 FORMAT(5X,*EQS=*,12F7.2)
 6030 FORMAT(5X,*SC,SCSTRT,SCSTP,LTSTOP=*,F7.2,3I7)
 6280 FORMAT(5X,20F6.2)
 7280 FORMAT(13F6.2)
      END
/EOR
      0
  0.   20.    4.    5.    5.    24.
  5.    1.  100.
      0    1  0.    0.     100     0  1936
 70.
 10.          1  9999
      3    8
     35
/EOR
/EOI


/JOB
A224,TL=200.
FAKE.
PDFV.
MNF(Y).
LGO,DN03A.
SAVE,DN03A.
EXIT.
SAVE,DN03A.
/EOR
      PROGRAM DENDR03(DNOUT,INPUT,OUTPUT,TAPE6=OUTPUT,TAPE7=DNOUT)
*
****          THIS PROGRAM SIMULATES THE RESPONSE OF A NEURON WITH A SINGLE
*             PASSIVE BIFURCATING DENDRITE ATTACHED TO THE SOMA TO A STEP
****          CURRENT INPUT.
*
      INTEGER SCSTRT,SCSTP
      REAL LBR,LSEG
      PARAMETER NDN=3,NCMPT=10,NSTEP=1000,STEP=1.,NDSPY=2,FD=1.,FS=1.
      DIMENSION G0(NDN,NCMPT),Z(NDN,NCMPT),V(NDN,NCMPT),U(NDN,NCMPT),
     *          GDD(3),IDB(10),IDC(10),PS(10,500)
*
****          THIS SECTION READS AND WRITES THE INPUT PARAMETERS
*
      READ 5000, IFLG1
```

```
      READ 5010, C,TTH,B,TGK,TMEM,THØ
      READ 5010, RS,RST,LBR
      READ 5020, SC,SCSTRT,SCSTP
      READ 5000, ((IDB(J),IDC(J)),J=1,NDSPY)
      READ 5000, LTSTOP
      RES=700000. $CON=1./(TMEM*10.**11)
*
      R=RST/2.**(2./3.) $AS=4.*3.14159*RS**2 $LSEG=LBR/FLOAT(NCMPT)
      AM1=2.*3.14159*RST*LSEG$AM2=2.*3.14159*R*LSEG$X=SQRT(RS**2-RST**2)
      RSD=RES*(LSEG/RST**2+ALOG((RS+X)/(RS-X))/RS)/2./3.14159*FS
      RDD1=RES*LSEG/3.14159/RST**2*FD $RDD2=RES*LSEG/3.14159/R**2*FD
      RTR=RES*LSEG*(1./RST**2+1./R**2)/2./3.14159*FD
      GDS=1./(RSD*CON*AS) $GSD=1./(RSD*CON*AM1)
      GDD(1)=1./(RDD1*CON*AM1) $GDD(2)=1./(RDD2*CON*AM2) $GDD(3)=GDD(2)
      GBT=1./(RTR*CON*AM1) $GTU=1./(RTR*CON*AM2) $GTA=1./(RDD2*CON*AM2)
      GLK=3.14159*R**2/AM2
*
      WRITE(6,6000) $WRITE(6,6005) NDN,NCMPT,NSTEP,STEP,NDSPY,RES,CON,
     *      FD,FS
      WRITE(6,6010) C,TTH,B,TGK,TMEM,THØ
      WRITE(6,6020) RS,RST,R,LBR,LSEG,AS,AM1,AM2
      WRITE(6,6025) RSD,RDD1,RDD2,RTR,GDS,GSD,GBT,GTU,GTA,
     *             (GDD(I),I=1,3),GLK
      WRITE(6,6030) SC,SCSTRT,SCSTP,LTSTOP
*
****          THIS SECTION INITIALIZES VARIABLES
*
      GØS=1.+GDS
      GØ(1,1)=1.+GSD+GDD(1)
      GØ(2,1)=1.+GTU+GTA+GDD(2)
      GØ(3,1)=1.+GTU+GTA+GDD(3)
      DO 110 I=1,NDN $DO 110 J=2,NCMPT-1
  110 GØ(I,J)=1.+2.*GDD(I)
      GØ(1,NCMPT)=1.+2.*GBT+GDD(1)
      GØ(2,NCMPT)=1.+GDD(2)+GLK
      GØ(3,NCMPT)=1.+GDD(3)+GLK
*
      DO 210 I=1,NDN $DO 210 J=1,NCMPT
  210 Z(I,J)=EXP(-GØ(I,J)*STEP/TMEM/FLOAT(NSTEP))
*
      US=0. $VS=0. $GK=0. $TH=THØ
      DGK=EXP(-STEP/TGK) $DTH=EXP(-STEP/TTH)
      DO 310 I=1,NDN $DO 310 J=1,NCMPT $U(I,J)=0.
  310 V(I,J)=0.
*
****          THIS SECTION UPDATES STATE VARIABLES AT EACH TIME STEP
*
      DO 1000 L=1,LTSTOP
      SCN=0. $IF(L.GE.SCSTRT.AND.L.LT.SCSTP) SCN=SC
      ZS=EXP(-(GØS+GK)*STEP/TMEM/FLOAT(NSTEP))
*
*
****              THIS SUB-SECTION PERFORMS FINE-GRAINED INTEGRATION
****              OF COMPARTMENT POTENTIALS.
*
      DO 4000 IS=1,NSTEP
      VS=US*ZS+(GDS*U(1,1)+SCN)*(1.-ZS)/GØS
      V(1,1)=U(1,1)*Z(1,1)+(GSD*US+GDD(1)*U(1,2))*(1.-Z(1,1))/GØ(1,1)

      V(2,1)=U(2,1)*Z(2,1)+(GTU*U(1,NCMPT)+GTA*U(3,1)+GDD(2)*U(2,2))*
     *        (1.-Z(2,1))/GØ(2,1)
      V(3,1)=U(3,1)*Z(3,1)+(GTU*U(1,NCMPT)+GTA*U(2,1)+GDD(3)*U(3,2))*
     *        (1.-Z(3,1))/GØ(3,1)
      DO 4010 I=1,NDN $DO 4010 J=2,NCMPT-1
 4010 V(I,J)=U(I,J)*Z(I,J)+GDD(I)*(U(I,J-1)+U(I,J+1))*(1.-Z(I,J))/
     *        GØ(I,J)
      V(1,NCMPT)=U(1,NCMPT)*Z(1,NCMPT)+(GBT*(U(2,1)+U(3,1))+
```

```
      *            GDD(1)*U(1,NCMPT-1))*(1.-Z(1,NCMPT))/GØ(1,NCMPT)
       V(2,NCMPT)=U(2,NCMPT)*Z(2,NCMPT)+GDD(2)*U(2,NCMPT-1)*
      *            (1.-Z(2,NCMPT))/GØ(2,NCMPT)
       V(3,NCMPT)=U(3,NCMPT)*Z(3,NCMPT)+GDD(3)*U(3,NCMPT-1)*
      *            (1.-Z(3,NCMPT))/GØ(3,NCMPT)
       DO 4020 I=1,NDN $DO 4020 J=1,NCMPT
 4020 U(I,J)=V(I,J) $US=VS
 4000 CONTINUE
*
       TH=TH*DTH+(THØ+C*VS)*(1.-DTH) $GK=GK*DGK
       S=Ø. $IF(VS.LT.TH) GO TO 4300
       S=1. $GK=GK+B
 4300 CONTINUE $PS(1,L)=VS+S*(5Ø.-VS)
       IF(IFLG1.GT.Ø) THEN
       WRITE(6,6280) VS,((V(I,J),J=1,NCMPT),I=1,NDN) $ENDIF
       DO 4500 I=1,NDSPY
 4500 PS(I+1,L)=V(IDB(I),IDC(I))
*
 1000 CONTINUE
*
**** THIS SECTION WRITES OUT ACTIVITY VARIABLES
*
       DO 1600 I=1,NDSPY+1 $WRITE(6,6002) $WRITE(7,6002)
       WRITE(6,6280) (PS(I,L),L=1,LTSTOP)
 1600 WRITE(7,7280) (PS(I,L),L=1,LTSTOP)
*
       STOP
**** THESE ARE THE FORMATS
*
 5000 FORMAT(12I6)
 5010 FORMAT(12F6.2)
 5020 FORMAT(F6.2,2I6)
 6000 FORMAT(1H1)
 6002 FORMAT(/)
 6005 FORMAT(5X,*NDN,NCMPT,NSTEP,STEP,NDSPY,RES,CON=*,3I7,F7.2,I7,
      *            2F9.1,2F7.2)
 6010 FORMAT(5X,*C,TTH,B,TGK,TMEM,THØ=*,6F7.2)
 6020 FORMAT(5X,*RS,RST,R,LBR,LSEG,AS,AM1,AM2=*,8F7.2)
 6025 FORMAT(5X,*RSD,RDD1,RDD2,RTR,GDS,GSD,GBT,GTU,GTA,GDDS,GLK=*,/,
      *            5X,13E10.2)
 6030 FORMAT(5X,*SC,SCSTRT,SCSTP,LTSTOP=*,F7.2,3I7)
 6280 FORMAT(5X,20F6.2)
 7280 FORMAT(13F6.2)
       END
/EOR
         Ø
  Ø.    20.     4.     5.     5.    24.
  5.     1.   100.
 10.           1  9999
      2        3      3      8
     35
/EOR
/EOI


/JOB
A224,CM=70000,TL=1200.
FAKE.
PDFV.
MNF(Y).
LGO,DN2Ø2.
SAVE,DN2Ø2.
EXIT.
SAVE,DN2Ø2.
/EOR
       PROGRAM DENDR21 (DNDR,INPUT,OUTPUT,TAPE6=OUTPUT,TAPE7=DNDR)
```

```
*
****        THIS PROGRAM SIMULATES THE ONGOING INPUT-OUTPUT ACTIVITY OF
*           A SINGLE NEURON WITH A PASSIVE DENDRITIC TREE OF ARBITRARY
*           MORPHOLOGY ACTIVATED BY AN ARBITRARY NUMBER OF SYNAPTIC INPUT
*           SYSTEMS WHICH ARE INDIVIDUALLY ADJUSTABLE. SUBROUTINE TREE
****        COMPUTES THE TOTAL RESISTANCE OF THE DENDRITIC TREE.
*
      INTEGER RG,TPCN,BTCN,DDN,CDN,SNTP,TYPE,SCSTRT,SCSTP
      REAL LBRNCH,LSEG
      PARAMETER NDN=30,NREG=4,NCMPT=10,NSTEPS=500,STEP=1.,
     *          NDSPY=2,SNTP=1,NFPOPS=1,NTGS=1,EK=-10.,FD=1.,FS=1.
      DIMENSION DDN(NDSPY),CDN(NDSPY),GDD(NREG),GRT(NREG),GTRB(NREG),
     *          EQ(SNTP),TG(SNTP),DG(SNTP),
     *          TYPE(NFPOPS),N(NFPOPS),NT(NFPOPS),NBC(NFPOPS),
     *          P(NFPOPS),STR(NFPOPS),LSTRT(NFPOPS),LSTP(NFPOPS),
     *          INSED1(NFPOPS),INSED2(NFPOPS),NTG(NFPOPS),
     *          IBC(NFPOPS,NDN),ITB(NFPOPS,NTGS),U(NDN,NCMPT),
     *          V(NDN,NCMPT),G(SNTP,NDN,NCMPT),GT(NDN,NCMPT),
     *          Z(NDN,NCMPT),GS(SNTP),SYN(NDN,NCMPT),PS(5,1000),
     *          NF(NFPOPS),IDF(100),G0(NDN,NCMPT),PT(NFPOPS,NTGS)
      COMMON/TREE/ RST,LBRNCH,LSEG,CON,RES,RG(NDN),TPCN(2,NDN),
     *          BTCN(2,NDN),RS,SC
*
****        THIS SECTION READS AND WRITES THE INPUT PARAMETERS
*
      READ 5000, IFLG1,IFLG2
      READ 5010, C,TTH,B,TGK,TMEM,TH0
      READ 5020, RS,NST,RST,LBRNCH
      DO 130 I=1,NDN
  130 READ 5000, J,RG(I),TPCN(1,I),TPCN(2,I),BTCN(1,I),BTCN(2,I)
      READ 5000, ((DDN(I),CDN(I)),I=1,NDSPY)
*
      RES=700000. $CON=1./(TMEM*10.**11) $LSEG=LBRNCH/FLOAT(NCMPT)
      AS=4.*3.14159*(RS**2) $R=RST $AM=2.*3.14159*R*LSEG
      RSD=RES*(LSEG/2./(R**2)+RS/((RS/SQRT(2.))**2))/3.14159*FS
      GDS=1./(RSD*CON*AS) $GSD=1./(RSD*CON*AM)
      RBB=RES*(LSEG/(R**2)+RS/((RS/SQRT(2.))**2))/3.14159*FS
      GBB=1./(RBB*CON*AM)
      DO 160 I=1,NREG
      IF(I.GT.1) R=R/(2.**(2./3.))
      AM=2.*3.14159*R*LSEG $RDD=RES*LSEG/3.14159/R**2*FD
      GDD(I)=1./(RDD*CON*AM)
      IF(I.LT.NREG) RTRB=RES*LSEG*(1./R**2+1./((R/(2.**(2./3.)))**2))
     *                                            /2./3.14159*FD
      RTRT=RES*LSEG*(1./R**2+1./((R*(2.**(2./3.)))**2))/2./3.14159*FD
      IF(I.LT.NREG) GTRB(I)=1./(RTRB*CON*AM)
      GTRT(I)=1./(RTRT*CON*AM)
      IF(I.EQ.NREG) GLK=3.14159*(R**2)/AM
  160 CONTINUE
*
      READ 5010, ((EQ(I),TG(I)),I=1,SNTP)
      DO 210 I=1,NFPOPS
      READ 5030, TYPE(I),N(I),NT(I),NBC(I),P(I),STR(I),LSTRT(I),LSTP(I),
     *           INSED1(I),INSED2(I),NTG(I)
      READ 5000, (IBC(I,J),J=1,NBC(I))
      DO 220 J=1,NTG(I)
  220 READ 5240, ITB(I,J),PT(I,J)
  210 CONTINUE
      READ 5250, SC,SCSTRT,SCSTP
      READ 5000, LTSTOP
*
      WRITE(6,6000)
      WRITE(6,6010) NDN,NREG,NCMPT,NSTEPS,STEP,RES,NDSPY,SNTP,
     *              NFPOPS,NTGS,IFLG1,IFLG2,FD,FS
      WRITE(6,6020) C,TTH,B,TGK,TMEM,TH0
      WRITE(6,6030) RS,NST,RST,LBRNCH
      WRITE(6,6040) $DO 310 I=1,NDN
```

```
310 WRITE(6,6050) I,RG(I),TPCN(1,I),TPCN(2,I),BTCN(1,I),BTCN(2,I)
    WRITE(6,6060) ((DDN(I),CDN(I)),I=1,NDSPY)
    WRITE(6,6070)
    WRITE(6,6075) GDS,GSD,GBB,((GTRT(I),GDD(I),GTRB(I)),I=1,NREG),GLK
    WRITE(6,6110) ((EQ(I),TG(I)),I=1,SNTP)
    WRITE(6,6120) $DO 320 I=1,NFPOPS
    WRITE(6,6130) TYPE(I),N(I),NT(I),NBC(I),P(I),STR(I),LSTRT(I),
    *              LSTP(I),INSED1(I),INSED2(I),NTG(I)
    WRITE(6,6075) (IBC(I,J),J=1,NBC(I))
    WRITE(6,6140)
320 WRITE(6,6150) ((ITB(I,J),PT(I,J)),J=1,NTG(I))
    WRITE(6,6160) SC,SCSTRT,SCSTP $WRITE(6,6170) LTSTOP
    IF(IFLG2.GT.0) CALL TREE $WRITE(6,6000)
*
****        THIS SECTION INITIALIZES VARIABLES
*
    G0S=1.+FLOAT(NST)*GDS
    DO 400 I=1,NDN $K=RG(I)
    IF(TPCN(1,I).LE.0) THEN
    G0(I,1)=1.+GSD+FLOAT(NST-1)*GBB+GDD(K) $ELSE
    IF(TPCN(2,I).EQ.0) G0(I,1)=1.+2.*GDD(K)
    IF(TPCN(2,I).GT.0) G0(I,1)=1.+GTRT(K)+2.*GDD(K) $ENDIF
    DO 410 J=2,NCMPT-1
410 G0(I,J)=1.+2.*GDD(K)
    IF(BTCN(1,I).LE.0) THEN $G0(I,NCMPT)=1.+GLK+GDD(K) $ELSE
    IF(BTCN(2,I).EQ.0) G0(I,NCMPT)=1.+2.*GDD(K)
    IF(BTCN(2,I).GT.0) G0(I,NCMPT)=1.+2.*GTRB(K)+GDD(K) $ENDIF
400 CONTINUE
*
    DO 500 I=1,NDN $DO 500 J=1,NCMPT
    U(I,J)=0. $V(I,J)=0. $DO 500 K=1,SNTP $G(K,I,J)=0.
500 CONTINUE $US=0. $VS=0. $GK=0. $TH=TH0
    DGK=EXP(-1./TGK) $DTH=EXP(-1./TTH)
    DO 510 K=1,SNTP $DG(K)=EXP(-STEP/TG(K))
510 GS(K)=0.
*
****        THIS SECTION UPDATES STATE VARIABLES AT EACH TIME STEP
*
    DO 1000 L=1,LTSTOP
    SCN=0. $IF(L.GE.SCSTRT.AND.L.LT.SCSTP) SCN=SC
*
****                UPDATING THE INPUT FIBERS
*
    DO 2000 I=1,NFPOPS
    NF(I)=0 $IF(L.LT.LSTRT(I).OR.L.GE.LSTP(I)) GO TO 2000
    DO 2900 J=1,N(I)
    CALL RANSET(J*L*INSED1(I))
    IF(RANF(0.).GT.P(I)) GO TO 2900
    NF(I)=NF(I)+1 $IDF(NF(I))=J
    DO 2800 K=1,NT(I)
    IDN=IBC(I,INT(RANF(0.)*FLOAT(NBC(I)))+1)
    IF(IDN.GT.0) THEN $ICN=INT(RANF(0.)*FLOAT(NCMPT))+1
    G(TYPE(I),IDN,ICN)=G(TYPE(I),IDN,ICN)+STR(I)
    ELSE $GS(TYPE(I))=GS(TYPE(I))+STR(I) $ENDIF
2800 CONTINUE
2900 CONTINUE
    DO 2600 J=1,NTG(I)
    CALL RANSET(J*L*INSED2(I))
    IF(RANF(0.).GT.PT(I,J)) GO TO 2600
    NF(I)=NF(I)+1 $IDF(NF(I))=J+N(I)
    IF(ITB(I,J).GT.0) THEN $ICN=INT(RANF(0.)*FLOAT(NCMPT))+1
    G(TYPE(I),ITB(I,J),ICN)=G(TYPE(I),ITB(I,J),ICN)+STR(I)
    ELSE $GS(TYPE(I))=GS(TYPE(I))+STR(I) $ENDIF
2600 CONTINUE
2000 CONTINUE
*
*
```

```
****                    UPDATING THE CELLS
*
      GSM=GK $SYNS=GK*EK $DO 3000 K=1,SNTP
      SYNS=SYNS+GS(K)*EQ(K)
 3000 GSM=GSM+GS(K) $GTS=G0S+GSM $ZS=EXP(-GTS*STEP/TMEM/FLOAT(NSTEPS))
      DO 3200 I=1,NDN $DO 3200 J=1,NCMPT $SYNG=0. $SYN(I,J)=0.
      DO 3210 K=1,SNTP
      SYN(I,J)=SYN(I,J)+G(K,I,J)*EQ(K)
 3210 SYNG=SYNG+G(K,I,J) $GT(I,J)=G0+SYNG
 3200 Z(I,J)=EXP(-GT(I,J)*STEP/TMEM/FLOAT(NSTEPS))
*
*
****          THIS SUB-SECTION PERFORMS FINE-GRAINED INTEGRATION
****          OF COMPARTMENT POTENTIALS.
*
      DO 4000 IS=1,NSTEPS
      DSC=0. $DO 4010 I=1,NST
 4010 DSC=DSC+U(I,1)*GDS
      VS=US*ZS+(DSC+SYNS+SCN)*(1.-ZS)/GTS
*
      SDP=0. $DO 4020 J=1,NST
 4020 SDP=SDP+U(J,1)
      DO 4100 I=1,NDN $K=RG(I)
*
      IF(TPCN(1,I).LE.0) THEN $SDT=GBB*(SDP-U(I,1))
      V(I,1)=U(I,1)*Z(I,1)+(GSD*US+SDT+GDD(K)*U(I,2)+SYN(I,1))*
     *                 (1.-Z(I,1))/GTS(I,1) $ELSE
      IF(TPCN(2,I).EQ.0) V(I,1)=U(I,1)*Z(I,1)+(GDD(K)*(U(TPCN(1,I)
     *    ,NCMPT)+U(I,2))+SYN(I,1))*(1.-Z(I,1))/GT(I,1)
      IF(TPCN(2,I).GT.0) V(I,1)=U(I,1)*Z(I,1)+(GTRT(K)*U(TPCN(1,I),
     *    NCMPT)+GDD(K)*(U(TPCN(2,I),1)+U(I,2))+SYN(I,1))*(1.-Z(I,1))/
     *    GT(I,1)
      ENDIF
*
      DO 4120 J=2,NCMPT-1
 4120 V(I,J)=U(I,J)*Z(I,J)+(GDD(K)*(U(I,J-1)+U(I,J+1))+SYN(I,J))*
     *       (1.-Z(I,J))/GT(I,J)
*
      IF(BTCN(1,I).LE.0) THEN
      V(I,NCMPT)=U(I,NCMPT)*Z(I,NCMPT)+(GDD(K)*U(I,NCMPT-1)+SYN(I,
     *    NCMPT))*(1.-Z(I,NCMPT))/GT(I,NCMPT)  $ELSE
      IF(BTCN(2,I).EQ.0) V(I,NCMPT)=U(I,NCMPT)*Z(I,NCMPT)+(GDD(K)*
     *    (U(BTCN(1,I),1)+U(I,NCMPT-1))+SYN(I,NCMPT))*(1.-Z(I,NCMPT))/
     *    GT(I,NCMPT)
      IF(BTCN(2,I).GT.0) V(I,NCMPT)=U(I,NCMPT)*Z(I,NCMPT)+(GTRB(K)*
     *    (U(BTCN(1,I),1)+U(BTCN(2,I),1))+GDD(K)*U(I,NCMPT-1)+SYN(I,
     *    NCMPT))*(1.-Z(I,NCMPT))/GT(I,NCMPT)   $ENDIF
*
 4100 CONTINUE
      US=VS $DO 4200 I=1,NDN $DO 4200 J=1,NCMPT
 4200 U(I,J)=V(I,J)
*
 4000 CONTINUE
*
      TH=TH*DTH+(TH0+C*VS)*(1.-DTH) $GK=GK*DGK
      S=0. $IF(VS.LT.TH) GO TO 4300
      S=1. $GK=GK+B*(1.-DGK)
 4300 CONTINUE $PS(1,L)=VS+S*(50.-VS)
*
      IF(IFLG1.GT.0) THEN $WRITE(6,6280) VS $DO 4400 I=1,NDN
 4400 WRITE(6,6280) (U(I,J),J=1,NCMPT) $ENDIF
      DO 4500 I=1,NDSPY
 4500 PS(I+1,L)=V(DDN(I),CDN(I))
*
      DO 4610 I=1,NDN $DO 4610 J=1,NCMPT $DO 4610 K=1,SNTP
 4610 G(K,I,J)=G(K,I,J)*DG(K)
 1000 CONTINUE
```

```
*
****          THIS SECTION WRITES OUT ACTIVITY VARIABLES
*
      DO 1600 I=1,NDSPY+1 $WRITE(6,6002)
 1600 WRITE(6,6280) (PS(I,L),L=1,LTSTOP)
      DO 1700 I=1,NDSPY+1 $WRITE(7,6002)
 1700 WRITE(7,7280) (PS(I,L),L=1,LTSTOP)
*
      STOP
****          THESE ARE THE FORMATS
*
 5000 FORMAT(12I6)
 5010 FORMAT(12F6.2)
 5020 FORMAT(F6.2,I6,2F6.2)
 5030 FORMAT(4I6,2F6.2,5I6)
 5240 FORMAT(I6,F6.2)
 5250 FORMAT(F6.2,2I6)
 6000 FORMAT(1H1)
 6002 FORMAT(/)
 6010 FORMAT(5X,*NDN,NREG,NCMPT,NSTEPS,STEP,RES,NDSPY,SNTP,NFPOPS,*
     *            ,*NTGS,IFLG1,ILFG2,FD,FS=*,/,10X,4I7,2F10.2,6I7,2F7.2)
 6020 FORMAT(5X,*C,TTH,B,TGK,TMEM,TH0,=*,6F7.2)
 6030 FORMAT(5X,*RS,NST,R,LBRNCH=*,F7.2,I7,2F7.2)
 6040 FORMAT(/,5X,*BR,RG,TPCNS,BTCNS=*)
 6050 FORMAT(5X,6I7)
 6060 FORMAT(5X,*DSP=*,2I7)
 6070 FORMAT(/,5X,*GDS,GSD,GBB,(GTRT-GDD-GTRB),GLK=*)
 6075 FORMAT(3X,16F8.2)
 6110 FORMAT(5X,*EQS,TGS=*,12F7.2)
 6120 FORMAT(5X,*SNT,N,NT,NBC,P,STR,LSTRT,LSTP,INSED1,INSED2,NTG,IBCS=*)
 6130 FORMAT(10X,4I7,2F7.2,6I7)
 6140 FORMAT(5X,*TARGETS:BRNCH,RATE=*)
 6150 FORMAT(5X,I7,F7.2)
 6160 FORMAT(5X,*SC,SCSTRT,SCSTP=*,F7.2,2I7)
 6170 FORMAT(5X,*LTSTOP=*,I7)
 6280 FORMAT(5X,20F6.2)
 7280 FORMAT(13F6.2)
      END
*
      SUBROUTINE TREE
*
****          THIS SUBROUTINE COMPUTES THE TOTAL RESISTANCE AND OTHER
****          QUANTITIES OF INTEREST FOR A BIFURCATING DENDRITIC TREE.
*
      INTEGER RG,TPCN,BTCN,USED
      REAL LBRNCH,LSEG,LC
      PARAMETER NDN=30,NREG=4
      COMMON/TREE/ RST,LBRNCH,LSEG,CON,RES,RG(NDN),TPCN(2,NDN),
     *             BTCN(2,NDN),RS,SC
      DIMENSION LC(NREG),RL(NREG),AD(NREG),USED(NDN),YN(NDN),NM(NDN)
*
      AS=4.*3.14159*RS**2
      WRITE(6,6102) RS,AS $WRITE(7,6102) RS,AS
      WRITE(6,6105) $WRITE(7,6105)
      DO 160 I=1,NREG
      IF(I.EQ.1) R=RST
      IF(I.GT.1) R=R/(2.**(2./3.))
      AM=2.*3.14159*R*LSEG
      LC(I)=SQRT(R/(2.*RES*CON))
      RL(I)=RES/(3.14159*R**2)
      AD(I)=(TANH(LBRNCH/LC(I))+R/2./LC(I))/
     *      ((1.+TANH(LBRNCH/LC(I))*R/2./LC(I))*LC(I)*RL(I))
      DCS=EXP(-LSEG/LC(I)) $DCB=EXP(-LBRNCH/LC(I))
      WRITE(6,6110) I,R,AM,RL(I),AD(I),LC(I),DCS,DCB
      WRITE(7,6110) I,R,AM,RL(I),AD(I),LC(I),DCS,DCB
 160  CONTINUE
*
```

```
       N=0  $DO 10  J=1,NDN
   10 USED(J)=0
       DO 100 I=1,NDN
       IF(BTCN(1,I).GT.0.OR.USED(I).GT.0) GO TO 100
       N=N+1 $NM(N)=I $YN(N)=AD(RG(I)) $USED(I)=1
       IF(TPCN(2,I).EQ.0) GO TO 90
       J=TPCN(2,I) $USED(J)=1 $YN(N)=YN(N)+AD(RG(J))
   90 CONTINUE $IF(TPCN(1,I).EQ.0) GO TO 100
       J=TPCN(1,I) $K=RG(J) $NM(N)=J
       YN(N)=(YN(N)+AD(K))/(1.+YN(N)*AD(K)*(LC(K)*RL(K))**2)
  100 CONTINUE        $NN=N
*
       DO 300 L=1,NREG-2    $NN=0
       DO 200 I=1,N  $IF(USED(NM(I)).GT.0) GO TO 200 $NN=NN+1
       IF(TPCN(2,NM(I)).EQ.0) GO TO 190
       YN(I)=YN(I)+YN(I+1) $USED(TPCN(2,NM(I)))=1
  190 J=TPCN(1,NM(I)) $K=RG(J) $NM(NN)=J
       YN(NN)=(YN(I)+AD(K))/(1.+YN(I)*AD(K)*(LC(K)*RL(K))**2)
  200 CONTINUE $N=NN
  300 CONTINUE
*
       YS=CON*4.*3.14159*RS**2    $YD=0.
       DO 400 I=1,NN
  400 YD=YD+YN(I)
       SIG=YD/YS $RTOT=1./(YS+YD) $RSMA=1./YS $RDN=1./YD $ESS=SC/(1.+SIG)
       WRITE(6,6120) SIG,YS,YD,RTOT,RSMA,RDN,ESS
       WRITE(7,6120) SIG,YS,YD,RTOT,RSMA,RDN,ESS
*
 6102 FORMAT(5X,*RS,AS=*,2F7.2)
 6105 FORMAT(/,5X,*RG,R,AM,RL,AD,LC,DC,DCB=:*)
 6110 FORMAT(5X,I7,2F7.2,2E12.2,F7.1,2F7.4)
 6120 FORMAT(5X,*SIG,YS,YD,RTOT,RSMA,RDN,ESS=*,/,5X,F7.2,5E10.2,F7.2)
*
       RETURN
       END
/EOR
      0       1
     0.    25.    400.    10.    11.    24.
     5.5      3  1.25   158.
      1      1      0      0      4      5
      2      1      0      0      6      7
      3      1      0      0      8      9
      4      2      1      5     10     11
      5      2      1      4     12      0
      6      2      2      7     13     14
      7      2      2      6     15      0
      8      2      3      9     16     17
      9      2      3      8     18      0
     10      3      4     11     19     20
     11      3      4     10     21      0
     12      2      5      0     22      0
     13      3      6     14     23     24
     14      3      6     13     25      0
     15      2      7      0     26      0
     16      3      8     17     27      0
     17      3      8     16     28      0
     18      2      9      0     29     30
     19      4     10     20      0      0
     20      4     10     19      0      0
     21      3     11      0      0      0
     22      2     12      0      0      0
     23      4     13     24      0      0
     24      4     13     23      0      0
     25      3     14      0      0      0
     26      2     15      0      0      0
     27      3     16      0      0      0
     28      3     17      0      0      0
```

```
    29      3     18     30      0      0
    30      3     18     29      0      0
     2      1     25      5
70.  3.
     1      0      0      1     0.     0.          0      0   1227   1936      1
     0
     0   0.
 3500.         1   9999
    15
/EOR
/EOI


/JOB
A224,CM=70000,TL=1200.
FAKE.
PDFV.
MNF(Y).
LGO,DN203.
SAVE,DN203.
EXIT.
SAVE,DN203.
/EOR
      PROGRAM DENDR31 (DNDR,INPUT,OUTPUT,TAPE6=OUTPUT,TAPE7=DNDR)
*
****      THIS PROGRAM SIMULATES THE ONGOING INPUT-OUTPUT ACTIVITY OF
*         A SINGLE NEURON WITH A DENDRITIC TREE OF ARBITRARY
*         MORPHOLOGY ACTIVATED BY AN ARBITRARY NUMBER OF SYNAPTIC INPUT
*         SYSTEMS WHICH ARE INDIVIDUALLY ADJUSTABLE. ACTIVE CALCIUM-
*         RELATED ACTIVE CONDUCTANCES ARE INCLUDED. SUBROUTINE TREE
****      COMPUTES THE TOTAL RESISTANCE OF THE DENDRITIC TREE.
*
      INTEGER RG,TPCN,BTCN,DDN,CDN,SNTP,TYPE,SCSTRT,SCSTP
      REAL LBRNCH,LSEG
      PARAMETER NDN=30,NREG=4,NCMPT=10,NSTEPS=500,STEP=1.,
     *   NDSPY=2,SNTP=1,NFPOPS=1,NTGS=1,EK=-10.,ECA=50.,FD=1.,FS=1.
      DIMENSION DDN(NDSPY),CDN(NDSPY),GDD(NREG),GTRT(NREG),GTRB(NREG),
     *         EQ(SNTP),TG(SNTP),DG(SNTP),
     *         TYPE(NFPOPS),N(NFPOPS),NT(NFPOPS),NBC(NFPOPS),
     *         P(NFPOPS),STR(NFPOPS),LSTRT(NFPOPS),LSTP(NFPOPS),
     *         INSED1(NFPOPS),INSED2(NFPOPS),NTG(NFPOPS),
     *         IBC(NFPOPS,NDN),ITB(NFPOPS,NTGS),U(NDN,NCMPT),
     *         V(NDN,NCMPT),G(SNTP,NDN,NCMPT),GT(NDN,NCMPT),
     *         Z(NDN,NCMPT),GS(SNTP),SYN(NDN,NCMPT),PS(5,1000),
     *         NF(NFPOPS),IDF(100),G0(NDN,NCMPT),PT(NFPOPS,NTGS),
     *         GCA(NDN,NCMPT),CA(NDN,NCMPT),GKD(NDN,NCMPT)
      COMMON/TREE/  RST,LBRNCH,LSEG,CON,RES,RG(NDN),TPCN(2,NDN),
     *              BTCN(2,NDN),RS,SC
*
****          THIS SECTION READS AND WRITES THE INPUT PARAMETERS
*
      READ 5000, IFLG1,IFLG2,IFLG3
      READ 5010, C,TTH,B,TGK,TMEM,TH0
      READ 5020, RS,NST,RST,LBRNCH
      READ 5010, THD,D,TGC,A,TCA,CA0,BD,TGKD
      DO 130 I=1,NDN
  130 READ 5000, J,RG(I),TPCN(1,I),TPCN(2,I),BTCN(1,I),BTCN(2,I)
      READ 5000, ((DDN(I),CDN(I)),I=1,NDSPY)
*
      RES=700000. $CON=1./(TMEM*10.**11) $LSEG=LBRNCH/FLOAT(NCMPT)
      AS=4.*3.14159*(RS**2) $R=RST $AM=2.*3.14159*R*LSEG
      RSD=RES*(LSEG/2./(R**2)+RS/((RS/SQRT(2.))**2))/3.14159*FS
      GDS=1./(RSD*CON*AS) $GSD=1./(RSD*CON*AM)
      RBB=RES*(LSEG/(R**2)+RS/((RS/SQRT(2.))**2))/3.14159*FS
      GBB=1./(RBB*CON*AM)
      DO 160 I=1,NREG
```

```
      IF(I.GT.1) R=R/(2.**(2./3.))
      AM=2.*3.14159*R*LSEG $RDD=RES*LSEG/3.14159/R**2*FD
      GDD(I)=1./(RDD*CON*AM)
      IF(I.LT.NREG) RTRB=RES*LSEG*(1./R**2+1./((R/(2.**(2./3.)))**2))
     *                      /2./3.14159*FD
      RTRT=RES*LSEG*(1./R**2+1./((R*(2.**(2./3.)))**2))/2./3.14159*FD
      IF(I.LT.NREG) GTRB(I)=1./(RTRB*CON*AM)
      GTRT(I)=1./(RTRT*CON*AM)
      IF(I.EQ.NREG) GLK=3.14159*(R**2)/AM
  160 CONTINUE
*
      READ 5010, ((EQ(I),TG(I)),I=1,SNTP)
      DO 210 I=1,NFPOPS
      READ 5030, TYPE(I),N(I),NT(I),NBC(I),P(I),STR(I),LSTRT(I),LSTP(I),
     *           INSED1(I),INSED2(I),NTG(I)
      READ 5000, (IBC(I,J),J=1,NBC(I))
      DO 220 J=1,NTG(I)
  220 READ 5240, ITB(I,J),PT(I,J)
  210 CONTINUE
      READ 5250, SC,SCSTRT,SCSTP
      READ 5000, LTSTOP
*
      WRITE(6,6000)
      WRITE(6,6010) NDN,NREG,NCMPT,NSTEPS,STEP,RES,NDSPY,SNTP,
     *              NFPOPS,NTGS,IFLG1,IFLG2,IFLG3,FD,FS
      WRITE(6,6020) C,TTH,B,TGK,TMEM,THØ
      WRITE(6,6030) RS,NST,RST,LBRNCH
      WRITE(6,6035) THD,D,TGC,A,TCA,CAØ,BD,TGKD
      WRITE(6,6040) $DO 310 I=1,NDN
  310 WRITE(6,6050) I,RG(I),TPCN(1,I),TPCN(2,I),BTCN(1,I),BTCN(2,I)
      WRITE(6,6060) ((DDN(I),CDN(I)),I=1,NDSPY)
      WRITE(6,6070)
      WRITE(6,6075) GDS,GSD,GBB,((GTRT(I),GDD(I),GTRB(I)),I=1,NREG),GLK
      WRITE(6,6110) ((EQ(I),TG(I)),I=1,SNTP)
      WRITE(6,6120) $DO 320 I=1,NFPOPS
      WRITE(6,6130) TYPE(I),N(I),NT(I),NBC(I),P(I),STR(I),LSTRT(I),
     *              LSTP(I),INSED1(I),INSED2(I),NTG(I)
      WRITE(6,6075) (IBC(I,J),J=1,NBC(I))
      WRITE(6,6140)
  320 WRITE(6,6150) ((ITB(I,J),PT(I,J)),J=1,NTG(I))
      WRITE(6,6160) SC,SCSTRT,SCSTP $WRITE(6,6170) LTSTOP
      IF(IFLG2.GT.Ø) CALL TREE $WRITE(6,6000)
*
****       THIS SECTION INITIALIZES VARIABLES
*
      GØS=1.+FLOAT(NST)*GDS
      DO 400 I=1,NDN $K=RG(I)
      IF(TPCN(1,I).LE.Ø) THEN
      GØ(I,1)=1.+GSD+FLOAT(NST-1)*GBB+GDD(K) $ELSE
      IF(TPCN(2,I).EQ.Ø) GØ(I,1)=1.+2.*GDD(K)
      IF(TPCN(2,I).GT.Ø) GØ(I,1)=1.+GTRT(K)+2.*GDD(K) $ENDIF
      DO 410 J=2,NCMPT-1
  410 GØ(I,J)=1.+2.*GDD(K)
      IF(BTCN(1,I).LE.Ø) THEN $GØ(I,NCMPT)=1.+GLK+GDD(K) $ELSE
      IF(BTCN(2,I).EQ.Ø) GØ(I,NCMPT)=1.+2.*GDD(K)
      IF(BTCN(2,I).GT.Ø) GØ(I,NCMPT)=1.+2.*GTRB(K)+GDD(K) $ENDIF
  400 CONTINUE
*
      DO 500 I=1,NDN $DO 500 J=1,NCMPT
      GCA(I,J)=Ø. $CA(I,J)=Ø. $GKD(I,J)=Ø.
      U(I,J)=Ø. $V(I,J)=Ø. $DO 500 K=1,SNTP $G(K,I,J)=Ø.
  500 CONTINUE $US=Ø. $VS=Ø. $GK=Ø. $TH=THØ
      DGK=EXP(-1./TGK) $DTH=EXP(-1./TTH)
      DGC=EXP(-STEP/TGC) $DCA=EXP(-STEP/TCA) $DGKD=EXP(-STEP/TGKD)
      DO 510 K=1,SNTP $DG(K)=EXP(-STEP/TG(K))
  510 GS(K)=Ø.
*
```

```
****        THIS SECTION UPDATES STATE VARIABLES AT EACH TIME STEP
*
      DO 1000 L=1,LTSTOP
      SCN=0. $IF(L.GE.SCSTRT.AND.L.LT.SCSTP) SCN=SC
*
****              UPDATING THE INPUT FIBERS
*
      DO 2000 I=1,NFPOPS
      NF(I)=0 $IF(L.LT.LSTRT(I).OR.L.GE.LSTP(I)) GO TO 2000
      DO 2900 J=1,N(I)
      CALL RANSET(J*L*INSED1(I))
      IF(RANF(0.).GT.P(I)) GO TO 2900
      NF(I)=NF(I)+1 $IDF(NF(I))=J
      DO 2800 K=1,NT(I)
      IDN=IBC(I,INT(RANF(0.)*FLOAT(NBC(I)))+1)
      IF(IDN.GT.0) THEN $ICN=INT(RANF(0.)*FLOAT(NCMPT))+1
      G(TYPE(I),IDN,ICN)=G(TYPE(I),IDN,ICN)+STR(I)
      ELSE $GS(TYPE(I))=GS(TYPE(I))+STR(I) $ENDIF
 2800 CONTINUE
 2900 CONTINUE
      DO 2600 J=1,NTG(I)
      CALL RANSET(J*L*INSED2(I))
      IF(RANF(0.).GT.PT(I,J)) GO TO 2600
      NF(I)=NF(I)+1 $IDF(NF(I))=J+N(I)
      IF(ITB(I,J).GT.0) THEN $ICN=INT(RANF(0.)*FLOAT(NCMPT))+1
      G(TYPE(I),ITB(I,J),ICN)=G(TYPE(I),ITB(I,J),ICN)+STR(I)
      ELSE $GS(TYPE(I))=GS(TYPE(I))+STR(I) $ENDIF
 2600 CONTINUE
 2000 CONTINUE
*
*
****              UPDATING THE CELLS
*
      GSM=GK $SYNS=GK*EK $DO 3000 K=1,SNTP
      SYNS=SYNS+GS(K)*EQ(K)
 3000 GSM=GSM+GS(K) $GTS=G0S+GSM $ZS=EXP(-GTS*STEP/TMEM/FLOAT(NSTEPS))
      DO 3200 I=1,NDN $DO 3200 J=1,NCMPT
      SYNG=GCA(I,J)+GKD(I,J) $SYN(I,J)=GCA(I,J)*ECA+GKD(I,J)*EK
      DO 3210 K=1,SNTP
      SYN(I,J)=SYN(I,J)+G(K,I,J)*EQ(K)
 3210 SYNG=SYNG+G(K,I,J) $GT(I,J)=G0(I,J)+SYNG
 3200 Z(I,J)=EXP(-GT(I,J)*STEP/TMEM/FLOAT(NSTEPS))
*
*
****              THIS SUB-SECTION PERFORMS FINE-GRAINED INTEGRATION
****              OF COMPARTMENT POTENTIALS.
*
      DO 4000 IS=1,NSTEPS
      DSC=0. $DO 4010 I=1,NST
 4010 DSC=DSC+U(I,1)*GDS
      VS=US*ZS+(DSC+SYNS+SCN)*(1.-ZS)/GTS
*
      SDP=0. $DO 4020 J=1,NST
 4020 SDP=SDP+U(J,1)
      DO 4100 I=1,NDN $K=RG(I)
*
      IF(TPCN(1,I).LE.0) THEN $SDT=GBB*(SDP-U(I,1))
      V(I,1)=U(I,1)*Z(I,1)+(GSD*US+SDT+GDD(K)*U(I,2)+SYN(I,1))*
     *                (1.-Z(I,1))/GT(I,1) $ELSE
      IF(TPCN(2,I).EQ.0) V(I,1)=U(I,1)*Z(I,1)+(GDD(K)*(U(TPCN(1,I)
     *      ,NCMPT)+U(I,2))+SYN(I,1))*(1.-Z(I,1))/GT(I,1)
      IF(TPCN(2,I).GT.0) V(I,1)=U(I,1)*Z(I,1)+(GTRT(K)*U(TPCN(1,I),
     *    NCMPT)+GDD(K)*(U(TPCN(2,I),1)+U(I,2))+SYN(I,1))*(1.-Z(I,1))/
     *    GT(I,1)
       ENDIF
*
      DO 4120 J=2,NCMPT-1
```

```
 4120 V(I,J)=U(I,J)*Z(I,J)+(GDD(K)*(U(I,J-1)+U(I,J+1))+SYN(I,J))*
      *          (1.-Z(I,J))/GT(I,J)
*
      IF(BTCN(1,I).LE.0) THEN
      V(I,NCMPT)=U(I,NCMPT)*Z(I,NCMPT)+(GDD(K)*U(I,NCMPT-1)+SYN(I,
      *          NCMPT))*(1.-Z(I,NCMPT))/GT(I,NCMPT)   $ELSE
      IF(BTCN(2,I).EQ.0) V(I,NCMPT)=U(I,NCMPT)*Z(I,NCMPT)+(GDD(K)*
      *          (U(BTCN(1,I),1)+U(I,NCMPT-1))+SYN(I,NCMPT))*(1.-Z(I,NCMPT))/
      *          GT(I,NCMPT)
      IF(BTCN(2,I).GT.0) V(I,NCMPT)=U(I,NCMPT)*Z(I,NCMPT)+(GTRB(K)*
      *          (U(BTCN(1,I),1)+U(BTCN(2,I),1))+GDD(K)*U(I,NCMPT-1)+SYN(I,
      *          NCMPT))*(1.-Z(I,NCMPT))/GT(I,NCMPT)   $ENDIF
*
 4100 CONTINUE
      US=VS $DO 4200 I=1,NDN $DO 4200 J=1,NCMPT
 4200 U(I,J)=V(I,J)
*
 4000 CONTINUE
*
      TH=TH*DTH+(TH0+C*VS)*(1.-DTH) $GK=GK*DGK
      S=0. $IF(VS.LT.TH) GO TO 4300
      S=1. $GK=GK+B*(1.-DGK)
 4300 CONTINUE $PS(1,L)=VS+S*(50.-VS)
*
      DO 4330 I=1,NDN $DO 4330 J=1,NCMPT
      GCA(I,J)=GCA(I,J)*DGC $IF(V(I,J).GE.THD) GCA(I,J)=GCA(I,J)+
      *                     D*(V(I,J)-THD)*(1.-DGC)
      CA(I,J)=CA(I,J)*DCA+A*GCA(I,J)*(1.-DCA)
      GKD(I,J)=GKD(I,J)*DGKD $IF(CA(I,J).GE.CA0) GKD(I,J)=GKD(I,J)+
      *                     BD*(1.-DGKD)
 4330 CONTINUE
*
      IF(IFLG3.GT.0) THEN $WRITE(6,6002) $DO 4410 I=1,NDN
      WRITE(6,6280) (GCA(I,J),J=1,NCMPT)
      WRITE(6,6280) (CA(I,J),J=1,NCMPT)
      WRITE(6,6280) (GKD(I,J),J=1,NCMPT)
 4410 CONTINUE $ENDIF
      IF(IFLG1.GT.0) THEN $WRITE(6,6280) VS $DO 4400 I=1,NDN
 4400 WRITE(6,6280) (U(I,J),J=1,NCMPT) $ENDIF
      DO 4500 I=1,NDSPY
 4500 PS(I+1,L)=V(DDN(I),CDN(I))
*
      DO 4610 I=1,NDN $DO 4610 J=1,NCMPT $DO 4610 K=1,SNTP
 4610 G(K,I,J)=G(K,I,J)*DG(K)
*
 1000 CONTINUE
*
****          THIS SECTION WRITES OUT ACTIVITY VARIABLES
*
      DO 1600 I=1,NDSPY+1 $WRITE(6,6002)
 1600 WRITE(6,6280) (PS(I,L),L=1,LTSTOP)
      DO 1700 I=1,NDSPY+1 $WRITE(7,6002)
 1700 WRITE(7,7280) (PS(I,L),L=1,LTSTOP)
*
      STOP
****          THESE ARE THE FORMATS
*
 5000 FORMAT(12I6)
 5010 FORMAT(12F6.2)
 5020 FORMAT(F6.2,I6,2F6.2)
 5030 FORMAT(4I6,2F6.2,5I6)
 5240 FORMAT(I6,F6.2)
 5250 FORMAT(F6.2,2I6)
 6000 FORMAT(1H1)
 6002 FORMAT(/)
 6010 FORMAT(5X,*NDN,NREG,NCMPT,NSTEPS,STEP,RES,NDSPY,SNTP,NFPOPS,*
      *          ,*NTGS,IFLG1,ILFG2,IFLG3,FD,FS=*,/,10X,4I7,2F10.2,
```

```
    *             7I7,2F7.2)
6020 FORMAT(5X,*C,TTH,B,TGK,TMEM,THØ,=*,6F7.2)
6030 FORMAT(5X,*RS,NST,R,LBRNCH=*,F7.2,I7,2F7.2)
6035 FORMAT(5X,*THD,D,TGC,A,TCA,CAØ,BD,TGKD=*,8F10.2)
6040 FORMAT(/,5X,*BR,RG,TPCNS,BTCNS=*)
6050 FORMAT(5X,6I7)
6060 FORMAT(5X,*DSP=*,2I7)
6070 FORMAT(/,5X,*GDS,GSD,GBB,(GTRT-GDD-GTRB),GLK=*)
6075 FORMAT(3X,16F8.2)
6110 FORMAT(5X,*EQS,TGS=*,12F7.2)
6120 FORMAT(5X,*SNT,N,NT,NBC,P,STR,LSTRT,LSTP,INSED1,INSED2,NTG,IBCS=*)
6130 FORMAT(10X,4I7,2F7.2,6I7)
6140 FORMAT(5X,*TARGETS:BRNCH,RATE=*)
6150 FORMAT(5X,I7,F7.2)
6160 FORMAT(5X,*SC,SCSTRT,SCSTP=*,F7.2,2I7)
6170 FORMAT(5X,*LTSTOP=*,I7)
6280 FORMAT(5X,20F6.2)
7280 FORMAT(13F6.2)
     END
*
     SUBROUTINE TREE
*
****      THIS SUBROUTINE COMPUTES THE TOTAL RESISTANCE AND OTHER
****      QUANTITIES OF INTEREST FOR A BIFURCATING DENDRITIC TREE.
*
     INTEGER RG,TPCN,BTCN,USED
     REAL LBRNCH,LSEG,LC
     PARAMETER NDN=30,NREG=4
     COMMON/TREE/ RST,LBRNCH,LSEG,CON,RES,RG(NDN),TPCN(2,NDN),
    *             BTCN(2,NDN),RS,SC
     DIMENSION LC(NREG),RL(NREG),AD(NREG),USED(NDN),YN(NDN),NM(NDN)
*
     AS=4.*3.14159*RS**2
     WRITE(6,6102) RS,AS $WRITE(7,6102) RS,AS
     WRITE(6,6105) $WRITE(7,6105)
     DO 160 I=1,NREG
     IF(I.EQ.1) R=RST
     IF(I.GT.1) R=R/(2.**(2./3.))
     AM=2.*3.14159*R*LSEG
     LC(I)=SQRT(R/(2.*RES*CON))
     RL(I)=RES/(3.14159*R**2)
     AD(I)=(TANH(LBRNCH/LC(I))+R/2./LC(I))/
    *        ((1.+TANH(LBRNCH/LC(I))*R/2./LC(I))*LC(I)*RL(I))
     DCS=EXP(-LSEG/LC(I)) $DCB=EXP(-LBRNCH/LC(I))
     WRITE(6,6110) I,R,AM,RL(I),AD(I),LC(I),DCS,DCB
     WRITE(7,6110) I,R,AM,RL(I),AD(I),LC(I),DCS,DCB
 160 CONTINUE
*
     N=0 $DO 10 J=1,NDN
  10 USED(J)=0
     DO 100 I=1,NDN
     IF(BTCN(1,I).GT.0.OR.USED(I).GT.0) GO TO 100
     N=N+1 $NM(N)=I $YN(N)=AD(RG(I)) $USED(I)=1
     IF(TPCN(2,I).EQ.0) GO TO 90
     J=TPCN(2,I) $USED(J)=1 $YN(N)=YN(N)+AD(RG(J))
  90 CONTINUE $IF(TPCN(1,I).EQ.0) GO TO 100
     J=TPCN(1,I) $K=RG(J) $NM(N)=J
     YN(N)=(YN(N)+AD(K))/(1.+YN(N)*AD(K)*(LC(K)*RL(K))**2)
 100 CONTINUE        $NN=N
*
     DO 300 L=1,NREG-2    $NN=0
     DO 200 I=1,N    $IF(USED(NM(I)).GT.0) GO TO 200 $NN=NN+1
     IF(TPCN(2,NM(I)).EQ.0) GO TO 190
     YN(I)=YN(I)+YN(I+1) $USED(TPCN(2,NM(I)))=1
 190 J=TPCN(1,NM(I)) $K=RG(J) $NM(NN)=J
     YN(NN)=(YN(I)+AD(K))/(1.+YN(I)*AD(K)*(LC(K)*RL(K))**2)
 200 CONTINUE  $N=NN
```

```
  300 CONTINUE
*
      YS=CON*4.*3.14159*RS**2    $YD=0.
      DO 400 I=1,NN
  400 YD=YD+YN(I)
      SIG=YD/YS $RTOT=1./(YS+YD) $RSMA=1./YS $RDN=1./YD $ESS=SC/(1.+SIG)
      WRITE(6,6120) SIG,YS,YD,RTOT,RSMA,RDN,ESS
      WRITE(7,6120) SIG,YS,YD,RTOT,RSMA,RDN,ESS
*
 6102 FORMAT(5X,*RS,AS=*,2F7.2)
 6105 FORMAT(/,5X,*RG,R,AM,RL,AD,LC,DC,DCB=:*)
 6110 FORMAT(5X,I7,2F7.2,2E12.2,F7.1,2F7.4)
 6120 FORMAT(5X,*SIG,YS,YD,RTOT,RSMA,RDN,ESS=*,/,5X,F7.2,5E10.2,F7.2)
*
      RETURN
      END
/EOR
      0     1     0
   0.   25.  400.   10.   11.   24.
   5.5      3  1.25  158.
  22.    0.     5.    2.   10.   30.   4000. 50.
         1    1     0     0     4     5
         2    1     0     0     6     7
         3    1     0     0     8     9
         4    2     1     5    10    11
         5    2     1     4    12     0
         6    2     2     7    13    14
         7    2     2     6    15     0
         8    2     3     9    16    17
         9    2     3     8    18     0
        10    3     4    11    19    20
        11    3     4    10    21     0
        12    2     5     0    22     0
        13    3     6    14    23    24
        14    3     6    13    25     0
        15    2     7     0    26     0
        16    3     8    17    27     0
        17    3     8    16    28     0
        18    2     9     0    29    30
        19    4    10    20     0     0
        20    4    10    19     0     0
        21    3    11     0     0     0
        22    2    12     0     0     0
        23    4    13    24     0     0
        24    4    13    23     0     0
        25    3    14     0     0     0
        26    2    15     0     0     0
        27    3    16     0     0     0
        28    3    17     0     0     0
        29    3    18    30     0     0
        30    3    18    29     0     0
         2    1    25     5
  70.  3.
         1    0     0     1   0.    0.        0     0  1227  1936     1
         0
         0  0.
 3500.      1  9999
        15
/EOR
/EOI


/JOB
A224,CM=110000,TL=1200.
FAKE.
PDFV.
```

```
MNF(Y).
LGO,DN507A.
SAVE,DN507A.
EXIT.
SAVE,DN507A.
/EOR
      PROGRAM DENDR51 (DNDR,INPUT,OUTPUT,TAPE6=OUTPUT,TAPE7=DNDR)
*
****      THIS PROGRAM SIMULATES THE ONGOING INPUT-OUTPUT ACTIVITY OF
*         A SINGLE NEURON WITH A DENDRITIC TREE OF ARBITRARY
*         MORPHOLOGY ACTIVATED BY AN ARBITRARY NUMBER OF SYNAPTIC INPUT
*         SYSTEMS WHICH ARE INDIVIDUALLY ADJUSTABLE. ACTIVE CALCIUM-
*         RELATED CONDUCTANCES ARE INCLUDED. SYNAPTIC INPUT ON SPINES
*         IS REPRESENTED BY AN EQUIVALENT SYNAPTIC CONDUCTANCE. SUB-
****      ROUTINE TREE COMPUTES THE RESISTANCE OF THE DENDRITIC TREE.
*
      INTEGER RG,TPCN,BTCN,DDN,CDN,SNTP,TYPE,SCSTRT,SCSTP
      REAL LBR,LSEG,LSK
      PARAMETER NDN=30,NST=3,NREG=4,NCMPT=10,NSTEPS=500,STEP=1.,
     *      NDSPY=2,SNTP=1,NFPOPS=1,NTGS=1,EK=-10.,ECA=50.,FD=1.,FS=1.
      DIMENSION GBB(NST,NST),GDS(NST),GTP(2,NDN),GDD(NDN),GBT(2,NDN),
     *      EQ(SNTP),TG(SNTP),DG(SNTP),
     *      DDN(NDSPY),CDN(NDSPY),G1(NDN),G2(NDN),
     *      TYPE(NFPOPS),N(NFPOPS),NT(NFPOPS),NBC(NFPOPS),
     *      P(NFPOPS),STR(NFPOPS),LSTRT(NFPOPS),LSTP(NFPOPS),
     *      INSED1(NFPOPS),INSED2(NFPOPS),NTG(NFPOPS),
     *      IBC(NFPOPS,NDN),ITB(NFPOPS,NTGS),U(NDN,NCMPT),
     *      V(NDN,NCMPT),G(SNTP,NDN,NCMPT),GT(NDN,NCMPT),
     *      Z(NDN,NCMPT),GS(SNTP),SYN(NDN,NCMPT),PS(5,1000),
     *      NF(NFPOPS),IDF(100),G0(NDN,NCMPT),PT(NFPOPS,NTGS),
     *      GCA(NDN,NCMPT),CA(NDN,NCMPT),GKD(NDN,NCMPT)
      COMMON/TREE/ RST,LSEG(NDN),AM(NDN),CON,RES,RG(NDN),TPCN(2,NDN),
     *      BTCN(2,NDN),RS,SC
*
****          THIS SECTION READS AND WRITES THE INPUT PARAMETERS
*
      READ 5000, IFLG1,IFLG2,IFLG3
      READ 5010, C,TTH,B,TGK,TMEM,THO
      READ 5010, RS,RST
      READ 5010,THD,D,TGC,A,TCA,CA0,BD,TGKD
      WRITE(6,6000)
      WRITE(6,6010) NDN,NST,NREG,NCMPT,NSTEPS,STEP,RES,NDSPY,SNTP,
     *      NFPOPS,NTGS,IFLG1,IFLG2,IFLG3,FD,FS
      WRITE(6,6020) C,TTH,B,TGK,TMEM,THO
      WRITE(6,6030) RS,RST
      WRITE(6,6035) THD,D,TGC,A,TCA,CA0,BD,TGKD
*
      RES=700000. $CON=1./(TMEM*10.**11) $AS=4.*3.14159*RS**2
      DO 150 I=1,NDN $DO 150 J=1,2 $GBT(J,I)=0.
  150 GTP(J,I)=0. $DO 152 I=1,NST $DO 152 J=1,NST
  152 GBB(I,J)=0.
      WRITE(6,6040)
      DO 160 I=1,NDN
      READ 5020, J,RG(I),LBR,TPCN(1,I),TPCN(2,I),BTCN(1,I),BTCN(2,I),
     *      NS,RH,RSK,LSK
      LSEG(I)=LBR/FLOAT(NCMPT) $R=RST/((2.**(2./3.))**(RG(I)-1))
      AH=3.14159*(4.*RH**2-RSK**2) $ASK=2.*3.14159*RSK*LSK
      AM(I)=2.*3.14159*R*LSEG(I)+FLOAT(NS)*(AH+ASK-3.14159*RSK**2)/
     *      FLOAT(NCMPT)
*
      IF(I.LE.NST) THEN $X=SQRT(RS**2-R**2)
     *RSD=RES*(LSEG(I)/R**2+ALOG((RS+X)/(RS-X))/RS)/2./3.14159*FS
      RSD=RES*(LSEG(I)/2./R**2+2./RS)/3.14159*FS
      GTP(1,I)=1./(RSD*CON*AM(I)) $GDS(I)=1./(RSD*CON*AS)
      IF(NST.GT.1.AND.I.EQ.NST) THEN $DO 158 J1=1,NST $DO 158 J2=1,NST
      RBB=RES*((LSEG(J1)+LSEG(J2))/2./R**2+2./RS)/3.14159*FS
  158 GBB(J1,J2)=1./(RBB*CON*AM(J2)) $ENDIF $ELSE
```

```
*
      IF(TPCN(2,I).EQ.0) THEN
      RTP1=RES*(LSEG(TPCN(1,I))+LSEG(I))/2./3.14159/R**2*FD
      GTP(1,I)=1./(RTP1*CON*AM(I))
      GBT(1,TPCN(1,I))=1./(RTP1*CON*AM(TPCN(1,I))) $ENDIF
      IF(TPCN(2,I).GT.0) THEN
      RTP1=RES*(LSEG(TPCN(1,I))/(R*2.**(2./3.))**2+LSEG(I)/R**2)/2./
     *          3.14159*FD
      GTP(1,I)=1./(RTP1*CON*AM(I)) $IT=TPCN(1,I)
      IF(GBT(1,IT).EQ.0) THEN $ID=1 $ELSE $ID=2 $ENDIF
      GBT(ID,IT)=1./(RTP1*CON*AM(IT))
      IT=TPCN(2,I) $IF(I.GT.IT) THEN
      RTP2=RES*(LSEG(IT)+LSEG(I))/2./3.14159/R**2*FD
      GTP(2,I)=1./(RTP2*CON*AM(I)) $GTP(2,IT)=1./(RTP2*CON*AM(IT))
      ENDIF $ENDIF $ENDIF
*
      RDD=RES*LSEG(I)/3.14159/R**2*FD $GDD(I)=1./(RDD*CON*AM(I))
*
      IF(BTCN(1,I).EQ.0) GBT(1,I)=3.14159*R**2/AM(I)
*
      IF(NS.GT.0) THEN
      X=SQRT(RH**2-RSK**2)
      RHS=RES*(LSK/RSK**2+ALOG((RH+X)/(RH-X))/RH)/2./3.14159
      RSKD=RES*((LSK/RSK**2+LSEG(I)/2./R**2)/3.14159+
     *          ARSIN(SQRT(R**2-RSK**2)/R)/LSEG(I))/2.
      AE=1./(RHS*CON*AH) $BE=1./(RHS*CON*ASK)
      CE=1./(RSKD*CON*ASK) $DE=1./(RSKD*CON*AM(I))
      G1(I)=DE*BE/(BE+CE) $G2(I)=AE*CE/(BE+CE)
      ELSE $G1(I)=10**10 $G2(I)=10**10 $ENDIF
      WRITE(6,6050) I,RG(I),LBR,TPCN(1,I),TPCN(2,I),BTCN(1,I),BTCN(2,I),
     *          NS,RH,RSK,LSK,
     *          GTP(1,I),GTP(2,I),GDD(I),GBT(1,I),GBT(2,I)
      IF(I.LE.NST) WRITE(6,6052) GDS(I)
      IF(I.EQ.NST) WRITE(6,6055) (((GBB(J1,J2)),J1=1,NST),J2=1,NST)
  160 CONTINUE
      READ 5000, ((DDN(I),CDN(I)),I=1,NDSPY)
      READ 5010, ((EQ(I),TG(I)),I=1,SNTP)
      DO 210 I=1,NFPOPS
      READ 5030, TYPE(I),N(I),NT(I),NBC(I),P(I),STR(I),LSTRT(I),LSTP(I),
     *          INSED1(I),INSED2(I),NTG(I)
      READ 5000, (IBC(I,J),J=1,NBC(I))
      DO 220 J=1,NTG(I)
  220 READ 5240, ITB(I,J),PT(I,J)
  210 CONTINUE
      READ 5250, SC,SCSTRT,SCSTP
      READ 5000, LTSTOP
*
      WRITE(6,6060) ((DDN(I),CDN(I)),I=1,NDSPY)
      WRITE(6,6110) ((EQ(I),TG(I)),I=1,SNTP)
      WRITE(6,6120) $DO 320 I=1,NFPOPS
      WRITE(6,6130) TYPE(I),N(I),NT(I),NBC(I),P(I),STR(I),LSTRT(I),
     *          LSTP(I),INSED1(I),INSED2(I),NTG(I)
      WRITE(6,6075) (IBC(I,J),J=1,NBC(I))
      WRITE(6,6140)
  320 WRITE(6,6150) ((ITB(I,J),PT(I,J)),J=1,NTG(I))
      WRITE(6,6160) SC,SCSTRT,SCSTP $WRITE(6,6170) LTSTOP
      IF(IFLG2.GT.0) CALL TREE $WRITE(6,6000)
*
****       THIS SECTION INITIALIZES VARIABLES
*
      G0S=1. $DO 402 J=1,NST
  402 G0S=G0S+GDS(J)
      DO 400 I=1,NDN
      IF(I.LE.NST) THEN $G0(I,1)=1.-GBB(I,I)+GTP(1,I)+GDD(I)
      DO 408 J1=1,NST
  408 G0(I,1)=G0(I,1)+GBB(J1,I) $ELSE
      G0(I,1)=1.+GTP(1,I)+GTP(2,I)+GDD(I) $ENDIF
```

```
      DO 410 J=2,NCMPT-1
  410 G0(I,J)=1.+2.*GDD(I)
      G0(I,NCMPT)=1.+GBT(1,I)+GBT(2,I)+GDD(I)
  400 CONTINUE
*
      DO 500 I=1,NDN $DO 500 J=1,NCMPT
      GCA(I,J)=0. $CA(I,J)=0. $GKD(I,J)=0.
      U(I,J)=0. $V(I,J)=0. $DO 500 K=1,SNTP $G(K,I,J)=0.
  500 CONTINUE $US=0. $VS=0. $GK=0. $TH=TH0
      DGK=EXP(-STEP/TGK) $DTH=EXP(-STEP/TTH)
      DGC=EXP(-STEP/TGC) $DCA=EXP(-STEP/TCA) $DGKD=EXP(-STEP/TGKD)
      DO 510 K=1,SNTP $DG(K)=EXP(-STEP/TG(K))
  510 GS(K)=0.
*
****          THIS SECTION UPDATES STATE VARIABLES AT EACH TIME STEP
*
      DO 1000 L=1,LTSTOP
      SCN=0. $IF(L.GE.SCSTRT.AND.L.LT.SCSTP) SCN=SC
*
****                 UPDATING THE INPUT FIBERS
*
      DO 2000 I=1,NFPOPS
      NF(I)=0 $IF(L.LT.LSTRT(I).OR.L.GE.LSTP(I)) GO TO 2000
      DO 2900 J=1,N(I)
      CALL RANSET(J*L*INSED1(I))
      IF(RANF(0.).GT.P(I)) GO TO 2900
      NF(I)=NF(I)+1 $IDF(NF(I))=J
      DO 2800 K=1,NT(I)
      IDN=IBC(I,INT(RANF(0.)*FLOAT(NBC(I)))+1)
      IF(IDN.GT.0) THEN $ICN=INT(RANF(0.)*FLOAT(NCMPT))+1
      G(TYPE(I),IDN,ICN)=G(TYPE(I),IDN,ICN)+STR(I)
      ELSE $GS(TYPE(I))=GS(TYPE(I))+STR(I) $ENDIF
 2800 CONTINUE
 2900 CONTINUE
      DO 2600 J=1,NTG(I)
      CALL RANSET(J*L*INSED2(I))
      IF(RANF(0.).GT.PT(I,J)) GO TO 2600
      NF(I)=NF(I)+1 $IDF(NF(I))=J+N(I)
      IF(ITB(I,J).GT.0) THEN $ICN=INT(RANF(0.)*FLOAT(NCMPT))+1
      G(TYPE(I),ITB(I,J),ICN)=G(TYPE(I),ITB(I,J),ICN)+STR(I)
      ELSE $GS(TYPE(I))=GS(TYPE(I))+STR(I) $ENDIF
 2600 CONTINUE
 2000 CONTINUE
*
*
****                 UPDATING THE CELLS
*
      GSM=GK $SYNS=GK*EK $DO 3000 K=1,SNTP
      SYNS=SYNS+GS(K)*EQ(K)
 3000 GSM=GSM+GS(K) $GTS=G0S+GSM $ZS=EXP(-GTS*STEP/TMEM/FLOAT(NSTEPS))
      DO 3200 I=1,NDN $DO 3200 J=1,NCMPT
      SYNG=GCA(I,J)+GKD(I,J) $SYN(I,J)=GCA(I,J)*ECA+GKD(I,J)*EK
      DO 3210 K=1,SNTP $GEQ=G1(I)*G(K,I,J)/(G2(I)+G(K,I,J))
      SYN(I,J)=SYN(I,J)+GEQ*EQ(K)
 3210 SYNG=SYNG+GEQ $GT(I,J)=G0(I,J)+SYNG
 3200 Z(I,J)=EXP(-GT(I,J)*STEP/TMEM/FLOAT(NSTEPS))
*
*
****             THIS SUB-SECTION PERFORMS FINE-GRAINED INTEGRATION
****             OF COMPARTMENT POTENTIALS.
*
      DO 4000 IS=1,NSTEPS
      DSC=0. $DO 4010 I=1,NST
 4010 DSC=DSC+U(I,1)*GDS(I)
      VS=US*ZS+(DSC+SYNS+SCN)*(1.-ZS)/GTS
*
      SDP=0. $DO 4020 J=1,NST
```

```
4020 SDP=SDP+U(J,1)
     DO 4100 I=1,NDN $K=RG(I)
*
     IF(I.LE.NST) THEN $SDT=-GBB(I,I)*U(I,1) $DO 4110 J1=1,NST
4110 SDT=SDT+GBB(J1,I)*U(J1,1)
     V(I,1)=U(I,1)*Z(I,1)+(GTP(1,I)*US+SDT+GDD(I)*U(I,2)+SYN(I,1))*
     *                      (1.-Z(I,1))/GT(I,1) $ELSE
     IF(TPCN(2,I).EQ.0) V(I,1)=U(I,1)*Z(I,1)+(GTP(1,I)*U(TPCN(1,I)
     *      ,NCMPT)+GDD(I)*U(I,2)+SYN(I,1))*(1.-Z(I,1))/GT(I,1)
     IF(TPCN(2,I).GT.0) V(I,1)=U(I,1)*Z(I,1)+(GTP(1,I)*U(TPCN(1,I),
     *    NCMPT)+GTP(2,I)*U(TPCN(2,I),1)+GDD(I)*U(I,2)+SYN(I,1))*
     *    (1.-Z(I,1))/GT(I,1)
     ENDIF
*
     DO 4120 J=2,NCMPT-1
4120 V(I,J)=U(I,J)*Z(I,J)+(GDD(I)*(U(I,J-1)+U(I,J+1))+SYN(I,J))*
     *      (1.-Z(I,J))/GT(I,J)
*
     IF(BTCN(1,I).LE.0) THEN
     V(I,NCMPT)=U(I,NCMPT)*Z(I,NCMPT)+(GDD(I)*U(I,NCMPT-1)+SYN(I,
     *    NCMPT))*(1.-Z(I,NCMPT))/GT(I,NCMPT)  $ELSE
     IF(BTCN(2,I).EQ.0) V(I,NCMPT)=U(I,NCMPT)*Z(I,NCMPT)+(GBT(1,I)*
     *    U(BTCN(1,I),1)+GDD(I)*U(I,NCMPT-1)+SYN(I,NCMPT))*
     *    (1.-Z(I,NCMPT))/GT(I,NCMPT)
     IF(BTCN(2,I).GT.0) V(I,NCMPT)=U(I,NCMPT)*Z(I,NCMPT)+(GBT(1,I)*
     *    U(BTCN(1,I),1)+GBT(2,I)*U(BTCN(2,I),1)+GDD(I)*U(I,NCMPT-1)+
     *    SYN(I,NCMPT))*(1.-Z(I,NCMPT))/GT(I,NCMPT)  $ENDIF
*
4100 CONTINUE
     US=VS $DO 4200 I=1,NDN $DO 4200 J=1,NCMPT
4200 U(I,J)=V(I,J)
*
4000 CONTINUE
*
     TH=TH*DTH+(TH0+C*VS)*(1.-DTH) $GK=GK*DGK
     S=0. $IF(VS.LT.TH) GO TO 4300
     S=1. $GK=GK+B*(1.-DGK)
4300 CONTINUE $PS(1,L)=VS+S*(50.-VS)
*
     DO 4330 I=1,NDN $DO 4330 J=1,NCMPT
     GCA(I,J)=GCA(I,J)*DGC $IF(V(I,J).GE.THD) GCA(I,J)=GCA(I,J)+
     *                      D*(V(I,J)-THD)*(1.-DGC)
     CA(I,J)=CA(I,J)*DCA+A*GCA(I,J)*(1.-DCA)
     GKD(I,J)=GKD(I,J)*DGKD $IF(CA(I,J).GE.CA0) GKD(I,J)=GKD(I,J)+
     *                      BD*(1.-DGKD)
4330 CONTINUE
*
     IF(IFLG1.GT.0) THEN $WRITE(6,6280) VS $DO 4400 I=1,NDN
4400 WRITE(6,6280) (U(I,J),J=1,NCMPT) $ENDIF
*
     IF(IFLG3.GT.0) THEN $WRITE(6,6002) $DO 4410 I=1,NDN
     WRITE(6,6280) (GCA(I,J),J=1,NCMPT)
     WRITE(6,6280) (CA(I,J),J=1,NCMPT)
     WRITE(6,6280) (GKD(I,J),J=1,NCMPT)
4410 CONTINUE $ENDIF
     DO 4500 I=1,NDSPY
4500 PS(I+1,L)=V(DDN(I),CDN(I))
*
     DO 4610 I=1,NDN $DO 4610 J=1,NCMPT $DO 4610 K=1,SNTP
4610 G(K,I,J)=G(K,I,J)*DG(K)
*
1000 CONTINUE
*
****          THIS SECTION WRITES OUT ACTIVITY VARIABLES
*
     DO 1600 I=1,NDSPY+1 $WRITE(6,6002)
1600 WRITE(6,6280) (PS(I,L),L=1,LTSTOP)
```

```
      DO 1700 I=1,NDSPY+1 $WRITE(7,6002)
 1700 WRITE(7,7280) (PS(I,L),L=1,LTSTOP)
*
      STOP
****          THESE ARE THE FORMATS
*
 5000 FORMAT(12I6)
 5010 FORMAT(12F6.2)
 5012 FORMAT(6F6.2,F6.0,F6.2)
 5020 FORMAT(2I6,F6.0,5I6,3F6.2)
 5030 FORMAT(4I6,2F6.2,5I6)
 5240 FORMAT(I6,F6.2)
 5250 FORMAT(F6.2,2I6)
 6000 FORMAT(1H1)
 6002 FORMAT(/)
 6010 FORMAT(5X,*NDN,NST,NREG,NCMPT,NSTEPS,STEP,RES,NDSPY,SNTP,NFPOPS,*
     *            ,*NTGS,IFLG1,ILFG2,IFLG3,FD,FS=*,/,10X,5I7,2F10.2,
     *          7I7,2F7.2)
 6020 FORMAT(5X,*C,TTH,B,TGK,TMEM,TH0,=*,6F7.2)
 6030 FORMAT(5X,*RS,RST=*,2F7.2)
 6035 FORMAT(5X,*THD,D,TGC,A,TCA,CA0,BD,TGKD=*,8F10.2)
 6040 FORMAT(/,5X,*BR,RG,L,TPCNS,BTCNS,NS,RH,RSK,LSK,GTPS,GDD,GBTS=:*)
 6050 FORMAT(5X,2I7,F7.0,5I7,8F8.2)
 6052 FORMAT(60X,*GDS=*,9F9.2)
 6055 FORMAT(35X,*GBBS=:*,9F9.2)
 6060 FORMAT(5X,*DSP=*,2I7)
 6075 FORMAT(3X,16F8.2)
 6110 FORMAT(5X,*EQS,TGS=*,12F7.2)
 6120 FORMAT(5X,*SNT,N,NT,NBC,P,STR,LSTRT,LSTP,INSED1,INSED2,NTG,IBCS=*)
 6130 FORMAT(10X,4I7,2F7.2,6I7)
 6140 FORMAT(5X,*TARGETS:BRNCH,RATE=*)
 6150 FORMAT(5X,I7,F7.2)
 6160 FORMAT(5X,*SC,SCSTRT,SCSTP=*,F7.2,2I7)
 6170 FORMAT(5X,*LTSTOP=*,I7)
 6280 FORMAT(5X,20F6.2)
 7280 FORMAT(13F6.2)
      END
*
      SUBROUTINE TREE
*
****          THIS SUBROUTINE COMPUTES THE TOTAL RESISTANCE AND OTHER
****          QUANTITIES OF INTEREST FOR A BIFURCATING DENDRITIC TREE.
*
      INTEGER RG,TPCN,BTCN,USED
      REAL LBR,LSEG,LC
      PARAMETER NDN=30,NREG=4,NCMPT=10
      COMMON/TREE/ RST,LSEG(NDN),AM(NDN),CON,RES,RG(NDN),TPCN(2,NDN),
     *             BTCN(2,NDN),RS,SC
      DIMENSION LC(NDN),RL(NDN),AD(NDN),USED(NDN),YN(NDN),NM(NDN)
*
      AS=4.*3.14159*RS**2
      WRITE(6,6102) RS,AS $WRITE(7,6102) RS,AS
      WRITE(6,6105) $WRITE(7,6105)
      DO 160 I=1,NDN
      R=RST/((2.**(2./3.))**(RG(I)-1)) $LBR=FLOAT(NCMPT)*LSEG(I)
      REQ=AM(I)/(2.*3.14159*LSEG(I))
      LC(I)=R/SQRT(2.*REQ*CON*RES)
      RL(I)=RES/(3.14159*R**2)
      AD(I)=(TANH(LBR/LC(I))+REQ/2./LC(I))/
     *        ((1.+TANH(LBR/LC(I))*REQ/2./LC(I))*LC(I)*RL(I))
      DCS=EXP(-LSEG(I)/LC(I)) $DCB=EXP(-LBR/LC(I))
      WRITE(6,6110) I,R,AM(I),RL(I),AD(I),LC(I),DCS,DCB
      WRITE(7,6110) I,R,AM(I),RL(I),AD(I),LC(I),DCS,DCB
  160 CONTINUE
*
      N=0 $DO 10 J=1,NDN
   10 USED(J)=0
```

```
        DO 100 I=1,NDN
        IF(BTCN(1,I).GT.0.OR.USED(I).GT.0) GO TO 100
        N=N+1  $NM(N)=I  $YN(N)=AD(I)  $USED(I)=1
        IF(TPCN(2,I).EQ.0) GO TO 90
        J=TPCN(2,I)  $USED(J)=1  $YN(N)=YN(N)+AD(J)
  90    CONTINUE  $IF(TPCN(1,I).EQ.0) GO TO 100
        J=TPCN(1,I)  $NM(N)=J
        YN(N)=(YN(N)+AD(J))/(1.+YN(N)*AD(J)*(LC(J)*RL(J))**2)
 100    CONTINUE       $NN=N
*
        DO 300 L=1,NREG-2    $NN=0
        DO 200 I=1,N  $IF(USED(NM(I)).GT.0) GO TO 200  $NN=NN+1
        IF(TPCN(2,NM(I)).EQ.0) GO TO 190
        YN(I)=YN(I)+YN(I+1)  $USED(TPCN(2,NM(I)))=1
 190    J=TPCN(1,NM(I))  $NM(NN)=J
        YN(NN)=(YN(I)+AD(J))/(1.+YN(I)*AD(J)*(LC(J)*RL(J))**2)
 200    CONTINUE  $N=NN
 300    CONTINUE
*
        YS=CON*4.*3.14159*RS**2   $YD=0.
        DO 400 I=1,NN
 400    YD=YD+YN(I)
        SIG=YD/YS  $RTOT=1./(YS+YD)  $RSMA=1./YS  $RDN=1./YD  $ESS=SC/(1.+SIG)
        WRITE(6,6120) SIG,YS,YD,RTOT,RSMA,RDN,ESS
        WRITE(7,6120) SIG,YS,YD,RTOT,RSMA,RDN,ESS
*
 6102   FORMAT(5X,*RS,AS=*,2F7.2)
 6105   FORMAT(/,5X,*RG,R,AM,RL,AD,LC,DCS,DCB=:*)
 6110   FORMAT(5X,I7,2F7.2,2E12.2,F7.1,2F7.4)
 6120   FORMAT(5X,*SIG,YS,YD,RTOT,RSMA,RDN,ESS=*,/,5X,F7.2,5E10.2,F7.2)
*
        RETURN
        END
/EOR
```

```
        0     1     0
   0.   25.  400.  10.   11.   24.
   5.5   1.25
  32.    0.    5.    2.    5.   20.  4000. 50.
      1    1  158.    0    0    4    5    0 0.   0.    0.
      2    1  158.    0    0    6    7    0 0.   0.    0.
      3    1  158.    0    0    8    9    0 0.   0.    0.
      4    2  158.    1    5   10   11  370  .145  .053  .375
      5    2  158.    1    4   12    0  370  .145  .053  .375
      6    2  158.    2    7   13   14  370  .145  .053  .375
      7    2  158.    2    6   15    0  370  .145  .053  .375
      8    2  158.    3    9   16   17  370  .145  .053  .375
      9    2  158.    3    8   18    0  370  .145  .053  .375
     10    3  158.    4   11   19   20  370  .166  .065  .375
     11    3  158.    4   10   21    0  370  .166  .065  .375
     12    2  158.    5    0   22    0  370  .145  .053  .375
     13    3  158.    6   14   23   24  370  .166  .065  .375
     14    3  158.    6   13   25    0  370  .166  .065  .375
     15    2  158.    7    0   26    0  370  .145  .053  .375
     16    3  158.    8   17   27    0  370  .166  .065  .375
     17    3  158.    8   16   28    0  370  .166  .065  .375
     18    2  158.    9    0   29   30  370  .145  .053  .375
     19    4  158.   10   20    0    0  370  .178  .067  .375
     20    4  158.   10   19    0    0  370  .178  .067  .375
     21    3  158.   11    0    0    0  370  .166  .065  .375
     22    2  158.   12    0    0    0  370  .145  .053  .375
     23    4  158.   13   24    0    0  370  .178  .067  .375
     24    4  158.   13   23    0    0  370  .178  .067  .375
     25    3  158.   14    0    0    0  370  .166  .065  .375
     26    2  158.   15    0    0    0  370  .145  .053  .375
     27    3  158.   16    0    0    0  370  .166  .065  .375
     28    3  158.   17    0    0    0  370  .166  .065  .375
     29    3  158.   18   30    0    0  370  .166  .065  .375
```

```
    30      3   158.    18    29    Ø     Ø    370   .166   .065   .375
     6      5    25     5
70.      3.
     1      Ø    Ø      1   Ø.    Ø.          Ø     Ø   1227   1936      1
     Ø
     Ø  Ø.
2500.      1  9999
    30
/EOR
/EOI


/JOB
A224,TL=6ØØ.
FAKE.
PDFV.
MNF(Y).
LGO,AXØØ3.
SAVE,AXØØ3.
EXIT.
SAVE,AXØØ3.
/EOR
      PROGRAM AXONØ1(AXOUT,INPUT,OUTPUT,TAPE6=OUTPUT,TAPE7=AXOUT)
*
****      THIS PROGRAM SIMULATES THE PROPAGATION OF ACTION POTENTIALS
*         ALONG A SINGLE NON-BIFURCATING AXON ACCORDING TO THE HODGKIN-
****      HUXLEY MODEL.
*
      INTEGER SCSTRT,SCSTP
      REAL LBR,LSEG,NØ,MØ,HØ
      PARAMETER NCMPT=10, NSTEP=1000, STEP=1.,NDSPY=2,GNØ=120.,GKØ=36.,
     *      GL=.3,NØ=.31767691,MØ=.05293249,HØ=.59612075
      DIMENSION GT(NCMPT),Z(NCMPT),V(NCMPT),U(NCMPT),IDC(10),
     *          PS(10,500),FN(NCMPT),FM(NCMPT),FH(NCMPT),ACT(NCMPT)
      COMMON GØ(NCMPT),GØS,DL
*
****      THIS SECTION READS AND WRITES THE INPUT PARAMETERS
*
      READ 5000, IFLG1
      READ 5010, RS,R,LBR
      READ 5020, SC,SCSTRT,SCSTP
      READ 5000, (IDC(J),J=1,NDSPY)
      READ 5000, LTSTOP
      RES=700000. $CON=GNØ*HØ*MØ**3+GKØ*NØ**4+GL
*
      AS=4.*3.14159*RS**2 $LSEG=LBR/FLOAT(NCMPT)
      AM=2.*3.14159*R*LSEG $X=SQRT(RS**2-R**2)
      RSD=RES*(LSEG/R**2+ALOG((RS+X)/(RS-X))/RS)/2./3.14159
      RDD=RES*LSEG/3.14159/R**2
      GDS=10.**11/(RSD*AS) $GSD=10.**11/(RSD*AM)
      GDD=10.**11/(RDD*AM) $GLK=CON*3.14159*R**2/AM
*
      WRITE(6,6000) $WRITE(6,6005) NCMPT,NSTEP,STEP,NDSPY,RES,CON
      WRITE(6,6020) RS,R,LBR,LSEG,AS,AM
      WRITE(6,6025) RSD,RDD,GDS,GSD,GDD,GLK
      WRITE(6,6030) SC,SCSTRT,SCSTP,LTSTOP
*
****      THIS SECTION INITIALIZES VARIABLES
*
      US=Ø. $VS=Ø. $SN=NØ $SM=MØ $SH=HØ
      DO 110 I=1,NCMPT $U(I)=Ø. $V(I)=Ø.
      FN(I)=NØ $FM(I)=MØ $FH(I)=HØ
  110 CONTINUE $DL=STEP/FLOAT(NSTEP)
*
      GØS=CON+GDS
      GØ(1)=CON+GSD+GDD
```

```
      DO 210 I=2,NCMPT-1
  210 GØ(I)=CON+2.*GDD
      GØ(NCMPT)=CON+GLK+GDD
*
****         THIS SECTION UPDATES STATE VARIABLES AT EACH TIME STEP
*
      DO 1000 L=1,LTSTOP
      SCN=Ø. $IF(L.GE.SCSTRT.AND.L.LT.SCSTP) SCN=SC
*
      DO 4000 IS=1,NSTEP
      CALL HHEQN(Ø,VS,SN,SM,SH,ACTS,GTS,ZS)
      DO 4005 J=1,NCMPT
 4005 CALL HHEQN(J,V(J),FN(J),FM(J),FH(J),ACT(J),GT(J),Z(J))
      VS=US*ZS+(ACTS+GDS*U(1)-SCN)*(1.-ZS)/GTS
      V(1)=U(1)*Z(1)+(ACT(1)+GSD*US+GDD*U(2))*(1.-Z(1))/GT(1)
      DO 4010 J=2,NCMPT-1
 4010 V(J)=U(J)*Z(J)+(ACT(J)+GDD*(U(J-1)+U(J+1)))*(1.-Z(J))/GT(J)
      V(NCMPT)=U(NCMPT)*Z(NCMPT)+(ACT(NCMPT)+GDD*U(NCMPT-1))*
     *           (1.-Z(NCMPT))/GT(NCMPT)
      DO 4020 J=1,NCMPT
 4020 U(J)=V(J) $US=VS
 4000 CONTINUE
*
      IF(IFLG1.GT.Ø) THEN
      WRITE(6,6280) VS,(V(J),J=1,NCMPT) $ENDIF
      PS(1,L)=VS $DO 4500 I=1,NDSPY
 4500 PS(I+1,L)=V(IDC(I))
*
 1000 CONTINUE
*
****         THIS SECTION WRITES OUT ACTIVITY VARIABLES
*
      DO 1600 I=1,NDSPY+1 $WRITE(6,6002) $WRITE(7,6002)
      WRITE(6,6280) (PS(I,L),L=1,LTSTOP)
 1600 WRITE(7,7280) (PS(I,L),L=1,LTSTOP)
*
      STOP
****         THESE ARE THE FORMATS
*
 5000 FORMAT(12I6)
 5010 FORMAT(2F6.2,F6.Ø)
 5020 FORMAT(F6.2,2I6)
 6000 FORMAT(1H1)
 6002 FORMAT(/)
 6005 FORMAT(5X,*NCMPT,NSTEP,STEP,NDSPY,RES,CON=*,2I7,F7.2,I7,2F10.2)
 6020 FORMAT(5X,*RS,R,LBR,LSEG,AS,AM=*,6F7.2)
 6025 FORMAT(5X,*RSD,RDD,GDS,GSD,GDD,GLK=*,6E12.2)
 6030 FORMAT(5X,*SC,SCSTRT,SCSTP,LTSTOP=*,F7.2,3I7)
 6280 FORMAT(5X,18F7.2)
 7280 FORMAT(13F6.2)
      END
*
      SUBROUTINE HHEQN(J,E,N,M,H,ACT,GT,Z)
      REAL N,M,H
      PARAMETER NCMPT=10,ENA=-115.,EK=+12.,EL=-10.6,GNØ=120.,GKØ=36.,
     *          GL=.3
      COMMON GØ(NCMPT),GØS,DL
*
      AN=.Ø1*(E+10.)/(EXP(E/10.+1.)-1.) $BN=.125*EXP(E/8Ø.)
      AM=.1*(E+25.)/(EXP(E/10.+2.5)-1.) $BM=4.*EXP(E/18.)
      AH=.Ø7*EXP(E/2Ø.) $BH=1./(EXP(E/10.+3.)+1.)
      Z=EXP(-DL*(AN+BN)) $N=N*Z+AN*(1.-Z)/(AN+BN)
      Z=EXP(-DL*(AM+BM)) $M=M*Z+AM*(1.-Z)/(AM+BM)
      Z=EXP(-DL*(AH+BH)) $H=H*Z+AH*(1.-Z)/(AH+BH)
      GNA=GNØ*H*M**3 $GK=GKØ*N**4
      ACT=GNA*ENA+GK*EK+GL*EL
      IF(J.EQ.Ø) THEN $GT=GØS+GNA+GK $ELSE
```

```
        GT=GØ(J)+GNA+GK $ENDIF $Z=EXP(-DL*GT)
        RETURN
        END
*
/EOR
        Ø
  5.       .12 1ØØØ.
  1Ø.        1 9999
        3    8
        8
/EOR
/EOI


/JOB
A224,CM=7ØØØØ,TL=2ØØ.
FAKE.
PDFV.
MNF(Y).
LGO.
/EOR
        PROGRAM HHNRNØ1 (INPUT,OUTPUT,TAPE6=OUTPUT)
*
****        THIS PROGRAM SIMULATES THE HODGKIN-HUXLEY MODEL OF THE
****        RESPONSE IN A NEURON PATCH TO A STIMULATING ELECTRODE.
*
        INTEGER SCSTRT,SCSTP
        REAL NØ,MØ,HØ
        PARAMETER STEP=1.,NSTEPS=5Ø,ENA=-115.,EK=12.,EL=-1Ø.6,GNØ=12Ø.,
     *          GKØ=36.,GL=.3,NØ=.31767691,MØ=.Ø5293249,HØ=.59612Ø75
        DIMENSION P(5ØØØ)
*
****        THIS SECTION READS AND WRITES THE INPUT PARAMETERS
*
        READ 5Ø1Ø, SC,SL,SCSTRT,SCSTP
        READ 5ØØØ, LTSTOP
*
        WRITE(6,6ØØØ) $WRITE(6,6Ø1Ø) SC,SL,SCSTRT,SCSTP
        WRITE(6,6Ø2Ø) LTSTOP
*
*
****        THIS SECTION INITIALIZES VARIABLES
*
        E=Ø. $FN=NØ $FM=MØ $FH=HØ
        DL=STEP/FLOAT(NSTEPS)
*
****        THIS SECTION UPDATES STATE VARIABLES AT EACH TIME STEP
*
        DO 1ØØØ L=1,LTSTOP
*
        SCN=Ø. $IF(L.GE.SCSTRT.AND.L.LT.SCSTP) SCN=SC
        DO 9ØØ I=1,NSTEPS
*
        AN=.Ø1*(E+1Ø.)/(EXP(E/1Ø.+1.)-1.) $BN=.125*EXP(E/8Ø.)
        AM=.1*(E+25.)/(EXP(E/1Ø.+2.5)-1.) $BM=4.*EXP(E/18.)
        AH=.Ø7*EXP(E/2Ø.) $BH=1./(EXP(E/1Ø.+3.)+1.)
        Z=EXP(-DL*(AN+BN)) $FN=FN*Z+AN*(1.-Z)/(AN+BN)
        Z=EXP(-DL*(AM+BM)) $FM=FM*Z+AM*(1.-Z)/(AM+BM)
        Z=EXP(-DL*(AH+BH)) $FH=FH*Z+AH*(1.-Z)/(AH+BH)
        GN=GNØ*FH*FM**3 $GK=GKØ*FN**4
*
        GTOT=GL+GN+GK $Z=EXP(-DL*GTOT)
        ACT=-SCN+GN*ENA+GK*EK+GL*EL
        E=E*Z+ACT*(1.-Z)/GTOT
        S=Ø. $IF(E.GE.4Ø.) S=1. $P((L-1)*NSTEPS+I)=E
  9ØØ CONTINUE
```

```
*
 1000 CONTINUE
*
****         THIS SECTION WRITES OUT ACTIVITY VARIABLES
*
      WRITE(6,6050) (P(L),L=1,LTSTOP)
*
****         THESE ARE THE FORMATS
*
      STOP
 5000 FORMAT(12I6)
 5010 FORMAT(2F6.2,2I6)
 6000 FORMAT(1H1)
 6010 FORMAT(5X,*SC,SL,SCSTRT,SCSTP=*,2F7.2,2I7)
 6020 FORMAT(5X,*LTSTOP=*,I7)
 6050 FORMAT(/,(2X,20F6.2))
      END
/EOR
 10.     0.        1  9999
     60
/EOR
/EOI


/JOB
A224,TL=600.
FAKE.
PDFV.
MNF(Y).
LGO,AX003.
SAVE,AX003.
EXIT.
SAVE,AX003.
/EOR
      PROGRAM AXON02(AXOUT,INPUT,OUTPUT,TAPE6=OUTPUT,TAPE7=AXOUT)
*
****         THIS PROGRAM SIMLUATES THE PROPAGATION OF ACTION POTENTIALS
*            ALONG A MYLINATED AXON WITH EQUALLY-SPACED NODES OF RANVIER
****         ACCORDING TO THE HODGKIN-HUXLEY MODEL.
*
      INTEGER SCSTRT,SCSTP
      REAL LBR,LSEG,LMYL,N0,M0,H0
      PARAMETER NCMPT=10, NSTEP=1000, STEP=1.,NDSPY=2,GN0=120.,GK0=36.,
     *        GL=.3,N0=.31767691,M0=.05293249,H0=.59612075
      DIMENSION GT(NCMPT),Z(NCMPT),V(NCMPT),U(NCMPT),IDC(10),
     *        PS(10,500),FN(NCMPT),FM(NCMPT),FH(NCMPT),ACT(NCMPT)
      COMMON G0(NCMPT),G0S,DL
*
****         THIS SECTION READS AND WRITES THE INPUT PARAMETERS
*
      READ 5000, IFLG1
      READ 5010, RS,R,LBR,FMYL
      READ 5020, SC,SCSTRT,SCSTP
      READ 5000, (IDC(J),J=1,NDSPY)
      READ 5000, LTSTOP
      RES=700000. $CON=GN0*H0*M0**3+GK0*N0**4+GL
*
      LSEG=LBR*(1.-FMYL)/FLOAT(NCMPT) $LMYL=LBR*FMYL/FLOAT(NCMPT)
      AS=4.*3.14159*RS**2
      AM=2.*3.14159*R*LSEG $X=SQRT(RS**2-R**2)
      RSD=RES*((LSEG+2.*LMYL)/R**2+ALOG((RS+X)/(RS-X))/RS)/2./3.14159
      RDD=RES*(LSEG+LMYL)/3.14159/R**2
      GDS=10.**11/(RSD*AS) $GSD=10.**11/(RSD*AM)
      GDD=10.**11/(RDD*AM) $GLK=CON*3.14159*R**2/AM
*
      WRITE(6,6000) $WRITE(6,6005) NCMPT,NSTEP,STEP,NDSPY,RES,CON
```

```
      WRITE(6,6020) RS,R,LBR,FMYL,LSEG,LMYL,AS,AM
      WRITE(6,6025) RSD,RDD,GDS,GSD,GDD,GLK
      WRITE(6,6030) SC,SCSTRT,SCSTP,LTSTOP
*
****      THIS SECTION INITIALIZES VARIABLES
*
      US=0. $VS=0. $SN=NØ $SM=MØ $SH=HØ
      DO 110 I=1,NCMPT $U(I)=0. $V(I)=0.
      FN(I)=NØ $FM(I)=MØ $FH(I)=HØ
  110 CONTINUE $DL=STEP/FLOAT(NSTEP)
*
      GØS=CON+GDS
      GØ(1)=CON+GSD+GDD
      DO 210 I=2,NCMPT-1
  210 GØ(I)=CON+2.*GDD
      GØ(NCMPT)=CON+GLK+GDD
*
****      THIS SECTION UPDATES STATE VARIABLES AT EACH TIME STEP
*
      DO 1000 L=1,LTSTOP
      SCN=0. $IF(L.GE.SCSTRT.AND.L.LT.SCSTP) SCN=SC
*
      DO 4000 IS=1,NSTEP
      CALL HHEQN(Ø,VS,SN,SM,SH,ACTS,GTS,ZS)
      DO 4005 J=1,NCMPT
 4005 CALL HHEQN(J,V(J),FN(J),FM(J),FH(J),ACT(J),GT(J),Z(J))
      VS=US*ZS+(ACTS+GDS*U(1)-SCN)*(1.-ZS)/GTS
      V(1)=U(1)*Z(1)+(ACT(1)+GSD*US+GDD*U(2))*(1.-Z(1))/GT(1)
      DO 4010 J=2,NCMPT-1
 4010 V(J)=U(J)*Z(J)+(ACT(J)+GDD*(U(J-1)+U(J+1)))*(1.-Z(J))/GT(J)
      V(NCMPT)=U(NCMPT)*Z(NCMPT)+(ACT(NCMPT)+GDD*U(NCMPT-1))*
     *              (1.-Z(NCMPT))/GT(NCMPT)
      DO 4020 J=1,NCMPT
 4020 U(J)=V(J) $US=VS
 4000 CONTINUE
*
      IF(IFLG1.GT.Ø) THEN
      WRITE(6,6280) VS,(V(J),J=1,NCMPT) $ENDIF
      PS(1,L)=VS $DO 4500 I=1,NDSPY
 4500 PS(I+1,L)=V(IDC(I))
*
 1000 CONTINUE
*
****      THIS SECTION WRITES OUT ACTIVITY VARIABLES
*
      DO 1600 I=1,NDSPY+1 $WRITE(6,6002) $WRITE(7,6002)
      WRITE(6,6280) (PS(I,L),L=1,LTSTOP)
 1600 WRITE(7,7280) (PS(I,L),L=1,LTSTOP)
*
      STOP
****      THESE ARE THE FORMATS
*
 5000 FORMAT(12I6)
 5010 FORMAT(F6.2,F6.4,F6.Ø,F6.4)
 5020 FORMAT(F6.2,2I6)
 6000 FORMAT(1H1)
 6002 FORMAT(/)
 6005 FORMAT(5X,*NCMPT,NSTEP,STEP,NDSPY,RES,CON=*,2I7,F7.2,I7,2F10.2)
 6020 FORMAT(5X,*RS,R,LBR,FMYL,LSEG,LMYL,AS,AM=*,8F7.2)
 6025 FORMAT(5X,*RSD,RDD,GDS,GSD,GDD,GLK=*,6E12.2)
 6030 FORMAT(5X,*SC,SCSTRT,SCSTP,LTSTOP=*,F7.2,3I7)
 6280 FORMAT(5X,18F7.2)
 7280 FORMAT(13F6.2)
      END
*
      SUBROUTINE HHEQN(J,E,N,M,H,ACT,GT,Z)
      REAL N,M,H
```

```
      PARAMETER NCMPT=10,ENA=-115.,EK=+12.,EL=-10.6,GN0=120.,GK0=36.,
     *          GL=.3
      COMMON G0(NCMPT),G0S,DL
*
      AN=.01*(E+10.)/(EXP(E/10.+1.)-1.)  $BN=.125*EXP(E/80.)
      AM=.1*(E+25.)/(EXP(E/10.+2.5)-1.)  $BM=4.*EXP(E/18.)
      AH=.07*EXP(E/20.)  $BH=1./(EXP(E/10.+3.)+1.)
      Z=EXP(-DL*(AN+BN))  $N=N*Z+AN*(1.-Z)/(AN+BN)
      Z=EXP(-DL*(AM+BM))  $M=M*Z+AM*(1.-Z)/(AM+BM)
      Z=EXP(-DL*(AH+BH))  $H=H*Z+AH*(1.-Z)/(AH+BH)
      GNA=GN0*H*M**3  $GK=GK0*N**4
      ACT=GNA*ENA+GK*EK+GL*EL
      IF(J.EQ.0) THEN $GT=G0S+GNA+GK $ELSE
      GT=G0(J)+GNA+GK $ENDIF $Z=EXP(-DL*GT)
      RETURN
      END
*
/EOR
      0
   5.    .1200 1000.  .9000
  10.        1  9999
      3     8
      8
/EOR
/EOI



/JOB
A224,TL=500.
FAKE.
PDFV.
MNF(Y).
LGO,AX004.
SAVE,AX004.
EXIT.
SAVE,AX004.
/EOR
      PROGRAM AXON03(AXOUT,INPUT,OUTPUT,TAPE6=OUTPUT,TAPE7=AXOUT)
*
****      THIS PROGRAM SIMULATES THE PROPAGATION OF ACTION POTENTIALS
*         ALONG AN AXON WITH A SINGLE BIFURCATION ACCORDING TO THE
****      HODGKIN-HUXLEY MODEL.
*
      INTEGER SCSTRT,SCSTP
      REAL LBR,LSEG,N0,M0,H0
      PARAMETER NAX=3,NCMPT=10,NSTEP=500,STEP=1.,NDSPY=2,GN0=120.,
     *          GK0=36.,GL=.3,N0=.31767691,M0=.05293249,H0=.59612
      DIMENSION GT(NAX,NCMPT),Z(NAX,NCMPT),V(NAX,NCMPT),U(NAX,NCMPT),
     *          GDD(3),IDB(10),IDC(10),PS(10,500),FN(NAX,NCMPT),
     *          FM(NAX,NCMPT),FH(NAX,NCMPT),ACT(NAX,NCMPT)
      COMMON G0(NAX,NCMPT),G0S,DL
*
****      THIS SECTION READS AND WRITES THE INPUT PARAMETERS
*
      READ 5000, IFLG1
      READ 5010, RS,RST,LBR
      READ 5020, SC,SCSTRT,SCSTP
      READ 5000, ((IDB(J),IDC(J)),J=1,NDSPY)
      READ 5000, LTSTOP
      RES=700000. $CON=GN0*H0*M0**3+GK0*N0**4+GL
*
      R=RST/2.**(2./3.) $AS=4.*3.14159*RS**2 $LSEG=LBR/FLOAT(NCMPT)
      AM1=2.*3.14159*RST*LSEG$AM2=2.*3.14159*R*LSEG$X=SQRT(RS**2-RST**2)
      RSD=RES*(LSEG/RST**2+ALOG((RS+X)/(RS-X))/RS)/2./3.14159
      RDD1=RES*LSEG/3.14159/RST**2 $RDD2=RES*LSEG/3.14159/R**2
      RTR=RES*LSEG*(1./RST**2+1./R**2)/2./3.14159
```

```
        GDS=10.**11/(RSD*AS) $GSD=10.**11/(RSD*AM1)
        GDD(1)=10.**11/(RDD1*AM1) $GDD(2)=10.**11/(RDD2*AM2)
        GDD(3)=GDD(2) $GBT=10.**11/(RTR*AM1)
        GTU=10.**11/(RTR*AM2) $GTA=10.**11/(RDD2*AM2)
        GLK=CON*3.14159*R**2/AM2
*
        WRITE(6,6000) $WRITE(6,6005) NAX,NCMPT,NSTEP,STEP,NDSPY,RES,CON
        WRITE(6,6020) RS,RST,R,LBR,LSEG,AS,AM1,AM2
        WRITE(6,6025) RSD,RDD1,RDD2,RTR,GDS,GSD,GBT,GTU,GTA,
     *                (GDD(I),I=1,3),GLK
        WRITE(6,6030) SC,SCSTRT,SCSTP,LTSTOP
*
****        THIS SECTION INITIALIZES VARIABLES
*
        US=0. $VS=0. $SN=N0 $SM=M0 $SH=H0
        DO 110 I=1,NAX $DO 110 J=1,NCMPT $U(I,J)=0. $V(I,J)=0.
        FN(I,J)=N0 $FM(I,J)=M0 $FH(I,J)=H0
  110 CONTINUE $DL=STEP/FLOAT(NSTEP)
*
        G0S=CON+GDS
        G0(1,1)=CON+GSD+GDD(1)
        G0(2,1)=CON+GTU+GTA+GDD(2)
        G0(3,1)=CON+GTU+GTA+GDD(3)
        DO 210 I=1,NAX $DO 210 J=2,NCMPT-1
  210 G0(I,J)=CON+2.*GDD(I)
        G0(1,NCMPT)=CON+2.*GBT+GDD(1)
        G0(2,NCMPT)=CON+GDD(2)+GLK
        G0(3,NCMPT)=CON+GDD(3)+GLK
*
****        THIS SECTION UPDATES STATE VARIABLES AT EACH TIME STEP
*
        DO 1000 L=1,LTSTOP
        SCN=0. $IF(L.GE.SCSTRT.AND.L.LT.SCSTP) SCN=SC
*
        DO 4000 IS=1,NSTEP
        CALL HHEQN(0,0,VS,SN,SM,SH,ACTS,GTS,ZS)
        DO 4005 I=1,NAX $DO 4005 J=1,NCMPT
 4005 CALL HHEQN(I,J,V(I,J),FN(I,J),FM(I,J),FH(I,J),ACT(I,J),
     *           GT(I,J),Z(I,J))
        VS=US*ZS+(ACTS+GDS*U(1,1)-SCN)*(1.-ZS)/GTS
        V(1,1)=U(1,1)*Z(1,1)+(ACT(1,1)+GSD*US+GDD(1)*U(1,2))*(1.-Z(1,1))/
     *         GT(1,1)
        V(2,1)=U(2,1)*Z(2,1)+(ACT(2,1)+GTU*U(1,NCMPT)+GTA*U(3,1)+GDD(2)*
     *         U(2,2))*(1.-Z(2,1))/GT(2,1)
        V(3,1)=U(3,1)*Z(3,1)+(ACT(3,1)+GTU*U(1,NCMPT)+GTA*U(2,1)+GDD(3)*
     *         U(3,2))*(1.-Z(3,1))/GT(3,1)
        DO 4010 I=1,NAX $DO 4010 J=2,NCMPT-1
 4010 V(I,J)=U(I,J)*Z(I,J)+(ACT(I,J)+GDD(I)*(U(I,J-1)+U(I,J+1)))*
     *       (1.-Z(I,J))/GT(I,J)
        V(1,NCMPT)=U(1,NCMPT)*Z(1,NCMPT)+(ACT(1,NCMPT)+GBT*(U(2,1)+U(3,1))
     *       +GDD(1)*U(1,NCMPT-1))*(1.-Z(1,NCMPT))/GT(1,NCMPT)
        V(2,NCMPT)=U(2,NCMPT)*Z(2,NCMPT)+(ACT(2,NCMPT)+GDD(2)*
     *           U(2,NCMPT-1))*(1.-Z(2,NCMPT))/GT(2,NCMPT)
        V(3,NCMPT)=U(3,NCMPT)*Z(3,NCMPT)+(ACT(3,NCMPT)+GDD(3)*
     *           U(3,NCMPT-1))*(1.-Z(3,NCMPT))/GT(3,NCMPT)
        DO 4020 I=1,NAX $DO 4020 J=1,NCMPT
 4020 U(I,J)=V(I,J) $US=VS
 4000 CONTINUE
*
        IF(IFLG1.GT.0) THEN
        WRITE(6,6280) VS,((V(I,J),J=1,NCMPT),I=1,NAX) $ENDIF
        PS(1,L)=VS $DO 4500 I=1,NDSPY
 4500 PS(I+1,L)=V(IDB(I),IDC(I))
*
 1000 CONTINUE
*
****        THIS SECTION WRITES OUT ACTIVITY VARIABLES
```

```
*
      DO 1600 I=1,NDSPY+1 $WRITE(6,6002) $WRITE(7,6002)
      WRITE(6,6280) (PS(I,L),L=1,LTSTOP)
 1600 WRITE(7,7280) (PS(I,L),L=1,LTSTOP)
*
      STOP
****        THESE ARE THE FORMATS
*
 5000 FORMAT(12I6)
 5010 FORMAT(2F6.2,F6.0)
 5020 FORMAT(F6.2,2I6)
 6000 FORMAT(1H1)
 6002 FORMAT(/)
 6005 FORMAT(5X,*NAX,NCMPT,NSTEP,STEP,NDSPY,RES,CON=*,3I7,F7.2,I7,2F9.1)
 6020 FORMAT(5X,*RS,RST,R,LBR,LSEG,AS,AM1,AM2=*,8F8.2)
 6025 FORMAT(5X,*RSD,RDD1,RDD2,RTR,GDS,GSD,GBT,GTU,GTA,GDDS,GLK=*,/,
     *            5X,13E10.2)
 6030 FORMAT(5X,*SC,SCSTRT,SCSTP,LTSTOP=*,F7.2,3I7)
 6280 FORMAT(5X,15F7.2)
 7280 FORMAT(13F6.2)
      END
*
      SUBROUTINE HHEQN(I,J,E,N,M,H,ACT,GT,Z)
      REAL N,M,H
      PARAMETER NAX=3,NCMPT=10,ENA=-115.,EK=+12.,EL=-10.6,GN0=120.,
     *          GK0=36.,GL=.3
      COMMON G0(NAX,NCMPT),G0S,DL
*
      AN=.01*(E+10.)/(EXP(E/10.+1.)-1.) $BN=.125*EXP(E/80.)
      AM=.1*(E+25.)/(EXP(E/10.+2.5)-1.) $BM=4.*EXP(E/18.)
      AH=.07*EXP(E/20.) $BH=1./(EXP(E/10.+3.)+1.)
      Q=EXP(-DL*(AN+BN)) $N=N*Q+AN*(1.-Q)/(AN+BN)
      Q=EXP(-DL*(AM+BM)) $M=M*Q+AM*(1.-Q)/(AM+BM)
      Q=EXP(-DL*(AH+BH)) $H=H*Q+AH*(1.-Q)/(AH+BH)
      GNA=GN0*H*M**3 $GK=GK0*N**4
      ACT=GNA*ENA+GK*EK+GL*EL
      IF(J.EQ.0) THEN $GT=G0S+GNA+GK $ELSE
      GT=G0(I,J)+GNA+GK $ENDIF $Z=EXP(-DL*GT)
      RETURN
      END
*
/EOR
      0
   5.       .12 1000.
  10.         1 9999
      2       3       3       8
     16
/EOR
/EOI



/JOB
A224,CM=110000,TL=1200.
FAKE.
PDFV.
MNF(Y).
LGO,AX007.
SAVE,AX007.
EXIT.
SAVE,AX007.
/EOR
      PROGRAM AXON04 (AXOUT,INPUT,OUTPUT,TAPE6=OUTPUT,TAPE7=AXOUT)
*
****        THIS PROGRAM SIMULATES THE PROPAGATION OF ACTION POTENTIALS
*           INTO AXON TERMINATION FIELDS OF ARBITRARY BIFURCATION PATTERNS
****        ACCORDING TO THE HODGKIN-HUXLEY MODEL.
```

```
*
      INTEGER TPCN,BTCN,DDN,CDN,SCSTRT,SCSTP
      REAL LBR,LSEG,N0,M0,H0
      PARAMETER NAX=11,NREG=4,NCMPT=10,NSTEPS=100,STEP=1.,NDSPY=2,
     *          GN0=120.,GK0=36.,GL=.3,N0=.31767691,
     *          M0=.05293249,H0=.59612075
      DIMENSION GTP(2,NAX),GDD(NAX),GBT(2,NAX),DDN(NDSPY),CDN(NDSPY),
     *          V(NAX,NCMPT),U(NAX,NCMPT),GT(NAX,NCMPT),
     *          Z(NAX,NCMPT),FN(NAX,NCMPT),FM(NAX,NCMPT),
     *          FH(NAX,NCMPT),PS(5,1000),ACT(NAX,NCMPT)
      COMMON/HH/G0(NAX,NCMPT),G0S,DL
      COMMON/TREE/ LSEG(NAX),AM(NAX),CON,RES,R(NAX),TPCN(2,NAX),
     *             BTCN(2,NAX),RS,SC
*
****       THIS SECTION READS AND WRITES THE INPUT PARAMETERS
*
      READ 5000, IFLG1,IFLG2
      READ 5010, RS
      WRITE(6,6000)
      WRITE(6,6010) NAX,NREG,NCMPT,NSTEPS,STEP,RES,NDSPY,
     *              IFLG1,IFLG2
      WRITE(6,6030) RS
*
      RES=700000. $CON=GN0*H0*M0**3+GK0*N0**4+GL
      AS=4.*3.14159*RS**2
      DO 150 I=1,NAX $DO 150 J=1,2 $GBT(J,I)=0.
  150 GTP(J,I)=0.
      WRITE(6,6040)
      DO 160 I=1,NAX
      READ 5020, J,R(I),LBR,TPCN(1,I),TPCN(2,I),BTCN(1,I),BTCN(2,I)
      LSEG(I)=LBR/FLOAT(NCMPT)
      AM(I)=2.*3.14159*R(I)*LSEG(I)
*
      IF(I.EQ.1) THEN $X=SQRT(RS**2-R(I)**2)
      RSD=RES*(LSEG(I)/R(I)**2+ALOG((RS+X)/(RS-X))/RS)/2./3.14159
      GTP(1,I)=10.**11/(RSD*AM(I)) $GDS=10.**11/(RSD*AS) $ELSE
*
      IF(TPCN(2,I).EQ.0) THEN
      RTP1=RES*(LSEG(TPCN(1,I))+LSEG(I))/2./3.14159/R(I)**2
      GTP(1,I)=10.**11/(RTP1*AM(I))
      GBT(1,TPCN(1,I))=10.**11/(RTP1*AM(TPCN(1,I))) $ENDIF
      IF(TPCN(2,I).GT.0) THEN
      RTP1=RES*(LSEG(TPCN(1,I))/R(TPCN(1,I))**2+LSEG(I)/R(I)**2)/
     *     2./3.14159
      GTP(1,I)=10.**11/(RTP1*AM(I))  $IT=TPCN(1,I)
      IF(GBT(1,IT).EQ.0) THEN $ID=1 $ELSE $ID=2 $ENDIF
      GBT(ID,IT)=10.**11/(RTP1*AM(IT))
      IT=TPCN(2,I) $IF(I.GT.IT) THEN
      RTP2=RES*(LSEG(IT)+LSEG(I))/2./3.14159/R(I)**2
      GTP(2,I)=10.**11/(RTP2*AM(I)) $GTP(2,IT)=10.**11/(RTP2*AM(IT))
      ENDIF $ENDIF $ENDIF
*
      RDD=RES*LSEG(I)/3.14159/R(I)**2 $GDD(I)=10.**11/(RDD*AM(I))
*
      IF(BTCN(1,I).EQ.0) GBT(1,I)=CON*3.14159*R(I)**2/AM(I)
*
      WRITE(6,6050) I,LBR,TPCN(1,I),TPCN(2,I),BTCN(1,I),BTCN(2,I),
     *              GTP(1,I),GTP(2,I),GDD(I),GBT(1,I),GBT(2,I)
      IF(I.EQ.1) WRITE(6,6052) GDS
  160 CONTINUE
      READ 5000, ((DDN(I),CDN(I)),I=1,NDSPY)
      READ 5250, SC,SCSTRT,SCSTP
      READ 5000, LTSTOP
*
      WRITE(6,6060) ((DDN(I),CDN(I)),I=1,NDSPY)
      WRITE(6,6160) SC,SCSTRT,SCSTP $WRITE(6,6170) LTSTOP
      IF(IFLG2.GT.0) CALL TREE $WRITE(6,6000)
```

```
*
*
****        THIS SECTION INITIALIZES VARIABLES
*
      US=0. $VS=0. $SN=N0 $SM=M0 $SH=H0
      DO 300 I=1,NAX $DO 300 J=1,NCMPT
      U(I,J)=0. $V(I,J)=0. $FN(I,J)=N0 $FM(I,J)=M0 $FH(I,J)=H0
  300 CONTINUE $DL=STEP/FLOAT(NSTEPS)
      G0S=CON+GDS
      DO 400 I=1,NAX
      IF(I.EQ.1) THEN $G0(I,1)=CON+GTP(1,I)+GDD(I) $ELSE
      G0(I,1)=CON+GTP(1,I)+GTP(2,I)+GDD(I) $ENDIF
      DO 410 J=2,NCMPT-1
  410 G0(I,J)=CON+2.*GDD(I)
      G0(I,NCMPT)=CON+GBT(1,I)+GBT(2,I)+GDD(I)
  400 CONTINUE
*
****        THIS SECTION UPDATES STATE VARIABLES AT EACH TIME STEP
*
      DO 1000 L=1,LTSTOP
      SCN=0. $IF(L.GE.SCSTRT.AND.L.LT.SCSTP) SCN=SC
*
      DO 4000 IS=1,NSTEPS
      CALL HHEQN(0,0,VS,SN,SM,SH,ACTS,GTS,ZS)
      DO 4005 I=1,NAX $DO 4005 J=1,NCMPT
 4005 CALL HHEQN(I,J,V(I,J),FN(I,J),FM(I,J),FH(I,J),ACT(I,J),
     *          GT(I,J),Z(I,J))
      VS=US*ZS+(-SCN+ACTS+GDS*U(1,1))*(1.-ZS)/GTS
*
      DO 4100 I=1,NAX
*
      IF(I.EQ.1) THEN
      V(I,1)=U(I,1)*Z(I,1)+(GTP(1,I)*US+GDD(I)*U(I,2)+ACT(I,1))*
     *              (1.-Z(I,1))/GT(I,1) $ELSE
      IF(TPCN(2,I).EQ.0) V(I,1)=U(I,1)*Z(I,1)+(GTP(1,I)*U(TPCN(1,I)
     *      ,NCMPT)+GDD(I)*U(I,2)+ACT(I,1))*(1.-Z(I,1))/GT(I,1)
      IF(TPCN(2,I).GT.0) V(I,1)=U(I,1)*Z(I,1)+(GTP(1,I)*U(TPCN(1,I),
     *      NCMPT)+GTP(2,I)*U(TPCN(2,I),1)+GDD(I)*U(I,2)+ACT(I,1))*
     *      (1.-Z(I,1))/GT(I,1)
        ENDIF
*
      DO 4120 J=2,NCMPT-1
 4120 V(I,J)=U(I,J)*Z(I,J)+(GDD(I)*(U(I,J-1)+U(I,J+1))+ACT(I,J))*
     *          (1.-Z(I,J))/GT(I,J)
*
      IF(BTCN(1,I).LE.0) THEN
      V(I,NCMPT)=U(I,NCMPT)*Z(I,NCMPT)+(GDD(I)*U(I,NCMPT-1)+ACT(I,
     *      NCMPT))*(1.-Z(I,NCMPT))/GT(I,NCMPT)  $ELSE
      IF(BTCN(2,I).EQ.0) V(I,NCMPT)=U(I,NCMPT)*Z(I,NCMPT)+(GBT(1,I)*
     *      U(BTCN(1,I),1)+GDD(I)*U(I,NCMPT-1)+ACT(I,NCMPT))*
     *      (1.-Z(I,NCMPT))/GT(I,NCMPT)
      IF(BTCN(2,I).GT.0) V(I,NCMPT)=U(I,NCMPT)*Z(I,NCMPT)+(GBT(1,I)*
     *      U(BTCN(1,I),1)+GBT(2,I)*U(BTCN(2,I),1)+GDD(I)*U(I,NCMPT-1)+
     *      ACT(I,NCMPT))*(1.-Z(I,NCMPT))/GT(I,NCMPT)   $ENDIF
*
 4100 CONTINUE
      US=VS $DO 4200 I=1,NAX $DO 4200 J=1,NCMPT
 4200 U(I,J)=V(I,J)
*
 4000 CONTINUE
*
      IF(IFLG1.GT.0) THEN $WRITE(6,6280) VS $DO 4400 I=1,NAX
 4400 WRITE(6,6280) (U(I,J),J=1,NCMPT) $ENDIF
      PS(1,L)=VS $DO 4500 I=1,NDSPY
 4500 PS(I+1,L)=V(DDN(I),CDN(I))
*
 1000 CONTINUE
```

```
*
****          THIS SECTION WRITES OUT ACTIVITY VARIABLES
*
       DO 1600 I=1,NDSPY+1 $WRITE(6,6002)
 1600 WRITE(6,6280) (PS(I,L),L=1,LTSTOP)
       DO 1700 I=1,NDSPY+1 $WRITE(7,6002)
 1700 WRITE(7,7280) (PS(I,L),L=1,LTSTOP)
*
       STOP
****          THESE ARE THE FORMATS
*
 5000 FORMAT(12I6)
 5010 FORMAT(12F6.2)
 5020 FORMAT(I6,F6.4,F6.0,4I6)
 5250 FORMAT(F6.2,2I6)
 6000 FORMAT(1H1)
 6002 FORMAT(/)
 6010 FORMAT(5X,*NAX,NREG,NCMPT,NSTEPS,STEP,RES,NDSPY,*
      *         ,*IFLG1,ILFG2=*,/,10X,4I7,2F10.2,3I7)
 6030 FORMAT(5X,*RS=*,F7.2)
 6040 FORMAT(/,5X,*BR,RG,L,TPCNS,BTCNS,NS,RH,RSK,LSK,GTPS,GDD,GBTS=:*)
 6050 FORMAT(5X,I7,F7.0,5I7,8F8.2)
 6052 FORMAT(5X,*GDS=*,9F9.2)
 6060 FORMAT(5X,*DSP=*,2I7)
 6160 FORMAT(5X,*SC,SCSTRT,SCSTP=*,F7.2,2I7)
 6170 FORMAT(5X,*LTSTOP=*,I7)
 6280 FORMAT(5X,15F7.2)
 7280 FORMAT(13F6.2)
       END
*
       SUBROUTINE HHEQN(I,J,E,N,M,H,ACT,GT,Z)
       REAL N,M,H
       PARAMETER NAX=11,NCMPT=10,ENA=-115.,EK=+12.,EL=-10.6,GN0=120.,
      *          GK0=36.,GL=.3
       COMMON/HH/ G0(NAX,NCMPT),G0S,DL
*
       AN=.01*(E+10.)/(EXP(E/10.+1.)-1.) $BN=.125*EXP(E/80.)
       AM=.1*(E+25.)/(EXP(E/10.+2.5)-1.) $BM=4.*EXP(E/18.)
       AH=.07*EXP(E/20.) $BH=1./(EXP(E/10.+3.)+1.)
       Q=EXP(-DL*(AN+BN)) $N=N*Q+AN*(1.-Q)/(AN+BN)
       Q=EXP(-DL*(AM+BM)) $M=M*Q+AM*(1.-Q)/(AM+BM)
       Q=EXP(-DL*(AH+BH)) $H=H*Q+AH*(1.-Q)/(AH+BH)
       GNA=GN0*H*M**3 $GK=GK0*N**4
       ACT=GNA*ENA+GK*EK+GL*EL
       IF(J.EQ.0) THEN $GT=G0S+GNA+GK $ELSE
       GT=G0(I,J)+GNA+GK $ENDIF $Z=EXP(-DL*GT)
       RETURN
       END
*
       SUBROUTINE TREE
       INTEGER TPCN,BTCN,USED
       REAL LBR,LSEG,LC
       PARAMETER NAX=11,NREG=4,NCMPT=10
       COMMON/TREE/ LSEG(NAX),AM(NAX),CON,RES,R(NAX),TPCN(2,NAX),
      *             BTCN(2,NAX),RS,SC
       DIMENSION LC(NAX),RL(NAX),AD(NAX),USED(NAX),YN(NAX),NM(NAX)
*
       AS=4.*3.14159*RS**2
       WRITE(6,6102) RS,AS $WRITE(7,6102) RS,AS
       WRITE(6,6105) $WRITE(7,6105)
       DO 160 I=1,NAX
       LBR=FLOAT(NCMPT)*LSEG(I)
       REQ=AM(I)/(2.*3.14159*LSEG(I))
       LC(I)=R(I)/SQRT(2.*REQ*CON*RES/10.**11)
       RL(I)=RES/(3.14159*R(I)**2)
       AD(I)=(TANH(LBR/LC(I))+REQ/2./LC(I))/
      *       ((1.+TANH(LBR/LC(I))*REQ/2./LC(I))*LC(I)*RL(I))
```

```
        DCS=EXP(-LSEG(I)/LC(I))  $DCB=EXP(-LBR/LC(I))
        WRITE(6,6110) I,R(I),AM(I),RL(I),AD(I),LC(I),DCS,DCB
        WRITE(7,6110) I,R(I),AM(I),RL(I),AD(I),LC(I),DCS,DCB
  160 CONTINUE
*
        N=0 $DO 10 J=1,NAX
   10 USED(J)=0
        DO 100 I=1,NAX
        IF(BTCN(1,I).GT.0.OR.USED(I).GT.0) GO TO 100
        N=N+1 $NM(N)=I $YN(N)=AD(I) $USED(I)=1
        IF(TPCN(2,I).EQ.0) GO TO 90
        J=TPCN(2,I) $USED(J)=1 $YN(N)=YN(N)+AD(J)
   90 CONTINUE $IF(TPCN(1,I).EQ.0) GO TO 100
        J=TPCN(1,I) $NM(N)=J
        YN(N)=(YN(N)+AD(J))/(1.+YN(N)*AD(J)*(LC(J)*RL(J))**2)
  100 CONTINUE       $NN=N
*
        DO 300 L=1,NREG-2   $NN=0
        DO 200 I=1,N  $IF(USED(NM(I)).GT.0) GO TO 200 $NN=NN+1
        IF(TPCN(2,NM(I)).EQ.0) GO TO 190
        YN(I)=YN(I)+YN(I+1) $USED(TPCN(2,NM(I)))=1
  190 J=TPCN(1,NM(I))  $NM(NN)=J
        YN(NN)=(YN(I)+AD(J))/(1.+YN(I)*AD(J)*(LC(J)*RL(J))**2)
  200 CONTINUE  $N=NN
  300 CONTINUE
*
        YS=CON/(10.**11)*4.*3.14159*RS**2   $YD=0.
        DO 400 I=1,NN
  400 YD=YD+YN(I)
        SIG=YD/YS $RTOT=1./(YS+YD) $RSMA=1./YS $RDN=1./YD $ESS=SC/(1.+SIG)
        WRITE(6,6120) SIG,YS,YD,RTOT,RSMA,RDN,ESS
        WRITE(7,6120) SIG,YS,YD,RTOT,RSMA,RDN,ESS
*
 6102 FORMAT(5X,*RS, AS=*,2F7.2)
 6105 FORMAT(/,5X,*RG,R,AM,RL,AD,LC,DCS,DCB=:*)
 6110 FORMAT(5X,I7,2F7.2,2E12.2,F7.1,2F7.4)
 6120 FORMAT(5X,*SIG,YS,YD,RTOT,RSMA,RDN,ESS=*,/,5X,F7.2,5E10.2,F7.2)
*
        RETURN
        END
/EOR
      0     1
   5.
       1 .1250   200.     0     0     2     3
       2 .0788   200.     1     3     4     0
       3 .0788   200.     1     2     5     6
       4 .0788   200.     2     0     7     8
       5 .0496   200.     3     6     9    10
       6 .0496   200.     3     5    11     0
       7 .0496   200.     4     8     0     0
       8 .0496   200.     4     7     0     0
       9 .0313   200.     5    10     0     0
      10 .0313   200.     5     9     0     0
      11 .0496   200.     6     0     0     0
       2     5     9     8
      10.     1  9999
      16
/EOR
/EOI


/JOB
KW67,CM=75000,TL=3000.
FAKE.
PDFV.
MNF(Y).
```

```
LGO.
/EOR
      PROGRAM HEART01 (INPUT,OUTPUT,HTOUT,TAPE5=INPUT,TAPE6=OUTPUT,
     *TAPE7=HTOUT)

***      THIS PROGRAM SIMULATES THE MCALLISTER-NOBLE MODEL
***      OF THE ELECTRICAL ACTIVITY IN A PURKINJE FIBER.
***      (ACCORDING TO THE EXPRESSIONS GIVEN IN TABLE 1.)

      DIMENSION P(2000)
      REAL M0,H0
      PARAMETER STEP=1.,NSTEPS=50,ENA=40.,EK=-110.,ESI=70.,
     *         ECL=-70.,GNA=150.,GSI=.8,GSIS=.04,GCL=2.5,
     *         GKX2=.385,GNAB=.105,GCLB=.01,M0=0.,D0=0.,
     *         Q0=0.,S0=0.,X10=0.,X20=0.,H0=1.,R0=1.,
     *         F0=1.,CM=10.

***      THIS SECTION READS THE INPUT PARAMETERS

      WRITE(6,666)
 666  FORMAT(2X,'INPUT LTSTOP :')
      READ 5000, LTSTOP

***      THIS SECTION INITIALIZES VARIABLES

      E=-50. $FM=M0 $FH=H0 $FD=D0 $FQ=Q0 $FS=S0 $FR=R0
     *$FX1=X10 $FX2=X20 $FF=F0
      DL=STEP/FLOAT(NSTEPS)

***      THIS SECTION UPDATES STATE VARIABLES AT EACH TIME STEP

      DO 1000 L=1,LTSTOP

      DO 900 I=1,NSTEPS

      AM=(E+47.)/(1.-EXP(-(E+47.)*.1)) $BM=9.86*EXP(-(E+47.)/17.86)
      AH=.000000113*EXP(-(E+10.)/5.43) $BH=2.5/(1.+EXP(-(E+10.)/12.2))
      AD=.002*(E+40.)/(1.-EXP(-(E+40.)*.1)) $BD=.02*EXP(-(E+40.)/11.26)
      AF=.0005*EXP(-(E+26.)/25.) $BF=.02/(1.+EXP(-(E+26.)/11.49))
      AQ=.008*E/(1.-EXP(-E*.1)) $BQ=.08*EXP(-E/11.26)
      AR=.000033*EXP(-E/17.) $BR=.033/(1.+EXP(-(E+30.)/8.))
      AS=.001*(E+52.)/(1.-EXP(-(E+52.)/5.))
      BS=.00005*EXP(-(E+52.)/14.93)
      AX1=.0005*EXP((E+50.)/12.1)/(1.+EXP((E+50.)/17.5))
      BX1=.0013*EXP(-(E+20.)/16.67)/(1.+EXP(-(E+20.)/25.))
      AX2=.000127/(1.+EXP(-(E+19.)/5.))
      BX2=.0003*EXP(-(E+20.)/16.67)/(1.+EXP(-(E+20.)/25.))
      D1=1./(1.+EXP(-.15*(E+40.)))

      Z=EXP(-DL*(AM+BM)) $FM=FM*Z+AM*(1.-Z)/(AM+BM)
      Z=EXP(-DL*(AH+BH)) $FH=FH*Z+AH*(1.-Z)/(AH+BH)
      Z=EXP(-DL*(AD+BD)) $FD=FD*Z+AD*(1.-Z)/(AD+BD)
      Z=EXP(-DL*(AF+BF)) $FF=FF*Z+AF*(1.-Z)/(AF+BF)
      Z=EXP(-DL*(AQ+BQ)) $FQ=FQ*Z+AQ*(1.-Z)/(AQ+BQ)
      Z=EXP(-DL*(AR+BR)) $FR=FR*Z+AR*(1.-Z)/(AR+BR)
      Z=EXP(-DL*(AS+BS)) $FS=FS*Z+AS*(1.-Z)/(AS+BS)
      Z=EXP(-DL*(AX1+BX1)) $FX1=FX1*Z+AX1*(1.-Z)/(AX1+BX1)
      Z=EXP(-DL*(AX2+BX2)) $FX2=FX2*Z+AX1*(1.-Z)/(AX2+BX2)
      GNA0=GNA*FM**3*FH $GK=GKX2*FX2 $GCA0=GSI*FD*FF
      GCA1=GSIS*D1 $GCL0=GCL*FQ*FR

      GTOT=GNA0+GK+GCA0+GCA1+GCL0+GCLB+GNAB
      Z=EXP(-DL*GTOT/CM)
      CX2=25.*FX2
      RK2=2.8*(EXP(.04*(E+110.))-1.)/(EXP(.08*(E+60.))+
     *EXP(.04*(E+60.)))
      RX1=1.2*(EXP(.04*(E+95.))-1.)/EXP(.04*(E+45.))
```

```
      FK2=(RK2/2.8)+.2*(E+30.)/(1.-EXP(-.04*(E+30.)))
      RECT=RK2*FS+RX1*FX1+FK2+CX2
      ACT=-RECT+(GNA0+GNAB)*ENA+(GCA0+GCA1)*ESI+(GCL0+GCLB)*ECL
      E=E*Z+ACT*(1.-Z)/GTOT
      P(L)=E

 900  CONTINUE

1000  CONTINUE

***          THIS SECTION WRITES OUT ACTIVITY VARIABLES
      WRITE(6,6050) (P(L),L=1,LTSTOP)
      WRITE(7,7050) (P(L),L=1,LTSTOP)

***          THESE ARE THE FORMATS
      STOP

5000  FORMAT(1I4)
6050  FORMAT(/,(2X,10F8.2))
7050  FORMAT(/,(2X,4F8.2))
      END
/EOR
1500
/EOR
/EOI




/JOB
A224,CM=70000,TL=200.
FAKE.
PDFV.
M77.
LGO,JN005.
SAVE,JN005.
EXIT.
SAVE,JN005.
/EOR
      PROGRAM JNCTN70 (JNOUT,INPUT,OUTPUT,TAPE7=JNOUT,TAPE6=OUTPUT)
*
****      THIS PROGRAM SIMULATES THE ACTIVATION OF A POOL OF NEURONS
****      BY AN INPUT FIBER SYSTEM. A SIMPLE GRAPHIC DISPLAY IS ADDED.
*
      PARAMETER NFPOPS=2,NFIB=50,NCLS=50,MCTP1=2,STEP=1.,
     *          NST=3,NRG=4,NRGP1=5
      DIMENSION T(NFPOPS),N(NFPOPS),P(NFPOPS),
     *          INSTRT(NFPOPS),INSTP(NFPOPS),INFSED(NFPOPS),
     *          NF(NFPOPS),IDFF(NFPOPS,NFIB),IDFC(NCLS),CHAR(2),
     *          ALINE(110),PER(NFPOPS),LSTM(NFPOPS),DCS(NFPOPS)
      COMMON G(NFPOPS,NCLS,NST,NRGP1,MCTP1)
      COMMON/SV/E(NCLS),TH(NCLS),S(NCLS),GK(NCLS),B,C,THO,TMEM,DCG,DCTH
     *         ,EQ(NFPOPS),V(NCLS,NST,NRG),GCA(NCLS,NST,NRG),
     *          CA(NCLS,NST,NRG),GKD(NCLS,NST,NRG),CDS,VRT,DGC,THD,D,
     *          DCA,A,DGKD,CA0,BD
      COMMON/TR/ NCT(NFPOPS),NT(NFPOPS,NRGP1),STR(NFPOPS,NRGP1),
     *           INSED(NFPOPS)
*
****          THIS SECTION READS AND WRITES THE INPUT PARAMETERS
*
      READ 5020, C,TTH,B,TGK,THO,TMEM,CDS,VRT
      READ 5020, THD,D,TGC,A,TCA,CA0,BD,TGKD
      DO 110 I=1,NFPOPS
      READ 5030, EQ(I),T(I),N(I),P(I),INSTRT(I),INSTP(I),INFSED(I)
  110 READ 5035, NCT(I),((NT(I,J),STR(I,J)),J=1,NRGP1),INSED(I)
      READ 5010, LTSTOP
*
```

```
      WRITE(6,6000)
      WRITE(6,6010) NFPOPS,NCLS,LTSTOP
      WRITE(6,6020) C,TTH,B,TGK,THO,TMEM,CDS,VRT
      WRITE(6,6022) THD,D,TGC,A,TCA,CA0,BD,TGKD
      DO 210 I=1,NFPOPS
      WRITE(6,6030) I,EQ(I),T(I),N(I),P(I),INSTRT(I),INSTP(I),INFSED(I)
  210 WRITE(6,6035) NCT(I),((NT(I,J),STR(I,J)),J=1,NRGP1),INSED(I)
      WRITE(6,6040)
*
****          THIS SECTION INITIALIZES VARIABLES
*
      DO 310 I=1,NCLS
      E(I)=0. $TH(I)=1. $S(I)=0. $GK(I)=0.
      DO 309 K=1,NFPOPS$DO 309 J=1,MCTP1$DO 309 II=1,NST$DO309 M=1,NRGP1
  309 G(K,I,II,M,J)=0.
  310 CONTINUE
      CHAR(1)=1H  $CHAR(2)=1H*
      DCTH=EXP(-STEP/TTH) $DCG=EXP(-STEP/TGK)
      DGC=EXP(-STEP/TGC) $DCA=EXP(-STEP/TCA) $DGKD=EXP(-STEP/TGKD)
      DO 320 I=1,NFPOPS
  320 DCS(I)=EXP(-STEP/T(I))
      DO 330 I=1,NCLS $DO 330 J=1,NST $DO 330 K=1,NRG
      V(I,J,K)=0. $GCA(I,J,K)=0. $CA(I,J,K)=0. $GKD(I,J,K)=0.
  330 CONTINUE
*
****          THIS SECTION UPDATES STATE VARIABLES AT EACH TIME STEP
*
      DO 1000 L=1,LTSTOP
      NFC=0 $DO 1010 I=1,110
 1010 ALINE(I)=CHAR(1)
*
      DO 2000 I=1,NFPOPS $NF(I)=0
      IF(L.LT.INSTRT(I).OR.L.GE.INSTP(I)) GO TO 2000
      CALL RANSET(INFSED(I)) $DO 1900 J=1,N(I)
      IF(RANF(0.).GT.P(I)) GO TO 1900 $CALL RANGET(INFSED(I))
      NF(I)=NF(I)+1 $IDFF(I,NF(I))=J $CALL TRNSMT(I,J)
      IF(I.EQ.1) ALINE(J)=CHAR(2)
      CALL RANSET(INFSED(I))
 1900 CONTINUE $CALL RANGET(INFSED(I))
 2000 CONTINUE
*
      DO 3000 I=1,NCLS
      CALL SVUPDT(I)
      IF(E(I).LT.TH(I)) GO TO 3000
      S(I)=1. $NFC=NFC+1 $IDFC(NFC)=I $ALINE(60+I)=CHAR(2)
 3000 CONTINUE
*
      WRITE(6,6050) L,NF(1),(ALINE(I),I=1,50),NFC,(ALINE(I),I=61,110)
      WRITE(7,7010) L,(NF(I),I=1,NFPOPS),NFC,((IDFF(I,J),J=1,NF(I)),
     *              I=1,NFPOPS),(IDFC(J),J=1,NFC)
*
      DO 4010 K=1,NFPOPS $DO 4010 I=1,NCLS
      DO 4010 M=1,NST $DO 4010 II=1,NRGP1
      G(K,I,M,II,1)=G(K,I,M,II,1)*DCS(K)+G(K,I,M,II,2)
      DO 4020 J=2,MCTP1-1 $G(K,I,M,II,J)=G(K,I,M,II,J+1)
 4020 CONTINUE $G(K,I,M,II,MCTP1)=0.
 4010 CONTINUE
*
 1000 CONTINUE
*
      STOP
****          THESE ARE THE FORMATS
*
 5010 FORMAT(12I6)
 5020 FORMAT(12F6.2)
 5030 FORMAT (2F6.2,I6,F6.2,5I6,F6.2,I6)
 5035 FORMAT(2I6,F6.2,I6,F6.2,I6,F6.2,I6,F6.2,I6,F6.2,I6)
```

```
6000 FORMAT(1H1)
6010 FORMAT(5X,*NFPOPS,NCLS,LTSTOP=*,3I6)
6020 FORMAT(5X,*C,TTH,B,TGK,THO,TMEM,CDS,VRT=*,8F8.2)
6022 FORMAT(5X,*THD,D,TGC,A,TCA,CA0,BD,TGKD=*,8F8.2)
6030 FORMAT(5X,*I,EQ,T,N,P,INSTRT,INSTP,INFSED,NCT,NT,STR,INSED=:*,
    *       I6,2F6.2,I6,F6.2,5I6,F6.2,I6)
6035 FORMAT(5X,*NCT,(NT,STR),INSED=*,2I6,F6.2,I6,F6.2,I6,F6.2,
    *                             I6,F6.2,I6)
6040 FORMAT(///)
6050 FORMAT(2X,2I4,2X,50A1,2X,I4,50A1)
7010 FORMAT(30I4)
     END
*
*
     SUBROUTINE SVUPDT(K)
*
*** THIS SUBROUTINE UPDATES THE STATE VARIABLES OF AN INDIVIDUAL NEURON
*
     PARAMETER NFPOPS=2,NCLS=50,MCTP1=2,STEP=1.,EK=-10,ECA=50.,
    *          NST=3,NRG=4,NRGP1=5
     DIMENSION U(NRG)
     COMMON G(NFPOPS,NCLS,NST,NRGP1,MCTP1)
     COMMON/SV/E(NCLS),TH(NCLS),S(NCLS),GK(NCLS),B,C,THO,TMEM,DCG,DCTH
    *          ,EQ(NFPOPS),V(NCLS,NST,NRG),GCA(NCLS,NST,NRG),
    *          CA(NCLS,NST,NRG),GKD(NCLS,NST,NRG),CDS,VRT,DGC,
    *          THD,D,DCA,A,DGKD,CA0,BD
*
     GS=0. $SYN=0. $DO 10 I=1,NFPOPS
     GS=GS+G(I,K,1,NRGP1,1) $SYN=SYN+G(I,K,1,NRGP1,1)*EQ(I)
  10 CONTINUE
     GK(K)=GK(K)*DCG+B*(1.-DCG)*S(K) $GTOT=1.+GK(K)+GS
     DCE=EXP(-GTOT*STEP/TMEM)
     E(K)=E(K)*DCE+(SYN+GK(K)*EK)*(1.-DCE)/GTOT $DM=0.
     DO 980 I=1,NST $DO 920 J=1,NRG $GT=1.+GCA(K,I,J)+GKD(K,I,J)
     SY=0. $DO 910 M=1,NFPOPS $GT=GT+G(M,K,I,J,1)
 910 SY=SY+G(M,K,I,J,1)*EQ(M) $Z=EXP(-GT*STEP/TMEM)
 920 V(K,I,J)=V(K,I,J)*Z+(SY+GCA(K,I,J)*ECA+GKD(K,I,J)*EK)*
    *         (1.-Z)/GT
*
     DM=DM+V(K,I,1)
     U(1)=(V(K,I,1)+V(K,I,2)+E(K)/CDS)/(2.+1./CDS)*(1.-VRT)+
    *     VRT*V(K,I,1)
     IF(NRG.GT.1) THEN $DO 930 J=2,NRG-1
 930 U(J)=(V(K,I,J-1)+V(K,I,J)+V(K,I,J+1))/3.*(1.-VRT)+
    *     VRT*V(K,I,J)
     U(NRG)=(V(K,I,NRG)+V(K,I,NRG-1))/2.*(1.-VRT)+VRT*V(K,I,NRG) $ENDIF
     DO 940 J=1,NRG
 940 V(K,I,J)=U(J)
*
     DO 950 J=1,NRG
     GCA(K,I,J)=GCA(K,I,J)*DGC
     IF(V(K,I,J).GE.THD) GCA(K,I,J)=GCA(K,I,J)+
    *          D*(V(K,I,J)-THD)*(1.-DGC)
     CA(K,I,J)=CA(K,I,J)*DCA+A*GCA(K,I,J)*(1.-DCA)
     GKD(K,I,J)=GKD(K,I,J)*DGKD
     IF(CA(K,I,J).GE.CA0) GKD(K,I,J)=GKD(K,I,J)+BD*(1.-DGKD)
 950 CONTINUE
*
 980 CONTINUE
*
     E(K)=(E(K)+CDS*DM)/(1.+CDS*FLOAT(NST))*(1.-VRT)+VRT*E(K)
     TH(K)=THO+(TH(K)-THO)*DCTH+C*E(K)*(1.-DCTH)
     S(K)=0.
     RETURN
     END
*
     SUBROUTINE TRNSMT(K,M)
```

```
      PARAMETER NFPOPS=2,NCLS=50,MCTP1=2,NST=3,NRG=4,NRGP1=5
      COMMON G(NFPOPS,NCLS,NST,NRGP1,MCTP1)
      COMMON/TR/ NCT(NFPOPS),NT(NFPOPS,NRGP1),STR(NFPOPS,NRGP1),
     *           INSED(NFPOPS)
*
****      THIS SUBROUTINE PROJECTS SYNAPTIC ACTIVATION FROM A SENDING
****      FIBER TO ITS TARGETED RECEIVING CELLS.
*
      CALL RANSET(M*INSED(K))
      DO 10 I=1,NRGP1 $DO 10 J=1,NT(K,I)
      NREC=INT(RANF(0.)*FLOAT(NCLS))+1
      KST=INT(RANF(0.)*FLOAT(NST))+1 $IF(I.EQ.NRGP1) KST=1
      KCT=INT(RANF(0.)*FLOAT(NCT(K)))+2
   10 G(K,NREC,KST,I,KCT)=G(K,NREC,KST,I,KCT)+STR(K,I)
*
      RETURN
      END
*
/EOR
  .00 20.   20.    5.   10.    5.    1.    .1
 32.    2.2 10.    2.   10.   20.   80.   15.
 70.    1.      50    .1      0   999  1227
      1    0  0.        0  0.        2  20.       2  20.       0  0.      4748
-10.    5.      50    .1     15    50  1936
      1    2 10.        0  0.        0  0.        0  0.        0  0.      3421
     60
/EOR
/EOI




/JOB
A224,CM=100000,TL=1000.
FAKE.
PDFV.
M77.
LGO,SYS713.
SAVE,SYS713.
EXIT.
SAVE,SYS713.
/EOR
      PROGRAM SYSTM70 (SYSTM,INPUT,OUTPUT,TAPE6=OUTPUT,TAPE7=SYSTM)
*
****      THIS PROGRAM SIMULATES THE ACTIVITY OF AN ARBITRARY SYSTEM OF
*         INTERCONNECTED NEURONAL POPULATIONS ACTIVATED BY AN ARBITRARY
*         NUMBER OF SPATIALLY-ORGANIZED INPUT FIBER SYSTEMS
****      ALL SYNAPTIC SYSTEMS ARE INDIVIDUALLY ADJUSTABLE.
*
      INTEGER TYPE,SNTP
      PARAMETER NTPOPS=5,NFPOPS=2,NCPOPS=3,NCLS=50,MCTP1=2,
     *          STEP=1.,SNTP=2,NSTMX=3,NRGMX=4,NRGP1M=5,NTGMX=5
      DIMENSION P(NFPOPS),INSTR(NFPOPS),INSTP(NFPOPS),INFSED(NFPOPS),
     *          IDF(NCLS),TTH(NCPOPS),TGK(NCPOPS),
     *          T(SNTP),DCS(SNTP),NF(NFPOPS),
     *          NTGR(NTPOPS),TGC(NCPOPS),TCA(NCPOPS),TGKD(NCPOPS)
      COMMON G(SNTP,NCPOPS,NCLS,NSTMX,NRGP1M,MCTP1),NST(NCPOPS),
     *        NRG(NCPOPS)
      COMMON/SV/E(NCPOPS,NCLS),TH(NCPOPS,NCLS),S(NCPOPS,NCLS),
     *          GK(NCPOPS,NCLS),B(NCPOPS),C(NCPOPS),TMEM(NCPOPS),
     *          THO(NCPOPS),DCG(NCPOPS),DCTH(NCPOPS),EQ(SNTP),
     *          V(NCPOPS,NCLS,NSTMX,NRGMX),GCA(NCPOPS,NCLS,NSTMX,
     *          NRGMX),CA(NCPOPS,NCLS,NSTMX,NRGMX),GKD(NCPOPS,NCLS,
     *          NSTMX,NRGMX),CDS(NCPOPS),VRT(NCPOPS),DGC(NCPOPS),
     *          THD(NCPOPS),D(NCPOPS),DCA(NCPOPS),A(NCPOPS),
     *          DGKD(NCPOPS),CA0(NCPOPS),BD(NCPOPS)
      COMMON/TR/NCT(NTPOPS,NTGMX),NT(NTPOPS,NTGMX),STR(NTPOPS,
     *          NTGMX),INSED(NTPOPS,NTGMX),
```

```
     *          N(NTPOPS),TYPE(NTPOPS,NTGMX),
     *          IRCP(NTPOPS,NTGMX),
     *          IRRG(NTPOPS,NTGMX)
*
****      THIS SECTION READS AND WRITES THE INPUT PARAMETERS
*
     READ 5110, IFLG1,IFLG2
     READ 5120, ((EQ(I),T(I)),I=1,SNTP)
     DO 110 I=1,NTPOPS $IF(I.LE.NCPOPS) THEN
     READ 5130, N(I),C(I),TTH(I),B(I),TGK(I),TMEM(I),THO(I),
     *          NTGR(I)
     READ 5135, NST(I),NRG(I),THD(I),D(I),TGC(I),A(I),TCA(I),CA0(I),
     *          BD(I),TGKD(I),CDS(I),VRT(I) $ELSE $II=I-NCPOPS
     READ 5140, N(I),P(II),INSTR(II),INSTP(II),INFSED(II),
     *          NTGR(I) $ENDIF
     DO 120 J=1,NTGR(I)
 120 READ 5150, IRCP(I,J),IRRG(I,J),TYPE(I,J),NCT(I,J),NT(I,J),
     *          STR(I,J),INSED(I,J)
 110 CONTINUE
     READ 5110, LTSTOP
*
     WRITE(6,6000) $WRITE(6,6005) NTPOPS,NFPOPS,NCPOPS,NCLS,MCTP1,
     *          NSTMX,NRGMX,NTGMX,IFLG1,IFLG2
     WRITE(6,6110) ((EQ(I),T(I)),I=1,SNTP)
     WRITE(6,6112) $WRITE(6,6114) $WRITE(6,6116) $WRITE(6,6118)
     DO 210 I=1,NTPOPS $IF(I.LE.NCPOPS) THEN
     WRITE(6,6130) N(I),C(I),TTH(I),B(I),TGK(I),TMEM(I),THO(I),
     *          NTGR(I)
     WRITE(6,6135) NST(I),NRG(I),THD(I),D(I),TGC(I),A(I),TCA(I),CA0(I),
     *          BD(I),TGKD(I),CDS(I),VRT(I) $ELSE $II=I-NCPOPS
     WRITE(6,6140) N(I),P(II),INSTR(II),INSTP(II),INFSED(II),
     *          NTGR(I) $ENDIF
     DO 220 J=1,NTGR(I)
 220 WRITE(6,6150) IRCP(I,J),IRRG(I,J),TYPE(I,J),NCT(I,J),NT(I,J),
     *          STR(I,J),INSED(I,J)
 210 CONTINUE
     WRITE(6,6160) LTSTOP
*
****      THIS SECTION INITIALIZES VARIABLES
*
     DO 310 I=1,NCPOPS
     DCTH(I)=EXP(-STEP/TTH(I)) $DCG(I)=EXP(-STEP/TGK(I))
     DO 310 J=1,NCLS
     E(I,J)=0. $TH(I,J)=THO(I) $S(I,J)=0. $GK(I,J)=0.
     DO 310 K=1,MCTP1 $DO 310 M=1,SNTP
     DO 310 II=1,NSTMX $DO 310 JJ=1,NRGP1M
     G(M,I,J,II,JJ,K)=0.
 310 CONTINUE
     DO 330 I=1,SNTP
 330 DCS(I)=EXP(-STEP/T(I))
     DO 340 I=1,NCPOPS $DO 340 J=1,NCLS
     DO 340 K=1,NST(I) $DO 340 M=1,NRG(I)
     V(I,J,K,M)=0. $GCA(I,J,K,M)=0. $CA(I,J,K,M)=0. $GKD(I,J,K,M)=0.
 340 CONTINUE
     DO 350 I=1,NCPOPS
     DGC(I)=EXP(-STEP/TGC(I)) $DCA(I)=EXP(-STEP/TCA(I))
 350 DGKD(I)=EXP(-STEP/TGKD(I))
*
****      THIS SECTION UPDATES STATE VARIABLES AT EACH TIME STEP
*
     DO 1000 L=1,LTSTOP
*
*
****              UPDATING THE CELLS
*
     DO 2000 I=1,NCPOPS
     NF(I)=0 $IF(IFLG1.GT.0) WRITE(7,7060) L,I
```

```
      DO 2900 J=1,N(I)
      CALL SVUPDT(I,J)
      IF(E(I,J).LT.TH(I,J)) GO TO 2850
      S(I,J)=1. $NF(I)=NF(I)+1 $IDF(NF(I))=J
      DO 2800 K=1,NTGR(I)
      CALL TRNSMT(I,K,J)
 2800 CONTINUE
 2850 CONTINUE
 2900 CONTINUE $IF(IFLG1.GT.0) WRITE(7,7060) (IDF(J),J=1,NF(I))
 2000 CONTINUE
*
****              UPDATING THE INPUT FIBERS
*
      DO 3000 I=1,NFPOPS
      II=I+NCPOPS $NF(II)=0 $IF(IFLG1.GT.0) WRITE(7,7060) L,II
      IF(L.LT.INSTR(I).OR.L.GE.INSTP(I)) GO TO 3000
      CALL RANSET(INFSED(I))
      DO 3900 J=1,N(II)
      IF(RANF(0.).GT.P(I)) GO TO 3850 $CALL RANGET(INFSED(I))
      NF(II)=NF(II)+1 $IDF(NF(II))=J
      DO 3800 K=1,NTGR(II)
      CALL TRNSMT(II,K,J)
 3800 CONTINUE $CALL RANSET(INFSED(I))
 3850 CONTINUE
 3900 CONTINUE $IF(IFLG1.GT.0) WRITE(7,7060) (IDF(J),J=1,NF(II))
      CALL RANGET(INFSED(I))
 3000 CONTINUE
*
      WRITE(6,6510) L,(NF(I),I=1,NTPOPS)
      WRITE(7,7510) L,(NF(I),I=1,NTPOPS)
*
      DO 1910 I=1,NCPOPS $DO 1910 J=1,NCLS $DO 1910 M=1,SNTP
      DO 1910 II=1,NSTMX $DO 1910 JJ=1,NRGP1M
      G(M,I,J,II,JJ,1)=G(M,I,J,II,JJ,1)*DCS(M)+G(M,I,J,II,JJ,2)
      DO 1920 K=2,MCTP1-1
 1920 G(M,I,J,II,JJ,K)=G(M,I,J,II,JJ,K+1) $G(M,I,J,II,JJ,MCTP1)=0.
 1910 CONTINUE
*
 1000 CONTINUE
*
****         THIS SECTION WRITES OUT ACTIVITY VARIABLES
*
*
      STOP
****             THESE ARE THE FORMATS
*
 5110 FORMAT(12I6)
 5120 FORMAT(12F6.2)
 5130 FORMAT(I6,6F6.2,I6)
 5135 FORMAT(2I6,10F6.2)
 5140 FORMAT(I6,F6.4,4I6)
 5150 FORMAT(5I6,F6.2,I6)
 6000 FORMAT(1H1)
 6005 FORMAT(5X,*NTPOPS,NFPOPS,NCPOPS,NCLS,MCTP1,NSTMX,NRGMX,NTGMX,IFLG1
     *,IFLG2=:*,10I7)
 6110 FORMAT(5X,*EQ,T=*,12F7.2)
 6112 FORMAT(//,5X,*HERE ARE, FOR EACH POPULATION:*,/,10X,*N,C,TTH,B,TG
     *K,TMEM,THO,NTGR*,/,20X,*AND*)
 6114 FORMAT(10X,*NST,NRG,THD,D,TGC,A,TCA,CA0,BD,TGKD,CDS,VRT*,/,20X,
     *     *OR*)
 6116 FORMAT(10X,*N,P,INSTR,INSTP,INFSED,NTGR*,/,20X,*AND*)
 6118 FORMAT(10X,*IRCP,IRRG,TYPE,NCT,NT,STR,INSED..FOR EACH TG*)
 6130 FORMAT(/,5X,I7,6F7.2,I7)
 6135 FORMAT(5X,2I7,10F7.2)
 6140 FORMAT(/,5X,I7,F7.4,4I7)
 6150 FORMAT(5I7,F7.2,I7)
 6160 FORMAT(5X,*LTSTOP=*,I7)
```

```
 6510 FORMAT(5X,15I7)
 7510 FORMAT(15I5)
 7060 FORMAT(20I4)
      END
*
      SUBROUTINE SVUPDT(KK,K)
*
*** THIS SUBROUTINE UPDATES THE STATE VARIABLES OF AN INDIVIDUAL NEURON
*   WITH SIMPLIFIED DENDRITIC TREE CONTAINING ACTIVE CALCIUM-RELATED
*** CONDUCTANCES.
*
      INTEGER SNTP
      PARAMETER SNTP=2,NCPOPS=3,NCLS=50,MCTP1=2,STEP=1.,EK=-10,ECA=50.,
     *          NSTMX=3,NRGMX=4,NRGP1M=5
      DIMENSION U(NRGMX)
      COMMON G(SNTP,NCPOPS,NCLS,NSTMX,NRGP1M,MCTP1),NST(NCPOPS),
     *          NRG(NCPOPS)
      COMMON/SV/E(NCPOPS,NCLS),TH(NCPOPS,NCLS),S(NCPOPS,NCLS),
     *          GK(NCPOPS,NCLS),B(NCPOPS),C(NCPOPS),TMEM(NCPOPS),
     *          THO(NCPOPS),DCG(NCPOPS),DCTH(NCPOPS),EQ(SNTP),
     *          V(NCPOPS,NCLS,NSTMX,NRGMX),GCA(NCPOPS,NCLS,NSTMX,
     *          NRGMX),CA(NCPOPS,NCLS,NSTMX,NRGMX),GKD(NCPOPS,NCLS,
     *          NSTMX,NRGMX),CDS(NCPOPS),VRT(NCPOPS),DGC(NCPOPS),
     *          THD(NCPOPS),D(NCPOPS),DCA(NCPOPS),A(NCPOPS),
     *          DGKD(NCPOPS),CA0(NCPOPS),BD(NCPOPS)
*
      GS=0. $SYN=0. $DO 10 I=1,SNTP
      GS=GS+G(I,KK,K,1,NRG(KK)+1,1)
      SYN=SYN+G(I,KK,K,1,NRG(KK)+1,1)*EQ(I)
   10 CONTINUE
      GK(KK,K)=GK(KK,K)*DCG(KK)+B(KK)*(1.-DCG(KK))*S(KK,K)
      GTOT=1.+GK(KK,K)+GS $DCE=EXP(-GTOT*STEP/TMEM(KK))
      E(KK,K)=E(KK,K)*DCE+(SYN+GK(KK,K)*EK)*(1.-DCE)/GTOT $DM=0.
      DO 980 I=1,NST(KK) $DO 920 J=1,NRG(KK)
      GT=1.+GCA(KK,K,I,J)+GKD(KK,K,I,J)
      SY=0. $DO 910 M=1,SNTP $GT=GT+G(M,KK,K,I,J,1)
  910 SY=SY+G(M,KK,K,I,J,1)*EQ(M) $Z=EXP(-GT*STEP/TMEM(KK))
  920 V(KK,K,I,J)=V(KK,K,I,J)*Z+(SY+GCA(KK,K,I,J)*ECA+GKD(KK,K,I,J)*EK)*
     *          (1.-Z)/GT
*
      DM=DM+V(KK,K,I,1)
      IF(NRG(KK).EQ.1) THEN
      U(1)=(V(KK,K,I,1)+E(KK,K)/CDS(KK))/(1.+1./CDS(KK))*
     *          (1.-VRT(KK))+VRT(KK)*V(KK,K,I,1) $ELSE
      U(1)=(V(KK,K,I,1)+V(KK,K,I,2)+E(KK,K)/CDS(KK))/(2.+1./CDS(KK))*
     *          (1.-VRT(KK))+VRT(KK)*V(KK,K,I,1)
      DO 930 J=2,NRG(KK)-1
  930 U(J)=(V(KK,K,I,J-1)+V(KK,K,I,J)+V(KK,K,I,J+1))/3.*(1.-VRT(KK))+
     *          VRT(KK)*V(KK,K,I,J)
      U(NRG(KK))=(V(KK,K,I,NRG(KK))+V(KK,K,I,NRG(KK)-1))/2.*
     *          (1.-VRT(KK))+VRT(KK)*V(KK,K,I,NRG(KK)) $ENDIF
      DO 940 J=1,NRG(KK)
  940 V(KK,K,I,J)=U(J)
*
      DO 950 J=1,NRG(KK)
      GCA(KK,K,I,J)=GCA(KK,K,I,J)*DGC(KK)
      IF(V(KK,K,I,J).GE.THD(KK)) GCA(KK,K,I,J)=GCA(KK,K,I,J)+
     *                D(KK)*(V(KK,K,I,J)-THD(KK))*(1.-DGC(KK))
      CA(KK,K,I,J)=CA(KK,K,I,J)*DCA(KK)+A(KK)*GCA(KK,K,I,J)*
     *          (1.-DCA(KK))
      GKD(KK,K,I,J)=GKD(KK,K,I,J)*DGKD(KK)
      IF(CA(KK,K,I,J).GE.CA0(KK)) GKD(KK,K,I,J)=GKD(KK,K,I,J)+
     *                BD(KK)*(1.-DGKD(KK))
  950 CONTINUE
*
  980 CONTINUE
*
```

```
      E(KK,K)=(E(KK,K)+CDS(KK)*DM)/(1.+CDS(KK)*FLOAT(NST(KK)))*
     *        (1.-VRT(KK))+VRT(KK)*E(KK,K)
      TH(KK,K)=THO(KK)+(TH(KK,K)-THO(KK))*DCTH(KK)+C(KK)*E(KK,K)*
     *        (1.-DCTH(KK))
      S(KK,K)=0.
      RETURN
      END
*
      SUBROUTINE TRNSMT(IS,L,K)
*
****      THIS SUBROUTINE PROJECTS SYNAPTIC ACTIVATION FROM A SENDING
****      FIBER TO ITS TARGETED RECEIVING CELLS.
*
      INTEGER TYPE,SNTP
      PARAMETER NTPOPS=5,NCPOPS=3,NCLS=50,MCTP1=2,SNTP=2,
     *          NSTMX=3,NRGMX=4,NRGP1M=5,NTGMX=5
      COMMON G(SNTP,NCPOPS,NCLS,NSTMX,NRGP1M,MCTP1),NST(NCPOPS),
     *          NRG(NCPOPS)
      COMMON/TR/NCT(NTPOPS,NTGMX),NT(NTPOPS,NTGMX),STR(NTPOPS,
     *          NTGMX),INSED(NTPOPS,NTGMX),
     *          N(NTPOPS),TYPE(NTPOPS,NTGMX),
     *          IRCP(NTPOPS,NTGMX),
     *          IRRG(NTPOPS,NTGMX)
*
      IR=IRCP(IS,L)
      CALL RANSET(K*INSED(IS,L))
      DO 100 J=1,NT(IS,L)
      NREC=INT(RANF(0.)*FLOAT(N(IR)))+1
      KST=INT(RANF(0.)*FLOAT(NST(IR)))+1
      IF(IRRG(IS,L).EQ.0) THEN
      KRG=INT(RANF(0.)*(NRG(IR)+1))+1
      ELSE $KRG=IRRG(IS,L) $ENDIF
      IF(KRG.EQ.NRG(IR)+1) KST=1
      KCT=INT(RANF(0.)*FLOAT(NCT(IS,L)))+2
      I=TYPE(IS,L)
      G(I,IR,NREC,KST,KRG,KCT)=G(I,IR,NREC,KST,KRG,KCT)+STR(IS,L)
  100 CONTINUE
*
      RETURN
      END
/EOR
      0       0
 70.    1.  -10.    5.
     50   0.   25.   20.     5.     5.    10.        2
      3    4  20.    2.2    5.     2.     5.    20.   100.    10.     8.        .1
      3    0    1     1     8    5.    1227
      2    0    1     1     8    5.    1936
     50   .5   25.   20.     5.     5.    10.        0
      3    4  20.    0.     5.     2.     5.    20.   100.    10.     8.        .1
     50   0.   25.   20.     5.     5.    10.        1
      3    4  60.    0.     5.     2.     5.    20.   100.    10.     8.        .1
      1    1    2     1    10   10.    3421
     50   .1         1   999  4165    1
      1    0    1     1    10    5.    8917
     50   .1        15    50  3315    1
      3    0    1     1    10    5.    6373
     80
/EOR
/EOI


/JOB
A224,CM=100000,TL=1000.
FAKE.
PDFV.
M77(PL=20000).
```

```
SETLIM,PR.
LGO,SYS713.
SAVE,SYS713.
EXIT.
SAVE,SYS713.
/EOR
      PROGRAM SYSTM71 (SYSTM,INPUT,OUTPUT,TAPE6=OUTPUT,TAPE7=SYSTM)
*
****    THIS PROGRAM SIMULATES THE ACTIVITY OF AN ARBITRARY SYSTEM OF
*       INTERCONNECTED SPATIALLY-ORGANIZED NEURAL NETWORKS ACTIVATED BY
*       AN ARBITRARY NUMBER OF SPATIALLY-ORGANIZED INPUT FIBER SYSTEMS.
****    ALL SYNAPTIC SYSTEMS ARE INDIVIDUALLY ADJUSTABLE.
*
      INTEGER TYPE,W,H,SNTP
      PARAMETER NTPOPS=5,NFPOPS=2,NCPOPS=3,NCLS=50,MCTP1=2,W=10,H=5,
     *       STEP=1.,SNTP=2,NSTMX=3,NRGMX=4,NRGP1M=5,NTGMX=5
      DIMENSION P(NFPOPS),INSTR(NFPOPS),INSTP(NFPOPS),INFSED(NFPOPS),
     *       IDF(NCLS),TTH(NCPOPS),TGK(NCPOPS),CHAR(2),ALINE(W,H),
     *       LINE(NTPOPS,W,H),T(SNTP),DCS(SNTP),NF(NTPOPS),
     *       NTGR(NTPOPS),TGC(NCPOPS),TCA(NCPOPS),TGKD(NCPOPS)
      COMMON G(SNTP,NCPOPS,NCLS,NSTMX,NRGP1M,MCTP1),NST(NCPOPS),
     *       NRG(NCPOPS)
      COMMON/SV/E(NCPOPS,NCLS),TH(NCPOPS,NCLS),S(NCPOPS,NCLS),
     *       GK(NCPOPS,NCLS),B(NCPOPS),C(NCPOPS),TMEM(NCPOPS),
     *       THO(NCPOPS),DCG(NCPOPS),DCTH(NCPOPS),EQ(SNTP),
     *       V(NCPOPS,NCLS,NSTMX,NRGMX),GCA(NCPOPS,NCLS,NSTMX,
     *       NRGMX),CA(NCPOPS,NCLS,NSTMX,NRGMX),GKD(NCPOPS,NCLS,
     *       NSTMX,NRGMX),CDS(NCPOPS),VRT(NCPOPS),DGC(NCPOPS),
     *       THD(NCPOPS),D(NCPOPS),DCA(NCPOPS),A(NCPOPS),
     *       DGKD(NCPOPS),CA0(NCPOPS),BD(NCPOPS)
      COMMON/TR/NCT(NTPOPS,NTGMX),NT(NTPOPS,NTGMX),STR(NTPOPS,
     *       NTGMX),INSED(NTPOPS,NTGMX),WTF(NTPOPS,NTGMX),
     *       HTF(NTPOPS,NTGMX),N(NTPOPS),TYPE(NTPOPS,NTGMX),
     *       NC(NTPOPS),NR(NTPOPS),IRCP(NTPOPS,NTGMX),
     *       IRRG(NTPOPS,NTGMX)
*
****        THIS SECTION READS AND WRITES THE INPUT PARAMETERS
*
      READ 5110, IFLG1,IFLG2
      READ 5120, ((EQ(I),T(I)),I=1,SNTP)
      DO 110 I=1,NTPOPS $IF(I.LE.NCPOPS) THEN
      READ 5130, N(I),C(I),TTH(I),B(I),TGK(I),TMEM(I),THO(I),
     *       NC(I),NR(I),NTGR(I)
      READ 5135, NST(I),NRG(I),THD(I),D(I),TGC(I),A(I),TCA(I),CA0(I),
     *       BD(I),TGKD(I),CDS(I),VRT(I) $ELSE $II=I-NCPOPS
      READ 5140, N(I),P(II),INSTR(II),INSTP(II),INFSED(II),
     *       NC(I),NR(I),NTGR(I) $ENDIF
      DO 120 J=1,NTGR(I)
  120 READ 5150, IRCP(I,J),IRRG(I,J),TYPE(I,J),NCT(I,J),NT(I,J),
     *          STR(I,J),INSED(I,J),WTF(I,J),HTF(I,J)
  110 CONTINUE
      READ 5110, LTSTOP
*
      WRITE(6,6000) $WRITE(6,6005) NTPOPS,NFPOPS,NCPOPS,NCLS,MCTP1,
     *       NSTMX,NRGMX,NTGMX,IFLG1,IFLG2
      WRITE(6,6110) ((EQ(I),T(I)),I=1,SNTP)
      WRITE(6,6112) $WRITE(6,6114) $WRITE(6,6116) $WRITE(6,6118)
      DO 210 I=1,NTPOPS $IF(I.LE.NCPOPS) THEN
      WRITE(6,6130) N(I),C(I),TTH(I),B(I),TGK(I),TMEM(I),THO(I),
     *       NC(I),NR(I),NTGR(I)
      WRITE(6,6135) NST(I),NRG(I),THD(I),D(I),TGC(I),A(I),TCA(I),CA0(I),
     *       BD(I),TGKD(I),CDS(I),VRT(I) $ELSE $II=I-NCPOPS
      WRITE(6,6140) N(I),P(II),INSTR(II),INSTP(II),INFSED(II),
     *       NC(I),NR(I),NTGR(I) $ENDIF
      DO 220 J=1,NTGR(I)
  220 WRITE(6,6150) IRCP(I,J),IRRG(I,J),TYPE(I,J),NCT(I,J),NT(I,J),
     *          STR(I,J),INSED(I,J),WTF(I,J),HTF(I,J)
```

```
 210 CONTINUE
     WRITE(6,6160) LTSTOP
*
**** THIS SECTION INITIALIZES VARIABLES
*
     CHAR(1)=1H  $CHAR(2)=1H*
     DO 310 I=1,NCPOPS
     DCTH(I)=EXP(-STEP/TTH(I))  $DCG(I)=EXP(-STEP/TGK(I))
     DO 310 J=1,NCLS
     E(I,J)=0.  $TH(I,J)=THO(I)  $S(I,J)=0.  $GK(I,J)=0.
     DO 310 K=1,MCTP1 $DO 310 M=1,SNTP
     DO 310 II=1,NSTMX $DO 310 JJ=1,NRGP1M
     G(M,I,J,II,JJ,K)=0.
 310 CONTINUE
     DO 320 I=1,NTPOPS $DO 320 J=1,W $DO 320 K=1,H
 320 LINE(I,J,K)=0
     DO 330 I=1,SNTP
 330 DCS(I)=EXP(-STEP/T(I))
     DO 340 I=1,NCPOPS $DO 340 J=1,NCLS
     DO 340 K=1,NST(I) $DO 340 M=1,NRG(I)
     V(I,J,K,M)=0.  $GCA(I,J,K,M)=0.  $CA(I,J,K,M)=0.  $GKD(I,J,K,M)=0.
 340 CONTINUE
     DO 350 I=1,NCPOPS
     DGC(I)=EXP(-STEP/TGC(I))  $DCA(I)=EXP(-STEP/TCA(I))
 350 DGKD(I)=EXP(-STEP/TGKD(I))
*
**** THIS SECTION UPDATES STATE VARIABLES AT EACH TIME STEP
*
     DO 1000 L=1,LTSTOP
*
*
**** UPDATING THE CELLS
*
     DO 2000 I=1,NCPOPS
     NF(I)=0 $IF(IFLG1.GT.0) WRITE(7,7060) L,I
     DO 2010 J=1,W $DO 2010 K=1,H
2010 ALINE(J,K)=CHAR(1)
     DO 2900 J=1,N(I)
     CALL SVUPDT(I,J)
     IF(E(I,J).LT.TH(I,J)) GO TO 2850
     S(I,J)=1.  $NF(I)=NF(I)+1 $IDF(NF(I))=J
     JX=INT(FLOAT(MOD((J-1),NC(I))+1)*FLOAT(W)/FLOAT(NC(I)))
     JY=INT(FLOAT((J-1)/NC(I)+1)*FLOAT(H)/FLOAT(NR(I)))
     ALINE(JX,JY)=CHAR(2)  $LINE(I,JX,JY)=LINE(I,JX,JY)+1
     DO 2800 K=1,NTGR(I)
     CALL TRNSMT(I,K,J)
2800 CONTINUE
2850 CONTINUE
2900 CONTINUE $IF(IFLG1.GT.0) WRITE(7,7060) (IDF(J),J=1,NF(I))
     IF(IFLG2.GT.0) THEN $WRITE(6,6122) $DO 2700 JY=1,H
2700 WRITE(6,6060) (ALINE(JX,H+1-JY),JX=1,W) $ENDIF
2000 CONTINUE
*
**** UPDATING THE INPUT FIBERS
*
     DO 3000 I=1,NFPOPS
     II=I+NCPOPS $NF(II)=0 $IF(IFLG1.GT.0) WRITE(7,7060) L,II
     IF(L.LT.INSTR(I).OR.L.GE.INSTP(I)) GO TO 3000
     IF(IFLG2.GT.0) THEN $DO 3010 J=1,W $DO 3010 K=1,H
3010 ALINE(J,K)=CHAR(1) $ENDIF
     CALL RANSET(INFSED(I))
     DO 3900 J=1,N(II)
     IF(RANF(0.).GT.P(I)) GO TO 3850 $CALL RANGET(INFSED(I))
     NF(II)=NF(II)+1 $IDF(NF(II))=J
     JX=INT(FLOAT(MOD((J-1),NC(I))+1)*FLOAT(W)/FLOAT(NC(I)))
     JY=INT(FLOAT((J-1)/NC(I)+1)*FLOAT(H)/FLOAT(NR(I)))
     ALINE(JX,JY)=CHAR(2)  $LINE(II,JX,JY)=LINE(II,JX,JY)+1
```

```
      DO 3800 K=1,NTGR(II)
      CALL TRNSMT(II,K,J)
 3800 CONTINUE $CALL RANSET(INFSED(I))
 3850 CONTINUE
 3900 CONTINUE $IF(IFLG1.GT.0) WRITE(7,7060) (IDF(J),J=1,NF(II))
      IF(IFLG2.GT.0) THEN $WRITE(6,6122) $DO 3700 JY=1,H
 3700 WRITE(6,6060) (ALINE(JX,H+1-JY),JX=1,W) $ENDIF
      CALL RANGET(INFSED(I))
 3000 CONTINUE
*
      WRITE(6,6510) L,(NF(I),I=1,NTPOPS)
      WRITE(7,7510) L,(NF(I),I=1,NTPOPS)
*
      DO 1910 I=1,NCPOPS $DO 1910 J=1,NCLS $DO 1910 M=1,SNTP
      DO 1910 II=1,NSTMX $DO 1910 JJ=1,NRGP1M
      G(M,I,J,II,JJ,1)=G(M,I,J,II,JJ,1)*DCS(M)+G(M,I,J,II,JJ,2)
      DO 1920 K=2,MCTP1-1
 1920 G(M,I,J,II,JJ,K)=G(M,I,J,II,JJ,K+1) $G(M,I,J,II,JJ,MCTP1)=0.
 1910 CONTINUE
*
 1000 CONTINUE
*
****        THIS SECTION WRITES OUT ACTIVITY VARIABLES
*
      DO 4700 I=1,NTPOPS $WRITE(6,6122) $DO 4710 JY=1,H
 4710 WRITE(6,6070) (LINE(I,JX,H+1-JY),JX=1,W)
 4700 CONTINUE
*
      STOP
****        THESE ARE THE FORMATS
*
 5110 FORMAT(12I6)
 5120 FORMAT(12F6.2)
 5130 FORMAT(I6,6F6.2,3I6)
 5135 FORMAT(2I6,10F6.2)
 5140 FORMAT(I6,F6.4,6I6)
 5150 FORMAT(5I6,F6.2,I6,2F6.2)
 6000 FORMAT(1H1)
 6005 FORMAT(5X,*NTPOPS,NFPOPS,NCPOPS,NCLS,MCTP1,NSTMX,NRGMX,NTGMX,IFLG1
     *,IFLG2=:*,10I7)
 6110 FORMAT(5X,*EQ,T=*,12F7.2)
 6112 FORMAT(//,5X,*HERE ARE, FOR EACH POPULATION:,*,/,10X,*N,C,TTH,B,TG
     *K,TMEM,THO,NC,NR,NTGR*,/,20X,*AND*)
 6114 FORMAT(10X,*NST,NRG,THD,D,TGC,A,TCA,CA0,BD,TGKD,CDS,VRT*,/,20X,
     *    *OR*)
 6116 FORMAT(10X,*N,P,INSTR,INSTP,INFSED,NC,NR,NTGR*,/,20X,*AND*)
 6118 FORMAT(10X,*IRCP,IRRG,TYPE,NCT,NT,STR,INSED,WTF,HTF..FOR EACH TG*)
 6122 FORMAT(//)
 6130 FORMAT(/,5X,I7,6F7.2,3I7)
 6135 FORMAT(5X,2I7,10F7.2)
 6140 FORMAT(/,5X,I7,F7.4,6I7)
 6150 FORMAT(5I7,F7.2,I7,2F7.2)
 6160 FORMAT(5X,*LTSTOP=*,I7)
 6060 FORMAT(5X,80A1)
 6070 FORMAT(2X,40I3)
 6510 FORMAT(5X,15I7)
 7510 FORMAT(15I5)
 7060 FORMAT(20I4)
      END
*
      SUBROUTINE SVUPDT(KK,K)
*
***   THIS SUBROUTINE UPDATES THE STATE VARIABLES OF AN INDIVIDUAL NEURON
*     WITH SIMPLIFIED DENDRITIC TREE CONTAINING ACTIVE CALCIUM-RELATED
***   CONDUCTANCES.
*
      INTEGER SNTP
```

```
      PARAMETER SNTP=2,NCPOPS=3,NCLS=50,MCTP1=2,STEP=1.,EK=-10,ECA=50.,
     *         NSTMX=3,NRGMX=4,NRGP1M=5
      DIMENSION U(NRGMX)
      COMMON G(SNTP,NCPOPS,NCLS,NSTMX,NRGP1M,MCTP1),NST(NCPOPS),
     *      NRG(NCPOPS)
      COMMON/SV/E(NCPOPS,NCLS),TH(NCPOPS,NCLS),S(NCPOPS,NCLS),
     *      GK(NCPOPS,NCLS),B(NCPOPS),C(NCPOPS),TMEM(NCPOPS),
     *      THO(NCPOPS),DCG(NCPOPS),DCTH(NCPOPS),EQ(SNTP),
     *      V(NCPOPS,NCLS,NSTMX,NRGMX),GCA(NCPOPS,NCLS,NSTMX,
     *      NRGMX),CA(NCPOPS,NCLS,NSTMX,NRGMX),GKD(NCPOPS,NCLS,
     *      NSTMX,NRGMX),CDS(NCPOPS),VRT(NCPOPS),DGC(NCPOPS),
     *      THD(NCPOPS),D(NCPOPS),DCA(NCPOPS),A(NCPOPS),
     *      DGKD(NCPOPS),CA0(NCPOPS),BD(NCPOPS)
*
      GS=0. $SYN=0. $DO 10 I=1,SNTP
      GS=GS+G(I,KK,K,1,NRG(KK)+1,1)
      SYN=SYN+G(I,KK,K,1,NRG(KK)+1,1)*EQ(I)
   10 CONTINUE
      GK(KK,K)=GK(KK,K)*DCG(KK)+B(KK)*(1.-DCG(KK))*S(KK,K)
      GTOT=1.+GK(KK,K)+GS $DCE=EXP(-GTOT*STEP/TMEM(KK))
      E(KK,K)=E(KK,K)*DCE+(SYN+GK(KK,K)*EK)*(1.-DCE)/GTOT $DM=0.
      DO 980 I=1,NST(KK) $DO 920 J=1,NRG(KK)
      GT=1.+GCA(KK,K,I,J)+GKD(KK,K,I,J)
      SY=0. $DO 910 M=1,SNTP $GT=GT+G(M,KK,K,I,J,1)
  910 SY=SY+G(M,KK,K,I,J,1)*EQ(M) $Z=EXP(-GT*STEP/TMEM(KK))
  920 V(KK,K,I,J)=V(KK,K,I,J)*Z+(SY+GCA(KK,K,I,J)*ECA+GKD(KK,K,I,J)*EK)*
     *      (1.-Z)/GT
*
      DM=DM+V(KK,K,I,1)
      IF(NRG(KK).EQ.1) THEN
      U(1)=(V(KK,K,I,1)+E(KK,K)/CDS(KK))/(1.+1./CDS(KK))*
     *      (1.-VRT(KK))+VRT(KK)*V(KK,K,I,1) $ELSE
      U(1)=(V(KK,K,I,1)+V(KK,K,I,2)+E(KK,K)/CDS(KK))/(2.+1./CDS(KK))*
     *      (1.-VRT(KK))+VRT(KK)*V(KK,K,I,1)
      DO 930 J=2,NRG(KK)-1
  930 U(J)=(V(KK,K,I,J-1)+V(KK,K,I,J)+V(KK,K,I,J+1))/3.*(1.-VRT(KK))+
     *      VRT(KK)*V(KK,K,I,J)
      U(NRG(KK))=(V(KK,K,I,NRG(KK))+V(KK,K,I,NRG(KK)-1))/2.*
     *      (1.-VRT(KK))+VRT(KK)*V(KK,K,I,NRG(KK)) $ENDIF
      DO 940 J=1,NRG(KK)
  940 V(KK,K,I,J)=U(J)
*
      DO 950 J=1,NRG(KK)
      GCA(KK,K,I,J)=GCA(KK,K,I,J)*DGC(KK)
      IF(V(KK,K,I,J).GE.THD(KK)) GCA(KK,K,I,J)=GCA(KK,K,I,J)+
     *      D(KK)*(V(KK,K,I,J)-THD(KK))*(1.-DGC(KK))
      CA(KK,K,I,J)=CA(KK,K,I,J)*DCA(KK)+A(KK)*GCA(KK,K,I,J)*
     *      (1.-DCA(KK))
      GKD(KK,K,I,J)=GKD(KK,K,I,J)*DGKD(KK)
      IF(CA(KK,K,I,J).GE.CA0(KK)) GKD(KK,K,I,J)=GKD(KK,K,I,J)+
     *                      BD(KK)*(1.-DGKD(KK))
  950 CONTINUE
*
  980 CONTINUE
*
      E(KK,K)=(E(KK,K)+CDS(KK)*DM)/(1.+CDS(KK)*FLOAT(NST(KK)))*
     *      (1.-VRT(KK))+VRT(KK)*E(KK,K)
      TH(KK,K)=THO(KK)+(TH(KK,K)-THO(KK))*DCTH(KK)+C(KK)*E(KK,K)*
     *      (1.-DCTH(KK))
      S(KK,K)=0.
      RETURN
      END
*
      SUBROUTINE TRNSMT(IS,L,K)
*
****      THIS SUBROUTINE PROJECTS SYNAPTIC ACTIVATION FROM A SENDING
****      FIBER TO ITS TARGETED RECEIVING CELLS.
```

```
*
      INTEGER TYPE,W,H,SNTP
      PARAMETER NTPOPS=5,NCPOPS=3,NCLS=50,MCTP1=2,W=10,H=5,SNTP=2,
     *         NSTMX=3,NRGMX=4,NRGP1M=5,NTGMX=5
      COMMON G(SNTP,NCPOPS,NCLS,NSTMX,NRGP1M,MCTP1),NST(NCPOPS),
     *       NRG(NCPOPS)
      COMMON/TR/NCT(NTPOPS,NTGMX),NT(NTPOPS,NTGMX),STR(NTPOPS,
     *         NTGMX),INSED(NTPOPS,NTGMX),WTF(NTPOPS,NTGMX),
     *         HTF(NTPOPS,NTGMX),N(NTPOPS),TYPE(NTPOPS,NTGMX),
     *         NC(NTPOPS),NR(NTPOPS),IRCP(NTPOPS,NTGMX),
     *         IRRG(NTPOPS,NTGMX)
*
      IR=IRCP(IS,L)
      XS=FLOAT(MOD(K,NC(IS)))*FLOAT(W)/FLOAT(NC(IS))
      IF(XS.EQ.0.) XS=FLOAT(W)
      YS=FLOAT(K/NC(IS)+1)*FLOAT(H)/FLOAT(NR(IS))
      CALL RANSET(K*INSED(IS,L))
      DO 100 J=1,NT(IS,L)
      XR=XS+(RANF(0.)-.5)*WTF(IS,L) $YR=YS+(RANF(0.)-.5)*HTF(IS,L)
      ICR=INT(FLOAT(NC(IR))*(XR-.5)/FLOAT(W))+1
      IRR=INT(FLOAT(NR(IR))*(YR-.5)/FLOAT(H))+1
      IF(ICR.LE.0) ICR=NC(IR)+ICR
      IF(ICR.GT.NC(IR)) ICR=ICR-NC(IR)
      IF(IRR.LE.0) IRR=NR(IR)+IRR
      IF(IRR.GT.NR(IR)) IRR=IRR-NR(IR)
      NREC=NC(IR)*(IRR-1)+ICR
      KST=INT(RANF(0.)*FLOAT(NST(IR)))+1
      IF(IRRG(IS,L).EQ.0) THEN
      KRG=INT(RANF(0.)*(NRG(IR)+1))+1
      ELSE $KRG=IRRG(IS,L) $ENDIF
      IF(KRG.EQ.NRG(IR)+1) KST=1
      KCT=INT(RANF(0.)*FLOAT(NCT(IS,L)))+2
      I=TYPE(IS,L)
      G(I,IR,NREC,KST,KRG,KCT)=G(I,IR,NREC,KST,KRG,KCT)+STR(IS,L)
  100 CONTINUE
*
      RETURN
      END
/EOR
      0       0
 70.    1.   -10.     5.
   50  0.   25.    20.     5.    5.    10.      10      5     2
    3     4 20.     2.2    5.    2.    5.    20.  100.   10.      8.       .1
    3     0    1     1     8 5.      1227  9.    5.
    2     0    1     1     8 5.      1936  9.    5.
   50  .5   25.    20.     5.    5.    10.      10      5     0
    3     4 20.     0.     5.    2.    5.    20.  100.   10.      8.       .1
   50  0.   25.    20.     5.    5.    10.      10      5     1
    3     4 60.     0.     5.    2.    5.    20.  100.   10.      8.       .1
    1     1    2     1    10 10.      3421 20.   10.
   50 .1        1   999  4165    10      5     1
    1     0    1     1    10 5.      8917 20.   10.
   50 .1       15    50  3315    10      5     1
    3     0    1     1    10 5.      6373 20.   10.
   80
/EOR
/EOI


/JOB
A224,CM=100000,TL=1000.
FAKE.
PDFV.
M77.
GET,STR000.
LGO,STR000,LN702,STR702.
```

```
SAVE,LN702,STR702.
EXIT.
SAVE,LN702,STR702.
/EOR
      PROGRAM LRSYS70 (ANIN,LRSYS,ANOUT,INPUT,OUTPUT,
     *           TAPE3=ANIN,TAPE6=OUTPUT,TAPE7=LRSYS,TAPE8=ANOUT)
*
****       THIS PROGRAM SIMULATES HEBBIAN LEARNING AT THE SYNAPSES OF AN
*          ARBITRARY SYSTEM OF INTERACTING NEURONAL POPULATIONS
*          ACTIVATED BY AN ARBITRARY NUMBER OF INPUT FIBER SYSTEMS.  ALL
****       SYNAPTIC SYSTEMS ARE INDIVIDUALLY ADJUSTABLE.
*
      INTEGER TYPE,ACIT,OVR,SNTP
      PARAMETER NTPOPS=5,NFPOPS=2,NCPOPS=3,NCLS=50,MCTP1=2,
     *     NJN=6,NLJN=1,NTR=20,NRNGE=10,MRNGE=1000,STEP=1.,SNTP=3,
     *     NSTMX=3,NRGMX=3,NRGP1M=4
      DIMENSION P(NFPOPS),INSTP(NFPOPS),INFSED(NFPOPS),NF(NTPOPS),
     *     IDF(NCLS),TTH(NCPOPS),TGK(NCPOPS),IFLG(NCPOPS),
     *     STRMX(NLJN),DLSTR(NLJN),LRN(NJN),IS(NJN),IR(NJN),
     *     T(SNTP),DCS(SNTP),INSTR(NFPOPS),LG1(NLJN),
     *     TGC(NCPOPS),TCA(NCPOPS),TGKD(NCPOPS)
      COMMON G(SNTP,NCPOPS,NCLS,NSTMX,NRGP1M,MCTP1),NST(NCPOPS),
     *     NRG(NCPOPS)
      COMMON/SV/E(NCPOPS,NCLS),TH(NCPOPS,NCLS),S(NCPOPS,NCLS),
     *     GK(NCPOPS,NCLS),B(NCPOPS),C(NCPOPS),TMEM(NCPOPS),
     *     THO(NCPOPS),DCG(NCPOPS),DCTH(NCPOPS),EQ(SNTP),
     *     V(NCPOPS,NCLS,NSTMX,NRGMX),GCA(NCPOPS,NCLS,NSTMX,
     *     NRGMX),CA(NCPOPS,NCLS,NSTMX,NRGMX),GKD(NCPOPS,NCLS,
     *     NSTMX,NRGMX),CDS(NCPOPS),VRT(NCPOPS),DGC(NCPOPS),
     *     THD(NCPOPS),D(NCPOPS),DCA(NCPOPS),A(NCPOPS),
     *     DGKD(NCPOPS),CA0(NCPOPS),BD(NCPOPS)
      COMMON/TR/TYPE(NJN),NT(NJN),NCT(NJN),STR0(NJN),
     *     INSED(NJN),LJNN(NTPOPS,NCPOPS),MLAG(NLJN),IRRG(NJN),
     *     ACIT(NLJN,NCLS,NRNGE),OVR(NLJN,MRNGE),IVR(NLJN),
     *     STR(NLJN,NCLS,NTR),N(NTPOPS),
     *     JNN(NTPOPS,NCPOPS),L
*
****       THIS SECTION READS AND WRITES THE INPUT PARAMETERS
*
      READ 5100, IFLG1,IFLG2
      READ 5105, ((EQ(I),T(I)),I=1,SNTP)
      DO 110 I=1,NCPOPS
      READ 5110, N(I),C(I),TTH(I),B(I),TGK(I),TMEM(I),THO(I)
  110 READ 5115, NST(I),NRG(I),THD(I),D(I),TGC(I),A(I),TCA(I),CA0(I),
     *     BD(I),TGKD(I),CDS(I),VRT(I)
      DO 120 I=1,NFPOPS $II=NCPOPS+I
  120 READ 5120, N(II),P(I),INSTR(I),INSTP(I),INFSED(I)
      DO 128 I=1,NCPOPS $IFLG(I)=0 $DO 128 J=1,NTPOPS
      JNN(J,I)=0 $LJNN(J,I)=0
  128 CONTINUE $J=0 $DO 130 I=1,NJN
      READ 5130, TYPE(I),LRN(I),IS(I),IR(I),IRRG(I),NT(I),NCT(I),STR0(I)
     *     ,INSED(I)
      JNN(IS(I),IR(I))=1
      IF(LRN(I).GT.0) THEN $J=J+1 $LJNN(IS(I),IR(I))=J $IFLG(IR(I))=1
      READ 5140, MLAG(J),STRMX(J),DLSTR(J) $ENDIF
  130 CONTINUE
      READ 5100, LTSTOP
*
      WRITE(6,6000) $WRITE(6,6100) NTPOPS,NFPOPS,NCPOPS,NCLS,MCTP1
      WRITE(6,6102) NJN,NLJN,NTR,NRNGE,MRNGE,IFLG1
      WRITE(6,6105) ((EQ(I),T(I)),I=1,SNTP)
      WRITE(6,6110) $DO 111 I=1,NCPOPS
      WRITE(6,6112) N(I),C(I),TTH(I),B(I),TGK(I),TMEM(I),THO(I)
  111 WRITE(6,6115) NST(I),NRG(I),THD(I),D(I),TGC(I),A(I),TCA(I),CA0(I),
     *     BD(I),TGKD(I),CDS(I),VRT(I)
      WRITE(6,6120) $DO 121 I=1,NFPOPS $II=NCPOPS+I
  121 WRITE(6,6122) N(II),P(I),INSTR(I),INSTP(I),INFSED(I)
```

```
      WRITE(6,6130) $J=0 $DO 131 I=1,NJN
      WRITE(6,6132) TYPE(I),LRN(I),IS(I),IR(I),IRRG(I),NT(I),NCT(I),
     *              STR0(I),INSED(I)
      IF(LRN(I).GT.0) THEN $J=J+1
      WRITE(6,6140) MLAG(J),STRMX(J),DLSTR(J) $ENDIF
  131 CONTINUE
      WRITE(6,6200) LTSTOP $WRITE(6,6000)
*
****          THIS SECTION INITIALIZES VARIABLES
*
      DO 310 I=1,NCPOPS
      DCTH(I)=EXP(-STEP/TTH(I)) $DCG(I)=EXP(-STEP/TGK(I))
      DO 310 J=1,NCLS
      E(I,J)=0. $TH(I,J)=THO(I) $S(I,J)=0. $GK(I,J)=0.
      DO 310 K=1,MCTP1 $DO 310 M=1,SNTP
      DO 310 II=1,NSTMX $DO 310 JJ=1,NRGP1M
      G(M,I,J,II,JJ,K)=0.
  310 CONTINUE
      DO 330 I=1,NLJN $IVR(I)=0 $LG1(I)=10000000*(MLAG(I)+1)
      DO 332 J=1,NCLS $DO 332 K=1,NRNGE
  332 ACIT(I,J,K)=0 $DO 334 K=1,MRNGE
  334 OVR(I,K)=0
  330 CONTINUE
      DO 340 I=1,SNTP
  340 DCS(I)=EXP(-STEP/T(I))
      DO 350 I=1,NCPOPS $DO 350 J=1,NCLS
      DO 350 K=1,NST(I) $DO 350 M=1,NRG(I)
      V(I,J,K,M)=0. $GCA(I,J,K,M)=0. $CA(I,J,K,M)=0. $GKD(I,J,K,M)=0.
  350 CONTINUE
      DO 360 I=1,NCPOPS
      DGC(I)=EXP(-STEP/TGC(I)) $DCA(I)=EXP(-STEP/TCA(I))
  360 DGKD(I)=EXP(-STEP/TGKD(I))
      IF(IFLG2.EQ.0) THEN $DO 320 I=1,NJN $IF(LRN(I).EQ.0) GO TO 320
      II=LJNN(IS(I),IR(I)) $DO 320 J=1,NCLS $DO 320 K=1,NTR
      STR(II,J,K)=STR0(I)
  320 CONTINUE $ELSE
      READ(3,8900) (((STR(I,J,K),K=1,NTR),J=1,NCLS),I=1,NLJN) $ENDIF
*
****          THIS SECTION UPDATES STATE VARIABLES AT EACH TIME STEP
*
      DO 1000 L=1,LTSTOP
*
*
****                  UPDATING THE CELLS
*
      DO 2000 I=1,NCPOPS
      NF(I)=0
      DO 2900 J=1,N(I)
      CALL SVUPDT(I,J)
      IF(E(I,J).LT.TH(I,J)) GO TO 2850
      S(I,J)=1. $NF(I)=NF(I)+1 $IDF(NF(I))=J
*
      IF(IFLG(I).EQ.0) GO TO 2699 $DO 2700 LS=1,NTPOPS
      IF(LJNN(LS,I).EQ.0) GO TO 2700 $LJ=LJNN(LS,I)
      DO 2500 K=1,NRNGE $IF(ACIT(LJ,J,K).EQ.0) GO TO 2499
      IF(ACIT(LJ,J,K).GT.LG1(LJ)) GO TO 2500
      ID=MOD(ACIT(LJ,J,K),10000000) $JC=ID/1000 $JT=ID-JC*1000
      STR(LJ,JC,JT)=STR(LJ,JC,JT)+DLSTR(LJ)*
     *                     (STRMX(LJ)-STR(LJ,JC,JT))
 2500 CONTINUE
 2499 CONTINUE $JUP=IVR(LJ) $DO 2600 K=1,JUP
      IF(OVR(LJ,K)/10000000000.NE.J) GO TO 2600
      IF(MOD(OVR(LJ,K),10000000000).GT.LG1(LJ)) GO TO 2600
      ID=MOD(OVR(LJ,K),10000000) $JC=ID/1000 $JT=ID-JC*1000
      STR(LJ,JC,JT)=STR(LJ,JC,JT)+DLSTR(LJ)*
     *                     (STRMX(LJ)-STR(LJ,JC,JT))
 2600 CONTINUE
```

```
2700 CONTINUE
*
2699 CONTINUE
     DO 2800 K=1,NCPOPS
     IF(JNN(I,K).GT.0) CALL TRNSMT(I,K,J)
2800 CONTINUE
2850 CONTINUE
2900 CONTINUE $IF(IFLG1.GT.0) WRITE(7,7060) (L,I,NF(I),
    *          (IDF(J),J=1,NF(I)))
2000 CONTINUE
*
****          UPDATING THE INPUT FIBERS
*
     DO 3000 I=1,NFPOPS
     II=I+NCPOPS $NF(II)=0
     IF(L.LT.INSTR(I).OR.L.GE.INSTP(I)) GO TO 3000
     CALL RANSET(INFSED(I))
     DO 3900 J=1,N(II)
     IF(RANF(0.).GT.P(I)) GO TO 3850 $CALL RANGET(INFSED(I))
     NF(II)=NF(II)+1 $IDF(NF(II))=J
     DO 3800 K=1,NCPOPS
     IF(JNN(II,K).GT.0) CALL TRNSMT(II,K,J)
3800 CONTINUE $CALL RANSET(INFSED(I))
3850 CONTINUE
3900 CONTINUE $IF(IFLG1.GT.0) WRITE(7,7060) (L,II,NF(II),
    *          (IDF(J),J=1,NF(II)))
     CALL RANGET(INFSED(I))
3000 CONTINUE
*
     WRITE(6,6510) L,(NF(I),I=1,NTPOPS)
     WRITE(7,7510) L,(NF(I),I=1,NTPOPS)
*
     DO 1910 I=1,NCPOPS $DO 1910 J=1,NCLS $DO 1910 M=1,SNTP
     DO 1910 II=1,NSTMX $DO 1910 JJ=1,NRGP1M
     G(M,I,J,II,JJ,1)=G(M,I,J,II,JJ,1)*DCS(M)+G(M,I,J,II,JJ,2)
     DO 1920 K=2,MCTP1-1
1920 G(M,I,J,II,JJ,K)=G(M,I,J,II,JJ,K+1) $G(M,I,J,II,JJ,MCTP1)=0.
1910 CONTINUE
     DO 1600 LJ=1,NLJN $JUP=IVR(LJ) $DO 1700 J=1,JUP
1710 IF(OVR(LJ,J).EQ.0) GO TO 1699
     IF(MOD(OVR(LJ,J),10000000000).GT.10000000) GO TO 1700
     IVR(LJ)=IVR(LJ)-1 $DO 1760 J1=J,IVR(LJ)
1760 OVR(LJ,J1)=OVR(LJ,J1+1) $OVR(LJ,IVR(LJ)+1)=0 $GO TO 1710
1700 OVR(LJ,J)=OVR(LJ,J)-10000000
1699 CONTINUE
     DO 1800 I=1,NCLS $DO 1850 J=1,NRNGE
1810 IF(ACIT(LJ,I,J).EQ.0) GO TO  1849
     IF(ACIT(LJ,I,J).GT.10000000) GO TO 1850 $DO 1860 J1=1,NRNGE-1
1860 ACIT(LJ,I,J1)=ACIT(LJ,I,J1+1) $ACIT(LJ,I,NRNGE)=0 $GO TO 1810
1850 ACIT(LJ,I,J)=ACIT(LJ,I,J)-10000000
1849 CONTINUE
1800 CONTINUE
1600 CONTINUE
*
1000 CONTINUE
*
****          THIS SECTION WRITES OUT ACTIVITY VARIABLES
*
     DO 4100 JS=1,NTPOPS $DO 4100 JR=1,NCPOPS
     IF(LJNN(JS,JR).EQ.0) GO TO 4100 $WRITE(6,6040)
     LJ=LJNN(JS,JR)
     WRITE(6,6900) ((STR(LJ,I,K),K=1,NTR),I=1,NCLS)
     WRITE(8,8900) ((STR(LJ,I,K),K=1,NTR),I=1,NCLS)
4100 CONTINUE
*
     STOP
****          THESE ARE THE FORMATS
```

```
*
 5100 FORMAT(12I6)
 5105 FORMAT(12F6.2)
 5110 FORMAT(I6,6F6.2)
 5115 FORMAT(2I6,10F6.2)
 5120 FORMAT(I6,F6.2,3I6)
 5130 FORMAT(7I6,F6.2,I6)
 5140 FORMAT(I6,2F6.2)
 6000 FORMAT(1H1)
 6040 FORMAT(///)
 6100 FORMAT(5X,*NTPOPS,NCPOPS,NFPOPS,NCLS,MCTP1=*,7I7)
 6102 FORMAT(5X,*NJN,NLJN,NTR,NRNGE,MRNGE,IFLG1=*,6I7)
 6105 FORMAT(5X,*EQ,T=*,12F7.2)
 6110 FORMAT(5X,*CELLS: N,C,TTH,B,TGK,TMEM,THO,NC,NR=*,
      *    /,*NST,NRG,THD,D,TGC,A,TCA,CA0,BD,TGKD,CDS,VRT=*,//)
 6112 FORMAT(9X,I7,6F7.2)
 6115 FORMAT(5X,2I7,10F7.2)
 6120 FORMAT(///,5X,*FIBERS: N,P,INSTR,INSTP,INFSED=*,//)
 6122 FORMAT(5X,I7,F7.2,3I7)
 6130 FORMAT(///,5X,*JNCTNS: TYPE,LRN,IS,IR,IRRG,NT,NCT,STR0,*
      *    ,*INSED=*,/10X,*MLAG,STRMX,DLSTR=*,//)
 6132 FORMAT(5X,7I7,F7.2,I7)
 6140 FORMAT(10X,I7,2F7.2)
 6200 FORMAT(5X,*LTSTOP=*,I7)
 6510 FORMAT(5X,15I7)
 6900 FORMAT(5X,20F6.3)
 7510 FORMAT(15I5)
 7060 FORMAT(20I4)
 8900 FORMAT(F6.3)
      END
*
      SUBROUTINE SVUPDT(KK,K)
*
***   THIS SUBROUTINE UPDATES THE STATE VARIABLES OF AN INDIVIDUAL NEURON
*     WITH SIMPLIFIED DENDRITIC TREE CONTAINING ACTIVE CALCIUM-RELATED
***   CONDUCTANCES.
*
      INTEGER SNTP
      PARAMETER SNTP=3,NCPOPS=3,NCLS=50,MCTP1=2,STEP=1.,EK=-10,ECA=50.,
      *          NSTMX=3,NRGMX=3,NRGP1M=4
      DIMENSION U(NRGMX)
      COMMON G(SNTP,NCPOPS,NCLS,NSTMX,NRGP1M,MCTP1),NST(NCPOPS),
      *       NRG(NCPOPS)
      COMMON/SV/E(NCPOPS,NCLS),TH(NCPOPS,NCLS),S(NCPOPS,NCLS),
      *         GK(NCPOPS,NCLS),B(NCPOPS),C(NCPOPS),TMEM(NCPOPS),
      *         THO(NCPOPS),DCG(NCPOPS),DCTH(NCPOPS),EQ(SNTP),
      *         V(NCPOPS,NCLS,NSTMX,NRGMX),GCA(NCPOPS,NCLS,NSTMX,
      *         NRGMX),CA(NCPOPS,NCLS,NSTMX,NRGMX),GKD(NCPOPS,NCLS,
      *         NSTMX,NRGMX),CDS(NCPOPS),VRT(NCPOPS),DGC(NCPOPS),
      *         THD(NCPOPS),D(NCPOPS),DCA(NCPOPS),A(NCPOPS),
      *         DGKD(NCPOPS),CA0(NCPOPS),BD(NCPOPS)
*
      GS=0. $SYN=0. $DO 10 I=1,SNTP
      GS=GS+G(I,KK,K,1,NRG(KK)+1,1)
      SYN=SYN+G(I,KK,K,1,NRG(KK)+1,1)*EQ(I)
   10 CONTINUE
      GK(KK,K)=GK(KK,K)*DCG(KK)+B(KK)*(1.-DCG(KK))*S(KK,K)
      GTOT=1.+GK(KK,K)+GS $DCE=EXP(-GTOT*STEP/TMEM(KK))
      E(KK,K)=E(KK,K)*DCE+(SYN+GK(KK,K)*EK)*(1.-DCE)/GTOT $DM=0.
      DO 980 I=1,NST(KK) $DO 920 J=1,NRG(KK)
      GT=1.+GCA(KK,K,I,J)+GKD(KK,K,I,J)
      SY=0. $DO 910 M=1,SNTP $GT=GT+G(M,KK,K,I,J,1)
  910 SY=SY+G(M,KK,K,I,J,1)*EQ(M) $Z=EXP(-GT*STEP/TMEM(KK))
  920 V(KK,K,I,J)=V(KK,K,I,J)*Z+(SY+GCA(KK,K,I,J)*ECA+GKD(KK,K,I,J)*EK)*
      *          (1.-Z)/GT
*
      DM=DM+V(KK,K,I,1)
```

```
      IF(NRG(KK).EQ.1) THEN
      U(1)=(V(KK,K,I,1)+E(KK,K)/CDS(KK))/(1.+1./CDS(KK))*
     *    (1.-VRT(KK))+VRT(KK)*V(KK,K,I,1) $ELSE
      U(1)=(V(KK,K,I,1)+V(KK,K,I,2)+E(KK,K)/CDS(KK))/(2.+1./CDS(KK))*
     *    (1.-VRT(KK))+VRT(KK)*V(KK,K,I,1)
      DO 930 J=2,NRG(KK)-1
930   U(J)=(V(KK,K,I,J-1)+V(KK,K,I,J)+V(KK,K,I,J+1))/3.*(1.-VRT(KK))+
     *    VRT(KK)*V(KK,K,I,J)
      U(NRG(KK))=(V(KK,K,I,NRG(KK))+V(KK,K,I,NRG(KK)-1))/2.*
     *    (1.-VRT(KK))+VRT(KK)*V(KK,K,I,NRG(KK)) $ENDIF
      DO 940 J=1,NRG(KK)
940   V(KK,K,I,J)=U(J)
*
      DO 950 J=1,NRG(KK)
      GCA(KK,K,I,J)=GCA(KK,K,I,J)*DGC(KK)
      IF(V(KK,K,I,J).GE.THD(KK)) GCA(KK,K,I,J)=GCA(KK,K,I,J)+
     *              D(KK)*(V(KK,K,I,J)-THD(KK))*(1.-DGC(KK))
      CA(KK,K,I,J)=CA(KK,K,I,J)*DCA(KK)+A(KK)*GCA(KK,K,I,J)*
     *    (1.-DCA(KK))
      GKD(KK,K,I,J)=GKD(KK,K,I,J)*DGKD(KK)
      IF(CA(KK,K,I,J).GE.CA0(KK)) GKD(KK,K,I,J)=GKD(KK,K,I,J)+
     *              BD(KK)*(1.-DGKD(KK))
950   CONTINUE
*
980   CONTINUE
*
      E(KK,K)=(E(KK,K)+CDS(KK)*DM)/(1.+CDS(KK)*FLOAT(NST(KK)))*
     *    (1.-VRT(KK))+VRT(KK)*E(KK,K)
      TH(KK,K)=THO(KK)+(TH(KK,K)-THO(KK))*DCTH(KK)+C(KK)*E(KK,K)*
     *    (1.-DCTH(KK))
      S(KK,K)=0.
      RETURN
      END
*
      SUBROUTINE TRNSMT(IS,IR,K)
*
****      THIS SUBROUTINE PROJECTS SYNAPTIC ACTIVATION FROM A SENDING
****      FIBER TO ITS TARGETED RECEIVING CELLS.
*
      INTEGER TYPE,ACIT,OVR,SNTP
      PARAMETER NTPOPS=5,NCPOPS=3,NCLS=50,MCTP1=2,
     *          NJN=6,NLJN=1,NTR=20,NRNGE=10,MRNGE=1000,SNTP=3,
     *          NSTMX=3,NRGP1M=4
      COMMON G(SNTP,NCPOPS,NCLS,NSTMX,NRGP1M,MCTP1),NST(NCPOPS),
     *    NRG(NCPOPS)
      COMMON/TR/TYPE(NJN),NT(NJN),NCT(NJN),STR0(NJN),
     *    INSED(NJN),LJNN(NTPOPS,NCPOPS),MLAG(NLJN),IRRG(NJN),
     *    ACIT(NLJN,NCLS,NRNGE),OVR(NLJN,MRNGE),IVR(NLJN),
     *    STR(NLJN,NCLS,NTR),N(NTPOPS),
     *    JNN(NTPOPS,NCPOPS),L
*
      JN=JNN(IS,IR)
      CALL RANSET(K*INSED(JN))
      DO 100 J=1,NT(JN)
      NREC=INT(RANF(0.)*FLOAT(N(IR)))+1
      KST=INT(RANF(0.)*FLOAT(NST(IR)))+1
      IF(IRRG(JN).EQ.0) THEN
      KRG=INT(RANF(0.)*(NRG(IR)+1))+1
      ELSE $KRG=IRRG(JN) $ENDIF
      IF(KRG.EQ.NRG(IR)+1) KST=1
      KCT=INT(RANF(0.)*FLOAT(NCT(JN)))+1
      IF(LJNN(IS,IR).GT.0) THEN $I=LJNN(IS,IR)
      DO 200 JR=1,NRNGE
      IF(ACIT(I,NREC,JR).EQ.0) GO TO 210
200   CONTINUE $IVR(I)=IVR(I)+1 $IF(IVR(I).GT.(MRNGE-1)) L=99999
      OVR(I,IVR(I))=J+1000*K+10000000*(KCT+MLAG(I))+10000000000*NREC
      GO TO 220
```

```
  210 ACIT(I,NREC,JR)=J+1000*K+10000000*(KCT+MLAG(I))
  220 G(TYPE(JN),IR,NREC,KST,KRG,KCT+1)=G(TYPE(JN),IR,NREC,KST,KRG,KCT+1
      *         )+STR(I,K,J)  $ELSE
        G(TYPE(JN),IR,NREC,KST,KRG,KCT+1)=G(TYPE(JN),IR,NREC,KST,KRG,KCT+1
      *         )+STRØ(JN)  $ENDIF
  100 CONTINUE
*
      RETURN
      END
/EOR
     Ø       Ø
  70.00    .50-10.00 30.00-10.00  2.00
     50    .00 20.00 20.00  5.00 11.00 24.00
      3      3 24.00  5.00  6.00 30.00  6.00 22.00 65.00  8.00  1.00  0.00
     10    .00 20.00  5.00  5.00  5.00 10.00
      3      2100.00  5.00  6.00 30.00  6.00 22.00 65.00  8.00  1.00  0.00
     50    .00 20.00 20.00  5.00 11.00 24.00
      3      3 24.00  5.00  6.00 30.00  6.00 22.00 65.00  8.00  1.00  0.00
     50    .05    300    999   2223
     50    .10      1    999   1027
      1      Ø      1      2      1      4      1    .20   1227
      1      Ø      1      3      2      2      1    .15   1936
      2      Ø      2      1      Ø      5      1    .30   1718
      1      Ø      3      1      1      4      1    .20   7250
      3      Ø      4      3      1      6      1   -.10   5231
      1      1      5      1      1     20      1   3.00   8413
      3  5.       .2
     30
/EOR
/EOI




/JOB
A224,CM=141000,TL=1000.
FAKE.
PDFV.
M77.
GET,STRØØØ.
LGO,STRØØØ,LN707,STR707.
SAVE,LN707,STR707.
EXIT.
SAVE,LN707,STR707.
/EOR
      PROGRAM LRSYS72 (ANIN,LRSYS,ANOUT,INPUT,OUTPUT,
     *          TAPE3=ANIN,TAPE6=OUTPUT,TAPE7=LRSYS,TAPE8=ANOUT)
*
****    THIS PROGRAM SIMULATES HEBBIAN LEARNING AT THE SYNAPSES OF AN
*       ARBITRARY SYSTEM OF INTERACTING SPATIALLY-ORGANIZED NEURAL NETS
*       ACTIVATED BY AN ARBITRARY NUMBER OF INPUT FIBER SYSTEMS.  ALL
****    SYNAPTIC SYSTEMS ARE INDIVIDUALLY ADJUSTABLE.
*
      INTEGER TYPE,W,H,ACIT,OVR,SNTP
      PARAMETER NTPOPS=5,NFPOPS=2,NCPOPS=3,NCLS=100,MCTP1=2,W=10,H=10,
     *      NJN=6,NLJN=1,NTR=20,NRNGE=10,MRNGE=1000,STEP=1.,SNTP=3,
     *      NSTMX=3,NRGMX=3,NRGP1M=4,NSTIM=10
      DIMENSION P(NFPOPS),INSTP(NFPOPS),INFSED(NFPOPS),NF(NTPOPS),
     *      IDF(NCLS),TTH(NCPOPS),TGK(NCPOPS),IFLG(NCPOPS),
     *      STRMX(NLJN),DLSTR(NLJN),LRN(NJN),IS(NJN),IR(NJN),
     *      T(SNTP),DCS(SNTP),INSTR(NFPOPS),LG1(NLJN),
     *      TGC(NCPOPS),TCA(NCPOPS),TGKD(NCPOPS)
      COMMON G(SNTP,NCPOPS,NCLS,NSTMX,NRGP1M,MCTP1),NST(NCPOPS),
     *      NRG(NCPOPS)
      COMMON/SV/E(NCPOPS,NCLS),TH(NCPOPS,NCLS),S(NCPOPS,NCLS),
     *      GK(NCPOPS,NCLS),B(NCPOPS),C(NCPOPS),TMEM(NCPOPS),
     *      THO(NCPOPS),DCG(NCPOPS),DCTH(NCPOPS),EQ(SNTP),
     *      V(NCPOPS,NCLS,NSTMX,NRGMX),GCA(NCPOPS,NCLS,NSTMX,
```

```
      *            NRGMX),CA(NCPOPS,NCLS,NSTMX,NRGMX),GKD(NCPOPS,NCLS,
      *            NSTMX,NRGMX),CDS(NCPOPS),VRT(NCPOPS),DGC(NCPOPS),
      *            THD(NCPOPS),D(NCPOPS),DCA(NCPOPS),A(NCPOPS),
      *            DGKD(NCPOPS),CA0(NCPOPS),BD(NCPOPS)
      COMMON/STIM/LOCATN(NSTIM)
      COMMON/TR/TYPE(NJN),NT(NJN),NCT(NJN),STR0(NJN),WTF(NJN),HTF(NJN),
      *            INSED(NJN),LJNN(NTPOPS,NCPOPS),MLAG(NLJN),IRRG(NJN),
      *            ACIT(NLJN,NCLS,NRNGE),OVR(NLJN,MRNGE),IVR(NLJN),
      *            STR(NLJN,NCLS,NTR),N(NTPOPS),NC(NTPOPS),NR(NTPOPS),
      *            JNN(NTPOPS,NCPOPS),L
      *
      ****       THIS SECTION READS AND WRITES THE INPUT PARAMETERS
      *
            READ 5100, IFLG1,IFLG2
            READ 5105, ((EQ(I),T(I)),I=1,SNTP)
            DO 110 I=1,NCPOPS
            READ 5110, N(I),C(I),TTH(I),B(I),TGK(I),TMEM(I),THO(I),
      *            NC(I),NR(I)
      110 READ 5115, NST(I),NRG(I),THD(I),D(I),TGC(I),A(I),TCA(I),CA0(I),
      *            BD(I),TGKD(I),CDS(I),VRT(I)
            DO 120 I=1,NFPOPS $II=NCPOPS+I
      120 READ 5120, N(II),P(I),INSTR(I),INSTP(I),INFSED(I),NC(II),NR(II)
            DO 128 I=1,NCPOPS $IFLG(I)=0 $DO 128 J=1,NTPOPS
            JNN(J,I)=0 $LJNN(J,I)=0
      128 CONTINUE $J=0 $DO 130 I=1,NJN
            READ 5130, TYPE(I),LRN(I),IS(I),IR(I),IRRG(I),NT(I),NCT(I),STR0(I)
      *            ,WTF(I),HTF(I),INSED(I)
            JNN(IS(I),IR(I))=I
            IF(LRN(I).GT.0) THEN $J=J+1 $LJNN(IS(I),IR(I))=J $IFLG(IR(I))=1
            READ 5140, MLAG(J),STRMX(J),DLSTR(J) $ENDIF
      130 CONTINUE
            READ 5100, ISTM,LSTIM,INTERV,LSTP,NSTIM1
            READ 5100, (LOCATN(I),I=1,NSTIM)
            READ 5100, LTSTOP
      *
            WRITE(6,6000) $WRITE(6,6100) NTPOPS,NFPOPS,NCPOPS,NCLS,MCTP1,W,H,
            WRITE(6,6102) NJN,NLJN,NTR,NRNGE,MRNGE,IFLG1
            WRITE(6,6105) ((EQ(I),T(I)),I=1,SNTP)
            WRITE(6,6110) $DO 111 I=1,NCPOPS
            WRITE(6,6112) N(I),C(I),TTH(I),B(I),TGK(I),TMEM(I),THO(I),
      *            NC(I),NR(I)
      111 WRITE(6,6115) NST(I),NRG(I),THD(I),D(I),TGC(I),A(I),TCA(I),CA0(I),
      *            BD(I),TGKD(I),CDS(I),VRT(I)
            WRITE(6,6120) $DO 121 I=1,NFPOPS $II=NCPOPS+I
      121 WRITE(6,6122) N(II),P(I),INSTR(I),INSTP(I),INFSED(I),
      *            NC(II),NR(II)
            WRITE(6,6130) $J=0 $DO 131 I=1,NJN
            WRITE(6,6132) TYPE(I),LRN(I),IS(I),IR(I),IRRG(I),NT(I),NCT(I),
      *            STR0(I),WTF(I),HTF(I),INSED(I)
            IF(LRN(I).GT.0) THEN $J=J+1
            WRITE(6,6140) MLAG(J),STRMX(J),DLSTR(J) $ENDIF
      131 CONTINUE
            WRITE(6,6150) ISTM,LSTIM,INTERV,LSTP,NSTIM,NSTIM1
            WRITE(6,6154) $WRITE(6,6155) (LOCATN(I),I=1,NSTIM)
            WRITE(6,6200) LTSTOP $WRITE(6,6000)
      *
      ****       THIS SECTION INITIALIZES VARIABLES
      *
            DO 310 I=1,NCPOPS
            DCTH(I)=EXP(-STEP/TTH(I)) $DCG(I)=EXP(-STEP/TGK(I))
            DO 310 J=1,NCLS
            E(I,J)=0. $TH(I,J)=THO(I) $S(I,J)=0. $GK(I,J)=0.
            DO 310 K=1,MCTP1 $DO 310 M=1,SNTP
            DO 310 II=1,NSTMX $DO 310 JJ=1,NRGP1M
            G(M,I,J,II,JJ,K)=0.
      310 CONTINUE
            DO 330 I=1,NLJN $IVR(I)=0 $LG1(I)=10000000*(MLAG(I)+1)
```

```
      DO 332 J=1,NCLS $DO 332 K=1,NRNGE
332   ACIT(I,J,K)=0 $DO 334 K=1,MRNGE
334   OVR(I,K)=0
330   CONTINUE
      DO 340 I=1,SNTP
340   DCS(I)=EXP(-STEP/T(I))
      DO 350 I=1,NCPOPS $DO 350 J=1,NCLS
      DO 350 K=1,NST(I) $DO 350 M=1,NRG(I)
      V(I,J,K,M)=0. $GCA(I,J,K,M)=0. $CA(I,J,K,M)=0. $GKD(I,J,K,M)=0.
350   CONTINUE
      DO 360 I=1,NCPOPS
      DGC(I)=EXP(-STEP/TGC(I)) $DCA(I)=EXP(-STEP/TCA(I))
360   DGKD(I)=EXP(-STEP/TGKD(I))
      IF(IFLG2.EQ.0) THEN $DO 320 I=1,NJN $IF(LRN(I).EQ.0) GO TO 320
      II=LJNN(IS(I),IR(I)) $DO 320 J=1,NCLS $DO 320 K=1,NTR
      STR(II,J,K)=STR0(I)
320   CONTINUE $ELSE
      READ(3,8900) (((STR(I,J,K),K=1,NTR),J=1,NCLS),I=1,NLJN) $ENDIF
      MARKER=0
*
****       THIS SECTION UPDATES STATE VARIABLES AT EACH TIME STEP
*
      DO 1000 L=1,LTSTOP
*
      IF(L.EQ.LSTIM.AND.L.LE.LSTP) CALL STIMULS(ISTM,LSTIM,INTERV,
     *                              NSTIM1,MARKER)
*
****                UPDATING THE CELLS
*
      DO 2000 I=1,NCPOPS
      NF(I)=0
      DO 2900 J=1,N(I)
      CALL SVUPDT(I,J)
      IF(E(I,J).LT.TH(I,J)) GO TO 2850
      S(I,J)=1. $NF(I)=NF(I)+1 $IDF(NF(I))=J
*
      IF(IFLG(I).EQ.0) GO TO 2699 $DO 2700 LS=1,NTPOPS
      IF(LJNN(LS,I).EQ.0) GO TO 2700 $LJ=LJNN(LS,I)
      DO 2500 K=1,NRNGE $IF(ACIT(LJ,J,K).EQ.0) GO TO 2499
      IF(ACIT(LJ,J,K).GT.LG1(LJ)) GO TO 2500
      ID=MOD(ACIT(LJ,J,K),10000000) $JC=ID/1000 $JT=ID-JC*1000
      STR(LJ,JC,JT)=STR(LJ,JC,JT)+DLSTR(LJ)*
     *                        (STRMX(LJ)-STR(LJ,JC,JT))
2500  CONTINUE
2499  CONTINUE $JUP=IVR(LJ) $DO 2600 K=1,JUP
      IF(OVR(LJ,K)/10000000000.NE.J) GO TO 2600
      IF(MOD(OVR(LJ,K),10000000000).GT.LG1(LJ)) GO TO 2600
      ID=MOD(OVR(LJ,K),10000000) $JC=ID/1000 $JT=ID-JC*1000
      STR(LJ,JC,JT)=STR(LJ,JC,JT)+DLSTR(LJ)*
     *                        (STRMX(LJ)-STR(LJ,JC,JT))
2600  CONTINUE
2700  CONTINUE
*
2699  CONTINUE
      DO 2800 K=1,NCPOPS
      IF(JNN(I,K).GT.0) CALL TRNSMT(I,K,J)
2800  CONTINUE
2850  CONTINUE
2900  CONTINUE $IF(IFLG1.GT.0) WRITE(7,7060) (L,I,NF(I),
     *            (IDF(J),J=1,NF(I)))
2000  CONTINUE
*
****                UPDATING THE INPUT FIBERS
*
      DO 3000 I=1,NFPOPS
      II=I+NCPOPS $NF(II)=0
      IF(L.LT.INSTR(I).OR.L.GE.INSTP(I)) GO TO 3000
```

```
      CALL RANSET(INFSED(I))
      DO 3900 J=1,N(II)
      IF(RANF(0.).GT.P(I)) GO TO 3850 $CALL RANGET(INFSED(I))
      NF(II)=NF(II)+1 $IDF(NF(II))=J
      DO 3800 K=1,NCPOPS
      IF(JNN(II,K).GT.0) CALL TRNSMT(II,K,J)
3800  CONTINUE $CALL RANSET(INFSED(I))
3850  CONTINUE
3900  CONTINUE $IF(IFLG1.GT.0) WRITE(7,7060) (L,II,NF(II),
     *          (IDF(J),J=1,NF(II)))
      CALL RANGET(INFSED(I))
3000  CONTINUE
*
      WRITE(6,6510) L,(NF(I),I=1,NTPOPS)
      WRITE(7,7510) L,(NF(I),I=1,NTPOPS)
*
      DO 1910 I=1,NCPOPS $DO 1910 J=1,NCLS $DO 1910 M=1,SNTP
      DO 1910 II=1,NSTMX $DO 1910 JJ=1,NRGP1M
      G(M,I,J,II,JJ,1)=G(M,I,J,II,JJ,1)*DCS(M)+G(M,I,J,II,JJ,2)
      DO 1920 K=2,MCTP1-1
1920  G(M,I,J,II,JJ,K)=G(M,I,J,II,JJ,K+1) $G(M,I,J,II,JJ,MCTP1)=0.
1910  CONTINUE
      DO 1600 LJ=1,NLJN $JUP=IVR(LJ) $DO 1700 J=1,JUP
1710  IF(OVR(LJ,J).EQ.0) GO TO 1699
      IF(MOD(OVR(LJ,J),10000000000).GT.10000000) GO TO 1700
      IVR(LJ)=IVR(LJ)-1 $DO 1760 J1=J,IVR(LJ)
1760  OVR(LJ,J1)=OVR(LJ,J1+1) $OVR(LJ,IVR(LJ)+1)=0 $GO TO 1710
1700  OVR(LJ,J)=OVR(LJ,J)-10000000
1699  CONTINUE
      DO 1800 I=1,NCLS $DO 1850 J=1,NRNGE
1810  IF(ACIT(LJ,I,J).EQ.0) GO TO 1849
      IF(ACIT(LJ,I,J).GT.10000000) GO TO 1850 $DO 1860 J1=1,NRNGE-1
1860  ACIT(LJ,I,J1)=ACIT(LJ,I,J1+1) $ACIT(LJ,I,NRNGE)=0 $GO TO 1810
1850  ACIT(LJ,I,J)=ACIT(LJ,I,J)-10000000
1849  CONTINUE
1800  CONTINUE
1600  CONTINUE
*
1000  CONTINUE
*
****         THIS SECTION WRITES OUT ACTIVITY VARIABLES
*
      DO 4100 JS=1,NTPOPS $DO 4100 JR=1,NCPOPS
      IF(LJNN(JS,JR).EQ.0) GO TO 4100 $WRITE(6,6040)
      JN=JNN(JS,JR)
      WRITE(6,6900) ((STR(LJ,I,K),K=1,NTR),I=1,NCLS)
      WRITE(8,8900) ((STR(LJ,I,K),K=1,NTR),I=1,NCLS)
4100  CONTINUE
*
      STOP
****         THESE ARE THE FORMATS
*
5100  FORMAT(12I6)
5105  FORMAT(12F6.2)
5110  FORMAT(I6,6F6.2,2I6)
5115  FORMAT(2I6,10F6.2)
5120  FORMAT(I6,F6.2,5I6)
5130  FORMAT(7I6,3F6.2,I6)
5140  FORMAT(I6,2F6.2)
6000  FORMAT(1H1)
6040  FORMAT(///)
6100  FORMAT(5X,*NTPOPS,NCPOPS,NFPOPS,NCLS,MCTP1,W,H=*,7I7)
6102  FORMAT(5X,*NJN,NLJN,NTR,NRNGE,MRNGE,IFLG1=*,6I7)
6105  FORMAT(5X,*EQ,T=*,12F7.2)
6110  FORMAT(5X,*CELLS: N,C,TTH,B,TGK,TMEM,THO,NC,NR=*,
     *    /,*NST,NRG,THD,D,TGC,A,TCA,CA0,BD,TGKD,CDS,VRT=*,//)
6112  FORMAT(9X,I7,6F7.2,2I7)
```

```
6115 FORMAT(5X,2I7,10F7.2)
6120 FORMAT(///,5X,*FIBERS: N,P,INSTR,INSTP,INFSED,NC,NR=*,//)
6122 FORMAT(5X,I7,F7.2,5I7)
6130 FORMAT(///,5X,*JNCTNS: TYPE,LRN,IS,IR,IRRG,NT,NCT,STRØ,WTF,HTF,*
    *      ,*INSED=*,/10X,*MLAG,STRMX,DLSTR=*,//)
6132 FORMAT(5X,7I7,3F7.2,I7)
6140 FORMAT(10X,I7,2F7.2)
6150 FORMAT(5X,*ISTM,LSTIM,INTERV,LSTP,NSTIM,NSTIM1=*,6I7)
6154 FORMAT(5X,*LOCATNS:*)
6155 FORMAT(10X,20I6)
6200 FORMAT(5X,*LTSTOP=*,I7)
6510 FORMAT(5X,15I7)
6900 FORMAT(5X,20F6.3)
7510 FORMAT(15I5)
7060 FORMAT(20I4)
8900 FORMAT(F6.3)
     END
*
     SUBROUTINE SVUPDT(KK,K)
*
*** THIS SUBROUTINE UPDATES THE STATE VARIABLES OF AN INDIVIDUAL NEURON
*   WITH SIMPLIFIED DENDRITIC TREE CONTAINING ACTIVE CALCIUM-RELATED
*** CONDUCTANCES.
*
     INTEGER SNTP
     PARAMETER SNTP=3,NCPOPS=3,NCLS=100,MCTP1=2,STEP=1.,EK=-10,ECA=50.,
    *          NSTMX=3,NRGMX=3,NRGP1M=4
     DIMENSION U(NRGMX)
     COMMON G(SNTP,NCPOPS,NCLS,NSTMX,NRGP1M,MCTP1),NST(NCPOPS),
    *        NRG(NCPOPS)
     COMMON/SV/E(NCPOPS,NCLS),TH(NCPOPS,NCLS),S(NCPOPS,NCLS),
    *        GK(NCPOPS,NCLS),B(NCPOPS),C(NCPOPS),TMEM(NCPOPS),
    *        THO(NCPOPS),DCG(NCPOPS),DCTH(NCPOPS),EQ(SNTP),
    *        V(NCPOPS,NCLS,NSTMX,NRGMX),GCA(NCPOPS,NCLS,NSTMX,
    *        NRGMX),CA(NCPOPS,NCLS,NSTMX,NRGMX),GKD(NCPOPS,NCLS,
    *        NSTMX,NRGMX),CDS(NCPOPS),VRT(NCPOPS),DGC(NCPOPS),
    *        THD(NCPOPS),D(NCPOPS),DCA(NCPOPS),A(NCPOPS),
    *        DGKD(NCPOPS),CAØ(NCPOPS),BD(NCPOPS)
*
     GS=0. $SYN=0. $DO 10 I=1,SNTP
     GS=GS+G(I,KK,K,1,NRG(KK)+1,1)
     SYN=SYN+G(I,KK,K,1,NRG(KK)+1,1)*EQ(I)
  10 CONTINUE
     GK(KK,K)=GK(KK,K)*DCG(KK)+B(KK)*(1.-DCG(KK))*S(KK,K)
     GTOT=1.+GK(KK,K)+GS $DCE=EXP(-GTOT*STEP/TMEM(KK))
     E(KK,K)=E(KK,K)*DCE+(SYN+GK(KK,K)*EK)*(1.-DCE)/GTOT $DM=0.
     DO 980 I=1,NST(KK) $DO 920 J=1,NRG(KK)
     GT=1.+GCA(KK,K,I,J)+GKD(KK,K,I,J)
     SY=0. $DO 910 M=1,SNTP $GT=GT+G(M,KK,K,I,J,1)
 910 SY=SY+G(M,KK,K,I,J,1)*EQ(M) $Z=EXP(-GT*STEP/TMEM(KK))
 920 V(KK,K,I,J)=V(KK,K,I,J)*Z+(SY+GCA(KK,K,I,J)*ECA+GKD(KK,K,I,J)*EK)*
    *          (1.-Z)/GT
*
     DM=DM+V(KK,K,I,1)
     IF(NRG(KK).EQ.1) THEN
     U(1)=(V(KK,K,I,1)+E(KK,K)/CDS(KK))/(1.+1./CDS(KK))*
    *      (1.-VRT(KK))+VRT(KK)*V(KK,K,I,1) $ELSE
     U(1)=(V(KK,K,I,1)+V(KK,K,I,2)+E(KK,K)/CDS(KK))/(2.+1./CDS(KK))*
    *      (1.-VRT(KK))+VRT(KK)*V(KK,K,I,1)
     DO 930 J=2,NRG(KK)-1
 930 U(J)=(V(KK,K,I,J-1)+V(KK,K,I,J)+V(KK,K,I,J+1))/3.*(1.-VRT(KK))+
    *      VRT(KK)*V(KK,K,I,J)
     U(NRG(KK))=(V(KK,K,I,NRG(KK))+V(KK,K,I,NRG(KK)-1))/2.*
    *          (1.-VRT(KK))+VRT(KK)*V(KK,K,I,NRG(KK)) $ENDIF
     DO 940 J=1,NRG(KK)
 940 V(KK,K,I,J)=U(J)
*
```

```
      DO 950 J=1,NRG(KK)
      GCA(KK,K,I,J)=GCA(KK,K,I,J)*DGC(KK)
      IF(V(KK,K,I,J).GE.THD(KK))  GCA(KK,K,I,J)=GCA(KK,K,I,J)+
     *                  D(KK)*(V(KK,K,I,J)-THD(KK))*(1.-DGC(KK))
      CA(KK,K,I,J)=CA(KK,K,I,J)*DCA(KK)+A(KK)*GCA(KK,K,I,J)*
     *            (1.-DCA(KK))
      GKD(KK,K,I,J)=GKD(KK,K,I,J)*DGKD(KK)
      IF(CA(KK,K,I,J).GE.CAØ(KK))  GKD(KK,K,I,J)=GKD(KK,K,I,J)+
     *                  BD(KK)*(1.-DGKD(KK))
  950 CONTINUE
*
  980 CONTINUE
*
      E(KK,K)=(E(KK,K)+CDS(KK)*DM)/(1.+CDS(KK)*FLOAT(NST(KK)))*
     *        (1.-VRT(KK))+VRT(KK)*E(KK,K)
      TH(KK,K)=THO(KK)+(TH(KK,K)-THO(KK))*DCTH(KK)+C(KK)*E(KK,K)*
     *        (1.-DCTH(KK))
      S(KK,K)=0.
      RETURN
      END
*
      SUBROUTINE TRNSMT(IS,IR,K)
*
****      THIS SUBROUTINE PROJECTS SYNAPTIC ACTIVATION FROM A SENDING
****      FIBER TO ITS TARGETED RECEIVING CELLS.
*
      INTEGER TYPE,W,H,ACIT,OVR,SNTP
      PARAMETER NTPOPS=5,NCPOPS=3,NCLS=100,MCTP1=2,W=10,H=10,
     *      NJN=6,NLJN=1,NTR=20,NRNGE=10,MRNGE=1000,SNTP=3,
     *      NSTMX=3,NRGP1M=4
      COMMON G(SNTP,NCPOPS,NCLS,NSTMX,NRGP1M,MCTP1),NST(NCPOPS),
     *      NRG(NCPOPS)
      COMMON/TR/TYPE(NJN),NT(NJN),NCT(NJN),STRØ(NJN),WTF(NJN),HTF(NJN),
     *      INSED(NJN),LJNN(NTPOPS,NCPOPS),MLAG(NLJN),IRRG(NJN),
     *      ACIT(NLJN,NCLS,NRNGE),OVR(NLJN,MRNGE),IVR(NLJN),
     *      STR(NLJN,NCLS,NTR),N(NTPOPS),NC(NTPOPS),NR(NTPOPS),
     *      JNN(NTPOPS,NCPOPS),L
*
      JN=JNN(IS,IR)
      XS=FLOAT(MOD(K,NC(IS)))*FLOAT(W)/FLOAT(NC(IS))
      IF(XS.EQ.Ø.)  XS=FLOAT(W)
      YS=FLOAT(K/NC(IS)+1)*FLOAT(H)/FLOAT(NR(IS))
      CALL RANSET(K*INSED(JN))
      DO 100 J=1,NT(JN)
      XR=XS+(RANF(Ø.)-.5)*WTF(JN)  $YR=YS+(RANF(Ø.)-.5)*HTF(JN)
      ICR=INT(FLOAT(NC(IR))*(XR-.5)/FLOAT(W))+1
      IRR=INT(FLOAT(NR(IR))*(YR-.5)/FLOAT(H))+1
      IF(ICR.LE.Ø)  ICR=NC(IR)+ICR
      IF(ICR.GT.NC(IR))  ICR=ICR-NC(IR)
      IF(IRR.LE.Ø)  IRR=NR(IR)+IRR
      IF(IRR.GT.NR(IR))  IRR=IRR-NR(IR)
      NREC=NC(IR)*(IRR-1)+ICR
      KST=INT(RANF(Ø.)*FLOAT(NST(IR)))+1
      IF(IRRG(JN).EQ.Ø) THEN
      KRG=INT(RANF(Ø.)*(NRG(IR)+1))+1
      ELSE  $KRG=IRRG(JN)  $ENDIF
      IF(KRG.EQ.NRG(IR)+1)  KST=1
      KCT=INT(RANF(Ø.)*FLOAT(NCT(JN)))+1
      IF(LJNN(IS,IR).GT.Ø)  THEN  $I=LJNN(IS,IR)
      DO 200 JR=1,NRNGE
      IF(ACIT(I,NREC,JR).EQ.Ø)  GO TO 210
  200 CONTINUE  $IVR(I)=IVR(I)+1  $IF(IVR(I).GT.(MRNGE-1))  L=99999
      OVR(I,IVR(I))=J+1000*K+10000000*(KCT+MLAG(I))+10000000000*NREC
      GO TO 220
  210 ACIT(I,NREC,JR)=J+1000*K+10000000*(KCT+MLAG(I))
  220 G(TYPE(JN),IR,NREC,KST,KRG,KCT+1)=G(TYPE(JN),IR,NREC,KST,KRG,KCT+1
     *      )+STR(I,K,J)  $ELSE
```

```
      G(TYPE(JN),IR,NREC,KST,KRG,KCT+1)=G(TYPE(JN),IR,NREC,KST,KRG,KCT+1
     *    )+STRØ(JN) $ENDIF
  100 CONTINUE
*
      RETURN
      END
*
      SUBROUTINE STIMULS(IR,LSTIM,INTERV,NSTIM1,MARKER)
*
****      THIS SUBROUTINE
      PARAMETER NSTIM=10,NCLS=100,NCPOPS=3
      COMMON/SV/E(NCPOPS,NCLS),TH(NCPOPS,NCLS)
      COMMON/STIM/LOCATN(NSTIM)
*
      DO 10 I=1,NSTIM1
      J=I+MARKER $E(IR,LOCATN(J))=TH(IR,LOCATN(J))+16.
   10 CONTINUE
      MARKER=MARKER+NSTIM1 $IF(MARKER.GE.NSTIM) MARKER=Ø
      LSTIM=LSTIM+INTERV
      RETURN
      END
*
/EOR
      Ø      Ø
  70.00    .50-10.00 30.00-10.00   2.00
    100    .00 20.00 20.00   5.00 11.00 24.00     10     10
      3      3 24.00   5.00   6.00 30.00   6.00 22.00 65.00   8.00     .50     .50
     10    .00 20.00   5.00   5.00   5.00 10.00      5      2
      3      2100.00   5.00   6.00 30.00   6.00 22.00 65.00   8.00     .50     .50
     50    .00 20.00 20.00   5.00 11.00 24.00     10      5
      3      3 24.00   5.00   6.00 30.00   6.00 22.00 65.00   8.00     .50     .50
     50    .05    300    999   2223     10      5
    100    .10      1    999   1027     10     10
      1      Ø      1      2      1      4      1     .20   2.00   2.00   1227
      1      Ø      1      3      2      2      1     .15   3.00   3.00   1936
      2      Ø      2      1      Ø      5      1     .30   3.00   3.00   1718
      1      Ø      3      1      1      4      1     .20   2.00   2.00    516
      3      Ø      4      3      1      6      1    -.10   3.00   3.00   7554
      1      1      5      1      1     20      1    3.00   3.00   3.00   8413
      3   5.         .2
      1      1      5     30     10
     63     64     65     66     53     56     43     44     45     46
     30
/EOR
/EOI



/JOB
A224,CM=140000,TL=1000.
FAKE.
PDFV.
M77.
GET,STR702.
LGO,STR702,LN703.
SAVE,LN703.
EXIT.
SAVE,LN703.
/EOR
      PROGRAM LDSYS73 (ANIN,LRSYS,INPUT,OUTPUT,
     *          TAPE3=ANIN,TAPE6=OUTPUT,TAPE7=LRSYS)
*
****     THIS PROGRAM SIMULATES HEBBIAN LEARNING AT THE SYNAPSES OF AN
*        ARBITRARY SYSTEM OF INTERACTING SPATIALLY-ORGANIZED NEURAL NETS
*        ACTIVATED BY AN ARBITRARY NUMBER OF INPUT FIBER SYSTEMS.   ALL
****     SYNAPTIC SYSTEMS ARE INDIVIDUALLY ADJUSTABLE.
*
```

```
      INTEGER TYPE,W,H,SNTP
      PARAMETER NTPOPS=5,NFPOPS=2,NCPOPS=3,NCLS=100,MCTP1=2,W=10,H=10,
     *          NJN=6,NLJN=1,NTR=20,STEP=1.,SNTP=3,
     *          NSTMX=3,NRGMX=3,NRGP1M=4,NSTIM=10
      DIMENSION P(NFPOPS),INSTP(NFPOPS),INFSED(NFPOPS),NF(NTPOPS),
     *          IDF(NCLS),TTH(NCPOPS),TGK(NCPOPS),IFLG(NCPOPS),
     *          LRN(NJN),IS(NJN),IR(NJN),
     *          T(SNTP),DCS(SNTP),INSTR(NFPOPS),
     *          TGC(NCPOPS),TCA(NCPOPS),TGKD(NCPOPS)
      COMMON G(SNTP,NCPOPS,NCLS,NSTMX,NRGP1M,MCTP1),NST(NCPOPS),
     *          NRG(NCPOPS)
      COMMON/SV/E(NCPOPS,NCLS),TH(NCPOPS,NCLS),S(NCPOPS,NCLS),
     *          GK(NCPOPS,NCLS),B(NCPOPS),C(NCPOPS),TMEM(NCPOPS),
     *          THO(NCPOPS),DCG(NCPOPS),DCTH(NCPOPS),EQ(SNTP),
     *          V(NCPOPS,NCLS,NSTMX,NRGMX),GCA(NCPOPS,NCLS,NSTMX,
     *          NRGMX),CA(NCPOPS,NCLS,NSTMX,NRGMX),GKD(NCPOPS,NCLS,
     *          NSTMX,NRGMX),CDS(NCPOPS),VRT(NCPOPS),DGC(NCPOPS),
     *          THD(NCPOPS),D(NCPOPS),DCA(NCPOPS),A(NCPOPS),
     *          DGKD(NCPOPS),CA0(NCPOPS),BD(NCPOPS)
      COMMON/STIM/LOCATN(NSTIM)
      COMMON/TR/TYPE(NJN),NT(NJN),NCT(NJN),STR0(NJN),WTF(NJN),HTF(NJN),
     *          INSED(NJN),LJNN(NTPOPS,NCPOPS),IRRG(NJN),
     *          STR(NLJN,NCLS,NTR),N(NTPOPS),NC(NTPOPS),NR(NTPOPS),
     *          JNN(NTPOPS,NCPOPS),L
*
****        THIS SECTION READS AND WRITES THE INPUT PARAMETERS
*
      READ 5100, IFLG1,IFLG2,IFLG3
      READ 5105, ((EQ(I),T(I)),I=1,SNTP)
      DO 110 I=1,NCPOPS
      READ 5110, N(I),C(I),TTH(I),B(I),TGK(I),TMEM(I),THO(I),
     *           NC(I),NR(I)
  110 READ 5115, NST(I),NRG(I),THD(I),D(I),TGC(I),A(I),TCA(I),CA0(I),
     *           BD(I),TGKD(I),CDS(I),VRT(I)
      DO 120 I=1,NFPOPS $II=NCPOPS+I
  120 READ 5120, N(II),P(I),INSTR(I),INSTP(I),INFSED(I),NC(II),NR(II)
      DO 128 I=1,NCPOPS $IFLG(I)=0 $DO 128 J=1,NTPOPS
      JNN(J,I)=0 $LJNN(J,I)=0
  128 CONTINUE $J=0 $DO 130 I=1,NJN
      READ 5130, TYPE(I),LRN(I),IS(I),IR(I),IRRG(I),NT(I),NCT(I),STR0(I)
     *           ,WTF(I),HTF(I),INSED(I)
      JNN(IS(I),IR(I))=I
      IF(LRN(I).GT.0) THEN $LJNN(IS(I),IR(I))=J $IFLG(IR(I))=1 $ENDIF
  130 CONTINUE
      READ 5100, ISTM,LSTIM,INTERV,LSTP,NSTIM1
      READ 5100, (LOCATN(I),I=1,NSTIM)
      READ 5100, LTSTOP
*
      WRITE(6,6000) $WRITE(6,6100) NTPOPS,NFPOPS,NCPOPS,NCLS,MCTP1,W,H,
      WRITE(6,6102) NJN,NLJN,NTR,IFLG1,IFLG2,IFLG3
      WRITE(6,6105) ((EQ(I),T(I)),I=1,SNTP)
      WRITE(6,6110) $DO 111 I=1,NCPOPS
      WRITE(6,6112) N(I),C(I),TTH(I),B(I),TGK(I),TMEM(I),THO(I),
     *           NC(I),NR(I)
  111 WRITE(6,6115) NST(I),NRG(I),THD(I),D(I),TGC(I),A(I),TCA(I),CA0(I),
     *           BD(I),TGKD(I),CDS(I),VRT(I)
      WRITE(6,6120) $DO 121 I=1,NFPOPS $II=NCPOPS+I
  121 WRITE(6,6122) N(II),P(I),INSTR(I),INSTP(I),INFSED(I),
     *           NC(II),NR(II)
      WRITE(6,6130) $J=0 $DO 131 I=1,NJN
      WRITE(6,6132) TYPE(I),LRN(I),IS(I),IR(I),IRRG(I),NT(I),NCT(I),
     *           STR0(I),WTF(I),HTF(I),INSED(I)
  131 CONTINUE
      WRITE(6,6150) ISTM,LSTIM,INTERV,LSTP,NSTIM,NSTIM1
      WRITE(6,6154) $WRITE(6,6155) (LOCATN(I),I=1,NSTIM)
      WRITE(6,6200) LTSTOP $WRITE(6,6000)
*
```

```
****          THIS SECTION INITIALIZES VARIABLES
*
      DO 310 I=1,NCPOPS
      DCTH(I)=EXP(-STEP/TTH(I)) $DCG(I)=EXP(-STEP/TGK(I))
      DO 310 J=1,NCLS
      E(I,J)=0. $TH(I,J)=THO(I) $S(I,J)=0. $GK(I,J)=0.
      DO 310 K=1,MCTP1 $DO 310 M=1,SNTP
      DO 310 II=1,NSTMX $DO 310 JJ=1,NRGP1M
      G(M,I,J,II,JJ,K)=0.
  310 CONTINUE
      DO 340 I=1,SNTP
  340 DCS(I)=EXP(-STEP/T(I))
      DO 350 I=1,NCPOPS $DO 350 J=1,NCLS
      DO 350 K=1,NST(I) $DO 350 M=1,NRG(I)
      V(I,J,K,M)=0. $GCA(I,J,K,M)=0. $CA(I,J,K,M)=0. $GKD(I,J,K,M)=0.
  350 CONTINUE
      DO 360 I=1,NCPOPS
      DGC(I)=EXP(-STEP/TGC(I)) $DCA(I)=EXP(-STEP/TCA(I))
  360 DGKD(I)=EXP(-STEP/TGKD(I))
      IF(IFLG2.EQ.0) THEN $DO 320 I=1,NJN $IF(LRN(I).EQ.0) GO TO 320
      II=LJNN(IS(I),IR(I)) $DO 320 J=1,NCLS $DO 320 K=1,NTR
      STR(II,J,K)=STR0(I)
  320 CONTINUE $ELSE
      READ(3,8900) (((STR(I,J,K),K=1,NTR),J=1,NCLS),I=1,NLJN) $ENDIF
      MARKER=0

****          THIS SECTION UPDATES STATE VARIABLES AT EACH TIME STEP
*
      DO 1000 L=1,LTSTOP
*
      IF(L.EQ.LSTIM.AND.L.LE.LSTP) CALL STIMULS(ISTM,LSTIM,INTERV,
     *                              NSTIM1,MARKER)
*
****                UPDATING THE CELLS
*
      DO 2000 I=1,NCPOPS
      NF(I)=0
      DO 2900 J=1,N(I)
      CALL SVUPDT(I,J)
      IF(E(I,J).LT.TH(I,J)) GO TO 2850
      S(I,J)=1. $NF(I)=NF(I)+1 $IDF(NF(I))=J
*
      DO 2800 K=1,NCPOPS
      IF(JNN(I,K).GT.0) CALL TRNSMT(I,K,J)
 2800 CONTINUE
 2850 CONTINUE
 2900 CONTINUE $IF(IFLG1.GT.0) WRITE(7,7060) (L,I,NF(I),
     *            (IDF(J),J=1,NF(I)))
 2000 CONTINUE
*
****                UPDATING THE INPUT FIBERS
*
      DO 3000 I=1,NFPOPS
      II=I+NCPOPS $NF(II)=0
      IF(L.LT.INSTR(I).OR.L.GE.INSTP(I)) GO TO 3000
      CALL RANSET(INFSED(I))
      DO 3900 J=1,N(II)
      IF(RANF(0.).GT.P(I)) GO TO 3850 $CALL RANGET(INFSED(I))
      NF(II)=NF(II)+1 $IDF(NF(II))=J
      DO 3800 K=1,NCPOPS
      IF(JNN(II,K).GT.0) CALL TRNSMT(II,K,J)
 3800 CONTINUE $CALL RANGET(INFSED(I))
 3850 CONTINUE
 3900 CONTINUE $IF(IFLG1.GT.0) WRITE(7,7060) (L,II,NF(II),
     *            (IDF(J),J=1,NF(II)))
      CALL RANGET(INFSED(I))
 3000 CONTINUE
```

```
*
      WRITE(6,6510) L,(NF(I),I=1,NTPOPS)
      WRITE(7,7510) L,(NF(I),I=1,NTPOPS)
*
      DO 1910 I=1,NCPOPS $DO 1910 J=1,NCLS $DO 1910 M=1,SNTP
      DO 1910 II=1,NSTMX $DO 1910 JJ=1,NRGP1M
      G(M,I,J,II,JJ,1)=G(M,I,J,II,JJ,1)*DCS(M)+G(M,I,J,II,JJ,2)
      DO 1920 K=2,MCTP1-1
 1920 G(M,I,J,II,JJ,K)=G(M,I,J,II,JJ,K+1) $G(M,I,J,II,JJ,MCTP1)=0.
 1910 CONTINUE
*
 1000 CONTINUE
*
****       THIS SECTION WRITES OUT ACTIVITY VARIABLES
*
      IF(IFLG3.GT.0) THEN $DO 4100 JS=1,NTPOPS $DO 4100 JR=1,NCPOPS
      IF(LJNN(JS,JR).EQ.0) GO TO 4100 $WRITE(6,6040)
      LJ=LJNN(JS,JR) $JN=JNN(JS,JR)
      WRITE(6,6900) ((STR(LJ,I,K),K=1,NT(JN)),I=1,N(JS))
 4100 CONTINUE $ENDIF
*
      STOP
****        THESE ARE THE FORMATS
*
 5100 FORMAT(12I6)
 5105 FORMAT(12F6.2)
 5110 FORMAT(I6,6F6.2,2I6)
 5115 FORMAT(2I6,10F6.2)
 5120 FORMAT(I6,F6.2,5I6)
 5130 FORMAT(7I6,3F6.2,I6)
 6000 FORMAT(1H1)
 6040 FORMAT(///)
 6100 FORMAT(5X,*NTPOPS,NCPOPS,NFPOPS,NCLS,MCTP1,W,H=*,7I7)
 6102 FORMAT(5X,*NJN,NLJN,NTR,IFLG1,IFLG2,IFLG3=*,6I7)
 6105 FORMAT(5X,*EQ,T=*,12F7.2)
 6110 FORMAT(5X,*CELLS: N,C,TTH,B,TGK,TMEM,THO,NC,NR=*,
     *  /,*NST,NRG,THD,D,TGC,A,TCA,CA0,BD,TGKD,CDS,VRT=*,//)
 6112 FORMAT(9X,I7,6F7.2,2I7)
 6115 FORMAT(5X,2I7,10F7.2)
 6120 FORMAT(///,5X,*FIBERS: N,P,INSTR,INSTP,INFSED,NC,NR=*,//)
 6122 FORMAT(5X,I7,F7.2,5I7)
 6130 FORMAT(///,5X,*JNCTNS: TYPE,LRN,IS,IR,IRRG,NT,NCT,STR0,WTF,HTF,*
     *        ,*INSED=*,//)
 6132 FORMAT(5X,7I7,3F7.2,I7)
 6150 FORMAT(5X,*ISTM,LSTIM,INTERV,LSTP,NSTIM,NSTIM1=*,6I7)
 6154 FORMAT(5X,*LOCATNS:*)
 6155 FORMAT(10X,20I6)
 6200 FORMAT(5X,*LTSTOP=*,I7)
 6510 FORMAT(5X,15I7)
 6900 FORMAT(5X,20F6.3)
 7510 FORMAT(15I5)
 7060 FORMAT(20I4)
 8900 FORMAT(F6.3)
      END
*
      SUBROUTINE SVUPDT(KK,K)
*
***  THIS SUBROUTINE UPDATES THE STATE VARIABLES OF AN INDIVIDUAL NEURON
*    WITH SIMPLIFIED DENDRITIC TREE CONTAINING ACTIVE CALCIUM-RELATED
***  CONDUCTANCES.
*
      INTEGER SNTP
      PARAMETER SNTP=3,NCPOPS=3,NCLS=100,MCTP1=2,STEP=1.,EK=-10,ECA=50.,
     *          NSTMX=3,NRGMX=3,NRGP1M=4
      DIMENSION U(NRGMX)
      COMMON G(SNTP,NCPOPS,NCLS,NSTMX,NRGP1M,MCTP1),NST(NCPOPS),
     *       NRG(NCPOPS)
```

```
      COMMON/SV/E(NCPOPS,NCLS),TH(NCPOPS,NCLS),S(NCPOPS,NCLS),
     *          GK(NCPOPS,NCLS),B(NCPOPS),C(NCPOPS),TMEM(NCPOPS),
     *          THO(NCPOPS),DCG(NCPOPS),DCTH(NCPOPS),EQ(SNTP),
     *          V(NCPOPS,NCLS,NSTMX,NRGMX),GCA(NCPOPS,NCLS,NSTMX,
     *          NRGMX),CA(NCPOPS,NCLS,NSTMX,NRGMX),GKD(NCPOPS,NCLS,
     *          NSTMX,NRGMX),CDS(NCPOPS),VRT(NCPOPS),DGC(NCPOPS),
     *          THD(NCPOPS),D(NCPOPS),DCA(NCPOPS),A(NCPOPS),
     *          DGKD(NCPOPS),CAØ(NCPOPS),BD(NCPOPS)
*
      GS=Ø. $SYN=Ø. $DO 1Ø I=1,SNTP
      GS=GS+G(I,KK,K,1,NRG(KK)+1,1)
      SYN=SYN+G(I,KK,K,1,NRG(KK)+1,1)*EQ(I)
   1Ø CONTINUE
      GK(KK,K)=GK(KK,K)*DCG(KK)+B(KK)*(1.-DCG(KK))*S(KK,K)
      GTOT=1.+GK(KK,K)+GS $DCE=EXP(-GTOT*STEP/TMEM(KK))
      E(KK,K)=E(KK,K)*DCE+(SYN+GK(KK,K)*EK)*(1.-DCE)/GTOT $DM=Ø.
      DO 98Ø I=1,NST(KK) $DO 92Ø J=1,NRG(KK)
      GT=1.+GCA(KK,K,I,J)+GKD(KK,K,I,J)
      SY=Ø. $DO 91Ø M=1,SNTP $GT=GT+G(M,KK,K,I,J,1)
  91Ø SY=SY+G(M,KK,K,I,J,1)*EQ(M) $Z=EXP(-GT*STEP/TMEM(KK))
  92Ø V(KK,K,I,J)=V(KK,K,I,J)*Z+(SY+GCA(KK,K,I,J)*ECA+GKD(KK,K,I,J)*EK)*
     *          (1.-Z)/GT
*
      DM=DM+V(KK,K,I,1)
      IF(NRG(KK).EQ.1) THEN
      U(1)=(V(KK,K,I,1)+E(KK,K)/CDS(KK))/(1.+1./CDS(KK))*
     *          (1.-VRT(KK))+VRT(KK)*V(KK,K,I,1) $ELSE
      U(1)=(V(KK,K,I,1)+V(KK,K,I,2)+E(KK,K)/CDS(KK))/(2.+1./CDS(KK))*
     *          (1.-VRT(KK))+VRT(KK)*V(KK,K,I,1)
      DO 93Ø J=2,NRG(KK)-1
  93Ø U(J)=(V(KK,K,I,J-1)+V(KK,K,I,J)+V(KK,K,I,J+1))/3.*(1.-VRT(KK))+
     *          VRT(KK)*V(KK,K,I,J)
      U(NRG(KK))=(V(KK,K,I,NRG(KK))+V(KK,K,I,NRG(KK)-1))/2.*
     *          (1.-VRT(KK))+VRT(KK)*V(KK,K,I,NRG(KK)) $ENDIF
      DO 94Ø J=1,NRG(KK)
  94Ø V(KK,K,I,J)=U(J)
*
      DO 95Ø J=1,NRG(KK)
      GCA(KK,K,I,J)=GCA(KK,K,I,J)*DGC(KK)
      IF(V(KK,K,I,J).GE.THD(KK)) GCA(KK,K,I,J)=GCA(KK,K,I,J)+
     *          D(KK)*(V(KK,K,I,J)-THD(KK))*(1.-DGC(KK))
      CA(KK,K,I,J)=CA(KK,K,I,J)*DCA(KK)+A(KK)*GCA(KK,K,I,J)*
     *          (1.-DCA(KK))
      GKD(KK,K,I,J)=GKD(KK,K,I,J)*DGKD(KK)
      IF(CA(KK,K,I,J).GE.CAØ(KK)) GKD(KK,K,I,J)=GKD(KK,K,I,J)+
     *          BD(KK)*(1.-DGKD(KK))
  95Ø CONTINUE
*
  98Ø CONTINUE
*
      E(KK,K)=(E(KK,K)+CDS(KK)*DM)/(1.+CDS(KK)*FLOAT(NST(KK)))*
     *          (1.-VRT(KK))+VRT(KK)*E(KK,K)
      TH(KK,K)=THO(KK)+(TH(KK,K)-THO(KK))*DCTH(KK)+C(KK)*E(KK,K)*
     *          (1.-DCTH(KK))
      S(KK,K)=Ø.
      RETURN
      END
*
      SUBROUTINE TRNSMT(IS,IR,K)
*
****      THIS SUBROUTINE PROJECTS SYNAPTIC ACTIVATION FROM A SENDING
****      FIBER TO ITS TARGETED RECEIVING CELLS.
*
      INTEGER TYPE,W,H,SNTP
      PARAMETER NTPOPS=5,NCPOPS=3,NCLS=1ØØ,MCTP1=2,W=1Ø,H=1Ø,
     *          NJN=6,NLJN=1,NTR=2Ø,NRNGE=1Ø,MRNGE=1ØØØ,SNTP=3,
     *          NSTMX=3,NRGP1M=4
```

```
      COMMON G(SNTP,NCPOPS,NCLS,NSTMX,NRGP1M,MCTP1),NST(NCPOPS),
     *        NRG(NCPOPS)
      COMMON/TR/TYPE(NJN),NT(NJN),NCT(NJN),STRØ(NJN),WTF(NJN),HTF(NJN),
     *        INSED(NJN),LJNN(NTPOPS,NCPOPS),IRRG(NJN),
     *        STR(NLJN,NCLS,NTR),N(NTPOPS),NC(NTPOPS),NR(NTPOPS),
     *        JNN(NTPOPS,NCPOPS),L
*
      JN=JNN(IS,IR)
      XS=FLOAT(MOD(K,NC(IS)))*FLOAT(W)/FLOAT(NC(IS))
      IF(XS.EQ.Ø.) XS=FLOAT(W)
      YS=FLOAT(K/NC(IS)+1)*FLOAT(H)/FLOAT(NR(IS))
      CALL RANSET(K*INSED(JN))
      DO 100 J=1,NT(JN)
      XR=XS+(RANF(Ø.)-.5)*WTF(JN) $YR=YS+(RANF(Ø.)-.5)*HTF(JN)
      ICR=INT(FLOAT(NC(IR))*(XR-.5)/FLOAT(W))+1
      IRR=INT(FLOAT(NR(IR))*(YR-.5)/FLOAT(H))+1
      IF(ICR.LE.Ø) ICR=NC(IR)+ICR
      IF(ICR.GT.NC(IR)) ICR=ICR-NC(IR)
      IF(IRR.LE.Ø) IRR=NR(IR)+IRR
      IF(IRR.GT.NR(IR)) IRR=IRR-NR(IR)
      NREC=NC(IR)*(IRR-1)+ICR
      KST=INT(RANF(Ø.)*FLOAT(NST(IR)))+1
      IF(IRRG(JN).EQ.Ø) THEN
      KRG=INT(RANF(Ø.)*(NRG(IR)+1))+1
      ELSE $KRG=IRRG(JN) $ENDIF
      IF(KRG.EQ.NRG(IR)+1) KST=1
      KCT=INT(RANF(Ø.)*FLOAT(NCT(JN)))+1
      IF(LJNN(IS,IR).GT.Ø) THEN $X=STR(LJNN(IS,IR),K,J)
      ELSE $X=STRØ(JN) $ENDIF
      G(TYPE(JN),IR,NREC,KST,KRG,KCT+1)=G(TYPE(JN),IR,NREC,KST,KRG,KCT+1
     *        )+X
  100 CONTINUE
*
      RETURN
      END
*
      SUBROUTINE STIMULS(IR,LSTIM,INTERV,NSTIM1,MARKER)
*
****      THIS SUBROUTINE
      PARAMETER NSTIM=10,NCLS=100,NCPOPS=3
      COMMON/SV/E(NCPOPS,NCLS),TH(NCPOPS,NCLS)
      COMMON/STIM/LOCATN(NSTIM)
*
      DO 10 I=1,NSTIM1
      J=I+MARKER $E(IR,LOCATN(J))=TH(IR,LOCATN(J))+3.
   10 CONTINUE
      MARKER=MARKER+NSTIM1 $IF(MARKER.GE.NSTIM) MARKER=Ø
      LSTIM=LSTIM+INTERV
      RETURN
      END
*
/EOR
        Ø      1      1
    70.00    .50-10.00 30.00-10.00   2.00
      100    .00 20.00 20.00   5.00 11.00 24.00     10     10
        3      3 24.00   5.00   6.00 30.00   6.00 22.00 65.00   8.00   1.00   0.00
       10    .00 20.00   5.00   5.00   5.00 10.00      5      2
        3      2100.00   5.00   6.00 30.00   6.00 22.00 65.00   8.00   1.00   0.00
       50    .00 20.00 20.00   5.00 11.00 24.00     10      5
        3      3 24.00   5.00   6.00 30.00   6.00 22.00 65.00   8.00   1.00   0.00
       50    .05    300    999   2223     10      5
      100    .10      1    999   1027     10     10
        1      Ø      1      2      1      4      1    .20   2.00   2.00   1227
        1      Ø      1      3      2      2      1    .15   3.00   3.00   1936
        2      Ø      2      1      Ø      5      1    .30   3.00   3.00   1718
        1      1      3      1      1      4      1    .20   2.00   2.00    516
        3      Ø      4      3      1      6      1   -.10   3.00   3.00   7554
```

```
      1        Ø       5        1        1      20       1    3.ØØ    3.ØØ    3.ØØ   8413
      1      999       1       3Ø      10
     63       64      65       66       53      56      43      44      45      46
     3Ø
/EOR
/EOI
```

Appendix 3

Programs for Multiple Embedding of Memory Traces (MLSYS41, LDSYS41, DSPLAY2, CMPTP1)

Programs MLSYS41 and LDSYS41 are extensions of LRSYS32 and LDSYS33 described in Chapter 20. They include the calculation of a Hamiltonian function for each constituent population, the automatic calculation of global statistics for temporal and spatial distributions of activity, and the production of topographic maps of activity. Program DSPLAY2 reads the output files produced by either MLSYS41 or LDSYS41 and produces movies, collections of constituent spike trains, and spike train correlation histograms (interval histogram, autocorrelation histogram, or cross-correlation histogram) with parameters defined by its READ statements. Program CMPTP1 produces the topographic difference map for two topographic distributions so as to compare them spatially.

```
/JOB
MX54,CM=143000,TL=2000.
FAKE.
PDFV.
M77.
GET,STR20C0.
LGO,STR20C0,MB20CD,MT20CD,TOP20CD,STR20CD.
REPLACE,MB20CD,MT20CD,TOP20CD,STR20CD.
EXIT.
REPLACE,MB20CD,MT20CD,TOP20CD,STR20CD.
/EOR
      PROGRAM MLSYS41 (STRIN,MBOUT,MTOUT,TOPOUT,STROUT,INPUT,OUTPUT,
     * TAPE3=STRIN,TAPE6=OUTPUT,TAPE7=MBOUT,TAPE8=MTOUT,TAPE9=TOPOUT,
     * TAPE10=STROUT)
*
****      THIS PROGRAM SIMULATES HEBBIAN-SELECTIVE LEARNING AT THE SYNAPSES
*         OF AN ARBITRARY SYSTEM OF INTERACTING SPATIALLY-ORGANIZED NEURAL
*         NETWORKS ACTIVATED BY AN ARBITRARY NUMBER OF INPUT FIBER SYSTEMS.
*         ALL SYNAPTIC SYSTEMS ARE INDIVIDUALLY ADJUSTABLE. SEPARATE TRANS-
****      MISSION AND MEMORY LINKS. ACIT,OVR PACKED.  JAN 29,1987
*
      INTEGER TYPE,W,H,ACIT,OVR,SNTP
      REAL MUT,MUS
      PARAMETER NTPOPS=3,NFPOPS=1,NCPOPS=2,NCLS=49,MCTP1=4,W=7,H=7,
     *         NJN=6,NLJN=4,NTR=25,NRNGE=17,MRNGE=5000,STEP=1.,SNTP=2,
     *         NSTIM=24,LCLS=1
      DIMENSION P(NFPOPS),INSTP(NFPOPS),INFSED(NFPOPS),NF(NTPOPS),
     *         IDF(NCLS),TTH(NCPOPS),TGK(NCPOPS),IFLG(NJN),LG1(NLJN),
     *         LRN(NJN),IS(NJN),IR(NJN),LINE(NTPOPS,W,H),MUT(NTPOPS),
     *         T(SNTP),DCS(SNTP),INSTR(NFPOPS),HP(NCPOPS),
     *         SIGT(NTPOPS),MUS(NTPOPS),SIGS(NTPOPS),CLS(NTPOPS),
     *         SLP(NTPOPS),STRS(NLJN),DLS(NLJN),SFT(NCPOPS),
     *         DLSF(NCPOPS)
      COMMON G(SNTP,NCPOPS,NCLS,MCTP1)
      COMMON/SV/E(NCPOPS,NCLS),TH(NCPOPS,NCLS),S(NCPOPS,NCLS),
     *         GK(NCPOPS,NCLS),B(NCPOPS),C(NCPOPS),DCG(NCPOPS),
     *         DCTH(NCPOPS),TMEM(NCPOPS),TH0(NCPOPS),EQ(SNTP),
     *         SF(NCPOPS,NCLS)
      COMMON/STIM/LOCATN(NSTIM)
      COMMON/TR/TYPE(NJN),NT(NJN),NCT(NJN),STR0(NJN),WTF(NJN),HTF(NJN),
     *       INSED(NJN),LJNN(NTPOPS,NCPOPS),MLAG(NLJN),
     *         STRF(NLJN),DLF(NLJN),
     *       ACIT(NLJN,NCLS,NRNGE),OVR(NLJN,MRNGE),IVR(NLJN),
     *       STR(NLJN,NCLS,NTR),N(NTPOPS),NC(NTPOPS),NR(NTPOPS),
     *         JNN(NTPOPS,NCPOPS),L
*
****      THIS SECTION READS AND WRITES THE INPUT PARAMETERS
*
      READ 5100, IFLG1,IFLG2
      READ 5105, ((EQ(I),T(I)),I=1,SNTP)
      DO 110 I=1,NCPOPS
  110 READ 5110, N(I),C(I),TTH(I),B(I),TGK(I),TMEM(I),TH0(I),
     *          NC(I),NR(I),SFT(I),DLSF(I)
      DO 120 I=1,NFPOPS $II=NCPOPS+I
  120 READ 5120, N(II),P(I),INSTR(I),INSTP(I),INFSED(I),NC(II),NR(II)
      DO 128 I=1,NCPOPS $IFLG(I)=0 $DO 128 J=1,NTPOPS
      JNN(J,I)=0 $LJNN(J,I)=0
  128 CONTINUE $J=0 $DO 130 I=1,NJN
      READ 5130, TYPE(I),LRN(I),IS(I),IR(I),NT(I),NCT(I),STR0(I),
     *          WTF(I),HTF(I),INSED(I)
      JNN(IS(I),IR(I))=I
      IF(LRN(I).GT.0) THEN $J=J+1 $LJNN(IS(I),IR(I))=J $IFLG(IR(I))=1
      ENDIF
  130 CONTINUE
*
      WRITE(7,6000) $WRITE(9,6001)
```

```
        WRITE(7,6001) $WRITE(7,6100) NTPOPS,NFPOPS,NCPOPS,NCLS,MCTP1,W,H,
        WRITE(7,6102) NJN,NLJN,NTR,NRNGE,MRNGE,IFLG1
        WRITE(7,6105) ((EQ(I),T(I)),I=1,SNTP)
        WRITE(7,6110) $DO 111 I=1,NCPOPS
    111 WRITE(7,6112) N(I),C(I),TTH(I),B(I),TGK(I),TMEM(I),TH0(I),
      *               NC(I),NR(I),SFT(I),DLSF(I)
        WRITE(7,6120) $DO 121 I=1,NFPOPS $II=NCPOPS+I
    121 WRITE(7,6122) N(II),P(I),INSTR(I),INSTP(I),INFSED(I),
      *               NC(II),NR(II)
        WRITE(7,6130) $J=0 $DO 131 I=1,NJN
        WRITE(7,6132) TYPE(I),LRN(I),IS(I),IR(I),NT(I),NCT(I),
      *               STR0(I),WTF(I),HTF(I),INSED(I)
    131 CONTINUE
*
        IF(IFLG2.EQ.0) THEN $DO 320 I=1,NJN $IF(LRN(I).EQ.0) GO TO 320
        II=LJNN(IS(I),IR(I)) $DO 320 J=1,NCLS $DO 320 K=1,NTR
        STR(II,J,K)=STR0(I)
    320 CONTINUE $DO 322 I=1,NCPOPS $DO 322 J=1,NCLS
    322 SF(I,J)=1. $ELSE
        READ(3,8900) ((SF(I,J),J=1,NCLS),I=1,NCPOPS)
        READ(3,8900) (((STR(I,J,K),K=1,NTR),J=1,NCLS),I=1,NLJN) $ENDIF
*
   8000 CONTINUE
        READ 5100, ISTM,LSTIM,INTERV,LSTP,NSTIM1
        IF(ISTM.EQ.0) GO TO 8500
        READ 5100, (LOCATN(I),I=1,NSTIM)
        READ 5105, ((SFT(J),DLSF(J)),J=1,NCPOPS)
        DO 8010 J=1,NLJN
        READ 5140, MLAG(J),STRS(J),DLS(J),STRF(J),DLF(J)
   8010 WRITE(7,6140) MLAG(J),STRS(J),DLS(J),STRF(J),DLF(J)
        READ 5100, LTSTOP
        WRITE(7,6150) ISTM,LSTIM,INTERV,LSTP,NSTIM,NSTIM1
        WRITE(7,6154) $WRITE(7,6155) (LOCATN(I),I=1,NSTIM)
        WRITE(7,6200) LTSTOP $WRITE(7,6000)
*
****           THIS SECTION INITIALIZES VARIABLES
*
        MARKER=0
        DO 310 I=1,NCPOPS
        DCTH(I)=EXP(-STEP/TTH(I)) $DCG(I)=EXP(-STEP/TGK(I))
        DO 310 J=1,NCLS
        E(I,J)=0. $TH(I,J)=TH0(I) $S(I,J)=0. $GK(I,J)=0.
        DO 310 K=1,MCTP1 $DO 310 M=1,SNTP
        G(M,I,J,K)=0.
    310 CONTINUE
        DO 330 I=1,NLJN $IVR(I)=0 $LG1(I)=10000000*(MLAG(I)+1)
        DO 332 J=1,NCLS $DO 332 K=1,NRNGE
    332 ACIT(I,J,K)=0 $DO 334 K=1,MRNGE
    334 OVR(I,K)=0
    330 CONTINUE
        DO 340 I=1,SNTP
    340 DCS(I)=EXP(-STEP/T(I))
        DO 350 I=1,NTPOPS $MUT(I)=0. $SIGT(I)=0.
        DO 350 J=1,W $DO 350 K=1,H
    350 LINE(I,J,K)=0
*
****           THIS SECTION UPDATES STATE VARIABLES AT EACH TIME STEP
*
        DO 1000 L=1,LTSTOP
*
        IF(ISTM.LE.NCPOPS.AND.L.EQ.LSTIM.AND.L.LE.LSTP)
      *   CALL STIMULS(ISTM,LSTIM,INTERV,NSTIM1,MARKER)
*
****                   UPDATING THE CELLS
*
        HT=0.
```

```
      DO 2000 I=1,NCPOPS
      NF(I)=0 $HP(I)=0.
      DO 2900 J=1,N(I)
      CALL SVUPDT(I,J)
      HP(I)=HP(I)+ABS(E(I,J)-TH(I,J))
      IF(E(I,J).LT.TH(I,J)) GO TO 2850
      SF(I,J)=SF(I,J)+DLSF(I)*(SFT(I)-SF(I,J))
      S(I,J)=1. $NF(I)=NF(I)+1 $IDF(NF(I))=J
      JX=INT(FLOAT(MOD((J-1),NC(I))+1)*FLOAT(W)/FLOAT(NC(I)))
      JY=INT(FLOAT((J-1)/NC(I)+1)*FLOAT(H)/FLOAT(NR(I)))
      LINE(I,JX,JY)=LINE(I,JX,JY)+1
*
      IF(IFLG(I).EQ.0) GO TO 2699 $DO 2700 LS=1,NTPOPS
      IF(LJNN(LS,I).EQ.0) GO TO 2700 $LJ=LJNN(LS,I)
      DO 2500 K=1,NRNGE $IF(ACIT(LJ,J,K).EQ.0) GO TO 2499
      IF(ACIT(LJ,J,K).GT.LG1(LJ)) GO TO 2500
      ID=MOD(ACIT(LJ,J,K),10000000) $JC=ID/1000 $JT=ID-JC*1000
      STR(LJ,JC,JT)=STR(LJ,JC,JT)+DLS(LJ)*
     *                (STRS(LJ)/SFT(I)-STR(LJ,JC,JT))
 2500 CONTINUE
 2499 CONTINUE $JUP=IVR(LJ) $DO 2600 K=1,JUP
      IF(OVR(LJ,K)/10000000000.NE.J) GO TO 2600
      IF(MOD(OVR(LJ,K),10000000000).GT.LG1(LJ)) GO TO 2600
      ID=MOD(OVR(LJ,K),10000000) $JC=ID/1000 $JT=ID-JC*1000
      STR(LJ,JC,JT)=STR(LJ,JC,JT)+DLS(LJ)*
     *                (STRS(LJ)/SFT(I)-STR(LJ,JC,JT))
 2600 CONTINUE
 2700 CONTINUE
*
 2699 CONTINUE
      DO 2800 K=1,NCPOPS
      IF(JNN(I,K).GT.0) CALL TRNSMT(I,K,J)
 2800 CONTINUE
 2850 CONTINUE
 2900 CONTINUE $IF(IFLG1.GT.0) WRITE(8,7060) (L,I,NF(I),
     *            (IDF(J),J=1,NF(I)))
      HT=HT+HP(I)
      MUT(I)=MUT(I)+FLOAT(NF(I)) $SIGT(I)=SIGT(I)+FLOAT(NF(I)**2)
 2000 CONTINUE
*
****                  UPDATING THE INPUT FIBERS
*
      DO 3000 I=1,NFPOPS
      II=I+NCPOPS $NF(II)=0
      IF(ISTM.EQ.II.AND.L.EQ.LSTIM.AND.L.LE.LSTP) THEN
      DO 3010 JJ=1,NSTIM1 $J=JJ+MARKER
      NF(II)=NF(II)+1 $IDF(NF(II))=LOCATN(J) $KK=LOCATN(J)-1
      JX=INT(FLOAT(MOD((KK),NC(II))+1)*FLOAT(W)/FLOAT(NC(II)))
      JY=INT(FLOAT((KK)/NC(II)+1)*FLOAT(H)/FLOAT(NR(II)))
      LINE(II,JX,JY)=LINE(II,JX,JY)+1
      DO 3810 K=1,NCPOPS
      IF(JNN(II,K).GT.0) CALL TRNSMT(II,K,LOCATN(J))
 3810 CONTINUE
 3010 CONTINUE
      MARKER=MARKER+NSTIM1 $IF(MARKER.GE.NSTIM) MARKER=0
      LSTIM=LSTIM+INTERV $ELSE
      IF(L.LT.INSTR(I).OR.L.GE.INSTP(I)) GO TO 3998
      CALL RANSET(INFSED(I))
      DO 3900 J=1,N(II)
      IF(RANF(0.).GT.P(I)) GO TO 3850 $CALL RANGET(INFSED(I))
      NF(II)=NF(II)+1 $IDF(NF(II))=J
      JX=INT(FLOAT(MOD((J-1),NC(II))+1)*FLOAT(W)/FLOAT(NC(II)))
      JY=INT(FLOAT((J-1)/NC(II)+1)*FLOAT(H)/FLOAT(NR(II)))
      LINE(II,JX,JY)=LINE(II,JX,JY)+1
      DO 3800 K=1,NCPOPS
      IF(JNN(II,K).GT.0) CALL TRNSMT(II,K,J)
```

```
3800 CONTINUE $CALL RANSET(INFSED(I))
3850 CONTINUE
3900 CONTINUE
     CALL RANGET(INFSED(I))
3998 CONTINUE $ENDIF
     IF(IFLG1.GT.0) WRITE(8,7060) (L,II,NF(II),(IDF(J),J=1,NF(II)))
     MUT(II)=MUT(II)+FLOAT(NF(II)) $SIGT(II)=SIGT(II)+FLOAT(NF(II)**2)
3000 CONTINUE
*
     WRITE(6,6510) L,(NF(I),I=1,NTPOPS)
     WRITE(7,7510) L,(NF(I),I=1,NTPOPS)
     WRITE(6,6511) L,HT,(HP(I),I=1,NCPOPS)
     WRITE(7,7511) L,HT,(HP(I),I=1,NCPOPS)
*
     DO 1910 I=1,NCPOPS $DO 1910 J=1,NCLS $DO 1910 M=1,SNTP
     G(M,I,J,1)=G(M,I,J,1)*DCS(M)+G(M,I,J,2) $DO 1920 K=2,MCTP1-1
1920 G(M,I,J,K)=G(M,I,J,K+1) $G(M,I,J,MCTP1)=0.
1910 CONTINUE
     DO 1600 LJ=1,NLJN $JUP=IVR(LJ) $DO 1700 J=1,JUP
1710 IF(OVR(LJ,J).EQ.0) GO TO 1699
     IF(MOD(OVR(LJ,J),10000000).GT.10000000) GO TO 1700
     IVR(LJ)=IVR(LJ)-1 $DO 1760 J1=J,IVR(LJ)
1760 OVR(LJ,J1)=OVR(LJ,J1+1) $OVR(LJ,IVR(LJ)+1)=0 $GO TO 1710
1700 OVR(LJ,J)=OVR(LJ,J)-10000000
1699 CONTINUE
     DO 1800 I=1,NCLS $DO 1850 J=1,NRNGE
1810 IF(ACIT(LJ,I,J).EQ.0) GO TO  1849
     IF(ACIT(LJ,I,J).GT.10000000) GO TO 1850 $DO 1860 J1=1,NRNGE-1
1860 ACIT(LJ,I,J1)=ACIT(LJ,I,J1+1) $ACIT(LJ,I,NRNGE)=0 $GO TO 1810
1850 ACIT(LJ,I,J)=ACIT(LJ,I,J)-10000000
1849 CONTINUE
1800 CONTINUE
1600 CONTINUE
*
1000 CONTINUE
*
**** THIS SECTION WRITES OUT ACTIVITY VARIABLES
*
     DO 4700 I=1,NTPOPS
     SIGT(I)=SQRT((FLOAT(LTSTOP)*SIGT(I)-MUT(I)**2)/
    *               FLOAT(LTSTOP*(LTSTOP-1)))
     MUT(I)=MUT(I)/FLOAT(LTSTOP) $CLS(I)=0. $SIGS(I)=0. $SLP(I)=0.
     DO 4705 J=1,W $DO 4705 K=1,H
     IF(J.GT.1.AND.J.LE.NC(I)-1.AND.K.GT.1.AND.K.LE.NR(I)-1) THEN
     DO 4703 JX=J-1,J+1 $DO 4703 JY=K-1,K+1
4703 SLP(I)=SLP(I)+ABS(FLOAT(LINE(I,JX,JY)-LINE(I,J,K))) $ENDIF
4705 SIGS(I)=SIGS(I)+FLOAT(LINE(I,J,K)**2)
     SLP(I)=SLP(I)/FLOAT(8*(W-1)*(H-1))
     SIGS(I)=SQRT((FLOAT(N(I))*SIGS(I)-(MUT(I)*FLOAT(LTSTOP))**2)/
    *              FLOAT(N(I)*(N(I)-1)))
     MUS(I)=MUT(I)*FLOAT(LTSTOP)/FLOAT(N(I))
     DO 4707 J=1+LCLS,NC(I)-LCLS $DO 4707 K=1+LCLS,NR(I)-LCLS $CLSP=0.
     DO 4706 JI=J-LCLS,J+LCLS $DO 4706 KI=K-LCLS,K+LCLS
4706 CLSP=CLSP+FLOAT(LINE(I,JI,KI))
4707 CLS(I)=CLS(I)+ABS(CLSP-((1.+2.*FLOAT(LCLS))**2)*MUS(I))
     CLS(I)=CLS(I)/FLOAT((NC(I)-2*LCLS)*(NR(I)-2*LCLS)*(1+2*LCLS)**2)
     WRITE(9,9010) MUT(I),SIGT(I),MUS(I),SIGS(I),SLP(I),CLS(I)
     WRITE(6,6040) $DO 4710 JY=1,H
     WRITE(9,6070) ((FLOAT(LINE(I,JX,H+1-JY))/MUS(I)),JX=1,W)
4710 WRITE(6,6070) ((FLOAT(LINE(I,JX,H+1-JY))/MUS(I)),JX=1,W)
4700 CONTINUE
*
     GO TO 8000
8500 CONTINUE
*
     WRITE(10,8900) ((SF(I,J),J=1,NCLS),I=1,NCPOPS)
```

```
      WRITE(6,6900) (((STR(I,J,K),K=1,NTR),J=1,NCLS),I=1,NLJN)
      WRITE(10,8900) (((STR(I,J,K),K=1,NTR),J=1,NCLS),I=1,NLJN)
*
      STOP
****         THESE ARE THE FORMATS
*
 5100 FORMAT(12I6)
 5105 FORMAT(12F6.2)
 5110 FORMAT(I6,6F6.2,2I6,2F6.2)
 5120 FORMAT(I6,F6.2,5I6)
 5130 FORMAT(6I6,3F6.2,I6)
 5140 FORMAT(I6,4F6.2)
 6000 FORMAT(1H1)
 6001 FORMAT(3X,*MB20CD*,//)
 6040 FORMAT(///)
 6070 FORMAT(2X,30F4.1)
 6100 FORMAT(5X,*NTPOPS,NFPOPS,NCPOPS,NCLS,MCTP1,W,H=*,7I7)
 6102 FORMAT(5X,*NJN,NLJN,NTR,NRNGE,MRNGE,IFLG1=*,6I7)
 6105 FORMAT(5X,*EQ,T=*,12F7.2)
 6110 FORMAT(5X,*CELLS: N,C,TTH,B,TGK,TMEM,THO,NC,NR,SFT,DLSF=*)
 6112 FORMAT(9X,I7,6F7.2,2I7,2F7.2)
 6120 FORMAT(///,5X,*FIBERS: N,P,INSTR,INSTP,INFSED,NC,NR=*,//)
 6122 FORMAT(5X,I7,F7.2,5I7)
 6130 FORMAT(///,5X,*JNCTNS: TYPE,LRN,IS,IR,NT,NCT,STR0,WTF,HTF,*
     *         ,*INSED=*,/10X,*MLAG,STRS,DLS,STRF,DLF=*,//)
 6132 FORMAT(5X,6I7,3F7.2,I7)
 6140 FORMAT(10X,I7,4F7.2,2F7.2)
 6150 FORMAT(5X,*ISTM,LSTIM,INTERV,LSTP,NSTIM,NSTIM1=*,6I7)
 6154 FORMAT(5X,*LOCATNS:*)
 6155 FORMAT(10X,20I6)
 6200 FORMAT(5X,*LTSTOP=*,I7)
 6510 FORMAT(5X,15I7)
 6511 FORMAT(25X,I10,10F10.0)
 6900 FORMAT(5X,20F6.3)
 7510 FORMAT(15I5)
 7511 FORMAT(20X,I10,6F10.0)
 7060 FORMAT(20I4)
 8900 FORMAT(F6.3)
 9010 FORMAT(8F8.2)
      END
*
      SUBROUTINE SVUPDT(I,K)
*
****         THIS SUBROUTINE UPDATES THE STATE VARIABLES OF AN INDIVIDUAL NEURON
*
      INTEGER SNTP
      PARAMETER NCPOPS=2,NCLS=49,MCTP1=4,STEP=1.,EK=-10.,SNTP=2
      COMMON G(SNTP,NCPOPS,NCLS,MCTP1)
      COMMON/SV/E(NCPOPS,NCLS),TH(NCPOPS,NCLS),S(NCPOPS,NCLS),
     *         GK(NCPOPS,NCLS),B(NCPOPS),C(NCPOPS),DCG(NCPOPS),
     *         DCTH(NCPOPS),TMEM(NCPOPS),TH0(NCPOPS),EQ(SNTP),
     *         SF(NCPOPS,NCLS)
*
      GS=0. $SYN=0. $DO 10 J=1,SNTP
      GS=GS+G(J,I,K,1) $SYN=SYN+G(J,I,K,1)*EQ(J)
   10 CONTINUE $GS=SF(I,K)*GS $SYN=SF(I,K)*SYN
      GK(I,K)=GK(I,K)*DCG(I)+B(I)*(1.-DCG(I))*S(I,K)
      GTOT=1.+GK(I,K)+GS $DCE=EXP(-GTOT*STEP/TMEM(I))
      E(I,K)=E(I,K)*DCE+(SYN+GK(I,K)*EK)*(1.-DCE)/
     *         GTOT
*     TH(I,K)=TH0(I)+(TH(I,K)-TH0(I))*DCTH(I)+C(I)*E(I,K)*(1.-DCTH(I))
      S(I,K)=0.
*
      RETURN
      END
*
```

```
      SUBROUTINE TRNSMT(IS,IR,K)
*
****      THIS SUBROUTINE PROJECTS SYNAPTIC ACTIVATION FROM A SENDING
****      FIBER TO ITS TARGETED RECEIVING CELLS.
*
      INTEGER TYPE,W,H,ACIT,OVR,SNTP
      PARAMETER NTPOPS=3,NCPOPS=2,NCLS=49,MCTP1=4,W=7,H=7,
     *      NJN=6,NLJN=4,NTR=25,NRNGE=17,MRNGE=5000,SNTP=2
      COMMON G(SNTP,NCPOPS,NCLS,MCTP1)
      COMMON/TR/TYPE(NJN),NT(NJN),NCT(NJN),STR0(NJN),WTF(NJN),HTF(NJN),
     *      INSED(NJN),LJNN(NTPOPS,NCPOPS),MLAG(NLJN),
     *      STRF(NLJN),DLF(NLJN),
     *      ACIT(NLJN,NCLS,NRNGE),OVR(NLJN,MRNGE),IVR(NLJN),
     *      STR(NLJN,NCLS,NTR),N(NTPOPS),NC(NTPOPS),NR(NTPOPS),
     *      JNN(NTPOPS,NCPOPS),L
*
      JN=JNN(IS,IR)
      XS=FLOAT(MOD(K,NC(IS)))*FLOAT(W)/FLOAT(NC(IS))
      IF(XS.EQ.0.) XS=FLOAT(W)
      YS=FLOAT((K-1)/NC(IS)+1)*FLOAT(H)/FLOAT(NR(IS))
      CALL RANSET(K*INSED(JN))
      DO 100 J=1,NT(JN)
      XR=XS+(RANF(0.)-.5)*WTF(JN) $YR=YS+(RANF(0.)-.5)*HTF(JN)
      ICR=INT(FLOAT(NC(IR))*(XR-.5)/FLOAT(W))+1
      IRR=INT(FLOAT(NR(IR))*(YR-.5)/FLOAT(H))+1
      IF(ICR.LE.0) ICR=NC(IR)+ICR
      IF(ICR.GT.NC(IR)) ICR=ICR-NC(IR)
      IF(IRR.LE.0) IRR=NR(IR)+IRR
      IF(IRR.GT.NR(IR)) IRR=IRR-NR(IR)
      NREC=NC(IR)*(IRR-1)+ICR
      KCT=INT(RANF(0.)*FLOAT(NCT(JN)))+1
      IF(LJNN(IS,IR).GT.0) THEN $I=LJNN(IS,IR)
      DO 200 JR=1,NRNGE
      IF(ACIT(I,NREC,JR).EQ.0) GO TO 210
200   CONTINUE $IVR(I)=IVR(I)+1 $IF(IVR(I).GT.(MRNGE-1)) L=99999
      OVR(I,IVR(I))=J+1000*K+10000000*(KCT+MLAG(I))+10000000000*NREC
      GO TO 220
210   ACIT(I,NREC,JR)=J+1000*K+10000000*(KCT+MLAG(I))
220   G(TYPE(JN),IR,NREC,KCT+1)=G(TYPE(JN),IR,NREC,KCT+1)+STR(I,K,J)
      STR(I,K,J)=STR(I,K,J)+DLF(I)*(STRF(I)-STR(I,K,J))
      ELSE
      G(TYPE(JN),IR,NREC,KCT+1)=G(TYPE(JN),IR,NREC,KCT+1)+STR0(JN)
      ENDIF
100   CONTINUE
*
      RETURN
      END
*
      SUBROUTINE STIMULS(IR,LSTIM,INTERV,NSTIM1,MARKER)
*
****      THIS SUBROUTINE
      PARAMETER NSTIM=24,NCLS=49,NCPOPS=2
      COMMON/SV/E(NCPOPS,NCLS),TH(NCPOPS,NCLS)
      COMMON/STIM/LOCATN(NSTIM)
*
      DO 10 I=1,NSTIM1
      J=I+MARKER $E(IR,LOCATN(J))=TH(IR,LOCATN(J))+3.
10    CONTINUE
      MARKER=MARKER+NSTIM1 $IF(MARKER.GE.NSTIM) MARKER=0
      LSTIM=LSTIM+INTERV
      RETURN
      END
*
/EOR
      1     1
70.      .01-10.     .01
```

```
    49   0.    999.99 20.    2.      5.00 10.           7       7     .2   1.
    49   0.    999.99 10.    2.      5.00 10.           7       7     .2   1.
    49   .01   999    000  2787      7       7
     1    1      1      1    25      3     .00  3.     3.     1227
     1    1      1      2    10      3     .00  5.     5.     1936
     2    1      2      2    25      3     .00  5.     5.     1130
     2    1      2      1    25      3     .00  5.     5.     1938
     1    0      3      1     1      1  99.     1.     1.     0604
     1    0      3      2     1      1  99.     1.     1.     1960
     3    1      1    100    12
     2    4      6      8    10     12      14      16      18      20     22     24
    26   28     30     32    34     36      38      40      42      44     46     48
    .2   .1     .2     .1
     2   .5     .2     .05   .05
     2   .5     .2     .05   .05
     2  2.     .00  4.       .1
     2  2.     .00  4.       .1
   100
     0
/EOR
/EOI
    40
     0
/EOR
/EOI
```

```
/JOB
MX54,CM=70000,TL=3000.
FAKE.
PDFV.
M77.
GET,STR10CD.
LGO,STR10CD,MRCD04,MTRCD04,TPRCD04.
REPLACE,MRCD04,MTRCD04,TPRCD04.
EXIT.
REPLACE,MRCD04,MTRCD04,TPRCD04.
/EOR
      PROGRAM LDSYS41 (STRIN,MROUT,MTROUT,TPR,INPUT,OUTPUT,
     * TAPE3=STRIN,TAPE6=OUTPUT,TAPE7=MROUT,TAPE8=MTROUT,TAPE9=TPR)
*
****      THIS PROGRAM SIMULATES HEBBIAN LEARNING AT THE SYNAPSES OF AN
*         ARBITRARY SYSTEM OF INTERACTING SPATIALLY-ORGANIZED NEURAL NETWORKS
*         ACTIVATED BY AN ARBITRARY NUMBER OF INPUT FIBER SYSTEMS.  ALL
*         SYNAPTIC SYSTEMS ARE INDIVIDUALLY ADJUSTABLE. SEPARATE TRANS-
****      MISSION AND MEMORY LINKS. ACIT,OVR PACKED.
*
      INTEGER TYPE,W,H,SNTP
      REAL MUT,MUS
      PARAMETER NTPOPS=3,NFPOPS=1,NCPOPS=2,NCLS=49,MCTP1=4,W=7,H=7,
     *          NJN=6,NLJN=4,NTR=25,STEP=1.,SNTP=2,LCLS=1,
     *          NSTIM=24
      DIMENSION P(NFPOPS),INSTP(NFPOPS),INFSED(NFPOPS),NF(NTPOPS),
     *          IDF(NCLS),TTH(NCPOPS),TGK(NCPOPS),IFLG(NJN),
     *          LRN(NJN),IS(NJN),IR(NJN),HP(NCPOPS),LINE(NTPOPS,W,H),
     *          T(SNTP),DCS(SNTP),INSTR(NFPOPS),MUT(NTPOPS),
     *          SIGT(NTPOPS),MUS(NTPOPS),SIGS(NTPOPS),CLS(NTPOPS),
     *          SLP(NTPOPS)
      COMMON G(SNTP,NCPOPS,NCLS,MCTP1)
      COMMON/SV/E(NCPOPS,NCLS),TH(NCPOPS,NCLS),S(NCPOPS,NCLS),
     *          GK(NCPOPS,NCLS),B(NCPOPS),C(NCPOPS),DCG(NCPOPS),
     *          DCTH(NCPOPS),TMEM(NCPOPS),TH0(NCPOPS),EQ(SNTP),
     *          SF(NCPOPS,NCLS)
      COMMON/STIM/LOCATN(NSTIM)
      COMMON/TR/TYPE(NJN),NT(NJN),NCT(NJN),STR0(NJN),WTF(NJN),HTF(NJN),
     *          INSED(NJN),LJNN(NTPOPS,NCPOPS),
     *          STR(NLJN,NCLS,NTR),N(NTPOPS),NC(NTPOPS),NR(NTPOPS),
     *          JNN(NTPOPS,NCPOPS),L
*
****      THIS SECTION READS AND WRITES THE INPUT PARAMETERS
*
      READ 5100, IFLG1,IFLG2,IFLG3
      READ 5105, ((EQ(I),T(I)),I=1,SNTP)
      DO 110 I=1,NCPOPS
  110 READ 5110, N(I),C(I),TTH(I),B(I),TGK(I),TMEM(I),TH0(I),
     *          NC(I),NR(I)
      DO 120 I=1,NFPOPS $II=NCPOPS+I
  120 READ 5120, N(II),P(I),INSTR(I),INSTP(I),INFSED(I),NC(II),NR(II)
      DO 128 I=1,NCPOPS $IFLG(I)=0 $DO 128 J=1,NTPOPS
      JNN(J,I)=0 $LJNN(J,I)=0
  128 CONTINUE $J=0 $DO 130 I=1,NJN
      READ 5130, TYPE(I),LRN(I),IS(I),IR(I),NT(I),NCT(I),STR0(I),
     *          WTF(I),HTF(I),INSED(I)
      JNN(IS(I),IR(I))=1
      IF(LRN(I).GT.0) THEN $J=J+1 $LJNN(IS(I),IR(I))=J $IFLG(IR(I))=1
      ENDIF
  130 CONTINUE
      READ 5100, ISTM,LSTIM,INTERV,LSTP,NSTIM1
      READ 5100, (LOCATN(I),I=1,NSTIM)
      READ 5100, LTSTOP
*
      WRITE(7,6000) $WRITE(9,6010)
      WRITE(7,6010) $WRITE(7,6100) NTPOPS,NFPOPS,NCPOPS,NCLS,MCTP1,W,H,
```

```
      WRITE(7,6102) NJN,NLJN,NTR,IFLG1,IFLG2,IFLG3
      WRITE(7,6105) ((EQ(I),T(I)),I=1,SNTP)
      WRITE(7,6110) $DO 111 I=1,NCPOPS
  111 WRITE(7,6112) N(I),C(I),TTH(I),B(I),TGK(I),TMEM(I),TH0(I),
     *              NC(I),NR(I)
      WRITE(7,6120) $DO 121 I=1,NFPOPS $II=NCPOPS+I
  121 WRITE(7,6122) N(II),P(I),INSTR(I),INSTP(I),INFSED(I),
     *              NC(II),NR(II)
      WRITE(7,6130) $J=0 $DO 131 I=1,NJN
      WRITE(7,6132) TYPE(I),LRN(I),IS(I),IR(I),NT(I),NCT(I),
     *              STR0(I),WTF(I),HTF(I),INSED(I)
  131 CONTINUE
      WRITE(7,6150) ISTM,LSTIM,INTERV,LSTP,NSTIM,NSTIM1
      WRITE(7,6154) $WRITE(7,6155) (LOCATN(I),I=1,NSTIM)
      WRITE(7,6200) LTSTOP $WRITE(7,6000)
*
****        THIS SECTION INITIALIZES VARIABLES
*
      MARKER=0
      DO 310 I=1,NCPOPS
      DCTH(I)=EXP(-STEP/TTH(I)) $DCG(I)=EXP(-STEP/TGK(I))
      DO 310 J=1,NCLS
      E(I,J)=0. $TH(I,J)=TH0(I) $S(I,J)=0. $GK(I,J)=0.
      DO 310 K=1,MCTP1 $DO 310 M=1,SNTP
      G(M,I,J,K)=0.
  310 CONTINUE
      DO 340 I=1,SNTP
  340 DCS(I)=EXP(-STEP/T(I))
      IF(IFLG2.EQ.0) THEN $DO 320 I=1,NJN $IF(LRN(I).EQ.0) GO TO 320
      II=LJNN(IS(I),IR(I)) $DO 320 J=1,NCLS $DO 320 K=1,NTR
      STR(II,J,K)=STR0(I)
  320 CONTINUE $DO 322 I=1,NCPOPS $DO 322 J=1,NCLS
  322 SF(I,J)=1. $ELSE
      READ(3,8900) ((SF(I,J),J=1,NCLS),I=1,NCPOPS)
      READ(3,8900) (((STR(I,J,K),K=1,NTR),J=1,NCLS),I=1,NLJN) $ENDIF
      DO 350 I=1,NTPOPS $MUT(I)=0. $SIGT(I)=0.
      DO 350 J=1,W $DO 350 K=1,H
  350 LINE(I,J,K)=0
*
****        THIS SECTION UPDATES STATE VARIABLES AT EACH TIME STEP
*
      DO 1000 L=1,LTSTOP
*
      IF(ISTM.LE.NCPOPS.AND.L.EQ.LSTIM.AND.L.LE.LSTP)
     *   CALL STIMULS(ISTM,LSTIM,INTERV,NSTIM1,MARKER)
*
****              UPDATING THE CELLS
*
      HT=0.
      DO 2000 I=1,NCPOPS
      NF(I)=0 $HP(I)=0.
      DO 2900 J=1,N(I)
      CALL SVUPDT(I,J)
      HP(I)=HP(I)+ABS(E(I,J)-TH(I,J))
      IF(E(I,J).LT.TH(I,J)) GO TO 2850
      S(I,J)=1. $NF(I)=NF(I)+1 $IDF(NF(I))=J
      JX=INT(FLOAT(MOD((J-1),NC(I))+1)*FLOAT(W)/FLOAT(NC(I)))
      JY=INT(FLOAT((J-1)/NC(I)+1)*FLOAT(H)/FLOAT(NR(I)))
      LINE(I,JX,JY)=LINE(I,JX,JY)+1
*
      DO 2800 K=1,NCPOPS
      IF(JNN(I,K).GT.0) CALL TRNSMT(I,K,J)
 2800 CONTINUE
 2850 CONTINUE
 2900 CONTINUE $IF(IFLG1.GT.0) WRITE(8,7060) (L,I,NF(I),
     *          (IDF(J),J=1,NF(I)))
```

```
      HT=HT+HP(I)
      MUT(I)=MUT(I)+FLOAT(NF(I)) $SIGT(I)=SIGT(I)+FLOAT(NF(I)**2)
2000 CONTINUE
*
****              UPDATING THE INPUT FIBERS
*
      DO 3000 I=1,NFPOPS
      II=I+NCPOPS $NF(II)=0
      IF(ISTM.EQ.II.AND.L.EQ.LSTIM.AND.L.LE.LSTP) THEN
      DO 3010 JJ=1,NSTIM1 $J=JJ+MARKER
      NF(II)=NF(II)+1 $IDF(NF(II))=LOCATN(J) $KK=LOCATN(J)-1
      JX=INT(FLOAT(MOD((KK),NC(II))+1)*FLOAT(W)/FLOAT(NC(II)))
TPOPS)JY=INT(FLOAT((KK)/NC(II)+1)*FLOAT(H)/FLOAT(NR(II)))
      WRITE(7,7510) L,(NF(I),I=1,NTPOPS)
      WRITE(6,6511) L,HT,(HP(I),I=1,NCPOPS)
      WRITE(7,7511) L,HT,(HP(I),I=1,NCPOPS)
*
      DO 1910 I=1,NCPOPS $DO 1910 J=1,NCLS $DO 1910 M=1,SNTP
      G(M,I,J,1)=G(M,I,J,1)*DCS(M)+G(M,I,J,2) $DO 1920 K=2,MCTP1-1
1920 G(M,I,J,K)=G(M,I,J,K+1) $G(M,I,J,MCTP1)=0.
1910 CONTINUE
*
1000 CONTINUE
*
****          THIS SECTION WRITES OUT ACTIVITY VARIABLES
*
      DO 4700 I=1,NTPOPS
      SIGT(I)=SQRT((FLOAT(LTSTOP)*SIGT(I)-MUT(I)**2)/
     *          FLOAT(LTSTOP*(LTSTOP-1)))
      MUT(I)=MUT(I)/FLOAT(LTSTOP) $CLS(I)=0. $SIGS(I)=0. $SLP(I)=0.
      DO 4705 J=1,W $DO 4705 K=1,H
      IF(J.GT.1.AND.J.LE.NC(I)-1.AND.K.GT.1.AND.K.LE.NR(I)-1) THEN
      DO 4703 JX=J-1,J+1 $DO 4703 JY=K-1,K+1
4703 SLP(I)=SLP(I)+ABS(FLOAT(LINE(I,JX,JY)-LINE(I,J,K))) $ENDIF
4705 SIGS(I)=SIGS(I)+FLOAT(LINE(I,J,K)**2)
      SLP(I)=SLP(I)/FLOAT(8*(W-1)*(H-1))
      SIGS(I)=SQRT((FLOAT(N(I))*SIGS(I)-(MUT(I)*FLOAT(LTSTOP))**2)/
     *          FLOAT(N(I)*(N(I)-1)))
      MUS(I)=MUT(I)*FLOAT(LTSTOP)/FLOAT(N(I))
      DO 4707 J=1+LCLS,NC(I)-LCLS $DO 4707 K=1+LCLS,NR(I)-LCLS $CLSP=0.
      DO 4706 JI=J-LCLS,J+LCLS $DO 4706 KI=K-LCLS,K+LCLS
4706 CLSP=CLSP+FLOAT(LINE(I,JI,KI))
4707 CLS(I)=CLS(I)+ABS(CLSP-((1.+2.*FLOAT(LCLS))**2)*MUS(I))
      CLS(I)=CLS(I)/FLOAT((NC(I)-2*LCLS)*(NR(I)-2*LCLS)*(1+2*LCLS)**2)
      WRITE(9,9010) MUT(I),SIGT(I),MUS(I),SIGS(I),SLP(I),CLS(I)
      WRITE(6,6040) $DO 4710 JY=1,H
      WRITE(9,6070) ((FLOAT(LINE(I,JX,H+1-JY))/MUS(I)),JX=1,W)
4710 WRITE(6,6070) ((FLOAT(LINE(I,JX,H+1-JY))/MUS(I)),JX=1,W)
4700 CONTINUE
      IF(IFLG3.GT.0) THEN
      WRITE(6,6900) ((SF(I,J),J=1,NCLS),I=1,NCPOPS)
      WRITE(6,6900) (((STR(I,J,K),K=1,NTR),J=1,NCLS),I=1,NLJN)
      ENDIF
*
      STOP
****           THESE ARE THE FORMATS
*
5100 FORMAT(12I6)
5105 FORMAT(12F6.2)
5110 FORMAT(I6,6F6.2,2I6)
5120 FORMAT(I6,F6.2,5I6)
5130 FORMAT(6I6,3F6.2,I6)
6000 FORMAT(1H1)
6010 FORMAT (3X,*MRCD04*,//)
6040 FORMAT(///)
6070 FORMAT(2X,30F4.1)
```

```
/JOB
MX54,CM=70000,TL=100.
FAKE.
PDFV.
MNF(Y).
GET,MTRCD01.
LGO,MTRCD01,MVECD01,SPTCD01,CRLCD01.
REPLACE,MVECD01,SPTCD01,CRLCD01.
EXIT.
REPLACE,MVECD01,SPTCD01,CRLCD01.
EXIT.
REPLACE,SPTCD01,CRLCD01.
EXIT.
REPLACE,CRLCD01.
/EOR
      PROGRAM DSPLAY2(MTOT,MVE,SPT,CRL,INPUT,OUTPUT,
     *        TAPE3=MTOT,TAPE6=OUTPUT,TAPE7=MVE,TAPE8=SPT,TAPE9=CRL)
*
      INTEGER POPTG,PT
      PARAMETER NTPOPS=3,LTSTOP=1000,NFMX=49,NC=7,NR=7,NTMX=17,
     *        NTT=9,NSPK=250,NBINS=31
*
      DIMENSION IDF(NFMX),CHAR(3),AFRAME(NC,NR),LIFE(NC,NR),
     *        NT(NTPOPS),ALINE(NFMX),PT(NTT),IDTT(NTT),IST(100),
     *        NSP(NTT),T(NTT,NSPK),X(NBINS),ST(NBINS),B(NBINS)
*
      CHAR(1)=1H   $CHAR(2)=1H*  $CHAR(3)=1H-
*
      READ 5010, IFLG1
*
      IF(IFLG1.EQ.1.OR.IFLG1.EQ.12.OR.IFLG1.EQ.13.OR.IFLG1.EQ.4) THEN
*
*******THIS SECTION PRODUCES THE MOVIE FILE, MVE******************
*
      READ 5010, POPTG,LSTP,LFTM
      DO 1010 JX=1,NC $DO 1010 JY=1,NR
      LIFE(JX,JY)=0
 1010 AFRAME(JX,JY)=CHAR(1)
*
      DO 1000 L=1,LSTP
*
      DO 900 I=1,NTPOPS
      READ(3,3060) LP,IP,NP,(IDF(J),J=1,NP)
      IF(IP.NE.POPTG) GO TO 900
      DO 910 J=1,NP $JJ=IDF(J)
      JX=MOD((JJ-1),NC)+1 $JY=INT(FLOAT((JJ-1)/NC+1))
      AFRAME(JX,JY)=CHAR(2) $LIFE(JX,JY)=LFTM
  910 CONTINUE
  900 CONTINUE
*
      WRITE(7,7010) L,POPTG,LFTM
*
      DO 1700 JY=1,NR $JYY=NR+1-JY
      WRITE(7,7020) (AFRAME(JX,JYY),JX=1,NC)
      DO 1690 JX=1,NC
      LIFE(JX,JYY)=LIFE(JX,JYY)-1
      IF(LIFE(JX,JYY).LE.0) AFRAME(JX,JYY)=CHAR(1)
 1690 CONTINUE
 1700 CONTINUE
*
 1000 CONTINUE $ENDIF
*
*
      IF(IFLG1.EQ.2.OR.IFLG1.EQ.12.OR.IFLG1.EQ.23.OR.IFLG1.EQ.4) THEN
      REWIND 3
      WRITE (8,8000)
```

```
*
*********THIS SECTION PRODUCES THE SPIKE TRAIN DISPLAY, SPT*********
*
      DO 2010 II=1,NTPOPS
      READ 5010, IFLG,LSTP
      IF(IFLG.EQ.0) GO TO 2010
*
*
      DO 2000 L=1,LSTP
*
      DO 2020 J=1,NFMX
 2020 ALINE(J)=CHAR(1)
      DO 1900 I=1,NTPOPS
      READ(3,3060) LP,IP,NP,(IDF(J),J=1,NP)
      IF(IP.EQ.II) THEN
      DO 1910 J=1,NFMX $DO 1910 K=1,NP
      IF(IDF(K).EQ.J) ALINE(J)=CHAR(3)
 1910 CONTINUE $ENDIF
 1900 CONTINUE
*
      WRITE(8,8010) L,(ALINE(J),J=1,NFMX)
      WRITE(6,6020) L,(ALINE(J),J=1,NFMX)
*
 2000 CONTINUE
      REWIND 3
 2010 CONTINUE $ENDIF
*
*
      IF(IFLG1.EQ.3.OR.IFLG1.EQ.13.OR.IFLG1.EQ.23.OR.IFLG1.EQ.4) THEN
      REWIND 3
      WRITE(9,9001)
*
***********THIS SECTION PRODUCES THE SPIKE TRAIN CORRELATIONS, CRL****
*
      READ 5010, NST,LSTP
      READ 5010, ((PT(I),IDTT(I)),I=1,NST)
      WRITE(9,9020) NST,((PT(I),IDTT(I)),I=1,NST)
*
      DO 3005 J=1,NST
 3005 NSP(J)=0
*
      DO 3090 L=1,LSTP
      DO 3080 I=1,NTPOPS
      READ(3,3060) LP,IP,NP,(IDF(J),J=1,NP)
      DO 3070 J=1,NST
      IF(PT(J).NE.IP) GO TO 3070
      DO 3050 K=1,NP
      IF(IDF(K).EQ.IDTT(J)) THEN
      NSP(J)=NSP(J)+1 $T(J,NSP(J))=FLOAT(LP) $ENDIF
 3050 CONTINUE
 3070 CONTINUE
 3080 CONTINUE
 3090 CONTINUE
*
    1 READ 6010, (KE,IT1,IT2,TLO,TUP,RLF,RRT,BSZ)
      IF(KE.EQ.0) GO TO 9000
      WRITE(6,6600) $WRITE(6,6610) (KE,IT1,IT2,TLO,TUP,RLF,RRT,BSZ)
      WRITE(9,9005) KE,IT1,IT2,TLO,TUP,RLF,RRT,BSZ
      X1=0. $DO 210 L=1,NBINS
  210 B(L)=0.
      IF(KE.EQ.1) GO TO 100
      IF(KE.EQ.2) GO TO 200
      IF(KE.EQ.3) GO TO 300
      GO TO 9000
*
  100 CONTINUE
```

```
*                    THIS DOES THE INTERVAL HISTOGRAM
*
      DO 1100 I=1,NSP(IT1)
      IF(T(IT1,I).LT.TLO) GO TO 1100
      IF(T(IT1,I).GT.TUP-RRT) GO TO 1300
      X1=X1+1. $L=INT((T(IT1,I+1)-T(IT1,I))/BSZ)+1 $B(L)=B(L)+1.
 1100 CONTINUE
 1300 CONTINUE
      GO TO 5000
*
  200 CONTINUE
*                    THIS DOES THE AUTO-CORRELATION HISTOGRAM
*
      DO 2100 I=1,NSP(IT1)
      IF(T(IT1,I).LT.TLO) GO TO 2100
      IF(T(IT1,I).GT.TUP-RRT) GO TO 2300
      X1=X1+1. $JL=I+1 $DO 2200 J=JL,NSP(IT1)
      TIN=T(IT1,J)-T(IT1,I) $IF(TIN.GT.RRT) GO TO 2100
      L=INT(TIN/BSZ)+1 $B(L)=B(L)+1.
 2200 CONTINUE
 2100 CONTINUE
 2300 CONTINUE
      GO TO 5000
*
  300 CONTINUE
*                    THIS DOES THE CROSS-CORRELATION HISTOGRAM
*
      JL=1 $DO 3100 I=1,NSP(IT1)
      IF(T(IT1,I).LT.TLO+RLF) GO TO 3100
      IF(T(IT1,I).GT.TUP-RRT) GO TO 3300
      X1=X1+1. $DO 3200 J=JL,NSP(IT2) $TIN=T(IT2,J)-T(IT1,I)
      IF(-TIN.GT.RLF) JL=J $IF(-TIN.GT.RLF) GO TO 3200
      IF(TIN.GT.RRT) GO TO 3100
      L=INT((TIN+RLF)/BSZ)+1 $B(L)=B(L)+1.
 3200 CONTINUE
 3100 CONTINUE
 3300 CONTINUE
*
 5000 CONTINUE
*
*                    THIS WRITES OUT THE NUMBERS
*
      X2=0. $DO 501 I=1,NSP(IT2)
      IF(T(IT2,I).GE.TLO.AND.T(IT2,I).LE.TUP) X2=X2+1.
  501 CONTINUE
      NB=(RLF+RRT)/BSZ $YMX=0. $XM=X1*X2*BSZ/(TUP-TLO)
      DO 5020 I=1,NB $X(I)=FLOAT(I)*BSZ-RLF
 5020 ST(I)=(B(I)-XM)/SQRT(B(I)*(X1-B(I))/(X1-1))
      WRITE(6,6620) (NB,X1,X2,XM)
      WRITE(6,6607) $WRITE(6,6630) (X(L),L=1,NB)
      WRITE(6,6605) $WRITE(6,6630) (B(L),L=1,NB)
      WRITE(6,6605) $WRITE(6,6630) (ST(L),L=1,NB)
      WRITE(9,9008) NB
      WRITE(9,9010) ((X(L),B(L)),L=1,NB)
      DO 5100 I=1,NB
      WRITE(9,9100) I,(CHAR(2),L=1,INT(B(I)))
 5100 CONTINUE
*
*
      GO TO 1
*
 9000 CONTINUE $ENDIF
*
      STOP
 5010 FORMAT(18I4)
 3060 FORMAT(20I4)
```

```
7010 FORMAT(3I4)
7020 FORMAT(80A1)
8000 FORMAT(1H1,//,5X,*SPTCD01*,/)
8010 FORMAT(I4,2X,50A1)
9001 FORMAT (1H1,//,5X,*CRLCD01*,/)
9005 FORMAT(//,2X,3I5,5F10.4)
9008 FORMAT(I4)
9010 FORMAT(8F8.2)
9020 FORMAT(20I4)
9100 FORMAT(3X,I2,100A1)
6020 FORMAT(I4,2X,50A2)
6010 FORMAT(3I5,5F10.4)
6600 FORMAT(1H1)
6605 FORMAT(/)
6607 FORMAT(//,7X,*HERE ARE X,B,AND ST:*,/)
6610 FORMAT(5X,*KE,IT1,IT2,TLO,TUP,RLF,RRT,BSZ=*,3I5,5F10.4)
6620 FORMAT(/,5X,*NB,N1,N2,NM=*,I5,3F7.1)
6630 FORMAT(15F8.2)
     END
/EOR
   2
   1 500
   1 500
   0 500
   1   1   1  13   1  23   1  25   1  37   1  39   1  41   1  45   1  47
```

2	1	1	25.	500.	0.0	15.	1.0000
2	2	2	25.	500.	0.0	15.	1.0000
2	3	3	25.	500.	0.0	15.	1.0000
2	4	4	25.	500.	0.0	15.	1.0000
2	5	5	25.	500.	0.0	15.	1.0000
2	6	6	25.	500.	0.0	15.	1.0000
2	7	7	25.	500.	0.0	15.	1.0000
2	8	8	25.	500.	0.0	15.	1.0000
2	9	9	25.	500.	0.0	15.	1.0000
3	1	2	25.	500.	15.	15.	1.
3	1	3	25.	500.	15.	15.	1.
3	1	4	25.	500.	15.	15.	1.
3	1	5	25.	500.	15.	15.	1.
3	1	6	25.	500.	15.	15.	1.
3	1	7	25.	500.	15.	15.	1.
3	1	8	25.	500.	15.	15.	1.
3	1	9	25.	500.	15.	15.	1.
3	2	3	25.	500.	15.	15.	1.
3	2	4	25.	500.	15.	15.	1.
3	2	5	25.	500.	15.	15.	1.
3	2	6	25.	500.	15.	15.	1.
3	2	7	25.	500.	15.	15.	1.
3	2	8	25.	500.	15.	15.	1.
3	2	9	25.	500.	15.	15.	1.
3	3	4	25.	500.	15.	15.	1.
3	3	5	25.	500.	15.	15.	1.
3	3	6	25.	500.	15.	15.	1.
3	3	7	25.	500.	15.	15.	1.
3	3	8	25.	500.	15.	15.	1.
3	3	9	25.	500.	15.	15.	1.
3	4	5	25.	500.	15.	15.	1.
3	4	6	25.	500.	15.	15.	1.
3	4	7	25.	500.	15.	15.	1.
3	4	8	25.	500.	15.	15.	1.
3	4	9	25.	500.	15.	15.	1.
3	5	6	25.	500.	15.	15.	1.
3	5	7	25.	500.	15.	15.	1.
3	5	8	25.	500.	15.	15.	1.
3	5	9	25.	500.	15.	15.	1.
3	6	7	25.	500.	15.	15.	1.
3	6	8	25.	500.	15.	15.	1.

```
/JOB
A224,CM=60000,TL=60.
FAKE.
PDFV.
GET,TPRDCX1,TPRDCY1.
MNF(Y).
LGO,TPRDCX1,TPRDCY1,DTPXY1.
REPLACE,DTPXY1.
EXIT.
REPLACE,DTPXY1.
/EOR
      PROGRAM CMPTP1 (TPIN1,TPIN2,CTPOUT,INPUT,OUTPUT,
     *          TAPE3=TPIN1,TAPE4=TPIN2,TAPE6=OUTPUT,TAPE7=CTPOUT)
*
      REAL MUT,MUS,MAD
      INTEGER W,H
      PARAMETER W=7,H=7,NBDF=40
      DIMENSION TP(2,W,H),N(NBDF),DF(W,H)
*
      READ 5010, BNSZ,LCLS
      NBINS=NBDF/2
*
      DO 3000 L=1,2
*
      DO 1000 I=1,3
*
      WRITE(7,7000) I
      IF(I.LE.2) THEN
      IF(I.EQ.1) READ(3,30UT,SIGT,MUS,SLP,CLS
      IF(I.EQ.2) READ(4,3010) MUT,SIGT,MUS,SIGS,SLP,CLS
      SIGT=SIGT/MUT $SIGS=SIGS/MUS $SLP=SLP/MUS $CLS=CLS/MUS
*
      DO 800 JY=1,H
      IF(I.EQ.1) READ(3,3070) (TP(I,JX,H+1-JY),JX=1,W)
      IF(I.EQ.2) READ(4,3070) (TP(I,JX,H+1-JY),JX=1,W)
  800 CONTINUE
*
      DO 810 NB=1,NBINS $N(NB)=0
      DO 805 JX=1,W $DO 805 JY=1,H
      IF(TP(I,JX,JY).GE.BNSZ*FLOAT(NB-1).AND.
     *   TP(I,JX,JY).LT.BNSZ*FLOAT(NB)) N(NB)=N(NB)+1
  805 CONTINUE
  810 CONTINUE
      WRITE(7,7010) (N(J),J=1,NBINS)
      WRITE(7,7020) MUS,SIGS,SLP,CLS,MUT,SIGT
*
      DO 820 JY=1,H
  820 WRITE(7,7070) (TP(I,JX,H+1-JY),JX=1,W) $ELSE
*
      DO 900 JX=1,W $DO 900 JY=1,H
  900 DF(JX,JY)=TP(1,JX,JY)-TP(2,JX,JY)
*
      DO 910 NB=1,NBDF $N(NB)=0
      DO 905 JX=1,W $DO 905 JY=1,H
      IF(DF(JX,JY).GE.BNSZ*FLOAT(NB-1-NBINS).AND.
     *   DF(JX,JY).LT.BNSZ*FLOAT(NB-NBINS)) N(NB)=N(NB)+1
  905 CONTINUE
  910 CONTINUE
      WRITE(7,7010) (N(J),J=1,NBDF)
*
      MAD=0. $SLD=0. $CLD=0.
      DO 950 JX=1,W $DO 950 JY=1,H
      MAD=MAD+ABS(DF(JX,JY))
      IF(JX.GT.1.AND.JX.LE.W-1.AND.JY.GT.1.AND.JY.LE.H-1) THEN
      DO 946 KX=JX-1,JX+1 $DO 946 KY=JY-1,JY+1
  946 SLD=SLD+ABS(DF(KX,KY)-DF(JX,JY)) $ENDIF
```

```
      IF(JX.GT.LCLS.AND.JX.LE.W-LCLS.AND.JY.GT.LCLS.AND.JY.LE.H-LCLS)
     *     THEN
      DO 948 KX=JX-LCLS,JX+LCLS $DO 948 KY=JY-LCLS,JY+LCLS
 948  CLD=CLD+ABS(DF(KX,KY)) $ENDIF
 950  CONTINUE
      MAD=MAD/FLOAT(W*H) $SLD=SLD/FLOAT(8*(W-1)*(H-1))
      CLD=CLD/FLOAT((W-2*LCLS)*(H-2*LCLS)*(1+2*LCLS)**2)
      WRITE(7,7020) MAD,SLD,CLD
*
      DO 920 JY=1,H
 920  WRITE(7,7070) (DF(JX,H+1-JY),JX=1,W) $ENDIF
*
1000  CONTINUE
3000  CONTINUE
*
      STOP
3010  FORMAT(8F8.2)
3070  FORMAT(2X,30F4.1)
5010  FORMAT(F4.2,I4)
7000  FORMAT(1H1,5X,*DTPXY1   *,/,5X,*I=*,I3)
7010  FORMAT(16I4)
7020  FORMAT(6F10.3)
7070  FORMAT(20F4.1)
      END
/EOR
 .2    1
/EOR
/EOI
```

Index